INTRODUCTION TO
ELASTICITY THEORY
FOR CRYSTAL DEFECTS

Second Edition

INTRODUCTION TO
ELASTICITY THEORY FOR CRYSTAL DEFECTS

Second Edition

Robert W Balluffi
Massachusetts Institute of Technology, USA

World Scientific

NEW JERSEY • LONDON • SINGAPORE • BEIJING • SHANGHAI • HONG KONG • TAIPEI • CHENNAI • TOKYO

Published by

World Scientific Publishing Co. Pte. Ltd.
5 Toh Tuck Link, Singapore 596224
USA office: 27 Warren Street, Suite 401-402, Hackensack, NJ 07601
UK office: 57 Shelton Street, Covent Garden, London WC2H 9HE

Library of Congress Cataloging-in-Publication Data
Names: Balluffi, R. W., author.
Title: Introduction to elasticity theory for crystal defects / Robert W. Balluffi, Massachusetts Institute of Technology.
Description: 2nd edition. | Singapore ; Hackensack, NJ : World Scientific, [2016] | 2016 | Includes bibliographical references and index.
Identifiers: LCCN 2015046960| ISBN 9789814749718 (hardcover ; alk. paper) | ISBN 9814749710 (hardcover ; alk. paper) | ISBN 9789814749725 (pbk. ; alk. paper) | ISBN 9814749729 (pbk. ; alk. paper)
Subjects: LCSH: Crystallography, Mathematical. | Elasticity. | Crystals--Defects. | Elastic analysis (Engineering)
Classification: LCC QD399 .B35 2016 | DDC 548/.7--dc23
LC record available at http://lccn.loc.gov/2015046960

British Library Cataloguing-in-Publication Data
A catalogue record for this book is available from the British Library.

Printed in Singapore

Preface to First Edition

A unified introduction to the theory of anisotropic elasticity for static defects in crystals is presented. The term "defects" is interpreted broadly to include defects of zero, one, two and three dimensionality: included are

- point defects (vacancies, self-interstitials, solute atoms, and small clusters of these species)
- line defects (dislocations)
- planar defects (homophase and heterophase interfaces)
- volume defects (inclusions and inhomogeneities)

The book is an outgrowth of a graduate course on "Defects in Crystals" offered by the author for many years at the Massachusetts Institute of Technology, and its purpose is to provide an introduction to current methods of solving defect elasticity problems through the use of anisotropic linear elasticity theory. Emphasis is put on methods rather than a wide range of applications and results. The theory generally allows multiple approaches to given problems, and a particular effort is made to formulate and compare alternative treatments.

Anisotropic linear elasticity is employed throughout. This is now practicable because of significant advances in the theory of anisotropic elasticity for crystal defects that have been made over the last thirty five years or so, including the development of anisotropic Green's functions for unit point forces in infinite spaces, half-spaces and joined dissimilar half-spaces. The use of anisotropic theory (rather than the simpler isotropic theory) is important, since even though the results obtained by employing the two approaches often agree to within 25% or so, there are many phenomena that depend entirely on elastic anisotropy. Unfortunately, however, the results obtained with the anisotropic theory are usually in the form of lengthy integrals that can be evaluated only by numerical methods and so lack transparency. To assist with this difficulty isotropic elasticity is employed in parallel treatments of many problems where sufficiently simple conditions are assumed so that tractable analytic solutions can be obtained that are more transparent physically. Treatments in the book where isotropic elasticity is employed are clearly distinguished to avoid confusion.

The results for the various defects are developed in a sequence of increasing complexity starting with their behavior in isolation in infinite homogeneous regions, where their elastic fields are derived along with, in many cases, corresponding elastic strain energies and induced volume changes. The treatment then progresses to interactions between the defects and imposed applied and internal stresses as well as the image stresses which arise when the defects are in finite homogeneous regions in the vicinity of interfaces. Finally, elastic interactions between the defects themselves are considered in terms of interaction energies and corresponding forces. Due to the breadth of the subject and the impossibility of including all important topics in detail, a selection is made of representative material. This should provide the reader with the background to master omitted topics.

The book is designed to be self-sufficient. Included is a preliminary chapter on the basic theory of linear elasticity that includes essentially all the elements of anisotropic and isotropic theory necessary to master the material that follows. A number of appendices is included containing other essentials. A particular effort has been made to write the book in a pedagogical manner useful for graduate students and workers in the field of materials science and engineering. Essentially all results are fully derived, as many intermediate steps as practicable are written out in full, and the use of the phrase "it can be shown" is avoided. Numerous exercises with solutions are provided, which in many cases expand the scope of the subject matter.

Requirements for use of the book are an undergraduate materials science familiarity with the structural aspects of the various defects and knowledge of linear algebra, vector calculus, and differential equations. To avoid long unwieldy expressions, the repeated index summation convention is employed. Consistent sign conventions are used, and introductory lists of the common symbols employed throughout the text are provided. To keep the notation as simple as possible, additional symbols are employed locally in various sections of the book and are identified in brief lists in the relevant chapters for the convenience of the reader.

Preface to Second Edition

A new chapter, *Defect self-interactions and self-forces*, has been added which introduces the important topic of the self-forces experienced by defects, such as dislocations and inclusions, which are extended in at least one dimension and whose self-energies therefore depend upon their shapes.

A considerable number of worked exercises has been added which expand the scope of the text and furnish further insights.

Numerous sections of the text have been rewritten, and/or expanded, to provide additional aspects and clarity. Again, as in the first edition, a particular effort has been made to include, and compare, different approaches to the solution of various problems.

Finally, typographical errors, that mysteriously escaped detection in proofing the first edition, have been corrected.

Acknowledgements

I am again particularly indebted to Professor David M. Barnett for permission to include his previously unpublished derivations of the anisotropic Green's functions for unit point forces in infinite spaces, half-spaces and joined elastically dissimilar half-spaces and for providing other valuable assistance. Professor Adrian Sutton again offered encouragement and advice. Professor John Hirth assisted with several questions. I am grateful to the Dept. of Materials Science and Engineering, Cornell University, for hospitality and support during the writing of this book.

Frequently Used Symbols

Roman

a: A	scalar quantities (light face)
a^* : A^*	complex conjugate of a or A
$\bar{a}$: $\bar{A}$	Fourier transform of a or A
a: **A**	vectors (bold face)
a_i : A_i	components of **a** or **A**
$\hat{\mathbf{a}}$	unit vector
$\lvert\mathbf{a}\rvert = a$	magnitude of **a**
$\underline{\mathbf{a}}$: $\underline{\mathbf{A}}$	second rank tensors (bold face, underlined)
a_{ij} : A_{ij}	components of $\underline{\mathbf{a}}$: $\underline{\mathbf{A}}$
$\underline{\underline{\mathbf{a}}}$: $\underline{\underline{\mathbf{A}}}$	fourth rank tensors (bold face and double underlined)
a_{ijkl} : A_{ijkl}	components of fourth rank tensor $\underline{\underline{\mathbf{A}}}$
[a]: [A]	matrices
a_i : A_i	elements of [a] or [A] if 1×3 or 3×1 matrix
a_{ij} : A_{ij}	elements of [a] or [A] if 3×3 matrix
a_{ijkl} : A_{ijkl}	elements of [a] or [A] if 9×9 matrix
$(ab)_{jk}$	notation used for element of matrix representing Christoffel tensor: defined by $(ab)_{jk} \equiv a_i C_{ijkl} b_l$ (employs curved brackets rather than the square brackets used for matrices elsewhere throughout book)
(ab)	matrix representing Christoffel tensor
$[A]^{-1}$	inverse of [A]
$[A]^T$	transpose of [A]
b	Burgers vector of dislocation
$\underline{\underline{\mathbf{C}}}$: C_{ijkl}	elastic stiffness tensor

e_{ijk}	alternator symbol: $e_{ijk} \equiv \mathbf{e}_i \cdot (\mathbf{e}_j \times \mathbf{e}_k)$
$\hat{\mathbf{e}}_i$	base unit vector of Cartesian, right-handed, orthogonal coordinate system
e	dilatation: (sum of the normal elastic strain components: $e = \varepsilon_{mm}$)
E	modulus of elasticity (or Young's modulus)
E	total elasto-mechanical energy, i.e., elastic strain energy plus potential energy of applied forces
F	force
f	force per unit length
$\mathcal{F}$	force per unit area
f	force density
H(x)	Heaviside step function: $H(x) = 0$ when $x < 0$; $H(x) = 1$ when $x > 0$
K	bulk elastic modulus
$\hat{\boldsymbol{l}} : \hat{l}_i$	unit directional vector: component of $\boldsymbol{l}$ (direction cosine)
N	number
n	number per unit volume (density)
$\hat{\mathbf{n}}$	unit vector normal to surface (taken to be positive for a closed surface when pointing outwards)
P	hydrostatic pressure (positive when compressive)
$\mathbf{k}$	Fourier transform vector
r	radius
r, θ, z	cylindrical coordinates (see Fig. A.1a)
r, θ, ϕ	spherical coordinates (see Fig. A.1b)
R	radius of curvature: distance between source point at $\mathbf{x}'$ and field point at $\mathbf{x}$
s	arc length along line: distance
S^E_{ijkl}	Eshelby tensor
$\mathcal{S}$	region of surface
S	surface area
$\hat{\mathcal{S}}$	surface of unit sphere
$\underline{\underline{\mathbf{S}}} : S_{ijkl}$	elastic compliance tensor
sgn(x)	$\text{sgn}(x) = 1$ if $x > 0$: $\text{sgn}(x) = -1$ if $x < 0$
$\hat{\mathbf{t}}$	unit vector tangent to dislocation

$\mathbf{T}$	traction vector
$\mathbf{u}$	elastic displacement
$\mathbf{u}^{\mathrm{T}}$	displacement associated with transformation strain
$\mathbf{u}^{\mathrm{tot}}$	total displacement ($\mathbf{u}^{\mathrm{tot}} = \mathbf{u} + \mathbf{u}^{\mathrm{T}}$)
$\mathcal{V}$	region of volume
V	volume
$\mathrm{W}:\mathrm{w}:\mathcal{w}$	elastic strain energy: elastic strain energy per unit volume: strain energy per unit length
$\mathcal{W}$	work
$\mathrm{x_1, x_2, x_3}$	Cartesian coordinates
$\mathbf{x} : \mathrm{x_i} : \mathrm{x}$	field vector in Cartesian coordinates: component of $\mathbf{x}$: magnitude of $\mathbf{x}$, i.e., $\mathrm{x} = \lvert\mathbf{x}\rvert = (\mathrm{x}_1^2 + \mathrm{x}_2^2 + \mathrm{x}_3^2)^{1/2}$
$\mathbf{x}' : \mathrm{x}'_\mathrm{i} : \mathrm{x}'$	source vector in Cartesian coordinates

Greek

δ_{ij}	Kronecker delta operator ($\delta_{\mathrm{ij}} = 1$ when i = j: $\delta_{\mathrm{ij}} = 0$ when i≠j)
$\delta(\mathbf{x} - \mathbf{x}_\circ)$	Dirac delta function
$\underline{\varepsilon}$: $\varepsilon_{\mathrm{ij}}$	elastic strain tensor: component of $\underline{\varepsilon}$
$\varepsilon_{\mathrm{ij}}^{\mathrm{T}}$	transformation strain
$\varepsilon_{\mathrm{ij}}^{\mathrm{T}^*}$	transformation strain of equivalent homogeneous inclusion
$\varepsilon_{\mathrm{ij}}^{\mathrm{tot}}$	total strain ($\varepsilon_{\mathrm{ij}}^{\mathrm{tot}} = \varepsilon_{\mathrm{ij}} + \varepsilon_{\mathrm{ij}}^{\mathrm{T}}$)
θ	sum of the normal stress components ($\theta \equiv \sigma_{\mathrm{mm}}$)
$\mathrm{r}, \theta, \mathrm{z}$	cylindrical coordinates (see Fig. A.1(a))
r, θ, ϕ	spherical coordinates (see Fig. A.1(b))
λ	Lame' elastic constant
μ	Lame' elastic constant (elastic shear modulus)
ν	Poisson's ratio
$\underline{\boldsymbol{\sigma}} : \sigma_{\mathrm{ij}}$	stress tensor: component of $\underline{\boldsymbol{\sigma}}$
Φ	potential energy of forces applied to body
ϕ	Newtonian potential
ψ	biharmonic potential
Ω	atomic volume

The ± Symbol

± a special symbol used throughout this book, and its relevant literature, which is employed in sums involving Stroh vectors possessing a summation index, such as α, that ranges from 1 to 6. It has the properties that $\pm = +$ when $\alpha = 1, 2, 3$ and $\pm = -$ when $\alpha = 4, 5, 6$. Hence, for example,

$$\sum_{\alpha=1}^{6} \pm \mathrm{A}_{s\alpha}\mathrm{A}_{k\alpha} = \mathrm{A}_{s1}\mathrm{A}_{k1} + \mathrm{A}_{s2}\mathrm{A}_{k2} + \mathrm{A}_{s3}\mathrm{A}_{k3} - \mathrm{A}_{s4}\mathrm{A}_{k4} - \mathrm{A}_{s5}\mathrm{A}_{k5} - \mathrm{A}_{s6}\mathrm{A}_{k6}$$

Its properties are therefore seen to be the same as those of the sign of the imaginary part of p_α, which, by convention (see Eq. (3.34), is positive when $\alpha = 1, 2, 3$ and negative when $\alpha = 4, 5, 6$.

Superscripts

D	defect
DIS	dislocation
IM	image
INC	inclusion
INH	inhomogeneity
LF	line force
M	matrix

Contents

Chapter 1

Introduction

1.1 Contents of Book

An introduction to the use of anisotropic linear elasticity in determining the static elastic properties of defects in crystals is presented. The defects possess different dimensionalities and span the defect spectrum, including:

- Point defects (vacancies, self-interstitials, solute atoms, and small clusters of these species),
- Line defects (dislocations),
- Planar defects (homophase and heterophase interfaces),
- Volume defects (inclusions and inhomogeneities).

To avoid confusion, an *inclusion* is defined as a misfitting region embedded within a larger constraining matrix, and, therefore, acts as a source of stress. It may be either *homogeneous* (if it possesses the same elastic properties as the matrix) or *inhomogeneous* (if its elastic properties differ). On the other hand, an *inhomogeneity* is simply an embedded region having different elastic constants but no misfit. A *homogeneous region* is one in which the elastic properties are uniform throughout.

Following the preliminaries of the present chapter, the book presents (Ch. 2) a concise account of the basic theory of anisotropic and isotropic linear elasticity, and, in addition, derivations of a number of special relationships needed throughout the text. This is followed by a review of general methods of solving defect elasticity problems (Ch. 3), derivations of useful Green's functions (Ch. 4), and the basic formulation of interactions between defects and imposed stress in the form of interaction energies and forces (Ch. 5). Then, (Chs. 6–15) attention is focused on individual defects in the following order: inclusions, inhomogeneities, point defects, dislocations and interfaces. In most cases the elastic field associated with the defect in an infinite homogeneous region is treated first. Then, the interaction of the defect with imposed stress is studied, and this sets the stage for analyzing the behavior of defects in finite homogeneous regions where interfaces and associated image stresses are present. Next (Ch. 16) a selection of interactions between various pairs of defects is analyzed, and finally (Ch. 17), the self-forces experienced by defects that are extended in at least one dimension, and whose energy depends upon their shape (configuration), are treated.

1.2 Sources

Important sources for the book include the pioneering work of J. D. Eshelby, especially Eshelby (1951, 1954, 1956, 1957, 1961), who invented imaginary cutting, straining and bonding operations to create defects in a manner that greatly expedites the analysis of their elastic properties. By applying potential theory to the results of these operations and using harmonic and biharmonic potentials and the divergence theorem (Gauss's theorem),[1] expressions for defect interaction energies and forces on defects are obtained in the form of integrals over surfaces enclosing the defects. This approach has connections with classic electrostatics and electromagnetism and produces an arsenal of general expressions that can be employed to treat specific defect problems. Other sources include the indispensable treatise of Bacon, Barnett and Scattergood (1979b), that demonstrated that the anisotropic elasticity theory can often be applied to defects with almost the same ease as isotropic theory, and the more recent book, *Elastic Strain Fields and Dislocation Mobility*, edited by V. L. Indenbom and J. Lothe (1992). Additional valuable sources include the books of Leibfried and Breuer (1978) on point defects, of Teodosiu (1982) on point defects and dislocations, of Hirth and Lothe (1982) on dislocations and of Mura (1987) on inclusions, dislocations and cracks. The book of Sutton and Balluffi (2006) provided a source for material on interfaces. Finally, many journal articles must be cited, especially those of J. Lothe and D. M. Barnett, dealing with the anisotropic theory.

[1] Eshelby has been quoted (Bilby 1990) as saying about this work, "amusing applications of the theorem of Gauss".

1.3 Symbols and Conventions

The symbols that are used most frequently are identified in the list before the main text. Components of vectors and tensors are generally indicated by subscripts, while the entities to which various quantities refer to are usually indicated by superscripts: the superscripts of most importance are also listed.

Cartesian coordinates and index notation, involving either Latin or Greek subscripts, are mainly employed. For Latin subscripts, the standard *repeated index summation convention* is employed. Here, any indexed quantity possessing a repeated subscript is automatically summed with respect to that subscript as it runs from 1 to 3 (unless specifically noted otherwise). For example,

$$\begin{aligned} \mathrm{x_{ii}} &= \mathrm{x_{11} + x_{22} + x_{33}}, \\ \mathrm{x_i x_i} &= \mathrm{x_1 x_1 + x_2 x_2 + x_3 x_3}, \\ \mathrm{x_{jk} y_k} &= \mathrm{x_{j1} y_1 + x_{j2} y_2 + x_{j3} y_3}, \end{aligned} \tag{1.1}$$

and

$$\mathrm{x_{ii} = x_{11} + x_{22} + x_{33} + x_{44} + x_{55} + x_{66}}. \quad (i = 1, 2 \ldots 6) \tag{1.2}$$

For Greek subscripts, this convention does not apply. Instead, summation of quantities with repeated Greek subscripts is not automatic but must be indicated explicitly by the usual summation symbol, e.g.,

$$\begin{aligned} \mathrm{x}_{\alpha\alpha} &\neq \mathrm{x_{11} + x_{22} + x_{33}}, \\ \sum_{\alpha=1}^{3} \mathrm{x}_{\alpha\alpha} &= \mathrm{x_{11} + x_{22} + x_{33}}. \end{aligned} \tag{1.3}$$

Cylindrical (r,θ,z) and spherical (r,θ, ϕ) orthogonal curvilinear coordinates are also employed, and basic elasticity formulae in these coordinates, rather than Cartesian coordinates, are presented in appendices A and G.

Complete descriptions of the elastic fields derived throughout the book, i.e., the displacements, strains and stresses, are normally not all presented together. Instead, to save space, results are presented in forms that can be used to obtain the complete descriptions relatively easily by employing standard relationships between the various quantities. For example, when only the displacement field is given, the corresponding strain field can be determined by simple differentiation, and then the stress field can be obtained using Hooke's law.

Unless noted otherwise, it can be assumed that the results presented throughout the book are valid for general anisotropic systems. Cases where isotropic elasticity is assumed are clearly identified to avoid any confusion.

1.4 On the Applicability of Linear Elasticity

Linear elasticity is an approximation that describes a homogeneous crystal as a uniform continuum in which the stress is proportional to the strain via constant elastic coefficients. For many defect applications this approximation is quite adequate. It is most reliable in regions that are large enough to span a significant number of atoms and where the atom displacements are small and consequently proportional to the forces exerted on them. With this assumption the effects of the displacements associated with the solution for one elastic displacement field on the solution for a superposed second displacement field can be neglected. The stresses and strains obtained as a solution of one boundary value problem can then be simply added to the solution of another problem involving other boundary conditions, i.e., linear superposition holds for both the boundary conditions and the solutions.

However, many of the defects of interest, such as point defects and dislocations, possess *core regions* of atomic dimensions where the atoms have undergone relatively large displacements out of the linear elastic range and find themselves in alien atomic environments. As discussed by Read (1953), such highly disturbed material, in which atoms are not surrounded by their usual neighbors and for which the linear continuum model breaks down, may be regarded as *bad material* in contrast to *good material* corresponding to defect-free crystalline material that is, at most, elastically strained. The core region of a vacancy, for example, consists of a small roughly spherical region of bad material centered on the vacant site where neighboring atoms have relaxed and undergone relatively large displacements. The core of a dislocation line consists of a long narrow cylindrical region of bad material, and the core of an incoherent grain boundary consists of a thin plate-like region of bad material in the transition region between the two adjoining bulk crystals.

In view of this, the displacement field of the defect can be broken down into the relatively small core region, where the linear theory cannot be applied, and the much larger surrounding matrix region where it serves as a good approximation. A quite reliable solution can then be obtained by employing a hybrid approach in which the displacements in the core are determined by means of atomistic calculations and are matched to the displacements in the adjoining bulk matrix region determined by using linear elasticity. Fortunately, such a complex calculation can be avoided in many situations by realizing that the displacements due to the defect generally decrease rapidly with distance into the matrix and, at distances several times the relevant core dimension, become insensitive to the detailed nature of the conditions at the core/matrix interface. An acceptable solution for the elastic field in the matrix region beyond a few core dimensions can then be obtained by the exclusive use of linear elasticity with the core described, at most, by a few simple parameters. Since the relevant core dimensions are relatively small, the regions that can be treated in this manner in bodies containing defects typically extend over essentially the entire body and have dimensions corresponding to length scales that are of major interest. This limitation is therefore not a major drawback under many circumstances. The difficulties of dealing with the large non-linear displacements at defect cores can be mitigated to a degree by employing *non-linear elasticity*, but this will not be considered in the present book.

A further complication with the use of linear elasticity occurs when abrupt step-like changes in bulk elastic constants are present in a system as, for example, at the interface between an inhomogeneous inclusion and the matrix. The assumption that the bulk elastic constants in the matrix and inclusion are truly constant right up to the interface is an approximation, since at small distances from the interface the elastic constants of the inclusion and matrix must be affected to at least some degree by their altered local environments, even under conditions when the atomic displacements are relatively small. This problem can be dealt with by employing *size-dependent elasticity*, where it is assumed that the elastic constants depend upon the local environment over a specified length scale (Eringen 2002, Sharma and Ganti 2003). This approach introduces additional complexities, however, and will not be considered in this book.

Chapter 2

Basic Elements of Linear Elasticity

2.1 Introduction

The basic elements of anisotropic linear elasticity are presented in concise form. First, the deformation of an elastically strained body is described in terms of the local displacements, strains and rotations that occur throughout the body. Requirements on the strains that ensure compatibility of the medium are then described. Next, the forces acting throughout the body are described in terms of surface tractions, body forces and stresses. Conditions for mechanical equilibrium are derived. The stresses and strains are then linearly coupled via elastic constants, and various stress-strain relationships are derived. Finally, the energy stored in an elastically strained medium is formulated. Elements of the theory for the special case when the medium is elastically isotropic are included[1] along with several additional items required for treating crystal defects.

References include: Love (1944); Sokolnikoff (1946); Muskhelishvili (1953); Nye (1957); Lekhnitskii (1963); Bacon, Barnett and Scattergood (1979b); Soutas-Little (1999); Hetnarski and Ignaczak (2004) and Asaro and Lubarda (2006).

[1]The theory of elasticity presented in this chapter holds for systems in which all displacements and strains are purely elastic. In Ch. 3 stress-free *transformation strains* are introduced as a means of mimicking crystal defects. For systems containing transformation strains the purely elastic formulation must therefore be modified as described in Sec. 3.6.

2.2 Elastic Displacement and Strain Tensor

2.2.1 *Straining versus rigid body rotation*

When a body is elastically deformed, points embedded in the body are generally displaced by differing degrees: local regions must therefore be strained (deformed) and also rotated in various ways. To analyze the connection between the displacements and the strains and rotations a Cartesian coordinate system is adopted having unit base vectors $\hat{\mathbf{e}}_i$ and coordinates $\mathbf{x}_i$. As illustrated in Fig. 2.1, a point initially at the position $\mathbf{P}^\circ = x_i^\circ \hat{\mathbf{e}}_i$ is then displaced by the vector $\mathbf{u}^\circ = u_i^\circ \hat{\mathbf{e}}_i$ to the position $\mathbf{P}'^\circ = x_i'^\circ \hat{\mathbf{e}}_i$, while a closely adjacent point initially at $\mathbf{P} = x_i \hat{\mathbf{e}}_i$ is displaced by $\mathbf{u} = u_i \hat{\mathbf{e}}_i$ to $\mathbf{P}' = x_i' \hat{\mathbf{e}}_i$. The difference between the initial positions is $\mathbf{A} = \mathbf{P} - \mathbf{P}^\circ$ and between the final positions $\mathbf{A}' = \mathbf{P}' - \mathbf{P}'^\circ$. The difference between the differences is then

$$\delta \mathbf{A} = \mathbf{A}' - \mathbf{A} = \mathbf{u} - \mathbf{u}^\circ. \tag{2.1}$$

The displacement of any point in the body is a function of its original position so that

$$\begin{aligned} u_i^\circ &= u_i^\circ(x_1^\circ, x_2^\circ, x_3^\circ), \\ u_i &= u_i(x_1, x_2, x_3), \end{aligned} \tag{2.2}$$

and by expanding Eq.(2.1) to first order around $(x_1^\circ, x_2^\circ, x_3^\circ)$,

$$dA_i = u_i - u_i^\circ = \left(u_i^\circ + \frac{\partial u_i}{\partial x_1} A_1 + \frac{\partial u_i}{\partial x_2} A_2 + \frac{\partial u_i}{\partial x_3} A_3 \right) - u_i^\circ = \frac{\partial u_i}{\partial x_j} A_j. \tag{2.3}$$

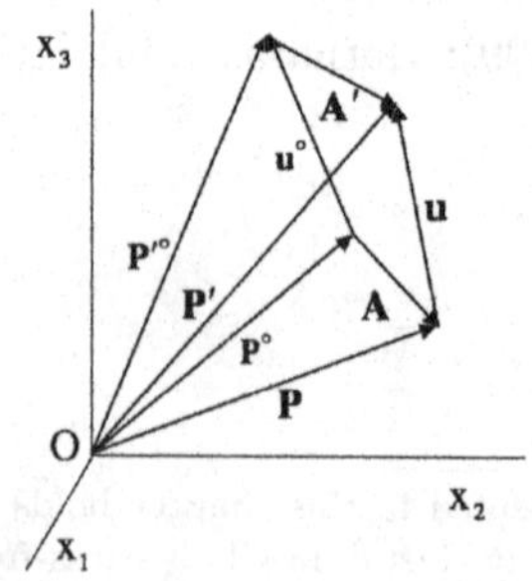

Fig. 2.1 Displacements $\mathbf{u}^\circ$ and $\mathbf{u}$ of points initially located at the vector positions $\mathbf{P}^\circ$ and $\mathbf{P}$, respectively.

Equation (2.3) can be written in the equivalent form

$$dA_i = \left[\frac{1}{2}\left(\frac{\partial u_i}{\partial x_j} + \frac{\partial u_j}{\partial x_i}\right) + \frac{1}{2}\left(\frac{\partial u_i}{\partial x_j} - \frac{\partial u_j}{\partial x_i}\right)\right] A_j, \tag{2.4}$$

where the quantities $\partial u_i/\partial x_j$ are termed *distortions.* Then, by introducing the symmetric quantity ε_{ij} and the skew-symmetric quantity ω_{ij}, defined by

$$\begin{aligned} \varepsilon_{ij} = \varepsilon_{ji} &\equiv \frac{1}{2}\left(\frac{\partial u_i}{\partial x_j} + \frac{\partial u_j}{\partial x_i}\right), \\ \omega_{ij} = -\omega_{ji} &\equiv \frac{1}{2}\left(\frac{\partial u_i}{\partial x_j} - \frac{\partial u_j}{\partial x_i}\right), \end{aligned} \tag{2.5}$$

Eq. (2.4) becomes

$$\delta A_i = \varepsilon_{ij}A_j + \omega_{ij}A_j, \tag{2.6}$$

which can be written in vector-tensor form as,[2]

$$\delta \mathbf{A} = \underline{\boldsymbol{\varepsilon}}\mathbf{A} + \underline{\boldsymbol{\omega}}\mathbf{A} = (\underline{\boldsymbol{\varepsilon}} + \underline{\boldsymbol{\omega}})\mathbf{A}, \tag{2.7}$$

or, in matrix form as,

$$[\delta A] = [\varepsilon][A] + [\omega][A] = \begin{bmatrix} \delta A_1 \\ \delta A_2 \\ \delta A_3 \end{bmatrix} = \begin{bmatrix} \varepsilon_{11} & \varepsilon_{12} & \varepsilon_{13} \\ \varepsilon_{12} & \varepsilon_{22} & \varepsilon_{23} \\ \varepsilon_{13} & \varepsilon_{23} & \varepsilon_{33} \end{bmatrix} \begin{bmatrix} A_1 \\ A_2 \\ A_3 \end{bmatrix} + \begin{bmatrix} 0 & -\omega_{21} & \omega_{13} \\ \omega_{21} & 0 & -\omega_{32} \\ -\omega_{13} & \omega_{32} & 0 \end{bmatrix} \begin{bmatrix} A_1 \\ A_2 \\ A_3 \end{bmatrix}. \tag{2.8}$$

As now shown, the symmetric portion of δA_i, $\varepsilon_{ij}A_j$, in Eq. (2.8) represents local straining (deformation) of the medium, while the skew-symmetric portion, $\omega_{ij}A_j$, represents local rigid body rotation.

[2]The quantities ε_{ij} and ω_{ij} are the components of the second rank tensors, $\underline{\boldsymbol{\varepsilon}}$ and $\underline{\boldsymbol{\omega}}$, respectively. A second rank tensor possesses nine components and maps one vector into another, as in Eq. (2.7) where $\mathbf{A}$ is linearly transformed into $\delta\mathbf{A}$ by the tensor $(\underline{\boldsymbol{\varepsilon}} + \underline{\boldsymbol{\omega}})$ (Nye 1957).

2.2.1.1 *Local straining and components of strain*

To reveal the effect of the $\underline{\varepsilon}$ tensor on $\mathbf{A}$ to produce the new vector, $\mathbf{A}'$, we write $\mathbf{A}' = \mathbf{A} + \delta\mathbf{A}$, where $\delta A_i = \varepsilon_{ij} A_j$, so that

$$\begin{bmatrix} A_1' \\ A_2' \\ A_3' \end{bmatrix} = \begin{bmatrix} A_1 \\ A_2 \\ A_3 \end{bmatrix} + \begin{bmatrix} \varepsilon_{11} & \varepsilon_{12} & \varepsilon_{13} \\ \varepsilon_{12} & \varepsilon_{22} & \varepsilon_{23} \\ \varepsilon_{13} & \varepsilon_{23} & \varepsilon_{33} \end{bmatrix} \begin{bmatrix} A_1 \\ A_2 \\ A_3 \end{bmatrix}$$

$$= \begin{bmatrix} (1+\varepsilon_{11})A_1 + \varepsilon_{12}A_2 + \varepsilon_{13}A_3 \\ \varepsilon_{12}A_1 + (1+\varepsilon_{22})A_2 + \varepsilon_{23}A_3 \\ \varepsilon_{13}A_1 + \varepsilon_{23}A_2 + (1+\varepsilon_{33})A_3 \end{bmatrix}. \tag{2.9}$$

Then, according to Eq. (2.9), if $\mathbf{A}$ lies along $\hat{\mathbf{e}}_1$, as illustrated in Fig. 2.2a, it will be transformed into the vector $\mathbf{A}' = (1+\varepsilon_{11})A\hat{\mathbf{e}}_1 + \varepsilon_{12}A\hat{\mathbf{e}}_2 + \varepsilon_{13}A\hat{\mathbf{e}}_3$ as shown in Fig. 2.2b. Dropping second order terms, its length will be increased by $\varepsilon_{11}A$ and it will be sheared in the direction $\hat{\mathbf{e}}_2$ by the distance $\varepsilon_{12}A$, and in the direction $\hat{\mathbf{e}}_3$ by $\varepsilon_{13}A$. Similar results will be obtained when $\mathbf{A}$ lies initially along $\hat{\mathbf{e}}_2$ or $\hat{\mathbf{e}}_3$.

The components ε_{11}, ε_{22} and ε_{33} are seen to be the fractional extensions of the local medium in the $\hat{\mathbf{e}}_1$, $\hat{\mathbf{e}}_2$ and $\hat{\mathbf{e}}_3$ directions respectively and are termed *normal strains.* On the other hand, as evident in Fig. 2.2, the quantity ε_{12} is a measure of the extent by which the local material is sheared through the angles $\phi_1 = \phi_2 = \varepsilon_{12}$ in the $\hat{\mathbf{e}}_2$ and $\hat{\mathbf{e}}_1$ directions, respectively. It can therefore be expressed in the form $\varepsilon_{12} = (\phi_1 + \phi_2)/2 = \phi_{12}/2$ where ϕ_{12} is the total angle of shear which converts the square cross section in Fig. 2.2 into a parallelogram. Similar results are obtained for ε_{13} and ε_{23}. The quantities ε_{12}, ε_{13} and ε_{23} are therefore identified as half the total shear

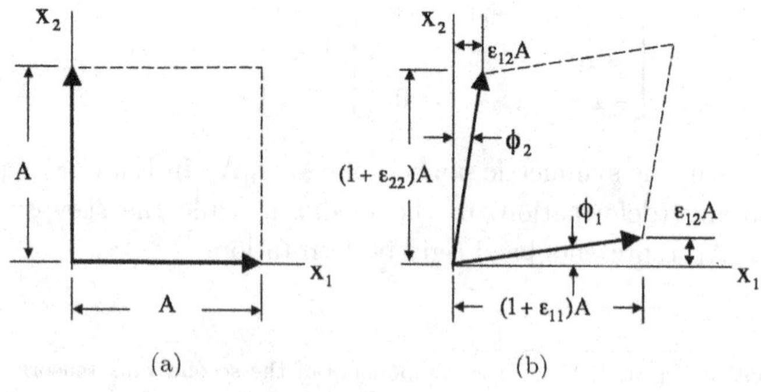

Fig. 2.2 Deformation of vector $\mathbf{A}$, lying initially along either x_1 or x_2. (a) Before strain. (b) After strain.

angles experienced by the local material in the $x_3 = 0$, $x_2 = 0$ and $x_1 = 0$ planes, respectively, and are termed *shear strains*.[3] The local deformation in the immediate vicinity of a point is therefore completely described by six independent components of $\underline{\varepsilon}$, i.e., the three normal strains, ε_{11}, ε_{22}, and ε_{33}, and the three shear strains, $\varepsilon_{12} = \varepsilon_{21}$, $\varepsilon_{13} = \varepsilon_{31}$ and $\varepsilon_{23} = \varepsilon_{32}$.

2.2.1.2 *Local rigid body rotation*

To reveal the effect of applying the skew-symmetric $\underline{\omega}$ tensor to $\mathbf{A}$, as in Eq. (2.8), we consider the general vector equation that yields the change in $\mathbf{A}$, i.e., $\delta\mathbf{A}$, owing to an infinitesimal right-handed rotation of $\mathbf{A}$ by the angle $\delta\theta$ around an axis parallel to the unit vector $\hat{\mathbf{w}}$. This can be written as

$$\delta\mathbf{A} = \delta\mathbf{w} \times \mathbf{A}, \tag{2.10}$$

where $\delta\mathbf{w}$ is an *infinitesimal rotation vector* given by

$$\delta\mathbf{w} = \delta\theta\hat{\mathbf{w}}. \tag{2.11}$$

The vector $\delta\mathbf{A}$ is perpendicular to $\mathbf{A}$ and, to first order, $(\mathbf{A} + \delta\mathbf{A}) \cdot (\mathbf{A} + \delta\mathbf{A}) = \mathbf{A} \cdot \mathbf{A}$. Therefore, $\mathbf{A}$ remains of constant length but is rotated by the angle $\delta\theta = |\delta\mathbf{w} \times \mathbf{A}|A^{-1}$. Then, writing out the expression for $\delta\mathbf{A}$ in full,

$$\begin{aligned}\delta\mathbf{A} = \delta\mathbf{w} \times \mathbf{A} &= (-\delta w_3 A_2 + \delta w_2 A_3)\hat{\mathbf{e}}_1 + (\delta w_3 A_1 - \delta w_1 A_3)\hat{\mathbf{e}}_2 \\ &\quad + (-\delta w_2 A_1 + \delta w_1 A_2)\hat{\mathbf{e}}_3\end{aligned} \tag{2.12}$$

or, alternatively,

$$\begin{bmatrix}\delta A_1 \\ \delta A_2 \\ \delta A_3\end{bmatrix} = \begin{bmatrix}0 & -\delta w_3 & \delta w_2 \\ \delta w_3 & 0 & -\delta w_1 \\ -\delta w_2 & \delta w_1 & 0\end{bmatrix}\begin{bmatrix}A_1 \\ A_2 \\ A_3\end{bmatrix}. \tag{2.13}$$

The rotation matrix in Eq. (2.13) and the $[\omega]$ matrix in Eq. (2.8), where

$$\begin{aligned}&\delta\omega_1 = \omega_{32} = \frac{1}{2}\left(\frac{\partial u_3}{\partial x_2} - \frac{\partial u_2}{\partial x_3}\right) \quad \delta\omega_2 = \omega_{13} = \frac{1}{2}\left(\frac{\partial u_1}{\partial x_3} - \frac{\partial u_3}{\partial x_1}\right) \\ &\delta\omega_3 = \omega_{21} = \frac{1}{2}\left(\frac{\partial u_2}{\partial x_1} - \frac{\partial u_1}{\partial x_2}\right)\end{aligned} \tag{2.14}$$

are seen to have the same form, thus confirming that the latter matrix indeed represents a rigid body rotation.

---

[3] The shear strain $\varepsilon_{ij} (i \neq j)$ employed in this book is a component of the strain tensor, $\underline{\varepsilon}$, and is equal to half the "engineering shear strain" which is often employed in the literature (e.g., Timoshenko and Goodier 1970) and is not the component of a tensor.

2.2.2 *Relationships for strain components*

2.2.2.1 *Transformation of strain components due to rotation of coordinate system*

When a strain tensor in a given coordinate system is known, it is often necessary to find the new form that the tensor will take when the coordinate system is rotated. This can be accomplished after first finding the relationship between a given vector displacement in the old system and in the new system.

Let $\mathbf{u} = u_i \hat{e}_i$ and $\mathbf{u}' = u'_i \hat{e}'_i$ represent the same displacement vector in the old and new coordinate systems, respectively. The components of the vector in the new system, in terms of its components in the old system, are then

$$u'_i = (u_1\hat{e}_1 + u_2\hat{e}_2 + u_3\hat{e}_3) \cdot \hat{e}'_i = u_j(\hat{e}_j \cdot \hat{e}'_i) = l_{ij}u_j \quad \text{or} \quad [u'], = [l][u], \tag{2.15}$$

where l_{ij} is the cosine of the angle between $\hat{e}'_i$ and $\hat{e}_j$. Conversely, the old components in terms of the new components are given by

$$u_i = (u'_1\hat{e}'_1 + u'_2\hat{e}'_2 + u'_3\hat{e}'_3) \cdot \hat{e}_i = u'_j(\hat{e}'_j \cdot \hat{e}_i) = l_{ji}u'_j \quad \text{or} \quad [u] = [l]^T[u']. \tag{2.16}$$

Solving Eq. (2.15) for [u],

$$[u] = [l]^{-1}[u'] \tag{2.17}$$

and, by comparing Eqs. (2.16) and (2.17),

$$[l]^T = [l]^{-1}. \tag{2.18}$$

Therefore,

$$[l][l]^T = [l]^T[l] = [I]. \tag{2.19}$$

Every column vector and row vector in $[l]$ is a unit vector, and every pair of column vectors and every pair of row vectors is orthogonal. $[l]$ is therefore termed a *unitary orthogonal matrix*.

The deformation of the old vector **A** in the old coordinate system and the new vector **A**′ in the new system by the strain tensor, will be of the respective forms

$$[\delta \mathrm{A}] = [\varepsilon][\mathrm{A}] \quad [\delta \mathrm{A}'] = [\varepsilon'][\mathrm{A}']. \tag{2.20}$$

The transformation matrix $[l]$, acting on the vector $\mathbf{u}$ in Eq. (2.15), can also be applied to the vectors $\delta\mathbf{A}'$ and $\mathbf{A}'$ in Eq. (2.20), and therefore

$$\begin{aligned} [\delta \mathrm{A}'] &= [\varepsilon'][\mathrm{A}'], \\ [l][\delta \mathrm{A}] &= [\varepsilon'][l][\mathrm{A}], \\ [\delta \mathrm{A}] &= [l]^{\mathrm{T}}[\varepsilon'][l][\mathrm{A}]. \end{aligned} \tag{2.21}$$

Then, comparing this result with Eq. (2.20), the strain tensors in the two systems are related by

$$[\varepsilon] = [l]^{\mathrm{T}}[\varepsilon'][l] \tag{2.22}$$

and, by inverting Eq. (2.22) by use of Eq. (2.18),

$$[\varepsilon'] = [l][\varepsilon][l]^{\mathrm{T}}. \tag{2.23}$$

The transformations given by Eqs. (2.22) and (2.23) may also be expressed in the component forms,

$$\varepsilon'_{\mathrm{ij}} = l_{\mathrm{im}} l_{\mathrm{jn}} \varepsilon_{\mathrm{mn}}, \tag{2.24}$$

$$\varepsilon_{\mathrm{ij}} = l_{\mathrm{mi}} l_{\mathrm{nj}} \varepsilon'_{\mathrm{mn}}. \tag{2.25}$$

All second rank tensors follow these transformation laws.

2.2.2.2 *Principal coordinate system for strain tensor*

Using the above results it is now shown that for any state of strain it is always possible to find a coordinate system, termed the *principal coordinate system*, that causes the strain tensor to take the simple diagonal matrix form[4]

$$[\tilde{\varepsilon}] = \begin{bmatrix} \tilde{\varepsilon}_{11} & 0 & 0 \\ 0 & \tilde{\varepsilon}_{22} & 0 \\ 0 & 0 & \tilde{\varepsilon}_{33} \end{bmatrix}, \tag{2.26}$$

[4] All quantities referred to a principal coordinate system in this section are distinguished by a tilde, as in Eq. (2.26).

where the diagonal elements are known as the *principal strains*. When the principal coordinate system is employed, and the strain tensor in the form of Eq. (2.26) is applied to various vectors in the medium, it is readily seen that a vector lying along any one of the three principal coordinate axes (i.e., the three *principal directions*) remains non-rotated and simply undergoes a fractional change in length corresponding to the principal strain along that axis. The principal directions are therefore special directions in which vectors embedded in the medium are simply changed in length (or scaled) by the corresponding principal strains and not rotated as the result of a general strain. The unit vectors along the three principal directions and the three principal strains correspond, respectively, to the *eigenvectors* and *eigenvalues* of the strain tensor.

To find the eigenvalues and eigenvectors of a general strain tensor, $\underline{\varepsilon}$, start by applying the tensor to a vector **A**, so that, following Eq. (2.20),

$$[\delta \mathrm{A}] = [\varepsilon][\mathrm{A}]. \tag{2.27}$$

The condition that **A** is simply scaled by the factor λ is then

$$[\varepsilon][\mathrm{A}] = \lambda[\mathrm{A}] \tag{2.28}$$

or, equivalently,

$$\begin{aligned}
(\varepsilon_{11} - \lambda)\mathrm{A}_1 + \varepsilon_{12}\mathrm{A}_2 + \varepsilon_{13}\mathrm{A}_3 &= 0,\\
\varepsilon_{12}\mathrm{A}_1 + (\varepsilon_{22} - \lambda)\mathrm{A}_2 + \varepsilon_{23}\mathrm{A}_3 &= 0,\\
\varepsilon_{13}\mathrm{A}_1 + \varepsilon_{23}\mathrm{A}_2 + (\varepsilon_{33} - \lambda)\mathrm{A}_3 &= 0.
\end{aligned} \tag{2.29}$$

For this set of simultaneous linear equations to have a non-trivial solution, the determinantal condition

$$\det \begin{bmatrix} (\varepsilon_{11} - \lambda) & \varepsilon_{12} & \varepsilon_{13} \\ \varepsilon_{12} & (\varepsilon_{22} - \lambda) & \varepsilon_{23} \\ \varepsilon_{13} & \varepsilon_{23} & (\varepsilon_{33} - \lambda) \end{bmatrix} = 0 \tag{2.30}$$

must be satisfied. The three roots $(\lambda_1, \lambda_2, \lambda_3)$ of the cubic equation obtained from the determinant are then the three eigenvalues of $\underline{\varepsilon}$. The corresponding eigenvectors can then be found by substituting the eigenvalues into Eq. (2.29).

As an example, we find the eigenvalues and eigenvectors of the tensor (in arbitrary units)

$$[\varepsilon] = \begin{bmatrix} 1 & 1 & 0 \\ 1 & 2 & 0 \\ 0 & 0 & 1 \end{bmatrix}. \tag{2.31}$$

The equation corresponding to Eq. (2.29) is

$$\begin{aligned} (1-\lambda)A_1 + A_2 &= 0, \\ A_1 + (2-\lambda)A_2 &= 0, \\ (1-\lambda)A_3 &= 0. \end{aligned} \tag{2.32}$$

The determinantal condition corresponding to Eq. (2.30) is

$$\det \begin{vmatrix} 1-\lambda & 1 & 0 \\ 1 & 2-\lambda & 0 \\ 0 & 0 & 1-\lambda \end{vmatrix} = 0. \tag{2.33}$$

The three eigenvalue solutions of the cubic equation obtained from Eq. (2.33) are

$$\lambda_1 = \frac{3-\sqrt{5}}{2} \quad \lambda_2 = 1 \quad \lambda_3 = \frac{3+\sqrt{5}}{2}. \tag{2.34}$$

Then, substituting these eigenvalues into Eq. (2.32) and solving for the corresponding eigenvectors, the following three orthogonal unit vectors are obtained which are suitable base vectors for a right-handed principal coordinate system:

$$\tilde{e}_1 = \begin{bmatrix} p \\ pq \\ 0 \end{bmatrix} \quad \tilde{e}_2 = \begin{bmatrix} 0 \\ 0 \\ 1 \end{bmatrix} \quad \tilde{e}_3 = \begin{bmatrix} r \\ rs \\ 0 \end{bmatrix}, \tag{2.35}$$

where $p = \sqrt{(5+\sqrt{5})/10}$, $q = (1-\sqrt{5})/2$, $r = \sqrt{(5-\sqrt{5})/10}$, and $s = (1+\sqrt{5})/2$.

Finally, we can use Eq. (2.23) to confirm that the strain tensor expressed in the principal coordinate system found above has the expected diagonal form. From Eq. (2.35), the required direction cosine matrix linking the new (principal) coordinate axes to the original coordinate axes is

$$[l] = \begin{bmatrix} p & pq & 0 \\ 0 & 0 & 1 \\ r & rs & 0 \end{bmatrix}, \tag{2.36}$$

and substituting this into Eq. (2.23), and employing Eq. (2.34),

$$\begin{aligned} [\tilde{\varepsilon}] &= [l] \begin{bmatrix} 1 & 1 & 0 \\ 1 & 2 & 0 \\ 0 & 0 & 1 \end{bmatrix} [l]^T = \begin{bmatrix} p & pq & 0 \\ 0 & 0 & 1 \\ r & rs & 0 \end{bmatrix} \begin{bmatrix} 1 & 1 & 0 \\ 1 & 2 & 0 \\ 0 & 0 & 1 \end{bmatrix} \begin{bmatrix} p & 0 & r \\ pq & 0 & rs \\ 0 & 1 & 0 \end{bmatrix} \\ &= \begin{bmatrix} (3-\sqrt{5})/2 & 0 & 0 \\ 0 & 1 & 0 \\ 0 & 0 & (3+\sqrt{5})/2 \end{bmatrix} = \begin{bmatrix} \lambda_1 & 0 & 0 \\ 0 & \lambda_2 & 0 \\ 0 & 0 & \lambda_3 \end{bmatrix} \end{aligned} \tag{2.37}$$

as anticipated.

2.2.2.3 *Strain ellipsoid*

The deformation associated with the strain in a local region sufficiently small that the strain is essentially uniform, can be readily visualized by employing the principal coordinate system. Imagine a small volume in this local region bounded by a spherical surface of radius R obeying

$$\tilde{x}_1^2 + \tilde{x}_2^2 + \tilde{x}_3^2 = R^2 \tag{2.38}$$

before the application of the strain. In the principal coordinate system the $\tilde{x}_i$ will be transformed by the strain (see Eq. (2.9)) according to

$$\begin{bmatrix} \tilde{x}'_1 \\ \tilde{x}'_2 \\ \tilde{x}'_3 \end{bmatrix} = \begin{bmatrix} \tilde{x}_1 \\ \tilde{x}_2 \\ \tilde{x}_3 \end{bmatrix} + \begin{bmatrix} \tilde{\varepsilon}_{11} & 0 & 0 \\ 0 & \tilde{\varepsilon}_{22} & 0 \\ 0 & 0 & \tilde{\varepsilon}_{33} \end{bmatrix} \begin{bmatrix} \tilde{x}_1 \\ \tilde{x}_2 \\ \tilde{x}_3 \end{bmatrix} = \begin{bmatrix} \tilde{x}_1(1+\tilde{\varepsilon}_{11}) \\ \tilde{x}_2(1+\tilde{\varepsilon}_{22}) \\ \tilde{x}_3(1+\tilde{\varepsilon}_{33}) \end{bmatrix}. \tag{2.39}$$

Substitution of Eq. (2.39) into Eq. (2.38) then yields

$$\frac{(\tilde{x}'_1)^2}{R^2(1+\tilde{\varepsilon}_{11})^2} + \frac{(\tilde{x}'_2)^2}{R^2(1+\tilde{\varepsilon}_{22})^2} + \frac{(\tilde{x}'_3)^2}{R^2(1+\tilde{\varepsilon}_{33})^2} = 1. \tag{2.40}$$

The initially spherical volume is therefore converted into an ellipsoid, termed the *strain ellipsoid,* possessing semiaxes $\mathrm{R}(1+\tilde{\varepsilon}_{11})$, $\mathrm{R}(1+\tilde{\varepsilon}_{22})$, and $\mathrm{R}(1+\tilde{\varepsilon}_{33})$. The local fractional volume change due to the strain, $\tilde{\mathrm{e}}$, is given by

$$\begin{aligned}\tilde{\mathrm{e}} &= \frac{\mathrm{V}'-\mathrm{V}}{\mathrm{V}} = \frac{\mathrm{R}(1+\tilde{\varepsilon}_{11})\mathrm{R}(1+\tilde{\varepsilon}_{22})\mathrm{R}(1+\tilde{\varepsilon}_{33})-\mathrm{R}^3}{\mathrm{R}^3} \\ &= \tilde{\varepsilon}_{11}+\tilde{\varepsilon}_{22}+\tilde{\varepsilon}_{33} \end{aligned} \tag{2.41}$$

which, with the aid of Eq. (2.5), can also be expressed as

$$\tilde{\mathrm{e}} = \frac{\partial \tilde{\mathrm{u}}_1}{\partial \mathrm{x}_1} + \frac{\partial \tilde{\mathrm{u}}_2}{\partial \mathrm{x}_2} + \frac{\partial \tilde{\mathrm{u}}_3}{\partial \mathrm{x}_3} = \tilde{\nabla}\cdot\tilde{\mathbf{u}}. \tag{2.42}$$

The fractional volume change, $\tilde{\mathrm{e}}$, or *cubical dilatation,* is therefore equal to the trace of the strain tensor, $[\tilde{\varepsilon}]$, or, alternatively, the divergence of the displacement vector. An important feature of the cubical dilatation is that it is invariant with respect to the choice of coordinate system. With the aid of Eq. (2.24), and since $[l]$ is a unitary orthogonal matrix, it is readily verified that $(\varepsilon_{11}+\varepsilon_{22}+\varepsilon_{33}) = \mathrm{e} = (\varepsilon'_{11}+\varepsilon'_{22}+\varepsilon'_{33}) = \mathrm{e}' = \nabla\cdot\mathbf{u} = \nabla\cdot\mathbf{u}'$ for any choice of orthogonal coordinate system.

2.2.2.4 *Strain compatibility*

So far, expressions for the six strains of the strain tensor have been obtained by using Eq. (2.5) which entails taking the derivatives of three displacements, each of which is a continuous function of $(\mathrm{x}_1,\mathrm{x}_2,\mathrm{x}_3)$. Consider now the inverse problem of finding the three displacements when the six strains are specified without the use of Eq. (2.5). It can be seen immediately that three "compatible" displacement functions may not be obtained by integrating the strains. Here, "compatible" means that, if it is imagined that the unstrained body is initially diced up into an ensemble of adjoining cubical volume elements and then subjected to six strains, the volume elements may not fit back together again when an attempt is made to rejoin them: i.e., various gaps and mismatches may remain. Clearly, certain restrictions on the six strains must exist to ensure compatibility.[5] The equations expressing these constraints are known as the *equations of compatibility.*

[5] An example of six strains that are continuous functions of $(\mathrm{x}_1,\mathrm{x}_2,0)$, and are compatible only under certain conditions, is given in Exercise 2.4.

To find these equations consider a point in the body at $\mathbf{P}' = x_i'\hat{\mathbf{e}}_i$ and a second point some distance away at $\mathbf{P}'' = x_i''\hat{\mathbf{e}}_i$ where the displacements are $\mathbf{u}'$ and $\mathbf{u}''$, respectively. If there are no gaps or mismatches in the medium after straining, the difference $\mathbf{u}'' - \mathbf{u}'$ should be expressible as a line integral of $d\mathbf{u}$ along a curve C from $\mathbf{P}'$ to $\mathbf{P}''$ that is independent of the path. This line integral can be written as

$$u_j'' - u_j' = \int_C du_j \tag{2.43}$$

and, since $du_j = (\partial u_j/\partial x_k)dx_k$, and $\partial u_j/\partial x_k = \varepsilon_{jk} + \omega_{jk}$,

$$u_j'' - u_j' = \int_C \frac{\partial u_j}{\partial x_k} dx_k = \int_C \varepsilon_{jk} dx_k + \int_C \omega_{jk} dx_k. \tag{2.44}$$

The last integral in Eq. (2.44) can be integrated by parts to obtain

$$\int_C \omega_{jk} dx_k = (\omega_{jk}'' x_k'' - \omega_{jk}' x_k') - \int_C x_k \frac{\partial \omega_{jk}}{\partial x_l} dx_l. \tag{2.45}$$

Putting Eq. (2.45) into Eq. (2.44),

$$u_j'' - u_j' = (\omega_{jk}'' x_k'' - \omega_{jk}' x_k') + \int_C \left(\varepsilon_{jl} - x_k \frac{\partial \omega_{jk}}{\partial x_l} \right) dx_l. \tag{2.46}$$

Then, by differentiating Eq. (2.5),

$$\frac{\partial \omega_{jk}}{\partial x_l} = \frac{\partial \varepsilon_{lj}}{\partial x_k} - \frac{\partial \varepsilon_{lk}}{\partial x_j}, \tag{2.47}$$

and substituting this into Eq. (2.46),

$$u_j'' - u_j' = (\omega_{jk}'' x_k'' - \omega_{jk}' x_k') + \int_C U_{jl} dx_l, \tag{2.48}$$

where

$$U_{jl} = \varepsilon_{jl} - x_k \left(\frac{\partial \varepsilon_{lj}}{\partial x_k} - \frac{\partial \varepsilon_{lk}}{\partial x_j} \right). \tag{2.49}$$

For $u_j'' - u_j'$, given by Eq. (2.48), to be independent of the path, $U_{jl} dx_l$ must be an exact differential, thus requiring

$$\frac{\partial U_{ji}}{\partial x_l} = \frac{\partial U_{jl}}{\partial x_i}. \tag{2.50}$$

Substitution of Eq. (2.49) into Eq. (2.50) then yields

$$\left[\frac{\partial\varepsilon_{ji}}{\partial x_l} - \delta_{kl}\left(\frac{\partial\varepsilon_{ij}}{\partial x_k} - \frac{\partial\varepsilon_{ki}}{\partial x_j}\right) - \frac{\partial\varepsilon_{jl}}{\partial x_i} + \delta_{ki}\left(\frac{\partial\varepsilon_{lj}}{\partial x_k} - \frac{\partial\varepsilon_{kl}}{\partial x_j}\right)\right]$$

$$-x_k\left[\frac{\partial^2\varepsilon_{ij}}{\partial x_l\partial x_k} - \frac{\partial^2\varepsilon_{ki}}{\partial x_l\partial x_j} - \frac{\partial^2\varepsilon_{lj}}{\partial x_i\partial x_k} + \frac{\partial^2\varepsilon_{kl}}{\partial x_i\partial x_j}\right] = 0. \tag{2.51}$$

The first term in square brackets in Eq. (2.51) vanishes, and since x_k can be varied independently, the second term in square brackets must also vanish. Therefore, the strains must satisfy

$$\frac{\partial^2\varepsilon_{ij}}{\partial x_l\partial x_k} - \frac{\partial^2\varepsilon_{ki}}{\partial x_l\partial x_j} - \frac{\partial^2\varepsilon_{lj}}{\partial x_i\partial x_k} + \frac{\partial^2\varepsilon_{kl}}{\partial x_i\partial x_j} = 0. \tag{2.52}$$

Equation (2.52) embodies $3^4 = 81$ equations, some of which are satisfied identically and others are repeated because of symmetries in the four-indice groupings. Detailed examination shows that only the following six equations, known as the *equations of compatibility*, need be retained:

$$2\frac{\partial^2\varepsilon_{23}}{\partial x_2\partial x_3} - \frac{\partial^2\varepsilon_{22}}{\partial x_3^2} - \frac{\partial^2\varepsilon_{33}}{\partial x_2^2} = C_{11} = 0,$$

$$\frac{\partial^2\varepsilon_{33}}{\partial x_1\partial x_2} + \frac{\partial}{\partial x_3}\left(-\frac{\partial\varepsilon_{23}}{\partial x_1} - \frac{\partial\varepsilon_{13}}{\partial x_2} + \frac{\partial\varepsilon_{12}}{\partial x_3}\right) = C_{12} = 0,$$

$$2\frac{\partial^2\varepsilon_{13}}{\partial x_1\partial x_3} - \frac{\partial^2\varepsilon_{11}}{\partial x_3^2} - \frac{\partial^2\varepsilon_{33}}{\partial x_1^2} = C_{22} = 0,$$

$$\frac{\partial^2\varepsilon_{22}}{\partial x_1\partial x_3} + \frac{\partial}{\partial x_2}\left(-\frac{\partial\varepsilon_{23}}{\partial x_1} + \frac{\partial\varepsilon_{13}}{\partial x_2} - \frac{\partial\varepsilon_{12}}{\partial x_3}\right) = C_{13} = 0,$$

$$2\frac{\partial^2\varepsilon_{12}}{\partial x_1\partial x_2} - \frac{\partial^2\varepsilon_{11}}{\partial x_2^2} - \frac{\partial^2\varepsilon_{22}}{\partial x_1^2} = C_{33} = 0,$$

$$\frac{\partial^2\varepsilon_{11}}{\partial x_2\partial x_3} + \frac{\partial}{\partial x_1}\left(\frac{\partial\varepsilon_{23}}{\partial x_1} - \frac{\partial\varepsilon_{13}}{\partial x_2} - \frac{\partial\varepsilon_{12}}{\partial x_3}\right) = C_{23} = 0. \tag{2.53}$$

Alternatively, the above six equations can be expressed more compactly as elements of a symmetric *incompatibility tensor*, $\underline{\mathbf{C}}$, having components

$$C_{ij} = \frac{\partial^2 \varepsilon_{ij}}{\partial x_m \partial x_m} + \frac{\partial^2 \varepsilon_{mm}}{\partial x_i \partial x_j} - \frac{\partial^2 \varepsilon_{im}}{\partial x_j \partial x_m} - \frac{\partial^2 \varepsilon_{jm}}{\partial x_i \partial x_m}$$

$$-\left(\frac{\partial^2 \varepsilon_{mm}}{\partial x_n \partial x_n} - \frac{\partial^2 \varepsilon_{mn}}{\partial x_m \partial x_n}\right)\delta_{ij} = -e_{ikp}e_{jlq}\frac{\partial^2 \varepsilon_{pq}}{\partial x_k \partial x_l} = 0. \quad (2.54)$$

2.3 Traction Vector, Stress Tensor and Body Forces

Consider a stressed body, $\mathcal{V}^{\circ}$, containing within it a region, $\mathcal{V}$, enclosed by an internal surface, $\mathcal{S}$, as in Fig. 2.3. Region $\mathcal{V}$ will generally be subjected to two types of forces: forces exerted on its surface $\mathcal{S}$ by the outside medium, and internal body forces due, for example, to gravitational or magnetic fields.

2.3.1 *Traction vector and components of stress*

The force, $\Delta\mathbf{F}$, experienced by an element of the surface $\mathcal{S}$, such as ΔS in Fig. 2.3, depends upon the local stress field and the inclination of the surface element as indicated by its positive unit normal, $\hat{\mathbf{n}}$,[6] i.e.,

$$\Delta\mathbf{F} = \Delta\mathbf{F}(\hat{\mathbf{n}}). \quad (2.55)$$

The *traction vector* acting on ΔS is then defined as

$$\mathbf{T}(\hat{\mathbf{n}}) \equiv \lim_{\Delta S \to 0} \frac{\Delta\mathbf{F}(\hat{\mathbf{n}})}{\Delta S}. \quad (2.56)$$

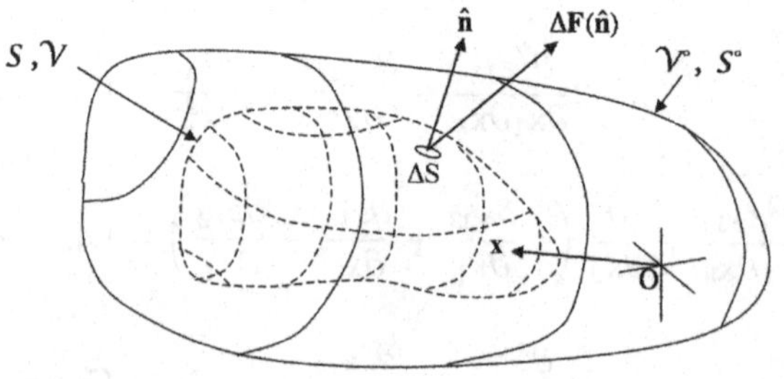

Fig. 2.3 Region, $\mathcal{V}$, (dashed) enclosed by the surface, $\mathcal{S}$, within larger stressed body, $\mathcal{V}^{\circ}$. Force $\Delta\mathbf{F}(\hat{\mathbf{n}})$ acts on surface element, ΔS, with positive unit normal $\hat{\mathbf{n}}$. Orthogonal coordinate system shown with field vector, $\mathbf{x}$.

[6]Throughout the present book the convention is adopted that positive unit normal vectors to closed surfaces point in the outward direction.

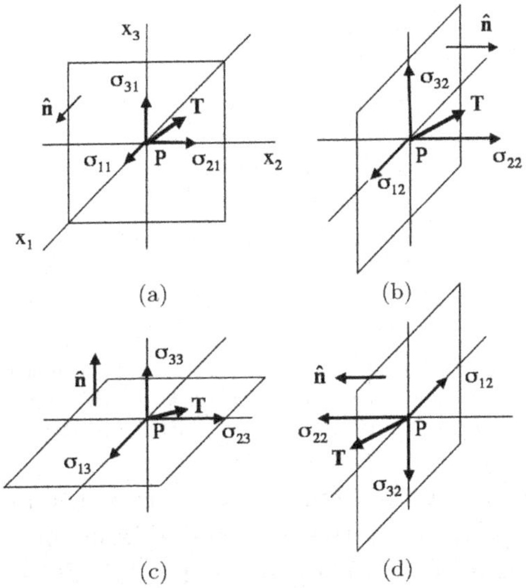

Fig. 2.4 (a–c) Three assumed traction vectors, $\mathbf{T}$, and nine components of stress, σ_{ij}, acting on the three surfaces with unit normal vectors, $\hat{\mathbf{n}} = \hat{\mathbf{e}}_1$, $\hat{\mathbf{n}} = \hat{\mathbf{e}}_2$ and $\hat{\mathbf{n}} = \hat{\mathbf{e}}_3$, respectively, at the point $P(\mathbf{x})$. Arrows indicate stress directions. (d) Same as (b) except that the outward unit normal vector to surface has been reversed.

Consider now the three planar surfaces corresponding to the three $x_i = \text{constant}_i$ planes passing through a point $P(\mathbf{x})$ within $\mathcal{V}$, illustrated in Fig. 2.4. Each of the three surfaces is subjected to a traction vector which, as shown below (Eq. (2.59), has components corresponding to three of the nine components, σ_{ij}, of a tensor, $\underline{\boldsymbol{\sigma}}$, termed the *stress tensor*, acting at the point P. The component normal to each plane is termed a *normal stress*, while the two components parallel to each plane are *shear stresses*. The accompanying arrow indicates its direction. The subscript i identifies the axis along which the stress is directed, and j identifies the axis that is normal to the plane on which it acts. A stress component acting on a plane whose outward normal is pointed along a positive axis direction is positive if it is directed along a positive axis direction. Otherwise, it is negative. When a component is acting on a plane whose outward normal is pointed along a negative axis direction the situation is reversed, and the component is positive when it is directed along a negative axis direction. A normal stress component is therefore positive when extensive and negative when compressive. All of the stress components indicated in Fig. 2.4 are seen to be positive and the normal stresses extensive.

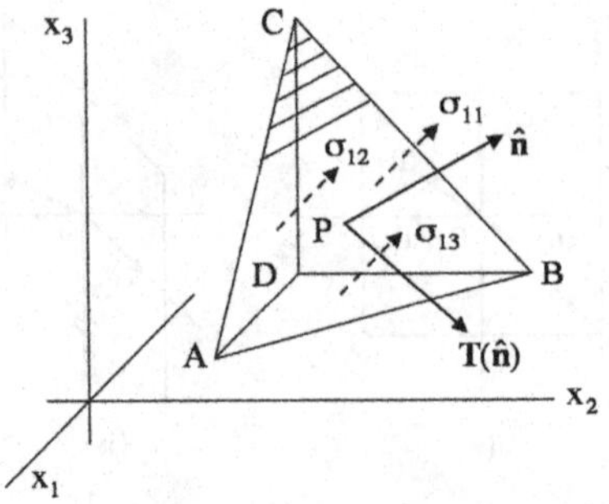

Fig. 2.5 Infinitesimal tetrahedral volume element embedded in stressed body at field point P(**x**). **T**($\hat{\mathbf{n}}$) is traction vector acting at P on face ABC with outward normal $\hat{\mathbf{n}}$. σ_{11}, σ_{12}, and σ_{13} stresses act on BCD, ACD and ABD faces, respectively.

To find the traction vector acting on a surface of arbitrary inclination at a point P in terms of the local stresses, consider the infinitesimal tetrahedral volume element centered on P illustrated in Fig. 2.5. The surface of interest is the front ABC face at the inclination $\hat{\mathbf{n}}$, and each of the three back faces is perpendicular to a coordinate axis. If the area of ABC is dS, the area of the back face with normal $\hat{\mathbf{e}}_1$ is $dS_1 = (\hat{\mathbf{n}} \cdot \hat{\mathbf{e}}_1)dS = \hat{n}_1 dS$. For mechanical equilibrium the sum of all forces acting on the tetrahedron parallel to the x_1 direction must vanish. The stress components anti-parallel to $\hat{\mathbf{e}}_1$ at the back faces are shown dashed in Fig. 2.5. The net force parallel to the x_1 direction is then

$$-\sigma_{11}\hat{n}_1 dS - \sigma_{12}\hat{n}_2 dS - \sigma_{13}\hat{n}_3 dS + T_1 dS = 0 \tag{2.57}$$

and, therefore,

$$T_1 = \sigma_{11}\hat{n}_1 + \sigma_{12}\hat{n}_2 + \sigma_{13}\hat{n}_3, \tag{2.58}$$

where all quantities refer to the point P(**x**). Similar results are obtained for the forces along x_2 and x_3, so that, in general,

$$T_i = \sigma_{ij}\hat{n}_j \quad \text{or} \quad [T] = [\sigma][\hat{n}] \tag{2.59}$$

and $\underline{\sigma}$ is seen as a second rank tensor mapping $\hat{\mathbf{n}}$ into **T**.

2.3.2 *Body forces*

Body forces, such as gravitational or magnetic forces, are generally represented by body force density distributions, $\mathbf{f}(x_1, x_2, x_3)$. Point forces can readily be represented by force distributions in the form of delta functions (see appendix D). The total body force imposed on $\mathcal{V}$ can then be expressed by the integral

$$F_i = \oiiint_{\mathcal{V}} f_i(\mathbf{x})dV. \tag{2.60}$$

2.3.3 *Relationships for stress components and body forces*

2.3.3.1 *Requirements for mechanical equilibrium*

The requirement of mechanical equilibrium imposes conditions on the stress components and body forces: the stress components are therefore not independent. Consider again the enclosed region $\mathcal{V}$ in Fig. 2.3 with body forces present. If $f_i(\mathbf{x})$ is the distribution of body force density, the total force on $\mathcal{V}$ along x_1 is

$$F_1 = \oiiint_{\mathcal{V}} f_1 dV + \oiint_{\mathcal{S}} T_1 dS, \tag{2.61}$$

where the first term sums the body forces over $\mathcal{V}$ and the second the tractions exerted on $\mathcal{S}$. Applying the divergence theorem to the surface integral,

$$\begin{aligned}\oiint_{\mathcal{S}} T_1 dS &= \oiint_{\mathcal{S}} (\sigma_{11}\hat{n}_1 + \sigma_{12}\hat{n}_2 + \sigma_{13}\hat{n}_3)dS \\ &= \oiiint_{\mathcal{V}} \left(\frac{\partial\sigma_{11}}{\partial x_1} + \frac{\partial\sigma_{12}}{\partial x_2} + \frac{\partial\sigma_{13}}{\partial x_3}\right)dV \end{aligned} \tag{2.62}$$

and putting the result into Eq. (2.61),

$$F_1 = \oiiint_{\mathcal{V}} \left(f_1 + \frac{\partial\sigma_{11}}{\partial x_1} + \frac{\partial\sigma_{12}}{\partial x_2} + \frac{\partial\sigma_{13}}{\partial x_3}\right)dV. \tag{2.63}$$

Since F_1 in Eq. (2.63) must vanish at equilibrium, and, since dV is arbitrary in $\mathcal{V}$, the condition

$$\frac{\partial\sigma_{11}}{\partial x_1} + \frac{\partial\sigma_{12}}{\partial x_2} + \frac{\partial\sigma_{13}}{\partial x_3} + f_1 = 0 \tag{2.64}$$

must apply. Similar considerations hold along x_2 and x_3, so that, in general,

$$\frac{\partial\sigma_{ij}(\mathbf{x})}{\partial x_j} + f_i(\mathbf{x}) = 0, \tag{2.65}$$

which is known as the *equation of equilibrium for the stresses and body forces.*

Further relationships between the stress components are obtained from the condition that the total moment due to the tractions on $\mathcal{S}$ and the distribution of body force density in $\mathcal{V}$ must vanish. Consider first the net moment around $\hat{e}_1$ given by

$$\oiiint_{\mathcal{V}} (f_3x_2 - f_2x_3)dV + \oiint_{\mathcal{S}} (T_3x_2 - T_2x_3)dS = 0. \qquad (2.66)$$

Using the divergence theorem and Eqs. (2.59) and (2.65) to convert the surface integral term in Eq. (2.66) to a volume integral,

$$\begin{aligned}\oiint_{\mathcal{S}} (T_3x_2 - T_2x_3)dS &= \oiiint_{\mathcal{V}} \left[\sigma_{32} - \sigma_{23} + x_2\left(\frac{\partial\sigma_{31}}{\partial x_1} + \frac{\partial\sigma_{32}}{\partial x_2} + \frac{\partial\sigma_{33}}{\partial x_3}\right)\right. \\ &\quad \left. - x_3\left(\frac{\partial\sigma_{21}}{\partial x_1} + \frac{\partial\sigma_{22}}{\partial x_2} + \frac{\partial\sigma_{23}}{\partial x_2}\right)\right] dV \\ &= \oiiint_{\mathcal{V}} (\sigma_{32} - \sigma_{23} - f_3x_2 + f_2x_3)dV. \end{aligned} \qquad (2.67)$$

Then, substituting Eq. (2.67) into Eq. (2.66),

$$\oiiint_{\mathcal{V}} (\sigma_{32} - \sigma_{23})dV = 0. \qquad (2.68)$$

Since dV is arbitrary within $\mathcal{V}$, the condition $\sigma_{32} = \sigma_{23}$ must be satisfied. Similar considerations of the moments about $\hat{e}_2$ and $\hat{e}_3$ show that $\sigma_{13} = \sigma_{31}$ and $\sigma_{12} = \sigma_{21}$ so that, in general,

$$\sigma_{ij} = \sigma_{ji}. \qquad (2.69)$$

The stress tensor must therefore be symmetric. This reduces the number of stress components which must be specified to define the state of stress at a point from nine to six, i.e., to the three normal stresses, σ_{11}, σ_{22} and σ_{33}, and the three shear stresses, $\sigma_{12} = \sigma_{21}$, $\sigma_{13} = \sigma_{31}$ and $\sigma_{23} = \sigma_{32}$.

2.3.3.2 *Transformation of stress components due to rotation of coordinate system*

Equation (2.59) shows that the matrix $[\sigma]$ maps one vector into another, thus establishing $\underline{\boldsymbol{\sigma}}$ as a second rank tensor. The transformation laws for the stress components are therefore of the same form as those for the strain components given by Eqs. (2.24) and (2.25), i.e.,

$$\sigma'_{ij} = l_{im}l_{jn}\sigma_{mn} \quad \sigma_{ij} = l_{mi}l_{nj}\sigma'_{mn}. \qquad (2.70)$$

If both sides of the two expressions in Eq. (2.70) are summed over the three normal stresses, it is found that, since $[l]$ is an orthogonal unitary matrix, the sum of the normal stresses, Θ, in each coordinate system is identical, i.e.,

$$\sigma'_{11} + \sigma'_{22} + \sigma'_{33} = \sigma_{11} + \sigma_{22} + \sigma_{33} = \Theta. \tag{2.71}$$

The result that Θ, the trace of the stress tensor, is invariant is analogous to the result in Sec. 2.2.2.3 that the sum of the normal strains, i.e. the cubical dilatation, is an invariant of the strain tensor.

2.3.3.3 *Principal coordinate system for stress tensor*

Since the stress tensor is a second rank tensor, a principal coordinate system can be found by using the same general procedure employed in Sec. 2.2.2.2 to obtain the principal coordinate system for the strain tensor. To obtain this coordinate system, start with Eq. (2.59) which yields the traction exerted on a plane with normal $\hat{\mathbf{n}}$ due to the stress tensor. In the principal coordinate system, where the stress tensor is diagonalized and the diagonal elements correspond to the principal stresses, the tractions on the three planes normal to the principal axes therefore consist only of the principal normal stresses, i.e. no shear stresses are exerted on these planes. Using Eq. (2.59), this condition is therefore

$$\begin{bmatrix} \mathrm{T}_1 \\ \mathrm{T}_2 \\ \mathrm{T}_3 \end{bmatrix} = \begin{bmatrix} \sigma_{11} & \sigma_{12} & \sigma_{13} \\ \sigma_{12} & \sigma_{22} & \sigma_{23} \\ \sigma_{13} & \sigma_{23} & \sigma_{33} \end{bmatrix} \begin{bmatrix} \hat{\mathrm{n}}_1 \\ \hat{\mathrm{n}}_2 \\ \hat{\mathrm{n}}_3 \end{bmatrix} = \begin{bmatrix} \lambda\hat{\mathrm{n}}_1 \\ \lambda\hat{\mathrm{n}}_2 \\ \lambda\hat{\mathrm{n}}_3 \end{bmatrix}, \tag{2.72}$$

or,

$$\begin{aligned} (\sigma_{11} - \lambda)\hat{\mathrm{n}}_1 + \sigma_{12}\hat{\mathrm{n}}_2 + \sigma_{13}\hat{\mathrm{n}}_3 &= 0, \\ \sigma_{12}\hat{\mathrm{n}}_1 + (\sigma_{22} - \lambda)\hat{\mathrm{n}}_2 + \sigma_{23}\hat{\mathrm{n}}_3 &= 0, \\ \sigma_{13}\hat{\mathrm{n}}_1 + \sigma_{23}\hat{\mathrm{n}}_2 + (\sigma_{33} - \lambda)\hat{\mathrm{n}}_3 &= 0, \end{aligned} \tag{2.73}$$

which may be compared to Eq. (2.29) for the strain tensor. Then, following the same procedure used for the strain tensor in Sec. 2.2.2.2, the three eigenvalues, λ_i, and corresponding eigenvectors can be obtained, and the principal coordinate system can be constructed by employing the eigenvectors. The stress tensor then assumes the diagonal form

$$[\tilde{\sigma}] = \begin{bmatrix} \lambda_1 & 0 & 0 \\ 0 & \lambda_2 & 0 \\ 0 & 0 & \lambda_3 \end{bmatrix} = \begin{bmatrix} \tilde{\sigma}_{11} & 0 & 0 \\ 0 & \tilde{\sigma}_{22} & 0 \\ 0 & 0 & \tilde{\sigma}_{33} \end{bmatrix} \tag{2.74}$$

and the tractions on the three planes normal to the principal axes correspond, respectively, to the three principal stresses, $\tilde{\sigma}_{\alpha\alpha}$.

In Exercise 2.9 it is shown that the principal coordinate systems for the stress and strain tensors are identical for isotropic systems but not for low-symmetry anisotropic systems.

2.4 Linear Coupling of Stress and Strain

2.4.1 *Stress as a function of strain*

For small elastic displacements it is expected that the deformation (strain) will vary in proportion to the force (stress) in accordance with Hooke's Law. The most general linear relationship for stress as a function of strain is then

$$\sigma_{ij} = C_{ijkl}\varepsilon_{kl} \tag{2.75}$$

which, when written in matrix form, appears as

$$\begin{bmatrix} \sigma_{11} \\ \sigma_{22} \\ \sigma_{33} \\ \sigma_{23} \\ \sigma_{31} \\ \sigma_{12} \\ \sigma_{32} \\ \sigma_{13} \\ \sigma_{21} \end{bmatrix} = \begin{bmatrix} C_{1111} & C_{1122} & C_{1133} & C_{1123} & C_{1131} & C_{1112} & C_{1132} & C_{1113} & C_{1121} \\ C_{2211} & C_{2222} & C_{2233} & C_{2223} & C_{2231} & C_{2212} & C_{2232} & C_{2213} & C_{2221} \\ C_{3311} & C_{3322} & C_{3333} & C_{3323} & C_{3331} & C_{3312} & C_{3332} & C_{3313} & C_{3321} \\ C_{2311} & C_{2322} & C_{2333} & C_{2323} & C_{2331} & C_{2312} & C_{2332} & C_{2313} & C_{2321} \\ C_{3111} & C_{3122} & C_{3133} & C_{3123} & C_{3131} & C_{3112} & C_{3132} & C_{3113} & C_{3121} \\ C_{1211} & C_{1222} & C_{1233} & C_{1223} & C_{1231} & C_{1212} & C_{1232} & C_{1213} & C_{1221} \\ C_{3211} & C_{3222} & C_{3233} & C_{3223} & C_{3231} & C_{3212} & C_{3232} & C_{3213} & C_{3221} \\ C_{1311} & C_{1322} & C_{1333} & C_{1323} & C_{1331} & C_{1312} & C_{1332} & C_{1313} & C_{1321} \\ C_{2111} & C_{2122} & C_{2133} & C_{2123} & C_{2131} & C_{2112} & C_{2132} & C_{2113} & C_{2121} \end{bmatrix}$$

$$\times \begin{bmatrix} \varepsilon_{11} \\ \varepsilon_{22} \\ \varepsilon_{33} \\ \varepsilon_{23} \\ \varepsilon_{31} \\ \varepsilon_{12} \\ \varepsilon_{32} \\ \varepsilon_{13} \\ \varepsilon_{21} \end{bmatrix} \tag{2.76}$$

The 81 constant coefficients, C_{ijkl}, are termed *elastic stiffnesses* and constitute the components of a fourth rank tensor, $\underline{\underline{C}}$, that maps the second rank strain tensor $\underline{\varepsilon}$ into the second rank stress tensor $\underline{\sigma}$.[7] The number of elastic stiffnesses that are required is lower than the above 81 because of existing symmetries. Since the stress and strain components are symmetric, it follows that the C_{ijkl} must have the symmetry properties

$$C_{ijkl} = C_{jikl} = C_{ijlk} = C_{jilk}. \tag{2.77}$$

A further symmetry property is obtained by examining the Helmholtz free energy, F, of an elastically strained body. The change of internal energy of any system due to a reversible exchange of heat and work with the environment at constant temperature is given by the combined first and second law of thermodynamics expression

$$dU = \delta Q - dW = TdS - d\mathcal{W} \tag{2.78}$$

where U = internal energy, Q = heat, $\mathcal{W}$ = work performed by the system, T = temperature and S = entropy. Equation (2.132) shows that the work (per unit volume) required to elastically strain a body homogeneously is $\sigma_{ij}d\varepsilon_{ij}$. Therefore, for a strained body of volume V, the work term in Eq. (2.78) is of the form

$$d\mathcal{W} = -V\sigma_{ij}d\varepsilon_{ij} \tag{2.79}$$

and the change in Helmholtz free energy per unit volume, at constant temperature, is therefore

$$df_T = du - Tds = \sigma_{ij}d\varepsilon_{ij} = C_{ijkl}\varepsilon_{kl}d\varepsilon_{ij} \tag{2.80}$$

after employing Eq. (2.75). Then, differentiating Eq. (2.80)

$$\frac{\partial^2 f_T}{\partial\varepsilon_{ij}\partial\varepsilon_{kl}} = C_{ijkl}. \tag{2.81}$$

---

[7]In general, a fourth rank tensor maps a second rank tensor into another second rank tensor (Nye 1957).

Since f_T is a state function, and therefore a perfect differential, the left side of Eq. (2.81) is unchanged and symmetrical with respect to the interchange of ij and kl. The symmetry relationship

$$C_{ijkl} = C_{klij} \tag{2.82}$$

must therefore apply. By applying the above symmetry properties of C_{ijkl}, and those of σ_{ij} and ε_{ij}, to Eq. (2.76), it reduces to

$$\begin{bmatrix} \sigma_{11} \\ \sigma_{22} \\ \sigma_{33} \\ \sigma_{23} \\ \sigma_{31} \\ \sigma_{12} \\ \sigma_{23} \\ \sigma_{31} \\ \sigma_{12} \end{bmatrix} = \begin{bmatrix} C_{1111} & C_{1122} & C_{1133} & C_{1123} & C_{1131} & C_{1112} & C_{1123} & C_{1131} & C_{1112} \\ C_{2211} & C_{2222} & C_{2233} & C_{2223} & C_{2231} & C_{2212} & C_{2223} & C_{2231} & C_{2212} \\ C_{3311} & C_{3322} & C_{3333} & C_{3323} & C_{3331} & C_{3312} & C_{3323} & C_{3331} & C_{3312} \\ C_{2311} & C_{2322} & C_{2333} & C_{2323} & C_{2331} & C_{2312} & C_{2323} & C_{2331} & C_{2312} \\ C_{3111} & C_{3122} & C_{3133} & C_{3123} & C_{3131} & C_{3112} & C_{3123} & C_{3131} & C_{3112} \\ C_{1211} & C_{1222} & C_{3233} & C_{1223} & C_{1231} & C_{1212} & C_{1223} & C_{1231} & C_{1212} \\ C_{2311} & C_{2322} & C_{2333} & C_{2323} & C_{2331} & C_{2312} & C_{2323} & C_{2331} & C_{2312} \\ C_{3111} & C_{3122} & C_{3133} & C_{3123} & C_{3131} & C_{3112} & C_{3123} & C_{3131} & C_{3112} \\ C_{1211} & C_{1222} & C_{1233} & C_{1223} & C_{1231} & C_{1212} & C_{1223} & C_{1231} & C_{1212} \end{bmatrix}$$

$$\times \begin{bmatrix} \varepsilon_{11} \\ \varepsilon_{22} \\ \varepsilon_{33} \\ \varepsilon_{23} \\ \varepsilon_{31} \\ \varepsilon_{12} \\ \varepsilon_{23} \\ \varepsilon_{31} \\ \varepsilon_{12} \end{bmatrix} \tag{2.83}$$

which contains only 21 independent elastic constants.

The transformation law for the components of a fourth rank tensor, such as $\underline{\underline{C}}$ in Eq. (2.83), due to a rotation of its coordinate system can now be found by using the transformation law for second rank tensors. Substituting Eq. (2.25) into Eq. (2.75),

$$\sigma_{gh} = C_{ghmn}\varepsilon_{mn} = C_{ghmn} l_{km} l_{ln} \varepsilon'_{kl}. \tag{2.84}$$

Then, multiplying Eq. (2.84) throughout by $l_{ig}l_{jh}$ and using Eq. (2.70),

$$l_{ig}l_{jh}\sigma_{gh} = l_{ig}l_{jh}C_{ghmn}l_{km}l_{ln}\varepsilon'_{kl} = \sigma'_{ij}. \tag{2.85}$$

However, in the new system

$$\sigma'_{ij} = C'_{ijkl}\varepsilon'_{kl} \tag{2.86}$$

and, comparing Eqs. (2.85) and (2.86), the transformation law is found to be

$$C'_{ijkl} = l_{ig}l_{jh}C_{ghmn}l_{km}l_{ln}. \tag{2.87}$$

An application of this law is carried out in Exercise 2.8.

Equation (2.83) can be written more compactly by eliminating redundancies and employing *contracted notation* (see Nye 1957) in which the C_{ijkl} components are written in the form C_{mn}, where m and n correspond to the paired indices ij and kl, respectively, according to the scheme

$$\begin{array}{lllllll} ij \text{ or } kl = & 11 & 22 & 33 & 23,32 & 13,31 & 12,21 \\ m \text{ or } n = & 1 & 2 & 3 & 4 & 5 & 6 \end{array} \tag{2.88}$$

and the σ_{ij} and ε_{ij} components, which appear in the 9×1 column matrices, are transformed according to

$$\begin{bmatrix} \sigma_{11} & \sigma_{12} & \sigma_{13} \\ \sigma_{12} & \sigma_{22} & \sigma_{23} \\ \sigma_{13} & \sigma_{23} & \sigma_{33} \end{bmatrix} \rightarrow \begin{bmatrix} \sigma_1 & \sigma_6 & \sigma_5 \\ \sigma_6 & \sigma_2 & \sigma_4 \\ \sigma_5 & \sigma_4 & \sigma_3 \end{bmatrix}$$

$$\begin{bmatrix} \varepsilon_{11} & \varepsilon_{12} & \varepsilon_{13} \\ \varepsilon_{12} & \varepsilon_{22} & \varepsilon_{23} \\ \varepsilon_{13} & \varepsilon_{23} & \varepsilon_{33} \end{bmatrix} \rightarrow \begin{bmatrix} \varepsilon_1 & \varepsilon_6/2 & \varepsilon_5/2 \\ \varepsilon_6/2 & \varepsilon_2 & \varepsilon_4/2 \\ \varepsilon_5/2 & \varepsilon_4/2 & \varepsilon_3 \end{bmatrix} \tag{2.89}$$

Therefore, for example, $C_{1122} \rightarrow C_{12}$, $C_{2312} \rightarrow C_{46}$, $\sigma_{33} \rightarrow \sigma_3$, $\sigma_{13} \rightarrow \sigma_5$, $\varepsilon_{22} \rightarrow \varepsilon_2$ and $\varepsilon_{21} \rightarrow \varepsilon_6/2$.

By applying these contraction rules to Eq. (2.83), we can write it in the reduced 6 × 6 form,

$$\begin{bmatrix}\sigma_1\\ \sigma_2\\ \sigma_3\\ \sigma_4\\ \sigma_5\\ \sigma_6\end{bmatrix} = \begin{bmatrix} C_{1111} & C_{1122} & C_{1133} & 2C_{1123} & 2C_{1131} & 2C_{1112}\\ C_{1122} & C_{2222} & C_{2233} & 2C_{2223} & 2C_{2231} & 2C_{2212}\\ C_{1133} & C_{2233} & C_{3333} & 2C_{3323} & 2C_{3331} & 2C_{3312}\\ C_{1123} & C_{2223} & C_{3323} & 2C_{2323} & 2C_{2331} & 2C_{2312}\\ C_{1131} & C_{2231} & C_{3331} & 2C_{2331} & 2C_{3131} & 2C_{3112}\\ C_{1112} & C_{2212} & C_{3312} & 2C_{2312} & 2C_{3112} & 2C_{1212}\end{bmatrix}\begin{bmatrix}\varepsilon_1\\ \varepsilon_2\\ \varepsilon_3\\ \varepsilon_4/2\\ \varepsilon_5/2\\ \varepsilon_6/2\end{bmatrix}$$

$$= \begin{bmatrix} C_{11} & C_{12} & C_{13} & C_{14} & C_{15} & C_{16}\\ C_{12} & C_{22} & C_{23} & C_{24} & C_{25} & C_{26}\\ C_{13} & C_{23} & C_{33} & C_{34} & C_{35} & C_{36}\\ C_{14} & C_{24} & C_{34} & C_{44} & C_{45} & C_{46}\\ C_{15} & C_{25} & C_{35} & C_{45} & C_{55} & C_{56}\\ C_{16} & C_{26} & C_{36} & C_{46} & C_{56} & C_{66}\end{bmatrix}\begin{bmatrix}\varepsilon_1\\ \varepsilon_2\\ \varepsilon_3\\ \varepsilon_4\\ \varepsilon_5\\ \varepsilon_6\end{bmatrix}. \qquad (2.90)$$

In the first expression, Eq. (2.83) has been consolidated, and the rules given by Eq. (2.89) have been applied. The second expression is then the result of applying the rules given by Eq. (2.88). Finally, this result shows that stress, as a linear function of strain, can be expressed, using contracted notation, in the relatively simple index form

$$\sigma_i = C_{ij}\varepsilon_j \quad (C_{ij} = C_{ji}) \quad (i, j = 1, 2 \ldots 6) \qquad (2.91)$$

in which, however, 21 independent elastic constants still remain.

The existence of symmetry elements in the crystalline medium to which the C_{ijkl} tensor applies further reduces the number of independent elastic constants in a manner that depends upon the extent of the symmetry. For triclinic crystals, which possess only a center of symmetry, or no symmetry at all, the number remains at 21, while for cubic crystals, with their relatively large number of symmetry elements, the number is reduced to three. As described by Nye (1957), the effects of symmetry can be determined in a systematic manner by transforming the coordinate system used for the elastic constant tensor according to each existing symmetry element operation and requiring that the tensor remain unchanged. For example, for cubic crystals referred to an orthogonal coordinate system based on the axes of the standard cubic unit cell, the tensor must be invariant to three-fold rotation of the coordinate system around the four diagonals of the cell. The procedure is tedious, and, since a detailed description is given by Nye

(1957), it will not be reproduced here. As an example, the result obtained in Eq. (2.90) reduces for the case of a cubic crystal, in a coordinate system corresponding to the cubic crystal axes, to the simpler form

$$\begin{bmatrix} \sigma_1 \\ \sigma_2 \\ \sigma_3 \\ \sigma_4 \\ \sigma_5 \\ \sigma_6 \end{bmatrix} = \begin{bmatrix} C_{11} & C_{12} & C_{12} & 0 & 0 & 0 \\ C_{12} & C_{11} & C_{12} & 0 & 0 & 0 \\ C_{12} & C_{12} & C_{11} & 0 & 0 & 0 \\ 0 & 0 & 0 & C_{44} & 0 & 0 \\ 0 & 0 & 0 & 0 & C_{44} & 0 \\ 0 & 0 & 0 & 0 & 0 & C_{44} \end{bmatrix} \begin{bmatrix} \varepsilon_1 \\ \varepsilon_2 \\ \varepsilon_3 \\ \varepsilon_4 \\ \varepsilon_5 \\ \varepsilon_6 \end{bmatrix}. \tag{2.92}$$

2.4.2 *Strain as a function of stress*

To obtain the strain as a linear function of stress, Eq. (2.75) is inverted so that

$$\varepsilon_{ij} = S_{ijkl}\sigma_{kl} \tag{2.93}$$

where the 81 S_{ijkl} coefficients are termed *elastic compliances.* Just as for the elastic stiffnesses, the S_{ijkl} compliances are components of a fourth rank tensor, and, therefore obey the same transformation law as the C_{ijkl} tensor components. The S_{ijkl} and C_{ijkl} tensors have the same symmetry properties, and a contracted 6×6 matrix equation for the strains, which contains 21 independent compliances, and is analogous to the result given by Eq. (2.91) for the stresses, can therefore be written (as shown in Exercise 2.6) in the form

$$\varepsilon_i = S_{ij}\sigma_j \quad (S_{ij} = S_{ji}) \quad (i, j = 1, 2 \ldots 6) \tag{2.94}$$

after using the contraction rules given by Eqs. (2.88) and (2.89) along with the additional rules,

$$\begin{aligned} S_{ijkl} &= S_{mn} \text{ if m and n are independently 1, 2, or 3} \\ 2S_{ijkl} &= S_{mn} \text{ if either (but not both) m or n is 4, 5, or 6} \\ 4S_{ijkl} &= S_{mn} \text{ if m and n are independently 4, 5, or 6} \end{aligned} \tag{2.95}$$

Crystal symmetry further reduces the number of independent S_{ij} matrix elements in the same manner as previously for the C_{ij} elements. For example, for a cubic crystal, in a coordinate system that again corresponds to the crystal cubic axes, we have

$$\begin{bmatrix} \varepsilon_1 \\ \varepsilon_2 \\ \varepsilon_3 \\ \varepsilon_4 \\ \varepsilon_5 \\ \varepsilon_6 \end{bmatrix} = \begin{bmatrix} S_{11} & S_{12} & S_{12} & 0 & 0 & 0 \\ S_{12} & S_{11} & S_{12} & 0 & 0 & 0 \\ S_{12} & S_{12} & S_{11} & 0 & 0 & 0 \\ 0 & 0 & 0 & S_{44} & 0 & 0 \\ 0 & 0 & 0 & 0 & S_{44} & 0 \\ 0 & 0 & 0 & 0 & 0 & S_{44} \end{bmatrix} \begin{bmatrix} \sigma_1 \\ \sigma_2 \\ \sigma_3 \\ \sigma_4 \\ \sigma_5 \\ \sigma_6 \end{bmatrix} \tag{2.96}$$

which may be compared with Eq. (2.92).

Relationships between the S_{ij} and C_{ij} matrix elements are readily found by first writing Eqs. (2.91) and (2.94) in the matrix forms

$$[\sigma] = [C][\varepsilon] \quad [\varepsilon] = [S][\sigma]. \tag{2.97}$$

Then, by substituting the first expression into the second we have,[8]

$$[C][S] = [S][C] = [I] \quad [S] = [C]^{-1} \quad [C] = [S]^{-1} \tag{2.98}$$

so that, in contracted matrix component form,

$$C_{ij}S_{jk} = S_{ij}C_{jk} = \delta_{ik} \quad (i, j, k = 1, 2 \ldots 6). \tag{2.99}$$

Using Eqs. (2.92), (2.96) and (2.98), the relationships between the three non-vanishing C_{ij} matrix elements and corresponding S_{ij} elements for a cubic crystal are then (see Exercise 2.5),

$$C_{11} = \frac{S_{11} + S_{12}}{(S_{11} - S_{12})(S_{11} + 2S_{12})} \quad C_{12} = \frac{-S_{12}}{(S_{11} - S_{12})(S_{11} + 2S_{12})}$$
$$C_{44} = \frac{1}{S_{44}} \tag{2.100}$$

and,

$$S_{11} = \frac{C_{11} + C_{12}}{(C_{11} - C_{12})(C_{11} + 2C_{12})} \quad S_{12} = \frac{-C_{12}}{(C_{11} - C_{12})(C_{11} + 2C_{12})}$$
$$S_{44} = \frac{1}{C_{44}}. \tag{2.101}$$

[8] Note that the 6 × 6 determinants of both [C] and [S] do not vanish, since the strain energy density (as given by Eq. (2.134)) is positive definite. No problems therefore arise in determining their inverses.

2.4.3 "Corresponding" elastic fields

If two elastic fields, A and B, have stresses and strains coupled by the same elastic constants in a homogeneous region, then

$$\sigma_{ij}^{A}\varepsilon_{ij}^{B} = \sigma_{ij}^{B}\varepsilon_{ij}^{A}. \tag{2.102}$$

This can be verified by substituting Hooke's law and using the symmetry property, $C_{ijmn} = C_{mnij}$, of the C_{ijmn} tensor, i.e.,

$$\sigma_{ij}^{A}\varepsilon_{ij}^{B} = C_{ijmn}\varepsilon_{mn}^{A}\varepsilon_{ij}^{B} = C_{mnij}\varepsilon_{mn}^{A}\varepsilon_{ij}^{B} = C_{ijmn}\varepsilon_{ij}^{A}\varepsilon_{mn}^{B} = \sigma_{ij}^{B}\varepsilon_{ij}^{A}. \tag{2.103}$$

If the ε_{ij}^{A} and ε_{ij}^{B} strains are both compatible, and, therefore, can be expressed by Eq. (2.5) (see Exercise 2.10), it follows that

$$\begin{aligned}
\sigma_{ij}^{A}\varepsilon_{ij}^{B} &= \sigma_{ij}^{A}\frac{1}{2}\left(\frac{\partial u_i^B}{\partial x_j} + \frac{\partial u_j^B}{\partial x_i}\right) = \sigma_{ij}^{A}\frac{\partial u_i^B}{\partial x_j} \\
\sigma_{ij}^{B}\varepsilon_{ij}^{A} &= \sigma_{ij}^{B}\frac{1}{2}\left(\frac{\partial u_i^A}{\partial x_j} + \frac{\partial u_j^A}{\partial x_i}\right) = \sigma_{ij}^{B}\frac{\partial u_i^A}{\partial x_j}
\end{aligned} \tag{2.104}$$

and, if the σ_{ij}^{A} and σ_{ij}^{B} stresses both obey the equations of equilibrium, i.e., Eq. (2.65),

$$\sigma_{ij}^{A}\frac{\partial u_i^B}{\partial x_j} = \frac{\partial}{\partial x_j}\left(\sigma_{ij}^{A}u_i^B\right) + f_i^A u_i^B \qquad \sigma_{ij}^{B}\frac{\partial u_i^A}{\partial x_j} = \frac{\partial}{\partial x_j}\left(\sigma_{ij}^{B}u_i^A\right) + f_i^B u_i^A. \tag{2.105}$$

Finally, if all of the above conditions are satisfied,

$$\begin{aligned}
\sigma_{ij}^{A}\varepsilon_{ij}^{B} &= \sigma_{ij}^{B}\varepsilon_{ij}^{A} = \sigma_{ij}^{A}\frac{\partial u_i^B}{\partial x_j} = \sigma_{ij}^{B}\frac{\partial u_i^A}{\partial x_j} \\
&= \frac{\partial}{\partial x_j}\left(\sigma_{ij}^{A}u_i^B\right) + f_i^A u_i^B = \frac{\partial}{\partial x_j}\left(\sigma_{ij}^{B}u_i^A\right) + f_i^B u_i^A
\end{aligned} \tag{2.106}$$

and the A and B fields are termed *corresponding fields.*

Further useful relationships involving corresponding A and B fields can be obtained by first forming the vector

$$v_j = \sigma_{ij}^{A}u_i^B - \sigma_{ij}^{B}u_i^A. \tag{2.107}$$

Then, if the A and B fields are corresponding fields everywhere in a region, $\mathcal{V}$, which is enclosed by the surface $\mathcal{S}$, and is embedded in a larger body, $\mathcal{V}^{\circ}$, as in Fig. 2.3, the integral relationship

$$\oiint_{\mathcal{S}} v_j \hat{n}_j dS = \oiint_{\mathcal{S}} (\sigma_{ij}^{A} u_i^{B} - \sigma_{ij}^{B} u_i^{A}) \hat{n}_j dS = 0 \tag{2.108}$$

is valid. This can be demonstrated by converting the surface integral to a volume integral and using Eqs. (2.65) and (2.106), so that

$$\begin{aligned} \oiint_{\mathcal{S}} (\sigma_{ij}^{A} u_i^{B} - \sigma_{ij}^{B} u_i^{A}) \hat{n}_j dS &= \oiiint_{\mathcal{V}} \frac{\partial}{\partial x_j} (\sigma_{ij}^{A} u_i^{B} - \sigma_{ij}^{B} u_i^{A}) dV \\ &= \oiiint_{\mathcal{V}} \left(\sigma_{ij}^{A} \frac{\partial u_i^{B}}{\partial x_j} - \sigma_{ij}^{B} \frac{\partial u_i^{A}}{\partial x_j} \right) dV = 0. \end{aligned} \tag{2.109}$$

Suppose next that $\mathcal{S}^{(2)}$ is a closed surface in $\mathcal{V}^{\circ}$, and $\mathcal{S}^{(1)}$ is a second closed surface, lying within $\mathcal{S}^{(2)}$, that can be obtained by continuously distorting $S^{(2)}$ without sweeping through any region of the body containing singularities where the A and B systems are not corresponding fields. Then,[9]

$$\begin{aligned} \oiint_{\mathcal{S}^{(2)}-\mathcal{S}^{(1)}} (\sigma_{ij}^{A} u_i^{B} - \sigma_{ij}^{B} u_i^{A}) \hat{n}_j dS &= \oiint_{\mathcal{S}^{(2)}} (\sigma_{ij}^{A} u_i^{B} - \sigma_{ij}^{B} u_i^{A}) \hat{n}_j dS \\ &\quad - \oiint_{\mathcal{S}^{(1)}} (\sigma_{ij}^{A} u_i^{B} - \sigma_{ij}^{B} u_i^{A}) \hat{n}_j dS = 0 \end{aligned} \tag{2.110}$$

where $\mathcal{S}^{(2)} - \mathcal{S}^{(1)}$ indicates integration over the surfaces $\mathcal{S}^{(2)}$ and $\mathcal{S}^{(1)}$ bounding the swept out volume $\mathcal{V}^{(2)} - \mathcal{V}^{(1)}$. Equation (2.110) can be validated by converting the surface integral to a volume integral and employing Eqs. (2.65) and (2.106), i.e.,

$$\begin{aligned} \oiint_{\mathcal{S}^{(2)}-\mathcal{S}^{(1)}} (\sigma_{ij}^{A} u_i^{B} - \sigma_{ij}^{B} u_i^{A}) \hat{n}_j dS &= \oiiint_{\mathcal{V}^{(2)}-\mathcal{V}^{(1)}} \frac{\partial}{\partial x_j} (\sigma_{ij}^{A} u_i^{B} - \sigma_{ij}^{B} u_i^{A}) dV \\ &= \oiiint_{\mathcal{V}^{(2)}-\mathcal{V}^{(1)}} \left(\sigma_{ij}^{A} \frac{\partial u_i^{B}}{\partial x_j} - \sigma_{ij}^{B} \frac{\partial u_i^{A}}{\partial x_j} \right) dV \\ &= 0. \end{aligned} \tag{2.111}$$

[9]Note that for Eq. (2.110), and also Eq. (2.113), to be valid it is only necessary that A and B be corresponding fields within the region between $\mathcal{S}^{(2)}$ and $\mathcal{S}^{(1)}$.

In addition, if $\mathcal{S}$ is again a closed surface enclosing a region, $\mathcal{V}$, containing A and B fields which are corresponding everywhere, the integral relationship

$$\oiint_{\mathcal{S}} \left(\frac{\partial \sigma_{ij}^{A}}{\partial x_l} u_i^{B} - \sigma_{ij}^{B} \frac{\partial u_i^{A}}{\partial x_l} \right) \hat{n}_j dS = 0 \tag{2.112}$$

is valid as shown below, and, consequently, the relationship

$$\oiint_{\mathcal{S}^{(2)}-\mathcal{S}^{(1)}} \left(\frac{\partial \sigma_{ij}^{A}}{\partial x_l} u_i^{B} - \sigma_{ij}^{B} \frac{\partial u_i^{A}}{\partial x_l} \right) \hat{n}_j dS = 0 \tag{2.113}$$

is also valid. Equation (2.112) can by verified by converting the surface integral to a volume integral and using Eq. (2.65) so that

$$\begin{aligned} \oiint_{\mathcal{S}} \left(\frac{\partial \sigma_{ij}^{A}}{\partial x_l} u_i^{B} - \sigma_{ij}^{B} \frac{\partial u_i^{A}}{\partial x_l} \right) \hat{n}_j dS &= \iiint_{\mathcal{V}} \frac{\partial}{\partial x_j} \left(\frac{\partial \sigma_{ij}^{A}}{\partial x_l} u_i^{B} - \sigma_{ij}^{B} \frac{\partial u_i^{A}}{\partial x_l} \right) dV \\ &= \iiint_{\mathcal{V}} \left(\frac{\partial \sigma_{ij}^{A}}{\partial x_l} \frac{\partial u_i^{B}}{\partial x_j} - \sigma_{ij}^{B} \frac{\partial^2 u_i^{A}}{\partial x_j \partial x_l} \right) dV. \end{aligned} \tag{2.114}$$

Then, substituting the relationships

$$\sigma_{ij}^{A} = C_{ijmn} \partial u_m^{A} / \partial x_n \quad \sigma_{ij}^{B} = C_{ijmn} \partial u_m^{B} / \partial x_n \tag{2.115}$$

into the last integral, and using Eq. (2.82), we obtain

$$\begin{aligned} &\iiint_{\mathcal{V}} \left(C_{ijmn} \frac{\partial^2 u_m^{A}}{\partial x_l \partial x_n} \frac{\partial u_i^{B}}{\partial x_j} - C_{ijmn} \frac{\partial u_m^{B}}{\partial x_n} \frac{\partial^2 u_i^{A}}{\partial x_j \partial x_l} \right) dV \\ &= \iiint_{\mathcal{V}} \left(C_{mnij} \frac{\partial^2 u_i^{A}}{\partial x_l \partial x_j} \frac{\partial u_m^{B}}{\partial x_n} - C_{ijmn} \frac{\partial u_m^{B}}{\partial x_n} \frac{\partial^2 u_i^{A}}{\partial x_j \partial x_l} \right) dV = 0 \end{aligned} \tag{2.116}$$

because of the symmetry property of the stiffness tensor,

$$C_{mnij} = C_{ijmn}. \tag{2.117}$$

2.4.4 *Stress-strain relationships and elastic constants for isotropic system*

The elastic constants needed to describe an isotropic medium can be found by requiring that the tensor components, C_{ijkl} and S_{ijkl}, be invariant to rotations of 45° around the coordinate axes (Nye 1957). The results show that the stress-strain relationships in matrix form then reduce to

$$\begin{bmatrix}\sigma_1\\ \sigma_2\\ \sigma_3\\ \sigma_4\\ \sigma_5\\ \sigma_6\end{bmatrix} = \begin{bmatrix} C_{11} & C_{12} & C_{12} & 0 & 0 & 0\\ C_{12} & C_{11} & C_{12} & 0 & 0 & 0\\ C_{12} & C_{12} & C_{11} & 0 & 0 & 0\\ 0 & 0 & 0 & (C_{11}-C_{12})/2 & 0 & 0\\ 0 & 0 & 0 & 0 & (C_{11}-C_{12})/2 & 0\\ 0 & 0 & 0 & 0 & 0 & (C_{11}-C_{12})/2 \end{bmatrix} \times \begin{bmatrix}\varepsilon_1\\ \varepsilon_2\\ \varepsilon_3\\ \varepsilon_4\\ \varepsilon_5\\ \varepsilon_6\end{bmatrix},$$

$$\begin{bmatrix}\varepsilon_1\\ \varepsilon_2\\ \varepsilon_3\\ \varepsilon_4\\ \varepsilon_5\\ \varepsilon_6\end{bmatrix} = \begin{bmatrix} S_{11} & S_{12} & S_{12} & 0 & 0 & 0\\ S_{12} & S_{11} & S_{12} & 0 & 0 & 0\\ S_{12} & S_{12} & S_{11} & 0 & 0 & 0\\ 0 & 0 & 0 & 2(S_{11}-S_{12}) & 0 & 0\\ 0 & 0 & 0 & 0 & 2(S_{11}-S_{12}) & 0\\ 0 & 0 & 0 & 0 & 0 & 2(S_{11}-S_{12}) \end{bmatrix} \times \begin{bmatrix}\sigma_1\\ \sigma_2\\ \sigma_3\\ \sigma_4\\ \sigma_5\\ \sigma_6\end{bmatrix}. \tag{2.118}$$

Then, by employing the relationship between [S] and [C] given by Eq. (2.98),

$$S_{11} = \frac{C_{11}+C_{12}}{C_{11}C_{12}+C_{11}^2-2C_{12}^2} \quad S_{12} = \frac{-C_{12}}{C_{11}C_{12}+C_{11}^2-2C_{12}^2}. \tag{2.119}$$

Therefore, only two independent elastic constants are required for an isotropic material.[10] These have been chosen in a variety of ways in the literature. A common choice is the use of the constants μ and ν which are related to the C_{ijkl} tensor by

$$C_{ijkl} = \mu\left[\delta_{ik}\delta_{jl} + \delta_{il}\delta_{jk} + \frac{2\nu}{1-2\nu}\delta_{ij}\delta_{kl}\right] = \mu(\delta_{ik}\delta_{jl} + \delta_{il}\delta_{jk}) + \lambda\delta_{ij}\delta_{kl} \tag{2.120}$$

so that

$$C_{11} = \frac{2\mu(1-\nu)}{1-2\nu} \quad C_{12} = \frac{2\mu\nu}{1-2\nu} \quad S_{11} = \frac{1}{2\mu(1+\nu)}$$
$$S_{12} = \frac{-\nu}{2\mu(1+\nu)}. \tag{2.121}$$

Using μ and ν, the stress-strain relationships for an isotropic material then appear as

$$\begin{aligned}
\sigma_{11} &= 2\mu\left[\frac{\nu}{(1-2\nu)}e + \varepsilon_{11}\right], & \sigma_{12} &= 2\mu\varepsilon_{12}\\
\sigma_{22} &= 2\mu\left[\frac{\nu}{(1-2\nu)}e + \varepsilon_{22}\right], & \sigma_{13} &= 2\mu\varepsilon_{13}\\
\sigma_{33} &= 2\mu\left[\frac{\nu}{(1-2\nu)}e + \varepsilon_{33}\right], & \sigma_{23} &= 2\mu\varepsilon_{23}\\
\varepsilon_{11} &= \frac{1}{2\mu(1+\nu)}[\sigma_{11} - \nu(\sigma_{22}+\sigma_{33})], & \varepsilon_{12} &= \frac{1}{2\mu}\sigma_{12}\\
\varepsilon_{22} &= \frac{1}{2\mu(1+\nu)}[\sigma_{22} - \nu(\sigma_{11}+\sigma_{33})], & \varepsilon_{13} &= \frac{1}{2\mu}\sigma_{13}\\
\varepsilon_{33} &= \frac{1}{2\mu(1+\nu)}[\sigma_{33} - \nu(\sigma_{11}+\sigma_{22})]. & \varepsilon_{23} &= \frac{1}{2\mu}\sigma_{23}
\end{aligned} \tag{2.122}$$

In this formulation, the constant, μ, that linearly couples shear stress to shear strain, is termed the *shear modulus.* The physical significance of the constant ν can be revealed by considering the case where only a normal

[10] Since a cubic crystal requires three constants, these results show explicitly that a cubic crystal is not elastically isotropic even though a second rank tensor property of a cubic crystal, such as the thermal conductivity, is isotropic (Nye 1957). Note that a comparison of Eqs. (2.118) and (2.92) indicates that an isotropic material can be regarded from an elastic properties stand point as having cubic symmetry with $2C_{44} = C_{11} - C_{12}$.

stress, for example, σ_{11}, is present. The accompanying strains are then $\varepsilon_{11} = \sigma_{11}/[2\mu(1+\nu)]$ and $\varepsilon_{22} = \varepsilon_{33} = -\nu\sigma_{11}/[2\mu(1+\nu)]$, and ν is identified as the ratio of the transverse strain to the normal strain, i.e., $\nu = |\varepsilon_{22}/\varepsilon_{11}| = |\varepsilon_{33}/\varepsilon_{11}|$ and is known as *Poisson's ratio.*

Other pairs of elastic constants, related to μ and ν, are often employed. An example is λ and μ, known as the *Lamé constants*, which are related to the C_{ijkl} tensor by Eq. (2.120) and couple stress and strain according to

$$\begin{aligned} \sigma_{ij} &= \lambda e \delta_{ij} + 2\mu\varepsilon_{ij}, \\ \varepsilon_{ij} &= -\frac{\lambda\Theta\delta_{ij}}{2\mu(3\lambda+2\mu)} + \frac{1}{2\mu}\sigma_{ij}. \end{aligned} \tag{2.123}$$

Still another choice is use of E (known as *Young's modulus*) and ν, in which case

$$\begin{aligned} \sigma_{ij} &= \frac{\nu E}{(1+\nu)(1-2\nu)} e\delta_{ij} + \frac{E}{1+\nu}\varepsilon_{ij}, \\ \varepsilon_{ij} &= -\frac{\nu}{E}\Theta\delta_{ij} + \frac{1+\nu}{E}\sigma_{ij}. \end{aligned} \tag{2.124}$$

Here, the physical significance of E can be revealed by again imposing only a normal stress such as σ_{11}. Using Eq. (2.124), the strains are then $\varepsilon_{11} = \sigma_{11}/E$ and $\varepsilon_{22} = \varepsilon_{33} = -\nu\sigma_{11}/E$, and E is identified as the elastic constant that couples normal stress to normal strain via the simple relationship $\sigma_{11} = E\varepsilon_{11}$.

The various constants involved in the above choices are all related, e.g.,

$$\begin{aligned} E &= \frac{\mu(3\lambda+2\mu)}{\lambda+\mu} \qquad \nu = \frac{\lambda}{2(\lambda+\mu)}, \\ \lambda &= \frac{E\nu}{(1+\nu)(1-2\nu)} \qquad \mu = \frac{E}{2(1+\nu)}. \end{aligned} \tag{2.125}$$

Another essential elastic constant is the *bulk modulus*, K, defined by

$$\frac{1}{K} = -\frac{1}{P}\frac{\delta V}{V} = -\frac{1}{P}e, \tag{2.126}$$

where $\delta V/V$ is the fractional volume change due to the application of hydrostatic pressure, P. In the case of pure hydrostatic pressure, the stresses are given by $\sigma_{ij} = -P\delta_{ij}$ since, by convention, P is regarded as positive when compressive. Putting these stresses into Eq. (2.122) we then have

$$P = -\frac{2\mu(1+\nu)}{3(1-2\nu)}e. \tag{2.127}$$

For the case of a general stress field, use of Eq. (2.122) shows that the quantity $2\mu(1+\nu)e/[3(1-2\nu)]$ in Eq. (2.127) is given by $\sigma_{mm}/3$ which is identical to the negative of the expression for the hydrostatic stress, P, when the stress field is purely hydrostatic. We may therefore regard the quantity $-\sigma_{mm}/3$ as the *hydrostatic part* of a general stress field. Note that when σ_{mm} is negative, and therefore compressive, the hydrostatic part of the stress field is positive, consistent with our convention.

Next, an expression for the bulk modulus can be obtained by putting Eq. (2.127) into Eq. (2.126), i.e.,

$$K = -\frac{P}{e} = \frac{2\mu(1+\nu)}{3(1-2\nu)}. \tag{2.128}$$

Finally, using previous expressions, additional relationships involving K are

$$3K = 3\lambda + 2\mu \quad \text{and} \quad \frac{3K}{3K+4\mu} = \frac{1+\nu}{3(1-\nu)}. \tag{2.129}$$

2.5 Elastic Strain Energy

When a volume element is elastically strained the forces acting on it perform work as it changes shape. If the process is carried out reversibly, the work is stored as potential energy and is recoverable if the forces are removed. This can be readily understood on an atomistic basis. During the straining the atoms (molecules) in the medium are displaced from their equilibrium positions and restoring forces between them develop. The work done by the applied forces against these forces is stored as elastic strain energy in the form of potential energy. The interatomic forces involved are conservative: when the applied forces are removed, the atoms return to their unstrained equilibrium positions, the interatomic restoring forces relax, and the stored potential energy is released. The elastic strain energy in a strained body is therefore the work required to produce the state of strain (and accompanying stress) throughout the body.

2.5.1 *General relationships*

Consider a large homogeneous body subjected to a general stress field caused by a distribution of surface tractions and a distribution of body forces. Then, focus on a region, $\mathcal{V}$, enclosed by the surface $\mathcal{S}$, that is embedded in the larger body. If the displacements of the elastic field are perturbed by δu_i the work done by the surface tractions acting on $\mathcal{V}$ and the body forces within $\mathcal{V}$ is given by

$$\delta \mathcal{W} = \oiint_{\mathcal{S}} T_i \delta u_i dS + \iiint_{\mathcal{V}} f_i \delta u_i dV. \tag{2.130}$$

This work must appear as an equivalent increase in the elastic strain energy within $\mathcal{V}$, corresponding to $\delta W = \delta \mathcal{W}$, and, therefore,

$$\begin{aligned} \delta W = \delta \mathcal{W} &= \oiint_{\mathcal{S}} \sigma_{ij} \hat{n}_j \delta u_i dS - \iiint_{\mathcal{V}} \frac{\partial \sigma_{ij}}{\partial x_j} \delta u_i dV \\ &= \iiint_{\mathcal{V}} \frac{\partial}{\partial x_j} \sigma_{ij} \delta u_i dV - \iiint_{\mathcal{V}} \frac{\partial \sigma_{ij}}{\partial x_j} \delta u_i dV \\ &= \iiint_{\mathcal{V}} \sigma_{ij} \frac{\partial}{\partial x_j} \delta u_i dV = \iiint_{\mathcal{V}} \sigma_{ij} \delta \varepsilon_{ij} dV \end{aligned} \tag{2.131}$$

after using Eqs. (2.130), (2.59), (2.65), the divergence theorem and Eq. (2.5). Inspection of Eq. (2.131) shows that we may then identify $\sigma_{ij}\delta\varepsilon_{ij}$ as the change in the strain energy density,

$$\delta w = \sigma_{ij} \delta \varepsilon_{ij} \tag{2.132}$$

caused by incremental changes in the displacements, δu_i, and corresponding strains, $\delta \varepsilon_{ij}$. (An alternative derivation of this relationship is carried out in Exercise 2.11.)

If it is now imagined that the body is initially stress-free and that the applied tractions and body forces are then brought to their final values, the corresponding stresses and strains will be linearly related as indicated in Fig. 2.6. Equation (2.132) can therefore be integrated to obtain the final

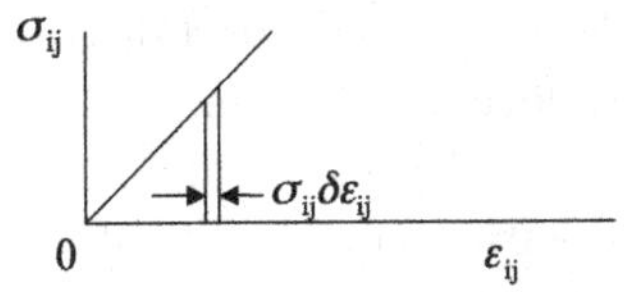

Fig. 2.6 Proportional relationship between stress and strain in elastically strained body.

strain energy density in the form

$$w = \frac{1}{2}\sigma_{ij}\varepsilon_{ij} = \frac{1}{2}\sigma_{ij}\frac{\partial u_i}{\partial x_j}, \tag{2.133}$$

where σ_{ij} and ε_{ij} are the final stresses and strains and use has been made of Eq. (2.5). Then, by introducing Eqs. (2.75) and (2.93),

$$w = \frac{1}{2}C_{ijkl}\varepsilon_{kl}\varepsilon_{ij} = \frac{1}{2}S_{ijkl}\sigma_{kl}\sigma_{ij}. \tag{2.134}$$

Finally, using contracted notation and Eqs. (2.91) and (2.94), we can also write

$$w = \frac{1}{2}\sigma_j\varepsilon_j = \frac{1}{2}C_{ij}\varepsilon_i\varepsilon_j = \frac{1}{2}S_{ij}\sigma_i\sigma_j. \quad (i, j = 1, 2 \ldots 6) \tag{2.135}$$

2.5.2 *Strain energy in isotropic systems*

The strain energy density in an isotropic system can be expressed as a function of either strain or stress in the equivalent forms

$$w = \mu\left(\frac{\nu}{1-2\nu}e^2 + \varepsilon_{ij}\varepsilon_{ij}\right) = \frac{1}{4\mu}\left(\sigma_{ij}\sigma_{ij} - \frac{\nu}{(1+\nu)}\Theta^2\right) \tag{2.136}$$

after using Eq. (2.133) and the stress-strain relationships in Sec. 2.4.4.

2.6 St.-Venant's Principle

St.-Venant's principle is a highly useful concept with relevance to many elasticity problems and can be stated as follows:

> *If a distribution of forces acting on some local portion of a body is replaced by a different distribution of forces, the effects of the two different distributions in regions sufficiently far removed from the region of application will be essentially the same, provided that the two distributions are statically equivalent, i.e., have the same resultant force and resultant moment.*

Therefore, for example, if a complicated distribution of forces is acting on a body in a relatively localized region, and if the detailed stress distribution in this region is not the main object of interest, the distribution of forces may be replaced with a single statically equivalent force in order to make it easier to find the elastic field throughout the body beyond this region. (Generally, the distance from the localized region where this approximation is satisfactory is of the order of the dimensions of the localized region.)

Exercises

2.1 Consider a body which is generally anisotropic and in the form of a rectangular parallelepiped defined by its three edge vectors **A**, **B** and **C** in the coordinate system illustrated in Fig. 2.7a. (a) Describe the change of body shape when (a) a uniform normal tensile stress, σ_{11}, is imposed, and (b) a uniform shear stress, σ_{12}, is imposed. For each case compare the results with those expected for an isotropic body.

Solution. The shape change can be analyzed by using the method employed in Sec. 2.2.1.1 in which the deformation of various vectors embedded in the body caused by a strain field is examined.

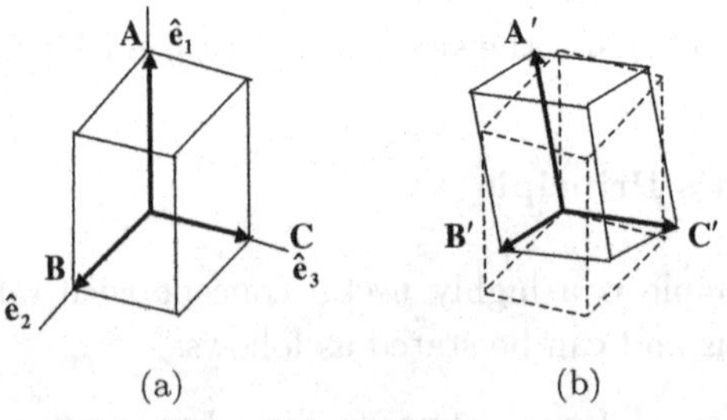

Fig. 2.7 (a) Rectangular parallelepiped defined by its edge vectors **A**, **B** and **C**. (b) Oblique parallelepiped resulting from changes in the lengths and directions of the **A**, **B** and **C** vectors in (a) due to uniform strains (exaggerated for purposes of illustration).

(a) When a pure normal stress, σ_{11}, is imposed, Eq. (2.94) shows that the resulting strain field is given by

$$\begin{bmatrix} \varepsilon_{11} \\ \varepsilon_{22} \\ \varepsilon_{33} \\ 2\varepsilon_{23} \\ 2\varepsilon_{13} \\ 2\varepsilon_{12} \end{bmatrix} = \sigma_{11} \begin{bmatrix} \mathrm{S}_{11} \\ \mathrm{S}_{12} \\ \mathrm{S}_{13} \\ \mathrm{S}_{14} \\ \mathrm{S}_{15} \\ \mathrm{S}_{16} \end{bmatrix}. \tag{2.137}$$

Therefore, using Eq. (2.9), the edge vectors $\mathbf{A} = \mathrm{A}\hat{\mathbf{e}}_1$, $\mathbf{B} = \mathrm{B}\hat{\mathbf{e}}_2$, and $\mathbf{C} = \mathrm{C}\hat{\mathbf{e}}_3$ are deformed by the strain field into

$$\mathbf{A}'(\sigma_{11}) = \mathrm{A} \begin{bmatrix} 1 + \mathrm{S}_{11}\sigma_{11} \\ \mathrm{S}_{16}\sigma_{11}/2 \\ \mathrm{S}_{15}\sigma_{11}/2 \end{bmatrix} \quad \mathbf{B}'(\sigma_{11}) = \mathrm{B} \begin{bmatrix} \mathrm{S}_{16}\sigma_{11}/2 \\ 1 + \mathrm{S}_{12}\sigma_{11} \\ \mathrm{S}_{14}\sigma_{11}/2 \end{bmatrix}$$

$$\mathbf{C}'(\sigma_{11}) = \mathrm{C} \begin{bmatrix} \mathrm{S}_{15}\sigma_{11}/2 \\ \mathrm{S}_{14}\sigma_{11}/2 \\ 1 + \mathrm{S}_{13}\sigma_{11} \end{bmatrix}. \tag{2.138}$$

When the body is isotropic, use of Eqs. (2.118) and (2.9) shows that the deformed vectors are, instead,

$$\mathbf{A}''(\sigma_{11}) = \mathrm{A} \begin{bmatrix} 1 + \mathrm{S}_{11}\sigma_{11} \\ 0 \\ 0 \end{bmatrix} \quad \mathbf{B}''(\sigma_{11}) = \mathrm{B} \begin{bmatrix} 0 \\ 1 + \mathrm{S}_{12}\sigma_{11} \\ 0 \end{bmatrix}$$

$$\mathbf{C}''(\sigma_{11}) = \mathrm{C} \begin{bmatrix} 0 \\ 0 \\ 1 + \mathrm{S}_{12}\sigma_{11} \end{bmatrix}. \tag{2.139}$$

(b) On the other hand, when a pure shear stress, σ_{12}, is imposed, Eq. (2.94) yields the strain field

$$\begin{bmatrix} \varepsilon_{11} \\ \varepsilon_{22} \\ \varepsilon_{33} \\ 2\varepsilon_{23} \\ 2\varepsilon_{13} \\ 2\varepsilon_{12} \end{bmatrix} = \begin{bmatrix} S_{16}\sigma_{12} \\ S_{26}\sigma_{12} \\ S_{36}\sigma_{12} \\ S_{46}\sigma_{12} \\ S_{56}\sigma_{12} \\ S_{66}\sigma_{12} \end{bmatrix}. \tag{2.140}$$

Then, using Eq. (2.9), **A**, **B** and **C** are deformed into the vectors

$$\mathbf{A}'(\sigma_{12}) = \mathrm{A}\begin{bmatrix} 1 + S_{16}\sigma_{12} \\ S_{66}\sigma_{12}/2 \\ S_{56}\sigma_{12}/2 \end{bmatrix} \quad \mathbf{B}'(\sigma_{12}) = \mathrm{B}\begin{bmatrix} S_{66}\sigma_{12}/2 \\ 1 + S_{26}\sigma_{12} \\ S_{46}\sigma_{12}/2 \end{bmatrix}$$

$$\mathbf{C}'(\sigma_{12}) = \mathrm{C}\begin{bmatrix} S_{56}\sigma_{12}/2 \\ S_{46}\sigma_{12}/2 \\ 1 + S_{36}\sigma_{12} \end{bmatrix}. \tag{2.141}$$

When the body is isotropic, Eqs. (2.118) and (2.9) show that the deformed vectors are, instead,

$$\mathbf{A}''(\sigma_{12}) = \mathrm{A}\begin{bmatrix} 1 \\ 2(S_{11} - S_{12})\sigma_{12} \\ 0 \end{bmatrix} \quad \mathbf{B}''(\sigma_{12}) = \mathrm{B}\begin{bmatrix} 2(S_{11} - S_{12})\sigma_{12} \\ 1 \\ 0 \end{bmatrix}$$

$$\mathbf{C}''(\sigma_{12}) = \mathrm{C}\begin{bmatrix} 0 \\ 0 \\ 1 \end{bmatrix}. \tag{2.142}$$

To summarize: when the body is generally anisotropic, Eqs. (2.138) and (2.141) show that under both the σ_{11} stress and the σ_{12} stress all three edge vectors change their lengths as well as their directions, as illustrated in Fig. 2.7b. Under either stress, the body therefore becomes an oblique parallelepiped with the spacings between its three sets of parallel faces changed by different degrees. On the other hand, when the body is isotropic the changes in shape are much simpler. Equation (2.139) shows that the σ_{11} stress simply changes the lengths of the edge vectors without rotating them, and the body remains a rectangular parallelepiped. However, the spacings between the three sets of parallel faces are changed.[11] Equation (2.142) shows that the σ_{12} stress causes the parallelepiped to undergo a simple ε_{12} shear of the type shown in projection in Fig. 2.2 where the spacings between its faces remains constant to first order.

2.2 Consider again the generally anisotropic rectangular parallelepiped of Exercise 2.1 and Fig. 2.7a. (a) Describe its shape change when it is subjected to hydrostatic pressure, and obtain an expression for its bulk modulus, K. (b) Compare the results with those expected for a body with cubic crystal symmetry.

Solution.

(a) We employ the same general procedure used in Exercise 2.1. Under hydrostatic pressure, $\sigma_{ij} = -P\delta_{ij}$, and Eq. (2.94) yields the strain field

$$\begin{bmatrix} \varepsilon_{11} \\ \varepsilon_{22} \\ \varepsilon_{33} \\ \varepsilon_{23} \\ \varepsilon_{13} \\ \varepsilon_{12} \end{bmatrix} = -P \begin{bmatrix} (S_{11}+S_{12}+S_{13}) \\ (S_{12}+S_{22}+S_{23}) \\ (S_{13}+S_{23}+S_{33}) \\ (S_{14}+S_{24}+S_{34})/2 \\ (S_{15}+S_{25}+S_{35})/2 \\ (S_{16}+S_{26}+S_{36})/2 \end{bmatrix}. \tag{2.143}$$

Then, using Eq. (2.9), the edge vectors $\mathbf{A} = A\hat{e}_1$, $\mathbf{B} = B\hat{e}_2$, and $\mathbf{C} = C\hat{e}_3$ are deformed by the strain field into

$$\mathbf{A}'(P) = A \begin{bmatrix} 1-(S_{11}+S_{12}+S_{13})P \\ -(S_{16}+S_{26}+S_{36})P/2 \\ -(S_{15}+S_{25}+S_{35})P/2 \end{bmatrix} \quad B \begin{bmatrix} -(S_{16}+S_{26}+S_{36})P/2 \\ 1-(S_{12}+S_{22}+S_{23})P \\ -(S_{14}+S_{24}+S_{34})P/2 \end{bmatrix}$$

$$\mathbf{C}'(P) = C \begin{bmatrix} -(S_{15}+S_{25}+S_{35})P/2 \\ -(S_{14}+S_{24}+S_{34})P/2 \\ 1-(S_{13}+S_{23}+S_{33})P \end{bmatrix}. \tag{2.144}$$

Next,

$$e = \varepsilon_{ii} = -[S_{11}+S_{22}+S_{33}+2(S_{23}+S_{13}+S_{12})]P \tag{2.145}$$

[11]More specifically, according to Eqs. (2.139) and (2.121), the fractional changes in the lengths of **A**, **B** and **C** are $\delta A/A = S_{11}\sigma_{11}$, and $\delta B/B = \delta C/C = S_{12}\sigma_{11} = -\nu S_{11}\sigma_{11}$, just as expected due to Poisson's ratio as described in Sec. 2.4.4.

and, therefore, using Eq. (2.126),

$$K = \frac{-P}{e} = \frac{1}{S_{11} + S_{22} + S_{33} + 2(S_{23} + S_{13} + S_{12})}. \tag{2.146}$$

(b) When the symmetry is cubic, Eq. (2.96) yields

$$\begin{bmatrix} \varepsilon_{11} \\ \varepsilon_{22} \\ \varepsilon_{33} \\ \varepsilon_{23} \\ \varepsilon_{13} \\ \varepsilon_{12} \end{bmatrix} = -P \begin{bmatrix} (S_{11} + 2S_{12}) \\ (S_{11} + 2S_{12}) \\ (S_{11} + 2S_{12}) \\ 0 \\ 0 \\ 0 \end{bmatrix} \tag{2.147}$$

and

$$\mathbf{A}''(P) = A \begin{bmatrix} 1 - (S_{11} + 2S_{12})P \\ 0 \\ 0 \end{bmatrix} \quad \mathbf{B}''(P) = B \begin{bmatrix} 0 \\ 1 - (S_{11} + 2S_{12})P \\ 0 \end{bmatrix}$$

$$\mathbf{C}''(P) = C \begin{bmatrix} 0 \\ 0 \\ 1 - (S_{11} + 2S_{12})P \end{bmatrix}. \tag{2.148}$$

Then,

$$e = 3(S_{11} + 2S_{12})P \tag{2.149}$$

and, with the help of Eq. (2.101),

$$K = -\frac{P}{e} = \frac{1}{3(S_{11} + 2S_{12})} = \frac{C_{11} + 2C_{12}}{3}.$$

To summarize: when the body is generally anisotropic, Eq. (2.144) shows that all three edge vectors undergo length changes as well as direction changes, and the body again becomes an oblique parallelepiped as illustrated in Fig. 2.7b. When the body has cubic symmetry, Eq. (2.148) shows that all three vectors undergo equal fractional length changes but do not change direction. The body, therefore, remains a rectangular parallelepiped.

2.3 Find expressions for the normal and tangential tractions acting on the plane

$$ax_1 + bx_2 + cx_3 + d = 0 \tag{2.150}$$

in a body subjected to a general stress system, σ_{ij}.

Solution. Following standard methods (Hildebrand 1949) the unit normal vector is obtained by first forming the function $g(x_1, x_2, x_3)$ given by,

$$g(x_1, x_2, x_3) = ax_1 + bx_2 + cx_3 + d. \tag{2.151}$$

Then,

$$\hat{\mathbf{n}} = \frac{\nabla g}{|\nabla g|} = \frac{\hat{\mathbf{e}}_1 a + \hat{\mathbf{e}}_2 b + \hat{\mathbf{e}}_3 c}{(a^2 + b^2 + c^2)^{1/2}}. \tag{2.152}$$

Using Eq. (2.59), the traction vector is then

$$\begin{bmatrix} T_1 \\ T_2 \\ T_3 \end{bmatrix} = \frac{1}{(a^2 + b^2 + c^2)^{1/2}} \begin{bmatrix} \sigma_{11} & \sigma_{12} & \sigma_{13} \\ \sigma_{12} & \sigma_{22} & \sigma_{23} \\ \sigma_{13} & \sigma_{23} & \sigma_{33} \end{bmatrix} \begin{bmatrix} a \\ b \\ c \end{bmatrix} \tag{2.153}$$

and the normal and tangential tractions are then, respectively,

$$\mathbf{T}(\text{norm}) = (\mathbf{T} \cdot \hat{\mathbf{n}})\hat{\mathbf{n}} \qquad \mathbf{T}(\tan) = (\hat{\mathbf{n}} \times \mathbf{T}) \times \hat{\mathbf{n}}. \tag{2.154}$$

2.4 A strain field is given by

$$[\varepsilon] = \begin{bmatrix} a(x_1^2 + x_2^2) & bx_1x_2 & 0 \\ bx_1x_2 & cx_1x_2 & 0 \\ 0 & 0 & 0 \end{bmatrix}. \tag{2.155}$$

Under what conditions on a, b and c is it compatible?

Solution. Substituting the components of $[\varepsilon]$ into Eq. (2.53), the conditions for compatibility are

$$\begin{gathered} C_{11} = C_{22} = C_{12} = C_{13} = C_{23} = 0, \\ C_{33} = 2b\frac{\partial^2(x_1x_2)}{\partial x_1 \partial x_2} - a\frac{\partial^2(x_1^2 + x_2^2)}{\partial x_2^2} - c\frac{\partial^2(x_1x_2)}{\partial x_1^2} = 0. \end{gathered} \tag{2.156}$$

Therefore, upon performing the differentiation, we must have

$$a = b \tag{2.157}$$

with c arbitrary.

2.5 Derive the relationships between the matrix elements of [C] and [S] for a cubic crystal given by Eq. (2.100).

Solution. The reciprocal relationship between [C] and [S], given by Eq. (2.98), when written out in full 6 × 6 form, is

$$\begin{bmatrix} C_{11} & C_{12} & C_{12} & 0 & 0 & 0 \\ C_{12} & C_{11} & C_{12} & 0 & 0 & 0 \\ C_{12} & C_{12} & C_{11} & 0 & 0 & 0 \\ 0 & 0 & 0 & C_{44} & 0 & 0 \\ 0 & 0 & 0 & 0 & C_{44} & 0 \\ 0 & 0 & 0 & 0 & 0 & C_{44} \end{bmatrix}$$

$$\times \begin{bmatrix} S_{11} & S_{12} & S_{12} & 0 & 0 & 0 \\ S_{12} & S_{11} & S_{12} & 0 & 0 & 0 \\ S_{12} & S_{12} & S_{11} & 0 & 0 & 0 \\ 0 & 0 & 0 & S_{44} & 0 & 0 \\ 0 & 0 & 0 & 0 & S_{44} & 0 \\ 0 & 0 & 0 & 0 & 0 & S_{44} \end{bmatrix}$$

$$= \begin{bmatrix} 1 & 0 & 0 & 0 & 0 & 0 \\ 0 & 1 & 0 & 0 & 0 & 0 \\ 0 & 0 & 1 & 0 & 0 & 0 \\ 0 & 0 & 0 & 1 & 0 & 0 \\ 0 & 0 & 0 & 0 & 1 & 0 \\ 0 & 0 & 0 & 0 & 0 & 1 \end{bmatrix}. \tag{2.158}$$

By carrying out the multiplication in Eq. (2.158), the three independent equations

$$\begin{aligned} C_{11}S_{11} + 2C_{12}S_{12} &= 1 \\ C_{12}S_{11} + C_{11}S_{12} + C_{12}S_{12} &= 0 \\ C_{44}S_{44} &= 1 \end{aligned} \tag{2.159}$$

are obtained. Solving these for C_{11}, C_{12} and C_{44} then yields Eq. (2.100).

2.6 Starting with Eq. (2.93), show how the strains given by Eq. (2.94) in contracted notation form are obtained by applying the contraction notation rules given by Eqs. (2.88), (2.89) and (2.95).

Solution. By writing Eq. (2.93) in full form, consolidating it, and then applying the rules given by Eq. (2.89),

$$\begin{bmatrix} \varepsilon_1 \\ \varepsilon_2 \\ \varepsilon_3 \\ \varepsilon_4/2 \\ \varepsilon_5/2 \\ \varepsilon_6/2 \end{bmatrix} = \begin{bmatrix} S_{1111} & S_{1122} & S_{1133} & 2S_{1123} & 2S_{1131} & 2S_{1112} \\ S_{1122} & S_{2222} & S_{2233} & 2S_{2223} & 2S_{2231} & 2S_{2212} \\ S_{1133} & S_{2233} & S_{3333} & 2S_{3323} & 2S_{3331} & 2S_{3312} \\ S_{1123} & S_{2223} & S_{3323} & 2S_{2323} & 2S_{2331} & 2S_{2312} \\ S_{1131} & S_{2231} & S_{3331} & 2S_{2331} & 2S_{3131} & 2S_{3112} \\ S_{1112} & S_{2212} & S_{3312} & 2S_{2312} & 2S_{3112} & 2S_{1212} \end{bmatrix} \times \begin{bmatrix} \sigma_1 \\ \sigma_2 \\ \sigma_3 \\ \sigma_4 \\ \sigma_5 \\ \sigma_6 \end{bmatrix}. \tag{2.160}$$

Therefore,

$$\begin{bmatrix} \varepsilon_1 \\ \varepsilon_2 \\ \varepsilon_3 \\ \varepsilon_4 \\ \varepsilon_5 \\ \varepsilon_6 \end{bmatrix} = \begin{bmatrix} S_{1111} & S_{1122} & S_{1133} & 2S_{1123} & 2S_{1131} & 2S_{1112} \\ S_{1122} & S_{2222} & S_{2233} & 2S_{2223} & 2S_{2231} & 2S_{2212} \\ S_{1133} & S_{2233} & S_{3333} & 2S_{3323} & 2S_{3331} & 2S_{3312} \\ 2S_{1123} & 2S_{2223} & 2S_{3323} & 4S_{2323} & 4S_{2331} & 4S_{2312} \\ 2S_{1131} & 2S_{2231} & 2S_{3331} & 4S_{2331} & 4S_{3131} & 4S_{3112} \\ 2S_{1112} & 2S_{2212} & 2S_{3312} & 4S_{2312} & 4S_{3112} & 4S_{1212} \end{bmatrix} \times \begin{bmatrix} \sigma_1 \\ \sigma_2 \\ \sigma_3 \\ \sigma_4 \\ \sigma_5 \\ \sigma_6 \end{bmatrix}. \tag{2.161}$$

Then, by applying the rules given by Eqs. (2.88) and (2.95)

$$\begin{bmatrix} \varepsilon_1 \\ \varepsilon_2 \\ \varepsilon_3 \\ \varepsilon_4 \\ \varepsilon_5 \\ \varepsilon_6 \end{bmatrix} = \begin{bmatrix} S_{11} & S_{12} & S_{13} & S_{14} & S_{15} & S_{16} \\ S_{12} & S_{22} & S_{23} & S_{24} & S_{25} & S_{26} \\ S_{13} & S_{23} & S_{33} & S_{34} & S_{35} & S_{36} \\ S_{14} & S_{24} & S_{34} & S_{44} & S_{45} & S_{46} \\ S_{15} & S_{25} & S_{35} & S_{45} & S_{55} & S_{56} \\ S_{16} & S_{26} & S_{36} & S_{46} & S_{56} & S_{66} \end{bmatrix} \begin{bmatrix} \sigma_1 \\ \sigma_2 \\ \sigma_3 \\ \sigma_4 \\ \sigma_5 \\ \sigma_6 \end{bmatrix} \tag{2.162}$$

and Eq. (2.94) follows.

2.7 Consider an internal stress field in a finite body, $\mathcal{V}^\circ$, that is free of body forces and whose surface, $\mathcal{S}^\circ$, is traction-free. Such a field would be produced, for example, by a defect, D, that is a source of internal stress, and is not mimicked by body forces (as described in Sec. 5.2.1.3). Show that the net average elastic dilatation associated with such a stress field vanishes, and that, therefore, the introduction of such a defect into the body will not change the body's overall volume. Hint: consider the stress components, σ_{ij}^D, first, and be aware of the relationship

$$\sigma_{ij}^D = \frac{\partial(\sigma_{ik}^D x_j)}{\partial x_k}. \tag{2.163}$$

Solution. First, we verify the expression given by Eq. (2.163) as follows:

$$\frac{\partial(\sigma_{ik}^D x_j)}{\partial x_k} = \sigma_{ik}^D \frac{\partial x_j}{\partial x_k} + x_j \frac{\partial \sigma_{ik}^D}{\partial x_k} = \sigma_{ik}^D \delta_{kj} + 0 = \sigma_{ij}^D, \tag{2.164}$$

where use has been made of Eq. (2.65). Then, using Eq. (2.164), the stress component, σ_{ij}^D, averaged over the body, is given by

$$\langle \sigma_{ij}^D \rangle = \frac{1}{\mathcal{V}^\circ} \oiiint_{\mathcal{V}^\circ} \sigma_{ij}^D dV = \frac{1}{\mathcal{V}^\circ} \oiiint_{\mathcal{V}^\circ} \frac{\partial(\sigma_{ik}^D x_j)}{\partial x_k} dV = \frac{1}{\mathcal{V}^\circ} \oiint_{\mathcal{S}^\circ} \sigma_{il}^D \hat{n}_l x_j dS \tag{2.165}$$

after employing the divergence theorem. Therefore, since $\sigma_{il}^D \hat{n}_l = 0$ at the traction-free surface,

$$\langle \sigma_{ij}^D \rangle = 0. \tag{2.166}$$

The averaged strain components, $\langle \varepsilon_{\mathrm{ij}}^{\mathrm{D}} \rangle$ therefore also vanish, since they are linearly related to the averaged stress components by

$$\langle \varepsilon_{\mathrm{ij}}^{\mathrm{D}} \rangle = \mathrm{S}_{\mathrm{ijk}l} \langle \sigma_{\mathrm{k}l}^{\mathrm{D}} \rangle. \tag{2.167}$$

The elastic dilatation, averaged over the body, therefore also vanishes. A demonstration of this result for the case edge and screw dislocations is given in Exercise 8.3.

2.8 The elastic stiffness constants for a crystal with cubic symmetry are given by the 6×6 matrix in Eq. (2.92) where the coordinate system corresponds to the cubic axes. Find the 6×6 matrix describing the new elastic constants when the crystal is referred to a new coordinate system obtained by rotating the system around its $\hat{\mathbf{e}}_1$ axis by $\pi/4$.

Solution. We employ the transformation law given by Eq. (2.87), i.e.,

$$\mathrm{C}'_{\mathrm{ijk}l} = l_{\mathrm{ig}} l_{\mathrm{jh}} \mathrm{C}_{\mathrm{ghmn}} l_{\mathrm{km}} l_{l\mathrm{n}}. \tag{2.168}$$

The [C] matrix in Eq. (2.168) must be in a 9×9 comformable format, and by comparing Eqs. (2.92) and (2.83) it is seen that [C] must therefore have the form

$$[\mathrm{C}] = \begin{bmatrix} \mathrm{C}_{11} & \mathrm{C}_{12} & \mathrm{C}_{12} & 0 & 0 & 0 & 0 & 0 & 0 \\ \mathrm{C}_{12} & \mathrm{C}_{11} & \mathrm{C}_{12} & 0 & 0 & 0 & 0 & 0 & 0 \\ \mathrm{C}_{12} & \mathrm{C}_{12} & \mathrm{C}_{11} & 0 & 0 & 0 & 0 & 0 & 0 \\ 0 & 0 & 0 & \mathrm{C}_{44} & 0 & 0 & \mathrm{C}_{44} & 0 & 0 \\ 0 & 0 & 0 & 0 & \mathrm{C}_{44} & 0 & 0 & \mathrm{C}_{44} & 0 \\ 0 & 0 & 0 & 0 & 0 & \mathrm{C}_{44} & 0 & 0 & \mathrm{C}_{44} \\ 0 & 0 & 0 & \mathrm{C}_{44} & 0 & 0 & \mathrm{C}_{44} & 0 & 0 \\ 0 & 0 & 0 & 0 & \mathrm{C}_{44} & 0 & 0 & \mathrm{C}_{44} & 0 \\ 0 & 0 & 0 & 0 & 0 & \mathrm{C}_{44} & 0 & 0 & \mathrm{C}_{44} \end{bmatrix}. \tag{2.169}$$

The rotation matrix is given by Eq. (2.15) in the form

$$[l] = \begin{bmatrix} 1 & 0 & 0 \\ 0 & \cos\theta & \sin\theta \\ 0 & -\sin\theta & \cos\theta \end{bmatrix} = \frac{1}{\sqrt{2}} \begin{bmatrix} \sqrt{2} & 0 & 0 \\ 0 & 1 & 1 \\ 0 & -1 & 1 \end{bmatrix}. \tag{2.170}$$

Then, substituting Eqs. (2.169) and (2.170) into Eq. (2.168), carrying out the tedious multiplication, and contracting the result back into a

6×6 format,

$$[C'] = \begin{bmatrix} C'_{11} & C'_{12} & C'_{13} & 0 & 0 & 0 \\ C'_{12} & C'_{22} & C'_{23} & 0 & 0 & 0 \\ C'_{13} & C'_{23} & C'_{33} & 0 & 0 & 0 \\ 0 & 0 & 0 & C'_{44} & 0 & 0 \\ 0 & 0 & 0 & 0 & C'_{55} & 0 \\ 0 & 0 & 0 & 0 & 0 & C'_{66} \end{bmatrix} \tag{2.171}$$

where

$$\begin{aligned} C'_{11} &= C_{11} \quad C'_{22} = C'_{33} = \frac{1}{2}(C_{11} + C_{12}) + C_{44} \\ C'_{12} &= C'_{13} = C_{12} \quad C'_{23} = \frac{1}{2}(C_{11} + C_{12}) - C_{44} \\ C'_{55} &= C'_{66} = C_{44} \quad C'_{44} = \frac{1}{2}(C_{11} - C_{12}) \end{aligned} \tag{2.172}$$

2.9. Show that the principal coordinate systems for the stress and strain tensors are identical for isotropic systems, but not for many low-symmetry anisotropic systems.

Solution. If the coordinate system is aligned along the principal axes of strain, the shear strains vanish. In an isotropic system, where the corresponding stresses are given by Eq. (2.123), i.e.,

$$\sigma_{ij} = \lambda e \delta_{ij} + 2\mu\varepsilon_{ij} \tag{2.173}$$

the shear stresses also vanish. The coordinate system must, therefore, be aligned with the principal axes of stress, and the two principal coordinate systems must be identical.

For a generally anisotropic system the shear strains again vanish when the coordinate system corresponds to the principal coordinate system for the strains. The corresponding stresses are given by Eq. (2.91), i.e.,

$$\sigma_i = C_{ij}\varepsilon_j \quad (C_{ij} = C_{ji}) \quad (i, j = 1, 2 \ldots 6) \tag{2.174}$$

and the shear stresses are then given by

$$\begin{aligned} \sigma_4 &= C_{14}\varepsilon_1 + C_{24}\varepsilon_2 + C_{34}\varepsilon_3 \\ \sigma_5 &= C_{15}\varepsilon_1 + C_{25}\varepsilon_2 + C_{35}\varepsilon_3 \\ \sigma_6 &= C_{16}\varepsilon_1 + C_{26}\varepsilon_2 + C_{36}\varepsilon_3 \end{aligned} \tag{2.175}$$

and do not generally vanish, since the relevant C_{ij} elastic constants do not all vanish for many anisotropic systems, e.g., the triclinic, monoclinic, tetragonal and trigonal systems (Nye 1957). The two principal coordinate systems are then not identical.

2.10 It has often been stated, e.g., Eshelby (1951), that for the strains to be derivable from the displacements according to Eq. (2.5) the strains must be compatible. Explain.

Solution. It is easily seen that Eq. (2.5), i.e.,

$$\varepsilon_{ij} = \frac{1}{2}\left(\frac{\partial u_i}{\partial x_j} + \frac{\partial u_j}{\partial x_i}\right)$$

is a solution for the strains in the compatibility equation given by Eq. (2.53). For example, substitution of Eq. (2.5) into the tensor component C_{11} in Eq. (2.53) yields

$$C_{11} = \frac{\partial^3 u_2}{\partial x_2 \partial x_3^2} + \frac{\partial^3 u_3}{\partial x_3 \partial x_2^2} - \frac{\partial^3 u_2}{\partial x_2 \partial x_3^2} - \frac{\partial^3 u_3}{\partial x_3 \partial x_2^2} = 0 \tag{2.176}$$

as required.

2.11 Derive Eq. (2.132), i.e.,

$$\delta w = \sigma_{ij} \delta \varepsilon_{ij} \tag{2.177}$$

for the incremental change in strain energy density due to an incremental increase in the strains, $\delta\varepsilon_{ij}$, by the alternative method of determining the work done by the stresses acting on a differential volume element $\delta V = dx_1 dx_2 dx_3$ when the strains are increased by $\delta\varepsilon_{ij}$.

Solution. The force in the direction $\hat{e}_1$ acting on the front ABCD face of the volume element in Fig. 2.8 is $F_1 = \sigma_{11}\delta x_2 \delta x_3$. When the strain, ε_{11}, is increased by $\delta\varepsilon_{11}$, the ABCD face is displaced relative to the back EFGH face in the direction $\hat{e}_1$ by the distance $\delta\varepsilon_{11}\delta x_1$, and the F_1 force therefore performs the work

$$\delta \mathcal{W}_1 = F_1 \delta\varepsilon_{11} \delta x_1 = \sigma_{11}\delta\varepsilon_{11}\delta x_1 \delta x_2 \delta x_3 = \sigma_{11}\delta\varepsilon_{11}\delta V. \tag{2.178}$$

The force in the direction $\hat{e}_2$ acting on the ABCD face is $F_2 = \sigma_{21}\delta x_2 \delta x_3$. When the strain ε_{21}, is increased by $\delta\varepsilon_{21}$, the ABCD face is displaced relative to the EFGH face in the direction $\hat{e}_2$ by the distance $\delta\varepsilon_{12}\delta x_1$ (see Fig. 2.2), and the F_2 force therefore performs the work

$$\delta \mathcal{W}_2 = F_2 \delta\varepsilon_{21} \delta x_1 = \sigma_{21}\delta\varepsilon_{21}\delta x_1 \delta x_2 \delta x_3 = \sigma_{21}\delta\varepsilon_{21}\delta V. \tag{2.179}$$

Similarly, the work done by the force acting in the direction $\hat{e}_3$ is $\delta\mathcal{W}_3 = \sigma_{31}\delta\varepsilon_{31}\delta V$. Similar expressions are obtained for the BFGC

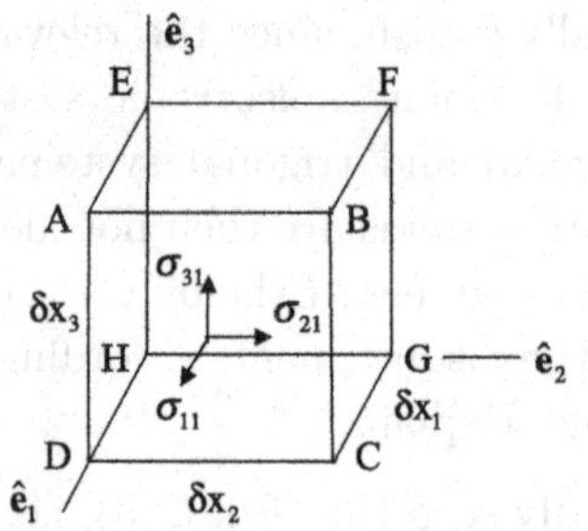

Fig. 2.8 Differential volume element $\delta V = \delta x_1 \delta x_2 \delta x_3$ embedded in stressed body. Stresses σ_{11}, σ_{21} and σ_{31} act on front face ABCD.

and EFBA faces, and the total work performed by the stress field on the volume element is therefore $\delta \mathcal{W} = \sigma_{ij} \delta \varepsilon_{ij} \delta V$. This work must appear as strain energy in the volume element, and, therefore, the corresponding increment of strain energy density in the element is

$$\delta w = \frac{\delta W}{\delta V} = \frac{\delta \mathcal{W}}{\delta V} = \sigma_{ij} \delta \varepsilon_{ij}. \tag{2.180}$$

2.12 Consider a boundary value problem in which a finite body $\mathcal{V}$, with surface $\mathcal{S}$, is elastically strained by the application of surface forces and body forces. The *Uniqueness Theorem* for solutions of boundary value problems states that:

> *The application of a given set of surface and body forces to a body produces only one (unique) stress/strain field.*

Prove that this is indeed true. Hint: Assume that two different solutions to the boundary value problem exist, and then prove, by considering their elastic strain energies, that they must be identical.

Solution. Assume that two different solutions, corresponding to a 1-field and a 2-field, exist, and consider the difference between these fields, which we term the Δ-field. Since the applied surface forces and body forces producing the 1-field and 2-field are identical, the Δ-field must be produced by vanishing surface and body forces. The elastic strain energy associated with such a field must therefore also vanish, since no forces would be available to perform work during its establishment (as described, for example, in the derivation of Eq. (2.133)).

We can therefore write for the Δ-field,

$$\oiiint_{\mathcal{V}} w^{\Delta}(\varepsilon_{ij}^{\Delta})dV = 0. \tag{2.181}$$

Here, the elastic strain energy density of the Δ-field, $w^{\Delta}(\varepsilon_{ij}^{\Delta})$, has been written as a function of $\varepsilon_{ij}^{\Delta}$ by virtue of Eq. (2.134). However, $w^{\Delta}(\varepsilon_{ij}^{\Delta})$ is a positive definite quantity, and Eq. (2.181) can, therefore, be satisfied only if $w^{\Delta}(\varepsilon_{ij}^{\Delta})$ vanishes everywhere in the body. But, this is only possible if the $\varepsilon_{ij}^{\Delta}$ vanish everywhere. We can therefore conclude that the Δ-field vanishes everywhere, and that the 1-field and 2-field are therefore identical and correspond to a single (unique) field.

2.13 Consider, again, the finite body in Exercise 2.12, which is strained by the application of surface forces and body forces. Show that *Clapeyron's formula*, i.e.,

$$\oiint_{\mathcal{S}} T_i u_i dS + \oiiint_{\mathcal{V}} f_i u_i dV = 2 \oiiint_{\mathcal{V}} w dV \tag{2.182}$$

applies (Asaro and Lumbarda 2006). Here, T_i, u_i and w are the final values of the surface tractions, displacements and strain energy density after the strain field has been established starting with the stress-free body.

Solution. By combining Eqs. (2.130) and (2.131),

$$\oiint_{\mathcal{S}} T_i \delta u_i dS + \oiiint_{\mathcal{V}} f_i \delta u_i dV = \oiiint_{\mathcal{V}} \sigma_{ij} \delta \varepsilon_{ij} dV. \tag{2.183}$$

As in the derivation of Eq.(2.133), if we now imagine that the body is initially stress-free and that the applied tractions, body forces and stresses are brought up to their final values, the corresponding displacements and strains will increase proportionally, and Eq. (2.183) can be integrated to obtain

$$\frac{1}{2} \oiint_{\mathcal{S}} T_i u_i dS + \frac{1}{2} \oiiint_{\mathcal{V}} f_i u_i dV = \frac{1}{2} \oiiint_{\mathcal{V}} \sigma_{ij} \varepsilon_{ij} dV. \tag{2.184}$$

Then, using Eq. (2.133), we obtain Clapeyron's formula,

$$\oiint_{\mathcal{S}} T_i u_i dS + \oiiint_{\mathcal{V}} f_i u_i dV = 2 \oiiint_{\mathcal{V}} w dV. \tag{2.185}$$

Chapter 3

Methods

3.1 Introduction

The main methods for treating the defect elasticity problems considered in this book are introduced. The chapter begins by reviewing the requirements that any solution for a defect elasticity problem must satisfy. A basic differential equation for the displacements whose solutions automatically satisfy these requirements is then formulated, and useful methods of solving it are described. These include its direct solution and various formalisms that expedite its solution under different conditions including the Fourier transform approach, the Green's function method, and the sextic and integral formalisms.

Following this, the transformation strain method, which is applicable to many defect problems, is described. Here, the defect is introduced in the form of an appropriate stress-free transformation strain. The resulting elastic field is then found by methods involving the use of Green's functions or Fourier transforms.

Next, the stress function method for solving problems is introduced, and the Airy stress function, which is applicable when plane strain conditions prevail, is described. Finally, the problem of finding solutions for defects in finite homogeneous regions bounded by interfaces, rather than in infinite regions, is outlined. The methods by which the boundary conditions at the interfaces, which are present in such cases, can be satisfied by the method of images and use of appropriate Green's functions are described.

3.2 Basic Field Equation for the Displacement

Any solution for a defect elasticity problem assuming linear elasticity must satisfy the following:

(1) The stresses must satisfy the equations of equilibrium, Eq. (2.65).
(2) The corresponding strains must satisfy the compatibility equations, Eq. (2.53), and, thereby, be derivable from a set of displacements according to Eq. (2.5).
(3) The elastic stresses and strains must be linearly coupled by suitable elastic constants and so satisfy Hooke's law.
(4) Conditions on the elastic field at any interfaces bounding the region containing the defect must be satisfied.

Because of this extensive set of requirements it is usually possible to find the solution for a given problem by multiple routes. In the great majority of cases in this book we shall solve defect problems by first finding the displacement field of the defect. Once this is known, the strains can be determined by simple differentiation of the displacements using Eq. (2.5), and the stresses can then be easily found using Hooke's law, e.g., Eq. (2.75). However, this procedure requires that the displacement field be of a form that will produce strains that obey the compatibility relationships, Eq. (2.53), and stresses that satisfy the equation of equilibrium, Eq. (2.65).

A basic differential equation for the displacements whose solutions are continuous functions and automatically satisfy the above requirements can be obtained by first substituting Eq. (2.5) into Eq. (2.75) and using the symmetry property $C_{ijlk} = C_{ijkl}$ so that

$$\sigma_{ij}(\mathbf{x}) = C_{ijkl}\varepsilon_{kl} = \frac{1}{2}C_{ijkl}\left[\frac{\partial u_k(\mathbf{x})}{\partial x_l} + \frac{\partial u_l(\mathbf{x})}{\partial x_k}\right] = C_{ijkl}\frac{\partial u_k(\mathbf{x})}{\partial x_l}. \tag{3.1}$$

Then, substituting this result into the equilibrium condition, Eq. (2.65), the desired basic equation is

$$\frac{\partial \sigma_{ij}(\mathbf{x})}{\partial x_j} + f_i(\mathbf{x}) = C_{ijkl}\frac{\partial^2 u_k(\mathbf{x})}{\partial x_j \partial x_l} + f_i(\mathbf{x}) = 0. \tag{3.2}$$

As will be seen, in many defect elasticity problems the defect can be represented by an effective force density distribution. Its elastic field can then be found by inserting the force density into Eq. (3.2) and solving for the $u_k(\mathbf{x})$.

For isotropic materials, Eq. (3.2) can be written in vector form by converting the elastic constants using Eq. (2.120) to obtain

$$\frac{\mu}{1-2\nu}\nabla(\nabla\cdot\mathbf{u})+\mu\nabla^2\mathbf{u}+\mathbf{f}=\frac{2\mu(1-\nu)}{1-2\nu}\nabla(\nabla\cdot\mathbf{u})-\mu\nabla\times(\nabla\times\mathbf{u})+\mathbf{f}=0 \tag{3.3}$$

where use has been made of the identity $\nabla^2\mathbf{u}=\nabla(\nabla\cdot\mathbf{u})-\nabla\times(\nabla\times\mathbf{u})$. Equation (3.3) is known as the *Navier equation.*

3.3 Fourier Transform Method

Equation (3.2) may be solved by the standard method of employing Fourier transforms (Sneddon, 1951). Here, the equation is first transformed into k-space using a Fourier transform (Appendix F). The equation is then solved in k-space, and the solution is transformed back into real space by means of an inverse transform.

Following this procedure, Eq. (3.2) is therefore transformed by applying Eq. (F.1), and after integrating the result by parts, the solution in k-space is

$$\mathrm{C_{ijk\mathit{l}}k_jk_\mathit{l}\bar{u}_k}(\mathbf{k})=\mathrm{\bar{f}_i}(\mathbf{k}). \tag{3.4}$$

At this point we introduce, for convenience, a contracted notation for the second rank tensor component, $\mathrm{C_{ijk\mathit{l}}k_jk_\mathit{l}}$, appearing in Eq. (3.4) by writing it in the form[1]

$$\mathrm{C_{ijk\mathit{l}}k_jk_\mathit{l}}\equiv\mathrm{(kk)_{ik}} \tag{3.5}$$

so that Eq. (3.4) appears more concisely as

$$\mathrm{(kk)_{ik}\bar{u}_k}(\mathbf{k})=\mathrm{\bar{f}_i}(\mathbf{k}) \tag{3.6}$$

or in matrix notation as[2]

$$\mathrm{(kk)[\bar{u}]}=\mathrm{[\bar{f}]}. \tag{3.7}$$

The solution of Eq. (3.7) is then (in matrix and component forms)

$$\mathrm{[\bar{u}]}=\mathrm{(kk)^{-1}[\bar{f}]}\quad \mathrm{\bar{u}_k}(\mathbf{k})=\mathrm{(kk)^{-1}_{ik}\bar{f}_i}(\mathbf{k}). \tag{3.8}$$

---

[1]Tensor components, such as $\mathrm{k_jC_{jik\mathit{l}}k_\mathit{l}}$, that are represented by the contracted notation $\mathrm{(kk)_{ik}}$, are known as components of *Christoffel stiffness tensors* and appear frequently throughout this book.

[2]For simplicity, matrices corresponding to Christoffel tensors are represented by the special notation (ab): i.e., curved brackets are employed rather than the square brackets that are employed elsewhere throughout the book to indicate matrices.

Finally, after performing the inverse transformation by substituting Eq. (3.8) into Eq. (F.2), the solution in real space, corresponding to the displacement field caused by a distribution of body force density, $f_i(\mathbf{x}')$, is

$$u_k(\mathbf{x}) = \frac{1}{(2\pi)^3}\int_{-\infty}^{\infty}\int_{-\infty}^{\infty}\int_{-\infty}^{\infty} dx_1' dx_2' dx_3'$$
$$\times \int_{-\infty}^{\infty}\int_{-\infty}^{\infty}\int_{-\infty}^{\infty} f_i(\mathbf{x}')e^{-i\mathbf{k}\cdot(\mathbf{x}-\mathbf{x}')}(kk)_{ik}^{-1} dk_1 dk_2 dk_3. \quad (3.9)$$

3.4 Green's Function Method

In many cases the solution for the displacement field can be expedited by the use of *Green's functions* (see, for example, Morse and Feshbach 1953). In considering the Green's function method it is instructive to review first the perhaps more familiar problem of finding the electrostatic potential due to a distribution of electrical charge density.

Suppose that it is desired to find the electrostatic potential at a field point $P(\mathbf{x})$ in an infinite homogeneous region due to a distribution of electrical charge density in a defined region such as $\mathcal{V}$ in Fig. 3.1. This can be accomplished by employing the classical expression for the potential produced at a field point, $P(\mathbf{x})$, by an electrical point charge q located at a source point, $Q(\mathbf{x}')$, in an infinite region which is given by

$$v(\mathbf{x}) = A\frac{q}{|\mathbf{x}-\mathbf{x}'|} \quad (3.10)$$

where $A = \text{constant}$. By virtue of this, the potential at $\mathbf{x}$ due to a distribution of charge density, $\rho(\mathbf{x}')$, in $\mathcal{V}$ is given by the integral

$$v(\mathbf{x}) = A \oiiint_{\mathcal{V}} \frac{\rho(\mathbf{x}')}{|\mathbf{x}-\mathbf{x}'|} dV'. \quad (3.11)$$

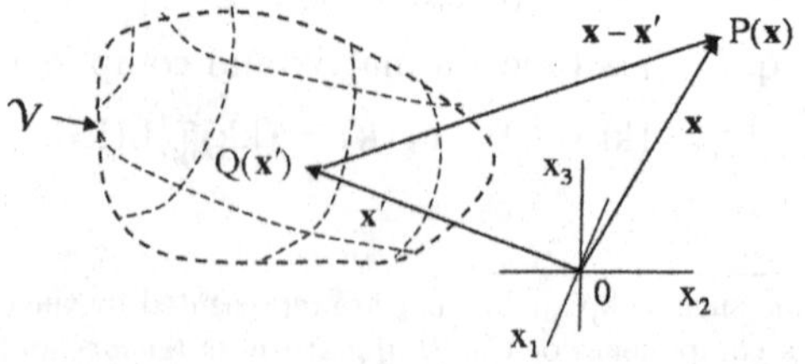

Fig. 3.1 Coordinate system used for finding potential at the field point $P(\mathbf{x})$ due to electrical charge density distributed at source points $Q(\mathbf{x}')$ distributed within the region $\mathcal{V}$ (dashed) which, in turn, is embedded in an infinite homogeneous region.

The potential given by Eq. (3.10) in the case of a unit point charge, is identified as a Green's function of the form

$$G(\mathbf{x}-\mathbf{x}') = A\frac{1}{|\mathbf{x}-\mathbf{x}'|} \tag{3.12}$$

and Eq. (3.11) can then be written as

$$v(\mathbf{x}) = \oiiint_{\mathcal{V}} G(\mathbf{x}-\mathbf{x}')\rho(\mathbf{x}')dV'. \tag{3.13}$$

Equation (3.13) is therefore a general expression that can be used to solve a wide range of electrostatic problems involving a charge distribution, $\rho(\mathbf{x}')$. If the charge density distribution in $\mathcal{V}$ corresponds to a point charge, q, located at $\mathbf{x}' = \mathbf{x}_\circ$, it can be represented by the delta function $\rho(\mathbf{x}') = q\delta(\mathbf{x}'-\mathbf{x}_\circ)$ (Appendix D). Equation (3.13) then becomes

$$v(\mathbf{x}) = A\oiiint_{\mathcal{V}} \frac{q\delta(\mathbf{x}'-\mathbf{x}^\circ)}{|\mathbf{x}-\mathbf{x}'|}dV' = A\frac{q}{|\mathbf{x}-\mathbf{x}_\circ|} \tag{3.14}$$

and Eq. (3.10), is recovered.

In many defect problems a solution for the displacements caused by a force density distribution (representing the defect) in an infinite homogeneous region is required. Therefore, by analogy with electrostatic problems, a Green's function corresponding to the displacement field produced by a unit point force in such a region is needed. If the Green's function is in the form of a second rank tensor, $\underline{\mathbf{G}}(\mathbf{x}-\mathbf{x}')$, whose component $G_{ij}(\mathbf{x}-\mathbf{x}')$ represents the displacement in the i direction at the field point $P(\mathbf{x})$ due to a unit point force applied in the j direction at the source point $Q(\mathbf{x}')$, the displacement $\mathbf{u}(\mathbf{x})$ produced at $\mathbf{x}$ by a point force $\mathbf{F}$ at $\mathbf{x}'$ will be of the form

$$u_i(\mathbf{x}) = G_{ij}(\mathbf{x}-\mathbf{x}')F_j \quad \text{or} \quad \begin{bmatrix} u_1 \\ u_2 \\ u_3 \end{bmatrix} = \begin{bmatrix} G_{11} & G_{12} & G_{13} \\ G_{21} & G_{22} & G_{23} \\ G_{31} & G_{32} & G_{33} \end{bmatrix} \begin{bmatrix} F_1 \\ F_2 \\ F_3 \end{bmatrix} \tag{3.15}$$

where the $\underline{\mathbf{G}}$ tensor maps the force vector into the displacement vector. As indicated, the Green's function in an infinite homogeneous medium is expected to be a function of only the vector difference $\mathbf{x}-\mathbf{x}'$, since in this case it should depend only upon the distance from the point force and the radial direction from the point force. The displacement field due to any distribution of force density, $f_i(\mathbf{x})$, within $\mathcal{V}$ must then possess the form

$$u_i(\mathbf{x}) = \oiiint_{\mathcal{V}} G_{ij}(\mathbf{x}-\mathbf{x}')f_j(\mathbf{x}')dV' \tag{3.16}$$

which is seen to be the counterpart to Eq. (3.13) for electrostatic problems.

To obtain a basic equation for finding such a Green's function, consider a point force F_k at $\mathbf{x}'$ in a region $\mathcal{V}$ enclosed by the surface $\mathcal{S}$. Mechanical equilibrium requires that

$$F_k + \oiint_{\mathcal{S}} \sigma_{kp}\hat{n}_p dS = 0 \tag{3.17}$$

where σ_{kp} is the stress due to F_k. Then, by use of Eq. (3.1) for σ_{kp}, Eq. (3.15), and the divergence theorem, Eq. (3.17) becomes

$$\begin{aligned} F_k &= -\oiint_{\mathcal{S}} C_{kpim}\frac{\partial u_i}{\partial x_m}\hat{n}_p dS = -\oiint_{\mathcal{S}} C_{kpim}\frac{\partial G_{ij}(\mathbf{x}-\mathbf{x}')}{\partial x_m}F_j\hat{n}_p dS \\ &= -\oiiint_{\mathcal{V}} C_{kpim}\frac{\partial^2 G_{ij}(\mathbf{x}-\mathbf{x}')}{\partial x_m \partial x_p}F_j dV. \end{aligned} \tag{3.18}$$

However, by using a delta function, the point force F_k at $\mathbf{x}'$ can be written in the alternative form

$$F_k = \oiiint_{\mathcal{V}} F_j\delta_{kj}\delta(\mathbf{x}-\mathbf{x}')dV \tag{3.19}$$

and substituting this into Eq. (3.18),

$$\oiiint_{\mathcal{V}} \left[C_{kpim}\frac{\partial^2 G_{ij}(\mathbf{x}-\mathbf{x}')}{\partial x_m \partial x_p} + \delta_{kj}\delta(\mathbf{x}-\mathbf{x}')\right]F_j dV = 0. \tag{3.20}$$

Then, since $F_j dV$ can be varied independently,

$$C_{kpim}\frac{\partial^2 G_{ij}(\mathbf{x}-\mathbf{x}')}{\partial x_m \partial x_p} + \delta_{kj}\delta(\mathbf{x}-\mathbf{x}') = 0. \tag{3.21}$$

Equation (3.21) is a useful equation for finding the Green's function, $G_{ij}(\mathbf{x}-\mathbf{x}')$, and is employed in Ch. 4 to find Green's functions for a point source in several different types of regions.

An expression for the Fourier transform of $G_{km}(\mathbf{x}-\mathbf{x}')$, which is also of use, is obtained by first comparing Eqs. (3.9) and (3.16) which shows that

$$G_{km}(\mathbf{x}-\mathbf{x}') = \frac{1}{(2\pi)^3}\int_{-\infty}^{\infty}\int_{-\infty}^{\infty}\int_{-\infty}^{\infty}(kk)^{-1}_{km}e^{-i\mathbf{k}\cdot(\mathbf{x}-\mathbf{x}')}dk_1dk_2dk_3. \tag{3.22}$$

Then, comparison of this result with Eq. (F.2) shows that $(kk)^{-1}_{km}$ can be identified as the Fourier transform of $G_{km}(\mathbf{x}-\mathbf{x}')$, i.e.,

$$\bar{G}_{km}(\mathbf{k}) = (kk)^{-1}_{km}. \tag{3.23}$$

In an alternative approach to obtain Eq. (3.23), the Fourier transform of Eq. (3.21) can be taken to obtain

$$C_{ijkl}\int_{-\infty}^{\infty}\int_{-\infty}^{\infty}\int_{-\infty}^{\infty}\frac{\partial^2 G_{km}(\mathbf{x}-\mathbf{x}')}{\partial x_j\partial x_l}e^{i\mathbf{k}\cdot(\mathbf{x}-\mathbf{x}')}dx_1dx_2dx_3$$

$$= -\delta_{im}\int_{-\infty}^{\infty}\int_{-\infty}^{\infty}\int_{-\infty}^{\infty}\delta(\mathbf{x}-\mathbf{x}')e^{i\mathbf{k}\cdot(\mathbf{x}-\mathbf{x}')}dx_1dx_2dx_3 = -\delta_{im}. \tag{3.24}$$

Then, by integrating by parts twice,

$$-k_jk_lC_{ijkl}\int_{-\infty}^{\infty}\int_{-\infty}^{\infty}\int_{-\infty}^{\infty}G_{km}(\mathbf{x}-\mathbf{x}')e^{i\mathbf{k}\cdot(\mathbf{x}-\mathbf{x}')}dx_1dx_2dx_3$$

$$= -k_jk_lC_{ijkl}\bar{G}_{km}(\mathbf{k}) = -\delta_{im} \tag{3.25}$$

and, solving for $\bar{G}_{km}(\mathbf{k})$, Eq. (3.23) is again obtained. The above results indicate that Eqs. (3.21), (3.22) or (3.23) can serve as starting points to obtain expressions for Green's functions: see, Teodosiu (1982), Mura (1987) and Exercise 4.7.

In Exercise 4.2, Eq. (3.21) is used to show that $G_{ij}(\mathbf{x}-\mathbf{x}')$ is a homogeneous function of degree -1 in the variable $(\mathbf{x}-\mathbf{x}')$ and therefore must have the form $G_{ij} = f_{ij}(\hat{l})/|\mathbf{x}-\mathbf{x}'|$, where $f_{ij}(\hat{l})$ is a function of the radial direction from the point force indicated by the unit vector $\hat{l}$.

3.5 Sextic and Integral Formalisms for Two-Dimensional Problems

We now develop the sextic and integral formalisms for treating two-dimensional elasticity problems. Two-dimensional cases arise when the elastic field is invariant when advancing along one dimension. If the

x_3 direction of a Cartesian coordinate system is taken in the invariant direction, the conditions

$$\frac{\partial}{\partial x_3} = 0 \quad u_1 = u_1(x_1, x_2) \quad u_2 = u_2(x_1, x_2) \quad u_3 = 0 \tag{3.26}$$

then apply, and the elastic field in every plane normal to x_3 is identical.

These formalisms provide essential tools for finding Green's functions for point body forces (Ch. 4) and solutions of Eq. (3.2) for the two-dimensional displacement fields of infinitely long straight dislocations and lines of force (Ch. 12). Further analysis shows that these solutions can then be used to determine the elastic fields of dislocation loops and various segmented dislocation structures in three dimensions (Ch. 12) as well as interfacial dislocations (Ch. 14).

The *sextic formalism* was pioneered by Stroh (1958, 1962) and the extended *integral formalism* was developed later by Barnett, Lothe, Malen and others (see, Bacon, Barnett and Scattergood 1979b). The sextic formulation is described first followed by the integral formulation. The source for most of the following analyses (Secs. 3.51. and 3.52) is the treatise of Bacon, Barnett and Scattergood (1979b).

3.5.1 *Sextic formalism*

3.5.1.1 *Basic formulation*

The approach is to solve Eq. (3.2) directly for the displacement. Two Cartesian coordinate systems are employed. The first, termed the *crystal system*, has unit base vectors $(\hat{\mathbf{e}}_1, \hat{\mathbf{e}}_2, \hat{\mathbf{e}}_3)$, and is attached to the crystal medium, preferably in an orientation for which the C_{ijkm} tensor assumes its simplest form. The second has unit base vectors $(\hat{\mathbf{m}}, \hat{\mathbf{n}}, \hat{\boldsymbol{\tau}})$ (see Fig. 3.2) and is oriented so that $\hat{\boldsymbol{\tau}}$ points in the invariant direction. When treating infinitely long straight dislocations and lines of force $\hat{\boldsymbol{\tau}}$ is then aligned with the defect. The $(\hat{\mathbf{m}}, \hat{\mathbf{n}}, \hat{\boldsymbol{\tau}})$ vectors can be readily referred to the crystal system, and the $(\hat{\mathbf{m}}, \hat{\mathbf{n}}, \hat{\boldsymbol{\tau}})$ system, along its attached line defect, can be rotated with respect to the crystal system in order to study effects due to elastic anisotropy.[3]

[3]Two coordinate systems, in which one is attached to the defect and the other to the crystal, are frequently employed in this manner throughout the book.

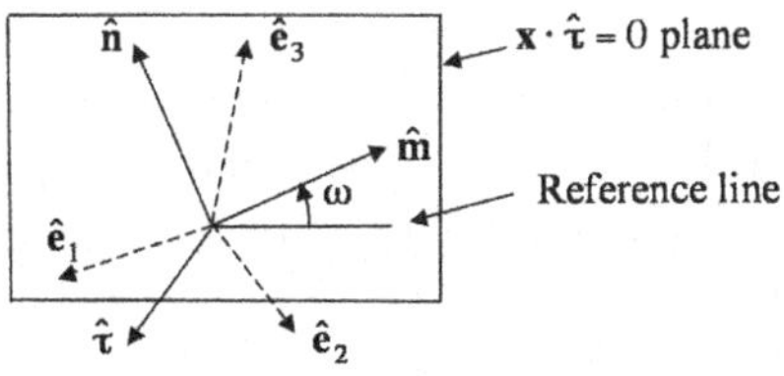

Fig. 3.2 Crystal $(\hat{\mathbf{e}}_1, \hat{\mathbf{e}}_2, \hat{\mathbf{e}}_3)$ and $(\hat{\mathbf{m}}, \hat{\mathbf{n}}, \hat{\boldsymbol{\tau}})$ coordinate systems. Base vectors $\hat{\mathbf{m}}$ and $\hat{\mathbf{n}}$ and reference line lie in $\mathbf{x} \cdot \hat{\boldsymbol{\tau}} = 0$ plane. $\hat{\boldsymbol{\tau}} = \hat{\mathbf{m}} \times \hat{\mathbf{n}}$.

A solution of Eq. (3.2) of the form

$$u_k = A_k f(\lambda) \tag{3.27}$$

is now assumed, where

$$\lambda = \hat{\mathbf{m}} \cdot \mathbf{x} + p\hat{\mathbf{n}} \cdot \mathbf{x} = \hat{m}_i x_i + p\hat{n}_i x_i \tag{3.28}$$

and A_k and p are constants. Note that this solution is two-dimensional and a function only of the position, $\mathbf{x}$, projected onto the $\mathbf{x} \cdot \hat{\boldsymbol{\tau}} = 0$ plane. Substitution of Eq. (3.27) into Eq. (3.2), with body forces absent, yields

$$C_{ijkm}\frac{\partial^2 u_k}{\partial x_i \partial x_m} = C_{ijkm}(\hat{m}_i + p\hat{n}_i)(\hat{m}_m + p\hat{n}_m)A_k\frac{d^2 f}{d\lambda^2} = 0 \tag{3.29}$$

with the help of the operator

$$\frac{\partial}{\partial x_i} = \frac{d}{d\lambda}\frac{\partial \lambda}{\partial x_i} = (\hat{m}_i + \hat{n}_i p)\frac{d}{d\lambda}. \tag{3.30}$$

Equation (3.27) is therefore a solution if the relationship

$$C_{ijkm}[\hat{n}_i\hat{n}_m p^2 + (\hat{m}_i\hat{n}_m + \hat{n}_i\hat{m}_m)p + \hat{m}_i\hat{m}_m]A_k = 0 \tag{3.31}$$

is satisfied. Equation (3.31) can be written in the alternative form

$$\left\{(\hat{n}\hat{n})_{jk}p^2 + [(\hat{m}\hat{n})_{jk} + (\hat{n}\hat{m})_{jk}]p + (\hat{m}\hat{m})_{jk}\right\}A_k = 0 \tag{3.32}$$

by employing the contracted Christoffel stiffness tensor notation introduced previously in the form of Eq. (3.5). Equation (3.32) is seen to constitute a classic eigenvalue problem that requires finding the values of p (i.e., eigenvalues) and A_k (i.e., eigenvectors) that satisfy it. A non-trivial solution can be found only if the determinantal equation

$$\det\left|(\hat{n}\hat{n})_{jk}p^2 + [(\hat{m}\hat{n})_{jk} + (\hat{n}\hat{m})_{jk}]p + (\hat{m}\hat{m})_{jk}\right| = 0 \tag{3.33}$$

is satisfied, and this leads to a sextic equation corresponding to a polynomial of the sixth degree in p which has six complex roots, i.e., the eigenvalues, $p_\alpha(\alpha = 1, 2, \ldots 6)$, which occur in complex conjugate pairs, since the coefficients of the polynomial are real.[4] The six eigenvalues can therefore always be assigned indices so that

$$p_{\alpha+3} = p_\alpha^*, \quad (\alpha = 1, 2, 3) \tag{3.34}$$

where the asterisk indicates a complex conjugate, and the first three eigenvalues, possessing subscripts 1, 2, 3, have positive imaginary parts, while the remaining three, possessing subscripts 4, 5, 6, have negative imaginary parts (see an example of this in Exercise 3.4). The corresponding six complex eigenvectors, obtained by substituting the six eigenvalues into Eq. (3.32), follow the same pattern, and, therefore,[5]

$$\begin{aligned} &A_{k\alpha}(k = 1, 2, 3 : \alpha = 1, 2, ...6) \\ &A_{k,\alpha+3} = A_{k\alpha}^*. \, (\alpha = 1, 2, 3) \end{aligned} \tag{3.35}$$

The most general solution based on Eq. (3.27) can now be written as a linear combination of the six eigenfunctions, $u_{k\alpha} = A_{k\alpha} f(\hat{\mathbf{m}} \cdot \mathbf{x} + p_\alpha \hat{\mathbf{n}} \cdot \mathbf{x})$, in the form

$$u_k = \sum_{\alpha=1}^{6} D_\alpha u_{k\alpha} = \sum_{\alpha=1}^{6} D_\alpha A_{k\alpha} f(\hat{\mathbf{m}} \cdot \mathbf{x} + p_\alpha \hat{\mathbf{n}} \cdot \mathbf{x}) \tag{3.36}$$

where the D_α are constants.

Additional vectors, $L_{j\alpha}$, which are related to the vectors $A_{j\alpha}$ by

$$L_{j\alpha} = -\hat{n}_i C_{ijkm}(\hat{m}_m + \hat{n}_m p_\alpha) A_{k\alpha} = -[(\hat{n}\hat{m})_{jk} + (\hat{n}\hat{n})_{jk} p_\alpha] A_{k\alpha} \tag{3.37}$$

and will be useful below in formulating the theory, are now introduced (Stroh 1962). An alternative expression for the $L_{j\alpha}$ vector is obtained by

[4] In addition, Eshelby, Read and Shockley (1953) have given a physical argument that the roots must be imaginary by showing that, otherwise, a condition of zero strain energy can be obtained in the presence of non-zero strain.

[5] In the double subscript notation introduced for a Stroh vector in Eq. (3.35), the Greek subscript identifies the corresponding eigenvalue whereas the Latin subscript indicates the vector component.

multiplying Eq. (3.37) throughout by p_α and adding it to Eq. (3.32), so that $L_{j\alpha}$ assumes the form

$$L_{j\alpha} = \frac{1}{p_\alpha}\hat{m}_i C_{ijkm}(\hat{m}_m + \hat{n}_m p_\alpha)A_{k\alpha} = \left[\frac{1}{p_\alpha}(\hat{m}\hat{m})_{jk} + (\hat{m}\hat{n})_{jk}\right] A_{k\alpha}. \quad (3.38)$$

So far, the magnitudes of the vectors $A_{i\alpha}$ and $L_{i\alpha}$ are undetermined, and they are therefore normalized according to

$$2A_{i\alpha}L_{i\alpha} = 1. \quad (3.39)$$

In addition, by virtue of Eqs. (3.34), (3.35), and (3.37),

$$L_{k,\alpha+3} = L^*_{k\alpha} \quad (\alpha = 1, 2, 3). \quad (3.40)$$

Having the above quantities, several further relationships are developed that will be needed when the general solution given by Eq. (3.36) is employed to find solutions for both long straight dislocations and lines of force running along $\hat{\tau}$.[6] The stresses derived from the displacements of Eq. (3.36) are

$$\sigma_{ij} = C_{ijkm}\frac{\partial u_k}{\partial x_m} = \sum_{\alpha=1}^{6} C_{ijkm}D_\alpha A_{k\alpha}\frac{df(\lambda_\alpha)}{d\lambda_\alpha}\frac{\partial \lambda_\alpha}{\partial x_m}$$

$$= \sum_{\alpha=1}^{6} C_{ijkm}D_\alpha A_{k\alpha}(\hat{m}_m + p_\alpha \hat{n}_m)\frac{df}{d\lambda_\alpha}, \quad (3.41)$$

where $\lambda_\alpha = \hat{\mathbf{m}} \cdot \mathbf{x} + p_\alpha \hat{\mathbf{n}} \cdot \mathbf{x}$. Multiplication of Eq. (3.41) throughout by $\hat{n}_i$ and use of Eq. (3.37), and multiplication of Eq. (3.41) throughout by $\hat{m}_i$ and use of Eq. (3.38), produces the two relationships,

$$\hat{n}_i\sigma_{ij} = -\sum_{\alpha=1}^{6} L_{j\alpha}D_\alpha\frac{df}{d\lambda_\alpha}, \quad (3.42)$$

$$\hat{m}_i\sigma_{ij} = \sum_{\alpha=1}^{6} p_\alpha L_{j\alpha}D_\alpha\frac{df}{d\lambda_\alpha}. \quad (3.43)$$

Next, the vector ψ_j defined by

$$\psi_j = \sum_{\alpha=1}^{6} D_\alpha L_{j\alpha} f(\hat{\mathbf{m}} \cdot \mathbf{x} + p_\alpha \hat{\mathbf{n}} \cdot \mathbf{x}) \quad (3.44)$$

is introduced. This quantity, which may be compared to the expression for the displacement given by Eq. (3.36), will be useful below in treating

[6] An expression for the elastic field generated by a line force is required in Ch. 14.

straight lines of force. Taking its derivative,

$$\frac{\partial \psi_j}{\partial x_i} = \sum_{\alpha=1}^{6} D_\alpha L_{j\alpha}(\hat{m}_i + p_\alpha \hat{n}_i)\frac{df}{d\lambda_\alpha}. \tag{3.45}$$

Then, using Eqs. (3.42), (3.43) and (3.45),

$$\hat{n}_i \sigma_{ij} = -m_i \frac{\partial \psi_j}{\partial x_i} \qquad \hat{m}_i \sigma_{ij} = \hat{n}_i \frac{\partial \psi_j}{\partial x_i}. \tag{3.46}$$

This result shows that ψ_j serves effectively as a stress function whose derivatives yield the stresses associated with the displacements given by Eq. (3.36).

The quantity ψ_j can now be used to formulate a condition for mechanical equilibrium in a region occupied by a line of force lying along the $\hat{\boldsymbol{\tau}}$ axis. Consider a cylindrical surface enclosing the line force, and let C be the closed curve corresponding to the intersection of the surface with a plane perpendicular to $\hat{\boldsymbol{\tau}}$. The force on a cylindrical segment of the surface bounded by C and C + dz, where z is measured along $\hat{\boldsymbol{\tau}}$, is then

$$dF_j = \left(\oint_C \sigma_{ij} \hat{\nu}_i ds \right) dz, \tag{3.47}$$

where $\hat{\nu}_i$ is the outward normal unit vector to C lying in the plane containing C, and s measures distance along C. Then,

$$\hat{\nu}_i = \hat{m}_i \cos\theta + \hat{n}_i \sin\theta, \tag{3.48}$$

where θ is the angle between $\hat{\boldsymbol{\nu}}$ and $\hat{\mathbf{m}}$ measured around the $\hat{\boldsymbol{\tau}}$ axis in a right-handed sense, and after substituting Eqs. (3.48) and (3.46) into Eq. (3.47),

$$dF_j = \left[\oint_C (-\hat{m}_i \sin\theta + \hat{n}_i \cos\theta) \frac{\partial \psi_j}{\partial x_i} ds \right] dz. \tag{3.49}$$

However, the tangent to C in the direction of increasing θ is given by

$$\hat{\rho}_i = -m_i \sin\theta + n_i \cos\theta = \frac{dx_i}{ds}, \tag{3.50}$$

so that Eq. (3.49) becomes

$$dF_j = \left(\oint_C \frac{\partial \psi_j}{\partial x_i} dx_i \right) dz = \Delta\psi_j dz. \tag{3.51}$$

Now, if a line force of strength per unit length, f_i, is present along the $\hat{\boldsymbol{\tau}}$ axis, mechanical equilibrium requires that $f_j dz + dF_j = 0$. Therefore, upon substituting Eq. (3.51) into this requirement,

$$\Delta\psi_j = -f_j. \tag{3.52}$$

Otherwise,

$$\Delta\psi_{\mathrm{j}} = 0. \tag{3.53}$$

Equations (3.52) and (3.53) will be of use in Sec. 12.3 when expressions for the elastic fields of straight lines of force are developed.

3.5.1.2 *Properties of the Stroh vectors and relationships between them*

We now obtain a number of essential relationships between the Stroh vectors, $\mathbf{A}_\alpha$ and $\mathbf{L}_\alpha$, that must exist by virtue of Eqs. (3.37) and (3.38). The first step is to write the six equations that are embodied in these two expressions in the 6×6 block matrix form

$$\begin{bmatrix} (\hat{n}\hat{m}) & [\mathrm{I}] \\ (\hat{m}\hat{m}) & [0] \end{bmatrix} \begin{bmatrix} [\mathrm{A}_\alpha] \\ [\mathrm{L}_\alpha] \end{bmatrix} = \mathrm{p}_\alpha \begin{bmatrix} -(\hat{n}\hat{n}) & [0] \\ -(\hat{m}\hat{n}) & [\mathrm{I}] \end{bmatrix} \begin{bmatrix} [\mathrm{A}_\alpha] \\ [\mathrm{L}_\alpha] \end{bmatrix} \tag{3.54}$$

which, when written out in full, appears as

$$\begin{bmatrix} (\hat{n}\hat{m})_{11} & (\hat{n}\hat{m})_{12} & (\hat{n}\hat{m})_{13} & 1 & 0 & 0 \\ (\hat{n}\hat{m})_{21} & (\hat{n}\hat{m})_{22} & (\hat{n}\hat{m})_{23} & 0 & 1 & 0 \\ (\hat{n}\hat{m})_{31} & (\hat{n}\hat{m})_{32} & (\hat{n}\hat{m})_{33} & 0 & 0 & 1 \\ (\hat{m}\hat{m})_{11} & (\hat{m}\hat{m})_{12} & (\hat{m}\hat{m})_{13} & 0 & 0 & 0 \\ (\hat{m}\hat{m})_{21} & (\hat{m}\hat{m})_{22} & (\hat{m}\hat{m})_{23} & 0 & 0 & 0 \\ (\hat{m}\hat{m})_{31} & (\hat{m}\hat{m})_{32} & (\hat{m}\hat{m})_{33} & 0 & 0 & 0 \end{bmatrix} \begin{bmatrix} \mathrm{A}_{1\alpha} \\ \mathrm{A}_{2\alpha} \\ \mathrm{A}_{3\alpha} \\ \mathrm{L}_{1\alpha} \\ \mathrm{L}_{2\alpha} \\ \mathrm{L}_{3\alpha} \end{bmatrix}$$

$$= -\mathrm{p}_\alpha \begin{bmatrix} (\hat{n}\hat{n})_{11} & (\hat{n}\hat{n})_{12} & (\hat{n}\hat{n})_{13} & 0 & 0 & 0 \\ (\hat{n}\hat{n})_{21} & (\hat{n}\hat{n})_{22} & (\hat{n}\hat{n})_{23} & 0 & 0 & 0 \\ (\hat{n}\hat{n})_{31} & (\hat{n}\hat{n})_{32} & (\hat{n}\hat{n})_{33} & 0 & 0 & 0 \\ (\hat{m}\hat{n})_{11} & (\hat{m}\hat{n})_{12} & (\hat{m}\hat{n})_{13} & -1 & 0 & 0 \\ (\hat{m}\hat{n})_{21} & (\hat{m}\hat{n})_{22} & (\hat{m}\hat{n})_{23} & 0 & -1 & 0 \\ (\hat{m}\hat{n})_{31} & (\hat{m}\hat{n})_{32} & (\hat{m}\hat{n})_{33} & 0 & 0 & -1 \end{bmatrix} \begin{bmatrix} \mathrm{A}_{1\alpha} \\ \mathrm{A}_{2\alpha} \\ \mathrm{A}_{3\alpha} \\ \mathrm{L}_{1\alpha} \\ \mathrm{L}_{2\alpha} \\ \mathrm{L}_{3\alpha} \end{bmatrix}. \tag{3.55}$$

Pre-multiplying Eq. (3.54) throughout by the matrix

$$\begin{bmatrix} (\hat{n}\hat{m})^{-1} & [0] \\ (\hat{m}\hat{n})(\hat{n}\hat{n})^{-1} & -[\mathrm{I}] \end{bmatrix} \tag{3.56}$$

then yields

$$[\mathrm{N}]\begin{bmatrix}[\mathrm{A}_\alpha]\\ [\mathrm{L}_\alpha]\end{bmatrix} = \mathrm{p}_\alpha \begin{bmatrix}[\mathrm{A}_\alpha]\\ [\mathrm{L}_\alpha]\end{bmatrix}, \tag{3.57}$$

where

$$[\mathrm{N}] = -\begin{bmatrix}(\hat{\mathrm{n}}\hat{\mathrm{n}})^{-1}(\hat{\mathrm{n}}\hat{\mathrm{m}}) & (\hat{\mathrm{n}}\hat{\mathrm{n}})^{-1}\\ (\hat{\mathrm{m}}\hat{\mathrm{n}})(\hat{\mathrm{n}}\hat{\mathrm{n}})^{-1}(\hat{\mathrm{n}}\hat{\mathrm{m}}) - (\hat{\mathrm{m}}\hat{\mathrm{m}}) & (\hat{\mathrm{m}}\hat{\mathrm{n}})(\hat{\mathrm{n}}\hat{\mathrm{n}})^{-1}\end{bmatrix}. \tag{3.58}$$

Equation (3.57) is seen to constitute an eigenvalue problem involving the 6-dimensional column vectors

$$\begin{bmatrix}[\mathrm{A}_\alpha]\\ [\mathrm{L}_\alpha]\end{bmatrix} \tag{3.59}$$

and the determinantal requirement

$$\det|[\mathrm{N}] - \mathrm{p}_\alpha[\mathrm{I}]| = 0. \tag{3.60}$$

Equation (3.57) is now employed to establish a number of properties of the $\mathbf{A}_\alpha$ and $\mathbf{L}_\alpha$ Stroh vectors and relationships between them.

Orthogonality of Stroh vectors

Consider two different sets of eigenvalues and Stroh vectors, identified by α and β, respectively, which obey Eq. (3.57), so that

$$[\mathrm{N}]\begin{bmatrix}[\mathrm{A}_\alpha]\\ [\mathrm{L}_\alpha]\end{bmatrix} = \mathrm{p}_\alpha \begin{bmatrix}[\mathrm{A}_\alpha]\\ [\mathrm{L}_\alpha]\end{bmatrix} \quad [\mathrm{N}]\begin{bmatrix}[\mathrm{A}_\beta]\\ [\mathrm{L}_\beta]\end{bmatrix} = \mathrm{p}_\beta \begin{bmatrix}[\mathrm{A}_\beta]\\ [\mathrm{L}_\beta]\end{bmatrix}. \tag{3.61}$$

Now, pre-multiply the first equation by the 1×6 row matrix $[[\mathrm{L}_\beta][\mathrm{A}_\beta]]$ and the second by $[[\mathrm{L}_\alpha][\mathrm{A}_\alpha]]$ and subtract the latter result from the former to obtain

$$\begin{aligned}&\mathrm{p}_\alpha[[\mathrm{L}_\beta][\mathrm{A}_\beta]]\begin{bmatrix}[\mathrm{A}_\alpha]\\ [\mathrm{L}_\alpha]\end{bmatrix} - \mathrm{p}_\beta[[\mathrm{L}_\alpha][\mathrm{A}_\alpha]]\begin{bmatrix}[\mathrm{A}_\beta]\\ [\mathrm{L}_\beta]\end{bmatrix}\\ &\quad = [[\mathrm{L}_\beta][\mathrm{A}_\beta]][\mathrm{N}]\begin{bmatrix}[\mathrm{A}_\alpha]\\ [\mathrm{L}_\alpha]\end{bmatrix} - [[\mathrm{L}_\alpha][\mathrm{A}_\alpha]]\,[\mathrm{N}]\begin{bmatrix}[\mathrm{A}_\beta]\\ [\mathrm{L}_\beta]\end{bmatrix}.\end{aligned} \tag{3.62}$$

To proceed with Eq. (3.62), the 6×6 [N] matrix, given by Eq. (3.58), is now developed further. First, write it in the block form

$$[\mathrm{N}] = -\begin{bmatrix}[\mathrm{R}] & [\mathrm{H}]\\ [\mathrm{F}] & [\mathrm{G}]\end{bmatrix} \tag{3.63}$$

where,

$$[R] = (\hat{n}\hat{n})^{-1}(\hat{n}\hat{m}),$$
$$[H] = (\hat{n}\hat{n})^{-1}, \tag{3.64}$$
$$[F] = (\hat{m}\hat{n})(\hat{n}\hat{n})^{-1}(\hat{n}\hat{m}) - (\hat{m}\hat{m}),$$
$$[G] = (\hat{m}\hat{n})(\hat{n}\hat{n})^{-1}.$$

However, the symmetry properties of the C_{ijkm} tensor show that

$$(\hat{n}\hat{m})_{jk} = \hat{n}_i C_{ijkm}\hat{m}_m = \hat{n}_i C_{kmij}\hat{m}_m = \hat{n}_i C_{mkji}\hat{m}_m = \hat{m}_m C_{mkji}\hat{n}_i = (\hat{m}\hat{n})_{kj} \tag{3.65}$$

or, alternatively,

$$(\hat{n}\hat{m}) = (\hat{m}\hat{n})^T. \tag{3.66}$$

Using this result and the standard transpose identity $[[A][B]]^T = [B]^T[A]^T$, it is found from Eq. (3.64) that

$$[R]^T = [(\hat{n}\hat{n})^{-1}(\hat{n}\hat{m})]^T = (\hat{n}\hat{m})^T[(\hat{n}\hat{n})^{-1}]^T = (\hat{m}\hat{n})(\hat{n}\hat{n})^{-1} = [G],$$
$$[H]^T = [(\hat{n}\hat{n})^{-1}]^T = (\hat{n}\hat{n})^{-1} = [H], \tag{3.67}$$
$$[F]^T = [(\hat{m}\hat{n})(\hat{n}\hat{n})^{-1}(\hat{n}\hat{m}) - (\hat{m}\hat{m})]^T = (\hat{n}\hat{m})^T[(\hat{n}\hat{n})^{-1}]^T(\hat{m}\hat{n})^T - (\hat{m}\hat{m})^T$$
$$= (\hat{m}\hat{n})(\hat{n}\hat{n})^{-1}(\hat{n}\hat{m}) - (\hat{m}\hat{m}) = [F].$$

Then putting these results into Eq. (3.62) and expanding its left side,

$$(p_\alpha - p_\beta)(L_{i\alpha}A_{i\beta} + L_{i\beta}A_{i\alpha}) = [[L_\beta][A_\beta]]\begin{bmatrix}[R] & [H]\\ [F] & [R]^T\end{bmatrix}\begin{bmatrix}[A_\alpha]\\ [L_\alpha]\end{bmatrix}$$
$$-[[L_\alpha][A_\alpha]]\begin{bmatrix}[R] & [H]\\ [F] & [R]^T\end{bmatrix}\begin{bmatrix}[A_\beta]\\ [L_\beta]\end{bmatrix}. \tag{3.68}$$

When the right-hand side of Eq. (3.68) is multiplied out, and the symmetric character of [H] and [F] is taken into account, it is found to vanish. Therefore,

$$(p_\alpha - p_\beta)(A_{i\alpha}L_{i\beta} + L_{i\alpha}A_{i\beta}) = 0 \tag{3.69}$$

and, if $\alpha \neq \beta$, and $p_\alpha \neq p_\beta$, the relationship

$$(A_{i\alpha}L_{i\beta} + L_{i\alpha}A_{i\beta}) = 0 \quad (\alpha \neq \beta) \tag{3.70}$$

must be valid. Then, by applying the normalization condition $2A_{i\alpha}L_{i\alpha} = 1$, expressed by Eq. (3.39), the general *orthogonality relationship* between the $\mathbf{A}_\alpha$ and $\mathbf{L}_\alpha$ Stroh vectors is given (in component and matrix forms) by

$$(A_{i\alpha}L_{i\beta} + L_{i\alpha}A_{i\beta}) = \delta_{\alpha\beta} \quad \text{or} \quad [A]^T[L] + [L]^T[A] = [I]. \tag{3.71}$$

Completeness of Stroh vectors

As now demonstrated, the Stroh vectors $\mathbf{A}_\alpha$ and $\mathbf{L}_\alpha$ must obey certain *completeness relationships.* Suppose that the solution of the following pair of simultaneous equations

$$\begin{aligned} \sum_{\alpha=1}^{6} A_{i\alpha}D_\alpha &= g_i \\ \sum_{\alpha=1}^{6} L_{i\alpha}D_\alpha &= h_i \end{aligned} \tag{3.72}$$

is sought for the complex constant, D_α, when the vectors $\mathbf{g}$ and $\mathbf{h}$ are arbitrary. By multiplying the first equation by $L_{i\beta}$ and the second by $A_{i\beta}$, and adding the results,

$$\sum_{\alpha=1}^{6} (A_{i\alpha}L_{i\beta} + L_{i\alpha}A_{i\beta})D_\alpha = g_iL_{i\beta} + h_iA_{i\beta}. \tag{3.73}$$

Then, substituting the orthogonality relation, Eq. (3.71), into Eq. (3.73),

$$\sum_{\alpha=1}^{6} \delta_{\alpha\beta}D_\alpha = D_\beta = g_iL_{i\beta} + h_iA_{i\beta}. \tag{3.74}$$

Next, substituting Eq. (3.74) back into Eq. (3.72)

$$\begin{aligned} \left(\sum_{\alpha=1}^{6} A_{i\alpha}L_{s\alpha}\right) g_s + \left(\sum_{\alpha=1}^{6} A_{i\alpha}A_{s\alpha}\right) h_s &= g_i, \\ \left(\sum_{\alpha=1}^{6} L_{i\alpha}L_{s\alpha}\right) g_s + \left(\sum_{\alpha=1}^{6} L_{i\alpha}A_{s\alpha}\right) h_s &= h_i. \end{aligned} \tag{3.75}$$

However, since $\mathbf{g}$ and $\mathbf{h}$ are arbitrary, and by virtue of Eqs. (3.35) and (3.40), the following completeness relationships must exist:

$$\sum_{\alpha=1}^{6} A_{i\alpha}A_{s\alpha} = \sum_{\alpha=1}^{3} A_{i\alpha}A_{s\alpha} + \sum_{\alpha=1}^{3} A^*_{i\alpha}A^*_{s\alpha} = 0 \quad \text{or}$$

$$[A][A]^T + [A^*][A^*]^T = [0], \tag{3.76}$$

$$\sum_{\alpha=1}^{6} \mathrm{L}_{\mathrm{i}\alpha}\mathrm{L}_{\mathrm{s}\alpha} = \sum_{\alpha=1}^{3} \mathrm{L}_{\mathrm{i}\alpha}\mathrm{L}_{\mathrm{s}\alpha} + \sum_{\alpha=1}^{3} \mathrm{L}^*_{\mathrm{i}\alpha}\mathrm{L}^*_{\mathrm{s}\alpha} = 0 \quad \text{or}$$

$$[\mathrm{L}][\mathrm{L}]^{\mathrm{T}} + [\mathrm{L}^*][\mathrm{L}^*]^{\mathrm{T}} = [0], \tag{3.77}$$

$$\sum_{\alpha=1}^{6} \mathrm{A}_{\mathrm{i}\alpha}\mathrm{L}_{\mathrm{s}\alpha} = \sum_{\alpha=1}^{3} \mathrm{A}_{\mathrm{i}\alpha}\mathrm{L}_{\mathrm{s}\alpha} + \sum_{\alpha=1}^{3} \mathrm{A}^*_{\mathrm{i}\alpha}\mathrm{L}^*_{\mathrm{s}\alpha} = \delta_{\mathrm{is}} \quad \text{or}$$

$$[\mathrm{A}][\mathrm{L}]^{\mathrm{T}} + [\mathrm{A}^*][\mathrm{L}^*]^{\mathrm{T}} = [\mathrm{I}]. \tag{3.78}$$

Invariance of Stroh vectors

Finally, it is shown that the Stroh vectors, $\mathbf{A}_\alpha$ and $\mathbf{L}_\alpha$, are independent of the angle ω that measures the inclination of the $\hat{\mathbf{m}}$ and $\hat{\mathbf{n}}$ base vectors of the $(\hat{\mathbf{m}}, \hat{\mathbf{n}}, \hat{\boldsymbol{\tau}})$ coordinate system in the $\mathbf{x} \cdot \hat{\boldsymbol{\tau}} = 0$ plane (Fig. 3.2). To obtain this important result it is convenient first to adopt the change in notation

$$\begin{bmatrix} [\mathrm{A}_\alpha] \\ [\mathrm{L}_\alpha] \end{bmatrix} \rightarrow [\zeta_\alpha] \tag{3.79}$$

so that Eq. (3.57) can be written more simply as

$$[\mathrm{N}][\zeta_\alpha] = \mathrm{p}_\alpha[\zeta_\alpha]. \tag{3.80}$$

Then, differentiation of Eq. (3.80) yields

$$\frac{\partial[\mathrm{N}]}{\partial\omega}[\zeta_\alpha] + [\mathrm{N}]\frac{\partial[\zeta_\alpha]}{\partial\omega} = [\zeta_\alpha]\frac{\partial \mathrm{p}_\alpha}{\partial\omega} + \mathrm{p}_\alpha\frac{\partial[\zeta_\alpha]}{\partial\omega}. \tag{3.81}$$

To formulate the derivative $\partial[\mathrm{N}]/\partial\omega$ in Eq. (3.81), Eqs. (3.63) and (3.67) are used to obtain

$$\begin{aligned} \frac{\partial[\mathrm{N}]}{\partial\omega} &= -\frac{\partial}{\partial\omega}\begin{bmatrix} [\mathrm{R}] & [\mathrm{H}] \\ [\mathrm{F}] & [\mathrm{R}]^{\mathrm{T}} \end{bmatrix} \\ &= -\frac{\partial}{\partial\omega}\begin{bmatrix} (\hat{\mathrm{n}}\hat{\mathrm{n}})^{-1}(\hat{\mathrm{n}}\hat{\mathrm{m}}) & (\hat{\mathrm{n}}\hat{\mathrm{n}})^{-1} \\ (\hat{\mathrm{m}}\hat{\mathrm{n}})(\hat{\mathrm{n}}\hat{\mathrm{n}})^{-1}(\hat{\mathrm{n}}\hat{\mathrm{m}}) - (\hat{\mathrm{m}}\hat{\mathrm{m}}) & (\hat{\mathrm{m}}\hat{\mathrm{n}})(\hat{\mathrm{n}}\hat{\mathrm{n}})^{-1} \end{bmatrix}. \end{aligned} \tag{3.82}$$

Then to determine the derivative of the [N] matrix in Eq (3.82), it is noted from Fig. 3.2, that

$$\frac{\partial\hat{\mathbf{m}}}{\partial\omega} = \hat{\mathbf{n}} \qquad \frac{\partial\hat{\mathbf{n}}}{\partial\omega} = -\hat{\mathbf{m}} \tag{3.83}$$

and, using this result,

$$\begin{aligned}\frac{\partial(\hat{m}\hat{m})_{jk}}{\partial\omega} &= \hat{m}_i C_{ijkm}\frac{\partial\hat{m}_m}{\partial\omega} + \hat{m}_m C_{ijkm}\frac{\partial\hat{m}_i}{\partial\omega} \\ &= \hat{m}_i C_{ijkm}\hat{n}_m + \hat{m}_m C_{ijkm}\hat{n}_i = (\hat{m}\hat{n})_{jk} + (\hat{n}\hat{m})_{jk} \end{aligned} \tag{3.84}$$

or, equivalently,

$$\frac{\partial(\hat{m}\hat{m})}{\partial\omega} = (\hat{m}\hat{n}) + (\hat{n}\hat{m}). \tag{3.85}$$

Similarly,

$$\begin{aligned}\frac{\partial(\hat{n}\hat{n})}{\partial\omega} &= -(\hat{n}\hat{m}) - (\hat{m}\hat{n}), \\ \frac{\partial(\hat{m}\hat{n})}{\partial\omega} &= (\hat{n}\hat{n}) - (\hat{m}\hat{m}), \\ \frac{\partial(\hat{n}\hat{m})}{\partial\omega} &= -(\hat{m}\hat{m}) + (\hat{n}\hat{n}). \end{aligned} \tag{3.86}$$

To find the required derivative, $\partial(\hat{n}\hat{n})^{-1}/\partial\omega$, the equality

$$(\hat{n}\hat{n})(\hat{n}\hat{n})^{-1} = [\mathrm{I}] \tag{3.87}$$

is differentiated to obtain

$$\frac{\partial(\hat{n}\hat{n})^{-1}}{\partial\omega} = (\hat{n}\hat{n})^{-1}\left[(\hat{m}\hat{n}) + (\hat{n}\hat{m})\right](\hat{n}\hat{n})^{-1} \tag{3.88}$$

after substitution from Eq. (3.86). Finally, substituting these results into Eq. (3.82),

$$\frac{\partial[\mathrm{N}]}{\partial\omega} = -\begin{bmatrix} [\mathrm{I}] + (\hat{n}\hat{n})^{-1}[(\hat{m}\hat{n}) + (\hat{n}\hat{m})](\hat{n}\hat{n})^{-1}(\hat{n}\hat{m}) & (\hat{n}\hat{n})^{-1}[(\hat{m}\hat{n}) + (\hat{n}\hat{m})](\hat{n}\hat{n})^{-1} \\ -(\hat{n}\hat{n})^{-1}(\hat{m}\hat{m}) & \\ -(\hat{m}\hat{m})(\hat{n}\hat{n})^{-1}(\hat{n}\hat{m}) - (\hat{m}\hat{n})(\hat{n}\hat{n})^{-1}(\hat{m}\hat{m}) & [\mathrm{I}] - (\hat{m}\hat{m})(\hat{n}\hat{n})^{-1} \\ +(\hat{m}\hat{n})(\hat{n}\hat{n})^{-1}[(\hat{m}\hat{n}) + (\hat{n}\hat{m})](\hat{n}\hat{n})^{-1}(\hat{n}\hat{m}) & +(\hat{m}\hat{n})(\hat{n}\hat{n})^{-1}[(\hat{m}\hat{n}) + (\hat{n}\hat{m})]](\hat{n}\hat{n})^{-1} \end{bmatrix}. \tag{3.89}$$

But, it can be confirmed by direct multiplication that Eq. (3.89) is given relatively simply by

$$\frac{\partial[\mathrm{N}]}{\partial\omega} = -\{[\mathrm{I}] + [\mathrm{N}][\mathrm{N}]\} \tag{3.90}$$

where [I] is now the 6×6 identity matrix. Then, post-multiplying this expression by $[\zeta_\alpha]$,

$$\frac{\partial[\mathrm{N}]}{\partial\omega}[\zeta_\alpha] = -[[\zeta_\alpha] + [\mathrm{N}][\mathrm{N}][\zeta_\alpha]]. \tag{3.91}$$

An expression for the quantity $[\mathrm{N}][\mathrm{N}][\zeta_\alpha]$ in Eq. (3.91) is next obtained by first multiplying Eq. (3.80) throughout by [N], i.e.,

$$[\mathrm{N}][\mathrm{N}][\zeta_\alpha] = \mathrm{p}_\alpha[\mathrm{N}][\zeta_\alpha] = \mathrm{p}_\alpha^2[\zeta_\alpha]. \tag{3.92}$$

Then, substituting Eq. (3.92) into Eq. (3.91) and substituting the result into Eq. (3.81),

$$-(1+\mathrm{p}_\alpha^2)[\zeta_\alpha] + [\mathrm{N}]\frac{\partial[\zeta_\alpha]}{\partial\omega} = [\zeta_\alpha]\frac{\partial\mathrm{p}_\alpha}{\partial\omega} + \mathrm{p}_\alpha\frac{\partial[\zeta_\alpha]}{\partial\omega}. \tag{3.93}$$

To proceed further with the development of Eq. (3.93), it is necessary to formulate several additional expressions. First, take the transposes of Eqs. (3.80) and (3.63) to obtain

$$[\zeta_\alpha]^{\mathrm{T}}[\mathrm{N}]^{\mathrm{T}} = \mathrm{p}_\alpha[\zeta_\alpha]^{\mathrm{T}} \tag{3.94}$$

and

$$[\mathrm{N}]^{\mathrm{T}} = -\begin{bmatrix} [\mathrm{R}]^{\mathrm{T}} & [\mathrm{F}]^{\mathrm{T}} \\ [\mathrm{H}]^{\mathrm{T}} & [\mathrm{G}]^{\mathrm{T}} \end{bmatrix} = -\begin{bmatrix} [\mathrm{R}]^{\mathrm{T}} & [\mathrm{F}] \\ [\mathrm{H}] & [\mathrm{R}] \end{bmatrix} \tag{3.95}$$

since $[\mathrm{G}] = [\mathrm{R}]^{\mathrm{T}}$, $[\mathrm{H}] = [\mathrm{H}]^{\mathrm{T}}$ and $[\mathrm{F}] = [\mathrm{F}]^{\mathrm{T}}$ according to Eq. (3.67). Now, introduce the block matrix [J] having the following properties:

$$[\mathrm{J}] = \begin{bmatrix} [\mathrm{O}] & [\mathrm{I}] \\ [\mathrm{I}] & [\mathrm{O}] \end{bmatrix} \quad [\mathrm{J}] = [\mathrm{J}]^{\mathrm{T}} \quad [\mathrm{J}][\mathrm{J}] = [\mathrm{I}]. \tag{3.96}$$

Then

$$[\mathrm{J}][\mathrm{N}]^{\mathrm{T}}[\mathrm{J}] = -\begin{bmatrix} [\mathrm{O}] & [\mathrm{I}] \\ [\mathrm{I}] & [\mathrm{O}] \end{bmatrix}\begin{bmatrix} [\mathrm{R}]^{\mathrm{T}} & [\mathrm{F}] \\ [\mathrm{H}] & [\mathrm{R}] \end{bmatrix}\begin{bmatrix} [\mathrm{O}] & [\mathrm{I}] \\ [\mathrm{I}] & [\mathrm{O}] \end{bmatrix} = [\mathrm{N}]. \tag{3.97}$$

Next, post-multiplying Eq. (3.94) by [J], and employing Eqs. (3.96) and (3.97),

$$\mathrm{p}_\alpha[\zeta_\alpha]^{\mathrm{T}}[\mathrm{J}] = [\zeta_\alpha]^{\mathrm{T}}[\mathrm{N}]^{\mathrm{T}}[\mathrm{J}] = [\zeta_\alpha]^{\mathrm{T}}[\mathrm{J}][\mathrm{N}]. \tag{3.98}$$

Also, it may be seen that, with the help of Eq. (3.39), that

$$[\zeta_\alpha]^T[J][\zeta_\alpha] = [[A_\alpha][L_\alpha]][J]\begin{bmatrix}[A_\alpha]\\ [L_\alpha]\end{bmatrix} = [[A_\alpha][L_\alpha]]\begin{bmatrix}[L_\alpha]\\ [A_\alpha]\end{bmatrix} = 2A_{i\alpha}L_{i\alpha} = 1. \tag{3.99}$$

Having these results, pre-multiply Eq. (3.93) by $[\zeta_\alpha]^T[J]$ to obtain

$$\begin{aligned}&-(1+p_\alpha^2)[\zeta_\alpha]^T[J][\zeta_\alpha] + [\zeta_\alpha]^T[J][N]\frac{\partial[\zeta_\alpha]}{\partial\omega}\\ &= [\zeta_\alpha]^T[J][\zeta_\alpha]\frac{\partial p_\alpha}{\partial\omega} + p_\alpha[\zeta_\alpha]^T[J]\frac{\partial[\zeta_\alpha]}{\partial\omega}.\end{aligned} \tag{3.100}$$

Then, using Eqs. (3.98) and (3.99), the differential equation for p_α as a function of ω is obtained in the simple form

$$\frac{\partial p_\alpha(\omega)}{\partial\omega} = -[1+p_\alpha^2(\omega)]. \tag{3.101}$$

Substitution of Eq. (3.101) back into Eq. (3.93) yields

$$[N]\frac{\partial[\zeta_\alpha]}{\partial\omega} = p_\alpha\frac{\partial[\zeta_\alpha]}{\partial\omega}. \tag{3.102}$$

Comparison of this result with the initial eigenvalue equation, Eq. (3.80), shows that the two equations are consistent with one another only if a relationship of the form

$$\frac{\partial[\zeta_\alpha]}{\partial\omega} = K(\omega)[\zeta_\alpha] \tag{3.103}$$

exists where $K(\omega)$ is a scalar complex function of ω. $K(\omega)$ is then determined by premultiplying Eq. (3.103) by $[\zeta_\alpha]^T[J]$ and using Eq. (3.99), to obtain

$$[\zeta_\alpha]^T[J]\frac{\partial[\zeta_\alpha]}{\partial\omega} = K(\omega)[\zeta_\alpha]^T[J][\zeta_\alpha] = K(\omega). \tag{3.104}$$

Then, using this result,

$$\begin{aligned}K(\omega) &= [\zeta_\alpha]^T[J]\frac{\partial[\zeta_\alpha]}{\partial\omega} = [[A_\alpha][L_\alpha]]\begin{bmatrix}[O] & [I]\\ [I] & [O]\end{bmatrix}\begin{bmatrix}\frac{\partial}{\partial\omega}[A_\alpha]\\ \frac{\partial}{\partial\omega}[L_\alpha]\end{bmatrix}\\ &= A_{i\alpha}\frac{\partial L_{i\alpha}}{\partial\omega} + L_{i\alpha}\frac{\partial A_{i\alpha}}{\partial\omega}.\end{aligned} \tag{3.105}$$

However, differentiation of the normalization equation, Eq. (3.39), yields

$$A_{i\alpha}\frac{\partial L_{i\alpha}}{\partial\omega} + L_{i\alpha}\frac{\partial A_{i\alpha}}{\partial\omega} = 0 \tag{3.106}$$

and comparison of Eqs. (3.106) and (3.105) shows that $\mathrm{K}(\omega) = 0$. Substitution of this result into Eq. (3.103) finally yields the results

$$\frac{\partial[\zeta_\alpha]}{\partial\omega} = 0 \quad \text{or} \quad \frac{\partial \mathrm{A}_{i\alpha}}{\partial\omega} = 0 \quad \frac{\partial \mathrm{L}_{i\alpha}}{\partial\omega} = 0. \tag{3.107}$$

This establishes the important result that the Stroh vectors are both invariant with respect to ω, i.e., rotation of the $(\hat{\mathbf{m}}, \hat{\mathbf{n}}, \hat{\boldsymbol{\tau}})$ coordinate system around the $\hat{\boldsymbol{\tau}}$ axis (see Fig. 3.2). However, they are dependent upon the direction of $\hat{\boldsymbol{\tau}}$ and the elastic constants C_{ijkm}. In contrast, the eigenvalues, p_α, are functions of ω as indicated in Eq. (3.101).

Sum rules for Stroh vectors

A number of useful *sum rules* for the Stroh vectors exists, which can be established with the help of the completeness relationships from Sec. 3.5.1.2.

(i) Starting with Eq. (3.37), multiplying it throughout by $\mathrm{A}_{s\alpha}$, and summing over α,

$$\sum_{\alpha=1}^{6} \mathrm{A}_{s\alpha}\mathrm{L}_{j\alpha} = -(\hat{n}\hat{m})_{jk} \sum_{\alpha=1}^{6} \mathrm{A}_{s\alpha}\mathrm{A}_{k\alpha} - (\hat{n}\hat{n})_{jk} \sum_{\alpha=1}^{6} p_\alpha \mathrm{A}_{s\alpha}\mathrm{A}_{k\alpha}. \tag{3.108}$$

Using the completeness relationships expressed by Eqs. (3.76) and (3.78), Eq. (3.108) is reduced to

$$-(\hat{n}\hat{n})_{jk} \sum_{\alpha=1}^{6} p_\alpha \mathrm{A}_{s\alpha}\mathrm{A}_{k\alpha} = \delta_{sj} \tag{3.109}$$

which, after multiplication throughout by $(\hat{n}\hat{n})^{-1}_{jk}$, takes the form

$$\sum_{\alpha=1}^{6} p_\alpha \mathrm{A}_{s\alpha}\mathrm{A}_{k\alpha} = -(\hat{n}\hat{n})^{-1}_{jk}\delta_{sj} = -(\hat{n}\hat{n})^{-1}_{sk}. \tag{3.110}$$

(ii) Multiplying Eq. (3.37) through by $\mathrm{L}_{s\alpha}$ and summing over α,

$$\sum_{\alpha=1}^{6} \mathrm{L}_{s\alpha}\mathrm{L}_{j\alpha} = -(\hat{n}\hat{m})_{jk} \sum_{\alpha=1}^{6} \mathrm{L}_{s\alpha}\mathrm{A}_{k\alpha} - (\hat{n}\hat{n})_{jk} \sum_{\alpha=1}^{6} p_\alpha \mathrm{L}_{s\alpha}\mathrm{A}_{k\alpha}. \tag{3.111}$$

By using the completeness relationships given by Eqs. (3.77) and (3.78), Eq. (3.111) is reduced to

$$(\hat{n}\hat{n})_{jk} \sum_{\alpha=1}^{6} p_\alpha \mathrm{L}_{s\alpha}\mathrm{A}_{k\alpha} = -(\hat{n}\hat{m})_{jk}\delta_{sk} = -(\hat{n}\hat{m})_{js} \tag{3.112}$$

which, after multiplication throughout by $(\hat{n}\hat{n})^{-1}_{jk}$, takes the form

$$\sum_{\alpha=1}^{6} p_\alpha L_{s\alpha} A_{k\alpha} = -(\hat{n}\hat{n})^{-1}_{kj}(\hat{n}\hat{m})_{js}. \tag{3.113}$$

(iii) Multiplying Eq. (3.38) throughout by $L_{s\alpha}$, summing over α, applying the completeness relationship Eq. (3.78), and using Eq. (3.113),

$$\begin{aligned}\sum_{\alpha=1}^{6} p_\alpha L_{s\alpha} L_{j\alpha} &= (\hat{m}\hat{m})_{js} + (\hat{m}\hat{n})_{jk} \sum_{\alpha=1}^{6} p_\alpha L_{s\alpha} A_{k\alpha} \\ &= (\hat{m}\hat{m})_{js} - (\hat{m}\hat{n})_{jk}(\hat{n}\hat{n})^{-1}_{kr}(\hat{n}\hat{m})_{rs}. \end{aligned} \tag{3.114}$$

(iv) Multiplying Eq. (3.38) throughout by $A_{s\alpha}$, summing over α, applying the completeness Eqs. (3.76) and (3.78), and multiplying throughout by $(\hat{m}\hat{m})^{-1}_{jk}$,

$$\sum_{\alpha=1}^{6} \frac{1}{p_\alpha} A_{s\alpha} A_{k\alpha} = (\hat{m}\hat{m})^{-1}_{jk}\delta_{sj} = (\hat{m}\hat{m})^{-1}_{sk}. \tag{3.115}$$

Additional sum rule expressions exist which are listed and referenced in Bacon, Barnett and Scattergood (1979b). Also, see Exercise 3.2.

Relationships involving the inverse matrix, $M_{\alpha i}$

In many cases equations involving the Stroh matrices, [A] and [L], arise in the sextic and integral formalisms that can be solved by employing the matrix, [M], which is inverse to [L] and possesses the following properties, expressed in both component and matrix notation:

$$M_{i,\alpha+3} = M^*_{i\alpha} \quad (\alpha = 1, 2, 3) \tag{3.116}$$

$$M_{\alpha k} L_{k\beta} = \delta_{\alpha\beta} \quad \sum_{\alpha=1}^{3} M_{i\alpha} L_{\alpha j} = \delta_{ij} \quad (\alpha, \beta = 1, 2, 3) \quad \text{or} \quad [M] = [L]^{-1} \tag{3.117}$$

$$M^*_{\alpha k} L^*_{k\beta} = \delta_{\alpha\beta} \quad \sum_{\alpha=1}^{3} M^*_{i\alpha} L^*_{\alpha j} = \delta_{ij} \quad (\alpha, \beta = 1, 2, 3) \quad \text{or} \quad [M^*] = [L^*]^{-1} \tag{3.118}$$

The completeness relationship between the [L]-type matrices given by Eq. (3.77) implies a similar relationship between the [M]-type matrices. Substituting the equalities $[L] = [M]^{-1}$ and $[L^*] = [M^*]^{-1}$ into Eq. (3.77),

$$[M]^{-1}[[M]^{-1}]^T + [M^*]^{-1}[[M^*]^{-1}]^T = [0]. \tag{3.119}$$

However, $[[M]^{-1}]^T = [[M^T]]^{-1}$, and, therefore,

$$[M]^{-1}[[M]^T]^{-1} + [M^*]^{-1}[[M^*]^T]^{-1} = [M]^T[M] + [M^*]^T[M^*] = [0] \quad (3.120)$$

which in component notation takes the form

$$\sum_{\alpha=1}^{3} M_{\alpha i}M_{\alpha s} + \sum_{\alpha=1}^{3} M^*_{\alpha i}M^*_{\alpha s} = \sum_{\alpha=1}^{6} M_{\alpha i}M_{\alpha s} = 0. \quad (3.121)$$

There are also several useful relationships between the M_{ij} and M^*_{ij} matrices and the Stroh matrices. The first is obtained by first substituting $[L] = [M]^{-1}$ and $[L]^T = [[M]^T]^{-1}$ into Eq. (3.71). Then, multiplying the result throughout by $[M]^T$ and post-multiplying by $[M]$,

$$[A][M] + [M]^T[A]^T = [M]^T[M] \quad \text{or}$$

$$\sum_{\alpha=1}^{3} (A_{s\alpha}M_{\alpha i} + A_{i\alpha}M_{\alpha s}) = \sum_{\alpha=1}^{3} M_{\alpha s}M_{\alpha i}. \quad (3.122)$$

Another is obtained by multiplying Eq. (3.77) throughout by $[M^*]$ and then post-multiplying by $[M]^T$ to produce

$$[M^*][L] + [L^*]^T[M]^T = [0] \quad \text{or} \quad M^*_{\alpha i}L_{i\beta} + L^*_{i\alpha}M_{\beta i} = 0. \quad (3.123)$$

Still another is obtained by first substituting $[L] = [M]^{-1}$ and $[L^*] = [M^*]^{-1}$ into Eqs. (3.71) and (3.78) to produce, respectively,

$$[A]^T[M]^{-1} + \left[[M]^{-1}\right]^T[A] = [I] \quad (3.124)$$

and

$$[A][[M]^{-1}]^T + [A^*][[M^*]^{-1}]^T = [I]. \quad (3.125)$$

Then, by post-multiplying Eq. (3.124) by $[M]$ and multiplying the result by $[M]^T$, and transposing Eq. (3.125) and multiplying the result by $[M]$ and then by $[M]^T$, respectively,

$$[M]^T[A]^T + [A][M] = [M]^T[M] \quad (3.126)$$

and

$$[M]^T[A]^T + [M]^T[M][M^*]^{-1}[A^*]^T = [M]^T[M]. \quad (3.127)$$

Subtraction of Eq. (3.127) from Eq. (3.126) then yields

$$[A][M] = [M]^T[M][M^*]^{-1}[A^*]^T \tag{3.128}$$

and substitution of Eq. (3.120) into Eq. (3.128) finally yields

$$[A][M] + [M^*]^T[A^*]^T = [0] \quad \text{or} \quad \sum_{\alpha=1}^{3}(A_{s\alpha}M_{\alpha i} + A^*_{i\alpha}M^*_{\alpha s}) = 0. \tag{3.129}$$

3.5.2 *Integral formalism*

The sextic formalism requires that the eigenvalue problem posed by Eq. (3.32) be solved. However, Barnett, Lothe, Malen and others (see, Bacon, Barnett and Scattergood 1979b), have used elements of the sextic formalism to construct an *integral formalism* that eliminates the necessity of solving this sometimes inconvenient problem. The essential quantities developed in the formalism are expressed in the form of well-behaved integrals that can be programmed efficiently for numerical calculations and are valid at the isotropic limit where the eigenvalues of the sextic formalism become degenerate (Nishioka and Lothe 1972). In the following, integral expressions are obtained for three quantities of the integral formalism that will be particularly useful in the treatment of dislocations and lines of force, namely the matrices Q_{sk}, S_{ks} and B_{js}.

3.5.2.1 *The matrices* Q_{sk}, S_{sk} *and* B_{sk}

As shown in Sec. 3.5.1.2, the vectors $\mathbf{A}_\alpha$ and $\mathbf{L}_\alpha$ are independent of ω in Fig. 3.2. A sum rule, such as Eq. (3.110), can therefore be integrated over the angular range $0 \le \omega \le 2\pi$ to obtain

$$\int_0^{2\pi}\sum_{\alpha=1}^{6} p_\alpha(\omega)A_{s\alpha}A_{k\alpha}d\omega = -\int_0^{2\pi}(\hat{n}\hat{n})^{-1}_{sk}d\omega = \sum_{\alpha=1}^{6}A_{s\alpha}A_{k\alpha}\int_0^{2\pi}p_\alpha(\omega)d\omega. \tag{3.130}$$

Now, the quantity $p_\alpha(\omega)$ in Eq. (3.130) is readily found by integrating Eq. (3.101) and is given by

$$p_\alpha(\omega) = \tan(\psi_\alpha - \omega) = -i\left\{\frac{1 - \exp[-i2(\psi_\alpha - \omega)]}{1 + \exp[-i2(\psi_\alpha - \omega)]}\right\} \tag{3.131}$$

since the constant of integration, ψ_α, is complex. Having this, the integral having the integrand $p_\alpha(\omega)$ in Eq. (3.130) is given by

$$\int_0^{2\pi} p_\alpha(\omega)d\omega = -i\int_0^{2\pi}\left\{\frac{1-\exp[-i2(\psi_\alpha-\omega)]}{1+\exp[-i2(\psi_\alpha-\omega)]}\right\}d\omega$$
$$=\begin{cases} 2\pi i & (\alpha=1,2,3)\\ -2\pi i & (\alpha=4,5,6)\end{cases} \qquad (3.132)$$

as shown by Bacon, Barnett and Scattergood (1979b) by means of contour integration on the complex plane. Substitution of Eq. (3.132) into Eq. (3.130) then yields the expression

$$i\sum_{\alpha=1}^{6}\pm A_{s\alpha}A_{k\alpha} = -\frac{1}{2\pi}\int_0^{2\pi}(\hat{n}\hat{n})^{-1}_{sk}d\omega \equiv Q_{sk} \qquad (3.133)$$

where the $\pm$ symbol has been employed. As indicated, Eq. (3.133) defines the Q_{ks} matrix, and it is seen that its evaluation involves an integration around a unit circle normal to the invariant direction.

Proceeding in the same fashion with the sum rules given by Eqs. (3.113) and 3.114), the matrices, S_{ks} and B_{js} are obtained in the respective forms

$$i\sum_{\alpha=1}^{6}\pm A_{k\alpha}L_{s\alpha} = -\frac{1}{2\pi}\int_0^{2\pi}(\hat{n}\hat{n})^{-1}_{kj}(\hat{n}\hat{m})_{js}d\omega \equiv S_{ks} \qquad (3.134)$$

and

$$\frac{i}{4\pi}\sum_{\alpha=1}^{6}\pm L_{s\alpha}L_{j\alpha} = \frac{1}{8\pi^2}\int_0^{2\pi}[(\hat{m}\hat{m})_{js}-(\hat{m}\hat{n})_{jk}(\hat{n}\hat{n})^{-1}_{kr}(\hat{n}\hat{m})_{rs}]d\omega \equiv B_{js}. \qquad (3.135)$$

The B_{js} matrix can be expressed in an alternative form by first writing Eq. (3.77) as

$$\sum_{\alpha=1}^{6}L_{s\alpha}L_{j\alpha} = \sum_{\alpha=1}^{3}L_{s\alpha}L_{j\alpha} + \sum_{\alpha=4}^{6}L_{s\alpha}L_{j\alpha} = 0. \qquad (3.136)$$

Then, substituting Eq. (3.136) into Eq. (3.135)

$$B_{js} = \frac{i}{4\pi}\left(\sum_{\alpha=1}^{3}L_{s\alpha}L_{j\alpha} - \sum_{\alpha=4}^{6}L_{s\alpha}L_{j\alpha}\right) = \frac{i}{2\pi}\sum_{\alpha=1}^{3}L_{s\alpha}L_{j\alpha}. \qquad (3.137)$$

Since the eigenvectors are independent of ω, each of the three matrices, Q_{sk}, S_{ks} and B_{sk} depends only upon the direction of $\hat{\mathbf{t}}$ and the elastic constants, C_{ijkm}. The Q_{sk} and B_{js} matrices are symmetric, and since the integrals are real, all three matrices are real. Furthermore, all integrals are well behaved and progress smoothly to the isotropic limit.

Various relationships also exist between the three matrices. For the pair [B] and [S],

$$
\begin{aligned}
B_{ij}S_{jk} + S_{ji}B_{jk} &= -\frac{1}{4\pi}\left[\sum_{\alpha=1}^{6} \pm L_{i\alpha}L_{j\alpha}\sum_{\beta=1}^{6} \pm A_{j\beta}L_{k\beta}\right. \\
&\qquad \left. + \sum_{\alpha=1}^{6} \pm A_{j\alpha}L_{i\alpha}\sum_{\beta=1}^{6} \pm L_{j\beta}L_{k\beta}\right] \\
&= -\frac{1}{4\pi}\sum_{\alpha=1}^{6}\sum_{\beta=1}^{6}[(\pm L_{i\alpha})(\pm L_{k\beta})(L_{j\alpha}A_{j\beta} + A_{j\alpha}L_{j\beta})] \\
&= -\frac{1}{4\pi}\sum_{\alpha=1}^{6}\sum_{\beta=1}^{6}[(\pm L_{i\alpha})(\pm L_{k\beta})\delta_{\alpha\beta}] \\
&= -\frac{1}{4\pi}\sum_{\alpha=1}^{6} L_{i\alpha}L_{k\alpha} = 0
\end{aligned} \tag{3.138}
$$

after using Eqs. (3.134), (3.135), (3.71), and (3.77). Using similar methods, for the pair [Q] and [S],

$$
Q_{ij}S_{kj} + S_{ij}Q_{jk} = 0 \tag{3.139}
$$

and, in addition, as shown in Exercise 3.3,

$$
4\pi B_{ij}Q_{jk} + S_{ji}S_{kj} = -\delta_{ik}. \tag{3.140}
$$

3.5.2.2 *The matrices* Q_{sk}, S_{sk} *and* B_{sk} *for isotropic systems*

The matrices Q_{sk}, S_{sk} and B_{sk} take relatively simple forms in isotropic systems and can be obtained by evaluating the integrals in Eqs. (3.133), (3.134) and (3.135). The first step is to determine the forms taken by the Christoffel stiffness tensors that appear in the integrands. By direct substitution of Eq. (2.120) for C_{ijkl}, the following results are obtained:

$$
\begin{aligned}
(\hat{n}\hat{n})_{jk} &= \hat{n}_i C_{ijkl}\hat{n}_l = \mu\left(\delta_{jk} + \frac{1}{1-2\nu}\hat{n}_j\hat{n}_k\right), \\
(\hat{n}\hat{n})^{-1}_{jk} &= \frac{1}{\mu}\left[\delta_{jk} - \frac{1}{2(1-\nu)}\hat{n}_j\hat{n}_k\right], \\
(\hat{n}\hat{m})_{jk} &= (\hat{m}\hat{n})_{kj} = \frac{2\mu\nu}{1-2\nu}\hat{n}_j\hat{m}_k + \mu\hat{n}_k\hat{m}_j.
\end{aligned} \tag{3.141}
$$

Q_{jk} can then be expressed as

$$Q_{jk} = -\frac{1}{2\pi}\int_0^{2\pi} (\hat{n}\hat{n})^{-1}_{jk} d\omega = -\frac{1}{2\pi\mu}\int_0^{2\pi}\left[\delta_{jk} - \frac{1}{2(1-\nu)}\hat{n}_j\hat{n}_k\right] d\omega. \tag{3.142}$$

For the integration, it is convenient to introduce a new $(\hat{\mathbf{M}}, \hat{\mathbf{N}}, \hat{\boldsymbol{\tau}})$ orthogonal coordinate system where $\hat{\mathbf{M}}$ lies along the reference line in Fig. 3.2, and the base vectors of the $(\hat{\mathbf{M}}, \hat{\mathbf{N}}, \hat{\boldsymbol{\tau}})$ and $(\hat{\mathbf{m}}, \hat{\mathbf{n}}, \hat{\boldsymbol{\tau}})$ systems are related by

$$\begin{aligned} \hat{\mathbf{m}} &= \cos\omega\hat{\mathbf{M}} + \sin\omega\hat{\mathbf{N}} & \hat{\mathbf{M}} &= \cos\omega\hat{\mathbf{m}} - \sin\omega\hat{\mathbf{n}} \\ \hat{\mathbf{n}} &= -\sin\omega\hat{\mathbf{M}} + \cos\omega\hat{\mathbf{N}} \quad \text{or} & \hat{\mathbf{N}} &= \sin\omega\hat{\mathbf{m}} + \cos\omega\hat{\mathbf{n}} \\ \hat{\boldsymbol{\tau}} &= \hat{\boldsymbol{\tau}} & \hat{\boldsymbol{\tau}} &= \hat{\boldsymbol{\tau}} \end{aligned} \tag{3.143}$$

Then, substituting Eq. (3.143) into Eq. (3.142),

$$\begin{aligned} Q_{jk} &= -\frac{1}{2\pi\mu}\int_0^{2\pi}\left[\delta_{jk} - \frac{1}{2(1-\nu)}(-\hat{M}_j\sin\omega + \hat{N}_j\cos\omega)\right. \\ &\quad \left.\times(-\hat{M}_k\sin\omega + \hat{N}_k\cos\omega)\right] d\omega \\ &= -\frac{\delta_{jk}}{\mu} + \frac{1}{4\mu(1-\nu)}(\hat{M}_j\hat{M}_k + \hat{N}_j\hat{N}_k) \\ &= -\frac{\delta_{jk}}{\mu} + \frac{1}{4\mu(1-\nu)}(\hat{m}_j\hat{m}_k + \hat{n}_j\hat{n}_k). \end{aligned} \tag{3.144}$$

If the $(\hat{\mathbf{m}}, \hat{\mathbf{n}}, \hat{\boldsymbol{\tau}})$ system is taken as the "new" system, and the crystal system as the "old" system, for a transformation of vector components, the transformation matrix for vector coordinates corresponding to $[l]$ in Eq. (2.15) is

$$[l] = \begin{bmatrix} \hat{m}_1 & \hat{m}_2 & \hat{m}_3 \\ \hat{n}_1 & \hat{n}_2 & \hat{n}_3 \\ \hat{\tau}_1 & \hat{\tau}_2 & \hat{\tau}_3 \end{bmatrix}. \tag{3.145}$$

Since $[l]$ is a unitary orthogonal matrix (see Sec. 2.2.2.1), $\hat{m}_j\hat{m}_k + \hat{n}_j\hat{n}_k + \hat{\tau}_j\hat{\tau}_k = \delta_{jk}$, and substituting this into Eq. (3.144),

$$Q_{jk} = -\frac{1}{4\mu(1-\nu)}[(3-4\nu)\delta_{jk} + \hat{\tau}_j\hat{\tau}_k]. \tag{3.146}$$

For an isotropic system, the $(\hat{\mathbf{m}}, \hat{\mathbf{n}}, \hat{\boldsymbol{\tau}})$ system can be replaced with a Cartesian system that can be oriented arbitrarily in the crystal, and, if it is aligned so that its $\hat{\tau}$ axis is parallel with $\hat{\mathbf{e}}_3$, the [Q] matrix assumes the form

$$[\mathrm{Q}] = -\frac{(3-4\nu)}{4\mu(1-\nu)}\begin{bmatrix} 1 & 0 & 0 \\ 0 & 1 & 0 \\ 0 & 0 & \dfrac{4(1-\nu)}{(3-4\mu)} \end{bmatrix} \qquad [\mathrm{S}] = -\frac{(1-2\nu)}{2(1-\nu)}\begin{bmatrix} 0 & 1 & 0 \\ -1 & 0 & 0 \\ 0 & 0 & 0 \end{bmatrix}$$

$$[\mathrm{B}] = \frac{\mu}{4\pi}\begin{bmatrix} \dfrac{1}{1-\nu} & 0 & 0 \\ 0 & \dfrac{1}{1-\nu} & 0 \\ 0 & 0 & 1 \end{bmatrix}. \tag{3.147}$$

The elements of the corresponding [S] and [B] matrices can be found in a similar manner, and are included in Eq. (3.147).

3.6 Elasticity Theory for Systems Containing Transformation Strains

Many of the defects in the present book can be mimicked by introducing a localized *transformation strain* which represents the defect and serves as the source of its elastic field. As will be seen, the use of transformation strains can be a powerful means for modeling inclusions (Ch. 6) as well as dislocations, which can be represented by highly localized transformation strains in the form of delta functions (Sec. 12.4.1.1). In addition to other applications they can be used treat the internal stresses caused by the differential thermal expansion that occurs in non-uniformly heated bodies.[7]

Until now, it has been assumed that all of the displacements and strains are purely elastic. However, as described below, this is not the case when transformation strains are introduced, and in the following the previous elastic theory is modified to produce a theory admitting both elastic and transformation displacements and strains.

For this section we employ the following notation:

$\varepsilon^{\mathrm{T}}_{\mathrm{ij}}$ = transformation strain

$\varepsilon_{\mathrm{ij}}$ = elastic strain

[7] Mura has made particular use of the transformation strain method for treating crystal defects, and his book (Mura 1987) should be consulted for many details and applications.

$\varepsilon_{ij}^{tot} = \varepsilon_{ij} + \varepsilon_{ij}^{T} =$ total strain

$\varepsilon_{ij}^{\mathcal{V}^D}, \varepsilon_{ij}^{M} =$ elastic strain in the region $\mathcal{V}^D$ and in the matrix, respectively

$\varepsilon_{ij}^{tot,\mathcal{V}^D}, \varepsilon_{ij}^{tot,M} =$ total strain in the region $\mathcal{V}^D$ and in the matrix, respectively

$\varepsilon_{ij}^{C,\mathcal{V}^D}, \varepsilon_{ij}^{C,M} =$ canceling elastic strain in the region $\mathcal{V}^D$ and the matrix, respectively

3.6.1 *Transformation strain formalism*

A transformation strain can be introduced into an initially stress-free homogeneous body by the following steps:

(1) Cut out of the body the local region, $\mathcal{V}^D$, that is to receive the transformation strain that will mimic the defect, D.

(2) Alter the size and shape of $\mathcal{V}^D$ without leaving residual stress by means of plastic deformation, phase transformation or the addition/removal of material. Describe the change in size and shape (which need not be uniform) by an effective transformation strain, $\varepsilon_{ij}^{T}(\mathbf{x})$.[8]

(3) Apply surface and body forces to $\mathcal{V}^D$ which cancel the transformation strain, $\varepsilon_{ij}^{T}(\mathbf{x})$. Then if

$$\sigma_{ij}^{T}(\mathbf{x}) = C_{ijkl}\varepsilon_{kl}^{T}(\mathbf{x}) \tag{3.148}$$

is the stress produced by an elastic strain, $\varepsilon_{ij}^{T}(\mathbf{x})$, the tractions and body forces in $\mathcal{V}^D$ at the end of this stage are

$$T_i(\mathbf{x}) = -\sigma_{ij}^{T}(\mathbf{x})\hat{n}_j \quad \text{and} \quad f_i(\mathbf{x}) = \partial\sigma_{ij}^{T}(\mathbf{x})/\partial x_j. \tag{3.149}$$

(4) Insert $\mathcal{V}^D$ back into its cavity in the unstressed body, which it now matches exactly, and bond it to the body.

(5) Cancel the surface tractions and body forces from step 3 by imposing equal and opposite forces which produce the *canceling* elastic strain fields, $\varepsilon_{ij}^{C,\mathcal{V}^D}(\mathbf{x})$ in the region V^D and $\varepsilon_{ij}^{C,M}(\mathbf{x})$ in the surrounding matrix.[9]

---

[8]Eshelby (1957, 1961) termed such a strain a *transformation strain*, and Mura (1987) has more recently called it an *eigenstrain*. However, since use of the latter term may conceivably imply an association with a classic eigenvalue problem, I favor the former term.

[9]The use of the superscript, C, (Eshelby 1957) seems appropriate, since the $\varepsilon_{ij}^{C}(\mathbf{x})$ strains arise from *canceling* forces. Also, the $\varepsilon_{ij}^{C}(\mathbf{x})$ strain in the matrix may be attributed to the *constraints* imposed by the matrix on the effects of the canceling forces on the $\mathcal{V}^D$ region.

In the final state the material within $\mathcal{V}^{\mathrm{D}}$ is *coherent* with respect to the surrounding matrix in the sense that all points on the surface of $\mathcal{V}^{\mathrm{D}}$ that were in registry initially are still in registry. The final elastic strains in $\mathcal{V}^{\mathrm{D}}$ and in the surrounding matrix, designated by $\varepsilon_{ij}^{\mathcal{V}^{\mathrm{D}}}$ and $\varepsilon_{ij}^{\mathrm{M}}$, respectively, and the corresponding elastic displacements, $u_i^{\mathcal{V}^{\mathrm{D}}}$ and u_i^{M}, are then

$$\begin{aligned} \varepsilon_{ij}^{\mathcal{V}^{\mathrm{D}}}(\mathbf{x}) &= \varepsilon_{ij}^{\mathrm{C},\mathcal{V}^{\mathrm{D}}}(\mathbf{x}) - \varepsilon_{ij}^{\mathrm{T}}(\mathbf{x}) \quad & u_i^{\mathcal{V}^{\mathrm{D}}}(\mathbf{x}) &= u_i^{\mathrm{C},\mathcal{V}^{\mathrm{D}}}(\mathbf{x}) - u_i^{\mathrm{T}}(\mathbf{x}) \\ \varepsilon_{ij}^{\mathrm{M}}(\mathbf{x}) &= \varepsilon_{ij}^{\mathrm{C,M}}(\mathbf{x}) & u_i^{\mathrm{M}}(\mathbf{x}) &= u_i^{\mathrm{C,M}}(\mathbf{x}) \end{aligned} \tag{3.150}$$

where $u_i^{\mathrm{T}}(\mathbf{x})$ is the *transformation displacement* that is associated with the transformation strain $\varepsilon_{ij}^{\mathrm{T}}(\mathbf{x})$.

The strains given by Eq. (3.150) are due to the incompatibility caused by the application of the transformation strain to the $\mathcal{V}^{\mathrm{D}}$ region which produced a misfit between the $\mathcal{V}^{\mathrm{D}}$ region and the surrounding matrix. Such residual strains are termed *internal strains*, since they are generated internally by the presence of the transformation strain in the absence of any applied forces. As discussed by Eshelby (1957), they can be formulated directly in terms of the incompatibility tensor given by Eq. (2.54).

To analyze bodies containing transformation strains, we follow Mura (1987) and employ a formalism that introduces the *total strain*, $\varepsilon_{ij}^{\mathrm{tot}}(\mathbf{x})$, defined as the sum of the elastic strain, $\varepsilon_{ij}(\mathbf{x})$, and the transformation strain, i.e.,

$$\varepsilon_{ij}^{\mathrm{tot}}(\mathbf{x}) = \varepsilon_{ij}(\mathbf{x}) + \varepsilon_{ij}^{\mathrm{T}}(\mathbf{x}). \tag{3.151}$$

Similarly, the corresponding *total displacement*, $u_i^{\mathrm{tot}}(\mathbf{x})$, is the sum of the elastic displacement $u_i(\mathbf{x})$ and transformation displacement $u_i^{\mathrm{T}}(\mathbf{x})$ i.e.,

$$u_i^{\mathrm{tot}}(\mathbf{x}) = u_i(\mathbf{x}) + u_i^{\mathrm{T}}(\mathbf{x}). \tag{3.152}$$

The total strain must be compatible everywhere, and the six total strains must therefore be derivable from three continuous total displacements (see Exercise 2.10) according to

$$\varepsilon_{ij}^{\mathrm{tot}} = \varepsilon_{ij} + \varepsilon_{ij}^{\mathrm{T}} = \frac{1}{2}\left(\frac{\partial u_i^{\mathrm{tot}}}{\partial x_j} + \frac{\partial u_j^{\mathrm{tot}}}{\partial x_i}\right). \tag{3.153}$$

Also, the stresses, σ_{ij}, must be related to the elastic strains by Hooke's law, and, therefore, with the use of Eq. (3.153),

$$\sigma_{ij} = C_{ijkl}(\varepsilon_{kl}^{\mathrm{tot}} - \varepsilon_{kl}^{\mathrm{T}}) = C_{ijkl}\frac{\partial u_k^{\mathrm{tot}}}{\partial x_l} - C_{ijkl}\varepsilon_{kl}^{\mathrm{T}} = C_{ijkl}\frac{\partial u_k^{\mathrm{tot}}}{\partial x_l} - \sigma_{ij}^{\mathrm{T}}. \tag{3.154}$$

By virtue of Eqs. (3.150)–(3.152), we then have

$$\begin{aligned} \varepsilon_{ij}^{tot,\mathcal{V}^D} &= (\varepsilon_{ij}^{C,\mathcal{V}^D} - \varepsilon_{ij}^{T}) + \varepsilon_{ij}^{T} = \varepsilon_{ij}^{C,\mathcal{V}^D} \qquad & \varepsilon_{ij}^{tot,M} &= \varepsilon_{ij}^{C,M} \\ u_i^{tot,\mathcal{V}^D} &= u_i^{C,\mathcal{V}^D} & u_i^{tot,M} &= u_i^{C,M} \end{aligned} \tag{3.155}$$

where $u_i^{tot,\mathcal{V}^D}$ and $u_i^{tot,M}$ are the total displacements in $\mathcal{V}^D$ and the matrix, respectively. The boundary conditions at the coherent $\mathcal{V}^D$/matrix interface that must be satisfied are continuity of the total displacements and continuity of the tractions, i.e.,

$$\begin{aligned} u_i^{tot,\mathcal{V}^D} &= u_i^{C,\mathcal{V}^D} = u_i^{tot,M} = u_i^{C,M}, \\ & \qquad\qquad\qquad\qquad\qquad (\text{on } \mathcal{V}^D/\text{matrix interface}) \\ \sigma_{ij}^{\mathcal{V}^D}\hat{n}_j &= (\sigma_{ij}^{C,\mathcal{V}^D} - \sigma_{ij}^{T})\hat{n}_j = \sigma_{ij}^{M}\hat{n}_j = \sigma_{ij}^{C,M}\hat{n}_j. \end{aligned} \tag{3.156}$$

Equation (3.155), with the above boundary conditions, shows that the total displacement field throughout a homogeneous body that is subjected to a transformation strain in a region, $\mathcal{V}^D$, is given by the canceling displacement field, i.e.,

$$u_i^{tot}(\mathbf{x}) = u_i^{C}(\mathbf{x}). \tag{3.157}$$

Next, differentiating Eq. (3.154) and substituting Eq. (2.65) in the absence of body forces,

$$C_{ijkl}\frac{\partial^2 u_k^{tot}(\mathbf{x})}{\partial x_j \partial x_l} - \frac{\partial \sigma_{ij}^{T}(\mathbf{x})}{\partial x_j} = \frac{\partial \sigma_{ij}(\mathbf{x})}{\partial x_j} = 0. \tag{3.158}$$

Comparison of Eqs. (3.158) and (3.2) shows the important result that the effect on the total displacement of the term, $-\partial\sigma_{ij}^{T}(\mathbf{x})/\partial x_j$, in a system containing a transformation strain and free of body force density, is identical to the effect of an equivalent body force density, f_i, in a system free of transformation strain.

Equations (3.154) and (3.158) can now be used to obtain a basic relationship between the Fourier amplitudes of the transformation strain and the stresses that they produce in a homogeneous system (Lothe 1992b). To obtain this, write the transformation strain as a wave with wave vector **k**, i.e.,

$$\varepsilon_{kl}^{T}(\mathbf{x}) = A_{kl}^{T}(\mathbf{k})e^{i\mathbf{k}\cdot\mathbf{x}}. \tag{3.159}$$

This will produce a wave of total displacement possessing the same wave vector, i.e.,

$$u_k^{tot}(\mathbf{x}) = A_k^{tot}(\mathbf{k})e^{i\mathbf{k}\cdot\mathbf{x}}. \tag{3.160}$$

The amplitude of the total displacement wave, $A_k^{tot}(\mathbf{k})$, is obtained by substituting Eqs. (3.160) and (3.159) (after employing Hooke's law) into Eq. (3.158) with the result

$$A_m^{tot}(\mathbf{k}) = -i(kk)_{mj}^{-1}k_iC_{ijkl}A_{kl}^T(\mathbf{k}). \tag{3.161}$$

Then, by substituting the above quantities into Eq. (3.154), the associated stress is obtained in the form

$$\sigma_{ij}(\mathbf{x}) = -C_{ijkl}^*(\hat{\mathbf{k}})A_{kl}^T(\mathbf{k})e^{i\mathbf{k}\cdot\mathbf{x}} = -C_{ijkl}^*(\hat{\mathbf{k}})\varepsilon_{kl}^T(\mathbf{x}). \tag{3.162}$$

where

$$C_{ijkl}^*(\hat{\mathbf{k}}) = C_{ijkl} - C_{ijpq}\hat{k}_s\hat{k}_q(\hat{k}\hat{k})_{pr}^{-1}C_{srkl}. \tag{3.163}$$

Then, taking the Fourier transform of Eq. (3.162),

$$\bar{\sigma}_{ij}(\mathbf{k}) = -C_{ijkl}^*(\hat{\mathbf{k}})\bar{\varepsilon}_{kl}^T(\mathbf{k}). \tag{3.164}$$

Equation (3.164) shows that the Fourier amplitudes of the transformation strain and the stresses that they produce are linearly coupled by the tensor, $C_{ijkl}^*(\hat{\mathbf{k}})$. This tensor, which is a function of the unit vector $\hat{\mathbf{k}}$, is known as the *planar elastic stiffness tensor* and, as shown in Exercise 3.1, it has the physical significance of describing Hooke's law under the condition that the planes in the elastic medium normal to $\hat{\mathbf{k}}$ experience a vanishing traction (see Lothe 1992b).

3.6.2 *Fourier transform solutions*

As noted in the previous section, a comparison of the basic field equation for systems with and without transformation strains, given by Eqs. (3.158) and (3.2) respectively, shows the equivalent roles played by the quantity, $-\partial\sigma_{ij}^T(\mathbf{x})/\partial x_j$, in the former case and the force density, $f_i(\mathbf{x})$, in the latter. Therefore, on the basis of this equivalence, the Fourier transform of $f_i(\mathbf{x})$, i.e., $\bar{f}_i(\mathbf{k})$, can be replaced with the Fourier transform of $-\partial\sigma_{ij}^T(\mathbf{x})/\partial x_j$, so that by virtue of Eq. (F.1),

$$\begin{aligned}\bar{f}_i(\mathbf{k}) &= -\int_{-\infty}^{\infty}\int_{-\infty}^{\infty}\int_{-\infty}^{\infty} \frac{\partial\sigma_{ij}^{T}(\mathbf{x})}{\partial x_j} e^{i\mathbf{k}\cdot\mathbf{x}} dx_1 dx_2 dx_3 \\ &= -C_{ijmn}\int_{-\infty}^{\infty}\int_{-\infty}^{\infty}\int_{-\infty}^{\infty} \frac{\partial\varepsilon_{mn}^{T}(\mathbf{x})}{\partial x_j} e^{i\mathbf{k}\cdot\mathbf{x}} dx_1 dx_2 dx_3 \\ &= -C_{ijmn}\int_{-\infty}^{\infty}\int_{-\infty}^{\infty}\int_{-\infty}^{\infty} \left[\frac{\partial(\varepsilon_{mn}^{T} e^{i\mathbf{k}\cdot\mathbf{x}})}{\partial x_j} - \varepsilon_{mn}^{T} e^{i\mathbf{k}\cdot\mathbf{x}} i k_j\right] dx_1 dx_2 dx_3 \\ &= iC_{ijmn}\int_{-\infty}^{\infty}\int_{-\infty}^{\infty}\int_{-\infty}^{\infty} \varepsilon_{mn}^{T} e^{i\mathbf{k}\cdot\mathbf{x}} k_j dx_1 dx_2 dx_3 - \oiint_{\mathcal{S}} \varepsilon_{mn}^{T} e^{i\mathbf{k}\cdot\mathbf{x}} \hat{n}_j dS \\ &= iC_{ijmn}\int_{-\infty}^{\infty}\int_{-\infty}^{\infty}\int_{-\infty}^{\infty} \varepsilon_{mn}^{T} e^{i\mathbf{k}\cdot\mathbf{x}} k_j dx_1 dx_2 dx_3. \end{aligned} \tag{3.165}$$

In this development, use has been made of the divergence theorem, and the expectation that the surface integral vanishes for all defect transformation strains employed in this book. Then, by substituting Eq. (3.165) into Eq. (3.8), and taking the inverse of the result using Eq. (F.2), the canceling displacement is given by

$$\begin{aligned}u_k^C(\mathbf{x}) = \frac{i}{(2\pi)^3} C_{ijmn} &\int_{-\infty}^{\infty}\int_{-\infty}^{\infty}\int_{-\infty}^{\infty} \varepsilon_{mn}^{T}(\mathbf{x}') dx_1' dx_2' dx_3' \\ \times &\int_{-\infty}^{\infty}\int_{-\infty}^{\infty}\int_{-\infty}^{\infty} e^{-i\mathbf{k}\cdot(\mathbf{x}-\mathbf{x}')} (kk)_{ik}^{-1} k_j dk_1 dk_2 dk_3. \end{aligned} \tag{3.166}$$

3.6.3 *Green's function solutions*

A useful differential equation for the canceling displacement $u_i^C(\mathbf{x})$ in a homogeneous body containing transformation strains in a region, $\mathcal{V}^D$, can also be obtained by employing Green's functions (Mura 1987). It is recalled that in the final step in the introduction of a transformation strain into a body in Sec. 3.6.1, the $u_i^C(\mathbf{x})$ displacement field is produced by applying surface tractions, $T_i = \sigma_{ij}^T \hat{n}_j$, and body forces, $f_i = -\partial\sigma_{ij}^T/\partial x_j$, to the $\mathcal{V}^D$ region which are just the opposite of those given by Eq. (3.149). Therefore, using Eq. (3.16),

$$u_i^C(\mathbf{x}) = -\oiiint_{\mathcal{V}^D} G_{ik}(\mathbf{x}-\mathbf{x}') \frac{\partial\sigma_{jk}^T(\mathbf{x}')}{\partial x_j'} dV' + \oiint_{\mathcal{S}^D} G_{ik}(\mathbf{x}-\mathbf{x}')\sigma_{jk}^T(\mathbf{x}')\hat{n}_j(\mathbf{x}') dS'. \tag{3.167}$$

Then, by employing the divergence theorem,

$$u_i^C(\mathbf{x}) = -\oiiint_{\mathcal{V}^D} G_{ik}(\mathbf{x}-\mathbf{x}')\frac{\partial\sigma_{jk}^T(\mathbf{x}')}{\partial x_j'}dV'$$
$$+\oiiint_{\mathcal{V}^D} \frac{\partial}{\partial x_j'}\left[G_{ik}(\mathbf{x}-\mathbf{x}')\sigma_{jk}^T(\mathbf{x}')\right]dV'$$
$$= \oiiint_{\mathcal{V}^D} \sigma_{jk}^T(\mathbf{x}')\frac{\partial G_{ik}(\mathbf{x}-\mathbf{x}')}{\partial x_j'}dV'. \qquad (3.168)$$

3.7 Stress Function Method for Isotropic Systems

As pointed out in Sec. 3.2, the stresses in a stable elastic field must satisfy the equations of equilibrium and also be associated with strains that satisfy the condition of compatibility. A wide variety of *stress functions* has been developed in the theory of elasticity that can be used to generate stresses that automatically satisfy these conditions, and are therefore useful for solving a variety of problems. For example, if the stresses generated by such functions can be matched to the existing boundary conditions, the elastic problem is solved immediately. Prominent stress functions include the Airy function, the Morera function, and the Maxwell function [Chou and Pagano (1967) and Hetnarski and Ignaczak (2004)].

We will have use for the Airy stress function, which is applicable for plane strain problems in isotropic systems when the elastic field is invariant along one dimension and the displacement along that dimension vanishes as is the case, for example, along an infinitely long straight edge dislocation. The Airy stress function for an isotropic system can be developed by first selecting $\hat{\mathbf{e}}_3$ as the invariant direction, so that, as already stated by Eq. (3.26),

$$\frac{\partial}{\partial x_3} = 0 \quad u_1 = u_1(x_1, x_2) \quad u_2 = u_2(x_1, x_2) \quad u_3 = 0 \qquad (3.169)$$

for the two-dimensional elastic field. The only non-vanishing strains and stresses are then ε_{11}, ε_{22}, ε_{12}, σ_{11}, σ_{22}, σ_{12} and $\sigma_{33} = \nu(\sigma_{11} + \sigma_{22})$. The equations of equilibrium, expressed in Cartesian and cylindrical coordinates by Eqs. (2.65) and (G.5), respectively, then appear, in the absence of body forces, as

$$\begin{aligned} &\frac{\partial \sigma_{11}}{\partial x_1} + \frac{\partial \sigma_{12}}{\partial x_2} = 0 \qquad && \frac{\partial \sigma_{rr}}{\partial r} + \frac{1}{r}\frac{\partial \sigma_{r\theta}}{\partial \theta} + \frac{\sigma_{rr} - \sigma_{\theta\theta}}{r} = 0 \\ & \qquad\qquad \text{or} && \\ &\frac{\partial \sigma_{12}}{\partial x_1} + \frac{\partial \sigma_{22}}{\partial x_2} = 0 \qquad && \frac{\partial \sigma_{r\theta}}{\partial r} + \frac{1}{r}\frac{\partial \sigma_{\theta\theta}}{\partial \theta} + \frac{2\sigma_{r\theta}}{r} = 0 \end{aligned} \quad . \tag{3.170}$$

These equations will be satisfied automatically if Airy stress functions, $\psi(x_1, x_2)$, or $\psi(r, \theta)$, can be found that satisfy, respectively

$$\begin{aligned} \sigma_{11} &= \frac{\partial^2 \psi}{\partial x_2^2} && \sigma_{rr} = \frac{1}{r}\frac{\partial \psi}{\partial r} + \frac{1}{r^2}\frac{\partial^2 \psi}{\partial \theta^2} \\ \sigma_{22} &= \frac{\partial^2 \psi}{\partial x_1^2} \qquad \text{or} && \sigma_{\theta\theta} = \frac{\partial^2 \psi}{\partial r^2} \\ \sigma_{12} &= -\frac{\partial^2 \psi}{\partial x_1 \partial x_2} && \sigma_{r\theta} = -\frac{\partial}{\partial r}\left(\frac{1}{r}\frac{\partial \psi}{\partial \theta}\right). \end{aligned} \tag{3.171}$$

However, compatibility of the corresponding strains must also be satisfied, and Eq. (2.53) therefore requires that the corresponding strains ε_{11}, ε_{22}, and ε_{12} satisfy

$$2\frac{\partial^2 \varepsilon_{12}}{\partial x_1 \partial x_2} - \frac{\partial^2 \varepsilon_{11}}{\partial x_2^2} - \frac{\partial^2 \varepsilon_{22}}{\partial x_1^2} = C_{33} = 0. \tag{3.172}$$

Then, converting the stresses in Eq. (3.171) to strains using Hooke's law and putting the results into Eq. (3.172), compatibility is achieved if

$$\frac{\partial^4 \psi}{\partial x_1^4} + 2\frac{\partial^4 \psi}{\partial x_2^2 \partial x_1^2} + \frac{\partial^4 \psi}{\partial x_2^4} = \nabla^4 \psi = \nabla^2(\nabla^2)\psi = 0. \tag{3.173}$$

The corresponding result in cylindrical coordinates, with the help of Eq. (A.3), is

$$\nabla^4 \psi = \nabla^2(\nabla^2)\psi = 0 \quad \text{where} \quad \nabla^2 \psi = \frac{1}{r}\frac{\partial}{\partial r}\left(r\frac{\partial \psi}{\partial r}\right) + \frac{1}{r^2}\frac{\partial^2 \psi}{\partial \theta^2}. \tag{3.174}$$

Therefore, if an Airy stress function is found that satisfies Eq. (3.173), or (3.174), the corresponding stresses can be obtained simply by use of Eq. (3.171).

3.8 Defects in Regions Bounded by Interfaces — Method of Image Stresses

Defects in real systems generally exist in finite regions bounded by interfaces. Any expressions for the elastic fields that they generate must therefore satisfy the boundary conditions that exist for these fields at the interfaces. For example, if the interface is a traction-free surface, the defect self-stress field, $\sigma_{ij}^{D}(\mathbf{x})$, must satisfy the condition $T_i^D = \sigma_{ij}^D \hat{n}_j = 0$ at the surface. In general, it will be shown in later chapters that for defects that are in large bodies and well away from any interfaces, the effect of the interfaces on the defect self-stress in the vicinity of the defect can usually be neglected. It can therefore be assumed that the self-stress field is the same as it would be if the defect were in an infinite body. However, when the defect is near an interface the effect of the interface becomes significant and must be taken into account.

The most direct method of dealing with problems where interface effects are significant is to treat them as boundary-value problems and seek solutions that simultaneously satisfy both the necessary equations of elasticity and the boundary conditions. This approach is often practicable, particularly when the problem has a high degree of symmetry and the boundary conditions can be expressed relatively simply (as, for example, in the case treated in Exercise 9.1). However, it is often more practicable to employ the method of *image stresses*.[10] Here, it is imagined that the defect is initially present in an infinite homogeneous region so that it possesses the elastic field that it would produce in an infinite body. Then, an additional solution, i.e., *an image solution*, is sought which, when added to the infinite body solution, produces a solution that satisfies both the equations of elasticity and the boundary conditions. The final stress field for the defect in the finite region, σ_{ij}^{D}, is therefore the sum

$$\sigma_{ij}^{D} = \sigma_{ij}^{D^{\infty}} + \sigma_{ij}^{D^{IM}} \tag{3.175}$$

where $\sigma_{ij}^{D^{\infty}}$ is the infinite-body stress, and $\sigma_{ij}^{D^{IM}}$ is the image stress. The advantage of this approach in many cases is that it is easier to formulate the $\sigma_{ij}^{D^{\infty}}$ and $\sigma_{ij}^{D^{IM}}$ solutions individually than a single direct solution which satisfies both the equations of elasticity and the boundary conditions.

To illustrate the image method explicitly, consider the case where a defect, D, is embedded in a finite homogeneous body, $\mathcal{V}^{\circ}$, with a traction-free surface $\mathcal{S}^{\circ}$. To obtain this situation, assume that D is initially embedded in an infinite homogeneous region, $\mathcal{V}^{\infty}$ as in Fig. 3.3a. Then (Fig. 3.3a), mark out the incipient finite region, $\mathcal{V}^{\circ}$, bounded by $\mathcal{S}^{\circ}$ (shown dashed)

[10]As shown in Sec. 5.3, the method of images is also useful for obtaining the force exerted on a defect lying in a finite homogeneous region.

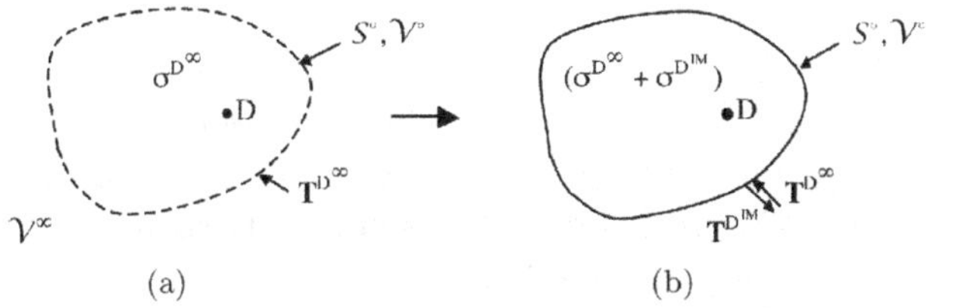

Fig. 3.3 (a) Defect, D, embedded in an infinite region, $\mathcal{V}^\infty$, where it produces the stress field $\sigma_{ij}^{D^\infty}$. Dashed line indicates an incipient region, $\mathcal{V}^\circ$, (bounded by $\mathcal{S}^\circ$ and subject to the tractions $T_i^{D^\infty}$) that is to become a free body with $\mathcal{S}^\circ$ traction-free. (b) Region $\mathcal{V}^\circ$ after removing it from infinite region and applying tractions $T_i^{D^{IM}} = -T_i^{D^\infty}$ to produce traction-free surface.

that D is to occupy with $\mathcal{S}^\circ$ traction-free. D is seen to produce the stress field $\sigma_{ij}^{D^\infty}$ throughout $\mathcal{V}^\infty$ and the traction, $T_i^{D^\infty}$, on $\mathcal{S}^\circ$. Next, remove $\mathcal{V}^\circ$ from $\mathcal{V}^\infty$ while at the same time applying tractions $T_i^{D^\infty}$ to its surface, $\mathcal{S}^\circ$, so that the stress field in $\mathcal{V}^\circ$ remains unchanged. Finally, apply image tractions, $T_i^{D^{IM}}$, on $\mathcal{S}^\circ$ that exactly cancel the tractions, $T_i^{D^\infty}$, thereby making $\mathcal{S}^\circ$ traction-free (Fig. 3.3b). The stress throughout $\mathcal{V}^\circ$ is then given be Eq. (3.175), where $\sigma_{ij}^{D^{IM}}$ is the stress produced by the tractions $T_i^{D^{IM}}$.

The above example of the image stress method for a defect in a finite homogeneous region possessing a traction-free surface is readily generalized to include cases where the region containing the defect is bonded to an elastically dissimilar region and the surface is no longer free, but instead, is subjected to elastic constraints.

The image stress will generally depend upon the position of the defect in the $\mathcal{V}^\circ$ region and the shape and size of the region, and, therefore, no general and tractable solutions exist for defect image stresses in finite homogeneous regions of arbitrary shapes. However, tractable solutions do exist for many relatively simple geometries possessing high degrees of symmetry. In a few special cases a suitable image stress for a defect corresponds to the stress field of the same defect (but of negative character) placed in an "image" position across an interface in a manner analogous to the electrostatic image fields produced by "image" electrical charges that are employed in classical electrostatics, e.g., Smythe (1950). [11]

[11] Also, in some cases, the placement of an "image defect" almost satisfies the prevailing boundary condition, and the discrepancy can be made up by adding an additional term that is relatively easy to formulate.

Exercises

3.1. Show that the planar elastic stiffness tensor, $C^*_{ijkl}(\hat{\mathbf{k}})$, given by Eq. (3.163), has the physical significance of describing Hooke's law under the condition that the planes in the elastic medium normal to $\hat{\mathbf{k}}$ experience vanishing traction.

Solution. Hooke's law, employing $C^*_{ijkl}(\hat{\mathbf{k}})$, has the form

$$\sigma_{ij} = C^*_{ijkl}(\hat{\mathbf{k}})\varepsilon_{kl}. \tag{3.176}$$

The condition for vanishing traction on planes normal to $\hat{\mathbf{k}}$ is then

$$T_i = \sigma_{ij}\hat{k}_j = C^*_{ijkl}(\hat{\mathbf{k}})\varepsilon_{kl}\hat{k}_j = 0 \tag{3.177}$$

after using Eq. (2.59). Then, substituting Eq. (3.163) into (3.177) the above condition assumes the form

$$\hat{k}_j C_{ijkl}\varepsilon_{kl} = \hat{k}_j C_{ijpq}\hat{k}_s\hat{k}_q(\hat{k}\hat{k})^{-1}_{pr}C_{srkl}\varepsilon_{kl}. \tag{3.178}$$

By invoking the symmetry properties of the C_{ijkl} tensor, it is then seen that the condition expressed by Eq. (3.178) is indeed satisfied.

$$\begin{aligned}\hat{k}_j C_{ijkl}\varepsilon_{kl} &= \hat{k}_j C_{ijpq}\hat{k}_s\hat{k}_q(\hat{k}\hat{k})^{-1}_{pr}C_{srkl}\varepsilon_{kl} = \hat{k}_j C_{jipq}\hat{k}_q\hat{k}_s(\hat{k}\hat{k})^{-1}_{pr}C_{srkl}\varepsilon_{kl}\\ &= \hat{k}_s(\hat{k}\hat{k})_{ip}(\hat{k}\hat{k})^{-1}_{pr}C_{srkl}\varepsilon_{kl} = \hat{k}_s\delta_{ir}C_{srkl}\varepsilon_{kl} = \hat{k}_s C_{sikl}\varepsilon_{kl}\\ &= \hat{k}_j C_{ijkl}\varepsilon_{kl}.\end{aligned} \tag{3.179}$$

3.2. Additional sum rules which involve the matrices Q_{sk}, S_{sk} and B_{sk} , and are related to those given in Sec. 3.5.1.2, can be formulated. Derive the following sum rule

$$\sum_{\alpha=1}^{6} \pm p_\alpha A_{r\alpha}L_{s\alpha} = i(nn)^{-1}_{rj}[4\pi B_{js} + (nm)_{jk}S_{ks}] \tag{3.180}$$

by starting with Eq. (3.37).

Solution First, multiply both sides of Eq. (3.37) throughout by $\pm L_{s\alpha}$, and sum over α from 1 to 6, to obtain

$$(nn)_{jk}\sum_{\alpha=1}^{6}\pm p_\alpha A_{k\alpha}L_{s\alpha} = -(nm)_{jk}\sum_{\alpha=1}^{6}\pm A_{k\alpha}L_{s\alpha} - \sum_{\alpha=1}^{6}\pm L_{j\alpha}L_{s\alpha}. \tag{3.181}$$

Then, solve for $\sum_{\alpha=1}^{6}\pm p_\alpha A_{k\alpha}L_{s\alpha}$ by multiplying Eq. (3.181) throughout by $(nn)_{rj}^{-1}$ so that

$$(nn)_{rj}^{-1}(nn)_{jk}\sum_{\alpha=1}^{6}\pm p_\alpha A_{k\alpha}L_{s\alpha}$$
$$= -(nn)_{rj}^{-1}\left[(nm)_{jk}\sum_{\alpha=1}^{6}\pm A_{k\alpha}L_{s\alpha} + \sum_{\alpha=1}^{6}\pm L_{j\alpha}L_{s\alpha}\right], \tag{3.182}$$

and, therefore

$$\delta_{rk}\sum_{\alpha=1}^{6}\pm p_\alpha A_{k\alpha}L_{s\alpha} = \sum_{\alpha=1}^{6}\pm p_\alpha A_{r\alpha}L_{s\alpha}$$
$$= -(nn)_{rj}^{-1}\left[(nm)_{jk}\sum_{\alpha=1}^{6}\pm A_{k\alpha}L_{s\alpha} + \sum_{\alpha=1}^{6}\pm L_{j\alpha}L_{s\alpha}\right], \tag{3.183}$$

Finally, substitution of Eqs. (3.134) and (3.135) into Eq. (3.183) yields Eq. (3.180).

3.3. Derive the relationship $4\pi B_{ij}Q_{jk} + S_{ji}S_{kj} = -\delta_{ik}$, which connects the matrices B_{ij}, Q_{ij} and S_{ij}, and is given by Eq. (3.140) in the text.

Solution Using Eqs. (3.133–3.135),

$$4\pi B_{ij}Q_{jk} + S_{ji}S_{kj} = -\left[\sum_{\alpha=1}^{6}\pm L_{i\alpha}L_{j\alpha}\right]\left[\sum_{\beta=1}^{6}\pm A_{j\beta}A_{k\beta}\right]$$
$$-\left[\sum_{\alpha=1}^{6}\pm A_{k\alpha}L_{j\alpha}\right]\left[\sum_{\beta=1}^{6}\pm A_{j\beta}L_{i\beta}\right]$$
$$= -\sum_{\alpha=1}^{6}\sum_{\beta=1}^{6}(\pm A_{k\beta})(\pm L_{i\alpha})(L_{j\alpha}A_{j\beta} + L_{j\beta}A_{j\alpha}) \tag{3.184}$$

Then, substituting Eqs. (3.71) and (3.78) into Eq. (3.184),

$$4\pi B_{ij}Q_{jk} + S_{ji}S_{kj} = -\sum_{\alpha=1}^{6}\sum_{\beta=1}^{6}(\pm A_{k\beta})(\pm L_{i\alpha})\delta_{\alpha\beta}$$

$$= -\sum_{\alpha=1}^{6} L_{i\alpha}A_{k\alpha} = -\delta_{ik}. \tag{3.185}$$

3.4. Consider the sixth-degree polynomial equation with real coefficients given by

$$p^6 - 2p^3 + 4 = 0. \tag{3.186}$$

Find its six roots and show that they occur in three conjugate pairs which can be indexed in the same manner as the roots in Eq. (3.34).

Solution Let

$$p^3 = q \tag{3.187}$$

Then, by substituting this into Eq. (3.186) we obtain the quadratic equation

$$q^2 - 2q + 4 = 0 \tag{3.188}$$

which possesses the two roots

$$\begin{aligned} q_1 &= 1 + \sqrt{-3} = 1 + i\sqrt{3} = 2e^{i\pi/3}, \\ q_2 &= 1 - \sqrt{-3} = 1 - i\sqrt{3} = 2e^{-i\pi/3}. \end{aligned} \tag{3.189}$$

Next, by substituting Eq. (3.189) into Eq. (3.187),

$$p^3 = 2e^{\pm i\pi/3}. \tag{3.190}$$

Using standard methods for finding the roots of complex numbers (Weir, Hass and Giordano 2000), Eq. (3.190) yields the following six roots, three that arise when the exponent is positive and three when it is negative:

$$p_1 = 2^{1/3}e^{i\pi/9} = 2^{1/3}\cos(\pi/9) + i2^{1/3}\sin(\pi/9),$$
$$p_2 = 2^{1/3}e^{i\pi 7/9} = -2^{1/3}\cos(2\pi/9) + i2^{1/3}\sin(2\pi/9),$$

$$
\begin{aligned}
p_3 &= 2^{1/3}e^{i\pi/9} = -2^{1/3}\sin(\pi/18) + i2^{1/3}\cos(\pi/18), \\
p_4 &= 2^{1/3}e^{-i\pi/9} = 2^{1/3}\cos(\pi/9) - i2^{1/3}\sin(\pi/9), \\
p_5 &= 2^{1/3}e^{-i\pi 7/9} = -2^{1/3}\cos(2\pi/9) - i2^{1/3}\sin(2\pi/9), \\
p_6 &= 2^{1/3}e^{i\pi/9} = -2^{1/3}\sin(\pi/18) - i2^{1/3}\cos(\pi/18).
\end{aligned}
\qquad (3.191)
$$

As may be seen, they consist of three conjugate pairs and are indexed in the same manner as the roots in Eq. (3.34).[12]

3.5. The basic differential equation for obtaining the point force Green's function, i.e., Eq. (3.21), was derived in the text by use of the definition of the Green's function given by Eq. (3.15). Now, reverse the procedure and verify that Eq. (3.15) can be obtained by use of Eq. (3.21).

Solution Starting with Eq. (3.21), an expression for $u_j(\mathbf{x})$ can be obtained by first multiplying this equation by $u_k(\mathbf{x}')$ and then integrating it over the region, $\mathcal{V}$, so that

$$
C_{kpim} \oiiint_{\mathcal{V}} \left[u_k(\mathbf{x}') \frac{\partial^2 G_{ij}(\mathbf{x}-\mathbf{x}')}{\partial x'_m \partial x'_p} + u_k(\mathbf{x}')\delta_{kj}\delta(\mathbf{x}-\mathbf{x}') \right] dV' = 0. \qquad (3.192)
$$

Then, solving for $u_j(\mathbf{x})$,

$$
u_j(\mathbf{x}) = -C_{kpim} \oiiint_{\mathcal{V}} u_k(\mathbf{x}') \frac{\partial^2 G_{ij}(\mathbf{x}-\mathbf{x}')}{\partial x'_m \partial x'_p} dV'. \qquad (3.193)
$$

However, we must introduce into the formulation for $u_j(\mathbf{x})$ the force which produces the displacement, and this can be done by employing the basic field equation, i.e., Eq. (3.2). By multiplying Eq. (3.2) by $G_{kj}(\mathbf{x}-\mathbf{x}')$ and then integrating the result over $\mathcal{V}$,

$$
\oiiint_{\mathcal{V}} \left[G_{kj}(\mathbf{x}-\mathbf{x}') C_{kpim} \frac{\partial^2 u_i(\mathbf{x}')}{\partial x'_m \partial x'_p} + G_{kj}(\mathbf{x}-\mathbf{x}') f_k(\mathbf{x}') \right] dV' = 0, \qquad (3.194)
$$

[12]Note that the sextic equation given by Eq. (3.186) is of a special form that can be solved analytically. In certain cases involving straight dislocations and lines of force, enough symmetry is present so that the relevant sextic equations can be solved analytically (Lothe 1992b). In such cases analytic expressions for their elastic fields can be obtained (Hirth and Lothe 1982). However, in the general case this is not possible, and resort must be made to numerical methods.

and adding this to Eq. (3.193), we obtain

$$u_j(\mathbf{x}) = \oiiint_{\mathcal{V}} G_{kj}(\mathbf{x}-\mathbf{x}')f_k(\mathbf{x}')dV' + C_{kpim}\oiiint_{\mathcal{V}} \times \left[G_{kj}(\mathbf{x}-\mathbf{x}')\frac{\partial^2 u_i(\mathbf{x}')}{\partial x'_m \partial x'_p} - u_k(\mathbf{x}')\frac{\partial^2 G_{ij}(\mathrm{x}-\mathrm{x}')}{\partial x'_m \partial x'_p}\right] dV'. \tag{3.195}$$

Equation (3.195) is seen to correspond to Eq. (3.16) except for the additional second term which vanishes as now shown. First, this term can be transformed according to

$$\begin{aligned} &C_{kpim}\oiiint_{\mathcal{V}}\left[G_{kj}(\mathbf{x}-\mathbf{x}')\frac{\partial^2 u_i(\mathbf{x}')}{\partial x'_m \partial x'_p} - u_k(\mathbf{x}')\frac{\partial^2 G_{ij}(\mathrm{x}-\mathrm{x}')}{\partial x'_m \partial x'_p}\right] dV' \\ &= C_{kpim}\oiiint_{\mathcal{V}} \frac{\partial}{\partial x'_p}\left[G_{kj}(\mathbf{x}-\mathbf{x}')\frac{\partial u_i(\mathbf{x}')}{\partial x'_m} - u_k(\mathbf{x}')\frac{\partial G_{ij}(\mathbf{x}-\mathbf{x}')}{\partial x'_m}\right] dV' \\ &= C_{kpim}\oiiint_{\mathcal{S}}\left[G_{kj}(\mathbf{x}-\mathbf{x}')\frac{\partial u_i(\mathbf{x}')}{\partial x'_m} - u_k(\mathbf{x}')\frac{\partial G_{ij}(\mathbf{x}-\mathbf{x}')}{\partial x'_m}\right] \hat{n}_p dS' \end{aligned} \tag{3.196}$$

by use of the divergence theorem and the relationships

$$\begin{aligned} C_{kpim}\frac{\partial u_i(\mathbf{x}')}{\partial x'_m}\frac{\partial G_{kj}(\mathbf{x}-\mathbf{x}')}{\partial x'_p} &= C_{imkp}\frac{\partial u_k(\mathbf{x}')}{\partial x'_p}\frac{\partial G_{ij}(\mathbf{x}-\mathbf{x}')}{\partial x'_m} \\ &= C_{kpim}\frac{\partial u_k(\mathbf{x}')}{\partial x'_p}\frac{\partial G_{ij}(\mathbf{x}-\mathbf{x}')}{\partial x'_m}. \end{aligned} \tag{3.197}$$

However, when $\mathcal{V}$ and $\mathcal{S}$ become very large and extend far beyond the force distribution, $f_k(\mathbf{x}')$, the surface integral given by Eq. (3.196) vanishes. Finally, if the force density distribution corresponds to a delta function point force of the form $f_k(\mathbf{x}') = F_k\delta(\mathbf{x}')$, and this is substituted into Eq. (3.195), with the second term set to zero,

$$u_j(\mathbf{x}) = \oiiint_{\mathcal{V}} G_{kj}(\mathbf{x}-\mathbf{x}')F_k\delta(\mathbf{x}')dV' = G_{kj}(\mathbf{x})F_k, \tag{3.198}$$

and we therefore recover Eq. (3.15).

3.6. Suppose an elastic field associated with a distribution of body force, $\mathbf{f}(\mathbf{x})$. If

$$\operatorname{div}\mathbf{f} = 0, \tag{3.199}$$

show that the displacement field must then obey the relationship

$$C_{ijkl}\frac{\partial^3 u_k}{\partial x_i \partial x_j \partial x_l} = 0. \tag{3.200}$$

On the other hand, if the system is isotropic, show that we must have

$$\frac{\partial}{\partial x_i}(\nabla^2 u_i) = 0. \tag{3.201}$$

Solution. By writing Eq. (3.199) in the form

$$\operatorname{div}\mathbf{f} = \frac{\partial f_i}{\partial x_i} = 0 \tag{3.202}$$

and substituting Eq. (3.2) into (3.202), we obtain

$$C_{ijkl}\frac{\partial^3 u_k}{\partial x_i \partial x_j \partial x_l} = 0. \tag{3.203}$$

For the isotropic system, take the divergence of Eq. (3.3), i.e.

$$\frac{\mu}{1-2\nu}\operatorname{div}\nabla(\nabla\cdot\mathbf{u}) + \mu\operatorname{div}\nabla^2\mathbf{u} + \operatorname{div}\mathbf{f} = 0. \tag{3.204}$$

Then, by substituting $\operatorname{div}\mathbf{f} = 0$ and the identities

$$\operatorname{div}\nabla(\nabla\cdot\mathbf{u}) = \operatorname{div}\nabla^2\mathbf{u} = \frac{\partial}{\partial x_i}(\nabla^2 u_i) \tag{3.205}$$

into Eq. (3.204),

$$\frac{\partial}{\partial x_i}(\nabla^2 u_i) = 0. \tag{3.206}$$

Alternatively, note that Eq. (3.206) could have been obtained from Eq. (3.203) by use of Eq. (2.120).

Chapter 4

Green's Functions for Unit Point Force

4.1 Introduction

The elastic Green's function for a point force was introduced in Ch. 3, and its general use in solving defect elasticity problems was described. Also, a basic differential equation for such a Green's function was derived in the form of Eq. (3.21). With the help of that equation, the elastic Green's functions for a unit point force is now obtained when it is present in three different types of regions: (1) an infinite homogeneous region, (2) a half-space with a planar traction-free surface, and (3) a half-space joined to an elastically dissimilar half-space along a planar interface. Following this, corresponding results for isotropic systems are derived.

For this chapter we employ the following notation:

G^{∞}_{km} = Green's function for point force in infinite homogeneous body

G_{km} = Green's function for point force in half-space with a planar traction-free surface

$G^{\mathrm{IM}}_{\mathrm{km}}$ = image Green's function for point force in half-space with a planar traction-free surface

$g^{(1)}_{\mathrm{km}}$ = Green's function in half space 1 for point force in half-space 1 joined to elastically dissimilar half-space 2 along planar interface

$g^{\mathrm{IM}^{(1)}}_{\mathrm{km}}$ = image Green's function in half space 1 for point force in half-space 1 joined to elastically dissimilar half-space 2 along planar interface

$g^{(2)}_{\mathrm{km}}$ = Green's function in half-space 2 for point force in half-space 1 joined to elastically dissimilar half-space 2 along planar interface

4.2 Green's Functions for Unit Point Force

The Green's functions for all three of the above regions are obtained following a treatment by Barnett (personal communication, 2007) which is constructed within the framework of the sextic formalism of Sec. 3.5.1,[1] The geometry employed for all three above regions is illustrated in Fig. 4.1. As in Sec. 3.5.1.1, two coordinate systems are employed. The first is the crystal system, already described in Sec. 3.5.1.1, while the second is the system illustrated in Fig. 4.1 employing the orthogonal base vectors, $(\hat{\mathbf{u}}, \hat{\mathbf{v}}, \hat{\mathbf{w}})$. The field point is at $\mathbf{x}$, ξ is the component of $\mathbf{x}$ along $\hat{\mathbf{w}}$ so that $\xi = \mathbf{x} \cdot \hat{\mathbf{w}}$, and the point force, $\mathbf{F}$, acts at the source point, $\mathbf{x}' = \xi_\circ \hat{\mathbf{w}}$.

A two-dimensional Fourier transform formulation is used in conjunction with Eq. (3.21) to find the Green's function. The problem is transformed into Fourier space where solutions are found and are then inversely transformed back to real space. Consistent with the geometry of Fig. 4.1, the two-dimensional transform of a function, $\mathrm{f_{km}}(\mathbf{x})$, and its inverse, will be written in the form (which may be compared with Eqs. (F.1) and (F.2)),

$$\begin{aligned} \bar{\mathrm{f}}_{\mathrm{km}}(\mathbf{k}, \xi = \mathbf{x} \cdot \hat{\mathbf{w}}) &= \int_{-\infty}^{\infty} \int_{-\infty}^{\infty} \mathrm{f_{km}}(\mathbf{x}) e^{i\mathbf{k}\cdot\mathbf{x}} d(\hat{\mathbf{u}} \cdot \mathbf{x}) d(\hat{\mathbf{v}} \cdot \mathbf{x}) \\ \mathrm{f_{km}}(\mathbf{x}) &= \frac{1}{4\pi^2} \int_{-\infty}^{\infty} \int_{-\infty}^{\infty} \bar{\mathrm{f}}_{\mathrm{km}}(\mathbf{k}, \xi = \mathbf{x} \cdot \hat{\mathbf{w}}) e^{-i\mathbf{k}\cdot\mathbf{x}} dk_1 dk_2 \end{aligned} \tag{4.1}$$

where the transform vector, $\hat{\mathbf{k}}$, lies in the plane normal to $\hat{\mathbf{w}}$ as indicated in Fig. 4.1. For the treatments of the two regions where planar interfaces are present, the base unit vector, $\hat{\mathbf{w}}$, is taken normal to the interface, whereas

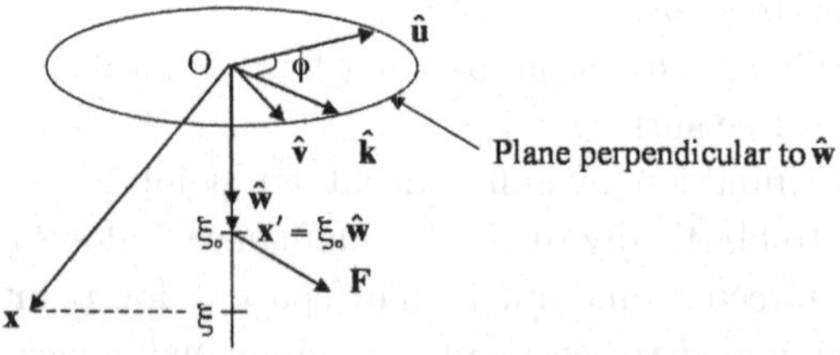

Fig. 4.1 Geometry for finding Green's function G^{∞}_{km} at field point $\mathbf{x}$, for point force $\mathbf{F}$, acting at source point $\mathbf{x}' = \xi_\circ \hat{\mathbf{w}}$.

[1] I am indebted to Prof. D. M. Barnett for permission to include his previously unpublished derivations of the Green's functions, G^{∞}_{km}, G_{km} and g_{km} in Secs. 4.2.1, 4.2.2 and 4.2.3, respectively.

for the infinite homogeneous region the direction of $\hat{\mathbf{w}}$ is arbitrary. In all cases the entire system (including the interface, if present) can therefore be rotated relative to the crystal system to study the effect of elastic anisotropy.[2]

The Green's function for the infinite homogeneous region is found first. Solutions are then found for the half-space and the joined half-spaces, by adding image terms to satisfy the boundary conditions at the interfaces.

4.2.1 *In infinite homogeneous region*

4.2.1.1 *Green's function*

Employing Eq. (4.1), the Fourier transform of G^{∞}_{km} and its inverse are first written (Barnett, personal communication, 2007) as

$$\begin{aligned}\bar{G}^{\infty}_{km}(k, \hat{\mathbf{w}} \cdot \mathbf{x}) &= \int_{-\infty}^{\infty}\int_{-\infty}^{\infty} G^{\infty}_{km}(\mathbf{x}) e^{i\mathbf{k}\cdot\mathbf{x}} d(\hat{\mathbf{u}} \cdot \mathbf{x}) d(\hat{\mathbf{v}} \cdot \mathbf{x}),\\ G^{\infty}_{km}(\mathbf{x}) &= \frac{1}{4\pi^2}\int_{-\infty}^{\infty}\int_{-\infty}^{\infty} \bar{G}^{\infty}_{km}(k, \hat{\mathbf{w}} \cdot \mathbf{x}) e^{-i\mathbf{k}\cdot\mathbf{x}} dk_1 dk_2.\end{aligned} \tag{4.2}$$

Since the point force is located at $(0, 0, \xi_\circ)$, Eq. (3.21) for the Green's function assumes the form

$$C_{ijkl}\frac{\partial^2 G^{\infty}_{km}}{\partial x_i \partial x_l} + \delta_{jm}\delta(\mathbf{x} \cdot \hat{\mathbf{u}})\delta(\mathbf{x} \cdot \hat{\mathbf{v}})\delta(\mathbf{x} \cdot \hat{\mathbf{w}} - \xi_\circ) = 0. \tag{4.3}$$

Then, by substituting the standard delta function relationship (Sneddon 1951),

$$\delta(\mathbf{x} \cdot \hat{\mathbf{u}})\delta(\mathbf{x} \cdot \hat{\mathbf{v}}) = \frac{1}{4\pi^2}\int_{-\infty}^{\infty}\int_{-\infty}^{\infty} e^{-i\mathbf{k}\cdot\mathbf{x}} dk_1 dk_2 \tag{4.4}$$

and Eq. (4.2) into Eq. (4.3), the equation

$$\int_{-\infty}^{\infty}\int_{-\infty}^{\infty}\left\{C_{ijkl}\frac{\partial^2}{\partial x_i \partial x_l}\left[\bar{G}^{\infty}_{km} e^{-i\mathbf{k}\cdot\mathbf{x}}\right] + \delta_{jm}\delta(\mathbf{x} \cdot \hat{\mathbf{w}} - \xi_\circ) e^{-i\mathbf{k}\cdot\mathbf{x}}\right\} dk_1 dk_2 = 0 \tag{4.5}$$

for the Fourier transform of the Green's function is obtained. Next, by writing $\mathbf{k} = k\hat{\mathbf{k}}$, and using $\xi = \hat{\mathbf{w}} \cdot \mathbf{x}$ and

$$\partial\bar{G}^{\infty}_{km}/\partial x_n = (\partial\bar{G}^{\infty}_{km}/\partial\xi)(\partial\xi/\partial x_n) = (\partial\bar{G}^{\infty}_{km}/\partial\xi)\hat{w}_n, \tag{4.6}$$

[2]In this respect, the $(\hat{\mathbf{u}}, \hat{\mathbf{v}}, \hat{\mathbf{w}})$ system is analogous to the $(\hat{\mathbf{m}}, \hat{\mathbf{n}}, \hat{\boldsymbol{\tau}})$ system of Sec. 3.5.1.1.

Eq. (4.5) can be put into the form

$$\int_{-\infty}^{\infty}\int_{-\infty}^{\infty}\left\{C_{ijkl}\left[\left(-ik\hat{k}_i+\hat{w}_i\frac{\partial}{\partial\xi}\right)\left(-ik\hat{k}_l+\hat{w}_l\frac{\partial}{\partial\xi}\right)\bar{G}^{\infty}_{km}\right]\right.$$
$$\left.+\delta_{jm}\delta(\xi-\xi_\circ)\right\}e^{-ik\hat{k}\cdot x}dk_1dk_2=0, \tag{4.7}$$

which is satisfied if

$$C_{ijkl}\left[\left(-ik\hat{k}_i+\hat{w}_i\frac{\partial}{\partial\xi}\right)\left(-ik\hat{k}_l+\hat{w}_l\frac{\partial}{\partial\xi}\right)\bar{G}^{\infty}_{km}\right]+\delta_{jm}\delta(\xi-\xi_\circ)=0. \tag{4.8}$$

Using the Christoffel stiffness tensor notation of Eq. (3.5), Eq. (4.8) then assumes the form

$$-k^2(\hat{k}\hat{k})_{jk}\bar{G}^{\infty}_{km}-ik[(\hat{k}\hat{w})_{jk}+(\hat{w}\hat{k})_{jk}]\frac{\partial\bar{G}^{\infty}_{km}}{\partial\xi}$$
$$+(\hat{w}\hat{w})_{jk}\frac{\partial}{\partial\xi}\left(\frac{\partial\bar{G}^{\infty}_{km}}{\partial\xi}\right)=-\delta_{jm}\delta(\xi-\xi_\circ). \tag{4.9}$$

The solution of Eq. (4.9) for $\bar{G}^{\infty}_{km}$ can be simplified by imposing on $\bar{G}^{\infty}_{km}$ the jump condition at the location of the point force,

$$\left|\frac{\partial\bar{G}^{\infty}_{km}}{\partial\xi}\right|_{\xi_\circ^+}-\left|\frac{\partial\bar{G}^{\infty}_{km}}{\partial\xi}\right|_{\xi_\circ^-}=-(\hat{w}\hat{w})^{-1}_{km}. \tag{4.10}$$

This incorporates the required delta function at the point force into the solution for $\bar{G}^{\infty}_{km}$, since the derivative of a Heaviside step function is a delta function at the step (see Table D.1). The delta function term in Eq. (4.9) can then be dropped leaving the more amenable homogeneous equation

$$-k^2(\hat{k}\hat{k})_{jk}\bar{G}^{\infty}_{km}-ik[(\hat{k}\hat{w})_{jk}+(\hat{w}\hat{k})_{jk}]\frac{\partial\bar{G}^{\infty}_{km}}{\partial\xi}+(\hat{w}\hat{w})_{jk}\frac{\partial}{\partial\xi}\left(\frac{\partial\bar{G}^{\infty}_{km}}{\partial\xi}\right)=0. \tag{4.11}$$

A solution of Eq. (4.11) of the form

$$\bar{G}^{\infty}_{km}=A_{k\alpha}B_{\alpha m}e^{-ikp_\alpha\xi} \tag{4.12}$$

is then assumed, which, after substitution into Eq. (4.11), yields

$$\left\{(\hat{w}\hat{w})_{jk}p_\alpha^2+[(\hat{k}\hat{w})_{jk}+(\hat{w}\hat{k})_{jk}]p_\alpha+(\hat{k}\hat{k})_{jk}\right\}A_{k\alpha}=0. \tag{4.13}$$

Equation (4.13) constitutes an eigenvalue problem that requires finding the eigenvalues, p_α, and corresponding eigenvectors, $A_{k\alpha}$. However, it is seen to be identical in form to Eq. (3.32) for the eigenvalue problem of

Ch. 3 (which determined the Stroh eigenvectors) with $\hat{\mathbf{w}}$ corresponding to $\hat{\mathbf{n}}$ and $\hat{\mathbf{k}}$ corresponding to $\hat{\mathbf{m}}$. The results in Ch. 3, involving the eigenvalues and eigenvectors in Eqs. (3.34) and (3.35), and the various Stroh vector expressions involving $\mathbf{A}_\alpha$ and $\mathbf{L}_\alpha$, are therefore valid for the present problem if $\hat{\mathbf{n}}$ is simply replaced by $\hat{\mathbf{w}}$ and $\hat{\mathbf{m}}$ replaced by $\hat{\mathbf{k}}$.

Having this result, a general series solution of the form

$$\begin{aligned} \bar{G}^{\infty}_{km} &= \sum_{\alpha=1}^{3} A^{*}_{k\alpha} B^{*}_{\alpha m} e^{-ikp^{*}_{\alpha}(\xi-\xi_{o})} \quad (\xi > \xi_{o}) \\ \bar{G}^{\infty}_{km} &= \sum_{\alpha=1}^{3} A_{k\alpha} B_{\alpha m} e^{-ikp_{\alpha}(\xi-\xi_{o})} \quad (\xi < \xi_{o}) \end{aligned} \tag{4.14}$$

is assumed where $B_{\alpha m}$ and $B^{*}_{\alpha m}$ are to be determined. Equation (4.14) satisfies the condition that $\bar{G}^{\infty}_{km}$ must vanish as $\xi \to \pm\infty$ because of the manner in which the positive and negative imaginary parts of the p_α and p^{*}_{α} eigenvalues appear in the two sums [see Eq. (3.34)]. Since $\bar{G}^{\infty}_{km}$ must be continuous at $\xi = \xi_o$, the condition

$$\sum_{\alpha=1}^{3} A_{k\alpha} B_{\alpha m} = \sum_{\alpha=1}^{3} A^{*}_{k\alpha} B^{*}_{\alpha m} \tag{4.15}$$

must be satisfied, and by substituting Eqs. (4.14) and (4.15), the condition given by Eq. (4.10) becomes

$$\sum_{\alpha=1}^{3} (p_\alpha A_{k\alpha} B_{\alpha m} - p^{*}_{\alpha} A^{*}_{k\alpha} B^{*}_{\alpha m}) = \frac{i}{k} (\hat{w}\hat{w})^{-1}_{km}. \tag{4.16}$$

Then, with the help of Eqs. (3.76) and (3.110) (with $\hat{\mathbf{n}}$ replaced by $\hat{\mathbf{w}}$ as discussed in the text following Eq. (4.13)), Eqs. (4.15) and (4.16) are satisfied when

$$B_{\alpha m} = -\frac{iA_{m\alpha}}{k} \quad B^{*}_{\alpha m} = \frac{iA^{*}_{m\alpha}}{k} \tag{4.17}$$

and, putting these results into Eq. (4.14),

$$\begin{aligned} \bar{G}^{\infty}_{km} &= \frac{i}{k} \sum_{\alpha=1}^{3} A^{*}_{k\alpha} A^{*}_{m\alpha} e^{-ikp^{*}_{\alpha}(\xi-\xi_{o})} \quad (\xi > \xi_{o}), \\ \bar{G}^{\infty}_{km} &= -\frac{i}{k} \sum_{\alpha=1}^{3} A_{k\alpha} A_{m\alpha} e^{-ikp_{\alpha}(\xi-\xi_{o})} \quad (\xi < \xi_{o}). \end{aligned} \tag{4.18}$$

Equation (4.18) is now inverted by first substituting it into Eq. (4.2) and setting $dk_1 dk_2 = kd\phi dk$ so that

$$G^{\infty}_{km}(\mathbf{x}) = -\frac{1}{4\pi^2 i}\int_0^{2\pi} d\phi \int_0^{\infty} dk \sum_{\alpha=1}^{3} A^*_{k\alpha}A^*_{m\alpha}e^{-ik[\hat{\mathbf{k}}\cdot\mathbf{x}+p^*_\alpha(\xi-\xi_\circ)]}, \quad (\xi > \xi_\circ)$$
$$G^{\infty}_{km}(\mathbf{x}) = \frac{1}{4\pi^2 i}\int_0^{2\pi} d\phi \int_0^{\infty} dk \sum_{\alpha=1}^{3} A_{k\alpha}A_{m\alpha}e^{-ik[\hat{\mathbf{k}}\cdot\mathbf{x}+p_\alpha(\xi-\xi_\circ)]}. \quad (\xi < \xi_\circ) \tag{4.19}$$

Then, after integrating over k,

$$G^{\infty}_{km}(\mathbf{x}) = \frac{1}{4\pi^2}\int_0^{2\pi} d\phi \sum_{\alpha=1}^{3} \frac{A^*_{k\alpha}A^*_{m\alpha}}{\hat{\mathbf{k}}\cdot\mathbf{x}+p^*_\alpha(\mathbf{x}\cdot\hat{\mathbf{w}}-\xi_\circ)}, \quad (\xi > \xi_\circ)$$
$$G^{\infty}_{km}(\mathbf{x}) = -\frac{1}{4\pi^2}\int_0^{2\pi} d\phi \sum_{\alpha=1}^{3} \frac{A_{k\alpha}A_{m\alpha}}{\hat{\mathbf{k}}\cdot\mathbf{x}+p_\alpha(\mathbf{x}\cdot\hat{\mathbf{w}}-\xi_\circ)}, \quad (\xi < \xi_\circ) \tag{4.20}$$

where ξ has been replaced by $\mathbf{x}\cdot\hat{\mathbf{w}}$.

Equation (4.20) can be expressed in a simpler form by recalling that the Green's function for a point force in an infinite homogeneous region with no interface present is a unique function of the vector difference $\mathbf{x}-\mathbf{x}'$, i.e., $G^{\infty}_{km} = G^{\infty}_{km}(\mathbf{x}-\mathbf{x}')$, as indicated previously in Eq. (3.15). The system in Fig. 4.1 can therefore be arranged so that $\hat{\mathbf{w}}$ is parallel to $\mathbf{x}$, and the source vector at $Q(\mathbf{x}')$ is given by $\mathbf{x}' = \xi_\circ\hat{\mathbf{w}}$, as illustrated in Fig. 4.2. Equation (4.20) still applies, but now

$$\hat{\mathbf{w}} = \frac{\mathbf{x}-\mathbf{x}'}{|\mathbf{x}-\mathbf{x}'|},$$
$$\hat{\mathbf{k}}\cdot\mathbf{x} = \hat{\mathbf{k}}\cdot\hat{\mathbf{w}}x = 0, \tag{4.21}$$
$$\mathbf{x}\cdot\hat{\mathbf{w}} - \xi_\circ = \hat{\mathbf{w}}\cdot(\mathbf{x}-\mathbf{x}') = |\mathbf{x}-\mathbf{x}'|.$$

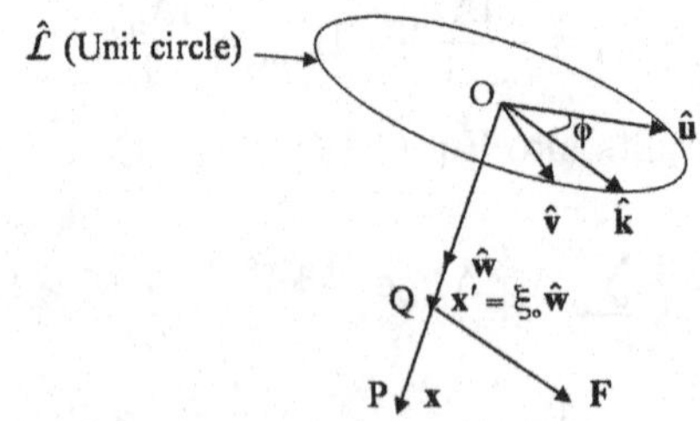

Fig. 4.2 Geometry for Eq. (4.22): $\hat{\mathbf{w}}$ is parallel to both $\mathbf{x}$ and $\mathbf{x}-\mathbf{x}'$: $\mathbf{x}' = \xi_\circ\hat{\mathbf{w}}$.

Therefore, substituting these relationships into Eq. (4.20),

$$
\begin{aligned}
G^{\infty}_{km}(\mathbf{x}-\mathbf{x}') &= \frac{1}{4\pi^2|\mathbf{x}-\mathbf{x}'|}\int_0^{2\pi} d\phi \sum_{\alpha=1}^{3} \frac{A^*_{k\alpha}A^*_{m\alpha}}{p^*_\alpha}, \quad (\xi > \xi_\circ) \\
G^{\infty}_{km}(\mathbf{x}-\mathbf{x}') &= \frac{1}{4\pi^2|\mathbf{x}-\mathbf{x}'|}\int_0^{2\pi} d\phi \sum_{\alpha=1}^{3} \frac{A_{k\alpha}A_{m\alpha}}{p_\alpha}. \quad (\xi < \xi_\circ)
\end{aligned}
\tag{4.22}
$$

The real parts of Eq. (4.22) can now be taken as the desired solution. According to Eq. (3.115), with $\hat{\mathbf{m}}$ replaced by $\hat{\mathbf{k}}$ as discussed in the text following Eq. (4.13), the real parts of the sums in Eq. (4.22) are given by

$$
\mathcal{R}e\sum_{\alpha=1}^{3}\frac{A_{k\alpha}A_{m\alpha}}{p_\alpha} = \mathcal{R}e\sum_{\alpha=1}^{3}\frac{A^*_{k\alpha}A^*_{m\alpha}}{p^*_\alpha} = \frac{1}{2}(\hat{k}\hat{k})^{-1}_{km}. \tag{4.23}
$$

Substituting these relationships into Eq. (4.22) then yields the expression

$$
G^{\infty}_{km}(\mathbf{x}-\mathbf{x}') = \frac{1}{8\pi^2|\mathbf{x}-\mathbf{x}'|}\int_0^{2\pi}(\hat{k}\hat{k})^{-1}_{km}d\phi. \tag{4.24}
$$

Recasting Eq. (4.24) as a line integral, $G^{\infty}_{km}(\mathbf{x}-\mathbf{x}')$ can also be written as

$$
G^{\infty}_{km}(\mathbf{x}-\mathbf{x}') = \frac{1}{8\pi^2|\mathbf{x}-\mathbf{x}'|}\oint_{\hat{\mathcal{L}}}(\hat{k}\hat{k})^{-1}_{km}ds. \tag{4.25}
$$

As is evident in Fig. 4.2, where $\hat{\mathbf{w}}$ is parallel to $(\mathbf{x}-\mathbf{x}')$, the integral in Eq. (4.25) corresponds to a line integral along s around the unit circle, $\hat{\mathcal{L}}$, traversed by the unit vector $\hat{\mathbf{k}}$ as it rotates in the plane perpendicular to $\hat{\mathbf{w}} = (\mathbf{x}-\mathbf{x}')/|\mathbf{x}-\mathbf{x}'|$. Using standard matrix algebra for the inversion, the integrand, $(\hat{k}\hat{k})^{-1}_{km}$, can be expressed in the form

$$
(\hat{k}\hat{k})^{-1}_{km} = \frac{\varepsilon_{ksj}\varepsilon_{mrw}(\hat{k}\hat{k})_{sr}(\hat{k}\hat{k})_{jw}}{2\varepsilon_{pgn}(\hat{k}\hat{k})_{1p}(\hat{k}\hat{k})_{2g}(\hat{k}\hat{k})_{3n}}. \tag{4.26}
$$

The above results for the Green's function are in agreement with results found elsewhere using other methods (Bacon, Barnett and Scattergood 1979b).[3] It must be emphasized that the evaluation of Eq. (4.25) does not require the sextic eigenvalue problem to be solved, and the integral

[3] Also, Eq. (4.24) is derived by an alternative method in Exercise 4.7.

can be readily evaluated numerically as discussed by Bacon, Barnett and Scattergood (1979b).

4.2.1.2 *Derivatives of the Green's function for an infinite homogeneous region*

Expressions for the spatial derivatives of G^{∞}_{km} are required frequently, and their determination, which is not straightforward, is now discussed. Following Bacon, Barnett and Scattergood (1979b), Eq. (4.25), which is expressed as a line integral around the unit circle $\hat{\mathcal{L}}$ in Fig. 4.2, is first rewritten (equivalently) as a surface integral of the form

$$G^{\infty}_{km}(\mathbf{x}-\mathbf{x}') = \frac{1}{8\pi^2|\mathbf{x}-\mathbf{x}'|} \oiint_{\hat{\mathcal{S}}} (\hat{k}\hat{k})^{-1}_{km}\delta(\hat{\mathbf{k}}\cdot\hat{\mathbf{w}})dS, \tag{4.27}$$

where the integration is over the unit sphere $\hat{\mathcal{S}}(|\hat{\mathbf{k}}|=1)$ but is subjected to a delta function that restricts the non-vanishing part of the integral to the unit circle, $\hat{\mathcal{L}}$, on which $\hat{\mathbf{k}}\cdot\hat{\mathbf{w}}=0$. Since $\hat{\mathbf{w}}=(\mathbf{x}-\mathbf{x}')/|\mathbf{x}-\mathbf{x}'|$,

$$\frac{\partial G^{\infty}_{km}(\mathbf{x}-\mathbf{x}')}{\partial x_i} = \frac{1}{8\pi^2}\left\{\frac{1}{|x-x'|}\oiint_{\hat{\mathcal{S}}}(\hat{k}\hat{k})^{-1}_{km}\frac{\partial\delta(\hat{\mathbf{k}}\cdot\hat{\mathbf{w}})}{\partial x_i} - \frac{(x_i-x_i')}{|\mathbf{x}-\mathbf{x}'|^3}\oiint_{\hat{\mathcal{S}}}(\hat{k}\hat{k})^{-1}_{km}\delta(\hat{\mathbf{k}}\cdot\hat{\mathbf{w}})\right\}dS. \tag{4.28}$$

Next, substituting the two equalities

$$\frac{\partial\delta(\hat{\mathbf{k}}\cdot\hat{\mathbf{w}})}{\partial x_i} = \frac{d\delta(\hat{\mathbf{k}}\cdot\hat{\mathbf{w}})}{d(\hat{\mathbf{k}}\cdot\hat{\mathbf{w}})}\frac{\partial(\hat{\mathbf{k}}\cdot\hat{\mathbf{w}})}{\partial x_i} \tag{4.29}$$

and

$$(\hat{\mathbf{k}}\cdot\hat{\mathbf{w}}) = \frac{\hat{k}_s(x_s-x_s')}{|\mathbf{x}-\mathbf{x}'|} \tag{4.30}$$

into Eq. (4.28), and using the properties of the delta function in Table D.1,

$$\frac{\partial G^{\infty}_{km}(\mathbf{x}-\mathbf{x}')}{\partial x_i} = \frac{1}{8\pi^2|\mathbf{x}-\mathbf{x}'|^2}\oiint_{\hat{\mathcal{S}}}(\hat{k}\hat{k})^{-1}_{km}\hat{k}_i\frac{d\delta(\hat{\mathbf{k}}\cdot\hat{\mathbf{w}})}{d(\hat{\mathbf{k}}\cdot\hat{\mathbf{w}})}dS. \tag{4.31}$$

Then, substituting the derivative of the delta function from appendix D, we obtain

$$\frac{\partial G^{\infty}_{km}(\mathbf{x}-\mathbf{x}')}{\partial x_i} = -\frac{1}{8\pi^2|\mathbf{x}-\mathbf{x}'|^2}\oiint_{\hat{\mathcal{S}}}\left\{\frac{\partial[(\hat{k}\hat{k})^{-1}_{km}\hat{k}_i]}{\partial(\hat{\mathbf{k}}\cdot\hat{\mathbf{w}})}\right\}_{\hat{\mathbf{k}}\cdot\hat{\mathbf{w}}=0} dS, \tag{4.32}$$

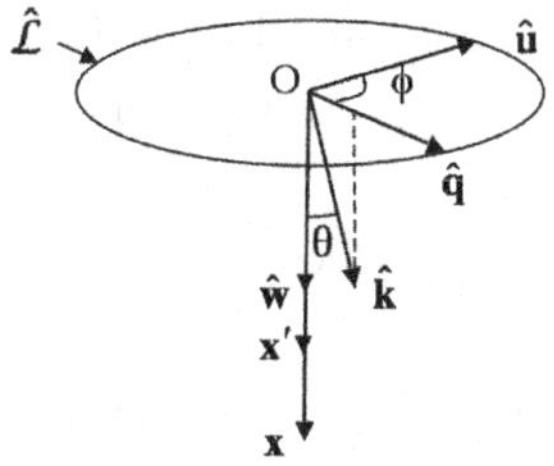

Fig. 4.3 Geometry used to obtain Eq. (4.34).

which can also be written as a line integral around the unit circle $\hat{\mathcal{L}}$ in the form

$$\frac{\partial G^{\infty}_{km}(\mathbf{x}-\mathbf{x}')}{\partial x_i} = -\frac{1}{8\pi^2|\mathbf{x}-\mathbf{x}'|^2}\oint_{\hat{\mathcal{L}}} \frac{\partial[(\hat{k}\hat{k})^{-1}_{km}\hat{k}_i]}{\partial(\hat{\mathbf{k}}\cdot\hat{\mathbf{w}})} ds. \tag{4.33}$$

The integral in Eq. (4.33) can be simplified by referring to Fig. 4.3 which shows the relevant parameters and introduces the unit vector, $\hat{\mathbf{q}}$, lying along the projection of $\hat{\mathbf{k}}$ on the plane of the unit circle. From the figure it is deduced that

$$\hat{\mathbf{k}} = \cos\theta\hat{\mathbf{w}} + \sin\theta\hat{\mathbf{q}} \quad \text{or} \quad \hat{k}_j = \hat{w}_j\cos\theta + \hat{q}_j\sin\theta. \tag{4.34}$$

Therefore,

$$\left[\frac{\partial\hat{k}_j}{\partial(\hat{\mathbf{k}}\cdot\hat{\mathbf{w}})}\right]_{\hat{\mathbf{k}}\cdot\hat{\mathbf{w}}=0} = \left[\frac{\partial\hat{k}_j}{\partial\cos\theta}\right]_{\theta=\pi/2} = \hat{w}_j \tag{4.35}$$

and using Eq. (4.35) to change variables in Eq. (4.33),

$$\frac{\partial G^{\infty}_{km}(\mathbf{x}-\mathbf{x}')}{\partial x_i} = -\frac{1}{8\pi^2|\mathbf{x}-\mathbf{x}'|^2}\oint_{\hat{\mathcal{L}}}\left[\hat{w}_i(\hat{k}\hat{k})^{-1}_{km} + \hat{w}_s\hat{k}_i\frac{\partial(\hat{k}\hat{k})^{-1}_{km}}{\partial\hat{k}_s}\right] ds. \tag{4.36}$$

An expression for the derivative in the integrand can then be obtained by first differentiating the identity $(\hat{k}\hat{k})^{-1}_{km}(\hat{k}\hat{k})_{mj} = \delta_{kj}$ so that

$$(\hat{k}\hat{k})^{-1}_{km}\frac{\partial(\hat{k}\hat{k})_{mj}}{\partial\hat{k}_s} + (\hat{k}\hat{k})_{mj}\frac{\partial(\hat{k}\hat{k})^{-1}_{km}}{\partial\hat{k}_s} = 0. \tag{4.37}$$

However, since, by definition, $(\hat{k}\hat{k})_{mj} = \hat{k}_i C_{imjn}\hat{k}_n$,

$$\frac{\partial(\hat{k}\hat{k})_{mj}}{\partial\hat{k}_s} = \hat{k}_n(C_{nmjs} + C_{smjn}) \tag{4.38}$$

and, after substituting Eq. (4.38) into Eq. (4.37) and multiplying the result throughout by $(\hat{k}\hat{k})^{-1}_{im}$, and using $(\hat{k}\hat{k})^{-1}_{im}(\hat{k}\hat{k})_{mj} = \delta_{ij}$,

$$\frac{\partial(\hat{k}\hat{k})^{-1}_{km}}{\partial \hat{k}_s} = -(\hat{k}\hat{k})^{-1}_{kp}(\hat{k}\hat{k})^{-1}_{jm}C_{qpjn}(\hat{k}_q\delta_{ns} + \hat{k}_n\delta_{qs}). \tag{4.39}$$

Finally, substituting Eq. (4.39) into Eq. (4.36),

$$\frac{\partial G^{\infty}_{km}(\mathbf{x}-\mathbf{x}')}{\partial x_i} = -\frac{1}{8\pi^2|\mathbf{x}-\mathbf{x}'|^2} \times \oint_{\hat{\mathcal{L}}} \left\{\hat{w}_i(\hat{k}\hat{k})^{-1}_{km} - \hat{k}_i(\hat{k}\hat{k})^{-1}_{kp}(\hat{k}\hat{k})^{-1}_{jm}[(\hat{k}\hat{w})_{pj} + (\hat{w}\hat{k})_{pj}]\right\} ds \tag{4.40}$$

where, as in the case of Eq. (4.25), the integral corresponds to a line integral along s around the unit circle, $\hat{\mathcal{L}}$, traversed by the unit vector $\hat{\mathbf{k}}$ as it rotates in the plane perpendicular to $\hat{\mathbf{w}} = (\mathbf{x}-\mathbf{x}')/|\mathbf{x}-\mathbf{x}'|$. Derivatives with respect to x'_i are then readily obtained using

$$\frac{\partial G^{\infty}_{km}(\mathbf{x}-\mathbf{x}')}{\partial x'_i} = -\frac{\partial G^{\infty}_{km}(\mathbf{x}-\mathbf{x}')}{\partial x_i}. \tag{4.41}$$

Higher order derivatives can be obtained by repeated differentiation and are described by Barnett (1972) and Bacon, Barnett and Scattergood (1979b). In general, it is found that the Nth derivative can be written in the form

$$\frac{\partial^N G^{\infty}_{km}}{\partial x_{s_1}\ldots\partial x_{s_N}} = \frac{(-1)^N}{8\pi^2|\mathbf{x}-\mathbf{x}'|^{N+1}}\oint_{\hat{\mathcal{L}}} f_N(\hat{\mathbf{w}},\hat{\mathbf{k}})ds. \tag{4.42}$$

In Exercise 4.3, Eq. (4.40) for the first derivative is derived by an alternative, and physically more transparent, approach.

4.2.2 *In half-space with planar free surface*

Having G^{∞}_{km}, the Green's function for a half-space with a planar traction-free surface, G_{km}, is now derived. This problem has been treated by Pan and Yuan (2000) and Barnett (personal communication, 2007), but we will follow Barnett in an extension of the preceding analysis.

Figure 4.1 still holds but with the origin and the unit circle now fixed in the surface plane. The procedure will be to assume that G^{∞}_{km} is valid everywhere in the half-space and then find an image term, G^{IM}_{km}, that vanishes as $\xi \to \infty$ and, when added to G^{∞}_{km}, produces a traction-free surface. The solution will therefore be of the form

$$G_{km} = G^{\infty}_{km} + G^{IM}_{km}(\mathbf{x}, \xi_{\circ}), \quad (\xi > 0) \tag{4.43}$$

where the image Green's function is a function of both the field vector, $\mathbf{x}$, and the position, $\xi_{\circ}$, of the source point along $\hat{\mathbf{w}}$. The corresponding transforms are then

$$\bar{G}_{km} = \bar{G}^{\infty}_{km} + \bar{G}^{IM}_{km}. \quad (\xi > 0) \tag{4.44}$$

An image transform solution of the form

$$\bar{G}^{IM}_{km} = \sum_{\alpha=1}^{3} A^{*}_{k\alpha} E^{*}_{\alpha m} e^{-ikp^{*}_{\alpha}\xi} \tag{4.45}$$

is therefore assumed which vanishes as $\xi \to \infty$. Next, the unknown, $E^{*}_{\alpha m}$, is determined by invoking the traction-free condition at the surface. Using Eq. (2.59), the traction at $\xi = 0$ that would be present if the region were infinite is

$$T^{\infty}_{j} = -(\sigma^{\infty}_{ij})_{\xi=0} \hat{w}_{i}, \tag{4.46}$$

where σ^{∞}_{ij} is the stress due to the point force in an infinite region which is obtained by using Eq. (3.15) in the form

$$\sigma^{\infty}_{ij} = C_{ijkl} \frac{\partial G^{\infty}_{km}}{\partial x_{l}} F_{m}. \tag{4.47}$$

Putting this into Eq. (4.46),

$$T^{\infty}_{j} = -\hat{w}_{i} C_{ijkl} \left(\frac{\partial G^{\infty}_{km}}{\partial x_{l}} \right)_{\xi=0} F_{m}. \tag{4.48}$$

Similarly, the traction associated with the image Green's function is

$$T^{IM}_{j} = -\hat{w}_{i} C_{ijkl} \left(\frac{\partial G^{IM}_{km}}{\partial x_{l}} \right)_{\xi=0} F_{m}. \tag{4.49}$$

The condition for a traction-free surface, i.e., $T^{\infty}_{j} + T^{IM}_{j} = 0$, is

$$\hat{w}_{i} C_{ijkl} \left(\frac{\partial G^{\infty}_{km}}{\partial x_{l}} + \frac{\partial G^{IM}_{km}}{\partial x_{l}} \right)_{\xi=0} F_{m} = 0 \tag{4.50}$$

and, since F_{m} is arbitrary,

$$\hat{w}_{i} C_{ijkl} \left(\frac{\partial G^{\infty}_{km}}{\partial x_{l}} + \frac{\partial G^{IM}_{km}}{\partial x_{l}} \right)_{\xi=0} = 0. \tag{4.51}$$

Next, substituting the inverse transform relationship given by Eq. (4.1) into Eq. (4.51), and using $\partial/\partial x_l = \hat{w}_l \partial/\partial\xi$,

$$\int_{-\infty}^{\infty}\int_{-\infty}^{\infty} \hat{w}_i C_{ijkl}\left(-ik\hat{k}_l + \hat{w}_l \frac{\partial}{\partial\xi}\right)(\bar{G}^{\infty}_{km} + \bar{G}^{IM}_{km})\, dk_1 dk_2 = 0, \quad (\xi = 0) \tag{4.52}$$

which is satisfied if

$$\hat{w}_i C_{ijkl}\left(-ik\hat{k}_l + \hat{w}_l \frac{\partial}{\partial\xi}\right)(\bar{G}^{\infty}_{km} + \bar{G}^{IM}_{km}) = 0. \quad (\xi = 0) \tag{4.53}$$

Finally, substituting Eqs. (4.18) (for $\xi < \xi_\circ$) and (4.45) into Eq. (4.53), and setting $\xi = 0$, the condition for a traction-free surface is

$$\sum_{\alpha=1}^{3} L^*_{j\alpha} E^*_{\alpha m} = \frac{i}{k} \sum_{\alpha=1}^{3} L_{j\alpha} A_{m\alpha} e^{ikp_\alpha \xi_\circ}. \tag{4.54}$$

Equation (4.54) can now be solved for $E^*_{\alpha m}$ by employing the matrix $M^*_{\beta j}$, which is the inverse of the matrix $L^*_{j\alpha}$ as indicated by Eq. (3.118). Multiplying Eq. (4.54) throughout by $M^*_{\beta j}$,

$$\sum_{\alpha=1}^{3} M^*_{\beta j} L^*_{j\alpha} E^*_{\alpha m} = \sum_{\alpha=1}^{3} \delta_{\beta\alpha} E^*_{\alpha m} = E^*_{\beta m} = \frac{i}{k} M^*_{\beta j} \sum_{\alpha=1}^{3} L_{j\alpha} A_{m\alpha} e^{ikp_\alpha \xi_\circ}. \tag{4.55}$$

The quantity $E^*_{\beta m}$, given by Eq. (4.55), is then substituted into Eq. (4.45), and after interchanging the α and β subscripts,

$$\bar{G}^{IM}_{km} = \frac{i}{k} \sum_{\alpha=1}^{3}\sum_{\beta=1}^{3} A^*_{k\alpha} M^*_{\alpha j} L_{j\beta} A_{m\beta} e^{-ik(p^*_\alpha \xi - p_\beta \xi_\circ)}. \tag{4.56}$$

Next, G^{IM}_{km} is obtained by an inverse transformation employing Eq. (4.1). Setting $dk_1 dk_2 = k d\phi dk$ (Fig. 4.4), and integrating,

$$\begin{aligned} G^{IM}_{km}(\mathbf{x}, \xi_\circ) &= \frac{i}{4\pi^2} \int_0^{2\pi} d\phi \sum_{\beta=1}^{3}\sum_{\alpha=1}^{3} A^*_{k\alpha} M^*_{\alpha j} L_{j\beta} A_{m\beta} \int_0^{\infty} e^{-ik[\hat{\mathbf{k}}\cdot\mathbf{x} + p^*_\alpha \xi - p_\beta \xi_\circ]} dk \\ &= \frac{1}{4\pi^2} \int_0^{2\pi} \sum_{\beta=1}^{3}\sum_{\alpha=1}^{3} \frac{A^*_{k\alpha} M^*_{\alpha j} L_{j\beta} A_{m\beta}}{[\hat{\mathbf{k}}\cdot\mathbf{x} + p^*_\alpha \mathbf{x}\cdot\hat{\mathbf{w}} - p_\beta \xi_\circ]} d\phi. \end{aligned} \tag{4.57}$$

which is of the functional form anticipated by Eq. (4.43).

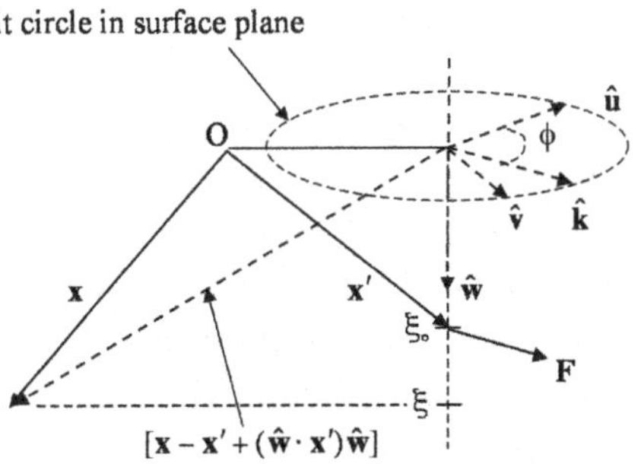

Fig. 4.4 Geometry for finding image Green's function, $G^{IM}_{km}(\mathbf{x}, \mathbf{x}')$ given by Eq. (4.58). Field point is at $\mathbf{x}$ and variable source point at $\mathbf{x}'$. Origin, at O, lies in surface plane containing unit circle. Dashed system corresponds to the construction in Fig. 4.1.

Equation (4.57) is valid for the coordinate system of Fig. 4.1 where the point of application of the force, $\mathbf{F}$ is restricted to lie along the vector $\hat{\mathbf{w}}$ at the position $\xi_\circ \hat{\mathbf{w}}$. However, it is more useful to have a formulation for G^{IM}_{km} at a field point, $\mathbf{x}$, in which $\mathbf{F}$ can be applied at a variable source point denoted, as usual, by the vector $\mathbf{x}'$. This can be accomplished by adopting the coordinate system shown in Fig. 4.4.

The coordinate transformation illustrated in Fig. 4.4 requires modification of the expression for G^{IM}_{km} given by Eq. (4.57). Comparison of Figs. 4.1 and 4.4, and consideration of the geometery in Fig. 4.4, shows that this requires the replacement of the vector $\mathbf{x}$ in Fig. 4.1 and Eq. (4.57) by the quantity $[\mathbf{x} - \mathbf{x}' + \xi_\circ \hat{\mathbf{w}}]$, where $\mathbf{x}$ and $\mathbf{x}'$ (in the replacement) are now the field vector and source vector, respectively, referred to the origin O in Fig. 4.4. Therefore, making the replacement and taking the real part as the solution, as previously in the case of G^{∞}_{km}, we obtain the image Green's function, as a function of the vectors $\mathbf{x}$ and $\mathbf{x}'$ in Fig. 4.4 in the form

$$G^{IM}_{km}(\mathbf{x}, \mathbf{x}') = \frac{1}{4\pi^2} \mathcal{R}_e \int_0^{2\pi} \sum_{\beta=1}^{3} \sum_{\alpha=1}^{3} \frac{A^*_{k\alpha} M^*_{\alpha j} L_{j\beta} A_{m\beta}}{[\hat{\mathbf{k}} \cdot (\mathbf{x} - \mathbf{x}') + p^*_\alpha \mathbf{x} \cdot \hat{\mathbf{w}} - p_\beta \mathbf{x}' \cdot \hat{\mathbf{w}}]} d\phi. \tag{4.58}$$

Note that Eq. (4.58) reverts to Eq. (4.57) in the special case where $\mathbf{x}' = \xi_\circ \hat{\mathbf{w}}$.

When evaluating Eq. (4.58), it must be recalled that Fig. 4.4 applies, the origin is in the surface, the unit vector $\hat{\mathbf{w}}$ is perpendicular to the surface, and in the integration over ϕ (with $\mathbf{x}$ and $\mathbf{x}'$ constant) the unit vector $\hat{\mathbf{k}}$ rotates over the range $0 \leq \phi \leq 2\pi$ in the surface plane. The integrand involves various Stroh eigenvalues p_α, Stroh vectors $A_{j\alpha}$ and $L_{j\alpha}$, and $[M^*]$ matrices, which are inverse with respect to corresponding $[L^*]$ matrices as indicated by Eq. (3.118). All of these quantities are functions of ϕ, and therefore must be determined by solving the Stroh eigenvalue problem after every incremental increase in ϕ during the integration. The necessary expressions for accomplishing this can be extracted from the Stroh sextic formalism presented in Sec. 3.5.1 with the unit vectors $\hat{\mathbf{n}}$ and $\hat{\mathbf{m}}$ replaced by $\hat{\mathbf{w}}$ and $\hat{\mathbf{k}}$, respectively, as described in the text following Eq. (4.13).

4.2.3 *In half-space joined to an elastically dissimilar half-space along planar interface*

This problem has also been treated by Pan and Yuan (2000) and Barnett (personal communication, 2007), and we again continue to follow Barnett. Figure 4.1 again applies initially with $\hat{\mathbf{w}}$ normal to the interface located in the $\mathbf{x} \cdot \hat{\mathbf{w}} = 0$ plane. Now, however, half-space 1 with elastic constants ${C_{ijkm}}^{(1)}$ occupies the $\xi > 0$ region, the point force $\mathbf{F}$ is located in half-space 1 at $\boldsymbol{\xi}_\circ = \xi_\circ \hat{\mathbf{w}}$, and half-space 2 with elastic constants $C^{(2)}_{ijkm}$ occupies the $\xi < 0$ region. To obtain a solution, it is assumed that the Green's function in half-space 1 consists of the Green's function for the point force in an infinite body, G^{∞}_{km}, plus an image term, $g^{IM^{(1)}}_{km}$, that vanishes as $\xi \to \infty$. A solution for half-space 2, denoted by $g^{(2)}_{km}$, is then assumed that vanishes as $\xi \to -\infty$ and matches the solution for half-space 1 at the interface. The solution is therefore of the form

$$g^{(1)}_{km} = G^{\infty}_{km} + g^{IM^{(1)}}_{km}, \qquad g^{(2)}_{km} = g^{(2)}_{km}, \tag{4.59}$$

or, in terms of corresponding transforms,

$$\bar{g}^{(1)}_{km}, = \overline{G}^{\infty}_{km} + \bar{g}^{IM^{(1)}}_{km}, \qquad \bar{g}^{(2)}_{km} = \bar{g}^{(2)}_{km}. \tag{4.60}$$

Using the expression for $\overline{G}_{km}^{\infty}$ given by Eq. (4.18), a solution of the form

$$\bar{g}_{km}^{(1)} = \frac{i}{k}\sum_{\alpha=1}^{3} A_{k\alpha}^{*(1)} A_{m\alpha}^{*(1)} e^{-ikp_{\alpha}^{*(1)}(\xi-\xi_\circ)} + \sum_{\alpha=1}^{3} A_{k\alpha}^{*(1)} H_{\alpha m}^{*} e^{-ikp_{\alpha}^{*(1)}\xi} \quad (\xi > \xi_\circ)$$

$$\bar{g}_{km}^{(1)} = -\frac{i}{k}\sum_{\alpha=1}^{3} A_{k\alpha}^{(1)} A_{m\alpha}^{(1)} e^{-ikp_{\alpha}^{(1)}(\xi-\xi_\circ)}$$

$$+\sum_{\alpha=1}^{3} A_{k\alpha}^{*(1)} H_{\alpha m}^{*} e^{-ikp_{\alpha}^{*(1)}\xi} \quad (0 < \xi < \xi_\circ) \tag{4.61}$$

$$\bar{g}_{km}^{(2)} = \sum_{\alpha=1}^{3} A_{k\alpha}^{(2)} F_{\alpha m} e^{-ikp_{\alpha}^{(2)}\xi} \quad (\xi < 0)$$

is assumed, with $H_{\alpha m}^{*}$ and $F_{\alpha m}$ to be determined. The solution vanishes as $\xi \to \pm\infty$, and it therefore remains to satisfy the boundary conditions at the interface. The transforms must be continuous across the interface, i.e., $(\bar{g}_{km}^{(1)} = \bar{g}_{km}^{(2)})_{\xi=0}$, so that

$$-\frac{i}{k}\sum_{\alpha=1}^{3} A_{k\alpha}^{(1)} A_{m\alpha}^{(1)} e^{ikp_{\alpha}^{(1)}\xi_\circ} = \sum_{\alpha=1}^{3} A_{k\alpha}^{(2)} F_{\alpha m} - \sum_{\alpha=1}^{3} A_{k\alpha}^{*(1)} H_{\alpha m}^{*}. \tag{4.62}$$

The tractions must also be continuous, and, therefore, using the same procedures that led to Eqs. (4.51) and (4.53),

$$\hat{w}_i \left(C_{ijkl}^{(1)} \frac{\partial g_{km}^{(1)}}{\partial x_l} - C_{ijkl}^{(2)} \frac{\partial g_{km}^{(2)}}{\partial x_l} \right)_{\xi=0} = 0 \tag{4.63}$$

and

$$\hat{w}_i \left(-ik\hat{k}_l + \hat{w}_l \frac{\partial}{\partial \xi} \right) \left(C_{ijkl}^{(1)} \bar{g}_{km}^{(1)} - C_{ijkl}^{(2)} \bar{g}_{km}^{(2)} \right)_{\xi=0} = 0. \tag{4.64}$$

Then, substituting Eq. (4.61) into Eq. (4.64) and using Eq. (3.37),

$$-\frac{i}{k}\sum_{\alpha=1}^{3} L_{k\alpha}^{(1)} A_{m\alpha}^{(1)} e^{ikp_{\alpha}^{(1)}\xi_\circ} = \sum_{\alpha=1}^{3} L_{k\alpha}^{(2)} F_{\alpha m} - \sum_{\alpha=1}^{3} L_{k\alpha}^{*(1)} H_{\alpha m}^{*}. \tag{4.65}$$

Equations (4.62) and (4.65) can now be used to solve for the two unknowns, $F^*_{\alpha m}$ and $H^*_{\alpha m}$. Using the matrix $M^{*(1)}_{\beta k}$, which is the inverse of $L^{*(1)}_{k\alpha}$ as indicated by Eq. (3.118), and the matrix $\mathcal{A}^{*(1)}_{\beta k}$, which is the inverse of $A^{*(1)}_{k\alpha}$ according to

$$\mathcal{A}^{*(1)}_{\beta k} A^{*(1)}_{k\alpha} = \delta_{\beta\alpha}, \quad (\alpha, \beta = 1, 2, 3) \tag{4.66}$$

multiplying Eq. (4.65) by $M^{*(1)}_{\beta k}$ and Eq. (4.62) by $\mathcal{A}^{*(1)}_{\beta k}$, and subtracting one from the other,

$$\begin{aligned} -\frac{i}{k}\sum_{\alpha=1}^{3}(\mathcal{A}^{*(1)}_{\beta k}A^{(1)}_{k\alpha} - M^{*(1)}_{\beta k}L^{(1)}_{k\alpha})A^{(1)}_{m\alpha}e^{ikp^{(1)}_{\alpha}\xi_{\circ}} \\ = \sum_{\alpha=1}^{3}(\mathcal{A}^{*(1)}_{\beta k}A^{(2)}_{k\alpha} - M^{*(1)}_{\beta k}L^{(2)}_{k\alpha})F_{\alpha m}. \end{aligned} \tag{4.67}$$

For convenience, the matrices $P_{\beta\alpha}$ and $R_{\beta\alpha}$ defined by

$$\begin{aligned} P_{\beta\alpha} &\equiv \mathcal{A}^{*(1)}_{\beta k}A^{(1)}_{k\alpha} - M^{*(1)}_{\beta k}L^{(1)}_{k\alpha} \\ R_{\beta\alpha} &\equiv \mathcal{A}^{*(1)}_{\beta k}A^{(2)}_{k\alpha} - M^{*(1)}_{\beta k}L^{(2)}_{k\alpha} \end{aligned} \tag{4.68}$$

are introduced so that Eq. (4.67) can be expressed more simply as

$$-\frac{i}{k}\sum_{\alpha=1}^{3}P_{\beta\alpha}A^{(1)}_{m\alpha}e^{ikp^{(1)}_{\alpha}\xi_{\circ}} = \sum_{\alpha=1}^{3}R_{\beta\alpha}F_{\alpha m}. \tag{4.69}$$

Then, to solve for $F_{\alpha m}$, the matrix $\mathcal{R}_{\gamma\beta}$, which is the inverse of $R_{\beta\alpha}$ according to

$$\sum_{\beta=1}^{3}\mathcal{R}_{\gamma\beta}R_{\beta\alpha} = \delta_{\gamma\alpha}, \quad (\gamma = 1, 2, 3) \tag{4.70}$$

is introduced, and by multiplying Eq. (4.69) throughout by $\mathcal{R}_{\gamma\beta}$, and summing over β,

$$F_{\gamma m} = -\frac{i}{k}\sum_{\beta=1}^{3}\sum_{\alpha=1}^{3}\mathcal{R}_{\gamma\beta}P_{\beta\alpha}A^{(1)}_{m\alpha}e^{ikp^{(1)}_{\alpha}\xi_{\circ}}. \tag{4.71}$$

The quantity $H^*_{\gamma m}$ is obtained in a similar manner. Starting with Eqs. (4.62) and (4.65) and employing the matrices $M^{(2)}_{\beta k}$ and $\mathcal{A}^{(2)}_{\beta k}$, which is the inverse of $A^{(2)}_{k\alpha}$ according to

$$\mathcal{A}^{(2)}_{\beta k} A^{(2)}_{k\alpha} = \delta_{\beta\alpha}, \quad (\alpha, \beta = 1, 2, 3) \tag{4.72}$$

and the matrices $U_{\beta\alpha}$ and $W_{\beta\alpha}$ defined by

$$\begin{aligned} U_{\beta\alpha} &\equiv \mathcal{A}^{(2)}_{\beta j} A^{(1)}_{j\alpha} - M^{(2)}_{\beta k} L^{(1)}_{j\alpha}, \\ W_{\beta\alpha} &\equiv \mathcal{A}^{(2)}_{\beta j} A^{*(1)}_{j\alpha} - M^{(2)}_{\beta k} L^{*(1)}_{j\alpha}, \end{aligned} \tag{4.73}$$

the expression

$$\frac{i}{k} \sum_{\alpha=1}^{3} U_{\beta\alpha} A^{(1)}_{m\alpha} e^{ikp^{(1)}_{\alpha}\xi_{\circ}} = \sum_{\alpha=1}^{3} W_{\beta\alpha} H^{*}_{\alpha m} \tag{4.74}$$

is obtained. Then, by multiplying this result throughout by the matrix, $\mathcal{W}_{\gamma\beta}$, which is the inverse of $W_{\beta\alpha}$ according to

$$\sum_{\beta=1}^{3} \mathcal{W}_{\gamma\beta} W_{\beta\alpha} = \delta_{\gamma\alpha} \quad (\gamma, \alpha = 1, 2, 3) \tag{4.75}$$

and summing over β, $H^{*}_{\gamma m}$ is obtained in the form

$$H^{*}_{\gamma m} = \frac{i}{k} \sum_{\beta=1}^{3} \sum_{\alpha=1}^{3} \mathcal{W}_{\gamma\beta} U_{\beta\alpha} A^{(1)}_{m\alpha} e^{ikp^{(1)}_{\alpha}\xi_{\circ}}. \tag{4.76}$$

Then, substituting the above expressions for $F_{\gamma m}$ and $H^{*}_{\gamma m}$ into Eq. (4.61),

$$\begin{aligned} \bar{g}^{IM^{(1)}}_{km} &= \frac{i}{k} \sum_{\alpha=1}^{3} \sum_{\gamma=1}^{3} A^{*(1)}_{k\gamma} \mathcal{W}_{\gamma j} U_{j\alpha} A^{(1)}_{m\alpha} e^{-ik[p^{*(1)}_{\gamma}\xi - p^{(1)}_{\alpha}\xi_{\circ}]}, \quad (0 < \xi) \\ \bar{g}^{(2)}_{km} &= -\frac{i}{k} \sum_{\alpha=1}^{3} \sum_{\gamma=1}^{3} A^{(2)}_{k\gamma} \mathcal{R}_{\gamma j} P_{j\alpha} A^{(1)}_{m\alpha} e^{-ik[p^{(2)}_{\gamma}\xi - p^{(1)}_{\alpha}\xi_{\circ}]}. \quad (\xi < 0) \end{aligned} \tag{4.77}$$

After inversely transforming Eq. (4.77) by the same method used to obtain Eq. (4.20), the image Green's function in half-space 1 and the solution in half-space 2, for a point force in half-space 1 at the source point $\xi_{\circ}\hat{\mathbf{w}}$ in the coordinate system of Fig. 4.1, are

$$\begin{aligned} g^{IM^{(1)}}_{km}(\mathbf{x}, \xi_{\circ}) &= \frac{1}{4\pi^2} \int_0^{2\pi} \sum_{\alpha=1}^{3} \sum_{\gamma=1}^{3} \frac{A^{*(1)}_{k\gamma} \mathcal{W}_{\gamma j} U_{j\alpha} A^{(1)}_{m\alpha}}{[\hat{\mathbf{k}} \cdot \mathbf{x} + p^{*(1)}_{\gamma} \mathbf{x} \cdot \hat{\mathbf{w}} - p^{(1)}_{\alpha} \xi_{\circ}]} d\phi, \quad (0 < \xi) \\ g^{(2)}_{km}(\mathbf{x}, \xi_{\circ}) &= -\frac{1}{4\pi^2} \int_0^{2\pi} \sum_{\alpha=1}^{3} \sum_{\gamma=1}^{3} \frac{A^{(2)}_{k\gamma} \mathcal{R}_{\gamma j} P_{j\alpha} A^{(1)}_{m\alpha}}{[\hat{\mathbf{k}} \cdot \mathbf{x} + p^{(2)}_{\gamma} \mathbf{x} \cdot \hat{\mathbf{w}} - p^{(1)}_{\alpha} \xi_{\circ}]} d\phi. \quad (\xi < 0) \end{aligned} \tag{4.78}$$

As in the case of the half-space image Green's function, i.e., Eq. (4.57), it is more useful to have the Green's functions given by Eq. (4.78) as functions of a field vector $\mathbf{x}$ and a general source vector, $\mathbf{x}'$. Therefore, by employing the same coordinate transformation used to obtain Eq. (4.58) from Eq. (4.57), the Green's functions as functions of the $\mathbf{x}$ and $\mathbf{x}'$ vectors in Fig. 4.4 are

$$
\begin{aligned}
g_{km}^{IM(1)}(\mathbf{x}, \mathbf{x}') &= \frac{1}{4\pi^2}\mathcal{R}_e \int_0^{2\pi} \sum_{\alpha=1}^{3}\sum_{\gamma=1}^{3} \\
&\quad \times \frac{A_{k\gamma}^{*(1)} \mathcal{W}_{\gamma j} U_{j\alpha} A_{m\alpha}^{(1)}}{[\hat{\mathbf{k}} \cdot (\mathbf{x} - \mathbf{x}') + p_\gamma^{*(1)} \mathbf{x} \cdot \hat{\mathbf{w}} - p_\alpha^{(1)} \mathbf{x}' \cdot \hat{\mathbf{w}}]} d\phi, \quad (\mathbf{x} \cdot \hat{\mathbf{w}} > 0) \\
g_{km}^{(2)}(\mathbf{x}, \mathbf{x}') &= -\frac{1}{4\pi^2}\mathcal{R}_e \int_0^{2\pi} \sum_{\alpha=1}^{3}\sum_{\gamma=1}^{3} \\
&\quad \times \frac{A_{k\gamma}^{(2)} \mathcal{R}_{\gamma j} P_{j\alpha} A_{m\alpha}^{(1)}}{[\hat{\mathbf{k}} \cdot (\mathbf{x} - \mathbf{x}') + p_\gamma^{(2)} \mathbf{x} \cdot \hat{\mathbf{w}} - p_\alpha^{(1)} \mathbf{x}' \cdot \hat{\mathbf{w}}]} d\phi. \quad (\mathbf{x} \cdot \hat{\mathbf{w}} < 0)
\end{aligned}
\tag{4.79}
$$

As is the case of Eq. (4.58), in the integration over ϕ (with $\mathbf{x}$ and $\mathbf{x}'$ constant) the unit vector $\hat{\mathbf{k}}$ again rotates over the range $0 \le \phi \le 2\pi$. The integrand again involves various Stroh eigenvalues p_α and Stroh vectors $A_{j\alpha}$ and $L_{j\alpha}$, and now, in addition, several "inverse type" matrices of the same general type as [M] which is the inverse of [L] as indicated by Eq. (3.117). All of these quantities are functions of ϕ, and therefore must be determined by solving the Stroh eigenvalue problem after every incremental increase in ϕ during the integration. The necessary expressions for accomplishing this can again be extracted from the Stroh sextic formalism presented in Sec. 3.5.1 with the unit vectors $\hat{\mathbf{n}}$ and $\hat{\mathbf{m}}$ replaced by $\hat{\mathbf{w}}$ and $\hat{\mathbf{k}}$, respectively, as described in the text following Eq. (4.13).

4.3 Green's Functions for Unit Point Force in Isotropic System

The Green's functions of the same three types analyzed in Sec. 4.2 are now determined for an isotropic system. The most complex of the three problems, i.e., the joined half-spaces problem, is first solved in a form that can be readily reduced to solutions for the half-space and the infinite region.

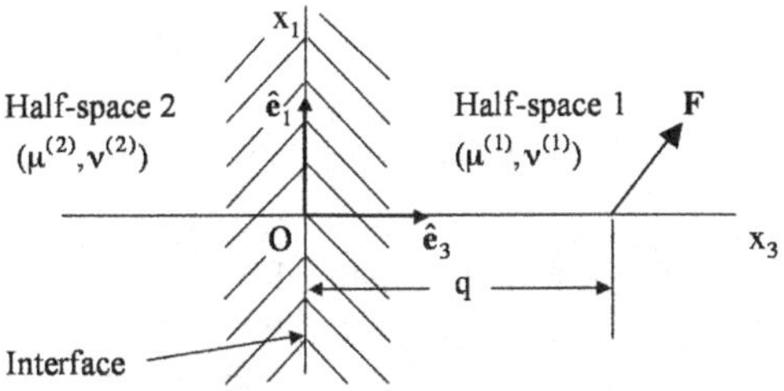

Fig. 4.5 Geometry for point force, **F**, applied at (0,0,q) in half-space 1 joined to elastically dissimilar half-space 2 along planar interface.

Following Rongved (1955), this is accomplished by employing a method originated by Mindlin (1953), where, instead of obtaining the Green's function directly as in the previous section, the displacements due to the point force are found first as solutions of the Navier equation with the help of *Papkovitch functions*. The corresponding Green's functions are then readily constructed.

4.3.1 *In half-space joined to elastically dissimilar half-space along planar interface*

As illustrated in Fig. 4.5, a Cartesian $(\hat{\mathbf{e}}_1, \hat{\mathbf{e}}_2, \hat{\mathbf{e}}_3)$ coordinate system is used with its origin in the interface and $\hat{\mathbf{e}}_3$ pointed into half-space 1, which occupies the region $x_3 > 0$ containing the point force, **F**, at (0, 0, q). Solutions are found for a point force that is normal, and then parallel, to the interface so that they can be readily combined to obtain the solution for a general force.

4.3.1.1 *Formulation of Papkovitch functions*

Displacement functions that obey the Navier equation in isotropic systems can generally be expressed in terms of the two Papkovitch functions **B** and β (Papkovitch 1932) and Mindlin (1936a,b 1953). The first step in deriving them is to divide the displacement field, $\mathbf{u}(\mathbf{x})$, into an irrotational part $\mathbf{u}^{\mathrm{irr}}(\mathbf{x})$, and a solenoidal part $\mathbf{u}^{\mathrm{sol}}(\mathbf{x})$, i.e.,[4]

$$\mathbf{u}(\mathbf{x}) = \mathbf{u}^{\mathrm{irr}}(\mathbf{x}) + \mathbf{u}^{\mathrm{sol}}(\mathbf{x}) = \nabla\phi + \nabla \times \mathbf{H}. \tag{4.80}$$

[4]A vector field can be generally expressed as the sum of two vector fields one of which is *solenoidal* and the other *irrotational* (e.g., Sokolnikoff and Redheffer 1958). A vector

Then, substituting Eq. (4.80) into the Navier equation, Eq. (3.3),

$$\nabla^2 \mathbf{B} = -\frac{\mathbf{f}}{\mu}, \tag{4.81}$$

where, **B** is the Papkovitch function defined by

$$\mathbf{B} = \alpha \nabla \phi + \nabla \times \mathbf{H} \quad \alpha \equiv \frac{2(1-\nu)}{1-2\nu}. \tag{4.82}$$

Then taking the divergence of **B**,

$$\nabla^2 \phi = \frac{1}{\alpha} \nabla \cdot \mathbf{B} \tag{4.83}$$

which has the solution

$$\phi = \frac{1}{2\alpha}(\mathbf{x} \cdot \mathbf{B} + \beta), \tag{4.84}$$

where β is the second Papkovitch function whose Laplacian is given by

$$\nabla^2 \beta = \frac{1}{\mu} \mathbf{x} \cdot \mathbf{f}. \tag{4.85}$$

Next, putting $\nabla \times \mathbf{H}$ from Eq. (4.82) and ϕ from Eq. (4.84) into Eq. (4.80),

$$\mathbf{u} = \mathbf{B} - \frac{1}{4(1-\nu)} \nabla(\mathbf{x} \cdot \mathbf{B} + \beta). \tag{4.86}$$

The displacement given by Eq. (4.86) is therefore a solution of the Navier equation expressed in terms of the Papkovitch functions, **B** and β, whose Laplacians are known from Eqs. (4.81) and (4.85).

A solution for the displacement field **u** in the present Green's function problem can therefore be found by determining the two Papkovitch functions consistent with the presence of the point force and the prevailing boundary conditions.

---

field, $\mathbf{u}^{\text{irr}}(\mathbf{x})$, is irrotational if $\mathbf{u}^{\text{irr}}(\mathbf{x}) = \nabla \phi(\mathbf{x})$ where $\phi(\mathbf{x})$ is a scalar function with continuous second derivatives. The function $\phi(\mathbf{x})$ is generally called the potential of the field and the field is called a potential field. A vector field, $\mathbf{u}^{\text{sol}}(\mathbf{x})$, is solenoidal if $\mathbf{u}^{\text{sol}}(\mathbf{x}) = \nabla \times \mathbf{H}$ where $\mathbf{H}(\mathbf{x})$ is a vector function with continuous second derivatives.

4.3.1.2 *Papkovitch functions for point force normal to interface*

The displacements and tractions must be continuous across the interface leading to the boundary conditions

$$\left.\begin{aligned} &u_i^{(1)} = u_i^{(2)}, \\ &T_i^{(1)} = \sigma_{ij}^{(1)}\hat{n}_j = T_i^{(2)} = \sigma_{ij}^{(2)}\hat{n}_j. \end{aligned}\right\} \quad (\text{on } x_3 = 0) \tag{4.87}$$

These are satisfied by the Papkovitch functions if

$$\begin{aligned} &B_1^{(1)} = B_2^{(1)} = 0, \quad (x_3 > 0) \\ &B_1^{(2)} = B_2^{(2)} = 0, \quad (x_3 < 0) \end{aligned} \tag{4.88}$$

and if, after using Eqs. (2.5) and (2.122), the following conditions are satisfied on $x_3 = 0$:

$$\begin{aligned}
(1-2\nu^{(1)})\frac{\partial B_3^{(1)}}{\partial x_1} - \frac{\partial^2 \beta^{(1)}}{\partial x_1 \partial x_3} &= \frac{\mu^{(2)}(1-\nu^{(1)})}{\mu^{(1)}(1-\nu^{(2)})} \\
&\quad \times \left[(1-2\nu^{(2)})\frac{\partial B_3^{(2)}}{\partial x_1} - \frac{\partial^2 \beta^{(2)}}{\partial x_1 \partial x_3}\right], \\
(1-2\nu^{(1)})\frac{\partial B_3^{(1)}}{\partial x_2} - \frac{\partial^2 \beta^{(1)}}{\partial x_2 \partial x_3} &= \frac{\mu^{(2)}(1-\nu^{(1)})}{\mu^{(1)}(1-\nu^{(2)})} \\
&\quad \times \left[(1-2\nu^{(2)})\frac{\partial B_3^{(2)}}{\partial x_2} - \frac{\partial^2 \beta^{(2)}}{\partial x_2 \partial x_3}\right], \\
2(1-\nu^{(1)})\frac{\partial B_3^{(1)}}{\partial x_3} - \frac{\partial^2 \beta^{(1)}}{\partial x_3^2} &= \frac{\mu^{(2)}(1-\nu^{(1)})}{\mu^{(1)}(1-\nu^{(2)})} \\
&\quad \times \left[2(1-\nu^{(2)})\frac{\partial B_3^{(2)}}{\partial x_3} - \frac{\partial^2 \beta^{(2)}}{\partial x_3^2}\right], \\
\frac{\partial \beta^{(1)}}{\partial x_1} &= \frac{(1-\nu^{(1)})}{(1-\nu^{(2)})}\frac{\partial \beta^{(2)}}{\partial x_1}, \\
\frac{\partial \beta^{(1)}}{\partial x_2} &= \frac{(1-\nu^{(1)})}{(1-\nu^{(2)})}\frac{\partial \beta^{(2)}}{\partial x_2}, \\
(3-4\nu^{(1)})B_3^{(1)} - \frac{\partial \beta^{(1)}}{\partial x_3} &= \frac{(1-\nu^{(1)})}{(1-\nu^{(2)})}\left[(3-4\nu^{(2)})B_3^{(2)} - \frac{\partial \beta^{(2)}}{\partial x_3}\right].
\end{aligned} \tag{4.89}$$

Consider next the point force, F_3 directed along x_3 as in Fig. 4.6b. Rather than describe it as usual with a delta function, we follow Rongved (1955) and formulate it, equivalently, as a force density distribution $f_3(\mathbf{x})$ that vanishes everywhere except within a small closed region, $\mathcal{R}$, where it produces the total force $F^{tot} = \oiiint_{\mathcal{R}} f_3(\mathbf{x})dV$. Then, taking the limit of the integral as $\mathcal{R} \to 0$, while maintaining F^{tot} constant,

$$F_3 = \mathrm{Lim}_{\mathcal{R}\to 0} \oiiint_{\mathcal{R}} f_3(\mathbf{x})dV. \tag{4.90}$$

The quantities $B_3^{(1)}$ and $\beta^{(1)}$ can now be related to the point force by employing Green's classical third identity in the form (Kellogg 1929),

$$\psi = -\frac{1}{4\pi} \oiint_{\mathcal{S}} \psi \mathbf{n} \cdot \nabla g dS - \frac{1}{4\pi} \oiiint_{\mathcal{V}} g \nabla^2 \psi dV. \tag{4.91}$$

Equation (4.91) states that a function $\psi(\mathbf{x})$ at any point $P(\mathbf{x})$, in a region $\mathcal{V}$, bounded by the surface $\mathcal{S}$, as illustrated in Fig. 4.6a, can be expressed in terms of its values on $\mathcal{S}$, its Laplacian, and the appropriate Green's function for the region denoted by g. Furthermore, the contribution to ψ made by the surface integral in Eq. (4.91) is harmonic,[5] while the contribution of the volume integral is generally not. To proceed, it is therefore convenient to divide each quantity $B_3^{(1)}$ and $\beta^{(1)}$ into harmonic and non-harmonic parts indicated by the subscripts $\mathcal{S}$ and $\mathcal{V}$, respectively, so that

$$\begin{aligned} B_3^{(1)} &= B_{\mathcal{S}}^{(1)} + B_{\mathcal{V}}^{(1)}, \\ \beta^{(1)} &= \beta_{\mathcal{S}}^{(1)} + \beta_{\mathcal{V}}^{(1)}. \end{aligned} \tag{4.92}$$

According to Eqs. (4.81) and (4.85),

$$\begin{aligned} \nabla^2 B_3^{(1)} &= -\frac{f_3}{\mu}, \\ \nabla^2 \beta^{(1)} &= \frac{\mathbf{x} \cdot \mathbf{f}}{\mu}, \end{aligned} \tag{4.93}$$

and, by employing these expressions in the volume integral term of Eq. (4.91), we obtain the non-harmonic parts of $B_3^{(1)}$ and $\beta^{(1)}$ in the forms

$$\begin{aligned} B_{\mathcal{V}}^{(1)} &= -\frac{1}{4\pi} \oiiint_{\mathcal{V}} g \nabla^2 B_3^{(1)} dV = \frac{1}{4\pi\mu} \oiiint_{\mathcal{V}} g f_3 dV, \\ \beta_{\mathcal{V}}^{(1)} &= -\frac{1}{4\pi} \oiiint_{\mathcal{V}} g \nabla^2 \beta^{(1)} dV = -\frac{1}{4\pi\mu} \oiiint_{\mathcal{V}} g f_3 x_3 dV. \end{aligned} \tag{4.94}$$

[5] A *harmonic function* obeys Laplace's equation.

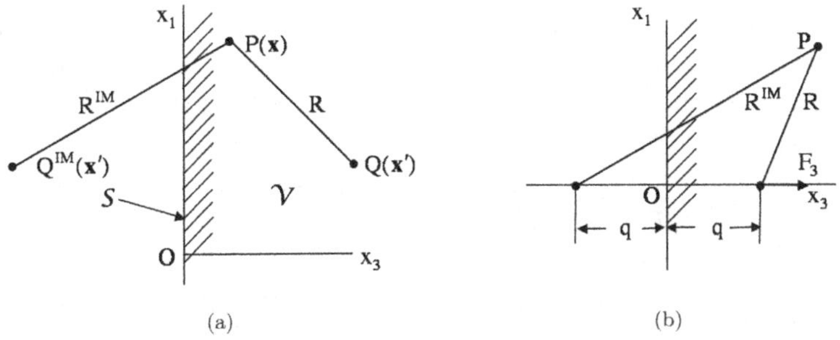

Fig. 4.6 (a) Half-space, $\mathcal{V}$, with planar free surface, $\mathcal{S}$, at $x_3 = 0$. P is the field point, Q is the source point, and Q^{IM} is the image of Q outside of $\mathcal{V}$. (b) Same general arrangement as (a) but with point force, F_3, directed along x_3, applied at the source point (0,0,q).

Next, the Green's function, g, required in Eq. (4.94) must be of the appropriate form for a half-space region $\mathcal{V}$, with a planar surface $\mathcal{S}$, and this can be found with the help of Fig. 4.6a which shows a *field point* $P(\mathbf{x})$ and a *source point* $Q(\mathbf{x}')$, at the distance R from P. An electrostatic potential argument (Kellogg 1929), in which a positive unit electrical charge is placed at Q, is then used to obtain g. Taking the potential at the distance R from a unit positive charge in the simple form 1/R, the Green's function is given by

$$g(P, Q) = v(P, Q) + \frac{1}{R}, \tag{4.95}$$

where 1/R is the potential produced at P by the charge at Q, and v(P,Q) is the potential at P produced by the electrical charge that is induced on a grounded conducting sheet having the same configuration as $\mathcal{S}$. The total potential along $\mathcal{S}$ must vanish because of the grounding, and this can be achieved by placing a unit negative charge at the image point, Q^{IM}, as shown in Fig. 4.6a. Therefore, $v(P, Q) = -1/R^{IM}$, where R^{IM} is the distance between the image source point Q^{IM} and the field point P and

$$g(P, Q) = \frac{1}{R} - \frac{1}{R^{IM}} \tag{4.96}$$

where

$$\begin{aligned} R &= [(x_1 - x_1')^2 + (x_2 - x_2')^2 + (x_3 - x_3')^2]^{1/2} \\ R^{IM} &= [(x_1 - x_1')^2 + (x_2 - x_2')^2 + (x_3 + x_3')^2]^{1/2} \end{aligned} \tag{4.97}$$

(see Mindlin (1953) and Rongved (1955)).

The quantities $B_{\mathcal{V}}^{(1)}$ and $\beta_{\mathcal{V}}^{(1)}$ for the point force F_3 at (0, 0, q), shown in Fig. 4.6b, can now be obtained by substituting Eq. (4.96) for g into Eq. (4.94), and then invoking Eq. (4.90) in order to carry out the volume integration, i.e.,

$$B_{\mathcal{V}}^{(1)} = \mathrm{Lim}_{\mathcal{R}\to 0}\frac{1}{4\pi\mu}\oiiint_{\mathcal{V}} gf_3 dV = \frac{F_3}{4\pi\mu}\left(\frac{1}{R} - \frac{1}{R^{IM}}\right),$$
$$\beta_{\mathcal{V}}^{(1)} = -\mathrm{Lim}_{\mathcal{R}\to 0}\frac{1}{4\pi\mu}\oiiint_{\mathcal{V}} gf_3 x_3 dV = -\frac{qF_3}{4\pi\mu}\left(\frac{1}{R} - \frac{1}{R^{IM}}\right), \tag{4.98}$$

where R and R^{IM} are now expressed by[6]

$$R = [x_1^2 + x_2^2 + (x_3 - q)^2)]^{1/2} \quad R^{IM} = [x_1^2 + x_2^2 + (x_3 + q)^2]^{1/2}. \tag{4.99}$$

With these results, the boundary conditions given by Eq. (4.89) can now be rewritten as functions of $B_{\mathcal{S}}^{(1)}$, $\beta_{\mathcal{S}}^{(2)}$, $B_3^{(2)}$ and $\beta^{(2)}$, i.e.,

$$(1 - 2\nu^{(1)})B_{\mathcal{S}}^{(1)} - \frac{\partial \beta_{\mathcal{S}}^{(1)}}{\partial x_3} - \frac{qF_3}{2\pi\mu^{(1)}}\frac{\partial}{\partial x_3}\left(\frac{1}{R^{IM}}\right)$$
$$= \frac{\mu^{(2)}(1-\nu^{(1)})}{\mu^{(1)}(1-\nu^{(2)})}\left[(1 - 2\nu^{(2)})B_3^{(2)} - \frac{\partial \beta^{(2)}}{\partial x_3}\right],$$

$$2(1-\nu^{(1)})\frac{\partial B_{\mathcal{S}}^{(1)}}{\partial x_3} - \frac{\partial^2 \beta_{\mathcal{S}}^{(1)}}{\partial x_3^2} - \frac{(1-\nu^{(1)})F_3}{\pi\mu^{(1)}}\frac{\partial}{\partial x_3}\left(\frac{1}{R^{IM}}\right)$$
$$= \frac{\mu^{(2)}(1-\nu^{(1)})}{\mu^{(1)}(1-\nu^{(2)})}\left[2(1-\nu^{(2)})\frac{\partial B_3^{(2)}}{\partial x_3} - \frac{\partial^2 \beta^{(2)}}{\partial x_3^2}\right],$$

$$\beta_{\mathcal{S}}^{(1)} = \frac{(1-\nu^{(1)})}{(1-\nu^{(2)})}\beta^{(2)}$$

$$(3 - 4\nu^{(1)})B_{\mathcal{S}}^{(1)} - \frac{\partial \beta_{\mathcal{S}}^{(1)}}{\partial x_3} - \frac{qF_3}{2\pi\mu^{(1)}}\frac{\partial}{\partial x_3}\left(\frac{1}{R^{IM}}\right)$$
$$= \frac{(1-\nu^{(1)})}{(1-\nu^{(2)})}\left[(3 - 4\nu^{(2)})B_3^{(2)} - \frac{\partial \beta^{(2)}}{\partial x_3}\right]. \tag{4.100}$$

[6]The symbol R is frequently employed throughout this book to denote the distance between a source point and a field point.

Equation (4.100) contains four relationships involving the four quantities $B_{\mathcal{S}}^{(1)}$, $\beta_{\mathcal{S}}^{(1)}$, $B_3^{(2)}$ and $\beta^{(2)}$. Solving these for $B_{\mathcal{S}}^{(1)}$ and $\beta_{\mathcal{S}}^{(1)}$, and substituting them into Eq. (4.92), the Papkovitch functions for half-space 1 are finally obtained (Rongved 1955) in the forms

$$B_1^{(1)} = B_2^{(1)} = 0,$$

$$B_3^{(1)} = \frac{F_3}{4\pi\mu^{(1)}}\left\{\frac{1}{R} + \frac{\mu^{(1)} - \mu^{(2)}}{\mu^{(1)} + (3-4\nu^{(1)})\mu^{(2)}}\left[\frac{(3-4\nu^{(1)})}{R^{IM}} + \frac{2q(x_3+q)}{(R^{IM})^3}\right]\right\},$$

$$\beta^{(1)} = -\frac{F_3}{4\pi\mu^{(1)}}\left\{\frac{q}{R} + \frac{\mu^{(1)} - \mu^{(2)}}{\mu^{(1)} + (3-4\nu^{(1)})\mu^{(2)}}\left[\frac{q(3-4\nu^{(1)})}{R^{IM}} - \frac{\{4(1-\nu^{(1)})\mu^{(1)}[(1-2\nu^{(1)})(3-4\nu^{(2)}) - 2\mu^{(2)}(\nu^{(1)}-\nu^{(2)})(\mu^{(1)}-\mu^{(2)})^{-1}]\}}{\mu^{(2)} + (3-4\nu^{(2)})\mu^{(1)}}\ln(R^{IM} + x_3 + q)\right]\right\}. \tag{4.101}$$

The Papkovitch functions for half-space 2 are found by solving the previous equations in a similar manner with the results (for details see Rongved (1955))

$$B_1^{(2)} = B_2^{(2)} = 0,$$

$$B_3^{(2)} = \frac{(1-\nu^{(2)})F_3}{\pi[\mu^{(2)} + (3-4\nu^{(2)})\mu^{(1)}]}\frac{1}{R},$$

$$\beta^{(2)} = \frac{(1-\nu^{(2)})F_3}{\pi[\mu^{(1)} + (3-4\nu^{(1)})\mu^{(2)}]} \times\left\{-\frac{q}{R} + \frac{\mu^{(1)}(1-2\nu^{(1)})(3-4\nu^{(2)}) - \mu^{(2)}(1-2\nu^{(2)})(3-4\nu^{(1)})}{\mu^{(2)} + (3-4\nu^{(2)})\mu^{(1)}} \times \ln(R - x_3 + q)\right\}. \tag{4.102}$$

4.3.1.3 *Papkovitch functions for point force parallel to interface*

The solution for a point force parallel to $\hat{\mathbf{e}}_1$ is obtained (Rongved 1955) by a generally similar procedure. However, it is lengthy, and we therefore shall not describe it here but merely present the final results for the Papkovitch functions: see Rongved (1955) for details.

$$
\begin{aligned}
B_1^{(1)} &= \frac{F_1}{4\pi\mu^{(1)}}\left[\frac{1}{R} + \left(\frac{\mu^{(1)} - \mu^{(2)}}{\mu^{(1)} + \mu^{(2)}}\right)\frac{1}{R^{IM}}\right], \\
B_2^{(1)} &= 0, \\
B_3^{(1)} &= \frac{(\mu^{(1)} - \mu^{(2)})F_1 x_1}{2\pi[\mu^{(1)} + \mu^{(2)}(3 - 4\nu^{(1)})]} \\
&\quad \times \left[-\frac{q}{\mu^{(1)}(R^{IM})^3} + \frac{1 - 2\nu^{(1)}}{(\mu^{(1)} + \mu^{(2)})(R^{IM} + x_3 + q)R^{IM}}\right], \\
\beta^{(1)} &= \frac{F_1 x_1}{2\pi(\mu^{(1)} + \mu^{(2)})[\mu^{(1)} + \mu^{(2)}(3 - 4\nu^{(1)})]} \\
&\quad \times \left[\frac{(1 - 2\nu^{(1)})(\mu^{(1)} - \mu^{(2)})q}{(R^{IM} + x_3 + q)R^{IM}} + \frac{A}{(R^{IM} + x_3 + q)}\right], \\
B_1^{(2)} &= \frac{F_1}{2\pi(\mu^{(1)} + \mu^{(2)})}\frac{1}{R}, \\
B_2^{(2)} &= 0, \\
B_3^{(2)} &= \frac{(1 - 2\nu^{(2)})(\mu^{(1)} - \mu^{(2)})F_1 x_1}{2\pi(\mu^{(1)} + \mu^{(2)})[\mu^{(2)} + \mu^{(1)}(3 - 4\nu^{(2)})](R - x_3 + q)R}, \\
\beta^{(2)} &= \frac{(1 - \nu^{(2)})F_1}{2\pi(1 - \nu^{(1)})(\mu^{(1)} + \mu^{(2)})[\mu^{(1)} + \mu^{(2)}(3 - 4\nu^{(1)})]} \\
&\quad \times \left\{\frac{x_1}{(R - x_3 + q)}\left[A + \frac{(\nu^{(1)} - \nu^{(2)})[\mu^{(1)} + \mu^{(2)}(3 - 4\nu^{(1)})]}{1 - \nu^{(2)}}\right]\right. \\
&\quad + \frac{q x_1}{R(R - x_3 + q)}\left[(1 - 2\nu^{(1)})(\mu^{(1)} - \mu^{(2)})\right. \\
&\quad \left.\left. + \frac{(\nu^{(1)} - \nu^{(2)})[\mu^{(1)} + \mu^{(2)}(3 - 4\nu^{(1)})]}{1 - \nu^{(2)}}\right]\right\},
\end{aligned}
\tag{4.103}
$$

with

$$A = \frac{\begin{array}{c}\{[\mu^{(2)}(3-4\nu^{(1)})(1-2\nu^{(2)}) - \mu^{(1)}(3-4\nu^{(2)})(1-2\nu^{(1)})] \\ \times (\mu^{(1)} - \mu^{(2)})(1-2\nu^{(1)}) - 2\mu^{(2)}(\nu^{(1)} - \nu^{(2)})[\mu^{(1)} + \mu^{(2)}(3-4\nu^{(1)})]\}\end{array}}{\mu^{(2)} + \mu^{(1)}(3-4\nu^{(2)})}. \tag{4.104}$$

4.3.1.4 *Determination of the displacements and the Green's function*

Knowing the Papkovitch functions, the displacements can be determined by use of Eq. (4.86) which, when expanded, expresses u_i in the form

$$u_i(F_j) = \frac{1}{4(1-\nu)}\left[(3-4\nu)B_i(F_j) - \left(x_m \frac{\partial B_m(F_j)}{\partial x_i} + \frac{\partial \beta(F_j)}{\partial x_i}\right)\right], \tag{4.105}$$

where $u_i(F_j)$ is the displacement in the direction i due to force applied in direction j, and $\mathbf{B}(F_j)$ and $\beta(F_j)$ are the Papkovitch functions associated with F_j. With this notation, the Green's function is

$$g_{ij} = u_i(F_j = 1) \tag{4.106}$$

and is therefore readily found by substituting the Papkovitch functions given by Eqs. (4.101–4.104) into Eq. (4.105) and employing Eq. (4.106).

In the following sections detailed results for the infinite body and the half-space with a planar free surface are written out for the reader's convenience. Results for the joined half-spaces, which can be readily obtained from the previous results in a similar manner, are omitted because of their unusual length. However, see Exercise 4.5.

4.3.2 *In infinite homogeneous region*

The Papkovitch functions for this case are obtained from the solution for the joined half-spaces by setting $\mu^{(1)} = \mu^{(2)}$, $\nu^{(1)} = \nu^{(2)}$ and $q = 0$ in Eqs. (4.101–4.104). The only non-vanishing functions are then

$$B_1^{(1)}(F_1) = B_1^{(2)}(F_1) = \frac{F_1}{4\pi\mu R} \quad B_2^{(1)}(F_2) = B_2^{(2)}(F_2) = \frac{F_2}{4\pi\mu R}$$

$$B_3^{(1)}(F_1) = B_3^{(2)}(F_3) = \frac{F_3}{4\pi\mu R} \tag{4.107}$$

where now

$$R = R^{IM} = (x_1^2 + x_2^2 + x_3^2)^{1/2} = |\mathbf{x}| = x. \tag{4.108}$$

Substituting these results into Eq. (4.105), the displacements due to F_1 are then

$$\begin{aligned}
u_1^{\infty}(F_1) &= \frac{F_1}{16\pi(1-\nu)\mu}\left[\frac{3-4\nu}{R} + \frac{x_1^2}{R^3}\right],\\
u_2^{\infty}(F_1) &= \frac{F_1}{16\pi(1-\nu)\mu}\frac{x_1x_2}{R^3},\\
u_3^{\infty}(F_1) &= \frac{F_1}{16\pi(1-\nu)\mu}\frac{x_1x_3}{R^3}.
\end{aligned} \tag{4.109}$$

Corresponding results for F_2^{∞} and F_3^{∞} are obtained by the cyclic exchange of indices. Using these results, and employing Eq. (4.106), the Green's function with the point force acting at the origin can be written (also see Exercises 4.1, 4.4 and 4.6) as

$$G_{ij}^{\infty}(\mathbf{x}) = \frac{1}{16\pi\mu(1-\nu)|\mathbf{x}|}\left[(3-4\nu)\delta_{ij} + \frac{x_ix_j}{|\mathbf{x}|^2}\right]. \tag{4.110}$$

Having this result, a useful expression for the displacement at $\mathbf{x}$ due to a general point force, $\mathbf{F}$, acting at a source point $\mathbf{x}'$ can be written. The displacement, by definition, is given by

$$\mathbf{u}^{\infty}(\mathbf{x}-\mathbf{x}') = G_{ij}^{\infty}(\mathbf{x}-\mathbf{x}')F_j\hat{\mathbf{e}}_i \tag{4.111}$$

and substituting Eq. (4.110) into Eq. (4.111), and moving the source point to $\mathbf{x}'$,

$$\mathbf{u}^{\infty}(\mathbf{x}-\mathbf{x}') = \frac{1}{16\pi\mu(1-\nu)}\left[\frac{3-4\nu}{|\mathbf{x}-\mathbf{x}'|}\mathbf{F} + \frac{(\mathbf{x}-\mathbf{x}')\cdot\mathbf{F}}{|\mathbf{x}-\mathbf{x}'|^3}(\mathbf{x}-\mathbf{x}')\right]. \tag{4.112}$$

4.3.3 *In half-space with planar free surface*

Here, the point force is at (0, 0, q) with the origin at the surface, and the Papkovitch functions can be obtained by setting $\mu^{(2)} = \nu^{(2)} = 0$ in Eqs. (4.101–4.104). Since the F_1 and F_2 quantities are related by symmetry, the only required non-vanishing quantities are then, after dropping the

(1) superscript,

$$B_1(F_1) = \frac{F_1}{4\pi\mu}\left(\frac{1}{R} + \frac{1}{R^{IM}}\right),$$

$$B_3(F_1) = \frac{F_1 x_1}{2\pi\mu}\left[\frac{1-2\nu}{(R^{IM}+x_3+q)R^{IM}} - \frac{q}{(R^{IM})^3}\right],$$

$$\beta(F_1) = \frac{F_1 x_1(1-2\nu)}{2\pi\mu}\left[\frac{q}{(R^{IM}+x_3+q)R^{IM}} - \frac{(1-2\nu)}{(R^{IM}+x_3+q)}\right], \quad (4.113)$$

$$B_3(F_3) = \frac{F_3}{4\pi\mu}\left[\frac{1}{R} + \frac{3-4\nu}{R^{IM}} + \frac{2q(x_3+q)}{(R^{IM})^3}\right],$$

$$\beta(F_3) = \frac{F_3}{4\pi\mu}\left[4(1-\nu)(1-2\nu)\ln(R^{IM}+x_3+q) - \frac{q}{R} - \frac{q(3-4\nu)}{R^{IM}}\right],$$

where $R = [x_1^2 + x_2^2 + (x_3-q)^2]^{1/2}$ and $R^{IM} = [x_1^2 + x_2^2 + (x_3+q)^2]^{1/2}$.

Putting these results into Eq. (4.105), and referring to Eq. (4.109), the displacement components u_i due to the force components F_j are

$$u_i(F_j) = u_i^{\infty}(F_j) + u_i^{IM}(F_j) = \frac{F_j}{A}\left[\frac{(3-4\nu)}{R} + \frac{x_i x_j}{R^3}\right] + u_i^{IM}(F_j) \quad (4.114)$$

i.e., the sum of the displacements expected in an infinite region plus associated image displacements. The latter are of the form

$$u_1^{IM}(F_1) = \frac{F_1}{A}\left[\frac{1}{R^{IM}} + \frac{4(1-\nu)(1-2\nu)}{(R^{IM}+x_3+q)} + \frac{(3-4\nu)x_1^2}{(R^{IM})^3} + \frac{2qx_3}{(R^{IM})^3}\right.$$
$$\left. - \frac{4(1-\nu)(1-2\nu)x_1^2}{R^{IM}(R^{IM}+x_3+q)^2} - \frac{6qx_3x_1^2}{(R^{IM})^5}\right],$$

$$u_2^{IM}(F_1) = \frac{F_1 x_1 x_2}{A}\left[\frac{(3-4\nu)}{(R^{IM})^3} - \frac{4(1-\nu)(1-2\nu)}{R^{IM}(R^{IM}+x_3+q)^2} - \frac{6qx_3}{(R^{IM})^5}\right],$$

$$u_3^{IM}(F_1) = \frac{F_1x_1}{A}\left[\frac{(3-4\nu)(x_3-q)}{(R^{IM})^3} + \frac{4(1-\nu)(1-2\nu)}{R^{IM}(R^{IM}+x_3+q)} - \frac{6qx_3(x_3+q)}{(R^{IM})^5}\right],$$

$$u_1^{IM}(F_2) = \frac{F_2x_1x_2}{A}\left[\frac{(3-4\nu)}{(R^{IM})^3} - \frac{4(1-\nu)(1-2\nu)}{R^{IM}(R^{IM}+x_3+q)^2} - \frac{6qx_3}{(R^{IM})^5}\right],$$

$$u_2^{IM}(F_2) = \frac{F_2}{A}\left[\frac{1}{R^{IM}} + \frac{4(1-\nu)(1-2\nu)}{(R^{IM}+x_3+q)} + \frac{(3-4\nu)x_2^2}{(R^{IM})^3} + \frac{2qx_3}{(R^{IM})^3}\right.$$
$$\left. - \frac{4(1-\nu)(1-2\nu)x_2^2}{R^{IM}(R^{IM}+x_3+q)^2} - \frac{6qx_3x_2^2}{(R^{IM})^5}\right],$$

$$u_3^{IM}(F_2) = \frac{F_2x_2}{A}\left[\frac{(3-4\nu)(x_3-q)}{(R^{IM})^3} + \frac{4(1-\nu)(1-2\nu)}{R^{IM}(R^{IM}+x_3+q)} - \frac{6qx_3(x_3+q)}{(R^{IM})^5}\right],$$

$$u_1^{IM}(F_3) = \frac{F_3x_1}{A}\left[-\frac{4(1-\nu)(1-2\nu)}{R^{IM}(R^{IM}+x_3+q)} + \frac{(3-4\nu)(x_3-q)}{(R^{IM})^3} + \frac{6qx_3(x_3+q)}{(R^{IM})^5}\right],$$

$$u_2^{IM}(F_3) = \frac{F_3x_2}{A}\left[-\frac{4(1-\nu)(1-2\nu)}{R^{IM}(R^{IM}+x_3+q)} + \frac{(3-4\nu)(x_3-q)}{(R^{IM})^3} + \frac{6qx_3(x_3+q)}{(R^{IM})^5}\right],$$

$$u_3^{IM}(F_3) = \frac{F_3}{A}\left[\frac{8(1-\nu)^2-(3-4\nu)}{R^{IM}} + \frac{(3-4\nu)(x_3+q)^2-2qx_3}{(R^{IM})^3}\right.$$
$$\left. + \frac{6qx_3(x_3+q)^2}{(R^{IM})^5}\right], \qquad (4.115)$$

where

$$A = 16\pi\mu(1-\nu) \quad R^{IM} = [x_1^2 + x_2^2 + (x_3+q)^2]^{1/2}, \qquad (4.116)$$

in agreement with Mindlin (1936).

The corresponding half-space image Green's functions are then readily found by substituting Eq. (4.115) into the basic relationship $G_{ij}^{IM} = u_i^{IM}(F_j = 1)$.

Exercises

4.1. Obtain Eq. (4.110) for the Green's function for a point force in an infinite isotropic region by starting with the general integral equation solution given by Eq. (4.25).

Solution. The first step is to rewrite Eq. (4.25) in terms of isotropic elastic constants. Substituting Eq.(3.141) into Eq. (4.25),

$$\begin{aligned} G^{\infty}_{km}(\mathbf{x}-\mathbf{x}') &= \frac{1}{8\pi^2\mu|\mathrm{x}-\mathrm{x}'|}\oint_{\hat{\mathcal{L}}}\left[\delta_{km}-\frac{1}{2(1-\nu)}\hat{k}_k\hat{k}_m\right]ds \\ &= \frac{1}{8\pi^2\mu|\mathbf{x}-\mathbf{x}'|}\left[2\pi\delta_{km}-\frac{1}{2(1-\nu)}\oint_{\hat{\mathcal{L}}}\hat{k}_k\hat{k}_m ds\right]. \end{aligned} \quad (4.117)$$

Equation (4.117) is referred to the $(\hat{\mathbf{e}}_1, \hat{\mathbf{e}}_2, \hat{\mathbf{e}}_3)$ crystal coordinate system, and the line integral is around a unit circle in the plane perpendicular to $\hat{\mathbf{w}} = (\mathbf{x}-\mathbf{x}')/|\mathbf{x}-\mathbf{x}'|$ (Fig. 4.2). Therefore, to simplify the integral, transform it to the $(\hat{\mathbf{u}}, \hat{\mathbf{v}}, \hat{\mathbf{w}})$ system, termed here the "new" system. Now, rotate the new system around $\hat{\mathbf{w}}$ so that $\hat{\mathbf{u}} = \hat{\mathbf{w}} \times \hat{\mathbf{e}}_3$, and the base vectors of the new and "old" (crystal) systems are related by

$$\begin{aligned} \hat{\mathbf{u}} &= \hat{\mathbf{w}} \times \hat{\mathbf{e}}_3 = \frac{\hat{w}_2\hat{\mathbf{e}}_1 - \hat{w}_1\hat{\mathbf{e}}_2}{\sqrt{\hat{w}_1^2+\hat{w}_2^2}}, \\ \hat{\mathbf{v}} &= \hat{\mathbf{w}} \times \hat{\mathbf{u}} = \frac{\hat{w}_1\hat{w}_3\hat{\mathbf{e}}_1 + \hat{w}_2\hat{w}_3\hat{\mathbf{e}}_2 - (\hat{w}_1^2+\hat{w}_2^2)\hat{\mathbf{e}}_3}{\sqrt{\hat{w}_1^2+\hat{w}_2^2}}, \\ \hat{\mathbf{w}} &= \hat{w}_1\hat{\mathbf{e}}_1 + \hat{w}_2\hat{\mathbf{e}}_2 + \hat{w}_3\hat{\mathbf{e}}_3. \end{aligned} \quad (4.118)$$

Then, using Eqs. (2.17) and (2.18), the $\hat{\mathbf{k}}$ vector in Eq. (4.117) will be transformed according to

$$[\hat{k}] = [l]^{T}[\hat{k}'] \quad \text{or} \quad \hat{k}_i = l_{ji}\hat{k}'_j \quad (4.119)$$

where the prime indicates the new system, and

$$[l]^T = \begin{bmatrix} l_{11} & l_{21} & l_{31} \\ l_{12} & l_{22} & l_{32} \\ l_{13} & l_{23} & l_{33} \end{bmatrix} = \begin{bmatrix} \hat{w}_2/\sqrt{\hat{w}_1^2+\hat{w}_2^2} & \hat{w}_1\hat{w}_2/\sqrt{\hat{w}_1^2+\hat{w}_2^2} & \hat{w}_1 \\ -\hat{w}_1/\sqrt{\hat{w}_1^2+\hat{w}_2^2} & \hat{w}_2\hat{w}_3/\sqrt{\hat{w}_1^2+\hat{w}_2^2} & \hat{w}_2 \\ 0 & -\sqrt{\hat{w}_1^2+\hat{w}_2^2} & \hat{w}_3 \end{bmatrix}. \quad (4.120)$$

In the line integral of Eq. (4.117), $\hat{\mathbf{k}}$ lies in the $\hat{\mathbf{w}} = 0$ plane, and using the above results to evaluate the integral,

$$\begin{aligned} \oint_{\hat{\mathcal{L}}}\hat{k}_k\hat{k}_m ds &= \oint_{\hat{\mathcal{L}}} l_{ik}\hat{k}'_i l_{jm}\hat{k}'_j ds \\ &= \int_0^{2\pi}[l_{1k}l_{1m}\cos^2\theta + (l_{1m}l_{2k} + l_{1k}l_{2m})\sin\theta\cos\theta \\ &\quad + l_{2k}l_{2m}\sin^2\theta]d\theta \\ &= (l_{1k}l_{1m} + l_{2k}l_{2m})\pi. \end{aligned} \quad (4.121)$$

Finally, substituting Eq. (4.121) into Eq. (4.117) and using Eq. (4.120) and $\hat{\mathbf{w}} = (\mathbf{x} - \mathbf{x}')/|\mathbf{x} - \mathbf{x}'|$, G^{∞}_{km} assumes the form

$$G^{\infty}_{ij}(\mathbf{x} - \mathbf{x}') = \frac{1}{16\pi\mu(1-\nu)|\mathbf{x} - \mathbf{x}'|}\left[(3 - 4\nu)\delta_{ij} + \frac{(x_i - x'_i)(x_j - x'_j)}{|\mathbf{x} - \mathbf{x}'|^2}\right] \tag{4.122}$$

in agreement with Eq. (4.110). Also, see Exercise 4.4.

4.2. Show that the Green's function for a point force in an infinite medium is a homogeneous function of degree −1 in the variable $(\mathbf{x} - \mathbf{x}')$ and therefore must be of the form

$$G^{\infty}_{ij} = \frac{f_{ij}(\hat{\boldsymbol{l}})}{|\mathbf{x} - \mathbf{x}'|}, \tag{4.123}$$

where $f_{ij}(\hat{\boldsymbol{l}})$ is a function of the unit vector $\hat{\boldsymbol{l}} = (\mathbf{x} - \mathbf{x}')/|\mathbf{x} - \mathbf{x}'|$.[7] It therefore falls off with distance from the point force as $|\mathbf{x} - \mathbf{x}'|^{-1}$ and is a function of direction according to $f_{ij}(\hat{\boldsymbol{l}})$. Hint: start with the basic Eq. (3.21).

Solution. Starting with Eq. (3.21), i.e.,

$$C_{kpim}\frac{\partial^2 G_{ij}(\mathbf{x} - \mathbf{x}')}{\partial x_m \partial x_p} + \delta_{kj}\delta(\mathbf{x} - \mathbf{x}') = 0 \tag{4.124}$$

and making the change of variable

$$\boldsymbol{\xi} = \mathbf{x} - \mathbf{x}', \tag{4.125}$$

$$C_{kpim}\frac{\partial^2 G^{\infty}_{ij}(\boldsymbol{\xi})}{\partial \xi_m \partial \xi_p} + \delta_{kj}\delta(\boldsymbol{\xi}) = 0\,. \tag{4.126}$$

Then, scaling $\boldsymbol{\xi}$ by setting

$$\boldsymbol{\xi} = \boldsymbol{\lambda}\mathbf{y} \tag{4.127}$$

where $\boldsymbol{\lambda}$ = constant, Eq. (4.126) becomes

$$\frac{1}{\boldsymbol{\lambda}^2}C_{kpim}\frac{\partial^2 G^{\infty}_{ij}(\boldsymbol{\lambda}\mathbf{y})}{\partial y_m \partial y_p} + \delta_{kj}\delta(\boldsymbol{\lambda}\mathbf{y}) = 0. \tag{4.128}$$

[7]This exercise was suggested by Prof. D. M. Barnett.

However, using the properties of the delta function from Eqs. (D.2) and (D.3),

$$\int_{-\infty}^{\infty}\int_{-\infty}^{\infty}\int_{-\infty}^{\infty}\delta(\mathbf{x})dx_1dx_2dx_3 = \lambda^3\int_{-\infty}^{\infty}\int_{-\infty}^{\infty}\int_{-\infty}^{\infty}\delta(\lambda\mathbf{y})dy_1dy_2dy_3$$
$$= \int_{-\infty}^{\infty}\int_{-\infty}^{\infty}\int_{-\infty}^{\infty}\delta(\mathbf{y})dy_1dy_2dy_3 = 1 \tag{4.129}$$

so that

$$\delta(\lambda\mathbf{y}) = \lambda^{-3}\delta(\mathbf{y}) \tag{4.130}$$

and, upon substituting this result into Eq. (4.128),

$$\lambda C_{kpim}\frac{\partial^2 G_{ij}^{\infty}(\lambda\mathbf{y})}{\partial y_m \partial y_p} + \delta_{kj}\delta(\mathbf{y}) = 0. \tag{4.131}$$

Then, in view of the functional form of Eq. (4.131), we can write

$$\lambda C_{kpim}\frac{\partial^2 G_{ij}^{\infty}[\lambda(\mathbf{x}-\mathbf{x}')]}{\partial x_m \partial x_p} + \delta_{kj}\delta(\mathbf{x}-\mathbf{x}') = 0. \tag{4.132}$$

Comparison of Eq. (4.132) with (4.124) shows that

$$G_{ij}^{\infty}[\lambda(\mathbf{x}-\mathbf{x}')] = \lambda^{-1}G_{ij}^{\infty}(\mathbf{x}-\mathbf{x}') \tag{4.133}$$

and therefore $G_{ij}^{\infty}(\mathbf{x}-\mathbf{x}')$ is indeed homogeneous of degree -1 in $(\mathbf{x}-\mathbf{x}')$.
Furthermore, since scaling $(\mathbf{x}-\mathbf{x}')$ by λ scales $G_{ij}^{\infty}(\mathbf{x}-\mathbf{x}')$ by λ^{-1}, it must be concluded that $G_{ij}^{\infty}(\mathbf{x}-\mathbf{x}')$ is of the form of Eq. (4.123) when the point force is located at $\mathbf{x}'$. As an example, note that the Green's function given by Eq. (4.110) in that case takes the form

$$G_{ij}^{\infty}(\mathbf{x}) = \frac{1}{16\pi\mu(1-\nu)|\mathbf{x}-\mathbf{x}'|}\left[(3-4\nu)\delta_{ij} + \frac{(x_i-x_i')(x_j-x_j')}{|\mathbf{x}-\mathbf{x}'|^2}\right]$$
$$= \frac{[(3-4\nu)\delta_{ij} + l_i l_j]}{16\pi\mu(1-\nu)}\frac{1}{|\mathbf{x}-\mathbf{x}'|}, \tag{4.134}$$

consistent with Eq. (4.123).

4.3. As shown by Lothe (1992b), the partial derivative of the Green's function with respect to the displacement of the field point, given by

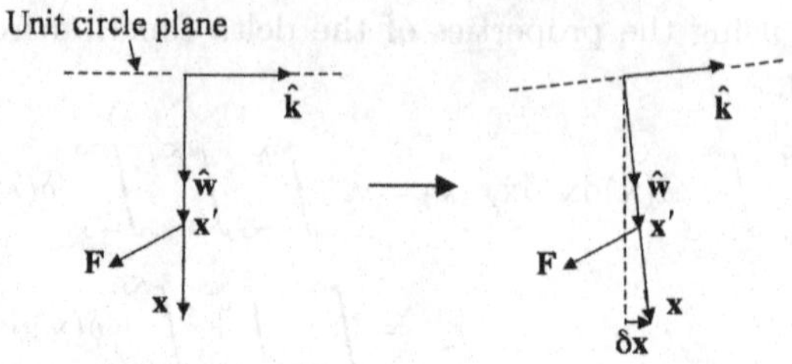

Fig. 4.7 Edge view of the tilt of the unit circle plane in the special case when the field point is displaced by the vector, $\delta\mathbf{x}$, in the direction of $\hat{\mathbf{k}}$.

Eq. (4.40), can be obtained by considering the tilt of the plane on which the unit circle integral is performed (Fig. 4.2) when the field point is displaced by the vector $\delta\mathbf{x}$. This tilt is illustrated (in exaggerated fashion) in Fig. 4.7 for the special case where $\delta\mathbf{x}$ is parallel to $\hat{\mathbf{k}}$. In the general case, the component of $\delta\mathbf{x}$ parallel to $\hat{\mathbf{k}}$ will produce a rotation of the vector $(\mathbf{x}-\mathbf{x}')$ corresponding to the angle $(\hat{\mathbf{k}}\cdot\delta\mathbf{x})/|\mathbf{x}-\mathbf{x}'|$ that will, in turn, cause $\hat{\mathbf{k}}$ to rotate so that

$$\delta\hat{\mathbf{k}} = -\frac{(\hat{\mathbf{k}}\cdot\delta\mathbf{x})\hat{\mathbf{w}}}{|\mathbf{x}-\mathbf{x}'|}. \tag{4.135}$$

Using this information, derive the derivative of the Green's function given by Eq. (4.40).

Solution. The derivative of $\mathrm{G}^{\infty}_{\mathrm{km}}$, using Eq. (4.25), is

$$\frac{\partial \mathrm{G}^{\infty}_{\mathrm{km}}(\mathbf{x}-\mathbf{x}')}{\partial \mathrm{x_i}} = \frac{1}{8\pi^2}\left[\frac{1}{|\mathbf{x}-\mathbf{x}'|}\oint_{\hat{\mathcal{L}}}\frac{\partial(\hat{\mathrm{k}}\hat{\mathrm{k}})^{-1}_{\mathrm{km}}}{\partial \mathrm{x_i}}\mathrm{ds}\right.$$
$$\left. + \oint_{\hat{\mathcal{L}}}(\hat{\mathrm{k}}\hat{\mathrm{k}})^{-1}_{\mathrm{km}}\mathrm{ds}\frac{\partial}{\partial \mathrm{x_i}}\left(\frac{1}{|\mathbf{x}-\mathbf{x}'|}\right)\right]. \tag{4.136}$$

An expression for the derivative of $(\hat{\mathrm{k}}\hat{\mathrm{k}})^{-1}_{\mathrm{km}}$ in the first integral of Eq. (4.136) can be obtained, using matrix and index notation, as follows:

$$(\hat{\mathrm{k}}\hat{\mathrm{k}})(\hat{\mathrm{k}}\hat{\mathrm{k}})^{-1} = [\mathrm{I}],$$

$$(\hat{\mathrm{k}}\hat{\mathrm{k}})\frac{\partial(\hat{\mathrm{k}}\hat{\mathrm{k}})^{-1}}{\partial \mathrm{x_i}} + \frac{\partial(\hat{\mathrm{k}}\hat{\mathrm{k}})}{\partial \mathrm{x_i}}(\hat{\mathrm{k}}\hat{\mathrm{k}})^{-1} = 0, \tag{4.137}$$

$$\frac{\partial(\hat{k}\hat{k})^{-1}}{\partial x_i} = -(\hat{k}\hat{k})^{-1}\frac{\partial(\hat{k}\hat{k})}{\partial x_i}(\hat{k}\hat{k})^{-1},$$

$$\frac{\partial(\hat{k}\hat{k})^{-1}_{km}}{\partial x_i} = -(\hat{k}\hat{k})^{-1}_{kp}\frac{\partial(\hat{k}\hat{k})_{pj}}{\partial x_i}(\hat{k}\hat{k})^{-1}_{jm},$$

where $\partial(\hat{k}\hat{k})_{pj}/\partial x_i$, is obtained by substituting Eq. (4.135) into Eq. (4.137), i.e.,

$$\frac{\partial(\hat{k}\hat{k})_{pj}}{\partial x_i} = \frac{\partial}{\partial x_i}(\hat{k}_l C_{lpjn}\hat{k}_n) = -\frac{\hat{k}_i}{|\mathbf{x}-\mathbf{x}'|}[(\hat{k}\hat{w})_{pj} + (\hat{w}\hat{k})_{pj}]. \tag{4.138}$$

In addition,

$$\frac{\partial}{\partial x_i}\left(\frac{1}{|\mathbf{x}-\mathbf{x}'|}\right) = -\frac{(x_i - x_i')}{|\mathbf{x}-\mathbf{x}'|^3} = -\frac{\hat{w}_i}{|\mathbf{x}-\mathbf{x}'|^2}. \tag{4.139}$$

Then, substitution of Eqs. (4.137)–(4.139) into Eq. (4.136) yields Eq. (4.40).

4.4 In Exercise 4.1 we obtained the Green's function for a point force in an infinite isotropic medium, Eq. (4.110), by integrating the expression for $G^{\infty}_{km}(\mathbf{x}-\mathbf{x}')$ for a general anisotropic medium, Eq. (4.25), after first rewriting it so that it pertained to an isotropic medium. Instead of using this approach, derive Eq. (4.110) by again starting with Eq. (4.25) and then employing the tensor, Q_{km}.

Solution According to Eq. (4.25), $G^{\infty}_{km}(\mathbf{x}-\mathbf{x}')$ is proportional to a line integral taken around the unit circle traversed by a unit vector as it rotates on a plane perpendicular to the unit vector $(\mathbf{x}-\mathbf{x}')/|\mathbf{x}-\mathbf{x}'|$. On the other hand, according to Eq. (3.133), the tensor Q_{km} is proportional to a line integral taken around the unit circle traversed by a unit vector as it rotates on a plane perpendicular to the unit vector $\hat{\boldsymbol{\tau}}$. The two line integrals will therefore be identical if $\hat{\boldsymbol{\tau}} = (\mathbf{x}-\mathbf{x}')/|\mathbf{x}-\mathbf{x}'|$. For an isotropic medium, Q_{km} has the form given by Eq. (3.146), and a comparison of Eqs. (3.133) and (3.146) shows that the line integral in Eq. (3.133) for an isotropic medium must be given by

$$I_{km} = \frac{\pi}{2\mu(1-\nu)}[(3-4\nu)\delta_{km} + \hat{\tau}_k\hat{\tau}_m]. \tag{4.140}$$

Therefore, setting $\hat{\boldsymbol{\tau}} = (\mathbf{x}-\mathbf{x}')/|\mathbf{x}-\mathbf{x}'|$ in Eq. (4.140), and substituting the resulting expression for the line integral into Eq. (4.25), we obtain Eq. (4.110), i.e.,

$$\begin{aligned} G^{\infty}_{km}(\mathbf{x}-\mathbf{x}') &= \frac{1}{8\pi^2|\mathbf{x}-\mathbf{x}'|}\oint_{\hat{\mathcal{L}}}(\hat{k}\hat{k})^{-1}_{km}ds \\ &= \frac{1}{16\pi\mu(1-\nu)|\mathbf{x}-\mathbf{x}'|} \\ &\quad\times\left[(3-4\nu)\delta_{km}+\frac{(x_k-x'_k)(x_m-x'_m)}{|x-x'|^2}\right]. \end{aligned} \tag{4.141}$$

4.5 For the joined half-spaces in Fig. 4.5 find an expression for the displacement field $u_1^{(2)}$ when the point force at $x_3 = q$ is a point force directed along x_3.

Solution According to Eq. (4.105), the desired displacement field then is given by

$$\begin{aligned} u_1^{(2)}(F_3) &= \frac{1}{4(1-\nu^{(2)})}\left[(3-4\nu^{(2)})B_1^{(2)}(F_3)\right. \\ &\quad\left. - x_m\frac{\partial B_m^{(2)}(F_3)}{\partial x_1} - \frac{\partial \beta^{(2)}(F_3)}{\partial x_1}\right] \end{aligned} \tag{4.142}$$

Next, substituting Eq. (4.102) into (4.142)

$$\begin{aligned} u_1^{(2)} &= \frac{x_1F_3}{4\pi R}\left\{\frac{1}{[\mu^{(2)}+(3-4\nu^{(2)})\mu^{(1)}]}\frac{x_3}{R^2} - \frac{1}{\mu^{(1)}+(3-4\nu^{(1)})\mu^{(2)}}\right. \\ &\quad\times\left[\frac{q}{R^2}+\frac{\mu^{(1)}(1-2\nu^{(1)})(3-4\nu^{(2)})-\mu^{(2)}(1-2\nu^{(2)})(3-4\nu^{(1)})}{\mu^{(2)}+(3-4\nu^{(2)})\mu^{(1)}}\right. \\ &\quad\left.\left.\times\frac{1}{(R-x_3-q)}\right]\right\}. \end{aligned} \tag{4.143}$$

4.6 The point force Green's function for an isotropic system given by Eq. (4.110) can be expressed in an alternative form that involves a differential operator, ϕ_{ij}, acting on the magnitude of the field vector and is therefore of the form

$$G^{\infty}_{ij}(\mathbf{x}) = \phi_{ij}|\mathbf{x}|. \tag{4.144}$$

Find the expression for ϕ_{ij}. Hint: consider the use of various derivatives of $|\mathbf{x}|$ in Eq. (4.110).

Solution Two derivatives of $|\mathbf{x}| = (x_1^2 + x_2^2 + x_3^2)^{1/2}$ are of the forms

$$\frac{\partial^2 |\mathbf{x}|}{\partial x_i \partial x_j} = \frac{1}{|\mathbf{x}|}\delta_{ij} - \frac{x_i x_i}{|\mathbf{x}|^3} \tag{4.145}$$

and

$$\nabla^2 |\mathbf{x}| = \frac{2}{|\mathbf{x}|}. \tag{4.146}$$

Then, substituting Eqs. (4.145) and (4.146) into Eq. (4.110),

$$\begin{aligned} G_{ij}^{\infty}(\mathbf{x}) &= \frac{1}{8\pi\mu}\left[\delta_{ij}\nabla^2|\mathbf{x}| - \frac{1}{2(1-\nu)}\frac{\partial^2|\mathbf{x}|}{\partial x_i \partial x_j}\right] \\ &= \frac{1}{8\pi\mu}\left[\delta_{ij}\nabla^2 - \frac{1}{2(1-\nu)}\frac{\partial^2}{\partial x_i \partial x_j}\right]|\mathbf{x}| \end{aligned} \tag{4.147}$$

and the differential operator therefore has the form

$$\phi_{ij} = \frac{1}{8\pi\mu}\left[\delta_{ij}\nabla^2 - \frac{1}{2(1-\nu)}\frac{\partial^2}{\partial x_i \partial x_j}\right]. \tag{4.148}$$

4.7 Derive Eq. (4.24) for the point force Green's function by starting with Eq. (3.23) and inverting it instead of by starting with Eq. (3.21) and proceeding as in the text in Sec. 4.2.1.1. Hint: be aware of the rule given by Eq. (12.100).

Solution Subsituting Eq. (3.23) into Eq. (F.4), the Green's function is

$$G_{km}(\mathbf{x}) = \frac{1}{8\pi^3}\int_{-\infty}^{\infty}\int_{-\infty}^{\infty}\int_{-\infty}^{\infty}(kk)_{km}^{-1}e^{i\mathbf{k}\cdot\mathbf{x}}dk_1dk_2dk_3. \tag{4.149}$$

However, by use of $\mathbf{k} = k\hat{\mathbf{k}}$ the amplitude of the above Fourier integral can be written as

$$(kk)_{km}^{-1} = k^{-2}(\hat{k}\hat{k})_{km}^{-1}, \tag{4.150}$$

and since k scales with the constant scaling factor, λ, as λk, $(kk)_{km}^{-1}$ scales as $\lambda^{-2}(kk)_{km}^{-1}$, and is therefore seen to be a homogeneous function of degree -2 in the variable k. The rule given by Eq. (12.100) then applies, and, therefore,

$$\int_{-\infty}^{\infty}\int_{-\infty}^{\infty}\int_{-\infty}^{\infty}(kk)_{km}^{-1}e^{i\mathbf{k}\cdot\mathbf{x}}dk_1dk_2dk_3 = \frac{\pi}{x}\int_0^{2\pi}(\hat{k}\hat{k})_{km}^{-1}d\phi \tag{4.151}$$

where the integral on the right hand side is around a unit circle normal to $\hat{\mathbf{w}}$ as illustrated in Fig. 4.2. Finally, substituting Eq. (4.151) into Eq. (4.149),

$$\mathrm{G_{km}}(\mathbf{x}) = \frac{1}{8\pi^2 \mathrm{x}} \int_0^{2\pi} (\hat{\mathrm{k}}\hat{\mathrm{k}})^{-1}_{\mathrm{km}} \mathrm{d}\phi \tag{4.152}$$

in agreement with Eq. (4.24).

Chapter 5

Interactions between Defects and Stress

5.1 Introduction

Defects can interact with stress in many different ways, and it is useful to begin by briefly identifying the types of interactions considered in the present book. The great majority of the defects under consideration, i.e., inclusions, point defects, dislocations and various interfaces containing discrete intrinsic dislocations, are sources of stress, and they will therefore interact elastically with *imposed stress* that may be *internal stress* due, for example, to the presence other defects, or *applied stress*, due to forces applied to the body.[1] When such a defect source of stress lies in a finite region bounded by an interface, an *image stress* is generated (Sec. 3.8) which then interacts with the defect. Thus, the interface acts, in a sense, as a source of stress. On the other hand, an inhomogeneity, which by itself is not a source of stress, perturbs an imposed stress field, and this perturbed field, in turn, interacts with the inhomogeneity (Sec. 5.4). The inhomogeneity may therefore be regarded as the indirect source of a stress with which it interacts. Also, if the displacement of a defect in a body causes the body to change shape, applied forces (which produce applied stresses) will be able to perform work, and thus interact with the defect (Sec. 15.3).

Finally, defects which are extended in at least one dimension, i.e., inclusions dislocations and interfaces, generally experience a *self-force* since the elastic self-strain energy of such a defect varies with its shape (configuration) (Sec. 17.1). A force therefore generally exists urging the defect to

[1]Throughout this book an *imposed stress* is regarded as any stress arising from a source independent of the defect. For examples, an *applied stress* is, therefore, classified as an imposed stress due to an applied force.

change its shape (configuration) in a direction that decreases its energy. We classify such interactions as defect *self-interactions.*

In summary, we have the following types of interactions:

(1) an interaction between a defect source of stress and imposed internal and applied stresses,
(2) an interaction between a defect source of stress and its image stress,
(3) an interaction between an imposed stress and an inhomogeneity,
(4) an interaction between imposed stress and a defect whose displacement in a body causes a change in body shape,
(5) a self-interaction experienced by a spatially extended defect.

This chapter focuses on basic formulations of interactions (1), (2) and (3). These are expressed in terms of *interaction energies* and corresponding *forces* which are derived in forms that will be of use in succeeding chapters devoted to specific defects. Interactions of types (4) and (5) are considered in Chs. 15 and 17, respectively.

The following notation is employed for this chapter:

ε_{ij}^{D} = strain field of defect in finite homogeneous region
$\varepsilon_{ij}^{D\infty}$ = strain field of defect in infinite homogeneous region
ε_{ij}^{DIM} = image strain field of defect in finite homogeneous region
ε_{ij}^{I} = imposed internal strain field, I
ε_{ij}^{A} = imposed applied strain field, A
ε_{ij}^{Q} = general imposed strain field, Q (either internal or applied)

5.2 Interaction Energies between a Defect Source of Stress and Various Stresses in Finite Homogeneous Body

Following the pioneering work of Eshelby (1956), consider an elastically homogeneous finite body $\mathcal{V}^{\circ}$, possessing a defect source of internal stress D, a second source of internal stress I, and a surface $\mathcal{S}^{\circ}$, subjected to tractions $\mathbf{T}^{A}$, as illustrated in Fig. 5.1.[2] As described in Sec. 3.8, the stress field of the defect, σ_{ij}^{D}, may be expressed as the sum of the field that it would produce

[2]Situations where the body is not homogeneous, as when the defect is an inhomogeneity or an inhomogeneous inclusion, are considered later.

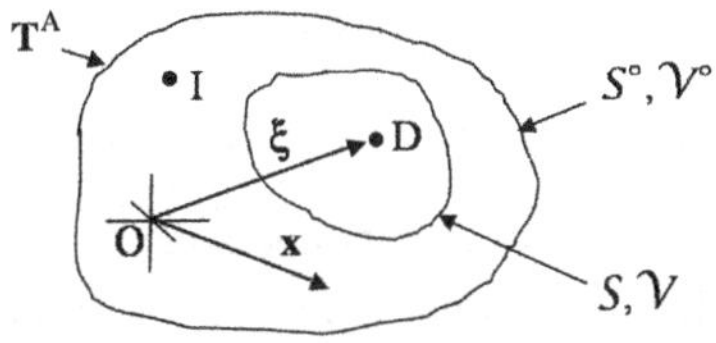

Fig. 5.1 Finite body $\mathcal{V}^\circ$ with surface $\mathcal{S}^\circ$, subjected to tractions, $\mathbf{T}^\mathrm{A}$, containing an embedded region, $\mathcal{V}$, enclosed by the surface, $\mathcal{S}$, that contains a defect source of stress, **D**. Also, a source of internal stress, I, lies in $\mathcal{V}^\circ - \mathcal{V}$. The field vector is $\mathbf{x}$, and the vector $\boldsymbol{\xi}$ indicates the defect position.

in an infinite body $\sigma_{ij}^{D^\infty}$ and its corresponding image field $\sigma_{ij}^{D^{IM}}$, so that $\sigma_{ij}^{D} = \sigma_{ij}^{D^\infty} + \sigma_{ij}^{D^{IM}}$ (as in Eq. (3.175)). Four distinguishable elastic fields are therefore present in this system, i.e., $\sigma_{ij}^{D^\infty}$, $\sigma_{ij}^{D^{IM}}$, σ_{ij}^{I} and σ_{ij}^{A}, where σ_{ij}^{I} is the internal stress due to the internal source, I, and σ_{ij}^{A} is the applied stress due to the surface tractions. The total stress, strain and displacement, are therefore

$$
\begin{aligned}
\sigma_{ij} &= \sigma_{ij}^{D^\infty} + \sigma_{ij}^{D^{IM}} + \sigma_{ij}^{I} + \sigma_{ij}^{A}, \\
\varepsilon_{ij} &= \varepsilon_{ij}^{D^\infty} + \varepsilon_{ij}^{D^{IM}} + \varepsilon_{ij}^{I} + \varepsilon_{ij}^{A}, \\
u_i &= u_i^{D^\infty} + u_i^{D^{IM}} + u_i^{I} + u_i^{A}.
\end{aligned}
\tag{5.1}
$$

Since this book is concerned only with the elasto-mechanical behavior of bodies under isothermal conditions, the total energy of the body, E, is taken to be the sum of the elastic strain energy in the body, W, plus the potential energy of any forces applied to the body, Φ, so that

$$
E = W + \Phi. \tag{5.2}
$$

Here, the potential energy is defined so that if an applied force, **F**, is displaced by $\delta\mathbf{u}$, and therefore performs the work $\mathcal{W} = \mathbf{F}\cdot\delta\mathbf{u}$, the change in potential energy is

$$
\delta\Phi = -\delta\mathcal{W} = -\mathbf{F}\cdot\delta\mathbf{u}. \tag{5.3}
$$

The interaction energy between a defect source of stress D and another stress in the system X, represented by $E_{int}^{D/X}$, is given by the difference between the energy when they coexist in the system, $E^{(D+X)}$ and the sum of the their energies when they exist separately, $(E^D + E^X)$, i.e.,

$$
E_{int}^{D/X} = E^{(D+X)} - \left(E^D + E^X\right). \tag{5.4}
$$

Alternatively, if $E^{D/X}$ is the energy change when the defect is added to the system in the presence of X, and E^D is the energy change when the defect is added by itself (as above), the interaction energy must be given by

$$E_{int}^{D/X} = E^{D/X} - E^{D}. \tag{5.5}$$

The equivalence of these two formulations is demonstrated in Exercise 5.6.

Since linear elasticity is assumed, the interactions between the various fields in the above system can be treated independently. The interaction energies of the defect, D, with the internal I field, the applied A field and its image IM field are, therefore, now considered in turn.

5.2.1 *Interaction energy with imposed internal stress*

For this interaction we assume that the surface tractions (applied forces) in Fig. 5.1 are absent, and only the D and I singularities are present.

5.2.1.1 *Defect represented by its stress field*

Since only strain energy is involved, the interaction energy between the D field and the I field, $E_{int}^{D/I}$, is then given by Eq. (5.4) in the form

$$\begin{aligned} E_{int}^{D/I} &= W_{int}^{D/I} = W^{D+I} - W^{D} - W^{I} \\ &= \frac{1}{2}\oiiint_{\mathcal{V}^\circ} (\sigma_{ij}^{D} + \sigma_{ij}^{I})(\varepsilon_{ij}^{D} + \varepsilon_{ij}^{I})\, dV \\ &\quad - \frac{1}{2}\oiiint_{\mathcal{V}^\circ} \sigma_{ij}^{D}\varepsilon_{ij}^{D} dV - \frac{1}{2}\oiiint_{\mathcal{V}^\circ} \sigma_{ij}^{I}\varepsilon_{ij}^{I} dV \\ &= \frac{1}{2}\oiiint_{\mathcal{V}^\circ} (\sigma_{ij}^{D}\varepsilon_{ij}^{I} + \sigma_{ij}^{I}\varepsilon_{ij}^{D})\, dV, \end{aligned} \tag{5.6}$$

where $W_{int}^{D/I}$ is the interaction strain energy, and Eq. (2.133) has been employed. Then, using Eq. (2.102),

$$E_{int}^{D/I} = \frac{1}{2}\oiiint_{\mathcal{V}^\circ} (\sigma_{ij}^{D}\varepsilon_{ij}^{I} + \sigma_{ij}^{I}\varepsilon_{ij}^{D})\, dV = \oiiint_{\mathcal{V}^\circ} \sigma_{ij}^{D}\varepsilon_{ij}^{I} dV = \oiiint_{\mathcal{V}^\circ} \sigma_{ij}^{I}\varepsilon_{ij}^{D} dV. \tag{5.7}$$

To integrate Eq. (5.5), the geometry indicated in Fig. 5.1 is employed (Eshelby 1956). The integral is written as the sum of two integrals, one over the region $\mathcal{V}$, containing D but excluding I, and the other over the region $(\mathcal{V}^\circ - \mathcal{V})$, containing I but excluding D, i.e.,

$$E_{int}^{D/I} = \oiiint_{\mathcal{V}} \sigma_{ij}^{D}\varepsilon_{ij}^{I} dV + \oiiint_{\mathcal{V}^\circ - \mathcal{V}} \sigma_{ij}^{I}\varepsilon_{ij}^{D} dV. \tag{5.8}$$

In this arrangement any strain incompatibilities associated with the sources of the D and I fields are avoided, and the D and I fields within the regions of integration are "corresponding fields", as described in Sec. 2.4.3. Therefore, substituting Eq. (2.106) into Eq. (5.8), and applying the divergence theorem, we have

$$\begin{aligned} E_{int}^{D/I} &= \oiiint_{\mathcal{V}} \frac{\partial}{\partial x_j}(\sigma_{ij}^{D} u_i^{I}) dV + \oiiint_{\mathcal{V}^{\circ}-\mathcal{V}} \frac{\partial}{\partial x_j}(\sigma_{ij}^{I} u_i^{D}) \\ &= \oiint_{\mathcal{S}} \sigma_{ij}^{D} u_i^{I} \hat{n}_j dS + \oiint_{\mathcal{S}^{\circ}} \sigma_{ij}^{I} u_i^{D} \hat{n}_j dS - \oiint_{\mathcal{S}} \sigma_{ij}^{I} u_i^{D} \hat{n}_j dS. \end{aligned} \tag{5.9}$$

The last surface integral is negative, since a normal vector to a closed surface directed inwards is taken to be negative. The penultimate surface integral vanishes since the tractions on $\mathcal{S}^{\circ}$ due to the internal stress, $\sigma_{ij}^{I}\hat{n}_j$, vanish, and Eq. (5.9) therefore takes the form

$$E_{int}^{D/I} = \oiint_{\mathcal{S}} \left(\sigma_{ij}^{D} u_i^{I} - \sigma_{ij}^{I} u_i^{D}\right) \hat{n}_j dS. \tag{5.10}$$

Equation (5.10) can be put into another form by setting $\sigma_{ij}^{D} = \sigma_{ij}^{D^{\infty}} + \sigma_{ij}^{D^{IM}}$ and $u_i^{D} = u_i^{D^{\infty}} + u_i^{D^{IM}}$ so that

$$\begin{aligned} E_{int}^{D/I} &= \oiint_{\mathcal{S}} \left(\sigma_{ij}^{D} u_i^{I} - \sigma_{ij}^{I} u_i^{D}\right) \hat{n}_j dS \\ &= \oiint_{\mathcal{S}} \left(\sigma_{ij}^{D^{\infty}} u_i^{I} - \sigma_{ij}^{I} u_i^{D^{\infty}}\right) \hat{n}_j dS \\ &\quad + \oiint_{\mathcal{S}} \left(\sigma_{ij}^{D^{IM}} u_i^{I} - \sigma_{ij}^{I} u_i^{D^{IM}}\right) \hat{n}_j dS \\ &= \oiint_{\mathcal{S}} \left(\sigma_{ij}^{D^{\infty}} u_i^{I} - \sigma_{ij}^{I} u_i^{D^{\infty}}\right) \hat{n}_j dS = E_{int}^{D^{\infty}/I}, \end{aligned} \tag{5.11}$$

where the integral containing image quantities vanishes since the D^{IM} and I fields are corresponding fields that obey Eq. (2.108) within $\mathcal{S}$ where the source of the I field is excluded, and we have introduced $E_{int}^{D^{\infty}/I}$ given by

$$E_{int}^{D^{\infty}/I} = \oiint_{\mathcal{S}} \left(\sigma_{ij}^{D^{\infty}} u_i^{I} - \sigma_{ij}^{I} u_i^{D^{\infty}}\right) \hat{n}_j dS = E_{int}^{D/I}. \tag{5.12}$$

Therefore, $E_{int}^{D/I} = E_{int}^{D^{\infty}/I}$, and the interaction energy between the defect and the I field takes several equivalent forms. In all cases, however, it is expressed in the form of an integral over a surface, $\mathcal{S}$, of arbitrary shape that encloses D but excludes I (Eshelby 1956).

5.2.1.2 *Defect represented by a transformation strain*

If the defect is mimicked by means of a transformation strain, ε_{ij}^{T}, Eq. (5.10) can be expressed in the form of a relatively simple volume integral (Eshelby 1956: Mura 1987). According to Sec. 3.6, the transformation strain is localized in a region, $\mathcal{V}^{D}$, embedded in the surrounding matrix, and, therefore, beginning with Eq. (5.10),

$$
\begin{aligned}
E_{int}^{D/I} &= \oiint_{\mathcal{S}} \left(\sigma_{ij}^{D} u_i^{I} - \sigma_{ij}^{I} u_i^{D}\right) \hat{n}_j dS \\
&= \oiint_{\mathcal{S}^D} \left[\sigma_{ij}^{D} u_i^{I} - \sigma_{ij}^{I} \left(u_i^{D} + u_i^{T}\right)\right] \hat{n}_j dS \\
&= \iiint_{\mathcal{V}^D} \frac{\partial}{\partial x_j} \left[\sigma_{ij}^{D} u_i^{I} - \sigma_{ij}^{I} (u_i^{D} + u_i^{T})\right] dV \\
&= \iiint_{\mathcal{V}^D} \left[\sigma_{ij}^{D} \frac{\partial u_i^{I}}{\partial x_j} - \sigma_{ij}^{I} \frac{\partial (u_i^{D} + u_i^{T})}{\partial x_j}\right] dV \\
&= \iiint_{\mathcal{V}^D} \left[\sigma_{ij}^{D} \varepsilon_{ij}^{I} - \sigma_{ij}^{I} \left(\varepsilon_{ij}^{D} + \varepsilon_{ij}^{T}\right)\right] dV \\
&= -\iiint_{\mathcal{V}^D} \sigma_{ij}^{I} \varepsilon_{ij}^{T} dV.
\end{aligned}
\tag{5.13}
$$

On the first line of Eq. (5.13), the surface $\mathcal{S}$ lies in the matrix outside of the $\mathcal{V}^{D}$ region, and all field quantities are elastic quantities as described in Sec. 3.6 (see Eq. (3.151)). On the second line the surface, $\mathcal{S}$, has been shrunk down to a surface $\mathcal{S}^{D}$ infinitesimally inside the $\mathcal{V}^{D}$ region. Since the $\mathcal{V}^{D}$ region contains a transformation strain, the matching conditions that must be satisfied across the interface between the $\mathcal{V}$ and $\mathcal{V}^{D}$ regions are then

$$
\begin{aligned}
\sigma_{ij}^{D}(\text{out})\hat{n}_j &= \sigma_{ij}^{D}(\text{in})\hat{n}_j, \\
\sigma_{ij}^{I}(\text{out})\hat{n}_j &= \sigma_{ij}^{I}(\text{in})\hat{n}_j, \\
u_i^{I}(\text{out}) &= u_i^{I}(\text{in}), \\
u_i^{D,tot}(\text{out}) &= u_i^{D,tot}(\text{in}) = u_i^{D}(\text{out}) = u_i^{D}(\text{in}) + u_i^{T},
\end{aligned}
\tag{5.14}
$$

where $\sigma_{ij}^{D}\hat{n}_j$, $\sigma_{ij}^{I}\hat{n}_j$, u_i^D and u_i^I (both in and out of $\mathcal{V}^D$) are elastic quantities, and the last condition satisfies the requirement that the total displacements, given by $u_i^{D,tot} = u_i^D + u_i^T$ (see Eq. (3.152)), must match across the interface. Then, in Eq. (5.13) the surface integral over $\mathcal{S}^D$ is converted to a volume integral. On the remaining lines, the integrand is differentiated, Eq. (2.65) is applied, and, finally, use is made of Eqs. (3.151) and (2.106). See also the related Eq. (7.59) derived in Exercise 7.5.

5.2.1.3 *Defect represented by body forces*

Now consider the case where the defect is mimicked by an array of point body forces $\mathbf{F}^{D(\alpha)}$ that produces the stress field σ_{ij}^{D}, as in the force multipole models of Sec. 10.3. In this case forces as well as strain energy are involved, and we must therefore include the potential energy of the forces and write the interaction energy as

$$E_{int}^{D/I} = W_{int}^{D/I} + \Phi_{int}^{D/I} = (W^{D+I} - W^{D} - W^{I}) + (\Phi^{D+I} - \Phi^{I}), \quad (5.15)$$

where $\Phi_{int}^{D/I}$ is the interaction potential energy. As indicated by Eq. (5.25), the interaction strain energy between the applied stress field due to the D body forces and the internal I stress field vanishes. The remaining interaction potential energy can then be obtained by imagining that the D field is first established in the body. Then, when the I field is introduced, the displacements associated with the I field will cause the body forces associated with the D field to be displaced by u_i^I. Their potential energy will therefore change by $-\sum_\alpha \mathbf{u}^I(\mathbf{x}^{(\alpha)}) \cdot \mathbf{F}^{D(\alpha)}$, where $\mathbf{x}^{(\alpha)}$ is the field vector to the point force $\mathbf{F}^{D(\alpha)}$. According to Eq. (5.15), the interaction energy is then

$$\begin{aligned} E_{int}^{D/I} &= \Phi_{int}^{D/I} = -\sum_{\alpha} \mathbf{u}^I(\mathbf{x}^{(\alpha)}) \cdot \mathbf{F}^{D(\alpha)}, \\ E_{int}^{D/I} &= -\iiint_{\mathcal{V}} \mathbf{u}^I(\mathbf{x}) \cdot \mathbf{f}^D(\mathbf{x}) dV, \end{aligned} \quad (5.16)$$

where the second expression holds when the forces are in the form of a continuous distribution in the region $\mathcal{V}$.

5.2.2 *Interaction energy with applied stress*

We now assume in Fig. 5.1 that the I singularity is absent and that only the defect and the applied tractions (forces) are present.

5.2.2.1 *Defect represented by its stress field*

Since both strain energy and the potential energy of applied forces are involved, the interaction energy is given by Eq. (5.15). The strain energy term takes the form obtained previously in Eq. (5.6), while the potential energy term is given by $-\oiint_{\mathcal{S}^\circ} \sigma_{ij}^{A} u_j^{D} \hat{n}_i dS$, which is the change in potential energy of the A system when the D field is added to the A system to form the D + A system. Therefore,

$$\begin{aligned} E_{int}^{D/A} &= \frac{1}{2}\oiiint_{\mathcal{V}^\circ} (\sigma_{ij}^{D}\varepsilon_{ij}^{A} + \sigma_{ij}^{A}\varepsilon_{ij}^{D}) dV - \oiint_{\mathcal{S}^\circ} \sigma_{ij}^{A} u_j^{D} \hat{n}_i dS \\ &= \oiiint_{\mathcal{V}} \sigma_{ij}^{D}\varepsilon_{ij}^{A} dV + \oiiint_{\mathcal{V}^\circ - \mathcal{V}} \sigma_{ij}^{A}\varepsilon_{ij}^{D} dV - \oiint_{\mathcal{S}^\circ} \sigma_{ij}^{A} u_j^{D} \hat{n}_i dS, \end{aligned} \tag{5.17}$$

where, as previously in Sec. 5.2.1.1, the volume integral is taken as the sum of integrals over the regions $\mathcal{V}$ and $\mathcal{V}^\circ - \mathcal{V}$, so that only corresponding fields are involved. Then, applying Eq. (2.106) and the divergence theorem,

$$\begin{aligned} E_{int}^{D/A} &= \oiiint_{\mathcal{V}} \frac{\partial}{\partial x_j}(\sigma_{ij}^{D} u_i^{A}) dV + \oiiint_{\mathcal{V}^\circ - \mathcal{V}} \frac{\partial}{\partial x_j}(\sigma_{ij}^{A} u_i^{D}) dV - \oiint_{\mathcal{S}^\circ} \sigma_{ij}^{A} u_i^{D} \hat{n}_j dS \\ &= \oiint_{\mathcal{S}} \sigma_{ij}^{D} u_i^{A} \hat{n}_j dS + \oiint_{\mathcal{S}^\circ} \sigma_{ij}^{A} u_i^{D} \hat{n}_j dS \\ &\quad - \oiint_{\mathcal{S}} \sigma_{ij}^{A} u_i^{D} \hat{n}_j dS - \oiint_{\mathcal{S}^\circ} \sigma_{ij}^{A} u_i^{D} \hat{n}_j dS \\ &= \oiint_{\mathcal{S}} (\sigma_{ij}^{D} u_i^{A} - \sigma_{ij}^{A} u_i^{D}) \hat{n}_j dS. \end{aligned} \tag{5.18}$$

Equation (5.18) can be put into another form by using the same procedure employed to obtain Eqs. (5.11) and Eq. (5.12), i.e.,

$$\begin{aligned} E_{int}^{D/A} &= \oiint_{\mathcal{S}} (\sigma_{ij}^{D} u_i^{A} - \sigma_{ij}^{A} u_i^{D}) \hat{n}_j dS \\ &= \oiint_{\mathcal{S}} \left(\sigma_{ij}^{D^\infty} u_i^{A} - \sigma_{ij}^{A} u_i^{D^\infty}\right) \hat{n}_j dS + \oiint_{\mathcal{S}} \left(\sigma_{ij}^{D^{IM}} u_i^{A} - \sigma_{ij}^{A} u_i^{D^{IM}}\right) \hat{n}_j dS \\ &= \oiint_{\mathcal{S}} \left(\sigma_{ij}^{D^\infty} u_i^{A} - \sigma_{ij}^{A} u_i^{D^\infty}\right) \hat{n}_j dS = E_{int}^{D^\infty/A}, \end{aligned} \tag{5.19}$$

where we have introduced $E_{int}^{D^\infty/A}$ given by

$$E_{int}^{D^\infty/A} = \oiint_{\mathcal{S}} \left(\sigma_{ij}^{D^\infty} u_i^A - \sigma_{ij}^A u_i^{D^\infty}\right) \hat{n}_j dS = E_{int}^{D/A}. \tag{5.20}$$

Therefore, $E_{int}^{D/A} = E_{int}^{D^\infty/A}$, and the interaction energy between the defect and the A field takes several equivalent forms.

To investigate the interaction strain energy between an applied field (represented here by the A field) and an internal field (represented by the D field), we first write Eq. (5.18) as

$$\begin{aligned} E_{int}^{D/A} &= \oiint_{\mathcal{S}} \sigma_{ij}^D u_i^A \hat{n}_j dS + \oiint_{\mathcal{S}^\circ} \sigma_{ij}^A u_i^D \hat{n}_j dS \\ &\quad - \oiint_{\mathcal{S}} \sigma_{ij}^A u_i^D \hat{n}_j dS - \oiint_{\mathcal{S}^\circ} \sigma_{ij}^A u_i^D \hat{n}_j dS \\ &= W_{int}^{D/A} + \Phi_{int}^{D/A}, \end{aligned} \tag{5.21}$$

where $W_{int}^{D/A}$ and $\Phi_{int}^{D/A}$ are identified, respectively, as

$$W_{int}^{D/A} = \oiint_{\mathcal{S}} \sigma_{ij}^D u_i^A \hat{n}_j dS + \oiint_{\mathcal{S}^\circ} \sigma_{ij}^A u_i^D \hat{n}_j dS - \oiint_{\mathcal{S}} \sigma_{ij}^A u_i^D \hat{n}_j dS \tag{5.22}$$

and

$$\Phi_{int}^{D/A} = -\oiint_{\mathcal{S}^\circ} \sigma_{ij}^A u_i^D \hat{n}_j dS. \tag{5.23}$$

Then, recognizing that $\mathcal{S}$ can be arbitrarily expanded, or contracted, as long as it does not sweep across a singularity, we expand $\mathcal{S}$ so that it is infinitesimally close to $\mathcal{S}^\circ$. Equation (5.22) then takes the form

$$\begin{aligned} W_{int}^{D/A} &= \oiint_{\mathcal{S}^\circ} \sigma_{ij}^D u_i^A \hat{n}_j dS + \oiint_{\mathcal{S}^\circ} \sigma_{ij}^A u_i^D \hat{n}_j dS - \oiint_{\mathcal{S}^\circ} \sigma_{ij}^A u_i^D \hat{n}_j dS \\ &= \oiint_{\mathcal{S}^\circ} \sigma_{ij}^D u_i^A \hat{n}_j dS = 0. \end{aligned} \tag{5.24}$$

after applying the condition $\sigma_{ij}^D \hat{n}_j = 0$ on $\mathcal{S}^\circ$.

This result establishes the important conclusion:

The interaction strain energy between an internal stress field, I, and an applied stress field, A, vanishes, i.e.,

$$W_{int}^{I/A} = 0 \tag{5.25}$$

5.2.2.2 *Defect represented by a transformation strain*

The interaction energy between a defect, represented by a transformation strain, and an internal stress, I, has been derived in the form of Eq. (5.13) after starting with the expression for the interaction energy given by Eq. (5.10). When the stress is the applied stress, A, the same type of derivation can be employed, starting in this case with Eq. (5.18), and leading to the result

$$\mathrm{E}_{\mathrm{int}}^{\mathrm{D/A}} = -\oiiint_{\mathcal{V}^{\mathrm{D}}} \sigma_{\mathrm{ij}}^{\mathrm{A}} \varepsilon_{\mathrm{ij}}^{\mathrm{T}} \mathrm{dV}, \tag{5.26}$$

which is of the same form as Eq. (5.13).

5.2.2.3 *Defect represented by body forces*

Since both forces and strain energy are involved, Eq. (5.15) applies, and

$$\mathrm{E}_{\mathrm{int}}^{\mathrm{D/A}} = \mathrm{W}_{\mathrm{int}}^{\mathrm{D/A}} + \Phi_{\mathrm{int}}^{\mathrm{D/A}}. \tag{5.27}$$

Then, as shown in Exercise 5.8, $\mathrm{W}_{\mathrm{int}}^{\mathrm{D/A}}$ vanishes, and $\Phi_{\mathrm{int}}^{\mathrm{D/A}}$ has the form of Eq. (5.16), after replacing I with A. $\mathrm{E}_{\mathrm{int}}^{\mathrm{D/A}}$ is then, finally, of the form

$$\begin{aligned} \mathrm{E}_{\mathrm{int}}^{\mathrm{D/A}} &= \Phi_{\mathrm{int}}^{\mathrm{D/A}} = -\sum_{\alpha} \mathbf{u}^{\mathrm{A}}(\mathbf{x}^{(\alpha)}) \cdot \mathbf{F}^{\mathrm{D}(\alpha)}, \\ \mathrm{E}_{\mathrm{int}}^{\mathrm{D/A}} &= -\oiiint_{\mathcal{V}} \mathbf{u}^{\mathrm{A}}(\mathbf{x}) \cdot \mathbf{f}^{\mathrm{D}}(\mathbf{x}) \mathrm{dV}, \end{aligned} \tag{5.28}$$

where the second expression holds when the forces are in the form of a continuous distribution in the region $\mathcal{V}$.

5.2.3 *Interaction energy with defect image stress*

5.2.3.1 *Defect represented by its stress field*

To determine the interaction energy between a defect and its image stress, $\mathrm{E}_{\mathrm{int}}^{\mathrm{D/D^{IM}}}$, we employ a finite body that contains only the defect stress field, $\sigma_{\mathrm{ij}}^{\mathrm{D}}$, and its image stress $\sigma_{\mathrm{ij}}^{\mathrm{D^{IM}}}$ as illustrated in Fig. 5.2.

The determination of $\mathrm{E}_{\mathrm{int}}^{\mathrm{D/D^{IM}}}$ is then similar to the determination of the interaction energy, $\mathrm{E}_{\mathrm{int}}^{\mathrm{D/A}}$, between a defect and the applied stress system, A, carried out in Sec. 5.2.2.1. Equation (5.17), again applies with A replaced

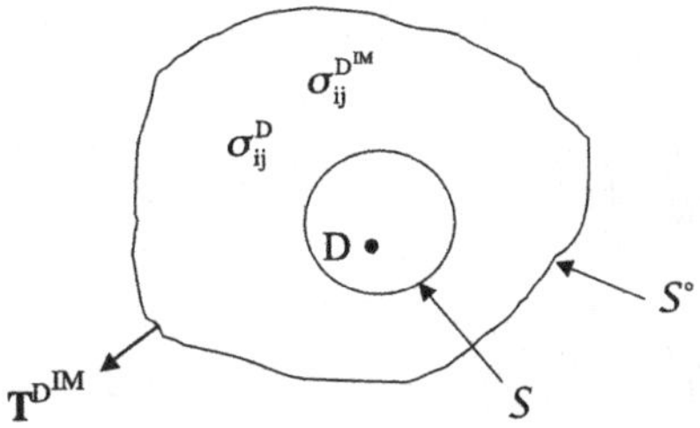

Fig. 5.2 Finite body containing σ_{ij}^{D} and $\sigma_{ij}^{D^{IM}}$ stresses due to presence of the defect, D, and the image tractions $T_i^{D^{IM}} = \sigma_{ij}^{D^{IM}}\hat{n}_j$. System is obtained by first adding the image stress through the application of the tractions $T_i^{D^{IM}}$ to $\mathcal{S}^{\circ}$, and then adding the defect to produce the σ_{ij}^{D} stress.

by D^{IM} so that the strain energy term is given by

$$\oiiint_{\mathcal{V}} \sigma_{ij}^{D}\varepsilon_{ij}^{D^{IM}}dV + \oiiint_{\mathcal{V}^{\circ}-\mathcal{V}} \sigma_{ij}^{D^{IM}}\varepsilon_{ij}^{D}dV \tag{5.29}$$

and the potential energy term, associated with the $T^{D^{IM}}$ tractions, is

$$-\oiint_{\mathcal{S}^{\circ}} \sigma_{ij}^{D^{IM}}u_i^{D}\hat{n}_j dS. \tag{5.30}$$

Therefore, by use of Eq. (2.106),

$$\begin{aligned} E_{int}^{D/D^{IM}} &= \oiiint_{\mathcal{V}} \frac{\partial}{\partial x_j}\left(\sigma_{ij}^{D}\varepsilon_{ij}^{D^{IM}}\right)dV \\ &\quad + \oiiint_{\mathcal{V}^{\circ}-\mathcal{V}} \frac{\partial}{\partial x_j}\left(\sigma_{ij}^{D^{IM}}\varepsilon_{ij}^{D}\right)dV - \oiint_{\mathcal{S}^{\circ}} \sigma_{ij}^{D^{IM}}u_i^{D}\hat{n}_j dS. \end{aligned} \tag{5.31}$$

Next, using the divergence theorem,

$$\begin{aligned} E_{int}^{D/D^{IM}} &= \oiint_{\mathcal{S}} \sigma_{ij}^{D}u_i^{D^{IM}}\hat{n}_j dS + \oiint_{\mathcal{S}^{\circ}} \sigma_{ij}^{D^{IM}}u_i^{D}\hat{n}_j dS \\ &\quad - \oiint_{\mathcal{S}} \sigma_{ij}^{D^{IM}}u_i^{D}\hat{n}_j dS - \oiint_{\mathcal{S}^{\circ}} \sigma_{ij}^{D^{IM}}u_i^{D}\hat{n}_j dS \\ &= \oiint_{\mathcal{S}} \left(\sigma_{ij}^{D}u_i^{D^{IM}} - \sigma_{ij}^{D^{IM}}u_i^{D}\right)\hat{n}_j dS. \end{aligned} \tag{5.32}$$

Finally, using the same procedure employed to obtain Eq. (5.19), we write

$$\begin{aligned}
E_{int}^{D/D^{IM}} &= \oiint_{S} \left(\sigma_{ij}^{D} u_i^{D^{IM}} - \sigma_{ij}^{D^{IM}} u_i^{D}\right)\hat{n}_j dS \\
&= \oiint_{S} \left(\sigma_{ij}^{D^\infty} u_i^{D^{IM}} - \sigma_{ij}^{D^{IM}} u_i^{D^\infty}\right)\hat{n}_j dS \\
&\quad + \oiint_{S} \left(\sigma_{ij}^{D^{IM}} u_i^{D^{IM}} - \sigma_{ij}^{D^{IM}} u_i^{D^{IM}}\right)\hat{n}_j dS \\
&= \oiint_{S} \left(\sigma_{ij}^{D^\infty} u_i^{D^{IM}} - \sigma_{ij}^{D^{IM}} u_i^{D^\infty}\right)\hat{n}_j dS = E_{int}^{D^\infty/D^{IM}}. \qquad (5.33)
\end{aligned}$$

so that $E_{int}^{D/D^{IM}} = E_{int}^{D^\infty/D^{IM}}$.

5.2.4 *Summary*

When the defect is represented by its stress field, the defect interaction energies with internal, applied, and image stress are given by Eqs. (5.11), (5.19) and (5.33), respectively. All are of the same general form and involve integration over a closed surface, $\mathcal{S}$, of arbitrary shape that encloses the defect but excludes the source of the imposed stress (which, in the case of an image stress can be regarded as the body surface).

When the defect is represented by a transformation strain, the defect interaction energies with internal and applied stress are given by Eqs. (5.13) and (5.26), respectively. They are of the same general form and involve integration of the product of the transformation strain and the imposed stress over the volume of the defect, i.e., the region containing the transformation strain.

When the defect is represented by an array of forces, as in Chap. 10, the defect interaction energies with internal and applied stress are given by Eqs. (5.16) and (5.28), respectively. They are of the same general form and consist of the sum of the products of the defect forces and the displacements caused by the imposed stress field.

5.3 Forces on A Defect Source of Stress in Finite Homogeneous Body

5.3.1 *General formulation*

We now turn to the forces imposed on defects by imposed internal, applied and image stresses. Consider again the defect, D, in the presence of the I, A, and D^{IM} fields in the finite homogeneous body shown in Fig. 5.1, where $\mathbf{x}$ is the usual field vector, and $\boldsymbol{\xi}$ is the vector indicating the position of the defect. The stresses and displacements of the elastic field of the defect must therefore be of the functional form

$$\sigma_{ij}^{D}(\mathbf{x},\boldsymbol{\xi}) = \sigma_{ij}^{D^{\infty}}(\mathbf{x}-\boldsymbol{\xi}) + \sigma_{ij}^{D^{IM}}(\mathbf{x},\boldsymbol{\xi}) \tag{5.34}$$

and

$$u_i^{D}(\mathbf{x},\boldsymbol{\xi}) = u_i^{D^{\infty}}(\mathbf{x}-\boldsymbol{\xi}) + u_i^{D^{IM}}(\mathbf{x},\boldsymbol{\xi}), \tag{5.35}$$

since the field properties of the defect in the finite body must depend upon the field vector, $\mathbf{x}$, as well as on $\boldsymbol{\xi}$ (which are independent of each other), while in an infinite body the field properties depend only upon the vector displacement from the defect, i.e., the vector difference $(\mathbf{x}-\boldsymbol{\xi})$. Therefore,

$$\frac{\partial\sigma_{ij}^{D^{\infty}}}{\partial x_{\alpha}} = -\frac{\partial\sigma_{ij}^{D^{\infty}}}{\partial\xi_{\alpha}} \tag{5.36}$$

and

$$\frac{\partial u_i^{D^{\infty}}}{\partial x_{\alpha}} = -\frac{\partial u_i^{D^{\infty}}}{\partial\xi_{\alpha}}. \tag{5.37}$$

The force exerted on the defect by any of these fields, X, in the α direction, $F_{\alpha}^{D/X}$, can be obtained by giving the defect an infinitesimal virtual displacement $\delta\xi_{\alpha}$ and determining the accompanying change in the total energy of the system. The force is then

$$F_{\alpha}^{D/X} = -\lim_{\delta\xi_{\alpha}\to 0}\frac{1}{\delta\xi_{\alpha}}(E'-E) = -\lim_{\delta\xi_{\alpha}\to 0}\frac{1}{\delta\xi_{\alpha}}\left(\frac{\partial E}{\partial\xi_{\alpha}}\delta\xi_{\alpha}\right) = -\frac{\partial E}{\partial\xi_{\alpha}}, \tag{5.38}$$

where the prime indicates the system after the displacement. Since linear elasticity applies, the total force if several fields are present, as in Fig. 5.1, is simply the sum of the individual forces, i.e.,

$$F_{\alpha} = F_{\alpha}^{D/I} + F_{\alpha}^{D/A} + F_{\alpha}^{D/D^{IM}}. \tag{5.39}$$

Alternatively, as demonstrated below, the force can be determined by finding the change in the interaction energy of the defect with the stress field, $E_{int}^{D/X}$, as the defect is displaced by $\delta\xi_\alpha$. The force is then defined by

$$F_\alpha^{D/X} = -\lim_{\delta\xi_\alpha \to 0} \frac{1}{\delta\xi_\alpha}\left(E_{int}^{\prime D/X} - E_{int}^{D/X}\right)$$

$$= -\lim_{\delta\xi_\alpha \to 0} \frac{1}{\delta\xi_\alpha}\left(\frac{\partial E_{int}^{D/X}}{\partial \xi_\alpha}\delta\xi_\alpha\right) = -\frac{\partial E_{int}^{D/X}}{\partial \xi_\alpha}. \tag{5.40}$$

5.3.2 *Force obtained from change of the system total energy*

5.3.2.1 *Forces obtained by use of the energy-momentum tensor*

Following Eshelby (1956), the total force exerted on the defect in Fig. 5.1 by the total stress field corresponding to the sum of the internal stress field I, the applied stress field A, and the defect stress field $D = D^\infty + D^{IM}$, is now obtained by applying Eq. (5.38) in a formulation that involves the *energy-momentum tensor*. However, it is simplest to start the analysis by omitting the internal stress, I, which will be added later. The first task is to determine the change in the total energy of the finite body when the defect is displaced by $\delta\xi_\alpha$, relative to a fixed point in the body, while keeping the boundary conditions at the surface due to the applied forces constant. This is accomplished in two steps: in step 1 every quantity ϕ associated with the elastic field of the defect throughout the body is replaced by $\phi - (\partial\phi/\partial x_\alpha)\delta\xi_\alpha$, and in step 2 the original boundary conditions at the surface are restored.

The change in strain energy in step 1 is therefore obtained, to first order, by integrating the changes in local strain energy density throughout the volume, and converting the result to a surface integral so that

$$\delta W^{(1)} = -\delta\xi_\alpha \oiiint_{\mathcal{V}^\circ} \frac{\partial w}{\partial \mathbf{x}_\alpha} dV = -\delta\xi_\alpha \oiint_{\mathcal{S}^\circ} w\hat{n}_\alpha dS. \tag{5.41}$$

Eshelby (1951; 1956) has pointed out that in the volume integral in Eq. (5.41) w may become infinite, or $\partial w/\partial x_\alpha$ may be undefined at the defect, thus causing a problem. However, an alternate procedure can be employed, equivalent to the step 1 procedure, in which the defect is displaced by moving the surface with respect to the defect rather than the defect relative to the surface. When the surface is displaced relative to the

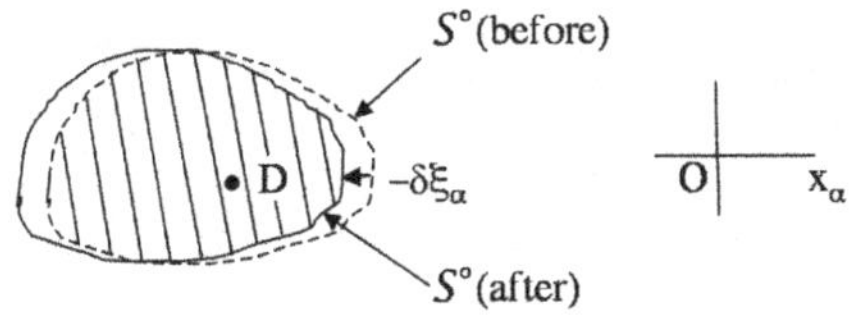

Fig. 5.3 Configuration before, and after, displacing the surface of the body relative to the embedded defect, D, by $-\delta\xi_\alpha$.

defect by $-\delta\xi_\alpha$, as illustrated in Fig. 5.3, $\delta W^{(1)}$ is obtained by integrating over each volume and taking the difference. The contributions from the overlapping hatched region then cancel. Since the defect will generally lie in this region, any problems with singularities in w due to the defect are effectively subtracted out, and Eq. (5.41) should therefore apply.

The strain energy is altered in step 2 because the surface tractions perform work on the body as they are adjusted to restore the boundary conditions. Again expanding to first order, the displacements and tractions at the surface at different stages can be written as

$$\begin{cases} u_i, & \text{(initially)} \\ u_i - \frac{\partial u_i}{\partial x_\alpha}\delta\xi_\alpha, & \text{(after step 1)} \\ u_i'. & \text{(finally, after step 2)} \end{cases}$$
$$\begin{cases} \sigma_{ij}\hat{n}_j, & \text{(initially)} \\ \left(\sigma_{ij} - \frac{\partial \sigma_{ij}}{\partial x_\alpha}\delta\xi_\alpha\right)\hat{n}_j, & \text{(after step 1)} \\ \left(\sigma_{ij} - \frac{\partial \sigma_{ij}}{\partial x_\alpha}\delta\xi_\alpha\right)\hat{n}_j + \frac{\partial \sigma_{ij}}{\partial x_\alpha}\delta\xi_\alpha\hat{n}_j = \sigma_{ij}\hat{n}_j. & \text{(finally, after step 2)} \end{cases} \tag{5.42}$$

The change in strain energy during step 2 is then obtained by integrating the work done by the tractions over $\mathcal{S}^\circ$. According to Eq. (5.42), the traction versus displacement curve during step 2 will appear as in Fig. 5.4, and the work performed is therefore

$$\begin{aligned} \delta W^{(2)} = \delta\mathcal{W}^{(2)} = \oiint_{\mathcal{S}^\circ} \Bigg[& \sigma_{ij}(u_i' - u_i) + \sigma_{ij}\frac{\partial u_i}{\partial x_\alpha}\delta\xi_{\alpha j} \\ & - (u_i' - u_i)\frac{\partial \sigma_{ij}}{\partial x_\alpha}\delta\xi_\alpha - \frac{\partial \sigma_{ij}}{\partial x_\alpha}\frac{\partial u_i}{\partial x_\alpha}(\delta\xi_\alpha)^2 \\ & + \frac{1}{2}(u_i' - u_i)\frac{\partial \sigma_{ij}}{\partial x_\alpha}\delta\xi_\alpha + \frac{1}{2}\frac{\partial \sigma_{ij}}{\partial x_\alpha}\frac{\partial u_i}{\partial x_\alpha}(\delta\xi_\alpha)^2 \Bigg] \hat{n}_j dS. \end{aligned} \tag{5.43}$$

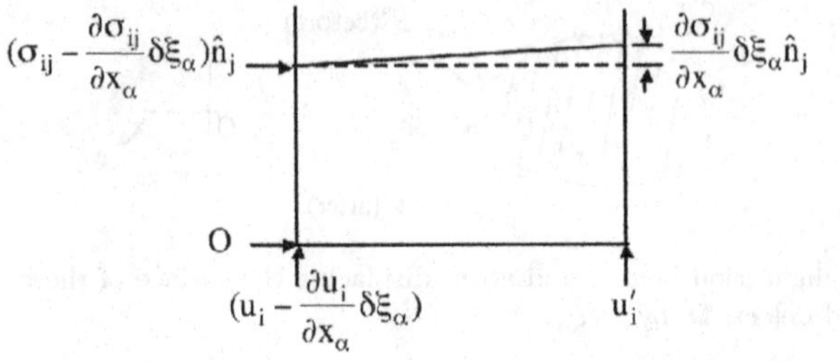

Fig. 5.4 Curve of traction versus displacement during step 2.

Then, recognizing that $(u_i' - u_i)$ is of order $\delta\xi_\alpha$, and dropping terms of order $(\delta\xi_\alpha)^2$,

$$\delta W^{(2)} = \oiint_{\mathcal{S}^\circ} \sigma_{ij}\left[(u_i' - u_i) + \frac{\partial u_i}{\partial x_\alpha}\delta\xi_\alpha\right]\hat{n}_j dS. \tag{5.44}$$

The change in potential energy of the tractions that occurs during steps 1 and 2 is given by

$$\delta\Phi^{(1)+(2)} = -\oiint_{\mathcal{S}^\circ} \sigma_{ij}(u_i' - u_i)\hat{n}_j dS. \tag{5.45}$$

and the change in total energy is therefore

$$\begin{aligned}\delta E = E' - E &= \delta W^{(1)} + \delta W^{(2)} + \delta\Phi^{(1)+(2)} \\ &= -\delta\xi_\alpha \oiint_{\mathcal{S}^\circ}\left(w\hat{n}_\alpha - \sigma_{ij}\frac{\partial u_i}{\partial x_\alpha}\hat{n}_j\right)dS. \end{aligned}\tag{5.46}$$

Using Eq. (5.38), the total force along x_α is then

$$F_\alpha = \oiint_{\mathcal{S}^\circ}\left(w\delta_{\alpha j} - \sigma_{ij}\frac{\partial u_i}{\partial x_\alpha}\right)\hat{n}_j dS. \tag{5.47}$$

A simpler expression is obtained by realizing that the above surface integral can be taken over any surface in the body, $\mathcal{S}$, that encloses D. This may be shown by first writing Eq. (5.47) in the form

$$F_\alpha = \oiint_{\mathcal{S}^\circ - \mathcal{S}}\left(w\delta_{\alpha j} - \sigma_{ij}\frac{\partial u_i}{\partial x_\alpha}\right)\hat{n}_j dS + \oiint_{\mathcal{S}}\left(w\delta_{\alpha j} - \sigma_{ij}\frac{\partial u_i}{\partial x_\alpha}\right)\hat{n}_j dS. \tag{5.48}$$

Then, applying the divergence theorem to the first term and setting it equal to ΔF_α,

$$\Delta F_\alpha = -\oiiint_{\mathcal{V}^\circ - \mathcal{V}} \frac{\partial}{\partial x_j}\left(w\delta_{\alpha j} - \sigma_{ij}\frac{\partial u_i}{\partial x_\alpha}\right) dV$$

$$= -\oiiint_{\mathcal{V}^\circ - \mathcal{V}} \left(\frac{\partial w}{\partial x_\alpha} - \sigma_{ij}\frac{\partial^2 u_i}{\partial x_j \partial x_\alpha}\right) dV \tag{5.49}$$

after using Eq. (2.65). The integrand in Eq. (5.49) over the region $\mathcal{V}^\circ - \mathcal{V}$, which is free of singularities, takes the form

$$\left(\frac{\partial w}{\partial x_\alpha} - \sigma_{ij}\frac{\partial^2 u_i}{\partial x_j \partial x_\alpha}\right) = \frac{1}{2}\left(\frac{\partial u_i}{\partial x_j}\frac{\partial \sigma_{ij}}{\partial x_\alpha} - \sigma_{ij}\frac{\partial^2 u_i}{\partial x_\alpha \partial x_j}\right) \tag{5.50}$$

after substituting Eq. (2.133), and vanishes when Eqs. (2.75) and (2.5) are substituted and the symmetry properties of the C_{ijkl} tensor are invoked. Therefore, $\Delta F_\alpha = 0$, and Eq. (5.48) can be written (Eshelby 1956) as

$$F_\alpha = \oiint_{\mathcal{S}} P_{j\alpha}\hat{n}_j dS, \tag{5.51}$$

where $P_{j\alpha}$, known as the *energy-momentum tensor* (Eshelby 1975), is given by

$$P_{j\alpha} = w\delta_{\alpha j} - \sigma_{ij}\frac{\partial u_i}{\partial x_\alpha}. \tag{5.52}$$

So far, the source of internal stress, I, in Fig. 5.1 has been omitted. However, this can be remedied on the basis of an argument by Eshelby (1956). Recall that the total energy was partitioned into the strain energy in the body and the potential energy of the forces producing the applied stress. However, it could equally well have been partitioned into the strain energy within $\mathcal{S}$ and the sum of the strain energy outside of $\mathcal{S}$ and the potential energy of the applied forces. By locating I outside of $\mathcal{S}$ and carrying out the derivation again, taking the "surface" of the body to be $\mathcal{S}$ and everything outside of $\mathcal{S}$ to be the agency producing "surface" forces on $\mathcal{S}$, the same result would have been obtained. Equation (5.51) is therefore valid when $\mathcal{S}$ encloses only D, as in Fig. 5.1, and the total stress, σ_{ij}, includes all stresses acting in the system.

Equation (5.51), involving the energy-momentum tensor, is quite general and applies to a wide range of defects. For example, it is demonstrated in Sec. 9.3 (see also Exercise 9.4) that it is valid for an inhomogeneity, even though an inhomogeneity, by itself, is not a source of stress.

The total force on the defect D in Fig. 5.1 can now be obtained by first writing Eq. (5.52) as

$$P_{j\alpha} = \frac{1}{2}\sigma_{ik}\frac{\partial u_i}{\partial x_k}\delta_{\alpha j} - \sigma_{ij}\frac{\partial u_i}{\partial x_\alpha}$$

with the help of Eq. (2.133), and then substituting the expressions for σ_{ij} and u_i given by Eq. (5.1). Then, after applying Eqs. (B.11) (Stokes' theorem) and (2.106), and putting the result into Eq. (5.51), the total force can be written as the sum

$$F_\alpha = F_\alpha^{D^\infty/D^\infty} + F_\alpha^{D^{IM}/D^{IM}} + F_\alpha^{I/I} + F_\alpha^{A/A} + F_\alpha^{D^\infty/D^{IM}} + F_\alpha^{D^\infty/I} + F_\alpha^{D^\infty/A} + F_\alpha^{D^{IM}/I} + F_\alpha^{D^{IM}/A} + F_\alpha^{I/A}, \tag{5.53}$$

where the diagonal and off-diagonal (cross) terms have the forms

$$F_\alpha^{X/X} = \frac{1}{2}\oiint_{\mathcal{S}}\left(\frac{\partial \sigma_{ij}^X}{\partial x_\alpha}u_i^X - \sigma_{ij}^X\frac{\partial u_i^X}{\partial x_\alpha}\right)\hat{n}_j dS \tag{5.54}$$

and

$$F_\alpha^{X/Y} = \oiint_{\mathcal{S}}\left(\frac{\partial \sigma_{ij}^X}{\partial x_\alpha}u_i^Y - \sigma_{ij}^Y\frac{\partial u_i^X}{\partial x_\alpha}\right)\hat{n}_j dS \tag{5.55}$$

respectively. However, as now seen, all terms in Eq. (5.53) vanish except the three cross terms $F_\alpha^{D^\infty/D^{IM}}$, $F_\alpha^{D^\infty/I}$ and $F_\alpha^{D^\infty/A}$. The integral representing the D^∞/D^∞ term is a constant independent of the choice of $\mathcal{S}$ by virtue of Eq. (2.113). However, it vanishes, since when $\mathcal{S}$ is expanded to infinity it is expected that $\sigma_{ij}^{D^\infty} \to 0$ at least as rapidly as x^{-2} in three dimensions and x^{-1} in two-dimensions. The $D^{IM}/D^{IM}, D^I/D^I, D^A/D^A, D^{IM}/D^I, D^{IM}/D^A$ and D^I/D^A terms vanish by virtue of Eq. (2.112) since in each case the two fields involved are corresponding fields throughout the volume enclosed by $\mathcal{S}$.

The total force given by Eq. (5.53) is therefore reduced to

$$F_\alpha = F_\alpha^{D^\infty/D^{IM}} + F_\alpha^{D^\infty/I} + F_\alpha^{D^\infty/A} \tag{5.56}$$

where (Eshelby 1956)

$$F_\alpha^{D^\infty/I} = F_\alpha^{D/I} = \oiint_{\mathcal{S}}\left(\frac{\partial \sigma_{ij}^{D^\infty}}{\partial x_\alpha}u_i^I - \sigma_{ij}^I\frac{\partial u_i^{D^\infty}}{\partial x_\alpha}\right)\hat{n}_j dS$$

$$= \oiint_{\mathcal{S}}\left(\frac{\partial \sigma_{ij}^D}{\partial x_\alpha}u_i^I - \sigma_{ij}^I\frac{\partial u_i^D}{\partial x_\alpha}\right)\hat{n}_j dS$$

$$F_\alpha^{D^\infty/A} = F_\alpha^{D/A} = \oiint_S \left(\frac{\partial \sigma_{ij}^{D^\infty}}{\partial x_\alpha} u_i^A - \sigma_{ij}^A \frac{\partial u_i^{D^\infty}}{\partial x_\alpha} \right) \hat{n}_j dS$$

$$= \oiint_S \left(\frac{\partial \sigma_{ij}^{D}}{\partial x_\alpha} u_i^A - \sigma_{ij}^A \frac{\partial u_i^{D}}{\partial x_\alpha} \right) \hat{n}_j dS$$

$$F_\alpha^{D^\infty/D^{IM}} = F_\alpha^{D/D^{IM}} = \oiint_S \left(\frac{\partial \sigma_{ij}^{D^\infty}}{\partial x_\alpha} u_i^{D^{IM}} - \sigma_{ij}^{D^{IM}} \frac{\partial u_i^{D^\infty}}{\partial x_\alpha} \right) \hat{n}_j dS$$

$$= \oiint_S \left(\frac{\partial \sigma_{ij}^{D}}{\partial x_\alpha} u_i^{D^{IM}} - \sigma_{ij}^{D^{IM}} \frac{\partial u_i^{D}}{\partial x_\alpha} \right) \hat{n}_j dS \qquad (5.57)$$

and, where the relationships

$$F_\alpha^{D^\infty/I} = F_\alpha^{D/I}$$
$$F_\alpha^{D^\infty/A} = F_\alpha^{D/A}$$
$$F_\alpha^{D^\infty/D^{IM}} = F_\alpha^{D/D^{IM}} \qquad (5.58)$$

are valid by virtue of Eq. (2.112).[3] For example, in the case of $F_\alpha^{D^\infty/I}$,

$$F_\alpha^{D^\infty/I} = \oiint_S \left(\frac{\partial \sigma_{ij}^{D^\infty}}{\partial x_\alpha} u_i^I - \sigma_{ij}^I \frac{\partial u_i^{D^\infty}}{\partial x} \right) \hat{n}_j dS$$

$$= \oiint_S \left(\frac{\partial \sigma_{ij}^{D}}{\partial x_\alpha} u_i^I - \sigma_{ij}^I \frac{\partial u_i^{D}}{\partial x_\alpha} \right) \hat{n}_j dS$$

$$- \oiint_S \left(\frac{\partial \sigma_{ij}^{D^{IM}}}{\partial x_\alpha} u_i^I - \sigma_{ij}^I \frac{\partial u_i^{D^{IM}}}{\partial x_\alpha} \right) \hat{n}_j dS$$

$$= \oiint_S \left(\frac{\partial \sigma_{ij}^{D}}{\partial x_\alpha} u_i^I - \sigma_{ij}^I \frac{\partial u_i^{D}}{\partial x_\alpha} \right) \hat{n}_j dS = F_\alpha^{D/I}, \qquad (5.59)$$

where the integral containing the image quantities vanishes because the D^{IM} and I fields are corresponding fields throughout $\mathcal{V}$.

[3]Note that the results for the forces involving the I, A and D^{IM} fields given by Eq. (5.57) form a pattern consistent with the corresponding results for the interaction energies involving these quantities given earlier by Eqs. (5.11), (5.19) and (5.53), respectively.

Finally, in the derivation of Eq. (5.55) we could, by virtue of Eq. (2.106), just as well have written the result with X and Y interchanged. Therefore,

$$\begin{aligned} F_\alpha^{X/Y} &= \oiint_S \left(\frac{\partial \sigma_{ij}^X}{\partial x_\alpha} u_i^Y - \sigma_{ij}^Y \frac{\partial u_i^X}{\partial x_\alpha} \right) \hat{n}_j dS \\ &= \oiint_S \left(\frac{\partial \sigma_{ij}^Y}{\partial x_\alpha} u_i^X - \sigma_{ij}^X \frac{\partial u_i^Y}{\partial x_\alpha} \right) \hat{n}_j dS = F_\alpha^{Y/X}. \end{aligned} \tag{5.60}$$

Confirmation of this relationship is obtained in Exercise 5.1.

5.3.2.2 *Forces obtained directly from changes in total system energy*

Instead of employing the energy-momentum approach, as previously, the force due to each type of stress can be considered individually and found by the use of Eq. (5.38) which involves determining directly the change in total energy as the defect is displaced (Eshelby 1951). This is demonstrated below by considering first the force due to an applied stress and then the force due to an image stress.

Force due to applied stress

Consider again the system in Fig. 5.1 and take into account only the interaction between the defect and the applied stress system, A.[4] When the defect is displaced by the distance $\delta\xi_\alpha$, the surface tractions perform the work

$$\delta\mathcal{W} = \oiint_{S^\circ} T_i^A \frac{\partial u_i^D}{\partial \xi_\alpha} \delta\xi_\alpha dS = \oiint_{S^\circ} \sigma_{ij}^A \frac{\partial u_i^D}{\partial \xi_\alpha} \delta\xi_\alpha \hat{n}_j dS \tag{5.61}$$

corresponding to a change in the potential energy of the system $\delta\Phi = -\delta\mathcal{W}$. Since the D field is internal, and the A field is applied, there is no interaction strain energy between them according to Eq. (5.25), and, therefore, $\delta E = \delta\Phi = -\delta\mathcal{W}$. Therefore, using this relationship and Eqs. (5.38) and (5.61),

$$F_\alpha^{D/A} = -\lim_{\delta\xi \to 0} \frac{\delta E}{\delta \xi_\alpha} = \oiint_{S^\circ} \sigma_{ij}^A \frac{\partial u_i^D}{\partial \xi_\alpha} \hat{n}_j dS. \tag{5.62}$$

[4]This is valid, of course, within the framework of linear elasticity.

The surface tractions due to the defect stress field must remain at zero as the defect is displaced, and the boundary conditions

$$\begin{aligned} &\sigma_{ij}^{D}\hat{n}_i = \sigma_{ij}^{D^\infty}\hat{n}_i + \sigma_{ij}^{D^{IM}}\hat{n}_i = 0 \\ &\frac{\partial \sigma_{ij}^{D}}{\partial \xi_\alpha}\hat{n}_i = \frac{\partial \sigma_{ij}^{D^\infty}}{\partial \xi_\alpha}\hat{n}_i + \frac{\partial \sigma_{ij}^{D^{IM}}}{\partial \xi_\alpha}\hat{n}_i = 0 \end{aligned} \qquad (\text{on } \mathcal{S}^\circ) \tag{5.63}$$

must therefore be satisfied on $\mathcal{S}^\circ$. The equation

$$\oiint_{\mathcal{S}^\circ} u_i^A \frac{\partial \sigma_{ij}^{D}}{\partial \xi_\alpha}\hat{n}_j dS = 0 \tag{5.64}$$

is therefore valid, and upon subtracting it from Eq. (5.62),

$$\begin{aligned} F_\alpha^{D/A} &= \oiint_{\mathcal{S}^\circ} \left(\sigma_{ij}^{A}\frac{\partial u_i^{D}}{\partial \xi_\alpha} - u_i^{A}\frac{\partial \sigma_{ij}^{D}}{\partial \xi_\alpha}\right)\hat{n}_j dS \\ &= \oiint_{\mathcal{S}^\circ - \mathcal{S}} \left(\sigma_{ij}^{A}\frac{\partial u_i^{D}}{\partial \xi_\alpha} - u_i^{A}\frac{\partial \sigma_{ij}^{D}}{\partial \xi_\alpha}\right)\hat{n}_j dS \\ &\quad + \oiint_{\mathcal{S}} \left(\sigma_{ij}^{A}\frac{\partial u_i^{D}}{\partial \xi_\alpha} - u_i^{A}\frac{\partial \sigma_{ij}^{D}}{\partial \xi_\alpha}\right)\hat{n}_j dS \\ &= \oiint_{\mathcal{S}} \left(\sigma_{ij}^{A}\frac{\partial u_i^{D}}{\partial \xi_\alpha} - u_i^{A}\frac{\partial \sigma_{ij}^{D}}{\partial \xi_\alpha}\right)\hat{n}_j dS, \end{aligned} \tag{5.65}$$

where $\mathcal{S}$ is a closed surface within the body that encloses the defect, and the integral over $\mathcal{S}^\circ - \mathcal{S}$ vanishes because of Eq. (5.113). Next, by expressing the D field in the form $D = D^\infty + D^{IM}$,

$$\begin{aligned} F_\alpha^{D/A} &= \oiint_{\mathcal{S}} \left(\sigma_{ij}^{A}\frac{\partial u_i^{D}}{\partial \xi_\alpha} - u_i^{A}\frac{\partial \sigma_{ij}^{D}}{\partial \xi_\alpha}\right)\hat{n}_j dS \\ &= \oiint_{\mathcal{S}} \left(\sigma_{ij}^{A}\frac{\partial u_i^{D^\infty}}{\partial \xi_\alpha} - u_i^{A}\frac{\partial \sigma_{ij}^{D^\infty}}{\partial \xi_\alpha}\right)\hat{n}_j dS \\ &\quad + \oiint_{\mathcal{S}} \left(\sigma_{ij}^{A}\frac{\partial u_i^{D^{IM}}}{\partial \xi_\alpha} - u_i^{A}\frac{\partial \sigma_{ij}^{D^{IM}}}{\partial \xi_\alpha}\right)\hat{n}_j dS. \end{aligned} \tag{5.66}$$

However, the last integral in Eq. (5.66) vanishes by virtue of Eq. (5.113) in which we can replace D by D^{IM} and then shrink $\mathcal{S}^{(1)}$ to disappearance and redesignate $\mathcal{S}^{(2)}$ as $\mathcal{S}$, since the D^{IM} and A fields are not accompanied by singularities and are corresponding fields throughout the body. Then, by applying Eq. (5.37),

$$F_\alpha^{D/A} = \oiint_{\mathcal{S}} \left(\sigma_{ij}^{A} \frac{\partial u_i^{D^\infty}}{\partial \xi_\alpha} - u_i^{A} \frac{\partial \sigma_{ij}^{D^\infty}}{\partial \xi_\alpha} \right) \hat{n}_j dS$$

$$= \oiint_{\mathcal{S}} \left(\frac{\partial \sigma_{ij}^{D^\infty}}{\partial x_\alpha} u_i^{A} - \sigma_{ij}^{A} \frac{\partial u_i^{D^\infty}}{\partial x_\alpha} \right) \hat{n}_j dS \tag{5.67}$$

in agreement with Eq. (5.57).

Force due to image stress

Following Eshelby (1951), start with the defect in an infinite homogeneous body and then cut out the region, $\mathcal{V}^\circ$, corresponding to the desired traction-free body containing the defect, while applying surface forces to maintain the initial stress field. Next, carry out the same operation but with the cutout region displaced relative to the location of the embedded defect by the distance $-\delta\xi_\alpha$. The defect in the latter body will therefore be displaced relative to the body surface, $\mathcal{S}^\circ$, by the distance $\delta\xi_\alpha$ as illustrated in Fig. 5.3. The elastic energy in the latter body (indicated by a prime) minus the energy in the former body, before any relaxation of the applied surface forces is allowed, is then

$$\delta W = W' - W = -\frac{1}{2} \oiint_{\mathcal{S}^\circ} \sigma_{ij}^{D^\infty} \varepsilon_{ij}^{D^\infty} \delta\xi_\alpha (\hat{\mathbf{e}}_\alpha \cdot \hat{\mathbf{n}}) dS$$

$$= -\frac{1}{2} \oiint_{\mathcal{S}^\circ} \sigma_{ij}^{D^\infty} \varepsilon_{ij}^{D^\infty} \delta\xi_\alpha \hat{n}_\alpha dS$$

$$= -\frac{1}{2} \delta\xi_\alpha \oiint_{\mathcal{S}^\circ} \sigma_{ik}^{D^\infty} \varepsilon_{ik}^{D^\infty} \delta_{\alpha j} \hat{n}_j dS. \tag{5.68}$$

When the surface forces are relaxed on the former body to produce a traction-free surface, an image displacement field $u_i^{D^{IM}}$ is introduced, causing a release of energy given by $(1/2) \oiint_{\mathcal{S}^\circ} \sigma_{ij}^{D^\infty} u_i^{D^{IM}} \hat{n}_j dS$. The corresponding energy release for the latter body is then

$$\frac{1}{2} \left(1 - \delta\xi_\alpha \frac{\partial}{\partial \xi_\alpha} \right) \oiint_{\mathcal{S}^\circ} \sigma_{ij}^{D^\infty} u_i^{D^{IM}} \hat{n}_j dS \tag{5.69}$$

and, by substituting these results into Eq. (5.38),

$$
\begin{aligned}
F_\alpha^{D^\infty/D^{IM}} &= -\lim_{\delta\xi_\alpha\to 0}\frac{1}{\delta\xi_\alpha}(E'-E)\\
&= \frac{1}{2}\left(\oiint_{S^\circ}\sigma_{ik}^{D^\infty}\varepsilon_{ik}^{D^\infty}\delta_{\alpha j}\hat{n}_j dS - \frac{\partial}{\partial\xi_\alpha}\oiint_{S^\circ}\sigma_{ij}^{D^\infty}u_i^{D^{IM}}\hat{n}_j dS\right)\\
&= \frac{1}{2}\oiint_{S^\circ}\left(\sigma_{ik}^{D^\infty}\frac{\partial u_i^{D^\infty}}{\partial x_k}\delta_{\alpha j} - \frac{\partial}{\partial\xi_\alpha}\sigma_{ij}^{D^\infty}u_i^{D^{IM}}\right)\hat{n}_j dS.
\end{aligned}
\tag{5.70}
$$

Then, by applying Stokes' theorem, Eq. (B.11), and Eqs. (5.37) and (5.63),

$$
\begin{aligned}
F_\alpha^{D^\infty/D^{IM}} &= \frac{1}{2}\oiint_{S^\circ}\left[\sigma_{ij}^{D^{IM}}\frac{\partial\left(u_i^{D^\infty}+u_i^{D^{IM}}\right)}{\partial\xi_\alpha}\right]\hat{n}_j dS\\
&\quad + \frac{1}{2}\oiint_{S^\circ}\left[\left(u_i^{D^\infty}+u_i^{D^{IM}}\right)\frac{\partial\sigma_{ij}^{D^{IM}}}{\partial\xi_\alpha}\right]\hat{n}_j dS.
\end{aligned}
\tag{5.71}
$$

However, the two integrals in Eq. (5.71) can be shown to be equal by demonstrating that their difference vanishes, i.e.,

$$
\begin{aligned}
&\oiint_{S^\circ}\left[\sigma_{ij}^{D^{IM}}\frac{\partial(u_i^{D^\infty}+u_i^{D^{IM}})}{\partial\xi_\alpha}\right]\hat{n}_j dS - \oiint_{S^\circ}\left[(u_i^{D^\infty}+u_i^{D^{IM}})\frac{\partial\sigma_{ij}^{D^{IM}}}{\partial\xi_\alpha}\right]\hat{n}_j dS\\
&= \oiint_{S^\circ}\left(-\frac{\partial\sigma_{ij}^{D^{IM}}}{\partial\xi_\alpha}u_i^{D^{IM}} + \sigma_{ij}^{D^{IM}}\frac{\partial u_i^{D^{IM}}}{\partial\xi_\alpha}\right)\hat{n}_j dS\\
&\quad + \oiint_{S^\circ}\left(-\frac{\partial\sigma_{ij}^{D^{IM}}}{\partial\xi_\alpha}u_i^{D^\infty} + \sigma_{ij}^{D^{IM}}\frac{\partial u_i^{D^\infty}}{\partial\xi_\alpha}\right)\hat{n}_j dS\\
&= \oiint_{S^\circ}\left(\frac{\partial\sigma_{ij}^{D^\infty}}{\partial\xi_\alpha}u_i^{D^\infty} - \sigma_{ij}^{D^\infty}\frac{\partial u_i^{D^\infty}}{\partial\xi_\alpha}\right)\hat{n}_j dS\\
&= \oiint_{S^\circ}\left(-\frac{\partial\sigma_{ij}^{D^\infty}}{\partial x_\alpha}u_i^{D^\infty} + \sigma_{ij}^{D^\infty}\frac{\partial u_i^{D^\infty}}{\partial x_\alpha}\right)\hat{n}_j dS = 0
\end{aligned}
\tag{5.72}
$$

In Eq. (5.72), the integral on the second line, which involves only the image field, vanishes, since there are no singularities in the body, and Eq. (5.113) then applies after replacing both D and A by D^{IM}, shrinking $\mathcal{S}^{(1)}$ to disappearance and expanding $\mathcal{S}^{(2)}$ to $\mathcal{S}^\circ$. The integrals on the fourth and fifth lines are then obtained by successively applying Eqs. (5.63) and (5.37) to the integral on the third line. Finally, the last integral vanishes because of the argument given previously in deriving Eq. (5.56). The two integrals in Eq. (5.71) have now been shown to be equal, and Eq. (5.71) can then be written as

$$F_{\alpha}^{D^{\infty}/D^{IM}} = \oiint_{S^{\circ}} \left[\sigma_{ij}^{D^{IM}} \frac{\partial \left(u_i^{D^{\infty}} + u_i^{D^{IM}} \right)}{\partial \xi_{\alpha}} \right] \hat{n}_j dS. \tag{5.73}$$

Now, the integral expression

$$\oiint_{S^{\circ}} \frac{\partial \left(\sigma_{ij}^{D^{\infty}} + \sigma_{ij}^{D^{IM}} \right)}{\partial \xi_{\alpha}} u_i^{D^{IM}} \hat{n}_j dS = 0 \tag{5.74}$$

is valid by virtue of Eq. (5.63), and it can then be subtracted from Eq. (5.73) to obtain

$$\begin{aligned}
F_{\alpha}^{D^{\infty}/D^{IM}} &= \oiint_{S^{\circ}} \left[-\frac{\partial \left(\sigma_{ij}^{D^{\infty}} + \sigma_{ij}^{D^{IM}} \right)}{\partial \xi_{\alpha}} u_i^{D^{IM}} + \sigma_{ij}^{D^{IM}} \frac{\partial \left(u_i^{D^{\infty}} + u_i^{D^{IM}} \right)}{\partial \xi_{\alpha}} \right] \hat{n}_j dS \\
&= \oiint_{S^{\circ}} \left(-\frac{\partial \sigma_{ij}^{D^{\infty}}}{\partial \xi_{\alpha}} u_i^{D^{IM}} + \sigma_{ij}^{D^{IM}} \frac{\partial u_i^{D^{\infty}}}{\partial \xi_{\alpha}} \right) \hat{n}_j dS \\
&\quad + \oiint_{S^{\circ}} \left(-\frac{\partial \sigma_{ij}^{D^{IM}}}{\partial \xi_{\alpha}} u_i^{D^{IM}} + \sigma_{ij}^{D^{IM}} \frac{\partial u_i^{D^{IM}}}{\partial \xi_{\alpha}} \right) \hat{n}_j dS \\
&= \oiint_{S^{\circ}} \left(\frac{\partial \sigma_{ij}^{D^{\infty}}}{\partial x_{\alpha}} u_i^{D^{IM}} - \sigma_{ij}^{D^{IM}} \frac{\partial u_i^{D^{\infty}}}{\partial x_{\alpha}} \right) \hat{n}_j dS \\
&= \oiint_{S^{\circ}-S} \left(\frac{\partial \sigma_{ij}^{D^{\infty}}}{\partial x_{\alpha}} u_i^{D^{IM}} - \sigma_{ij}^{D^{IM}} \frac{\partial u_i^{D^{\infty}}}{\partial x_{\alpha}} \right) \hat{n}_j dS \\
&\quad + \oiint_{S} \left(\frac{\partial \sigma_{ij}^{D^{\infty}}}{\partial x_{\alpha}} u_i^{D^{IM}} - \sigma_{ij}^{D^{IM}} \frac{\partial u_i^{D^{\infty}}}{\partial x_{\alpha}} \right) \hat{n}_j dS \\
&= \oiint_{S} \left(\frac{\partial \sigma_{ij}^{D^{\infty}}}{\partial x_{\alpha}} u_i^{D^{IM}} - \sigma_{ij}^{D^{IM}} \frac{\partial u_i^{D^{\infty}}}{\partial x_{\alpha}} \right) \hat{n}_j dS
\end{aligned} \tag{5.75}$$

for the image force in agreement with Eq. (5.57) obtained previously through use of the energy-momentum tensor. In Eq. (5.75), the integral on the third line also appears in Eq. (5.72) and vanishes for reasons explained there. Finally, the last four lines are obtained by applying Eqs. (5.37) and (2.113), respectively.

5.3.3 *Force obtained from change of the interaction energy*

As indicated by Eq. (5.40), the force exerted on a defect by an imposed stress can also be obtained by determining the rate at which its interaction energy with the stress varies as the defect is displaced. This is readily demonstrated in the case of an imposed internal stress, I, which remains fixed as the defect is displaced.

The interaction energy is given by Eq. (5.11), i.e.,

$$E_{int}^{D^{\infty}/I} = \oiint_{\mathcal{S}} (\sigma_{ij}^{D^{\infty}} u_i^I - \sigma_{ij}^I u_i^{D^{\infty}})\hat{n}_j dS \tag{5.76}$$

and when the defect is displaced by $\delta\xi_\alpha$ the incremental changes in the defect stress and displacement are given by

$$\sigma_{ij}^{D'^{\infty}} - \sigma_{ij}^{D^{\infty}} = -\frac{\partial\sigma_{ij}^{D^{\infty}}}{\partial x_\alpha}\delta\xi_\alpha, \qquad u_i^{D'^{\infty}} - u_i^{D^{\infty}} = -\frac{\partial u_i^{D^{\infty}}}{\partial x_\alpha}\delta\xi_\alpha. \tag{5.77}$$

Then, using Eqs. (5.76) and (5.77), the incremental change in interaction energy due to the displacement is

$$\delta E_{int}^{D/I} = \left(E_{int}^{D'/I} - E_{int}^{D/I}\right) = -\oiint_{\mathcal{S}} \left(\frac{\partial\sigma_{ij}^{D^{\infty}}}{\partial x_\alpha} u_i^I - \sigma_{ij}^I \frac{\partial u_i^{D^{\infty}}}{\partial x_\alpha}\right)\delta\xi_\alpha \hat{n}_j dS. \tag{5.78}$$

Finally, putting Eq. (5.78) into Eq. (5.40),

$$F_\alpha^{D^{\infty}/I} = \oiint_{\mathcal{S}} \left(\frac{\partial\sigma_{ij}^{D^{\infty}}}{\partial x_\alpha} u_i^I - \sigma_{ij}^I \frac{\partial u_i^{D^{\infty}}}{\partial x_\alpha}\right)\hat{n}_j dS \tag{5.79}$$

in agreement with Eq. (5.57), which was obtained by determining the rate at which the total system energy changes as the defect is displaced.

5.3.4 *Summary*

When the defect is represented by its stress field, the force exerted on it by imposed internal stress, applied stresses or image stresses can be determined by use of the energy-momentum tensor given by Eq. (5.51). The results lead to the equations included under Eq. (5.57) which are all of the same general form and involve integration over a closed surface, $\mathcal{S}$, of arbitrary shape that encloses the defect but excludes the source of the imposed stress (which, in the case of an image stress can be regarded as the body surface).

The same results can be obtained by determining directly the rate of change of the total system energy as the defect is displaced according to Eq. (5.38) or, alternatively, the rate of change of the interaction energy between the defect and the imposed stress as the defect is displaced according to Eq. (5.40).

5.4 Interaction Energy and Force Between an Inhomogeneity and Imposed Stress

To formulate the interaction energy between an inhomogeneity and an imposed stress field Q, start with a homogeneous body with the Q field already present, and mark out the region intended for the inhomogeneity. Then convert the elastic constants in that region from C^{M}_{ijkl} to C^{INH}_{ijkl}. The changes in the elastic constants will perturb the Q field, resulting in an interaction energy between the inhomogeneity and the Q field that can be expressed as

$$E^{INH/Q}_{int} = E^{Q'} - E^{Q} = (W^{Q'} + \Phi^{Q'}) - (W^{Q} + \Phi^{Q}), \tag{5.80}$$

where $E^{Q'}$ is the total energy of the system with the perturbed Q field (indicated by Q′) present, and E^{Q} is the total energy of the Q field before the introduction of the inhomogeneity. If there is a gradient in the Q field, the resulting force on the inhomogeneity is given by the rate of change of the total energy of the system as the inhomogeneity is displaced as expressed by Eq. (5.38), i.e.,

$$F^{INH/Q}_{\alpha} = -\lim_{\delta\xi_\alpha \to 0} \frac{1}{\delta\xi_\alpha}\left(\frac{\partial E^{Q'}}{\partial \xi_\alpha}\delta\xi_\alpha\right) = -\frac{\partial E^{Q'}}{\partial \xi_\alpha}. \tag{5.81}$$

On the other hand, the force should also be given by the corresponding rate of change in the interaction energy, $E^{INH/Q}_{int}$, according to Eq. (5.40). Therefore, using Eq. (5.80),

$$\begin{aligned} F^{INH/Q}_{\alpha} &= -\lim_{\delta\xi_\alpha \to 0} \frac{1}{\delta\xi_\alpha}\left(\frac{\partial E^{INH/Q}_{int}}{\partial \xi_\alpha}\delta\xi_\alpha\right) = -\frac{\partial E^{INH/Q}_{int}}{\partial \xi_\alpha} \\ &= -\frac{\partial (E^{Q'} - E^{Q})}{\partial \xi_\alpha} = -\frac{\partial E^{Q'}}{\partial \xi_\alpha}. \end{aligned} \tag{5.82}$$

As anticipated, both expressions are seen to yield the same result, since E^{Q} is independent of the position of the inhomogeneity.

The force on a inhomogeneity can also be obtained by employing the Eshelby energy-momentum tensor. As is demonstrated in Sec. 9.3, the force exerted on an inhomogeneity by an imposed applied stress field, A, is given by Eq. (9.53) in the form of the energy-momentum tensor. Using Eq. (9.53), the force takes the form

$$F_\alpha^{INH/A} = \oiint_{S^\circ} \left(w^{A'} \delta_{j\alpha} - \sigma_{ij}^{A'} \frac{\partial u_i^{A'}}{\partial x_\alpha} \right) \hat{n}_j dS$$

$$= \oiint_{S^\circ} \left(\frac{1}{2} \sigma_{ik}^{A'} \frac{\partial u_i^{A'}}{\partial x_k} \delta_{j\alpha} - \sigma_{ij}^{A'} \frac{\partial u_i^{A'}}{\partial x_\alpha} \right) \hat{n}_j dS, \qquad (5.83)$$

where A′ is the A field after it has been perturbed by the presence of the inhomogeneity, and use has been made of Eq. (2.133) for the strain energy density. Then, using Eq. (B.11) for the term containing the delta function

$$F_\alpha^{INH/A} = \frac{1}{2} \oiint_{S^\circ} \left(\frac{\partial \sigma_{ij}^{A'}}{\partial x_\alpha} u_i^{A'} - \sigma_{ij}^{A'} \frac{\partial u_i^{A'}}{\partial x_\alpha} \right) \hat{n}_j dS. \qquad (5.84)$$

The force given by Eq. (5.84) is seen to be expressed entirely in terms of the A′ field. We may, therefore, conclude that the inhomogeneity, which is not a source of stress, perturbs the imposed A field by itself, and the resulting perturbed field, i.e., the A′ field, then exerts a force on the inhomogeneity.

Exercises

5.1. Confirm Eq. (5.60), i.e., $F_\alpha^{X/Y} = F_\alpha^{Y/X}$, by applying Stokes' theorem, the condition of equilibrium, and the rules for corresponding fields.

Solution. First, write the difference, $F_\alpha^{X/Y} - F_\alpha^{Y/X}$, which takes the form

$$F_\alpha^{X/Y} - F_\alpha^{Y/X} = \oiint_S \left(\frac{\partial \sigma_{ij}^X}{\partial x_\alpha} u_i^Y - \sigma_{ij}^Y \frac{\partial u_i^X}{\partial x_\alpha} \right) \hat{n}_j dS$$

$$- \oiint_S \left(\frac{\partial \sigma_{ij}^Y}{\partial x_\alpha} u_i^X - \sigma_{ij}^X \frac{\partial u_i^Y}{\partial x_\alpha} \right) \hat{n}_j dS$$

$$= \oiint_S \left[\sigma_{ij}^X \frac{\partial u_i^Y}{\partial x_\alpha} + \frac{\partial \sigma_{ij}^X}{\partial x_\alpha} u_i^Y - \left(\sigma_{ij}^Y \frac{\partial u_i^X}{\partial x_\alpha} + \frac{\partial \sigma_{ij}^Y}{\partial x_\alpha} u_i^X \right) \right] \hat{n}_j dS. \qquad (5.85)$$

Then, upon application of Stokes' theorem, i.e., Eq. (B.11), and Eq. (2.106),

$$F_\alpha^{X/Y} - F_\alpha^{Y/X} = \oiint_S \left(\sigma_{ij}^X \frac{\partial u_i^Y}{\partial x_j} - \sigma_{ij}^Y \frac{\partial u_i^X}{\partial x_j} \right) \hat{n}_\alpha dS = 0. \qquad (5.86)$$

5.2. Demonstrate that the force exerted on a defect source of stress by an applied field, A, given by Eq. (5.57), vanishes when A is uniform throughout the body.

Solution. First use Eq. (5.60) to rewrite Eq. (5.57) as

$$F_{\alpha}^{D^\infty/A} = \oiint_{\mathcal{S}} \left(\frac{\partial \sigma_{ij}^{D^\infty}}{\partial x_\alpha} u_i^A - \sigma_{ij}^A \frac{\partial u_i^{D^\infty}}{\partial x_\alpha} \right) \hat{n}_j dS$$

$$= \oiint_{\mathcal{S}} \left(\frac{\partial \sigma_{ij}^{A}}{\partial x_\alpha} u_i^{D^\infty} - \sigma_{ij}^{D^\infty} \frac{\partial u_i^{A}}{\partial x_\alpha} \right) \hat{n}_j dS. \tag{5.87}$$

The first term in the second integral then vanishes, since $\partial \sigma_{ij}^A/\partial x_\alpha = 0$, and the second term also vanishes since

$$\oiint_{\mathcal{S}} \sigma_{ij}^{D^\infty} \frac{\partial u_i^A}{\partial x_\alpha} \hat{n}_j dS = \frac{\partial u_i^A}{\partial x_\alpha} \oiint_{\mathcal{S}} \sigma_{ij}^{D^\infty} \hat{n}_j dS = 0. \tag{5.88}$$

Here, $\partial u_i^A/\partial x_\alpha$ is a constant, and is taken out of the integrand. The remaining integral then vanishes since it represents the net force exerted on the closed surface $\mathcal{S}$ by the defect stress field, which must vanish to satisfy equilibrium.

5.3. Consider the relatively simple case of a single defect, D, lying in a body with a traction-free surface, $\mathcal{S}^\circ$. The defect then experiences an image force that can be written as

$$F_{\alpha}^{D/D^{IM}} = \frac{1}{2} \oiint_{\mathcal{S}^\circ} \frac{\partial \sigma_{ij}^{D}}{\partial x_\alpha} u_i^D \hat{n}_j dS. \tag{5.89}$$

(a) Derive Eq. (5.89) starting with the Eq. (5.47) which involves the energy-momentum tensor directly. (b) Alternatively, derive Eq. (5.89) by starting with Eq. (5.57).

Solution (a) Since $\sigma_{ij}^D \hat{n}_j = 0$ on $\mathcal{S}^\circ$, Eq. (5.47), with the help of Eq. (2.133), reduces to

$$F_{\alpha}^{D/D^{IM}} = \oiint_{\mathcal{S}^\circ} w \hat{n}_\alpha dS = \frac{1}{2} \oiint_{\mathcal{S}^\circ} \sigma_{ij}^{D} \frac{\partial u_i^D}{\partial x_j} \hat{n}_\alpha dS. \tag{5.90}$$

Then, applying Eq. (B.11) (Stokes' theorem) and the condition $\sigma_{ij}^D \hat{n}_j = 0$ on $\mathcal{S}^\circ$, we obtain Eq. (5.89).

(b) Equations (5.57) and (5.60) can be used to write

$$F_\alpha = F_\alpha^{D/D^{IM}} = \frac{1}{2} \oiint_{\mathcal{S}^\circ} \left[\left(\frac{\partial \sigma_{ij}^{D^\infty}}{\partial x_\alpha} u_i^{D^{IM}} - \sigma_{ij}^{D^{IM}} \frac{\partial u_i^{D^\infty}}{\partial x_\alpha} \right) \right.$$

$$\left. + \left(\frac{\partial \sigma_{ij}^{D^{IM}}}{\partial x_\alpha} u_i^{D^\infty} - \sigma_{ij}^{D^\infty} \frac{\partial u_i^{D^{IM}}}{\partial x_\alpha} \right) \right] \hat{n}_j dS. \tag{5.91}$$

But, by virtue of the discussion following Eq. (5.55), we have the equality

$$\frac{1}{2} \oiint_{\mathcal{S}^\circ} \left[\left(\frac{\partial \sigma_{ij}^{D^\infty}}{\partial x_\alpha} u_i^{D^\infty} - \sigma_{ij}^{D^\infty} \frac{\partial u_i^{D^\infty}}{\partial x_\alpha} \right) \right.$$

$$\left. + \left(\frac{\partial \sigma_{ij}^{D^{IM}}}{\partial x_\alpha} u_i^{D^{IM}} - \sigma_{ij}^{D^{IM}} \frac{\partial u_i^{D^{IM}}}{\partial x_\alpha} \right) \right] \hat{n}_j dS = 0 \tag{5.92}$$

and, upon adding this to Eq. (5.91), we have, finally

$$F_\alpha^{D/D^{IM}} = \frac{1}{2} \oiint_{\mathcal{S}^\circ} \left(\frac{\partial \sigma_{ij}^{D}}{\partial x_\alpha} u_i^{D} - \sigma_{ij}^{D} \frac{\partial u_i^{D}}{\partial x_\alpha} \right) \hat{n}_j dS = \frac{1}{2} \oiint_{\mathcal{S}^\circ} \frac{\partial \sigma_{ij}^{D}}{\partial x_\alpha} u_i^{D} \hat{n}_j dS \tag{5.93}$$

since $\sigma_{ij}^{D} \hat{n}_j = 0$ on $\mathcal{S}^\circ$.

5.4. Suppose a body, $\mathcal{V}^\circ$, which possesses a traction-free surface $\mathcal{S}^\circ$, and contains two distinguishable defects D1 and D2, each represented by a localized body force density distribution. Then, by virtue of Eq. (5.28), the interaction energy between them can be expressed by either of the two forms

$$E_{int}^{D1/D2} = - \oiiint_{\mathcal{V}} u_i^{D2}(\mathbf{x}) f_i^{D1}(\mathbf{x}) dV \tag{5.94}$$

and

$$E_{int}^{D1/D2} = - \oiiint_{\mathcal{V}} u_i^{D1}(\mathbf{x}) f_i^{D2}(\mathbf{x}) dV \tag{5.95}$$

depending upon which defect is regarded as producing the displacement field. Explain, in simple physical terms, the equal validity of these two expressions.

Solution If we imagine that D2 is inserted into the system in the presence of D1, the integral given by Eq. (5.94) represents the work done by the force distribution of D1 as it is displaced by the displacement field of D2. Conversely, if we imagine that D1 is inserted in the presence of D2 the integral given by Eq. (5.95) represents the work done by the force distribution of D2 as it is displaced by the displacement field of D1. The two interaction energies must therefore be identical, since they will not depend upon the order in which the defects are introduced, and, therefore

$$\oiiint_{\mathcal{V}} u_i^{D2}(\mathbf{x}) f_i^{D1}(\mathbf{x}) dV = \oiiint_{\mathcal{V}} u_i^{D1}(\mathbf{x}) f_i^{D2}(\mathbf{x}) dV. \tag{5.96}$$

As shown in the following Exercise 5.5, this result can be established more formally by applying *Betti's reciprocity theorem.*

5.5. Betti's reciprocity theorem states (Sokolnikoff 1946) that: *If a region of a body, $\mathcal{V}$, with surface $\mathcal{S}$, is subjected to two systems of body and surface forces, then the work that would be done by the forces of the first system during the displacements associated with the second system is equal to the work that would be done by the forces of the second system during the displacements associated with the first system.*

(a) Give a proof of the above theorem. (b) Show that the result given by Eq. (5.96) in Exercise 5.4 can be obtained by a simple application of the theorem. Hint: start by considering a body containing two such systems, i.e., A and B, and consider the following equation which can be written solely on the basis of the divergence theorem and Eqs. (2.65) and (2.5):

$$\begin{aligned}\oiint_{\mathcal{S}} \sigma_{ij}^{A} u_i^{B} \hat{n}_j dS &= \oiiint_{\mathcal{V}} \frac{\partial}{\partial x_j} \left(\sigma_{ij}^{A} u_i^{B}\right) dV \\ &= \oiiint_{\mathcal{V}} \sigma_{ij}^{A} \varepsilon_{ij}^{B} dV - \oiiint_{\mathcal{V}} f_i^{A} u_i^{B} dV. \end{aligned} \tag{5.97}$$

Solution (a) Equation (5.97) can also be written with A and B interchanged, i.e.,

$$\begin{aligned}\oiint_{\mathcal{S}} \sigma_{ij}^{B} u_i^{A} \hat{n}_j dS &= \oiiint_{\mathcal{V}} \frac{\partial}{\partial x_j} (\sigma_{ij}^{B} u_i^{A}) dV \\ &= \oiiint_{\mathcal{V}} \sigma_{ij}^{B} \varepsilon_{ij}^{A} dV - \oiiint_{\mathcal{V}} f_i^{B} u_i^{A} dV. \end{aligned} \tag{5.98}$$

However, by virtue of Eq. (2.102),

$$\oiiint_{\mathcal{V}} \sigma_{ij}^{A} \varepsilon_{ij}^{B} dV = \oiiint_{\mathcal{V}} \sigma_{ij}^{B} \varepsilon_{ij}^{A} dV. \tag{5.99}$$

Then, by combining Eqs. (5.97–5.99),

$$\oiiint_{\mathcal{V}} f_i^A u_i^B dV + \oiint_{\mathcal{S}} \sigma_{ij}^A u_i^B \hat{n}_j dS$$
$$= \oiiint_{\mathcal{V}} f_i^B u_i^A dV + \oiint_{\mathcal{S}} \sigma_{ij}^B u_i^A \hat{n}_j dS \qquad (5.100)$$

which establishes Betti's theorem.

(b) For the stress systems in Exercise 5.4, $\mathcal{V} \to \mathcal{V}^\circ$, $\mathcal{S} \to \mathcal{S}^\circ$, and the surface tractions vanish. Therefore, Betti's theorem, given by Eq. (5.100), yields Eq. (5.96).

5.6. According to Eq. (5.5), the interaction energy between a defect, D, and a stress system, X, must be the difference between the energy change $E^{D/X}$ when the defect is added to the system in the presence of the X field, and the energy change E^D when the defect is added to the system by itself, i.e.,

$$E_{int}^{D/X} = E^{D/X} - E^D. \qquad (5.101)$$

Show that this is consistent with the formulation for the interaction energy given by Eq. (5.4) where the interaction energy is taken as the difference between total energy of the body, $E^{(D+X)}$, when the defect and X fields are present together, and the sum of the energies of their individual fields, i.e.,

$$E_{int}^{D/X} = E^{(D+X)} - (E^D + E^X). \qquad (5.102)$$

Solution This can be done very simply by showing that the two formulations are equivalent. By equating the two expression, we obtain

$$E^{(D+X)} = E^X + E^{D/X} \qquad (5.103)$$

which states that the energy of the body when the D and X fields are present together is equal to its energy when the X field is introduced by itself first and the D field is then introduced in the presence of the X field. This is most certainly true, and the consistency therefore holds.

5.7. The interaction energy between a defect and its image stress (as seen in Fig. 5.2) is given in the text by Eq. (5.32), i.e.,

$$E_{int}^{D/D^{IM}} = \oiint_{\mathcal{S}} \left(\sigma_{ij}^D u_i^{D^{IM}} - \sigma_{ij}^{D^{IM}} u_i^D \right) \hat{n}_j dS. \qquad (5.104)$$

This expression was obtained by employing the basic formulation for the interaction energies given at the beginning of the chapter by

Eq. (5.4). Show that a result that is consistent with Eq. (5.104) can be obtained by employing the alternative basic formulation for interaction energies given by Eq. (5.5).

Solution. Equation (5.5), for the present system, is of the form

$$E_{int}^{D/D^{IM}} = E^{D/D^{IM}} - E^{D}. \tag{5.105}$$

When the defect is added in the presence of the image stress,

$$\begin{aligned} W^{D/D^{IM}} &= W^{D}, \\ \Phi^{D/D^{IM}} &= -\oiint_{\mathcal{S}^{\circ}} \sigma_{ij}^{D^{IM}} u_i^{D} \hat{n}_j dS, \end{aligned} \tag{5.106}$$

since the interaction energy between the applied image stress and the internal defect stress vanishes because of Eq. (5.25), and the potential energy of the $\sigma_{ij}^{D^{IM}} \hat{n}_j$ tractions acting on the surface $\mathcal{S}^{\circ}$ (see Fig. 5.2) changes, as indicated, because of the displacement of the surface $\mathcal{S}^{\circ}$ due to the displacements u_i^{D} which occur when the defect is introduced. Therefore, substituting Eq. (5.106) into Eq. (5.105), and recognizing that $E^{D} = W^{D}$,

$$E_{int}^{D/D^{IM}} = W^{D} - \oiint_{\mathcal{S}^{\circ}} \sigma_{ij}^{D^{IM}} u_i^{D} \hat{n}_j dS + W^{D} = -\oiint_{\mathcal{S}^{\circ}} \sigma_{ij}^{D^{IM}} u_i^{D} \hat{n}_j dS. \tag{5.107}$$

To show that this result is consistent with Eq. (5.104) we can, by virtue of Eq. (2.110), expand the surface $\mathcal{S}$ in Eq. (5.104) so that it is infinitesimally close to $\mathcal{S}^{\circ}$ without changing the value of the integral. Then, since $\sigma_{ij}^{D} \hat{n}_j = 0$ on $\mathcal{S}^{\circ}$, we obtain

$$E_{int}^{D/D^{IM}} = \oiint_{\mathcal{S}} \left(\sigma_{ij}^{D} u_i^{D^{IM}} - \sigma_{ij}^{D^{IM}} u_i^{D} \right) \hat{n}_j dS = -\oiint_{\mathcal{S}^{\circ}} \sigma_{ij}^{D^{IM}} u_i^{D} \hat{n}_j dS, \tag{5.108}$$

which is identical to the result given by Eq. (5.107).

5.8. Derive Eq. (5.28) for the interaction energy between an applied stress and a defect represented by an array of body forces.

Solution. Both strain energy and potential energy are involved so Eq. (5.15) applies. If it is imagined that the defect is introduced into the body first, the potential energy of the defect forces will change by $-\sum_{\alpha} \mathbf{u}^{\mathbf{A}}(\mathbf{x}^{(\alpha)}) \cdot \mathbf{F}^{\mathbf{D}(\alpha)}$ when the A field is then introduced, and the interaction potential energy is therefore,

$$\Phi_{\mathrm{int}}^{\mathrm{D/A}} = -\sum_{\alpha} \mathbf{u}^{\mathrm{A}}(\mathbf{x}^{(\alpha)}) \cdot \mathbf{F}^{\mathrm{D}(\alpha)}. \tag{5.109}$$

The interaction strain energy is of the form of Eq. (5.22), i.e.,

$$\mathrm{W}_{\mathrm{int}}^{\mathrm{D/A}} = \oiint_{\mathcal{S}} \sigma_{\mathrm{ij}}^{\mathrm{D}} \mathrm{u}_{\mathrm{i}}^{\mathrm{A}} \hat{\mathrm{n}}_{\mathrm{j}} \mathrm{dS} + \oiint_{\mathcal{S}^{\circ}} \sigma_{\mathrm{ij}}^{\mathrm{A}} \mathrm{u}_{\mathrm{i}}^{\mathrm{D}} \hat{\mathrm{n}}_{\mathrm{j}} \mathrm{dS} - \oiint_{\mathcal{S}} \sigma_{\mathrm{ij}}^{\mathrm{A}} \mathrm{u}_{\mathrm{i}}^{\mathrm{D}} \hat{\mathrm{n}}_{\mathrm{j}} \mathrm{dS}. \tag{5.110}$$

However, the surface $\mathcal{S}$ can be expanded so that it is infinitesimally close to the surface, $\mathcal{S}^{\circ}$, without changing any of the integrals by virtue of Eq. (2.110). Therefore,

$$\mathrm{W}_{\mathrm{int}}^{\mathrm{D/A}} = \oiint_{\mathcal{S}^{\circ}} \sigma_{\mathrm{ij}}^{\mathrm{D}} \mathrm{u}_{\mathrm{i}}^{\mathrm{A}} \hat{\mathrm{n}}_{\mathrm{j}} \mathrm{dS} + \oiint_{\mathcal{S}^{\circ}} \sigma_{\mathrm{ij}}^{\mathrm{A}} \mathrm{u}_{\mathrm{i}}^{\mathrm{D}} \hat{\mathrm{n}}_{\mathrm{j}} \mathrm{dS} - \oiint_{\mathcal{S}^{\circ}} \sigma_{\mathrm{ij}}^{\mathrm{A}} \mathrm{u}_{\mathrm{i}}^{\mathrm{D}} \hat{\mathrm{n}}_{\mathrm{j}} \mathrm{dS} \tag{5.111}$$

and $\mathrm{W}_{\mathrm{int}}^{\mathrm{D/A}}$ vanishes since $\sigma_{\mathrm{ij}}^{\mathrm{D}} \hat{\mathrm{n}}_{\mathrm{j}} = 0$ on $\mathcal{S}^{\circ}$. Then,

$$\mathrm{E}_{\mathrm{int}}^{\mathrm{D/A}} = \mathrm{W}_{\mathrm{int}}^{\mathrm{D/A}} + \Phi_{\mathrm{int}}^{\mathrm{D/A}} = -\sum_{\alpha} \mathbf{u}^{\mathrm{A}}(\mathbf{x}^{(\alpha)}) \cdot \mathbf{F}^{\mathrm{D}(\alpha)}. \tag{5.112}$$

5.9. Consider the system, shown in Fig. 5.5, which consists of a defect source of stress, D, and an applied stress system, A, in a finite body. The closed surface $\mathcal{S}^{(1)}$ encloses D, and an additional closed surface $\mathcal{S}^{(2)}$ encloses $\mathcal{S}^{(1)}$. The stress due to the defect is then a function of both the field vector, $\mathbf{x}$, and the position of the defect in the body, indicated by the vector $\boldsymbol{\xi}$, so that $\sigma_{\mathrm{ij}}^{\mathrm{D}} = \sigma_{\mathrm{ij}}^{\mathrm{D}}(\mathbf{x}, \boldsymbol{\xi})$ and $\mathrm{u}_{\mathrm{i}}^{\mathrm{D}} = \mathrm{u}_{\mathrm{i}}^{\mathrm{D}}(\mathbf{x}, \boldsymbol{\xi})$. Now, prove that

$$\oiint_{\mathcal{S}^{(2)}-\mathcal{S}^{(1)}} \left(\frac{\partial \sigma_{\mathrm{ij}}^{\mathrm{D}}}{\partial \xi_l} \mathrm{u}_{\mathrm{i}}^{\mathrm{A}} - \sigma_{\mathrm{ij}}^{\mathrm{A}} \frac{\partial \mathrm{u}_{\mathrm{i}}^{\mathrm{D}}}{\partial \xi_l} \right) \hat{\mathrm{n}}_{\mathrm{j}} \mathrm{dS} = 0, \tag{5.113}$$

which may be compared to Eq. (2.113).

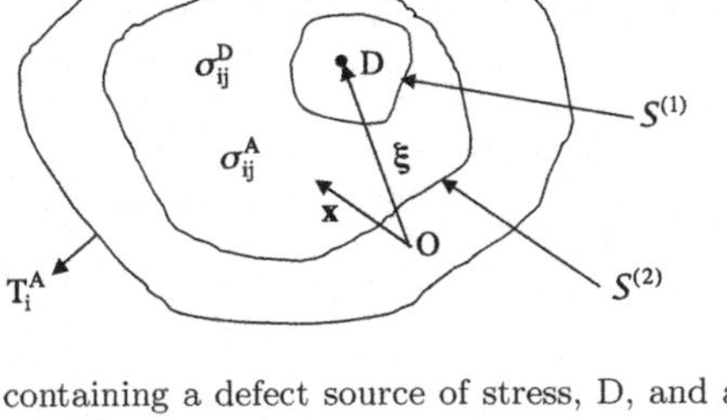

Fig. 5.5 Finite body containing a defect source of stress, D, and an applied stress, A. Position of defect indicated by vector $\boldsymbol{\xi}$.

Solution. The proof follows along the lines of the verification of Eq. (2.112). The D and A fields are corresponding fields within the region between $\mathcal{S}^{(1)}$ and $\mathcal{S}^{(2)}$, since no singularities are present there. We can, therefore, convert the surface integral into a volume integral according to

$$\oiint_{\mathcal{S}^{(2)}-\mathcal{S}^{(1)}} \left(\frac{\partial \sigma_{ij}^{D}}{\partial \xi_l} u_i^{A} - \sigma_{ij}^{A} \frac{\partial u_i^{D}}{\partial \xi_l} \right) \hat{n}_j dS$$

$$= \oiiint_{\mathcal{V}^{(2)}-\mathcal{V}^{(1)}} \frac{\partial}{x_j} \left(\frac{\partial \sigma_{ij}^{D}}{\partial \xi_l} u_i^{A} - \sigma_{ij}^{A} \frac{\partial u_i^{D}}{\partial \xi_l} \right) dV$$

$$= \oiiint_{\mathcal{V}^{(2)}-\mathcal{V}^{(1)}} \left(\frac{\partial \sigma_{ij}^{D}}{\partial \xi_l} \frac{\partial u_i^{A}}{\partial x_j} - \sigma_{ij}^{A} \frac{\partial^2 u_i^{D}}{\partial \xi_l x_j} \right) dV \quad (5.114)$$

where $\mathcal{V}^{(2)}$ and $\mathcal{V}^{(1)}$ are the regions enclosed by $\mathcal{S}^{(2)}$ and $\mathcal{S}^{(1)}$, respectively, and Eq. (2.65) has been employed. Then, substituting

$$\sigma_{ij}^{D} = C_{ijmn} \partial u_m^{D} / \partial x_n \qquad \sigma_{ij}^{A} = C_{ijmn} \partial u_m^{A} / \partial x_n \quad (5.115)$$

into Eq. (5.114), and using Eq. (2.82), we have the result

$$\oiiint_{\mathcal{V}^{(2)}-\mathcal{V}^{(1)}} \left(C_{ijmn} \frac{\partial^2 u_m^{D}}{\partial \xi_l \partial x_n} \frac{\partial u_i^{A}}{\partial x_j} - C_{ijmn} \frac{\partial u_m^{A}}{\partial x_n} \frac{\partial^2 u_i^{D}}{\partial x_j \partial \xi_l} \right) dV$$

$$= \oiiint_{\mathcal{V}^{(2)}-\mathcal{V}^{(1)}} \left(C_{ijmn} \frac{\partial^2 u_m^{D}}{\partial \xi_l \partial x_n} \frac{\partial u_i^{A}}{\partial x_j} - C_{mnij} \frac{\partial u_i^{A}}{\partial x_j} \frac{\partial^2 u_m^{D}}{\partial x_n \partial \xi_l} \right) dV = 0. \quad (5.116)$$

Chapter 6

Inclusions in Infinite Homogeneous Regions

6.1 Introduction

The elastic properties of various types of inclusions in infinite homogeneous regions are determined. Expressions for their elastic fields and strain energies are obtained by treating them as defects produced by the introduction of transformation strains as outlined in Sec. 3.6.

The problem of a coherent, homogeneous inclusion of arbitrary shape and transformation strain is treated first. This is followed by treatments of coherent inclusions with ellipsoidal shapes (which in limiting cases can be spheres, thin-disks or needles) and various transformation strains. Next, the more complicated problem of treating coherent inhomogeneous inclusions by Eshelby's *equivalent homogeneous inclusion* method is described. Finally, the elastic fields and strain energies of incoherent ellipsoidal inclusions produced by coherent → incoherent transitions are considered.

Since the inclusions are characterized throughout the chapter by transformation strains, the special elasticity theory for systems containing transformation strains formulated in Sec. 3.6 is used.

The following notation is employed for this chapter:

$\varepsilon_{ij}^{INC}, \varepsilon_{ij}^{M}$ = elastic strain due to inclusion in inclusion and matrix, respectively
$\varepsilon_{ij}^{C,INC}, \varepsilon_{ij}^{C,M}$ = canceling strain in inclusion and matrix, respectively

6.2 Characterization of Inclusions

An inclusion is characterized by five features, i.e., its *shape* and *volume*, its *misfit* relative to the matrix, its *inhomogeneity*, and its degree of *coherence* with respect to the matrix. To characterize an inclusion with respect to these features we imagine producing it by the following procedure consistent with that described in Sec. 3.6.1 for introducing general transformation strains:

(1) Start with a stress-free matrix with elastic constants, C^{M}_{ijkl}, mark out the region intended for the desired inclusion, and cut it out of the matrix. This establishes its shape and volume.
(2) For inhomogeneous inclusions, change the elastic constants of the cut-out region to C^{INC}_{ijkl}. This establishes its degree of inhomogeneity.
(3) Subject the cut out region to a transformation strain, $\varepsilon^{T}_{ij}(\mathbf{x})$, to establish its *misfit.* Then, apply forces to restore it to its original shape and volume. This produces an elastic strain, $-\varepsilon^{T}_{ij}(\mathbf{x})$, and a corresponding stress, $\sigma^{T}_{ij}(\mathbf{x}) = -C^{INC}_{ijmn}\varepsilon^{T}_{mn}$.
(4) Insert the cut-out region back into the matrix cavity, while maintaining the applied forces, and bond it to the matrix.
(5) Then, cancel these forces by applying a distribution of equal and opposite forces which produces a canceling strain field, $\varepsilon^{C,INC}_{ij}(\mathbf{x})$, in the inclusion and a corresponding field, $\varepsilon^{C,M}_{ij}(\mathbf{x})$, in the matrix, thus establishing a state of pure internal stress in the body.
(6) The inclusion at this stage is regarded as *coherent,* since all points on opposite sides of the inclusion/matrix interface in registry after the marking out operation in step 1 are still in registry. However, the strain energy due to the inclusion can generally be reduced by changing the shape and/or volume of either the inclusion or the cavity in the matrix that it occupies. This can be accomplished by diffusional transport of material or by plastic deformation as described in Sec. 6.5.1 (Fig. 6.6). These processes generally destroy the coherence producing an *incoherent* inclusion via a *coherent → incoherent transition.* The degree to which this occurs establishes the final degree of coherence.

6.3 Coherent Inclusions

For a coherent homogeneous inclusion produced by the outlined procedure, the elastic displacement, elastic strain, and stress in the inclusion and matrix, are given by

$$
\begin{aligned}
u_i^{INC} &= u_i^{C,INC} - u_i^T \quad u_i^M = u_i^{C,M} \\
\varepsilon_{ij}^{INC} &= \varepsilon_{ij}^{C,INC} - \varepsilon_{ij}^T \quad \varepsilon_{ij}^M = \varepsilon_{ij}^{C,M} \\
\sigma_{ij}^{INC} &= \sigma_{ij}^{C,INC} - \sigma_{ij}^T \quad \sigma_{ij}^M = \sigma_{ij}^{C,M}
\end{aligned}
\tag{6.1}
$$

in terms of quantities introduced in Sec. 3.6 for systems containing transformation strains. The corresponding boundary conditions at the coherent inclusion/matrix interface are then

$$
\left.
\begin{aligned}
u_i^{INC,tot} &= u_i^{INC} + u_i^T = u_i^M, \\
\sigma_{ij}^{INC}\hat{n}_j &= \sigma_{ij}^M \hat{n}_j.
\end{aligned}
\right. \quad (\text{on } \mathcal{S}^{INC})
\tag{6.2}
$$

The central problem is the determination of the $u_i^C(\mathbf{x})$ strains in both the inclusion and matrix, since once these are known, all other elastic field quantities can be determined by use of Eq. (2.5), Hooke's law and Eq. (6.1).

6.3.1 *Elastic field of homogeneous inclusion by Fourier transform method*

6.3.1.1 *Arbitrary shape and* ε_{ij}^T

For an inclusion containing the transformation strain, $\varepsilon_{mn}^T(\mathbf{x}')$, and embedded in an infinite homogeneous matrix, Eq. (3.166) can be rewritten in the equivalent differential form

$$
\begin{aligned}
u_i^C(\mathbf{x}) = &-\frac{1}{(2\pi)^3} C_{jlmn} \frac{\partial}{\partial \mathbf{x}_l} \oiiint_{V^{INC}} \varepsilon_{mn}^T(\mathbf{x}') dx_1' dx_2' dx_3' \\
&\times \int_{-\infty}^{\infty}\int_{-\infty}^{\infty}\int_{-\infty}^{\infty} e^{-i\mathbf{k}\cdot(\mathbf{x}-\mathbf{x}')} (kk)_{ij}^{-1} dk_1 dk_2 dk_3.
\end{aligned}
\tag{6.3}
$$

or, by changing $\mathbf{k}$ to $-\mathbf{k}$, as

$$
\begin{aligned}
u_i^C(\mathbf{x}) = &-\frac{1}{(2\pi)^3} C_{jlmn} \frac{\partial}{\partial \mathbf{x}_l} \oiiint_{V^{INC}} \varepsilon_{mn}^T(\mathbf{x}') dx_1' dx_2' dx_3' \\
&\times \int_{-\infty}^{\infty}\int_{-\infty}^{\infty}\int_{-\infty}^{\infty} e^{i\mathbf{k}\cdot(\mathbf{x}-\mathbf{x}')} (kk)_{ij}^{-1} dk_1 dk_2 dk_3.
\end{aligned}
\tag{6.4}
$$

Following Mura (1987), the integration over $\mathbf{k}$-space, can then be performed using spherical coordinates k, θ, ϕ (Fig. A.1b) so that

$dk_1dk_2dk_3 = k^2 dk \sin\phi d\phi d\theta = k^2 dk d\hat{S}(\hat{\mathbf{k}})$, where $d\hat{S}(\hat{\mathbf{k}})$ represents a differential element of area on the surface of the unit sphere, $|\hat{k} = 1|$. This introduces two integrations, i.e., one over the unit sphere and the other over $0 < k \leq \infty$. Then, Eq. (6.4), can be written as

$$u_i^C(\mathbf{x}) = -\frac{1}{(2\pi)^3} C_{jlmn} \frac{\partial}{\partial x_l} \oiiint_{V^{INC}} \varepsilon_{mn}^T(\mathbf{x}') dx_1' dx_2' dx_3' \times \int_0^{\infty} k^2 dk \oiint_{\hat{S}} e^{i\mathbf{k}\cdot(\mathbf{x}-\mathbf{x}')} (kk)_{ij}^{-1} d\hat{S}(\hat{\mathbf{k}}) \tag{6.5}$$

and Eq. (6.3) can be written as

$$u_i^C(\mathbf{x}) = -\frac{1}{(2\pi)^3} C_{jlmn} \frac{\partial}{\partial x_l} \oiiint_{V^{INC}} \varepsilon_{mn}^T(\mathbf{x}') dx_1' dx_2' dx_3' \times \int_{-\infty}^{0} k^2 dk \oiint_{\hat{S}} e^{i\mathbf{k}\cdot(\mathbf{x}-\mathbf{x}')} (kk)_{ij}^{-1} d\hat{S}(\hat{\mathbf{k}}). \tag{6.6}$$

by changing $-\mathbf{k}$ to $\mathbf{k}$. Then, adding Eqs. (6.5) and (6.6), and introducing $\mathbf{k} = k\hat{\mathbf{k}}$ so that $(kk)_{ij}^{-1} = k^{-2}(\hat{k}\hat{k})_{ij}^{-1}$, and dividing by two,

$$u_i^C(\mathbf{x}) = -\frac{1}{2(2\pi)^3} C_{jlmn} \frac{\partial}{\partial x_l} \oiiint_{V^{INC}} \varepsilon_{mn}^T(\mathbf{x}') dx_1' dx_2' dx_3' \times \int_{-\infty}^{\infty} e^{ik\hat{\mathbf{k}}\cdot(\mathbf{x}-\mathbf{x}')} dk \oiint_{\hat{S}} (\hat{k}\hat{k})_{ij}^{-1} d\hat{S}(\hat{\mathbf{k}}). \tag{6.7}$$

Next, by substituting the standard delta function expression (Sneddon 1951),

$$\int_{-\infty}^{\infty} e^{ik[\hat{\mathbf{k}}\cdot(\mathbf{x}-\mathbf{x}')]} dk = 2\pi\delta[\hat{\mathbf{k}}\cdot(\mathbf{x}-\mathbf{x}')] \tag{6.8}$$

into Eq. (6.7), and performing the differentiation with respect to x_l with the help of the relationship

$$\frac{\partial\delta[\hat{\mathbf{k}}\cdot(\mathbf{x}-\mathbf{x}')]}{\partial x_l} = \frac{\partial\delta[\hat{\mathbf{k}}\cdot(\mathbf{x}-\mathbf{x}')]}{\partial[\hat{\mathbf{k}}\cdot(\mathbf{x}-\mathbf{x}')]} \frac{\partial[\hat{\mathbf{k}}\cdot(\mathbf{x}-\mathbf{x}')]}{\partial x_l} = \hat{\mathbf{k}}_l \delta'[\hat{\mathbf{k}}\cdot(\mathbf{x}-\mathbf{x}')], \tag{6.9}$$

where the prime indicates differentiation of the delta function with respect to its argument,

$$u_i^C(\mathbf{x}) = -\frac{1}{8\pi^2} C_{jlmn} \oiiint_{V^{INC}} \varepsilon_{mn}^T(\mathbf{x}') dx_1' dx_2' dx_3' \times \oiint_{\hat{S}} \hat{k}_l (\hat{k}\hat{k})_{ij}^{-1} \delta'[\hat{\mathbf{k}}\cdot(\mathbf{x}-\mathbf{x}')] d\hat{S}(\hat{\mathbf{k}}). \tag{6.10}$$

Equation (6.10) is quite general and involves one integration over the surface of the unit sphere, $\hat{\mathcal{S}}$, in **k**-space and a second in Cartesian space over the inclusion, whose shape and transformation strain have yet to be specified.

6.3.1.2 *Ellipsoidal shape and arbitrary* $\varepsilon^{\mathrm{T}}_{\mathrm{ij}}$

Equation (6.10) can now be formulated for an inclusion of ellipsoidal shape but arbitrary transformation strain. This is of special interest, since the ellipsoidal shape can be readily varied to serve as a good approximation for the shapes that inclusions often take in real materials. For example, if the a_i are the principal axes, the shape is a sphere when $a_1 = a_2 = a_3$, a thin circular disk when $a_1 \ll a_2 = a_3$, and a long thin cylinder (needle) when $a_1 \gg a_2 = a_3$.

First, the principal axes of the ellipsoid are taken (Mura 1987) parallel to the base vectors of the (x_1, x_2, x_3) coordinate system,[1] so that the condition

$$\frac{x'^2_1}{a_1^2} + \frac{x'^2_2}{a_2^2} + \frac{x'^2_3}{a_3^2} \leq 1 \tag{6.11}$$

is imposed on Eq. (6.10). Then, by making the changes of variable

$$\begin{aligned} x_\alpha &= a_\alpha y_\alpha \\ x'_\alpha &= a_\alpha y'_\alpha \end{aligned} \tag{6.12}$$

the ellipsoidal inclusion is transformed into a unit sphere when the new (y_1, y_2, y_3) and (y'_1, y'_2, y'_3) coordinates are employed. It is also useful, for purposes of integrating Eq. (6.10) over the inclusion volume and the surface $\hat{\mathcal{S}}$, to introduce the additional quantities illustrated in Fig. 6.1. The vector $\boldsymbol{\zeta}$ is defined by $\zeta_\alpha = a_\alpha \hat{k}_\alpha$, and, therefore, the unit vector $\hat{\boldsymbol{\zeta}}$ in the figure possesses the components $\hat{\zeta}_\alpha = a_\alpha \hat{k}_\alpha / \zeta$, where $\zeta = |\boldsymbol{\zeta}|$. The unit vectors $\hat{\mathbf{m}}$ and $\hat{\mathbf{n}}$ are orthogonal to each other and to $\hat{\boldsymbol{\zeta}}$, and OA lies along the projection of r in the plane containing $\hat{\mathbf{m}}$ and $\hat{\mathbf{n}}$. An element of inclusion volume is then $a_1 a_2 a_3 r d\psi dr dz$. Relationships involving the various

[1]This axis system generally differs from the crystal axis system, and, by rotating it relative to the crystal system the effect of varying the crystal orientation of the inclusion relative to that of the crystal matrix can be studied.

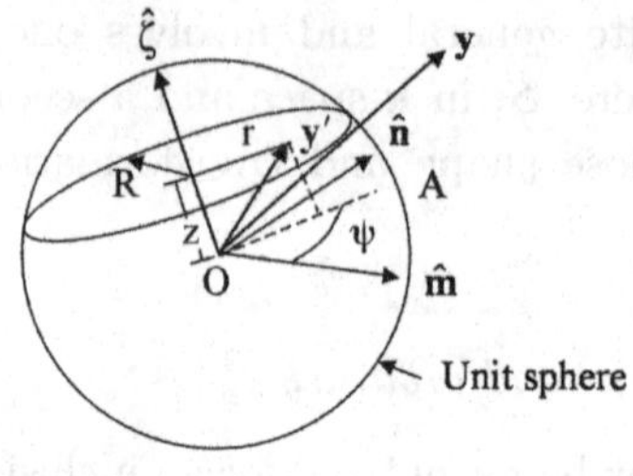

Fig. 6.1 Geometry for integrating over the inclusion volume.

quantities are therefore,

$$\begin{aligned}
\mathrm{dx}'_1\mathrm{dx}'_2\mathrm{dx}'_3 &= \mathrm{a_1a_2a_3dy'_1dy'_2dy'_3} = \mathrm{a_1a_2a_3rdrd}\psi\mathrm{dz},\\
\zeta_\alpha &= \mathrm{a}_\alpha\hat{\mathrm{k}}_\alpha,\\
\hat{\boldsymbol{\zeta}} &= \boldsymbol{\zeta}/\zeta,\\
\hat{\zeta}_\alpha &= \zeta_\alpha/\zeta = \mathrm{a}_\alpha\hat{\mathrm{k}}_\alpha/\zeta,\\
\hat{\boldsymbol{\zeta}}\cdot\mathbf{y} &= \hat{\mathbf{k}}\cdot\mathbf{x}/\zeta,\\
\mathrm{z} &= \hat{\boldsymbol{\zeta}}\cdot\mathbf{y}' = \left[\hat{\mathrm{k}}_1\mathrm{x}'_1 + \hat{\mathrm{k}}_2\mathrm{x}'_2 + \hat{\mathrm{k}}_3\mathrm{x}'_3\right]/\zeta = \hat{\mathbf{k}}\cdot\mathbf{x}'/\zeta,\\
\hat{\mathbf{k}}\cdot\mathbf{x} &= \hat{\mathrm{k}}_1\mathrm{a}_1\mathrm{y}_1 + \hat{\mathrm{k}}_2\mathrm{a}_2\mathrm{y}_2 + \hat{\mathrm{k}}_3\mathrm{a}_3\mathrm{y}_3 = \boldsymbol{\zeta}\cdot\mathbf{y} = \zeta\hat{\boldsymbol{\zeta}}\cdot\mathbf{y},\\
\mathrm{R} &= (1-\mathrm{z}^2)^{1/2}, \qquad (6.13)
\end{aligned}$$

where it is noted that the quantities y, z, r and R are dimensionless, whereas ζ has the dimensions of length.

In addition, the differential element of area, $\mathrm{d}\hat{\mathrm{S}}(\hat{\mathbf{k}})$, on the surface of the unit sphere in **k**-space associated with the unit vector $\hat{\mathbf{k}}$ in Eq. (6.10), is transformed to an element of area, $\mathrm{d}\hat{\mathrm{S}}(\hat{\boldsymbol{\zeta}})$, associated with the new unit vector $\hat{\boldsymbol{\zeta}}$. The relationship between these two quantities can be obtained by writing them in the respective forms $\mathrm{d}\hat{\mathrm{S}}(\hat{\mathbf{k}}) = |\mathrm{d}\hat{\mathbf{k}}^{(1)}\times\mathrm{d}\hat{\mathbf{k}}^{(2)}|$ and $\mathrm{d}\hat{\mathrm{S}}(\hat{\boldsymbol{\zeta}}) = |\mathrm{d}\hat{\boldsymbol{\zeta}}^{(1)}\times\mathrm{d}\hat{\boldsymbol{\zeta}}^{(2)}|$. Then, denoting the unit base vectors for $\hat{\mathbf{k}}$ and $\hat{\boldsymbol{\zeta}}$ respectively by $\hat{\mathbf{k}}^\circ_\alpha$ and $\hat{\boldsymbol{\zeta}}^\circ_\alpha$, and using $\hat{\zeta}_\alpha = \mathrm{a}_\alpha\hat{\mathrm{k}}_\alpha/\zeta$, it is found by writing out the vector products that the components of $\mathrm{d}\hat{\mathbf{k}}^{(1)}\times\mathrm{d}\hat{\mathbf{k}}^{(2)}$ and $\mathrm{d}\hat{\boldsymbol{\zeta}}^{(1)}\times\mathrm{d}\hat{\boldsymbol{\zeta}}^{(2)}$ are in the ratio $\mathrm{a_1a_2a_3}/\zeta^3$. Therefore,

$$\mathrm{d}\hat{\mathrm{S}}(\hat{\boldsymbol{\zeta}}) = \frac{\mathrm{a_1a_2a_3}}{\zeta^3}\mathrm{d}\hat{\mathrm{S}}(\hat{\mathbf{k}}). \qquad (6.14)$$

Substituting the relevant quantities introduced above into Eq. (6.10) and using the relationship

$$\delta'(\zeta\hat{\boldsymbol{\zeta}}\cdot\mathbf{y} - \zeta\mathrm{z}) = -\frac{1}{\zeta}\frac{\partial\delta(\zeta\hat{\boldsymbol{\zeta}}\cdot\mathbf{y} - \zeta\mathrm{z})}{\partial\mathrm{z}}, \qquad (6.15)$$

we obtain the expression

$$u_i^C(\mathbf{x}) = \frac{1}{8\pi^2} C_{jlmn} \int_0^R r\mathrm{d}r \int_0^{2\pi} \mathrm{d}\psi \int_{-1}^{1} \mathrm{d}z$$

$$\times \oiint_{\hat{\mathcal{S}}} \varepsilon_{mn}^T(\mathbf{x}')\hat{k}_l(\hat{k}\hat{k})_{ij}^{-1}\zeta^2 \frac{\partial\delta(\zeta\hat{\boldsymbol{\zeta}}\cdot\mathbf{y} - \zeta z)}{\partial z}\mathrm{d}\hat{S}(\hat{\boldsymbol{\zeta}}) \tag{6.16}$$

where $R = (1 - z^2)^{1/2}$ must be a real quantity.

Next, the derivative of the delta function in Eq. (6.16) can be eliminated by integrating by parts with respect to z to obtain

$$u_i^C(\mathbf{x}) = \frac{1}{8\pi^2} C_{jlmn} \Bigg| \int_0^{2\pi} \mathrm{d}\psi \int_0^R r\mathrm{d}r$$

$$\times \oiint_{\hat{\mathcal{S}}} \varepsilon_{mn}^T(\mathbf{x}')\hat{k}_l(\hat{k}\hat{k})_{ij}^{-1}\zeta^2\delta(\zeta\hat{\boldsymbol{\zeta}}\cdot\mathbf{y} - \zeta z)\mathrm{d}\hat{S}(\hat{\boldsymbol{\zeta}})\Bigg|_{z=-1}^{z=1}$$

$$-\frac{1}{8\pi^2} C_{jlmn} \int_{-1}^{1} \delta(\zeta\hat{\boldsymbol{\zeta}}\cdot\mathbf{y} - \zeta z)\frac{\partial}{\partial z}\left[\int_0^{2\pi} \mathrm{d}\psi \int_0^R r\mathrm{d}r\right.$$

$$\left.\times \oiint_{\hat{\mathcal{S}}} \varepsilon_{mn}^T(\mathbf{x}')\hat{k}_l(\hat{k}\hat{k})_{ij}^{-1}\zeta^2\mathrm{d}\hat{S}(\hat{\boldsymbol{\zeta}})\right]\mathrm{d}z. \tag{6.17}$$

Further treatment depends upon whether $\mathbf{y}$ is inside, or outside, the unit sphere, i.e., whether the field point, $\mathbf{x}$, is inside, or outside, the inclusion.

Elastic field inside inclusion

When $\mathbf{y}$ is inside the unit sphere, inspection of Fig. 6.1 shows that a value of z exists for all possible orientations of $\hat{\boldsymbol{\zeta}}$ which satisfy the condition

$$\hat{\boldsymbol{\zeta}}\cdot\mathbf{y} - z = 0, \qquad -1 \le z \le 1. \tag{6.18}$$

The delta function in Eq. (6.17) is then non-vanishing and $R = (1 - z^2)^{1/2}$ is real, as required. The first term in Eq. (6.17) vanishes, and the derivative with respect to z of the relevant quantities in the second term, can be expressed as

$$\frac{\partial}{\partial z}\left[\varepsilon^T \int_0^R r\mathrm{d}r\right] = \frac{\partial\varepsilon^T}{\partial z}\int_0^R r\mathrm{d}r - z\left[\varepsilon^T\right]_{r=R} \tag{6.19}$$

by applying Leibniz's rule. Then, upon substituting this expression into Eq. (6.17),

$$u_i^C(\mathbf{x}) = -\frac{1}{8\pi^2}C_{jlmn}\left\{\int_{-1}^{1}dz\int_0^{2\pi}d\psi\int_0^R rdr\right.$$

$$\times \oint\!\!\!\oint_{\hat{\mathcal{S}}} \frac{\partial\varepsilon^T_{mn}(\mathbf{x}')}{\partial z}\hat{k}_l(\hat{k}\hat{k})^{-1}_{ij}\delta(\zeta\hat{\boldsymbol{\zeta}}\cdot\mathbf{y}-\zeta z)\zeta^2 d\hat{S}(\hat{\boldsymbol{\zeta}})$$

$$-\int_{-1}^{1}dz\int_0^{2\pi}d\psi$$

$$\left.\times \oint\!\!\!\oint_{\hat{\mathcal{S}}} z\left[\varepsilon^T_{mn}(\mathbf{x}')\right]_{r=R}\hat{k}_l(\hat{k}\hat{k})^{-1}_{ij}\delta(\zeta\hat{\boldsymbol{\zeta}}\cdot\mathbf{y}-\zeta z)\zeta^2 d\hat{S}(\hat{\boldsymbol{\zeta}})\right\}. \quad (6.20)$$

The integration with respect to z is carried out using the following property of the delta function (Appendix D):

$$\zeta\int_{-1}^{1}\delta(\zeta\hat{\boldsymbol{\zeta}}\cdot\mathbf{y}-\zeta z)dz = \zeta\zeta^{-1}\int_{-1}^{1}\delta(\hat{\boldsymbol{\zeta}}\cdot\mathbf{y}-z)dz = \int_{-1}^{1}\delta(\hat{\boldsymbol{\zeta}}\cdot\mathbf{y}-z)dz = 1. \quad (6.21)$$

Substitution of this expression into Eq. (6.20) yields (Mura 1987),

$$u_i^C(\mathbf{x}) = -\frac{1}{8\pi^2}C_{jlmn}\int_0^{2\pi}d\psi\left\{\int_0^R rdr\frac{\partial\varepsilon^T_{mn}(\mathbf{x}')}{\partial z} - z\left[\varepsilon^T_{mn}(\mathbf{x}')\right]_{r=R}\right\}_{z=\hat{\boldsymbol{\zeta}}\cdot\mathbf{y}}$$

$$\times \oint\!\!\!\oint_{\hat{\mathcal{S}}}\hat{k}_l(\hat{k}\hat{k})^{-1}_{ij}\zeta d\hat{S}(\hat{\boldsymbol{\zeta}}), \quad (6.22)$$

where $\hat{\mathcal{S}}$ represents the entire surface of the unit sphere, and the vector $\mathbf{x}'$ must satisfy the condition $\hat{\mathbf{k}}\cdot(\mathbf{x}-\mathbf{x}')=0$ to satisfy the condition $z=\hat{\boldsymbol{\zeta}}\cdot\mathbf{y}$ as is readily demonstrated by the use of Eq. (6.13), i.e.,

$$\hat{\boldsymbol{\zeta}}\cdot\mathbf{y}-z = \frac{\hat{\mathbf{k}}\cdot\mathbf{x}}{\zeta}-\frac{\hat{\mathbf{k}}\cdot\mathbf{x}'}{\zeta} = \frac{1}{\zeta}\hat{\mathbf{k}}\cdot(\mathbf{x}-\mathbf{x}') = 0. \quad (6.23)$$

Further details are given by Mura (1987).

Elastic field outside inclusion

When $\mathbf{y}$ is outside the inclusion, inspection of Fig. 6.1 shows that the condition given by Eq. (6.18) can be satisfied only when $\hat{\boldsymbol{\zeta}}$ impinges on the surface of the unit sphere in the region outside the shaded polar caps illustrated

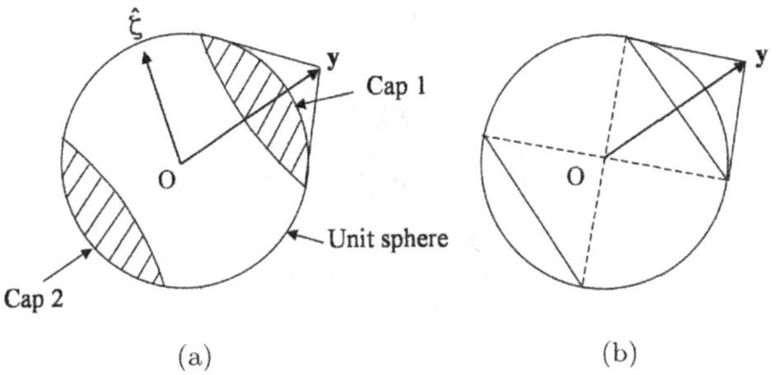

Fig. 6.2 (a) Region of the surface of the unit sphere (i.e., the region between the shaded polar caps 1 and 2) where the condition given by Eq. (6.18) is satisfied when $\mathbf{y}$ is outside the unit sphere. (b) Corresponding cross-section through center of unit sphere containing O and the vector $\mathbf{y}$.

in Fig. 6.2a. Then, Eq. (6.22) again applies (Mura 1987), but with the integration over the unit sphere restricted to the region between the polar caps where Eq. (6.18) is satisfied, i.e.,

$$\mathrm{u}_{\mathrm{i}}^{\mathrm{C,M}}(\mathbf{x}) = -\frac{1}{8\pi^2}\mathrm{C}_{\mathrm{j}l\mathrm{mn}}\int_0^{2\pi}\mathrm{d}\psi\left\{\int_0^{\mathrm{R}}\mathrm{rdr}\frac{\partial\varepsilon_{\mathrm{mn}}^{\mathrm{T}}(\mathbf{x}')}{\partial\mathrm{z}} - \mathrm{z}\left[\varepsilon_{\mathrm{mn}}^{\mathrm{T}}(\mathbf{x}')\right]_{\mathrm{r=R}}\right\}_{\mathrm{z}=\hat{\boldsymbol{\zeta}}\cdot\mathbf{y}}$$
$$\times\iint_{\hat{\mathcal{S}}^*}\hat{\mathrm{k}}_l(\hat{\mathrm{k}}\hat{\mathrm{k}})_{\mathrm{ij}}^{-1}\zeta\mathrm{d}\hat{\mathrm{S}}(\hat{\boldsymbol{\zeta}}), \qquad (6.24)$$

where the notation $\hat{\mathcal{S}}^*$ indicates the surface region between the polar caps.

Mura (1987) and Mura and Cheng (1977) have derived companion expressions for the derivatives of $\mathrm{u}_{\mathrm{i}}^{\mathrm{C}}(\mathbf{x})$, i.e., distortions, which are useful for calculating elastic strains. Further developments and details are given by Mura (1987).

6.3.1.3 *Ellipsoidal shape and uniform* $\varepsilon_{\mathrm{ij}}^{\mathrm{T}}$

Elastic field inside inclusion

When $\varepsilon_{\mathrm{mn}}^{\mathrm{T}}$ is uniform, Eq. (6.22) for $\mathbf{x}$ inside the inclusion reduces to

$$\mathrm{u}_{\mathrm{i}}^{\mathrm{C,INC}}(\mathbf{x}) = \frac{1}{4\pi}\mathrm{C}_{\mathrm{j}l\mathrm{mn}}\varepsilon_{\mathrm{mn}}^{\mathrm{T}}\oiint_{\hat{\mathcal{S}}}(\hat{\mathbf{k}}\cdot\mathbf{x})\hat{\mathrm{k}}_l(\hat{\mathrm{k}}\hat{\mathrm{k}})_{\mathrm{ij}}^{-1}\mathrm{d}\hat{\mathrm{S}}(\hat{\boldsymbol{\zeta}}) \qquad (6.25)$$

after employing Eq. (6.13). To expedite the surface integration over $\hat{\mathcal{S}}$, spherical coordinates $(\mathrm{r}=1,\theta,\phi)$, illustrated in Fig. 6.3, are now introduced. An element of area on the unit sphere is given by

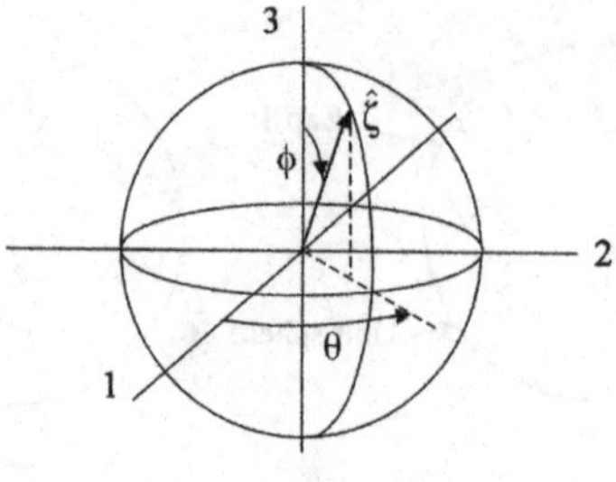

Fig. 6.3 Vector $\hat{\boldsymbol{\zeta}}$ in spherical coordinate system, $(\mathrm{r} = 1, \theta, \phi)$.

$d\hat{S}(\hat{\boldsymbol{\zeta}}) = \sin\phi d\phi d\theta = d\hat{\zeta}_3 d\theta$, and, therefore,

$$u_i^{C,INC}(\mathbf{x}) = \frac{1}{4\pi} C_{jlmn} \varepsilon_{mn}^{T} x_k \oiint_{\hat{\mathcal{S}}} \hat{k}_k \hat{k}_l (\hat{k}\hat{k})_{ij}^{-1} d\hat{\zeta}_3 d\theta$$

$$= \frac{1}{4\pi} C_{jlmn} \varepsilon_{mn}^{T} x_k \int_{-1}^{1} d\hat{\zeta}_3 \int_0^{2\pi} P_{ijkl}(\hat{\mathbf{k}}) d\theta, \qquad (6.26)$$

where

$$P_{ijkl}(\hat{\mathbf{k}}) \equiv \hat{k}_k \hat{k}_l (\hat{k}\hat{k})_{ij}^{-1}. \qquad (6.27)$$

An expression for $\varepsilon_{ij}^{C,INC}$ is obtained by using Eqs. (6.26) and (2.5) with the result

$$\varepsilon_{ij}^{C,INC} = \frac{1}{8\pi} C_{pqmn} \varepsilon_{mn}^{T} \int_{-1}^{1} d\hat{\zeta}_3 \int_0^{2\pi} \left[P_{ipjq}(\hat{\mathbf{k}}) + P_{jpiq}(\hat{\mathbf{k}})\right] d\theta, \qquad (6.28)$$

which can be written in the form

$$\varepsilon_{ij}^{C,INC} = S_{ijmn}^{E} \varepsilon_{mn}^{T}, \qquad (6.29)$$

where

$$S_{ijmn}^{E} = \frac{1}{8\pi} C_{pqmn} \int_{-1}^{1} d\hat{\zeta}_3 \int_0^{2\pi} \left[P_{ipjq}(\hat{\mathbf{k}}) + P_{jpiq}(\hat{\mathbf{k}})\right] d\theta. \qquad (6.30)$$

The fourth rank tensor, S_{ijmn}^{E}, that linearly couples the canceling strain, $\varepsilon_{ij}^{C,INC}$, with the transformation strain in Eq. (6.29), is known as the *Eshelby tensor*. This tensor is seen to be independent of x_i which establishes the important result that the canceling strain is uniform throughout an ellipsoidal inclusion when the transformation strain is uniform. Then, the final elastic strain in the inclusion, ε_{ij}^{INC}, is also uniform by virtue of Eq. (6.1).

Elastic field outside inclusion

In the matrix, Eq. (6.24) takes the form

$$u_i^{C,M}(\mathbf{x}) = \frac{1}{4\pi} C_{jlmn}\varepsilon_{mn}^T x_k \int_{-s(x)}^{s(x)} d\hat{\zeta}_3 \int_0^{2\pi} P_{ijkl}(\hat{\mathbf{k}}) d\theta, \tag{6.31}$$

where s(x) and −s(x) are the limits of $\hat{\zeta}_3$ which depend upon the distance of the field point $\mathbf{x}$ from the inclusion. Inspection of Fig. 6.2 shows that these limits decrease as x (and, therefore, y) increases, thereby causing $u_i^{C,M}(\mathbf{x})$ to decrease as x increases, as we would expect.

Finally, the displacements in the inclusion and matrix are obtained by substituting Eqs. (6.26) and (6.31) and the known u_i^T into Eq. (6.1). Mura (1987) has given results for a number of cases where the above equations have been integrated for crystals of different symmetries with limiting ellipsoidal shapes, i.e., spheres, thin-disks and needles.

In Exercise 6.1, Eqs. (6.26) and (6.31) are used to find the displacement fields in the inclusion and matrix for a homogeneous spherical inclusion with $\varepsilon_{ij}^T = \varepsilon^T \delta_{ij}$ in an isotropic system.

6.3.1.4 *Ellipsoidal shape and arbitrary* ε_{ij}^T

When the transformation strain is non-uniform, it can be represented (Asaro and Barnett 1975) by a general polynomial of the form

$$\begin{aligned}\varepsilon_{mn}^T(\mathbf{x}') &= \varepsilon_{mn}^T + \sum_\alpha \varepsilon_{mn\alpha}^T \left(\frac{x'_\alpha}{a_\alpha}\right) + \sum_\alpha \sum_\beta \varepsilon_{mn\alpha\beta}^T \left(\frac{x'_\alpha}{a_\alpha}\right)\left(\frac{x'_\beta}{a_\beta}\right) + \cdots\cdots \\ &= \varepsilon_{mn}^T + \varepsilon_{mnp}^T y'_p + \varepsilon_{mnpq}^T y'_p y'_q + \cdots\cdots \end{aligned} \tag{6.32}$$

From Fig. (6.1), the vector $\mathbf{y}'$ can be written as

$$y'_i = z\hat{\zeta}_i + r[\cos\psi \hat{m}_i + \sin\psi \hat{n}_i] \tag{6.33}$$

and using the binomial theorem, quantities such as $(y'_1)^{\alpha_1}$ can then be written as

$$(y'_1)^{\alpha_1} = \sum_{\beta_1=0}^{\alpha_1} \frac{\alpha_1!}{\beta_1!(\alpha_1-\beta_1)!} \left(z\hat{\zeta}_1\right)^{\alpha_1-\beta_1} r^{\beta_1}[\cos\psi \hat{m}_1 + \sin\psi \hat{n}_1]. \tag{6.34}$$

Since all equations are linear with respect to ε_{ij}^T and its derivatives, the displacement field due to each term retained in Eq. (6.32) can be determined and then simply summed to obtain the final result. The required canceling displacement field associated with each term can be obtained by employing Eq. (6.22) or (6.24), and integrating over both ψ and the unit sphere.

Consider the relatively simple case where the transformation strain is a polynomial of the first degree in the form $\varepsilon^{T}_{mn}(\mathbf{x}') = \varepsilon^{T}_{mnp}y'_{p}$. Then, using Eqs. (6.32) and (6.33),

$$\frac{\partial \varepsilon^{T}_{mn}(\mathbf{x}')}{\partial z} = \varepsilon^{T}_{mnp}\hat{\zeta}_{p}. \tag{6.35}$$

Next, substituting this expression and Eq. (6.33) into Eq. (6.22),

$$u^{C}_{i}(\mathbf{x}) = -\frac{\varepsilon^{T}_{mnp}}{8\pi^{2}}\int_{0}^{2\pi} d\psi \left\{ \int_{0}^{R} r dr \hat{\zeta}_{p} - z[z\hat{\zeta}_{p} + R[\cos\psi \hat{m}_{p} + \sin\psi \hat{n}_{p}] \right\}_{z=\hat{\boldsymbol{\zeta}}\cdot\mathbf{y}}$$
$$\times \oiint_{\hat{\mathcal{S}}} C_{jlmn}\hat{k}_{l}(\hat{k}\hat{k})^{-1}_{ij}\zeta d\hat{S}(\hat{\boldsymbol{\zeta}}). \tag{6.36}$$

Then, after performing the first two integrations,

$$u^{C}_{i}(\mathbf{x}) = -\frac{\varepsilon^{T}_{mnp}}{8\pi}\oiint_{\hat{\mathcal{S}}}\left[1 - 3(\hat{\boldsymbol{\zeta}}\cdot \mathrm{y})^{2}\right]\hat{\zeta}_{p}C_{jlmn}\hat{k}_{l}(\hat{k}\hat{k})^{-1}_{ij}\zeta d\hat{S}(\hat{\boldsymbol{\zeta}}). \tag{6.37}$$

Next, substituting $\hat{\boldsymbol{\zeta}}\cdot\mathbf{y} = (\hat{\mathbf{k}}\cdot\mathbf{x})\zeta^{-1}$, from Eq. (6.13), into Eq. (6.37), and using Eq. (2.5), the corresponding canceling strain in the inclusion, $\varepsilon^{C,INC}_{kl}$, is given by

$$\varepsilon^{C,INC}_{kl} = \frac{3}{8\pi}\varepsilon^{T}_{qnp}\left\{\oiint_{\hat{\mathcal{S}}} C^{INC}_{jmqn}\hat{\zeta}_{p}\hat{k}_{m}\hat{k}_{s}[\hat{k}_{l}(\hat{k}\hat{k})^{-1}_{kj} + \hat{k}_{k}(\hat{k}\hat{k})^{-1}_{lj}]\zeta^{-1}d\hat{S}(\hat{\boldsymbol{\zeta}})\right\}x_{s} \tag{6.38}$$

or, alternatively,

$$\varepsilon^{C,INC}_{kl} = \varepsilon^{T}_{qnp}\Gamma_{qnpkls}x_{s} \tag{6.39}$$

where

$$\Gamma_{qnpkls} = \frac{3}{8\pi}\oiint_{\hat{\mathcal{S}}} C^{INC}_{jmqn}\hat{\zeta}_{p}\hat{k}_{m}\hat{k}_{s}\left[\hat{k}_{l}(\hat{k}\hat{k})^{-1}_{kj} + \hat{k}_{k}(\hat{k}\hat{k})^{-1}_{lj}\right]\zeta^{-1}d\hat{S}(\hat{\boldsymbol{\zeta}}). \tag{6.40}$$

Equation (6.39) shows that the canceling strain in the inclusion is a polynomial of the first degree and, therefore, since the transformation strain is also a polynomial of the first degree, the strain in the inclusion is a polynomial of the first degree. In Exercise 6.2 it is shown that when the transformation strain is a second degree polynomial, the strain in the inclusion is also a second degree polynomial. As shown by Asaro and Barnett (1975) and Mura (1987) this result, i.e., that the inclusion strain is a polynomial of the same degree as the transformation strain polynomial, is true for polynomial strains of any degree. However, the proofs are lengthy and will not be presented here. These results, therefore, lead to the *polynomial theorem*:

> *The canceling strain field within an inclusion possessing a transformation strain that is a polynomial of degree M in the* $\mathbf{x}_i$ *coordinates, is itself a polynomial of degree M in the* $\mathbf{x}_i$ *coordinates.*

This will be of use in the formulation of the equivalent homogenous inclusion method for treating inhomogeneous inclusions in Sec. 6.3.2.2.

6.3.2 *Elastic field of inhomogeneous ellipsoidal inclusion*

6.3.2.1 *Uniform* ε_{ij}^{T}

Having solutions for homogeneous ellipsoidal inclusions, solutions for corresponding inhomogeneous ellipsoidal inclusions can now be found by employing the *equivalent homogeneous inclusion method* of Eshelby (1961). This involves the construction of a fictitious homogeneous inclusion that is "equivalent" to the corresponding actual inhomogeneous inclusion. The solution for this equivalent homogeneous inclusion, which can be found by using our previous methods for solving homogeneous inclusion problems, then provides the desired solution for the inhomogeneous inclusion.

First, the inhomogeneous inclusion is constructed by the procedure described in Sec. 6.2, i.e.:

(1) Cut out of the matrix the region intended for the inclusion.
(2) Change its elastic constants from C_{ijkl}^{M} to C_{ijkl}^{INC}.
(3) Subject it to the uniform transformation strain, ε_{ij}^{T}.
(4) Impose on it the elastic strain, $-\varepsilon_{ij}^{T}$, by applying suitable forces.
(5) Incorporate it back in its original cavity.
(6) Apply forces that cancel the forces of step 4 and generate the elastic strain $\varepsilon_{ij}^{C,INC}$, in the inclusion and the accompanying $\varepsilon_{ij}^{C,M}$ strain in the matrix.

The final elastic strain in the inclusion is therefore given by

$$\varepsilon_{ij}^{INC} = \varepsilon_{ij}^{C,INC} - \varepsilon_{ij}^{T} \tag{6.41}$$

and in the matrix by

$$\varepsilon_{ij}^{M} = \varepsilon_{ij}^{C,M}, \tag{6.42}$$

and the final size and shape of the inclusion differ from its initial size and shape by the strain $\varepsilon_{ij}^{C,INC}$.

Next, the equivalent homogeneous inclusion is constructed as follows:

(1) Cut it out of the matrix with the same size and shape as the corresponding inhomogeneous inclusion.
(2) Subject it to the transformation strain, $\varepsilon_{ij}^{T^*}$.[2]
(3) Subject it to the elastic strain, $-\varepsilon_{ij}^{T^*}$, by applying suitable forces.
(4) Incorporate it back in its cavity.
(5) Apply forces that cancel the forces of step 3 and generate the elastic strain, $\varepsilon_{ij}^{C,INC^*}$, in the inclusion and ε_{ij}^{C,M^*} in the matrix.

The final elastic strains in the equivalent homogeneous inclusion and the matrix are then

$$\varepsilon_{ij}^{INC^*} = \varepsilon_{ij}^{C,INC^*} - \varepsilon_{ij}^{T^*}, \qquad \varepsilon_{ij}^{M^*} = \varepsilon_{ij}^{C,M^*}, \tag{6.43}$$

and the size and shape of the inclusion differ from its initial size and shape by $\varepsilon_{ij}^{C,INC^*}$. Now, if the two inclusions are under the same stress and have the same final size and shape, they can be interchanged without disturbing the matrix, i.e., they will be "equivalent". This requirement will be met if the two conditions

$$\sigma_{ij}^{INC^*} = \sigma_{ij}^{INC} \qquad \text{and} \qquad \varepsilon_{ij}^{C,INC^*} = \varepsilon_{ij}^{C,INC} \tag{6.44}$$

are satisfied. To satisfy the equal stress condition, the relationship

$$\sigma_{ij}^{C,INC^*} - \sigma_{ij}^{T^*} = \sigma_{ij}^{C,INC} - \sigma_{ij}^{T} \tag{6.45}$$

must be satisfied. Then, using Hooke's law, and substituting the equal strain condition, $\varepsilon_{ij}^{C,INC^*} = \varepsilon_{ij}^{C,INC}$,

$$C_{ijkl}^{M}\left(\varepsilon_{kl}^{C,INC^*} - \varepsilon_{kl}^{T^*}\right) = C_{ijkl}^{INC}\left(\varepsilon_{kl}^{C,INC^*} - \varepsilon_{kl}^{T}\right). \tag{6.46}$$

However, since the transformation strain, $\varepsilon_{kl}^{T^*}$, is uniform, and the equivalent homogeneous inclusion is uniform, the relationship $\varepsilon_{kl}^{C,INC^*} = S_{klmn}^{E}\varepsilon_{mn}^{T^*}$ given by Eq. (6.29) is valid, and substituting this into Eq. (6.46),

$$\left[\left(C_{ijkl}^{INC} - C_{ijkl}^{M}\right)S_{klmn}^{E} + C_{ijmn}^{M}\right]\varepsilon_{mn}^{T^*} = C_{ijmn}^{INC}\varepsilon_{mn}^{T}. \tag{6.47}$$

Then, by introducing the tensor Y_{ijmn}^{INC} defined by

$$Y_{ijmn}^{INC} \equiv \left(C_{ijkl}^{INC} - C_{ijkl}^{M}\right)S_{klmn}^{E} + C_{ijmn}^{M}. \tag{6.48}$$

---

[2] Quantities associated with the equivalent homogeneous inclusion are indicated by an asterisk.

the solution for $\varepsilon^{T^*}_{mn}$ can be written in the matrix form

$$\left[\varepsilon^{T^*}\right] = \left[Y^{INC}\right]^{-1}\left[C^{INC}\right]\left[\varepsilon^{T}\right]. \tag{6.49}$$

The problem is now solved. The elastic strain in the inhomogeneous inclusion is first written as

$$\varepsilon^{INC}_{ij} = \varepsilon^{C,INC}_{ij} - \varepsilon^{T}_{ij} = \varepsilon^{C,INC^*}_{ij} - \varepsilon^{T}_{ij} = S^{E}_{ijkl}\varepsilon^{T^*}_{kl} - \varepsilon^{T}_{ij} \tag{6.50}$$

after using Eqs. (6.1), (6.44) and (6.29). Then, the transformation strain of the equivalent homogeneous inclusion, $\varepsilon^{T^*}_{mn}$, given by Eq. (6.49), is substituted into Eq. (6.50) to obtain

$$\left[\varepsilon^{INC}\right] = \left(\left[S^{E}\right]\left[Y^{INC}\right]^{-1}\left[C^{INC}\right] - [I]\right)\left[\varepsilon^{T}\right]. \tag{6.51}$$

Finally, the elastic field in the matrix, which is the same as the corresponding field in the matrix surrounding the equivalent homogeneous inclusion, is found by employing the methods described in Sec. 6.3.1 for homogeneous inclusions.

A complete solution for the inclusion and matrix displacement fields in the case of an inhomogeneous spherical inclusion with a uniform transformation strain in an isotropic system is presented in Sec. 6.4.3.1.

6.3.2.2 *Non-uniform* ε^{T}_{ij} *represented by polynomial*

When the transformation strain of the inhomogeneous inclusion, ε^{T}_{ij}, is non-uniform it can be represented generally by a polynomial in x_i as in Eq. (6.32). Since all of the relevant equations are linear, each term in the polynomial can be substituted by itself for ε^{T}_{kl} in Eq. (6.46). If the polynomial is of degree M, and the transformation strain of the equivalent homogeneous inclusion, $\varepsilon^{T^*}_{kl}$, is also taken to be a polynomial of degree M, then, by the polynomial theorem on p. 185, the corresponding canceling strain, $\varepsilon^{C,INC^*}_{kl}$, in Eq. (6.46) will also be a polynomial of degree M. Then, by collecting the coefficients of common quantities in x_i that appear in Eq. (6.46), a set of simultaneous linear equations can be developed that yield the required coefficients in the assumed polynomial expression for $\varepsilon^{T^*}_{kl}$.

Consider now this method for the case of a transformation strain given by the first order polynomial

$$\varepsilon^{T}_{mn} = \sum_{\alpha}\varepsilon^{T}_{mn\alpha}\left(\frac{x'_{\alpha}}{a_{\alpha}}\right) = \varepsilon^{T}_{mnp}y'_{p}. \tag{6.52}$$

Following the above procedure, writing $\varepsilon_{kl}^{T^*}$ in the same polynomial form as Eq. (6.52), and using Eqs. (6.52) and (6.39), Eq. (6.46) takes the form

$$C_{ijkl}^{M}\left[\varepsilon_{qnp}^{T^*}\sum_{\alpha}\Gamma_{qnpkl\alpha}x'_{\alpha}-\sum_{\alpha}\left(\frac{\varepsilon_{kl\alpha}^{T^*}}{a_{\alpha}}\right)x'_{\alpha}\right]$$

$$=C_{ijkl}^{INC}\left[\varepsilon_{qnp}^{T^*}\sum_{\alpha}\Gamma_{qnpkl\alpha}x'_{\alpha}-\sum_{\alpha}\left(\frac{\varepsilon_{kl\alpha}^{T}}{a_{\alpha}}\right)x'_{\alpha}\right]. \tag{6.53}$$

Because of the symmetry requirement, $\varepsilon_{kl\alpha}^{T^*}=\varepsilon_{lk\alpha}^{T^*}$, there are 18 unknown $\varepsilon_{kl\alpha}^{T^*}$ coefficients to be determined. Since the x'_{α} are independent, Eq. (6.53) provides 3 equations corresponding to $\alpha=1,2,3$, and each of these provide 6 equations corresponding to $ij=11,22,33,12,13,23$. There are therefore 18 linear equations available to solve for the 18 unknown polynomial coefficients. Having solved for the transformation strain in the equivalent homogeneous inclusion, $\varepsilon_{kl}^{T^*}$, the equivalent homogeneous inclusion problem and corresponding inhomogeneous problem can be solved using the methods of Sec. 6.3.1.

6.3.3 *Strain energy*

The strain energy due to an inclusion in an infinite homogeneous region can be found by considering step 4 of the procedure for producing the inclusion described in Sec. 6.2. At the point where the inclusion has been put back into its cavity and the applied forces have not yet been removed, the inclusion is subjected to the elastic strain, $-\varepsilon_{ij}^{T}(\mathbf{x})$, and corresponding stress, $-\sigma_{ij}^{T}(\mathbf{x})$, and the matrix is strain-free. The strain energy in the entire system is then

$$W'=\frac{1}{2}\oiiint_{V^{INC}}\sigma_{ij}^{T}\varepsilon_{ij}^{T}dV. \tag{6.54}$$

When the distribution of applied force is now cancelled, we can imagine that this occurs by a process in which the forces are simply allowed to relax to zero thus causing the canceling displacement, u_i^C, to reach its maximum (Eshelby 1961). Since these quantities are linearly coupled, and the distribution of applied force is

$$dF_i=-\sigma_{ij}^{T}\hat{n}_j dS, \tag{6.55}$$

the energy released is

$$\Delta W=\frac{1}{2}\oiint_{S^{INC}}\sigma_{ij}^{T}u_i^{C,INC}\hat{n}_j dS=\frac{1}{2}\oiiint_{V^{INC}}\sigma_{ij}^{T}\varepsilon_{ij}^{C,INC}dV \tag{6.56}$$

after using the divergence theorem. The energy remaining is then the final strain energy in the entire system, given by the difference between W' and ΔW, i.e.,

$$W^{INC} = W' - \Delta W = \frac{1}{2}\oiiint_{V^{INC}} \sigma_{ij}^{T}\left(\varepsilon_{ij}^{T} - \varepsilon_{ij}^{C,INC}\right) dV$$

$$= -\frac{1}{2}\oiiint_{V^{INC}} \sigma_{ij}^{T}\varepsilon_{ij}^{INC} dV = -\frac{1}{2}\oiiint_{V^{INC}} \sigma_{ij}^{INC}\varepsilon_{ij}^{T} dV \quad (6.57)$$

after use of Eqs. (6.1) and (2.102).

We therefore have the remarkably simple result that the strain energy depends only upon the transformation strain and the stress within the inclusion. This result is quite general and applies to both homogeneous and inhomogeneous inclusions with arbitrary shapes and transformation strains in both general and isotropic systems. An alternative derivation of Eq. (6.57) is carried out in Exercise 6.5.

6.4 Coherent Inclusions in Isotropic Systems

6.4.1 *Elastic field of homogeneous inclusion by Fourier transform method*

The preceding equations, which were obtained using the Fourier transform method, can be converted to corresponding equations valid for an isotropic system by use of Eqs. (2.120) and (3.141). They can then be solved subject to the prevailing boundary conditions. For convenience, several factors that appear frequently in the previous equations for general systems, along with their corresponding isotropic forms, are presented here:

$$(\hat{k}\hat{k})_{ij} = (\hat{k}_k C_{kijl}\hat{k}_l) \rightarrow \mu\left[\delta_{ij} + \frac{1}{1-2\nu}\hat{k}_i\hat{k}_j\right],$$

$$(\hat{k}\hat{k})^{-1}_{ij} = (\hat{k}_k C_{kijl}\hat{k}_l)^{-1} \rightarrow \frac{1}{\mu}\left[\delta_{ij} + \frac{1}{2(1-\nu)}\hat{k}_i\hat{k}_j\right], \quad (6.58)$$

$$C_{jlmn}\varepsilon^{T}_{mn}\hat{k}_l(\hat{k}\hat{k})^{-1}_{ij} \rightarrow 2\varepsilon^{T}_{im}\hat{k}_m - \frac{1}{1-\nu}\hat{k}_i\left[\hat{k}_m\hat{k}_n\varepsilon^{T}_{mn} - \nu\varepsilon^{T}_{kk}\right]. \quad (6.59)$$

This procedure is used in Exercise 6.1 where, in an isotropic system, the displacement field of a spherical inclusion with the uniform transformation strain $\varepsilon^{T}_{ij} = \delta_{ij}\varepsilon^{T}$ is found by use of Eqs. (6.26) and (6.31) after converting them to forms valid for isotropic systems.

6.4.2 *Elastic field of homogeneous inclusion by Green's function method*

As shown in Sec. 4.3.2, relatively simple analytical expressions exist for the point force Green's functions for an infinite isotropic body. The Green's function method is, therefore, an attractive method for finding further expressions for the displacement field of inclusions in isotropic systems as first shown by Eshelby (1957). In the following, Eshelby's method, which also involves elements of classical potential theory, is applied to inclusions with uniform transformation strains. Many of the results can be expressed in relatively simple analytical forms and provide insights into the general elastic behavior of inclusions.

6.4.2.1 *Arbitrary shape and uniform* ε^{T}_{ij}

Starting with Eq. (3.168), and substituting the Green's function applicable for an isotropic system given by Eq. (4.110) while using $\partial G_{ij}(\mathbf{x}-\mathbf{x}')/\partial x'_l = -\partial G_{ij}(\mathbf{x}-\mathbf{x}')/\partial x_l$,

$$\mathbf{u}^{C}_{i}(\mathbf{x}) = -\frac{1}{4\pi\mu^{M}}\sigma^{T}_{ik}\frac{\partial\phi(\mathbf{x})}{\partial x_k} + \frac{1}{16\pi\mu^{M}(1-\nu^{M})}\sigma^{T}_{jk}\frac{\partial^3\psi(\mathbf{x})}{\partial x_i\partial x_j\partial x_k}, \tag{6.60}$$

where

$$\phi(\mathbf{x}) \equiv \oiiint_{\mathcal{V}^{INC}} \frac{dV'}{|\mathbf{x}-\mathbf{x}'|}, \qquad \psi(\mathbf{x}) \equiv \oiiint_{\mathcal{V}^{INC}} |\mathbf{x}-\mathbf{x}'|dV', \tag{6.61}$$

The functions $\phi(\mathbf{x})$ and $\psi(\mathbf{x})$ are well-known potentials termed, respectively, the *Newtonian potential* (Kellogg 1929; MacMillan 1930) and the *biharmonic potential* (of attracting mass of unit density distributed over the volume $\mathcal{V}^{INC}$): they are related by

$$\begin{aligned}\nabla^2\psi(\mathbf{x}) &= \nabla^2 \oiiint_{\mathcal{V}^{INC}} |\mathbf{x}-\mathbf{x}'|dV' = \oiiint_{\mathcal{V}^{INC}} \nabla^2|\mathbf{x}-\mathbf{x}'|dV' \\ &= 2\oiiint_{\mathcal{V}^{INC}} \frac{1}{|\mathbf{x}-\mathbf{x}'|}dV' = 2\phi(\mathbf{x}).\end{aligned} \tag{6.62}$$

Also, using Eqs. (6.61) and (D.6),

$$\nabla^2\phi(\mathbf{x}) \equiv \oiiint_{\mathcal{V}^{INC}} \nabla^2\frac{1}{|\mathbf{x}-\mathbf{x}'|}dV' = -4\pi\oiiint_{\mathcal{V}^{INC}} \delta(\mathbf{x}-\mathbf{x}')dV' = -4\pi. \tag{6.63}$$

Therefore,

$$\nabla^2\phi(\mathbf{x}) = \begin{cases} -4\pi & \text{inside } \mathcal{V}^{\mathrm{INC}} \\ 0 & \text{outside } \mathcal{V}^{\mathrm{INC}} \end{cases} \qquad \nabla^4\psi(\mathbf{x}) = \begin{cases} -8\pi & \text{inside } \mathcal{V}^{\mathrm{INC}} \\ 0 & \text{outside } \mathcal{V}^{\mathrm{INC}} \end{cases}. \tag{6.64}$$

Equation (6.64) follows from Eqs. (6.62) and (6.63) and the fact that $\mathbf{x}'$ is restricted to $\mathcal{V}^{\mathrm{INC}}$ in Eq. (6.63).

Equation (6.60) is quite general, is valid for regions inside as well as outside the inclusion, and is restricted only by the assumption of a uniform transformation strain. In the following, several useful results for the elastic fields in the inclusion and matrix are first obtained by the use of Eq. (6.60) without requiring its complete solution (Eshelby 1957).

Dilatation in inclusion and matrix

The cubical dilatation, $e^{\mathrm{C}}(\mathbf{x})$, in both the inclusion and matrix, can be found directly without knowledge of ψ. Using Eqs. (6.60) and (6.62),

$$\begin{aligned} e^{\mathrm{C}}(\mathbf{x}) = \frac{\partial u_i^{\mathrm{C}}}{\partial x_i} &= -\frac{1}{4\pi\mu^{\mathrm{M}}}\sigma_{jk}^{\mathrm{T}}\frac{\partial^2\phi}{\partial x_j\partial x_k} + \frac{1}{16\pi\mu^{\mathrm{M}}(1-\nu^{\mathrm{M}})}\sigma_{jk}^{\mathrm{T}}\frac{\partial^4\psi}{\partial x_i^2\partial x_j\partial x_k} \\ &= -\frac{1}{4\pi\mu^{\mathrm{M}}}\sigma_{jk}^{\mathrm{T}}\frac{\partial^2\phi}{\partial x_j\partial x_k} + \frac{1}{16\pi\mu^{\mathrm{M}}(1-\nu^{\mathrm{M}})}\sigma_{jk}^{\mathrm{T}}\frac{\partial^2}{\partial x_j\partial x_k}\left(\nabla^2\psi\right) \\ &= -\frac{(1-2\nu^{\mathrm{M}})}{8\pi\mu^{\mathrm{M}}(1-\nu^{\mathrm{M}})}\sigma_{jk}^{\mathrm{T}}\frac{\partial^2\phi}{\partial x_j\partial x_k}, \end{aligned} \tag{6.65}$$

which is seen to be independent of ψ. This result is used in Exercise 6.3 to find $e^{\mathrm{C,INC}}$ and $e^{\mathrm{C,M}}$ for an inclusion with the transformation strain $\varepsilon_{ij}^{\mathrm{T}} = \delta_{ij}\varepsilon^{\mathrm{T}}$.

Elastic field in matrix at inclusion/matrix interface

For an inclusion with a uniform transformation strain, the elastic field in the matrix is generally more difficult to determine than in the inclusion. However, by using general potential theory, the elastic field in the matrix at the inclusion/matrix interface can be readily found from knowledge of the field at a directly opposite point across the interface just inside the inclusion. This is useful, since the stress concentration in the matrix directly adjoining the inclusion, where it is generally at a maximum, is often of special interest.

According to classical potential theory (Poincare 1899), the second derivatives of any potential $\mathrm{v}(\mathbf{x})$ obeying Poisson's equation, $\nabla^2\mathrm{v}(\mathbf{x}) = -4\pi\rho(\mathbf{x})$, as in Eq. (D.5), undergo discontinuous jumps whenever $\mathbf{x}$ crosses a surface $\mathcal{S}$ separating two regions of different density ρ.[3] When the potential is the Newtonian potential ϕ, and $\mathcal{S}$ encloses a region $\mathcal{V}$, and $\mathbf{x}$ crosses $\mathcal{S}$ from the inside to the outside at a point where the inclination of $\mathcal{S}$ is indicated by $\hat{\mathbf{n}}$, the jump is given by

$$\frac{\partial^2\phi^{\mathrm{IN}}}{\partial x_i \partial x_j} - \frac{\partial^2\phi^{\mathrm{OUT}}}{\partial x_i \partial x_j} = -4\pi\hat{n}_i\hat{n}_j(\rho^{\mathrm{IN}} - \rho^{\mathrm{OUT}}) = -4\pi\hat{n}_i\hat{n}_j \tag{6.66}$$

when the mass densities inside and outside $\mathcal{V}$ are unity and zero, respectively. By differentiating Eq. (6.62) for the biharmonic potential, we have

$$\nabla^2\left(\frac{\partial^2\psi}{\partial x_k \partial x_l}\right) = 2\frac{\partial^2\phi}{\partial x_k \partial x_l}. \tag{6.67}$$

Comparison of this with Poisson's equation shows that the quantity $\partial^2\psi/(\partial x_k \partial x_l)$ behaves as a potential produced by an effective density $\rho' = -[1/(2\pi)][\partial^2\phi/(\partial x_k \partial x_l)]$. Therefore, substituting these results into Eq. (6.66),

$$\begin{aligned}\frac{\partial^4\psi^{\mathrm{IN}}}{\partial x_i \partial x_j \partial x_k \partial x_l} &- \frac{\partial^4\psi^{\mathrm{OUT}}}{\partial x_i \partial x_j \partial x_k \partial x_l} \\ &= -4\pi\hat{n}_i\hat{n}_j(\rho'^{\mathrm{IN}} - \rho'^{\mathrm{OUT}}) \\ &= 2\hat{n}_i\hat{n}_j\left[\frac{\partial^2\phi^{\mathrm{IN}}}{\partial x_k \partial x_l} - \frac{\partial^2\phi^{\mathrm{OUT}}}{\partial x_k \partial x_l}\right] = 2\hat{n}_i\hat{n}_j(-4\pi\hat{n}_k\hat{n}_l) = -8\pi\hat{n}_i\hat{n}_j\hat{n}_k\hat{n}_l\end{aligned} \tag{6.68}$$

Next, the canceling strain is obtained by applying Eq. (2.5) to the general solution for u_i^{C} given by Eq. (6.60), i.e.,

$$\varepsilon_{il}^{\mathrm{C}} = \frac{\sigma_{jk}^{\mathrm{T}}}{16\pi\mu^{\mathrm{M}}(1-\nu^{\mathrm{M}})}\frac{\partial^4\psi}{\partial x_i \partial x_j \partial x_k \partial x_l} - \frac{1}{8\pi\mu^{\mathrm{M}}}\left(\sigma_{ik}^{\mathrm{T}}\frac{\partial^2\phi}{\partial x_k \partial x_l} + \sigma_{lk}^{\mathrm{T}}\frac{\partial^2\phi}{\partial x_k \partial x_i}\right). \tag{6.69}$$

Then, using Eqs. (6.66) and (6.68) to evaluate the derivatives in Eq. (6.69), the difference between the canceling strains in the matrix and the inclusion

[3] ρ is a generalized density corresponding to mass density in the case of a Newtonian potential and electrical charge density in the case of an electrostatic potential.

directly across the inclusion/matrix interface is

$$\left(\varepsilon_{il}^{C,M} - \varepsilon_{il}^{C,INC}\right)_{INC/M} = \frac{1}{2\mu^M}\left[\frac{\sigma_{jk}^T}{(1-\nu^M)}\hat{n}_i\hat{n}_j\hat{n}_k\hat{n}_l - \left(\sigma_{ik}^T\hat{n}_k\hat{n}_l + \sigma_{lk}^T\hat{n}_i\hat{n}_k\right)\right]. \tag{6.70}$$

Elastic field in matrix far from inclusion

By virtue of Eq. (6.1), this problem can be solved directly by finding an expression for the canceling displacements, u_i^C, that occur in step 5 of the procedure described in Sec. 6.2. Using Eq. (3.16) and Fig. 3.1, and the Green's function given by Eq. (4.110), the canceling displacement at $\mathbf{x}$ due to the application of the distribution of body force, $dF_j = \sigma_{jk}^T\hat{n}_k dS$, that is applied in this step is

$$u_i^C(\mathbf{x}) = \frac{1}{16\pi\mu^M(1-\nu^M)}\oiint_{\mathcal{S}^{INC}}\left[\frac{(3-4\nu)}{|\mathbf{x}-\mathbf{x}'|}\delta_{ij} + \frac{(x_i - x_i')(x_j - x_j')}{|\mathbf{x}-\mathbf{x}'|^3}\right]\sigma_{jk}^T\hat{n}_k dS'. \tag{6.71}$$

Then, using the divergence theorem,

$$\begin{aligned} u_i^C(\mathbf{x}) &= \frac{\sigma_{jk}^T}{16\pi\mu^M(1-\nu^M)}\iiint_{V^{INC}}\frac{1}{|\mathbf{x}-\mathbf{x}'|^2}f_{ijk}(\hat{\boldsymbol{l}})dV' \\ &= \frac{\varepsilon_{jk}^T}{8\pi(1-\nu^M)}\iiint_{V^{INC}}\frac{1}{|\mathbf{x}-\mathbf{x}'|^2}g_{ijk}(\hat{\boldsymbol{l}})dV', \end{aligned} \tag{6.72}$$

where $\hat{\boldsymbol{l}}$ is the unit directional vector

$$\hat{\boldsymbol{l}} = (\mathbf{x}-\mathbf{x}')/|\mathbf{x}-\mathbf{x}'| \tag{6.73}$$

and f_{ijk} and g_{ijk} are given by

$$\begin{aligned} f_{ijk} &= (1-2\nu)(\delta_{ij}\hat{l}_k + \delta_{ik}\hat{l}_j) - \delta_{jk}\hat{l}_i + 3\hat{l}_i\hat{l}_j\hat{l}_k, \\ g_{ijk} &= (1-2\nu)(\delta_{ij}\hat{l}_k + \delta_{ik}\hat{l}_j - \delta_{jk}\hat{l}_i) + 3\hat{l}_i\hat{l}_j\hat{l}_k. \end{aligned} \tag{6.74}$$

Equation (6.72) is valid both inside and outside the inclusion. At distances $|\mathbf{x}-\mathbf{x}'|$ in the matrix from the inclusion that are large compared with its size, f_{ijk} and g_{ijk} become essentially constant over the integrations. In this case

$$u_i^{C,M}(\mathbf{x}) = \frac{\sigma_{jk}^{T} f_{ijk} V^{INC}}{16\pi\mu^{M}(1-\nu^{M})} \frac{1}{|\mathbf{x}-\mathbf{x}'|^2} = \frac{\varepsilon_{jk}^{T} g_{ijk} V^{INC}}{8\pi(1-\nu^{M})} \frac{1}{|\mathbf{x}-\mathbf{x}'|^2}, \quad (6.75)$$

and the displacement is seen to fall off with distance from the inclusion as $|\mathbf{x}-\mathbf{x}'|^{-2}$.

6.4.2.2 *Ellipsoidal shape and uniform* ε_{ij}^{T}

Let us now consider homogeneous inclusions of ellipsoidal shape in isotropic systems following Eshelby (1957; 1961) and Mura (1987). Fortunately, the potentials associated with ellipsoidal bodies are well known (Kellogg 1929: MacMillan 1930). These have different forms inside and outside the inclusion, and the elastic fields in the inclusion and matrix are therefore treated separately.

Elastic field inside inclusion

The canceling displacement within the inclusion, $u_i^{C,INC}(\mathbf{x})$, is obtained (Eshelby 1957) by integrating Eq. (6.72) over the volume of the inclusion subject to the condition of Eq. (6.11). With the origin at the center of the ellipsoid, the geometry is shown in Fig. 6.4.

The conical volume element with apex angle, $d\omega$, has its vertex at $\mathbf{x}$, its axis along the unit vector, $\hat{\boldsymbol{l}}$, and intersects the inclusion surface at the distance $\rho(\hat{\boldsymbol{l}})$. The differential volume element, located at $\mathbf{x}'$, has the volume $dV' = |\mathbf{x}'-\mathbf{x}|^2 d|\mathbf{x}'-\mathbf{x}| d\omega$. For convenience, the positive direction of $\hat{\boldsymbol{l}}$ (see Eq. (6.73)) has been reversed and now points from the field point to the surface of the ellipsoid. This has required a change of the sign of g_{ijk}, since, from Eq. (6.74), g_{ijk} is an odd function of $\hat{\boldsymbol{l}}$. Equation (6.72) is first integrated along $|\mathbf{x}-\mathbf{x}'|$ and then over ω by integrating over the surface,

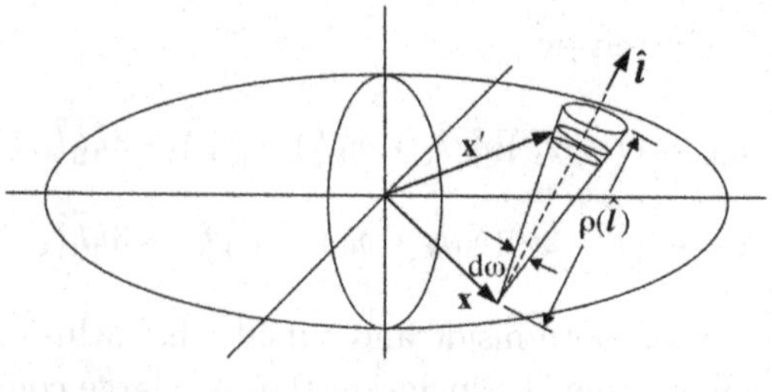

Fig. 6.4 Geometry for integration of Eq. (6.72) over ellipsoidal inclusion.

$\hat{\mathcal{S}}$, of a unit sphere centered on the field point at $\mathbf{x}$. Therefore,

$$
\begin{aligned}
u_i^{C,INC}(\mathbf{x}) &= \frac{\varepsilon_{jk}^T}{8\pi(1-\nu^M)} \oiiint_{V^{INC}} \frac{1}{|\mathbf{x}-\mathbf{x}'|^2} g_{ijk}(\hat{\boldsymbol{l}}) dV' \\
&= -\frac{\varepsilon_{jk}^T}{8\pi(1-\nu^M)} \oiint_{\hat{\mathcal{S}}} \int_0^{\rho(\hat{\boldsymbol{l}})} d|\mathbf{x}-\mathbf{x}'| g_{ijk}(\hat{\boldsymbol{l}}) d\hat{S}(\hat{\boldsymbol{l}}) \\
&= -\frac{\varepsilon_{jk}^T}{8\pi(1-\nu^M)} \oiint_{\hat{\mathcal{S}}} \rho(\hat{\boldsymbol{l}}) g_{ijk}(\hat{\boldsymbol{l}}) d\hat{S}(\hat{\boldsymbol{l}}).
\end{aligned} \tag{6.76}
$$

An expression for $\rho(\hat{\boldsymbol{l}})$ in the ellipsoidal body can be found by writing

$$
\frac{X_1^2}{a_1^2} + \frac{X_2^2}{a_2^2} + \frac{X_3^2}{a_3^2} = 1, \tag{6.77}
$$

where $\mathbf{X} = \mathbf{x} + \rho\hat{\boldsymbol{l}}$, so that

$$
\frac{(x_1+\rho\hat{l}_1)^2}{a_1^2} + \frac{(x_2+\rho\hat{l}_2)^2}{a_2^2} + \frac{(x_3+\rho\hat{l}_3)^2}{a_3^2} = 1. \tag{6.78}
$$

Then, solving this equation for ρ, and taking the positive root,

$$
\rho(\hat{\boldsymbol{l}}) = -\frac{f}{g} + \sqrt{\frac{f^2}{g^2} + \frac{e}{g}}, \tag{6.79}
$$

where

$$
e = 1 - \frac{x_i^2}{a_i^2}, \qquad f = \frac{\hat{l}_1 x_1}{a_1^2} + \frac{\hat{l}_2 x_2}{a_2^2} + \frac{\hat{l}_3 x_3}{a_3^2}, \qquad g = \frac{\hat{l}_i^2}{a_i^2}. \tag{6.80}
$$

Next, substituting Eqs. (6.79) and (6.80) and the quantities

$$
\lambda_1 = \frac{\hat{l}_1}{a_1^2} \qquad \lambda_2 = \frac{\hat{l}_2}{a_2^2} \qquad \lambda_3 = \frac{\hat{l}_3}{a_3^2} \tag{6.81}
$$

into Eq. (6.76), the displacements $\mathbf{u}_i^{C,INC}(\mathbf{x})$ are given by

$$
\begin{aligned}
\mathbf{u}_i^{C,INC}(\mathbf{x}) &= -\frac{\varepsilon_{jk}^T}{8\pi(1-\nu^M)} \oiint_{\hat{\mathcal{S}}} \rho(\hat{\boldsymbol{l}}) g_{ijk}(\hat{\boldsymbol{l}}) d\hat{S}(\hat{\boldsymbol{l}}) \\
&= \frac{x_m \varepsilon_{jk}^T}{8\pi(1-\nu^M)} \oiint_{\hat{\mathcal{S}}} \frac{\lambda_m g_{ijk}}{g} d\hat{S}(\hat{\boldsymbol{l}}),
\end{aligned} \tag{6.82}
$$

and the corresponding strains by

$$\varepsilon_{il}^{C,INC}(\mathbf{x}) = \frac{\varepsilon_{jk}^{T}}{16\pi(1-\nu^M)} \oiint_{\hat{S}} \frac{\lambda_i g_{ljk} + \lambda_l g_{ijk}}{g} d\hat{S}(\hat{\boldsymbol{l}}). \tag{6.83}$$

Note that the square root term in Eq. (6.79) has been dropped since it is an even function of $\hat{\boldsymbol{l}}$.

Using Eqs. (6.81) and (6.74), the surface integral in Eq. (6.83) can be broken down into a sum of integrals that are of the forms (Eshelby 1957) given by

$$\begin{aligned}
I_1 &= \oiint_{\hat{S}} \frac{\hat{l}_1^2}{a_1^2}\frac{d\hat{S}(\hat{\boldsymbol{l}})}{g} = 2\pi a_1 a_2 a_3 \int_0^\infty \frac{du}{(a_1^2+u)\Delta(u)},\\
I_{11} &= \oiint_{\hat{S}} \frac{\hat{l}_1^4}{a_1^4}\frac{d\hat{S}(\hat{\boldsymbol{l}})}{g} = 2\pi a_1 a_2 a_3 \int_0^\infty \frac{du}{(a_1^2+u)^2\Delta(u)},\\
I_{12} &= 3\oiint_{\hat{S}} \frac{\hat{l}_1^2\hat{l}_2^2}{a_1^2 a_2^2}\frac{d\hat{S}(\hat{\boldsymbol{l}})}{g} = 2\pi a_1 a_2 a_3 \int_0^\infty \frac{du}{(a_1^2+u)(a_2^2+u)\Delta(u)},\\
\Delta(u) &= (a_1^2+u)^{1/2}(a_2^2+u)^{1/2}(a_3^2+u)^{1/2},
\end{aligned} \tag{6.84}$$

where the additional non-vanishing expressions can be obtained by the cyclic interchange of (1,2,3), (a_1, a_2, a_3) and $(\hat{l}_1, \hat{l}_2, \hat{l}_3)$.

The I_i and I_{ij} quantities in Eq. (6.84) are connected by several relationships. Using Eqs. (6.84) and (6.80),

$$I_1 + I_2 + I_3 = \oiint_{\hat{S}} \left[\frac{\hat{l}_1^2}{a_1^2} + \frac{\hat{l}_2^2}{a_2^2} + \frac{\hat{l}_3^2}{a_3^2}\right] \frac{d\hat{S}(\hat{\boldsymbol{l}})}{g} = \oiint_{\hat{S}} d\hat{S}(\hat{\boldsymbol{l}}) = 4\pi. \tag{6.85}$$

Also,

$$\begin{aligned}
3a_1^2 I_{11} + a_2^2 I_{12} + a_3^2 I_{13} &= \oiint_{\hat{S}} \left[3a_1^2\frac{\hat{l}_1^4}{a_1^4} + 3a_2^2\frac{\hat{l}_1^2\hat{l}_2^2}{a_1^2a_2^2} + 3a_3^2\frac{\hat{l}_1^2\hat{l}_3^2}{a_1^2a_3^2}\right] \frac{d\hat{S}(\hat{\boldsymbol{l}})}{g}\\
&= 3\oiint_{\hat{S}} \frac{\hat{l}_1^2}{a_1^2}\frac{d\hat{S}(\hat{\boldsymbol{l}})}{g} = 3I_1.
\end{aligned} \tag{6.86}$$

Furthermore, starting with the integral equation for I_{12}, and decomposing the integrand into partial fractions,

$$
\begin{aligned}
I_{12} &= 2\pi a_1 a_2 a_3 \int_0^\infty \frac{du}{(a_1^2+u)(a_2^2+u)\Delta(u)} \\
&= 2\pi a_1 a_2 a_3 \int_0^\infty \frac{1}{(a_2^2-a_1^2)}\left[\frac{1}{(a_1^2+u)} - \frac{1}{(a_2^2+u)}\right]\frac{du}{\Delta(u)} \\
&= \frac{1}{(a_2^2-a_1^2)}(I_1 - I_2) \qquad (6.87)
\end{aligned}
$$

and, by using all three of the above relationships,

$$3I_{11} + I_{12} + I_{13} = \frac{4\pi}{a_1^2}. \qquad (6.88)$$

Using the above relationships and their cyclic variants, all I_i and I_{ij} can be expressed in terms of I_1 and I_3, which, in turn, are given (when $a_1 > a_2 > a_3$) by

$$
\begin{aligned}
I_1 &= \frac{4\pi a_1 a_2 a_3}{(a_1^2-a_2^2)(a_1^2-a_3^2)^{1/2}}[F(k,\theta) - E(k,\theta)], \\
I_3 &= \frac{4\pi a_1 a_2 a_3}{(a_2^2-a_3^2)(a_1^2-a_3^2)^{1/2}}\left(\frac{a_2(a_1^2-a_3^2)^{1/2}}{a_1 a_3} - E(k,\theta)\right),
\end{aligned}
\qquad (6.89)
$$

where F and E are the standard elliptic integrals of the first and second kind (Gradshteyn and Ryzhik 1980)

$$
\begin{aligned}
&F(k,\theta) = \int_0^\theta \frac{d\alpha}{\sqrt{1-k^2\sin^2\alpha}} \qquad E(k,\theta) = \int_0^\theta \sqrt{1-k^2\sin^2\alpha}\,d\alpha \\
&(0<k<1) \qquad k^2 = \frac{(a_1^2-a_2^2)}{a_1^2-a_3^2} \qquad \theta = \sin^{-1}\left(1-\frac{a_3^2}{a_1^2}\right)^{1/2}
\end{aligned}
\qquad (6.90)
$$

As pointed out by Eshelby (1961), the condition $a_1 > a_2 > a_3$, that ensures that $0 < k < 1$, is not actually necessary. If it is not satisfied, the elliptic integrals can be transformed into new elliptic integrals, $F(\theta', k')$ and $E(\theta', k')$, where $0 < k' < 1$ (Byrd and Friedman 1954).

Finally, having the above results, Eq. (6.83) can be put into the same form as Eq. (6.29) for the anisotropic case, i.e.,

$$\varepsilon_{ij}^{C,INC} = S_{ijkl}^{E}\varepsilon_{kl}^{T}, \tag{6.91}$$

where

$$S_{ijkl}^{E} = \frac{1}{16\pi(1-\nu^{M})} \oiint_{\hat{S}} \frac{\lambda_i g_{jkl} + \lambda_j g_{ikl}}{g} d\hat{S}(\hat{l}) \tag{6.92}$$

is the Eshelby tensor for an isotropic system having the symmetry properties

$$S_{ijkl}^{E} = S_{jikl}^{E} = S_{ijlk}^{E}. \tag{6.93}$$

Again, as in the case of Eq. (6.29), the important result is obtained that the canceling strain in the inclusion is uniform when the transformation strain is uniform. Using the previous expressions for g_{ijk} and λ_i and also Eq. (6.84),

$$\begin{aligned}
S_{1111}^{E} &= \frac{1}{8\pi(1-\nu^{M})}\left[3a_1^2 I_{11} + \left(1-2\nu^{M}\right) I_1\right],\\
S_{1122}^{E} &= \frac{1}{8\pi(1-\nu^{M})}\left[a_2^2 I_{12} - \left(1-2\nu^{M}\right) I_1\right],\\
S_{1133}^{E} &= \frac{1}{8\pi(1-\nu^{M})}\left[a_3^2 I_{13} - \left(1-2\nu^{M}\right) I_1\right],\\
S_{1212}^{E} &= \frac{a_1^2 + a_2^2}{16\pi(1-\nu^{M})}\left[I_{12} + \left(1-2\nu^{M}\right)(I_1 + I_2)\right],
\end{aligned} \tag{6.94}$$

where all additional non-zero tensor components can be obtained by cyclic interchange of (1,2,3). The symmetry properties resemble those of the elastic constant tensor, C_{ijkl}, except that in general, $S_{ijkl}^{E} \neq S_{klij}^{E}$. There is therefore no coupling between unlike shear strains and between shear strains and normal strains, and there are twelve non-zero independent S_{ijkl}^{E} components: nine of these couple the normal transformation strains and normal canceling strains, i.e., $S_{1111}^{E}, S_{1122}^{E}, S_{1133}^{E}, S_{2211}^{E}, S_{2222}^{E}, S_{2233}^{E}, S_{3311}^{E}, S_{3322}^{E}$ and S_{3333}^{E}, and three couple the corresponding shear transformation strains and shear canceling strains, i.e., $S_{1212}^{E}, S_{1313}^{E}$ and S_{2323}^{E}.

The quantities in Eq. (6.94) can be determined for various ellipsoidal shapes, and expressions for the S_{ijkl}^{E} tensor for ellipsoids of revolution as a function of their shape (as determined by their principal axes) are given in Appendix H. An expression for S_{1212}^{E} for a spherical inclusion is derived in Exercise 6.11. Additional expressions are given by Mura (1987).

The elastic field in the inclusion can now be obtained by using the inclusion shape and volume and transformation strain as inputs. The coefficients I_1 and I_3 are first calculated using Eq. (6.89), and the S^E_{ijkl} are determined using Eq. (6.94). Then, the $\varepsilon^{C,INC}_{ij}$ strains are determined by using Eq. (6.91) in matrix form and employing the index contraction rules of Eqs. (2.88) and (2.89) to construct the matrix representing the Eshelby tensor. Therefore,

$$\left[\varepsilon^{C,INC}\right] = \left[S^E\right]\left[\varepsilon^T\right] \tag{6.95}$$

which in full form appears as

$$\begin{bmatrix} \varepsilon_1^{C,INC} \\ \varepsilon_2^{C,INC} \\ \varepsilon_3^{C,INC} \\ \varepsilon_4^{C,INC} \\ \varepsilon_5^{C,INC} \\ \varepsilon_6^{C,INC} \end{bmatrix} = \begin{bmatrix} S^E_{11} & S^E_{12} & S^E_{13} & 0 & 0 & 0 \\ S^E_{21} & S^E_{22} & S^E_{23} & 0 & 0 & 0 \\ S^E_{31} & S^E_{32} & S^E_{33} & 0 & 0 & 0 \\ 0 & 0 & 0 & 2S^E_{44} & 0 & 0 \\ 0 & 0 & 0 & 0 & 2S^E_{55} & 0 \\ 0 & 0 & 0 & 0 & 0 & 2S^E_{66} \end{bmatrix} \begin{bmatrix} \varepsilon_1^T \\ \varepsilon_2^T \\ \varepsilon_3^T \\ \varepsilon_4^T \\ \varepsilon_5^T \\ \varepsilon_6^T \end{bmatrix}. \tag{6.96}$$

The elastic strain in the inclusion, ε^{INC}_{ij}, is finally obtained from Eq. (6.1), and the stresses are obtained via Hooke's law. This method is employed in Exercise 6.7, to obtain σ^{INC}_{11} and σ^{INC}_{12} for a thin-disk inclusion.

Elastic field outside inclusion

This problem has been treated in several ways by Eshelby (1959, 1961) and Mura (1987). Following Eshelby, and employing elements of potential theory, an expression for $u^{C,M}_i(\mathbf{x})$ solely in terms of the well-known Newtonian potential $\phi(\mathbf{x})$ is now obtained.

The general expression for $u^C_i(\mathbf{x})$ given by Eq. (6.60) contains both $\phi(\mathbf{x})$ and $\psi(\mathbf{x})$. However, $\psi(\mathbf{x})$ can be eliminated by following a procedure that starts by introducing the function f_{ij} that is related to ϕ and ψ by

$$f_{ij} = x_i \frac{\partial \phi}{\partial x_j} - \frac{\partial^2 \psi}{x_i x_j}. \tag{6.97}$$

The Laplacian of f_{ij}, given by

$$\nabla^2 f_{ij} = x_i \frac{\partial(\nabla^2 \phi)}{\partial x_j} + 2\frac{\partial^2 \phi}{\partial x_i \partial x_j} - \frac{\partial^2(\nabla^2 \psi)}{\partial x_i \partial x_j} \tag{6.98}$$

is seen to vanish both inside and outside the inclusion by virtue of Eqs. (6.62) and (6.64), and f_{ij} is therefore *harmonic* in both regions. Furthermore, its normal derivative at the $\mathcal{S}^{INC}$ surface, i.e., $\nabla f_{ij} \cdot \hat{n}$, undergoes a discontinuous jump on passing through $\mathcal{S}^{INC}$.[4] Making use of

$$\hat{n} = \left(\frac{x_1}{a_1^2 h}, \frac{x_2}{a_2^2 h}, \frac{x_3}{a_3^2 h} \right) \qquad h^2 = \frac{x_1^2}{a_1^4} + \frac{x_2^2}{a_2^4} + \frac{x_3^2}{a_3^4} \tag{6.99}$$

and the fact that the first derivatives of ϕ and the third derivatives of ψ are continuous across $\mathcal{S}^{INC}$, the jump is given by

$$(\nabla f_{ij} \cdot \hat{n})^{IN} - (\nabla f_{ij} \cdot \hat{n})^{OUT} = -4\pi x_i \hat{n}_j. \tag{6.100}$$

According to classical potential theory, Poincare (1899), f_{ij} must therefore be the harmonic potential of a layer of density $x_i \hat{n}_j$ distributed on $\mathcal{S}^{INC}$.

Next, compare f_{12} with the function g_{12} given by

$$g_{12} = \frac{a_1^2}{(a_1^2 - a_2^2)} \left(x_1 \frac{\partial \phi}{\partial x_2} - x_2 \frac{\partial \phi}{\partial x_1} \right). \tag{6.101}$$

By using the same methods, it is found that g_{12} is harmonic both inside and outside of $\mathcal{S}^{INC}$, and its normal derivative undergoes a discontinuous jump on passing through $\mathcal{S}^{INC}$ which is identical to the jump for f_{12}. Furthermore, both functions are continuous across $\mathcal{S}^{INC}$ and vanish at infinity. Both quantities must then be harmonic potentials of the same surface distribution of density and so must be identical: therefore,

$$x_1 \frac{\partial \phi}{\partial x_2} - \frac{\partial^2 \psi}{\partial x_1 \partial x_2} = \frac{a_1^2}{(a_1^2 - a_2^2)} \left(x_1 \frac{\partial \phi}{\partial x_2} - x_2 \frac{\partial \phi}{\partial x_1} \right). \tag{6.102}$$

Similar expressions for $i, j = 1, 3$ and $i, j = 2, 3$ hold so that

$$\begin{aligned}
\frac{\partial^2 \psi}{\partial x_1 \partial x_2} &= \frac{a_1^2}{a_1^2 - a_2^2} \frac{\partial \phi}{\partial x_1} x_2 + \frac{a_2^2}{a_2^2 - a_1^2} \frac{\partial \phi}{\partial x_2} x_1, \\
\frac{\partial^2 \psi}{\partial x_2 \partial x_3} &= \frac{a_2^2}{a_2^2 - a_3^2} \frac{\partial \phi}{\partial x_2} x_3 + \frac{a_3^2}{a_3^2 - a_2^2} \frac{\partial \phi}{\partial x_3} x_2, \\
\frac{\partial^2 \psi}{\partial x_3 \partial x_1} &= \frac{a_3^2}{a_3^2 - a_1^2} \frac{\partial \phi}{\partial x_3} x_1 + \frac{a_1^2}{a_1^2 - a_3^2} \frac{\partial \phi}{\partial x_1} x_3.
\end{aligned} \tag{6.103}$$

All remaining required derivatives of ψ in terms of derivatives of ϕ can now be obtained through use of the above expressions. For example, by

[4]See discussion and analysis preceding Eq. (6.68).

differentiating Eq. (6.62),

$$\frac{\partial(\nabla^2\psi)}{\partial x_1} = \frac{\partial^3\psi}{\partial x_1^3} + \frac{\partial^3\psi}{\partial x_1 \partial x_2^2} + \frac{\partial^3\psi}{\partial x_1 \partial x_3^2} = 2\frac{\partial\phi}{\partial x_1}, \tag{6.104}$$

and, therefore,

$$\frac{\partial^3\psi}{\partial x_1^3} = 2\frac{\partial\phi}{\partial x_1} - \frac{\partial}{\partial x_2}\left(\frac{\partial^2\psi}{\partial x_1 \partial x_2}\right) - \frac{\partial}{\partial x_3}\left(\frac{\partial^2\psi}{\partial x_1 \partial x_3}\right). \tag{6.105}$$

Also,

$$\frac{\partial^3\psi}{\partial x_1^2 \partial x_2} = \frac{\partial}{\partial x_1}\left(\frac{\partial^2\psi}{\partial x_1 \partial x_2}\right). \tag{6.106}$$

The derivatives of ψ in Eq. (6.60) are now eliminated by substituting the above expressions, and the desired expression for u_1^C in the matrix is obtained as a function of ϕ in the form

$$\begin{aligned} u_1^{C,M} = \frac{1}{8\pi(1-\nu^M)} \Bigg\{ & \frac{\varepsilon_{22}^T - \varepsilon_{11}^T}{a_1^2 - a_2^2}\frac{\partial}{\partial x_2}\left(a_1^2 x_2 \frac{\partial\phi}{\partial x_1} - a_2^2 x_1 \frac{\partial\phi}{\partial x_2}\right) \\ & + \frac{\varepsilon_{33}^T - \varepsilon_{11}^T}{a_3^2 - a_1^2}\frac{\partial}{\partial x_3}\left(a_3^2 x_1 \frac{\partial\phi}{\partial x_3} - a_1^2 x_3 \frac{\partial\phi}{\partial x_1}\right) \\ & - 2\left[\left(1-\nu^M\right)\varepsilon_{11}^T + \nu^M\left(\varepsilon_{22}^T + \varepsilon_{33}^T\right)\right]\frac{\partial\phi}{\partial x_1} \\ & - 4(1-\nu^M)\left(\varepsilon_{12}^T\frac{\partial\phi}{\partial x_2} + \varepsilon_{13}^T\frac{\partial\phi}{\partial x_3}\right) + \frac{\partial\Gamma}{\partial x_1}\Bigg\}, \end{aligned} \tag{6.107}$$

where

$$\begin{aligned} \Gamma = & \frac{2\varepsilon_{12}^T}{a_1^2 - a_2^2}\left(a_1^2 x_2 \frac{\partial\phi}{\partial x_1} - a_2^2 x_1 \frac{\partial\phi}{\partial x_2}\right) \\ & + \frac{2\varepsilon_{23}^T}{a_2^2 - a_3^2}\left(a_2^2 x_3 \frac{\partial\phi}{\partial x_2} - a_3^2 x_2 \frac{\partial\phi}{\partial x_3}\right) \\ & + \frac{2\varepsilon_{31}^T}{a_3^2 - a_1^2}\left(a_3^2 x_1 \frac{\partial\phi}{\partial x_3} - a_1^2 x_3 \frac{\partial\phi}{\partial x_1}\right). \end{aligned} \tag{6.108}$$

Corresponding expressions for $u_2^{C,M}$ and $u_3^{C,M}$ are obtained by cyclic interchange of (1,2,3) and (a_1, a_2, a_3).

Now, the Newtonian potential external to the ellipsoidal body, i.e., ϕ^M, has the well-known form (Kellogg 1929; MacMillan 1930)

$$\phi^M = \frac{2\pi a_1 a_2 a_3}{l^3}\left\{\left[l^2 - \frac{x_1^2}{k^2} + \frac{x_2^2}{k^2}\right]F(\theta, k)\right.$$
$$+ \left[\frac{x_1^2}{k^2} - \frac{x_2^2}{k^2(1-k^2)} + \frac{x_3^2}{(1-k^2)}\right]E(\theta, k)$$
$$\left. + \frac{l}{1-k^2}\left[\frac{C}{AB}x_2^2 - \frac{B}{AC}x_3^2\right]\right\} \qquad (6.109)$$

where $a_1^2 > a_2^2 > a_3^2$,

$$A = (a_1^2 + \lambda)^{1/2} \quad B = (a_2^2 + \lambda)^{1/2} \quad C = (a_3^2 + \lambda)^{1/2}$$
$$l = (a_1^2 - a_3^2)^{1/2} \quad k^2 = \frac{a_1^2 - a_2^2}{a_1^2 - a_3^2} \quad \theta = \sin^{-1}\left(\frac{l}{A}\right), \qquad (6.110)$$

and λ is the largest root of the equation

$$\frac{x_1^2}{(a_1^2+\lambda)} + \frac{x_2^2}{(a_2^2+\lambda)} + \frac{x_3^2}{(a_3^2+\lambda)} = 1. \qquad (6.111)$$

Therefore, $u_i^{C,M}$ is finally obtained by substituting Eq. (6.109) into Eq. (6.107). The substitution and differentiations are lengthy, and detailed results will not be presented here. Since $\lambda = \lambda(\mathbf{x})$ and $\theta = \theta(\lambda) = \theta(\mathbf{x})$, the differentiations can be helped by using the relationships

$$\frac{\partial F}{\partial \lambda} = -\frac{l}{2ABC} \qquad \frac{\partial E}{\partial \lambda} = -\frac{lB}{2A^3C} \qquad h^2 = \frac{x_1^2}{A^4} + \frac{x_2^2}{B^4} + \frac{x_3^2}{C^4}$$
$$\frac{\partial \lambda}{\partial x_1} = \frac{2x_1}{A^2h^2} \qquad \frac{\partial \lambda}{\partial x_2} = \frac{2x_2}{B^2h^2} \qquad \frac{\partial \lambda}{\partial x_3} = \frac{2x_3}{C^2h^2}. \qquad (6.112)$$

The first relationship in the first row, for example, is obtained by multiplying $\partial F/\partial\theta = A/B$ by $\partial\theta/\partial\lambda = -l/(2A^2C)$ while the first relationship in the second row is obtained by differentiating Eq. (6.111). Further aspects of determining the elastic field in the matrix are discussed by Eshelby (1959; 1961) and Mura (1987).

6.4.3 *Elastic field of inhomogeneous ellipsoidal inclusion with uniform ε_{ij}^{T}*

With the help of Eq. (2.120), the general result given by Eq. (6.47) reduces for an isotropic system to

$$\left(\lambda^{M} S^{E}_{mmkl}\varepsilon^{T^*}_{kl}\delta_{ij} + 2\mu^{M} S^{E}_{ijkl}\varepsilon^{T^*}_{kl}\right) - \left(\lambda^{M} e^{T^*}\delta_{ij} + 2\mu^{M}\varepsilon^{T^*}_{ij}\right)$$
$$= \left(\lambda^{INC} S^{E}_{mmkl}\varepsilon^{T^*}_{kl}\delta_{ij} + 2\mu^{INC} S^{E}_{ijkl}\varepsilon^{T^*}_{kl}\right) - \left(\lambda^{INC} e^{T}\delta_{ij} + 2\mu^{INC}\varepsilon^{T}_{ij}\right) \tag{6.113}$$

with S^{E}_{ijkl} given by Eq. (6.92). Equation (6.113) breaks down into separate expressions for the shear and normal transformation strains, respectively. When $i \neq j$,

$$\left(\mu^{INC} - \mu^{M}\right) S^{E}_{ijkl}\varepsilon^{T^*}_{kl} + \mu^{M}\varepsilon^{T^*}_{ij} = \mu^{INC}\varepsilon^{T}_{ij}, \qquad (i \neq j) \tag{6.114}$$

which yields the solution for the shear transformation strains in the equivalent homogeneous inclusion in the form

$$\varepsilon^{T^*}_{\alpha\beta} = \frac{\mu^{INC}}{2(\mu^{INC} - \mu^{M})S^{E}_{\alpha\beta\alpha\beta} + \mu^{M}}\varepsilon^{T}_{\alpha\beta} \qquad (\alpha \neq \beta) \tag{6.115}$$

after invoking the properties of the S^{E}_{ijkl} tensor described in the text following Eq. (6.94), and employing Greek indices to avoid the index summation convention.

When $i = j$, the three normal transformation strains for the equivalent homogeneous inclusion are obtained as the solution of the three simultaneous linear equations represented by

$$\left(\lambda^{INC} - \lambda^{M}\right) S^{E}_{mmkl}\varepsilon^{T^*}_{kl} + 2(\mu^{INC} - \mu^{M})S^{E}_{ijkl}\varepsilon^{T^*}_{kl}$$
$$+ \lambda^{M} e^{T^*} + 2\mu^{M}\varepsilon^{T^*}_{ij} \qquad (i = j) \tag{6.116}$$
$$= \lambda^{INC} e^{T} + 2\mu^{INC}\varepsilon^{T}_{ij}.$$

The solution of the three equations represented by Eq. (6.116) is expedited by writing them in matrix form using the contracted notation given by Eqs. (2.88) and (2.89), i.e.,

$$M_{ij}\varepsilon^{T^*}_{j} = N_{ij}\varepsilon^{T}_{j}, \qquad [M]\left[\varepsilon^{T^*}\right] = [N]\left[\varepsilon^{T}\right], \tag{6.117}$$

or

$$\begin{bmatrix} M_{11} & M_{12} & M_{13} \\ M_{21} & M_{22} & M_{23} \\ M_{31} & M_{32} & M_{33} \end{bmatrix} \begin{bmatrix} \varepsilon_1^{T^*} \\ \varepsilon_2^{T^*} \\ \varepsilon_3^{T^*} \end{bmatrix} = \begin{bmatrix} N_{11} & N_{12} & N_{13} \\ N_{21} & N_{22} & N_{23} \\ N_{31} & N_{32} & N_{33} \end{bmatrix} \begin{bmatrix} \varepsilon_1^{T} \\ \varepsilon_2^{T} \\ \varepsilon_3^{T} \end{bmatrix}, \qquad (6.118)$$

where

$$M_{ij} = (\lambda^{INC} - \lambda^{M})(S_{1j}^{E} + S_{2j}^{E} + S_{3j}^{E}) + 2(\mu^{INC} - \mu^{M})S_{ij}^{E} + \lambda^{M} + 2\mu^{M}\delta_{ij}$$

$$N_{ij} = \lambda^{INC} + 2\mu^{INC}\delta_{ij}. \qquad (6.119)$$

The normal transformation strains for the equivalent homogeneous inclusion are then

$$[\varepsilon^{T^*}] = [M]^{-1}[N][\varepsilon^{T}] \qquad (i = j). \qquad (6.120)$$

Having the results given by Eqs. (6.115) and (6.120), the remainder of the solution is carried out using the procedure described in the text following Eq. (6.49).

6.4.3.1 *Spherical inclusion with* $\varepsilon_{ij}^{T} = \varepsilon^{T}\delta_{ij}$

The equivalent homogeneous inclusion method described above can be illustrated by applying it to the relatively simple case of an inhomogeneous spherical inclusion of radius R with the uniform transformation strains $\varepsilon_{ij}^{T} = \varepsilon^{T}\delta_{ij}$. The only non-vanishing transformation strains are therefore the three equal normal strains, and using the values of S_{ijkl}^{E} given by Eq. (H.4) for a sphere, the M_{ij} matrix elements in Eq. (6.118) are

$$M_{11} = M_{22} = M_{33}$$

$$= (\lambda^{INC} - \lambda^{M})\frac{1+\nu^{M}}{3(1-\nu^{M})} + 2(\mu^{INC} - \mu^{M})\frac{7-5\nu^{M}}{15(1-\nu^{M})} + \lambda^{M} + 2\mu^{M},$$

$$M_{12} = M_{21} = M_{13} = M_{31} = M_{23} = M_{32}$$

$$= (\lambda^{INC} - \lambda^{M})\frac{1+\nu^{M}}{3(1-\nu^{M})} + 2(\mu^{INC} - \mu^{M})\frac{5\nu^{M}-1}{15(1-\nu^{M})} + \lambda^{M}. \qquad (6.121)$$

Then, using Eq. (6.120), ε^{T^*} is obtained in the form

$$\varepsilon^{T^*} = \frac{(3K^M + 4\mu^M)K^{INC}}{(3K^{INC} + 4\mu^M)K^M}\varepsilon^T$$

$$= \frac{3\mu^{INC}(1+\nu^{INC})(1-\nu^M)}{(1+\nu^M)[\mu^{INC}(1+\nu^{INC}) + 2\mu^M(1-2\nu^{INC}]}\varepsilon^T. \quad (6.122)$$

Substitution of Eq. (6.122) into Eq. (6.50) then yields

$$\varepsilon_{11}^{INC} = \varepsilon_{22}^{INC} = \varepsilon_{33}^{INC} = -\frac{2\mu^M(1-2\nu^{INC})}{\mu^{INC}(1+\nu^{INC}) + 2\mu^M(1-2\nu^{INC})}\varepsilon^T$$

$$= -\frac{4\mu^M}{3K^{INC} + 4\mu^M}\varepsilon^T, \quad (6.123)$$

leading to the corresponding displacement

$$u_i^{INC} = -\frac{4\mu^M}{3K^{INC} + 4\mu^M}\varepsilon^T x_i. \quad (6.124)$$

The displacements in the matrix are given by $u_i^M = u_i^{C,M}$ according to Eq. (6.1), and, therefore, can be obtained from Eq. (6.107) which requires an expression for the potential exterior to the spherical inclusion, i.e., ϕ^M. This has the known form (MacMillan 1930),

$$\phi^M(x) = \frac{4\pi R^3}{3x} \quad (6.125)$$

and, therefore, making the substitution,

$$u_i^{C,M} = u_i^M = \frac{(1+\nu^M)\varepsilon^{T^*}V^{INC}}{4\pi(1-\nu^M)}\frac{x_i}{x^3} = \frac{9K^M\varepsilon^{T^*}V^{INC}}{4\pi(3K^M + 4\mu^M)}\frac{x_i}{x^3}. \quad (6.126)$$

Then, substituting Eq. (6.122) into Eq. (6.126),

$$u_i^M = c\frac{x_i}{x^3}, \quad (6.127)$$

where the constant c is a measure of the 'strength' of the inclusion as a center of dilatation given by

$$c = \frac{9K^{INC}}{4\pi(3K^{INC} + 4\mu^{M})} V^{INC} \varepsilon^{T}$$

$$= \frac{3\mu^{INC}(1 + \nu^{INC})}{4\pi[\mu^{INC}(1 + \nu^{INC}) + 2\mu^{M}(1 - 2\nu^{INC})]} V^{INC} \varepsilon^{T}. \tag{6.128}$$

Finally, since the above solutions are spherically symmetric, they can be expressed simply in spherical coordinates as

$$u_r^{INC}(r) = -\frac{4\mu^{M}}{(3K^{INC} + 4\mu^{M})} \varepsilon^{T} r,$$

$$u_r^{M}(r) = c\frac{1}{r^2} = \frac{9K^{INC}}{4\pi(3K^{INC} + 4\mu^{M})} V^{INC} \varepsilon^{T} \frac{1}{r^2},$$

$$u_\theta = u_\phi = 0. \tag{6.129}$$

In Exercise 6.8, the results given by Eq. (6.129) are obtained by using an alternative method of solution in which the Navier equation is solved directly in both the inclusion and matrix as a boundary value problem.

6.4.4 *Strain energy*

6.4.4.1 *Homogeneous, or inhomogeneous, with arbitrary shape and ε_{ij}^{T}*

The general expression for the strain energy due to a coherent homogeneous or inhomogeneous inclusion of arbitrary shape and transformation strain given by Eq. (6.57), and derived in Sec. 6.3.3 without reference to any elastic constants, applies to both anisotropic and isotropic systems.

6.4.4.2 *Homogeneous with ellipsoidal shape and uniform ε_{ij}^{T} in isotropic system*

The dependence of the inclusion strain energy on its homogeneity, shape and transformation strain is a topic of considerable interest, since it affects the kinetics and morphology of many phase transformations where new phases are produced in the form of inclusions. Examples include precipitation and martensitic transformations (Balluffi, Allen and Carter 2005). It is also important in determining the thermal stability and morphology of many microstructures. This dependence is particularly easy to study for coherent, homogeneous and ellipsoidal inclusions with uniform transformation strains in isotropic systems. In such cases σ_{ij}^{INC} in Eq. (6.57) is uniform, and, therefore,

$$W = -\frac{1}{2}\sigma_{ij}^{INC}\varepsilon_{ij}^{T}V^{INC}, \quad V^{INC} = \frac{4}{3}\pi a_1 a_2 a_3. \tag{6.130}$$

The stresses in the inclusion required in Eq. (6.130) for inclusions of various ellipsoidal shapes are then found by substituting values of the Eshelby tensor from Appendix H into Eq. (6.96) to find the relevant $\varepsilon_{ij}^{C,INC}$ strains in terms of the transformation strains. Then, Eq. (6.1) is used to find the elastic strains, and the stresses are finally obtained by use of Hooke's law. A calculation of this type for a thin-disk inclusion is carried out in Exercise 6.7.

The strain energies calculated in the above manner for homogeneous spherical, thin-disk and needle-shaped inclusions subjected to general uniform transformation strains in isotropic systems are as follows:

Sphere ($a_1 = a_2 = a_3 = a$)

$$\begin{aligned} W = \frac{2\mu^M}{15(1-\nu^M)}V^{INC}\Big\{&4\left[\left(\varepsilon_{11}^T\right)^2 + \left(\varepsilon_{22}^T\right)^2 + \left(\varepsilon_{33}^T\right)^2\right] \\ &+ (5\nu^M + 1)\left(\varepsilon_{11}^T\varepsilon_{22}^T + \varepsilon_{11}^T\varepsilon_{33}^T + \varepsilon_{22}^T\varepsilon_{33}^T\right) \\ &+(7 - 5\nu^M)\left[(\varepsilon_{12}^T)^2 + (\varepsilon_{13}^T)^2 + (\varepsilon_{23}^T)^2\right]\Big\}. \end{aligned} \tag{6.131}$$

Thin-disk ($a_1 = a_2, a_3 \to 0$)

$$W = \frac{\mu^M V^{INC}}{1-\nu^M}\left\{\left[\left(\varepsilon_{11}^T\right)^2 + \left(\varepsilon_{22}^T\right)^2\right] + 2\nu^M\varepsilon_{11}^T\varepsilon_{22}^T + 2\left(1-\nu^M\right)\left(\varepsilon_{12}^T\right)^2\right\}. \tag{6.132}$$

Needle ($a_1 = a_2, a_3 \to \infty$)

$$\begin{aligned} W = \frac{\mu^M V^{INC}}{2(1-\nu^M)}\Bigg\{&\frac{3}{4}\left[\left(\varepsilon_{11}^T\right)^2 + \left(\varepsilon_{22}^T\right)^2\right] + 2\left(\varepsilon_{33}^T\right)^2 + \frac{1}{2}\varepsilon_{11}^T\varepsilon_{22}^T \\ &+ 2\nu^M\left[\varepsilon_{22}^T\varepsilon_{33}^T + \varepsilon_{11}^T\varepsilon_{33}^T\right] + \left(\varepsilon_{12}^T\right)^2 + 2\left(1-\nu^M\right)\left[\left(\varepsilon_{13}^T\right)^2 + \left(\varepsilon_{23}^T\right)^2\right]\Bigg\} \end{aligned} \tag{6.133}$$

where, in the needle case, W and V^{INC} are measured per unit inclusion length.

In Exercise 6.6, it is shown that the strain energy due to a homogeneous inclusion of arbitrary ellipsoidal shape having the uniform transformation strain $\varepsilon_{ij}^{T} = \varepsilon^{T}\delta_{ij}$ is given by

$$W = \frac{2\mu^{M}K^{M}}{(3K^{M} + 4\mu^{M})}(e^{T})^{2}V^{INC} = \frac{2\mu^{M}(1+\nu^{M})}{9(1-\nu^{M})}(e^{T})^{2}V^{INC} \quad (6.134)$$

and so is independent of its shape (as long as it remains ellipsoidal).

6.4.4.3 *Inhomogeneous with ellipsoidal shape and* $\varepsilon_{ij}^{T} = \varepsilon^{T}\delta_{ij}$

Strain energies for spherical and thin-disk inhomogeneous inclusions with $\varepsilon_{ij}^{T} = \varepsilon^{T}\delta_{ij}$ in isotropic systems are as follows:

Sphere ($a_1 = a_2 = a_3 = a$)

This case has spherical symmetry with $\sigma_{11}^{INC} = \sigma_{22}^{INC} = \sigma_{33}^{INC}$, and Eq. (6.130) becomes

$$W = -\frac{1}{2}\sigma_{ij}^{INC}\varepsilon_{ij}^{T}V^{INC} = -\frac{3}{2}\sigma_{11}^{INC}\varepsilon^{T}V^{INC}. \quad (6.135)$$

The stresses in the inhomogeneous inclusion and its equivalent homogeneous inclusion are equal, and therefore, since $\sigma_{11}^{INC} = 3K^{M}\varepsilon_{11}^{M}$, and ε_{11}^{INC} is given by Eq. (6.123) with $K^{INC} \rightarrow K^{M}$ and $\varepsilon^{T} \rightarrow \varepsilon^{T^*}$,

$$\sigma_{11}^{INC} = -\frac{12\mu^{M}K^{M}}{3K^{M} + 4\mu^{M}}\varepsilon^{T^*}. \quad (6.136)$$

Then, substituting Eq. (6.122) into Eq. (6.136), and Eq. (6.136) into Eq. (6.135),

$$W = \frac{18K^{INC}\mu^{M}}{3K^{INC} + 4\mu^{M}}V^{INC}(\varepsilon^{T})^{2}$$

$$= \frac{6\mu^{INC}\mu^{M}(1+\nu^{INC})}{[\mu^{INC}(1+\nu^{INC}) + 2\mu^{M}(1-2\nu^{INC})]}V^{INC}(\varepsilon^{T})^{2}. \quad (6.137)$$

Note that Eq. (6.137) reduces to Eq. (6.134) for a corresponding homogeneous inclusion with $\varepsilon_{ij}^{T} = \varepsilon^{T}\delta_{ij}$ when $\mu^{INC} \rightarrow \mu^{M}$ and $\nu^{INC} \rightarrow \nu^{M}$. It is shown in Exercise 6.9 that Eq. (6.137) can also be obtained by directly summing the strain energy in the inclusion and in the matrix.

Thin-disk ($a_1 = a_2, a_3 \rightarrow 0$)

With $\varepsilon_{ij}^{T} = \varepsilon^{T}\delta_{ij}$ and $(a_1 = a_2, a_3 \to 0)$, Eq. (6.130) takes the form

$$W = -\frac{1}{2}\sigma_{ij}^{INC}\varepsilon_{ij}^{T}V^{INC} = -\frac{1}{2}\left[\sigma_{11}^{INC} + \sigma_{22}^{INC} + \sigma_{33}^{INC}\right]\varepsilon^{T}V^{INC}. \tag{6.138}$$

The σ_{ij}^{INC} normal stresses are obtained from Eq. (6.182) after replacing the given ε_{i}^{T} strains by the $\varepsilon_{i}^{T^*}$ strains of the equivalent homogeneous inclusion and are of the forms,

$$\sigma_{11}^{INC} = \sigma_{22}^{INC} = -\frac{2\mu^{M}(1+\nu^{M})}{(1-\nu^{M})}\varepsilon_{11}^{T^*}, \quad \sigma_{33}^{INC} = 0. \tag{6.139}$$

The $\varepsilon_{ij}^{T^*}$ strains are then obtained as a function of the given ε_{ij}^{T} strains by using Eq. (6.120) with the M_{ij} given by

$$M_{33} = \lambda^{INC} + 2\mu^{INC},$$

$$M_{31} = M_{32} = \frac{\lambda^{INC}\nu^{M}}{1-\nu^{M}} + \frac{\lambda^{M}(1-2\nu^{M})}{1-\nu^{M}} + \frac{2\mu^{INC}\nu^{M}}{1-\nu^{M}} - \frac{2\mu^{M}\nu^{M}}{1-\nu^{M}},$$

$$M_{13} = M_{23} = \lambda^{INC},$$

$$M_{12} = M_{21} = \frac{\lambda^{INC}\nu^{M}}{1-\nu^{M}} + \frac{\lambda^{M}(1-2\nu^{M})}{1-\nu^{M}}, \qquad M_{11} = M_{22} = M_{21} + 2\mu^{M}. \tag{6.140}$$

Then, after some algebra,

$$\begin{aligned}\varepsilon_{33}^{T^*} &= \frac{[\mu^{M}(1+\nu^{M}) - 2\mu^{INC}\nu^{M}](1+\nu^{INC})}{\mu^{M}(1+\nu^{M})(1-\nu^{INC})}\varepsilon^{T},\\ \varepsilon_{11}^{T^*} &= \varepsilon_{22}^{T^*} = \frac{\mu^{INC}}{\mu^{M}}\frac{(1-\nu^{M})}{(1+\nu^{M})}\frac{(1+\nu^{INC})}{(1-\nu^{INC})}\varepsilon^{T}.\end{aligned} \tag{6.141}$$

Putting the above results into Eq. (6.139), and substituting the result into Eq. (6.138),

$$W = \frac{2\mu^{INC}(1+\nu^{INC})}{(1-\nu^{INC})}V^{INC}(\varepsilon^{T})^{2} \tag{6.142}$$

in agreement with Barnett (1971). Again, as for the spherical inclusion treated above, the expression for the inhomogeneous inclusion, i.e., Eq. (6.142), reduces to the expression for the corresponding homogeneous inclusion, i.e., Eq. (6.132), when $\mu^{INC} \to \mu^{M}$, $\nu^{INC} \to \nu^{M}$, and $\varepsilon_{ij}^{T} = \varepsilon^{T}\delta_{ij}$ for the homogeneous inclusion.

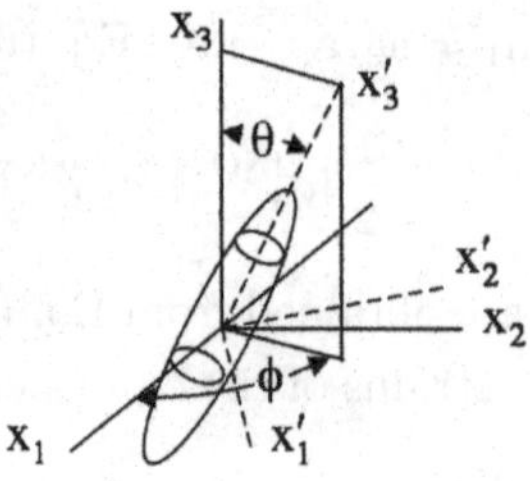

Fig. 6.5 Coordinate systems for studying strain energies of various inhomogeneous ellipsoidal inclusions.

6.4.5 *Further results*

Further effects of varying the inclusion shape, transformation strain and the elastic constants of the inclusion, or matrix, can be conveniently studied (Kato, Fujii and Onaka 1996c) by employing the two coordinate systems illustrated in Fig. 6.5 where the (x_1, x_2, x_3) system is the crystal system fixed to the matrix, and the (x'_1, x'_2, x'_3) system is fixed to the inclusion with the x'_3 axis coinciding with the a_3 axis of the ellipsoid. The x'_2 axis is maintained in the $x_3 = 0$ plane, and rotation of the inclusion with respect to the matrix is obtained by varying the angles θ and ϕ. The transformation strains are specified in the (x_1, x_2, x_3) system in the form of the three principal strains, $\tilde{\varepsilon}^T_{11}, \tilde{\varepsilon}^T_{22}$ and $\tilde{\varepsilon}^T_{33}$. The matrix of direction cosines describing the rotation of the inclusion relative to the matrix is

$$l_{ij} = \begin{bmatrix} \cos\theta\cos\phi & \cos\theta\sin\phi & -\sin\theta \\ -\sin\phi & \cos\phi & 0 \\ \sin\theta\cos\phi & \sin\theta\sin\phi & \cos\theta \end{bmatrix} \tag{6.143}$$

and using Eq. (2.24), the transformation strains in the inclusion, expressed in the (x'_1, x'_2, x'_3) system, are

$$\varepsilon'^{T}_{ij}(\theta, \phi) = \begin{bmatrix} \cos\theta\cos\phi & \cos\theta\sin\phi & -\sin\theta \\ -\sin\phi & \cos\phi & 0 \\ \sin\theta\cos\phi & \sin\theta\sin\phi & \cos\theta \end{bmatrix} \begin{bmatrix} \tilde{\varepsilon}^T_{11} & 0 & 0 \\ 0 & \tilde{\varepsilon}^T_{22} & 0 \\ 0 & 0 & \tilde{\varepsilon}^T_{33} \end{bmatrix}$$

$$\times \begin{bmatrix} \cos\theta\cos\phi & -\sin\phi & \sin\theta\cos\phi \\ \cos\theta\sin\phi & \cos\phi & \sin\theta\sin\phi \\ -\sin\theta & 0 & \cos\theta \end{bmatrix}. \tag{6.144}$$

Finally, the ratio, $f \equiv \mu^{INC}/\mu^{M}$, is introduced, and it is assumed that $\nu^{INC} = \nu^{M}$.

With this arrangement the strain energy due to spherical, thin-disk and needle inclusions can be studied as a function of the above variables. The first step is finding the transformation strains of the equivalent homogeneous inclusion expressed in the (x_1', x_2', x_3') inclusion coordinate system.

For a sphere these are given by

$$\varepsilon'^{T^*}_{11} = f(1-v)\left[\frac{10\varepsilon'^{T}_{11} - 5(\varepsilon'^{T}_{22} + \varepsilon'^{T}_{33})}{2f(4-5\nu) + (7-5\nu)} + \frac{\varepsilon'^{T}_{11} + \varepsilon'^{T}_{22} + \varepsilon'^{T}_{33}}{f(1+\nu) + 2(1-2\nu)}\right],$$
$$\varepsilon'^{T^*}_{12} = \frac{15f(1-\nu)}{2f(4-5\nu) + (7-5\nu)}\varepsilon'^{T}_{12}, \tag{6.145}$$

with the remaining transformation strains obtained by cyclic interchange of the indices. For an ellipsoidal thin-disk $(a_1 = a_2, a_3 \to 0)$,

$$\varepsilon'^{T^*}_{11} = f\varepsilon'^{T}_{11}, \qquad \varepsilon'^{T^*}_{22} = f\varepsilon'^{T}_{22}, \qquad \varepsilon'^{T^*}_{33} = \frac{(1-f)\nu}{1-\nu}\left(\varepsilon'^{T}_{11} + \varepsilon'^{T}_{22}\right) + \varepsilon'^{T}_{33},$$
$$\varepsilon'^{T^*}_{12} = f\varepsilon'^{T}_{12}, \qquad \varepsilon'^{T^*}_{13} = \varepsilon'^{T}_{13}, \qquad \varepsilon'^{T^*}_{23} = \varepsilon'^{T}_{23}, \tag{6.146}$$

and for an ellipsoidal needle $(a_1 = a_2, a_3 \to \infty)$,

$$\varepsilon'^{T^*}_{11} = f(1-\nu)\left\{\frac{[f(5-4\nu) + (3-4\nu)]\,\varepsilon'^{T}_{11} + (f-1)(1-4\nu)\varepsilon'^{T}_{22}}{[f(3-4\nu)+1]\,[f + (1-2\nu)]}\right\}$$
$$+ \frac{f(1-f)\nu\varepsilon'^{T}_{33}}{f + (1-2\nu)},$$
$$\varepsilon'^{T^*}_{22} = f(1-\nu)\left\{\frac{[f(5-4\nu) + (3-4\nu)]\,\varepsilon'^{T}_{22} + (f-1)(1-4\nu)\varepsilon'^{T}_{11}}{[f(3-4\nu)+1]\,[f + (1-2\nu)]}\right\}$$
$$+ \frac{f(1-f)\nu\varepsilon'^{T}_{33}}{f + (1-2\nu)},$$
$$\varepsilon'^{T^*}_{33} = f\varepsilon'^{T}_{33}, \qquad \varepsilon'^{T^*}_{12} = \frac{4f(1-\nu)}{f(3-4\nu)+1}\varepsilon'^{T}_{12}, \qquad \varepsilon'^{T^*}_{13} = \frac{2f}{1+f}\varepsilon'^{T}_{13},$$
$$\varepsilon'^{T^*}_{23} = \frac{2f}{1+f}\varepsilon'^{T}_{23}. \tag{6.147}$$

The above expressions for the sphere and thin-disk agree with Eqs. (6.122) and (6.141), respectively, when $\varepsilon'^{T}_{ij} = \varepsilon^{T}\delta_{ij}$, $\mu^{INC}/\mu^{M} = f$, and $\nu^{INC} = \nu^{M} = \nu$. Using these results, inclusion strain energies can be calculated using Eq. (6.130) after evaluating σ^{INC}_{ij} by the equivalent homogeneous inclusion method.

Many calculations of the elastic energy of various types of inclusions have been published. See, for example, Barnett (1971), Barnett, Lee, Aaronson and Russell (1974), Mura (1987), Onaka, Fujii and Kato (1995), Kato and Fujii (1994) and Kato, Fujii and Onaka (1996a 1996b 1996c). The results are complex and generally involve cumbersome expressions that must be evaluated numerically.

6.5 Coherent → Incoherent Transitions in Isotropic Systems

6.5.1 *General formulation*

The inclusions considered so far have been elastically *coherent* in the sense described in Sec. 6.2, and we now consider *coherent → incoherent transitions* in which the strain energy is reduced, by changing the shape or volume (or both) of either the embedded inclusion or the matrix cavity accommodating the inclusion, and in the process destroying the initial coherence across the inclusion/matrix interface. As mentioned in Sec. 6.2, this can be accomplished by diffusional transport or plastic deformation mechanisms which can be either *conservative* or *non-conservative* depending upon whether they produce a net gain (or loss) of mass to (or from) the inclusion or the matrix immediately adjoining the inclusion. Examples of conservative processes are shown in Fig. 6.6. Figures 6.6a,b show the misfit between the inclusion and its cavity immediately after the inclusion has undergone a transformation strain. Figure 6.6c shows how the shape misfit can be subsequently reduced by conservative diffusional transport within the inclusion which changes the shape of the inclusion but not its volume. Alternatively, this can be accomplished by diffusion along the inclusion/matrix interface or by plastic deformation within the inclusion as in Fig. 6.6d. Conservative diffusional and plastic deformation processes that change the shape of the cavity can also occur in the matrix in the immediate vicinity of the cavity surface. Since the inclusion and matrix volumes remain constant in these mechanisms, the original volume misfit (i.e., the original transformation strain dilatation, e^{T}) remains unchanged. At the limit,

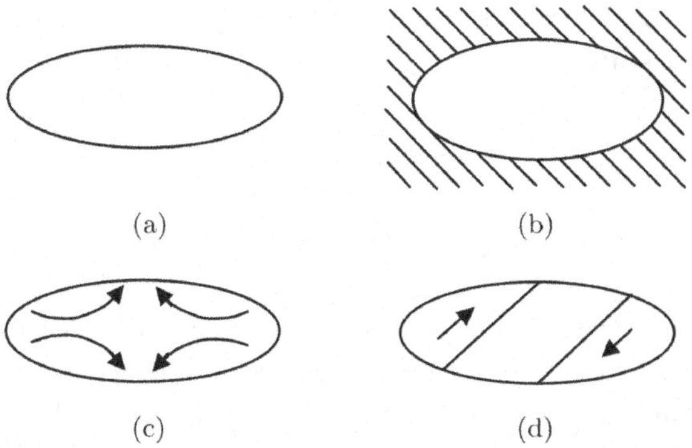

Fig. 6.6 (a) and (b) Misfitting inclusion and corresponding cavity in matrix directly after transformation strain; (c) and (d) Diffusional transport and plastic deformation mechanisms, respectively, within inclusion, that will conservatively decrease the misfit between the inclusion and cavity in (a) and (b).

the shape misfit can be eliminated completely, producing an incoherent inclusion in a state of hydrostatic pressure inherited from its original volume misfit.

Reducing the volume misfit requires a non-conservative process such as the long range diffusional transport of atoms between the inclusion/matrix interface acting as a net source (or sink) and sinks (or sources) in the matrix far from the inclusion, or, alternatively, by plastic deformation in the matrix by a mechanism such as prismatic dislocation punching (Balluffi, Allen and Carter 2005).

In principle, all shape and volume misfit can be eliminated by the above conservative and non-conservative mechanisms thus converting an initial inhomogeneous inclusion into a simple inhomogeneity. However, the extent to which this actually occurs depends upon a host of conditions too extensive to discuss here. In the present context such a complete transition is of relatively little interest, and we therefore focus attention on conservative transitions and prove formally that in such cases an initially coherent inhomogeneous inclusion achieves minimum elastic energy by eliminating its original deviatoric strain (appendix J), or, equivalently, its deviatoric stress, and reverting to a state of hydrostatic pressure. Expressions for the final pressure and elastic energy reached are then obtained in terms of the original transformation strain dilatation, e^T.

According to Eq. (6.130), the elastic energy due to the inclusion in its initial coherent state is

$$\mathrm{W} = -\frac{1}{2}\left(\sigma_{11}^{\mathrm{INC}}\varepsilon_{11}^{\mathrm{T}} + \sigma_{22}^{\mathrm{INC}}\varepsilon_{22}^{\mathrm{T}} + \sigma_{33}^{\mathrm{INC}}\varepsilon_{33}^{\mathrm{T}} + 2\sigma_{12}^{\mathrm{INC}}\varepsilon_{12}^{\mathrm{T}} + 2\sigma_{13}^{\mathrm{INC}}\varepsilon_{13}^{\mathrm{T}} + 2\sigma_{23}^{\mathrm{INC}}\varepsilon_{23}^{\mathrm{T}}\right)\mathrm{V}^{\mathrm{INC}}. \tag{6.148}$$

Making changes in the misfit of an embedded coherent inclusion by the processes illustrated in Fig. 6.6 is equivalent to making changes in the transformation strains assigned to the inclusion at the onset. We can therefore adopt the transformation strains as variables and minimize W with respect to these in order to find a new set of transformation strains that minimizes the energy. Since there are no physical restraints on changing the transformation shear strains in a conservative transition, W can be minimized by simply setting them to zero. However, the requirement of constant inclusion volume demands that the transformation strain dilatation be held constant. The problem of minimizing W is therefore reduced to the formal problem of minimizing the function

$$\mathrm{W}\left(\varepsilon_{11}^{\mathrm{T}}, \varepsilon_{22}^{\mathrm{T}}, \varepsilon_{33}^{\mathrm{T}}\right) = -\frac{1}{2}\left(\sigma_{11}^{\mathrm{INC}}\varepsilon_{11}^{\mathrm{T}} + \sigma_{22}^{\mathrm{INC}}\varepsilon_{22}^{\mathrm{T}} + \sigma_{33}^{\mathrm{INC}}\varepsilon_{33}^{\mathrm{T}}\right)\mathrm{V}^{\mathrm{INC}} \tag{6.149}$$

with respect to the variables $\varepsilon_{11}^{\mathrm{T}}$, $\varepsilon_{22}^{\mathrm{T}}$ and $\varepsilon_{33}^{\mathrm{T}}$ subject to the constraint

$$\varepsilon_{11}^{\mathrm{T}} + \varepsilon_{22}^{\mathrm{T}} + \varepsilon_{33}^{\mathrm{T}} = \varepsilon'^{\mathrm{T}}_{11} + \varepsilon'^{\mathrm{T}}_{22} + \varepsilon'^{\mathrm{T}}_{33} = \mathrm{e}^{\mathrm{T}} = \text{constant}, \tag{6.150}$$

where the primes indicate the new minimum energy quantities. This problem is solved below, for various ellipsoidal inclusion shapes in an isotropic system, by employing Lagrange multipliers.

6.5.2 *Inhomogeneous sphere*

Consider first a spherical ($a_1 = a_2 = a_3 = a$) inhomogeneous inclusion. Following Kato, Fujii and Onaka (1996c), set $\mu^{\mathrm{INC}}/\mu^{\mathrm{M}} \equiv \mathrm{f}$, and make the relatively modest approximation that $\nu^{\mathrm{INC}} = \nu^{\mathrm{M}} = \nu$. The stress $\sigma_{\mathrm{ij}}^{\mathrm{INC}}$ required in Eq. (6.149) is then obtained by using Eq. (2.123) and (6.44) with $\varepsilon_{\mathrm{ij}}^{\mathrm{INC}} = \mathrm{S}_{\mathrm{ijk}l}^{\mathrm{E}}\varepsilon_{kl}^{\mathrm{T}^*} - \varepsilon_{\mathrm{ij}}^{\mathrm{T}}$. The transformation strains, $\varepsilon_{\mathrm{ij}}^{\mathrm{T}^*}$, for the equivalent homogeneous inclusion, are given as functions of the $\varepsilon_{\mathrm{ij}}^{\mathrm{T}}$ by Eq. (6.145), and the S^{E}_{ijkl} are given by Eq. (H.4). Using these relationships,

$$\sigma^{INC}_{11} = \lambda^{INC}\frac{(1+\nu)a-(1-\nu)c}{(1-\nu)c}e^{T}$$
$$+2\mu^{INC}\left[\frac{4(4-5\nu)ca+(1+\nu)ab-3(1-\nu)bc}{3(1-\nu)bc}\varepsilon^{T}_{11}\right.$$
$$\left.+\frac{(1+\nu)ab-2(4-5\nu)ac}{3(1-\nu)bc}(\varepsilon^{T}_{22}+\varepsilon^{T}_{33})\right], \tag{6.151}$$

where

$$a = f(1-\nu), \quad b = 2f(4-5\nu)+(7-5\nu), \quad c = f(1+\nu)+2(1-2\nu). \tag{6.152}$$

σ^{INC}_{22} and σ^{INC}_{33} are obtained by cyclic interchange. Substitution of Eq. (6.151) into Eq. (6.149) therefore yields W as a function of the ε^{T}_{ij}. The minimum of $W(\varepsilon^{T}_{11},\varepsilon^{T}_{22},\varepsilon^{T}_{33})$ under the constraint given by Eq. (6.150) is now obtained by introducing the Lagrange multiplier, $\lambda_{\circ}$, and requiring that

$$\frac{\partial W}{\partial\varepsilon^{T}_{11}} = \lambda_{\circ}\frac{\partial g}{\partial\varepsilon^{T}_{11}}, \qquad \frac{\partial W}{\partial\varepsilon^{T}_{22}} = \lambda_{\circ}\frac{\partial g}{\partial\varepsilon^{T}_{22}}, \qquad \frac{\partial W}{\partial\varepsilon^{T}_{33}} = \lambda_{\circ}\frac{\partial g}{\partial\varepsilon^{T}_{33}},$$
$$g\left(\varepsilon^{T}_{11},\varepsilon^{T}_{22},\varepsilon^{T}_{33}\right) = \varepsilon^{T}_{11}+\varepsilon^{T}_{22}+\varepsilon^{T}_{33}-e^{T}. \tag{6.153}$$

Solving the above equations for the new transformation strains, and calculating the new stresses and strain energy, it is found that in the final minimum energy state

$$\sigma'^{INC}_{11} = \sigma'^{INC}_{22} = \sigma'^{INC}_{33} = -\frac{\lambda_{\circ}}{V^{INC}} = -\frac{4f(1+\nu)}{3[f(1+\nu)+2(1-2\nu)]}\mu^{M}e^{T},$$
$$(\nu = \nu^{INC} = \nu^{M}),$$
$$W'(\text{sphere}) = \frac{\lambda_{\circ}}{2}e^{T} = \frac{2f(1+\nu)}{3[f(1+\nu)+2(1-2\nu)]}\mu^{M}(e^{T})^{2}V^{INC},$$
$$\varepsilon'^{T}_{11} = \varepsilon'^{T}_{22} = \varepsilon'^{T}_{33} = \frac{e^{T}}{3}. \tag{6.154}$$

in agreement with results obtained by Kato, Fujii and Onaka (1996c) using a different route. The inclusion has therefore minimized its energy by adopting a state in which the shear transformation strains have vanished, and the total conserved dilatational transformation strain, e^{T}, is partitioned into three equal normal transformation strains, thereby producing a state of hydrostatic stress. The strain energy, W', is proportional to $(e^T)^2$ and, since $f = \mu^{INC}/\mu^M$, is seen to be identical to the expression given by Eq. (6.137) for the strain energy of a coherent inclusion having the same transformation strains. In this case the elastic states of the two inclusions, are the same. For the incoherent inclusion, that state was reached via a coherent → incoherent transition, while for the coherent inclusion no transition occurred.

6.5.3 *Inhomogeneous thin-disk*

Using the same methods as before, the results obtained in Exercise 6.10 for an incoherent ellipsoidal thin-disk ($a_3 \to 0, a_1 = a_2$) are

$$\begin{aligned} &\sigma'^{INC}_{11} = \sigma'^{INC}_{22} = \sigma'^{INC}_{33} = 0, \qquad (\nu = \nu^{INC} = \nu^M) \\ &W'(\text{thin} - \text{disk}) = 0, \qquad \varepsilon'^T_{11} = \varepsilon'^T_{22} = 0, \qquad \varepsilon'^T_{33} = e^T. \end{aligned} \tag{6.155}$$

In this case the inclusion has eliminated the initial shear transformation strains and has concentrated all of the conserved transformation strain dilatation in the direction normal to the broad face of the disk. The thin disk geometry allows this to occur without the development of stress. All stresses, as well as the strain energy, therefore vanish.

6.5.4 *Inhomogeneous needle*

Finally, for an incoherent ellipsoidal needle ($a_1 = a_2, a_3 \to \infty$), it is found by the above methods that

$$\begin{aligned} \sigma'^{INC}_{11} = \sigma'^{INC}_{22} = \sigma'^{INC}_{33} &= -\frac{2f(1+\nu)}{2f(1+\nu)+3(1-2\nu)}\mu^M e^T, \\ &(\nu = \nu^{INC} = \nu^M), \\ W'(\text{needle}) = -\frac{1}{2}\sigma'^{INC}_{11} V^{INC} e^T &= \frac{f(1+\nu)}{2f(1+\nu)+3(1-2\nu)} V^{INC} \mu^M (e^T)^2. \end{aligned} \tag{6.156}$$

In this case the inclusion has eliminated all shear transformation strains and shear stresses and produced a state of hydrostatic stress. Again, the strain energy is proportional to the square of the conserved transformation strain dilatation, e^T.

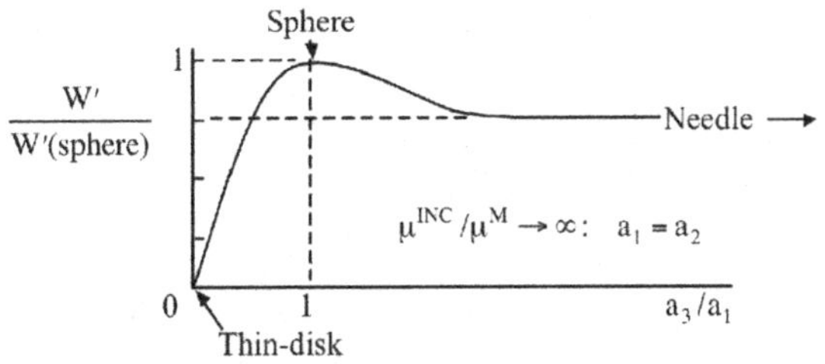

Fig. 6.7 Schematic plot of relative strain energy of infinitely stiff incoherent ellipsoidal inclusion in isotropic system as a function of its eccentricity, a_3/a_1.

These results indicate that for fully incoherent inclusions the strain energy varies with inclusion shape in the sequence

$$W'(\text{thin}-\text{disk}) < W'(\text{needle}) < W'(\text{sphere}). \qquad (6.157)$$

In the limit when $f = \mu^{INC}/\mu^{M} \to \infty$, and all of the strain energy resides in the matrix, the energies of the differently shaped incoherent inclusions have the relative values

$$W'(\text{thin}-\text{disk}) : W'(\text{needle}) : W'(\text{sphere}) = 0 : \frac{3}{4} : 1, \qquad (6.158)$$

in agreement with the classic results obtained for this case by Nabarro (1940) and illustrated schematically in Fig. 6.7.

Exercises

6.1. Find the elastic displacement field, in the inclusion and matrix, of a coherent homogeneous spherical inclusion of radius a with *a* uniform transformation strain $\varepsilon^{T}_{ij} = \delta_{ij}\varepsilon^{T}$ in an isotropic system by starting with Eq. (6.26), to treat the inclusion, and (6.31), to treat the matrix.

Solution. Since the displacement field has spherical symmetry, it suffices to determine the field as a function of distance along $\hat{\mathbf{e}}_3$. Therefore, with $\mathbf{x}$ fixed along $\hat{\mathbf{e}}_3$, and after substituting $\varepsilon^{T}_{ij} = \delta_{ij}\varepsilon^{T}$, Eq. (6.26) takes the form

$$u_3^{C,INC}(x_3) = \frac{1}{2}(C_{jl11} + C_{jl22} + C_{jl33})\varepsilon^{T} x_3 \int_{-1}^{1} d\hat{\zeta}_3 \hat{k}_3 \hat{k}_l (\hat{k}\hat{k})_{3j}, \qquad (6.159)$$

Then, after using Eq. (6.58),

$$u_3^{C,INC}(x_3) = \frac{1}{2\mu}(C_{jl11} + C_{jl22} + C_{jl33})\varepsilon^T x_3 \times \int_{-1}^{1} d\hat{\zeta}_3 \hat{k}_3 \hat{k}_l \left[\delta_{3j} - \frac{1}{2(1-\nu)}\hat{k}_3\hat{k}_j\right] \quad (6.160)$$

and after using Eq. (6.13),

$$u_3^{C,INC}(x_3) = \frac{1}{2\mu}(C_{jl11} + C_{jl22} + C_{jl33})\varepsilon^T x_3 \times \int_{-1}^{1} \hat{\zeta}_3 \hat{\zeta}_l \left[\delta_{3j} - \frac{1}{2(1-\nu)}\hat{\zeta}_3\hat{\zeta}_j\right] d\hat{\zeta}_3 \quad (6.161)$$

and after using Eq. (2.120),

$$u_3^{C,INC}(x_3) = \frac{(1+\nu)}{2(1-\nu)}\varepsilon^T x_3 \int_{-1}^{1} \hat{\zeta}_3^2 d\hat{\zeta}_3 = \frac{(1+\nu)}{3(1-\nu)}\varepsilon^T x_3 \quad (6.162)$$

and after using Eq. (6.1), with $u_3^T = \varepsilon^T x_3$,

$$u_3^{INC}(x_3) = u_3^{C,INC}(x_3) - u_3^T = -\frac{2(1-2\nu)}{3(1-\nu)}\varepsilon^T x_3, \quad (6.163)$$

which agrees with the displacement for this inclusion given by Eq. (6.124).

Turning now to the matrix, if Eq. (6.31) is treated in the same way as Eq. (6.26) to arrive at Eq. (6.162),

$$u_3^{C,M}(x_3) = \frac{(1+\nu)}{2(1-\nu)}\varepsilon^T x_3 \int_{-s(x)}^{s(x)} \hat{\zeta}_3^2 d\hat{\zeta}_3. \quad (6.164)$$

The limits for $\hat{\zeta}_3$ required for the integral in Eq. (6.164) are found with the help of Figs. 6.2b and 6.3. Since **x** and, therefore **y**, are parallel to $\hat{\mathbf{e}}_3$, component $\hat{\zeta}_3$ can be measured along **y**, and using Fig. 6.2b, it is readily seen that the limits on $\hat{\zeta}_3$ are

$$s = \pm\frac{1}{y} = \pm\frac{a}{x}. \quad (6.165)$$

Then, substituting these into Eq. (6.164) and using Eq. (6.1),

$$u_3^M(x_3) = u_3^{C,M}(x_3) = \frac{(1+\nu)}{2(1-\nu)}\varepsilon^T x_3 \int_{-a/x}^{a/x} \hat{\zeta}_3^2 d\hat{\zeta}_3 = \frac{(1+\nu)}{3(1-\nu)} a^3 \varepsilon^T \frac{x_3}{x^3}, \tag{6.166}$$

in agreement with Eq. (6.126) for the case of a homogeneous inclusion.

6.2. Consider a coherent homogeneous inclusion with a non-uniform transformation strain given by the second degree polynomial

$$\varepsilon_{mn}^T(\mathbf{x}') = \varepsilon_{mnpq}^T y_p' y_q'. \tag{6.167}$$

Show that the final strain in the inclusion, ε_{mn}^{INC}, is also a second degree polynomial

Solution. By differentiating Eq. (6.167) after substituting Eq. (6.33),

$$\frac{\partial \varepsilon_{mn}^T}{\partial z} = \varepsilon_{mnpq}^T \Big\{ 2z\hat{\zeta}_p\hat{\zeta}_q + r\Big[\Big(\hat{\zeta}_q\hat{m}_p + \hat{\zeta}_p\hat{m}_q\Big)\cos\psi + \Big(\hat{\zeta}_q\hat{n}_p + \hat{\zeta}_p\hat{n}_q\Big)\sin\psi\Big]\Big\}, \tag{6.168}$$

and then substituting this expression and Eq. (6.167) into Eq. (6.22) and integrating over r and ψ,

$$u_i^{C,INC} = -\frac{\varepsilon_{mnpq}^T}{8\pi^2} \oiint_{\hat{\mathcal{S}}} \Big\{ \Big[\hat{\zeta}_p\hat{\zeta}_q - \pi(\hat{m}_p\hat{m}_q + \hat{n}_p\hat{n}_q)\Big] z - \Big[2\hat{\zeta}_p\hat{\zeta}_q - \pi(\hat{m}_p\hat{m}_q + \hat{n}_p\hat{n}_q)\Big] z^3 \Big\}_{z=\hat{\boldsymbol{\zeta}}\cdot\mathbf{y}} C_{jlmn}\hat{k}_l(\hat{k}\hat{k})_{ij}^{-1}\zeta d\hat{S}(\hat{\boldsymbol{\zeta}}). \tag{6.169}$$

Then, employing $\hat{\boldsymbol{\zeta}}\cdot\mathbf{y} = (\hat{\mathbf{k}}\cdot\mathbf{x})\zeta^{-1}$ from Eq. (6.13), and differentiating Eq. (6.169),

$$\frac{\partial u_i^{C,INC}}{\partial x_k} = -\frac{\varepsilon_{mnpq}^T}{8\pi^2} \oiint_{\hat{\mathcal{S}}} \Big\{ \Big[\hat{\zeta}_p\hat{\zeta}_q - \pi(\hat{m}_p\hat{m}_q + \hat{n}_p\hat{n}_q)\Big] - \Big[6\hat{\zeta}_p\hat{\zeta}_q - 3\pi(\hat{m}_p\hat{m}_q + \hat{n}_p\hat{n}_q)\Big]\zeta^{-2}(\hat{\mathbf{k}}\cdot\mathbf{x})^2 \Big\} \times C_{jlmn}\hat{k}_l\hat{k}_k(\hat{k}\hat{k})_{ij}^{-1} d\hat{S}(\hat{\boldsymbol{\zeta}}). \tag{6.170}$$

Inspection of Eq. (6.170) shows that by virtue of Eq. (2.5), $\varepsilon^{C,INC}_{mn}$ will be a second degree polynomial in x_i. Then, according to Eq. (6.1), since ε^{T}_{mn} is a second degree polynomial, ε^{INC}_{mn} is also a second degree polynomial.

6.3. Using Eq. (6.65), show that the dilatations, $e^{C,INC}$ and $e^{C,M}$, for a coherent homogeneous inclusion of arbitrary shape in an isotropic system are given by

$$e^{C,INC} = \frac{(1+\nu^M)}{3(1-\nu^M)} e^T, \quad e^{C,M} = 0 \tag{6.171}$$

when $\varepsilon^{T}_{jk} = \varepsilon^{T}\delta_{jk}$.

Solution. By applying Hooke's law,

$$\sigma^{T}_{jk} = \lambda^M e^T \delta_{jk} + 2\mu^M \varepsilon^{T}_{jk} = \left(\frac{3\lambda^M + 2\mu^M}{3}\right) e^T \delta_{jk}, \tag{6.172}$$

and by substituting Eq. (6.172) into Eq. (6.65),

$$e^C = -\frac{1}{4\pi}\frac{(1+\nu^M)}{3(1-\nu^M)} e^T \nabla^2 \phi. \tag{6.173}$$

Then, use of Eq. (6.64) produces the required results. The dilatation therefore vanishes in the matrix and is uniform throughout the inclusion, a result that is verified, for example, in Sec. 6.4.3.1 for the case of a spherical inclusion.

6.4. Show that, for a coherent homogeneous inclusion of arbitrary shape with $\varepsilon^{T}_{ij} = \varepsilon^{T}\delta_{ij}$ in an isotropic system, Eq. (6.70) produces the same result for the difference in dilatation, $e^{C,M} - e^{C,INC}$, across the inclusion/matrix interface as predicted by the results of Exercise 6.3.

Solution. The transformation stress, σ^{T}_{jk}, required to evaluate $e^{C,M} - e^{C,INC}$ using Eq. (6.70) is again given by Eq. (6.172). Substituting this into Eq. (6.70), and performing the required sums,

$$e^{C,INC} - e^{C,M} = \frac{(1+\nu^M)}{3(1-\nu^M)} e^T \tag{6.174}$$

in agreement with the result obtained for $e^{C,M} - e^{C,INC}$ using Eq. (6.171).

6.5. The strain energy due to a coherent homogeneous inclusion of arbitrary shape and transformation strain is given by Eq. (6.57) where it is obtained by determining the change in the elastic energy of the entire system that occurs as the canceling displacement u_i^C is imposed. Obtain the same result by determining the strain energy as the sum of the strain energies in the inclusion and matrix taken separately.

Solution. The final strain energy in the inclusion is given by

$$W^{INC} = \frac{1}{2} \oiiint_{V^{INC}} \sigma_{ij}^{INC} \varepsilon_{ij}^{INC} dV. \tag{6.175}$$

The final strain energy in the matrix corresponds to the work done on it during step 5 of Sec. 6.2 when the forces $dF_i = \sigma_{ij}^T \hat{n}_j dS$ are applied to $\mathcal{S}^{INC}$ to produce the displacement $u_i^{C,M}$. Since the displacements increase linearly with the forces, the work is

$$\mathcal{W}^M = \frac{1}{2} \oiint_{\mathcal{S}^{INC}} u_i^{C,M} \sigma_{ij}^{C,M} \hat{n}_j dS = \frac{1}{2} \oiint_{\mathcal{S}^{INC}} u_i^{C,INC} \sigma_{ij}^{C,INC} \hat{n}_j dS, \tag{6.176}$$

where the positive direction of $\hat{\mathbf{n}}$ is taken in this case towards the inclusion, and $\sigma_{ij}^{C,M} \hat{n}_j = \sigma_{ij}^{INC} \hat{n}_j$ and $u_i^{C,M} = u_i^{C,INC}$ on $\mathcal{S}^{INC}$. Then, reversing the direction of $\hat{\mathbf{n}}$ in the surface integral in Eq. (6.176) and converting the integral to a volume integral,

$$\mathcal{W}^M = W^M = -\frac{1}{2} \oiiint_{V^{INC}} \sigma_{ij}^{INC} \varepsilon_{ij}^{C,INC} dV. \tag{6.177}$$

The total strain energy due to the inclusion is then, from Eqs. (6.175), (6.177) and (6.1),

$$\begin{aligned} W = W^{INC} + W^M &= -\frac{1}{2} \oiiint_{V^{INC}} \sigma_{ij}^{INC} (\varepsilon_{ij}^{C,INC} - \varepsilon_{ij}^{INC}) dV \\ &= -\frac{1}{2} \oiiint_{V^{INC}} \sigma_{ij}^{INC} \varepsilon_{ij}^{T} dV \end{aligned} \tag{6.178}$$

in agreement with Eq. (6.57).

6.6. Show that the strain energy due to a coherent homogeneous inclusion of arbitrary ellipsoidal shape in an isotropic system is given by

$$W = \frac{2\mu^M K^M}{(3K^M + 4\mu^M)} (e^T)^2 V^{INC} = \frac{2\mu^M (1 + \nu^M)}{9(1 - \nu^M)} (e^T)^2 V^{INC} \tag{6.179}$$

when the transformation strain is $\varepsilon_{ij}^T = \varepsilon^T \delta_{ij}$. Its strain energy is therefore independent of its particular ellipsoidal shape (i.e., spherical, thin-disk, or needle).

Solution. Equation (6.130) for the strain energy becomes

$$W = -\frac{1}{2}\left(\sigma_{11}^{INC} + \sigma_{22}^{INC} + \sigma_{33}^{INC}\right)\varepsilon^{T}V^{INC} = -\frac{3}{2}K^{M}e^{INC}\varepsilon^{T}V^{INC}. \tag{6.180}$$

Using Eq. (6.1), $e^{INC} = e^{C,INC} - e^{T}$, and the result given by Eq. (6.171) in Exercise 6.3 that $e^{C,INC} = (1+\nu^{M})e^{T}/[3(1-\nu^{M})]$ for a homogeneous inclusion of arbitrary shape and uniform transformation strain, $\varepsilon_{ij}^{T} = \varepsilon^{T}\delta_{ij}$, Eq. (6.180) takes the form of Eq. (6.179).

6.7. In an isotropic system find expressions for the stresses in a coherent homogeneous thin-disk ellipsoidal inclusion with a uniform transformation strain, ε_{ij}^{T}, in the limit where its thickness tends to zero.

Solution. Substitute values of the Eshelby tensor from Eq. (H.6) into Eq. (6.96) to find the $\varepsilon_{ij}^{C,INC}$ strains in terms of the transformation strains. Then use Eq. (6.1) to obtain the elastic strains in the inclusion given by

$$\varepsilon_{11}^{INC} = -\varepsilon_{11}^{T}, \quad \varepsilon_{22}^{INC} = -\varepsilon_{22}^{T}, \quad \varepsilon_{33}^{INC} = \frac{\nu^{M}}{1-\nu^{M}}(\varepsilon_{11}^{T} + \varepsilon_{22}^{T}),$$

$$\varepsilon_{12}^{INC} = -\varepsilon_{12}^{T}, \quad \varepsilon_{13}^{INC} = \varepsilon_{23}^{INC} = 0. \tag{6.181}$$

Finally, substitute these strains into Eq. (2.122) to obtain the stresses

$$\sigma_{11}^{INC} = -2\mu^{M}\left[\frac{\nu^{M}}{1-\nu^{M}}(\varepsilon_{11}^{T} + \varepsilon_{22}^{T}) + \varepsilon_{11}^{T}\right],$$

$$\sigma_{22}^{INC} = -2\mu^{M}\left[\frac{\nu^{M}}{1-\nu^{M}}(\varepsilon_{11}^{T} + \varepsilon_{22}^{T}) + \varepsilon_{22}^{T}\right],$$

$$\sigma_{12}^{INC} = -2\mu^{M}\varepsilon_{12}^{T}, \quad \sigma_{33}^{INC} = \sigma_{13}^{INC} = \sigma_{23}^{INC} = 0. \tag{6.182}$$

6.8. Instead of using the equivalent homogeneous inclusion method to obtain the results given by Eq. (6.129) for a spherical inhomogeneous inclusion of radius R with transformation strain $\varepsilon_{ij}^{T} = \varepsilon^{T}\delta_{ij}$ in an isotropic system, derive the same results by solving the Navier equation directly.

Solution. Treat the problem as a boundary value problem requiring the direct solution of the Navier equation in both the matrix and inclusion subject to the prevailing boundary conditions. The system contains transformation strains, and we therefore employ the formulation of elasticity theory presented in Sec. 3.6. The Navier equation for an isotropic system without transformation strains is given by Eq. (3.3), and a comparison of Eqs. (3.2) and (3.158) shows that the corresponding Navier equation for an isotropic system with transformation strains must therefore be

$$\frac{2\mu(1-\nu)}{1-2\nu}\nabla(\nabla\cdot\mathbf{u}^{\mathrm{tot}})-\mu\nabla\times(\nabla\times\mathbf{u}^{\mathrm{tot}})-\hat{\mathbf{e}}_{\mathrm{i}}\frac{\partial\sigma_{\mathrm{ij}}^{\mathrm{T}}}{\partial\mathrm{x_j}}=0. \quad (6.183)$$

Then, since $\varepsilon_{\mathrm{ij}}^{\mathrm{T}}=\varepsilon^{\mathrm{T}}\delta_{\mathrm{ij}}$,

$$\frac{2\mu(1-\nu)}{1-2\nu}\nabla(\nabla\cdot\mathbf{u}^{\mathrm{tot}})-\mu\nabla\times(\nabla\times\mathbf{u}^{\mathrm{tot}})=0. \quad (6.184)$$

The problem is spherically symmetric, and $\mathbf{u}^{\mathrm{tot}}$ is therefore radial and a function of r only. Therefore, employing spherical coordinates, $\nabla\times\mathbf{u}^{\mathrm{tot}}=0$, and

$$\nabla(\nabla\cdot\mathbf{u}^{\mathrm{tot}})=\frac{\partial}{\partial\mathrm{r}}\left\{\frac{1}{\mathrm{r}^2}\frac{\partial}{\partial\mathrm{r}}\left[\mathrm{r}^2\mathrm{u}_{\mathrm{r}}^{\mathrm{tot}}(\mathrm{r})\right]\right\}=0. \quad (6.185)$$

The general solutions of Eq. (6.185) in the inclusion and matrix are then, respectively,

$$\mathrm{u}_{\mathrm{r}}^{\mathrm{tot,INC}}(\mathrm{r})=\mathrm{u}_{\mathrm{r}}^{\mathrm{INC}}(\mathrm{r})+\varepsilon^{\mathrm{T}}\mathrm{r}=\frac{\mathrm{c}_1^{\mathrm{INC}}}{\mathrm{r}^2}+\mathrm{c}_2^{\mathrm{INC}}\mathrm{r},$$

$$\mathrm{u}_{\mathrm{r}}^{\mathrm{tot,M}}(\mathrm{r})=\mathrm{u}_{\mathrm{r}}^{\mathrm{M}}(\mathrm{r})=\frac{\mathrm{c}_1^{\mathrm{M}}}{\mathrm{r}^2}+\mathrm{c}_2^{\mathrm{M}}\mathrm{r}. \quad (6.186)$$

In the matrix a bound is put on $\mathrm{u}^{\mathrm{M}}(\mathrm{r})$ by requiring that $\mathrm{c}_2^{\mathrm{M}}=0$. Then, using Eqs. (G.10) and (G.11), the elastic displacements, stresses and strains in the matrix are

$$\begin{aligned}
&\mathrm{u}_{\mathrm{r}}^{\mathrm{M}}=\mathrm{c}_1^{\mathrm{M}}\frac{1}{\mathrm{r}^2}, && \mathrm{u}_{\theta}^{\mathrm{M}}=0, && \mathrm{u}_{\phi}^{\mathrm{M}}=0,\\
&\varepsilon_{\mathrm{rr}}^{\mathrm{M}}=-2\mathrm{c}_1^{\mathrm{M}}\frac{1}{\mathrm{r}^3}, && \varepsilon_{\theta\theta}^{\mathrm{M}}=\mathrm{c}_1^{\mathrm{M}}\frac{1}{\mathrm{r}^3}, && \varepsilon_{\phi\phi}^{\mathrm{M}}=\mathrm{c}_1^{\mathrm{M}}\frac{1}{\mathrm{r}^3},\\
&\sigma_{\mathrm{rr}}^{\mathrm{M}}=-4\mu^{\mathrm{M}}\mathrm{c}_1^{\mathrm{M}}\frac{1}{\mathrm{r}^3}, && \sigma_{\theta\theta}^{\mathrm{M}}=2\mu^{\mathrm{M}}\mathrm{c}_1^{\mathrm{M}}\frac{1}{\mathrm{r}^3}, && \sigma_{\phi\phi}^{\mathrm{M}}=2\mu^{\mathrm{M}}\mathrm{c}_1^{\mathrm{M}}\frac{1}{\mathrm{r}^3}.
\end{aligned} \quad (6.187)$$

The dilatation $\mathrm{e}^{\mathrm{M}}=\varepsilon_{\mathrm{rr}}^{\mathrm{M}}+\varepsilon_{\phi\phi}^{\mathrm{M}}+\varepsilon_{\theta\theta}^{\mathrm{M}}$ is therefore seen to vanish throughout the matrix. When $\mathrm{c}_1^{\mathrm{M}}$ is positive, for example, each differential volume element (Fig. G.1b) contracts in the radial r direction and expands in the θ and ϕ circumferential directions without producing any volume change.

In the inclusion, a singularity at the origin is avoided by requiring that $c_1^{INC} = 0$. Then, using Eqs. (6.186) and (3.152),

$$u_r^{tot,INC}(r) = u_r^{INC}(r) + u_r^T(r) = c_2^{INC} r = u_r^{INC}(r) + \varepsilon^T r \tag{6.188}$$

and using Eqs. (G.10) and (G.11), the elastic strains and stresses are

$$\begin{aligned} \varepsilon_{rr}^{INC} &= \varepsilon_{\phi\phi}^{INC} = \varepsilon_{\theta\theta}^{INC} = c_2^{INC} - \varepsilon^T, \\ \sigma_{rr}^{INC} &= \sigma_{\phi\phi}^{INC} = \sigma_{\theta\theta}^{INC} = 3K^{INC}\left(c_2^{INC} - \varepsilon^T\right). \end{aligned} \tag{6.189}$$

The remaining constants, c_2^{INC} and c_1^M, can now be obtained by invoking the boundary conditions at the inclusion/matrix interface. The radial stresses in the matrix and inclusion must match across the interface, i.e., $\sigma_{rr}^{INC}(R) = \sigma_{rr}^M(R)$, and therefore

$$3K^{INC}(c_2^{INC} - \varepsilon^T) = -\frac{4\mu^M c_1^M}{R^3}. \tag{6.190}$$

The total radial displacements in the inclusion and matrix must also match, so that

$$c_2^{INC} R = c_1^M \frac{1}{R^2}. \tag{6.191}$$

Then, solving Eqs. (6.190) and (6.191) for the two constants, the elastic displacements are

$$\begin{aligned} u_r^{INC}(r) &= u_r^{tot,INC}(r) - \varepsilon^T r = (c_2^{INC} - \varepsilon^T) r = -\frac{4\mu^M}{(3K^{INC} + 4\mu^M)}\varepsilon^T r \\ u_r^M(r) &= \frac{c_1^M}{r^2} = \frac{9K^{INC}V^{INC}}{4\pi(3K^{INC} + 4\mu^M)}\varepsilon^T \frac{1}{r^2} \end{aligned} \tag{6.192}$$

in agreement with Eq. (6.129).

6.9. In Sec. 6.4.4.3, the strain energy due to a spherical coherent inhomogeneous inclusion with a uniform transformation strain, $\varepsilon_{ij}^T = \varepsilon^T \delta_{ij}$, in an isotropic system was found in the form of Eq. (6.137) with the help of the general Eq. (6.130). Obtain Eq. (6.137) by an alternative approach that uses the stresses and strains associated with such an inclusion and calculates the strain energy as the sum of the strain energies in the inclusion and matrix.

Solution. In spherical coordinates the displacements in the inclusion and matrix are given by Eq. (6.129). Using Eq. (G.10), the strains are therefore

$$\varepsilon_{rr}^{INC} = -\frac{4\mu^{M}\varepsilon^{T}}{3K^{INC}+4\mu^{M}}, \quad \sigma_{rr}^{INC} = 3K^{INH}\varepsilon_{rr}^{INC},$$

$$\varepsilon_{rr}^{M} = -\frac{2c}{r^3}, \quad \sigma_{rr}^{M} = -\frac{4\mu^{M}c}{r^3},$$

$$\varepsilon_{\theta\theta}^{INC} = \varepsilon_{\phi\phi}^{INC} = \varepsilon_{rr}^{INC}, \quad \sigma_{\theta\theta}^{INC} = \sigma_{\phi\phi}^{INC} = \sigma_{rr}^{INC},$$

$$\varepsilon_{\theta\theta}^{M} = \varepsilon_{\phi\phi}^{M} = \frac{c}{r^3}, \quad \sigma_{\theta\theta}^{M} = \sigma_{\phi\phi}^{M} = \frac{2\mu^{M}c}{r^3}. \tag{6.193}$$

The strain energy in the inclusion is then

$$\begin{aligned} W^{INC} &= \frac{1}{2}\oiiint_{V^{INC}} \left(\sigma_{rr}^{INC}\varepsilon_{rr}^{INC} + \sigma_{\theta\theta}^{INC}\varepsilon_{\theta\theta}^{INC} + \sigma_{\phi\phi}^{INC}\varepsilon_{\phi\phi}^{INC}\right) dV \\ &= \frac{72K^{INC}(\mu^{M})^2 V^{INC}(\varepsilon^{T})^2}{(3K^{INC}+4\mu^{M})^2} \end{aligned} \tag{6.194}$$

and the strain energy in the matrix (when the inclusion radius is a) is

$$\begin{aligned} W^{M} &= \frac{1}{2}\int_a^{\infty} (\sigma_{rr}^{M}\varepsilon_{rr}^{M} + \sigma_{\theta\theta}^{M}\varepsilon_{\theta\theta}^{M} + \sigma_{\phi\phi}^{M}\varepsilon_{\phi\phi}^{M})4\pi r^2 dr \\ &= 24\pi\mu^{M}c^2\int_a^{\infty}\frac{dr}{r^3} = \frac{32\pi^2\mu^{M}c^2}{3V^{INC}}. \end{aligned} \tag{6.195}$$

The total strain energy is therefore

$$W = W^{INC} + W^{M} = \frac{18K^{INC}\mu^{M}}{3K^{INC}+4\mu^{M}}V^{INC}\left(\varepsilon^{T}\right)^2 \tag{6.196}$$

in agreement with Eq. (6.137).

6.10. Derive the results given by Eq. (6.155) for the final minimum energy state of an incoherent ellipsoidal thin-disk inhomogeneous inclusion in an isotropic system.

Solution. Using the same procedure used to obtain Eq. (6.151),

$$\sigma_{11}^{INC} = a(\varepsilon_{11}^{T} + \varepsilon_{22}^{T}) - b\varepsilon_{11}^{T}, \qquad \sigma_{22}^{INC} = a(\varepsilon_{11}^{T} + \varepsilon_{22}^{T}) - b\varepsilon_{22}^{T},$$

$$\sigma_{33}^{INC} = 0, \qquad a = \frac{\lambda^{INC} f(2\nu - 1)}{1-\nu}, \qquad b = 2f\mu^{INC}. \tag{6.197}$$

The inclusion strain energy, from Eq. (6.130), is then

$$W = \frac{\mu^{INC} f}{1-\nu}\left[\left(\varepsilon_{11}^{T}\right)^2 + 2\nu\varepsilon_{11}^{T}\varepsilon_{22}^{T} + \left(\varepsilon_{22}^{T}\right)^2\right] V^{INC}. \tag{6.198}$$

Equation (6.153) applies in this case, and therefore

$$\frac{2\mu^{INC} f V^{INC}}{1-\nu}(\varepsilon_{11}^{T} + \nu\varepsilon_{22}^{T}) = \lambda_{\circ}$$

$$\frac{2\mu^{INC} f V^{INC}}{1-\nu}(\varepsilon_{22}^{T} + \nu\varepsilon_{11}^{T}) = \lambda_{\circ}$$

$$\lambda_{\circ} = 0$$

$$\varepsilon_{11}^{T} + \varepsilon_{22}^{T} + \varepsilon_{33}^{T} = e^{T} = \text{constant}. \tag{6.199}$$

Solving Eq. (6.199) simultaneously for ε'^{T}_{11}, ε'^{T}_{22} and ε'^{T}_{33}, then allows the calculation of the results given by Eq. (6.155).

6.11. The S_{66}^{E} component of the Eshelby tensor for a spherical homogeneous inclusion in an isotropic system is given by Eq. (H.4) as

$$S_{66}^{E} = \frac{(4-5\nu)}{15(1-\nu^{M})}. \tag{6.200}$$

Derive this relationship. Hint: the Newtonian and biharmonic potentials (see Eq. (6.61)) at a point inside a sphere of radius R are given by

$$\phi = 2\pi\left(R^2 - \frac{r^2}{3}\right)$$

$$\psi = \frac{2\pi R^2}{3}r^2 - \frac{\pi}{15}r^4 + \pi R^4 \tag{6.201}$$

with the origin at the sphere center.

Solution. Assume a spherical homogeneous inclusion where the only non-zero transformation strain is uniform and given by ε_{12}^{T}. Then, according to Eq. (6.91) and our contracted tensor notation,

$$\varepsilon_{12}^{C,INC} = S_{1212}^{E}\varepsilon_{12}^{T} = S_{66}^{E}\varepsilon_{12}^{T}. \tag{6.202}$$

The strain $\varepsilon_{12}^{C,INC}$ in Eq. (6.202) can be found by employing Eq. (6.60). Substituting

$$\sigma_{ik}^{T} = 2\mu^{M}\varepsilon_{ik}^{T} \tag{6.203}$$

and the transformation strain ε_{12}^{T} into Eq. (6.60) yields

$$u_i^{C,INC} = -\frac{1}{2\pi}\varepsilon_{i2}^{T}\frac{\partial\phi}{\partial x_2} + \frac{1}{8\pi(1-\nu^{M})}\varepsilon_{12}^{T}\frac{\partial^3\psi}{\partial x_i\partial x_1\partial x_2}. \tag{6.204}$$

Then, using Eq. (6.201) after substituting $r = (x_1^2 + x_2^2 + x_3^2)^{1/2}$,

$$\frac{\partial\phi}{\partial x_2} = -\frac{4\pi}{3}x_2,$$

$$\frac{\partial^3\psi}{\partial x_i\partial x_1\partial x_2} = -\frac{8\pi}{15}\frac{\partial(x_1x_2)}{\partial x_i}. \tag{6.205}$$

Substituting these results into Eq. (6.204) then yields

$$u_i^{C,INC} = \frac{2}{3}\varepsilon_{i2}^{T}x_2 - \frac{1}{15(1-\nu^{M})}\varepsilon_{12}^{T}\frac{\partial(x_1x_2)}{\partial x_i}. \tag{6.206}$$

Then, using Eq. (2.5),

$$\varepsilon_{12}^{C,INC} = \frac{1}{2}\left(\frac{\partial u_1^{C,INC}}{\partial x_2} + \frac{\partial u_2^{C,INC}}{\partial x_1}\right) = \frac{(4-5\nu^{M})}{15(1-\nu^{M})}\varepsilon_{12}^{T}. \tag{6.207}$$

By comparing Eq. (6.207) with (6.202), we therefore have

$$S_{66}^{E} = \frac{4-5\nu^{M}}{15(1-\nu^{M})}. \tag{6.208}$$

Chapter 7

Interactions Between Inclusions and Imposed Stress

7.1 Introduction

Interaction energies and forces between homogeneous inclusions and imposed stress are treated first using the basic expressions developed in Ch. 5. Then, attention is shifted to inhomogeneous inclusions where the treatment is more complicated, since the inhomogeneity associated with such inclusions perturbs the imposed stress field. It therefore becomes necessary to determine the total field produced by the imposed stress, the inhomogeneity associated with the inclusion and the transformation strain (misfit) of the inclusion. This problem is treated using results obtained in Chs. 6 and 9 for ellipsoidal inhomogeneous inclusions and inhomogeneities, respectively, by use of Eshelby's equivalent homogeneous inclusion method.

7.2 Interactions between Inclusions and Imposed Stress in Isotropic Systems

7.2.1 *Homogeneous inclusion*

As shown by Eqs. (5.13) and (5.26), the interaction energy between a homogeneous inclusion of arbitrary shape and transformation strain and a general imposed stress system, Q, in a large but finite body, is given by

$$\mathrm{E}_{\mathrm{int}}^{\mathrm{INC/Q}} = -\oiiint_{\mathcal{V}^{\mathrm{INC}}} \sigma_{\mathrm{ij}}^{\mathrm{Q}} \varepsilon_{\mathrm{ij}}^{\mathrm{T}} \mathrm{dV}, \qquad (7.1)$$

where the Q stress can be either internal or applied.[1]

[1]It is also readily shown that Eq. (7.1) is valid for both finite and infinite bodies.

In the simple case when the transformation strain is the dilatation, $\varepsilon_{ij}^{T} = \varepsilon^{T}\delta_{ij}$, and σ_{ij}^{Q} is uniform throughout $\mathcal{V}^{INC}$, Eq. (7.1) yields

$$E_{int}^{INC/Q} = -\sigma_{ij}^{Q}\varepsilon_{ij}^{T}V^{INC} = -\sigma_{mm}^{Q}\varepsilon^{T}V^{INC} = -\frac{1}{3}\sigma_{mm}^{Q}\Delta V^{INC}, \qquad (7.2)$$

where $-\sigma_{mm}^{Q}/3$ is the hydrostatic part of the Q stress field (as described in the text following Eq. (2.127)), and $\Delta V^{INC} = 3V^{INC}\varepsilon^{T}$ is the misfit volume of the inclusion due to the transformation strain. The interaction energy is thus the familiar "work of expansion" against hydrostatic pressure, and takes this form because the inclusion with its isotropic transformation strain interacts only with the normal stress components of the σ_{ij}^{Q} stress tensor. Note that when the inclusion is oversize (so that ΔV^{INC} is positive), and σ_{mm}^{Q} is compressive (so that σ_{mm}^{Q} is negative), $E_{int}^{INC/Q}$ is positive, as would be expected. In Exercise 7.2, Eq. (7.2) is obtained for the above inclusion by mimicking it as a force density distribution, as described in Sec. 5.2.1.3, and employing Eq. (5.28).

When σ_{ij}^{Q} varies throughout the body, and gradients are, therefore, present, the inclusion will experience a force that, as demonstrated below, can be obtained to first order by employing either Eq. (5.57) in the form

$$F_{l}^{INC/Q} = F_{l}^{INC^{\infty}/Q} = \oiint_{S}\left(\frac{\partial\sigma_{ij}^{Q}}{\partial x_{l}}u_{i}^{INC^{\infty},M} - \sigma_{ij}^{INC^{\infty},M}\frac{\partial u_{i}^{Q}}{\partial x_{l}}\right)\hat{n}_{j}dS, \qquad (7.3)$$

where $u_{i}^{INC^{\infty},M}$ is the displacement in the matrix produced by the inclusion in an infinite body, or, alternatively by Eq. (5.40) as expressed by

$$F_{l}^{INC/Q} = -\frac{\partial E_{int}^{INC/Q}}{\partial\xi_{l}} = -\lim_{\delta\xi_{l}\to 0}\frac{1}{\delta\xi_{l}}\left(E_{int}^{\prime INC/Q} - E_{int}^{INC/Q}\right). \qquad (7.4)$$

Consider first the approach based on Eq. (7.3) which requires the integration of the Q field and INC$^{\infty}$ field properties over the surface $\mathcal{S}$. To illustrate its use, we assume, for simplicity, that the inclusion is spherical with $\varepsilon_{ij}^{T} = \varepsilon^{T}\delta_{ij}$. The inclusion is centered at the origin, and $\mathcal{S}$ is shrunk down so that it is a sphere of radius R in the matrix infinitesimally outside the inclusion/matrix interface. For the integration over $\mathcal{S}$ in Eq. (7.3), the inclusion is placed at the origin, and the derivatives $\partial u_{i}^{Q}/\partial x_{l}$ and $\partial\sigma_{ij}^{Q}/\partial x_{l}$

in Eq. (7.3) are expanded around the origin to first order, i.e.,

$$\begin{aligned}\frac{\partial u_i^Q}{\partial x_l} &= \left(\frac{\partial u_i^Q}{\partial x_l}\right)_{0,0,0} + \left(\frac{\partial^2 u_i^Q}{\partial x_m \partial x_l}\right)_{0,0,0} x_m,\\ \frac{\partial \sigma_{ij}^Q}{\partial x_l} &= \left(\frac{\partial \sigma_{ij}^Q}{\partial x_l}\right)_{0,0,0} + \left(\frac{\partial^2 \sigma_{ij}^Q}{\partial x_m \partial x_l}\right)_{0,0,0} x_m.\end{aligned} \tag{7.5}$$

Then, substituting Eq. (7.5) into Eq. (7.3),

$$\begin{aligned}F_l^{INC/Q} = &\left(\frac{\partial \sigma_{ij}^Q}{\partial x_l}\right)_{0,0,0} \oiint_{\mathcal{S}} u_i^{INC^\infty,M} \hat{n}_j dS\\ &+ \left(\frac{\partial^2 \sigma_{ij}^Q}{\partial x_m \partial x_l}\right)_{0,0,0} \oiint_{\mathcal{S}} x_m u_i^{INC^\infty,M} \hat{n}_j dS\\ &- \left(\frac{\partial u_i^Q}{\partial x_l}\right)_{0,0,0} \oiint_{\mathcal{S}} \sigma_{ij}^{INC^\infty,M} \hat{n}_j dS\\ &- \left(\frac{\partial^2 u_i^Q}{\partial x_m \partial x_l}\right)_{0,0,0} \oiint_{\mathcal{S}} x_m \sigma_{ij}^{INC^\infty,M} \hat{n}_j dS.\end{aligned} \tag{7.6}$$

The third integral vanishes since the net traction on $\mathcal{S}$ must vanish. The remaining three integrals can be evaluated by switching to spherical coordinates (Fig. A.1b) where $x_1 = R\cos\theta\sin\phi$, $x_2 = R\sin\theta\sin\phi$, $x_3 = R\cos\phi$, $\hat{n}_j = x_j/R$ and $dS = R^2 \sin\phi d\phi d\theta$. The displacement field of the inclusion in an infinite matrix is given by Eq. (6.127), and, therefore,

$$u_i^{INC^\infty,M}(\mathbf{x}) = c\frac{x_i}{x^3}, \tag{7.7}$$

where $c = 3K^M V^{INC} e^T/[4\pi(3K^M + 4\mu^M)]$. The stress $\sigma_{ij}^{INC^\infty,M}$ required in Eq. (7.6) can then be obtained by use of Eqs. (7.7), (2.5) and (2.122). Then, performing the integrations,

$$\begin{aligned}&\oiint_{\mathcal{S}} u_i^{INC^\infty,M} \hat{n}_j dS = \frac{4\pi c}{3}\delta_{ij}, \quad \oiint_{\mathcal{S}} x_m u_i^{INC^\infty,M} \hat{n}_j dS = 0,\\ &\oiint_{\mathcal{S}} x_m \sigma_{ij}^{INC^\infty,M} \hat{n}_j dS = -\frac{16\pi\mu^M c}{3}\delta_{im},\end{aligned} \tag{7.8}$$

and substituting these results into Eq. (7.6),

$$F_l^{INC/Q} = \frac{4\pi c}{3}\left[\left(\frac{\partial\sigma_{ii}^Q}{\partial x_l}\right)_{0,0,0} + 4\mu^M\left(\frac{\partial^2 u_i^Q}{\partial x_l \partial x_i}\right)_{0,0,0}\right]. \qquad (7.9)$$

To evaluate the second term in Eq. (7.9) we use Eq. (2.5) and write

$$\left(\frac{\partial^2 u_i^Q}{\partial x_l \partial x_i}\right)_{0,0,0} = \left[\frac{\partial}{\partial x_l}\left(\frac{\partial u_i^Q}{\partial x_i}\right)\right]_{0,0,0} = \left(\frac{\partial e^Q}{\partial x_l}\right)_{0,0,0}. \qquad (7.10)$$

Then, since $e^Q = \sigma_{mm}^Q/(3K^M)$ according to Eq. (2.122),

$$\left(\frac{\partial^2 u_i^Q}{\partial x_l \partial x_i}\right)_{0,0,0} = \left(\frac{\partial e^Q}{\partial x_l}\right)_{0,0,0} = \frac{1}{3K^M}\left(\frac{\partial\sigma_{mm}^Q}{\partial x_l}\right)_{0,0,0}. \qquad (7.11)$$

Substitution of Eq. (7.11) into Eq. (7.9) then yields

$$F_l^{INC/Q} = \frac{4\pi c}{3}\left(1 + \frac{4\mu^M}{3K^M}\right)\left(\frac{\partial\sigma_{mm}^Q}{\partial x_l}\right)_{0,0,0}. \qquad (7.12)$$

and after substituting for c, the force takes the form

$$F_l^{INC/Q} = V^{INC}\varepsilon^T\left(\frac{\partial\sigma_{mm}^Q}{\partial x_l}\right)_{0,0,0} = \Delta V^{INC}\left[\frac{\partial}{\partial x_l}\left(\frac{\sigma_{mm}^Q}{3}\right)\right]_{0,0,0} \qquad (7.13)$$

which is seen to be proportional to the gradient of the hydrostatic part of the Q stress field at the inclusion. According to Eq. (7.13), if the inclusion is oversize, so that ΔV^{INC} is positive, and if σ_{mm}^Q is extensive, so that it is positive, and if $(\partial\sigma_{mm}^Q/\partial x_l)_{0,0,0}$ is also positive, then $F_l^{INC,Q}$ will be positive and the inclusion will be urged in the direction of higher σ_{mm}^Q, as would be expected.

As demonstrated in Exercise 7.7, the $F_l^{INC/Q}$ force given by Eq. (7.13) can also be obtained by starting with Eq. (7.4) rather than Eq. (7.3).

7.2.2 *Inhomogeneous ellipsoidal inclusion*

The relatively simple treatments used in the previous section for homogeneous inclusions cannot be used for an inhomogeneous inclusion because the inhomogeneity associated with the inclusion perturbs the imposed stress. To analyze this situation, we focus on ellipsoidal inhomogeneous inclusions with uniform transformation strains and compositions. The Eshelby equivalent homogeneous inclusion method (Sec. 6.3.2) is then employed to find first the elastic fields in bodies when the imposed stress is an applied stress, denoted by σ_{ij}^{A}, that would be uniform in the absence of the inclusion. Having these, the interaction energy between the imposed field and the inclusion is formulated.

The following notation is employed for this section:

ε_{ij}^{A} = uniform strain that the forces applied to the body surface would produce in the absence of the inhomogeneous inclusion.
$\varepsilon_{ij}^{INC,INC}, \varepsilon_{ij}^{INC,M}$ = strain in the inclusion and matrix, respectively, associated with an inhomogeneous inclusion in the absence of the A field.
$\varepsilon_{ij}^{A'}$ = the A strain perturbed by the inhomogeneity associated with the inhomogeneous inclusion.
$\varepsilon_{ij}^{A',INC}, \varepsilon_{ij}^{A',M}$ = A′ strain in the inclusion and matrix, respectively.
$\varepsilon_{ij}^{INC} = \varepsilon_{ij}^{(INC+A'),INC} = \varepsilon_{ij}^{INC,INC} + \varepsilon_{ij}^{A',INC}$ = strain in inhomogeneous inclusion in presence of A field.
$\varepsilon_{ij}^{M} = \varepsilon_{ij}^{(INC+A'),M} = \varepsilon_{ij}^{INC,M} + \varepsilon_{ij}^{A',M}$ = strain in matrix due to inhomogeneous inclusion

7.2.2.1 *Elastic field in body containing inclusion and imposed stress*

The system, consisting of the inclusion and imposed A field, is constructed as follows:

(1) Start with the stress-free homogeneous body and introduce the inhomogeneous inclusion.
(2) Apply forces to the body that would produce a uniform applied strain field, ε_{ij}^{A}, throughout the body in the absence of the inclusion.

The final strain field is then the sum of the field associated with the initial inhomogeneous inclusion (termed the INC field and represented by $\varepsilon_{ij}^{INC,INC}$ and $\varepsilon_{ij}^{INC,M}$) and the A′ field (represented by $\varepsilon_{ij}^{A',INC}$ and $\varepsilon_{ij}^{A',M}$). It is assumed throughout this chapter that the body is relatively large and the inclusion sufficiently small and distant from the body surface so that image strains can be neglected as discussed in Sec. 8.2.1 [see Eq. (8.3)]. The INC and A′ fields in the inclusion are therefore assumed to be the same as if the body were infinite. As determined in Ch. 6 for ellipsoidal inclusions, the

INC field in the inclusion is therefore uniform and given by Eq. (6.51), and as determined in Ch. 9, the A′ field in an ellipsoidal inclusion is uniform and given by Eq. (9.11), after setting $C^{INH}_{ijkl} = C^{INC}_{ijkl}$.[2] Therefore, we have the relation

$$[\varepsilon^{A',INC}] = \{[S^E][Y^{INC}]^{-1}([C^M] - [C^{INC}]) + [I]\}[\varepsilon^A] \tag{7.14}$$

and the strain field in the present inhomogeneous inclusion subjected to the imposed A field is

$$[\varepsilon^{(INC+A'),INC}] = [\varepsilon^{INC,INC}] + [\varepsilon^{A',INC}] \tag{7.15}$$

or, after substituting Eqs. (6.51) and (7.14),

$$\begin{aligned}[\varepsilon^{(INC+A'),INC}] &= \left([S^E][Y^{INC}]^{-1}[C^{INC}] - [I]\right)[\varepsilon^T] \\ &\quad + \left\{[S^E][Y^{INC}]^{-1}\left([C^M] - [C^{INC}]\right) + [I]\right\}[\varepsilon^A]. \end{aligned} \tag{7.16}$$

In Exercise 7.4 it is demonstrated that, instead of using the results in Eqs. (6.51) and (9.11) to obtain Eq. (7.16), it can be obtained directly by employing the equivalent homogeneous inclusion method under conditions where the bodies containing the inhomogeneous inclusion and the equivalent homogeneous inclusion are each subjected to the applied forces which produce the A field.

7.2.2.2 *Interaction energy between inclusion and imposed stress*

Since the total elastic field in the body containing the inhomogeneous inclusion and imposed A field consists of the sum of the INC and A′ fields, the interaction energy between the inclusion and the A field is then the difference between the total energy when both fields are present, i.e., $E^{INC+A'}$, and the sum of the energies of the individual INC and A fields, i.e., $E^{INC} + E^A$. Therefore,

$$E^{INC/A}_{int} = E^{INC+A'} - E^{INC} - E^A. \tag{7.17}$$

Now, according to Eq. (5.80), the interaction energy between the inhomogeneity associated with the inhomogeneous inclusion and the A field,

---

[2]These strain fields for the special case of a spherical inclusion with $\varepsilon^T_{ij} = \varepsilon^T \delta_{ij}$ and $\varepsilon^A_{ij} = \varepsilon^A \delta_{ij}$ in an isotropic system can be obtained from Eq. (6.129) or Eq. (8.41), for $\varepsilon^{INC,INC}_{ij}$, and Eq. (9.59) for $\varepsilon^{A',INC}_{ij}$.

$E_{int}^{INH/A}$, is expressible in the form

$$E_{int}^{INH/A} = E^{A'} - E^{A} \tag{7.18}$$

and is formulated in detail in Ch. 9 [see Eq. (9.26)]. Therefore, by introducing this quantity into Eq. (7.17),

$$E_{int}^{INC/A} = E^{INC+A'} - E^{INC} + E_{int}^{INH/A} - E^{A'}. \tag{7.19}$$

Since the interaction strain energy between the internal INC stress field and the applied A′ field vanishes according to Eq. (5.25),

$$\begin{aligned} E^{INC+A'} &= W^{INC} + W^{A'} + \Phi^{INC+A'} \\ E^{INC} &= W^{INC} \\ E^{A'} &= W^{A'} + \Phi^{A'} \end{aligned} \tag{7.20}$$

and upon substituting these expressions into Eq. (7.19), the interaction energy is given by

$$\begin{aligned} E_{int}^{INC/A} &= E_{int}^{INH/A} + \Phi^{INC+A'} - \Phi^{A'} = E_{int}^{INH/A} \\ &\quad - \oiint_{\mathcal{S}^\circ} \sigma_{ij}^{A'}(u_i^{INC,tot+A'} - u_i^{A'})\hat{n}_j dS \\ &= E_{int}^{INH/A} - \oiint_{\mathcal{S}^\circ} \sigma_{ij}^{A'} u_i^{INC,tot} \hat{n}_j dS. \end{aligned} \tag{7.21}$$

A further expression for the interaction energy, $E_{int}^{INC/A}$, can be obtained by converting the surface integral in Eq. (7.21) to a volume integral by the following procedure which employs the transformation strain formalism of Sec. 3.6, the divergence theorem, and Eqs. (2.5), (2.65) and (2.75):

$$\begin{aligned} \oiint_{\mathcal{S}^\circ} \sigma_{ij}^{A'} u_i^{INC,tot} \hat{n}_j dS &= \oiiint_{\mathcal{V}^\circ} \frac{\partial}{\partial x_j}(\sigma_{ij}^{A'} u_i^{INC,tot}) dV \\ &= \oiiint_{\mathcal{V}^\circ} \sigma_{ij}^{A'} \frac{\partial u_i^{INC,tot}}{\partial x_j} dV \\ &= \oiiint_{\mathcal{V}^{INC}} \sigma_{ij}^{A'} \frac{\partial u_i^{INC,tot}}{\partial x_j} dV \\ &\quad + \oiiint_{\mathcal{V}^\circ - \mathcal{V}^{INC}} \sigma_{ij}^{A'} \frac{\partial u_i^{INC,tot}}{\partial x_j} dV. \end{aligned} \tag{7.22}$$

The penultimate integral in Eq. (7.22) then takes the form

$$
\begin{aligned}
\oiiint_{\mathcal{V}^{\rm INC}} \sigma_{ij}^{A'} \frac{\partial u_i^{\rm INC,tot}}{\partial x_j} dV &= \oiiint_{\mathcal{V}^{\rm INC}} C_{ijkl}^{\rm INC} \frac{\partial u_k^{A'}}{\partial x_l} \frac{\partial u_i^{\rm INC,tot}}{\partial x_j} dV \\
&= \oiiint_{\mathcal{V}^{\rm INC}} C_{klij}^{\rm INC} \frac{\partial u_i^{\rm INC,tot}}{\partial x_j} \frac{\partial u_k^{A'}}{\partial x_l} dV \\
&= \oiiint_{\mathcal{V}^{\rm INC}} C_{klij}^{\rm INC} (\varepsilon_{ij}^{\rm INC} + \varepsilon_{ij}^{T}) \frac{\partial u_k^{A'}}{\partial x_l} dV \\
&= \oiiint_{\mathcal{V}^{\rm INC}} \sigma_{kl}^{\rm INC} \frac{\partial u_k^{A'}}{\partial x_l} dV \\
&\quad + \oiiint_{\mathcal{V}^{\rm INC}} C_{klij}^{\rm INC} \varepsilon_{ij}^{T} \frac{\partial u_k^{A'}}{\partial x_l} dV \\
&= \oiiint_{\mathcal{V}^{\rm INC}} \frac{\partial}{\partial x_j} (\sigma_{ij}^{\rm INC} u_i^{A'}) dV \\
&\quad + \oiiint_{\mathcal{V}^{\rm INC}} C_{ijkl}^{\rm INC} \frac{\partial u_k^{A'}}{\partial x_l} \varepsilon_{ij}^{T} dV \\
&= \oiint_{\mathcal{S}^{\rm INC}} \sigma_{ij}^{\rm INC} u_i^{A'} \hat{n}_j dS + \oiiint_{\mathcal{V}^{\rm INC}} \sigma_{ij}^{A'} \varepsilon_{ij}^{T} dV
\end{aligned}
\tag{7.23}
$$

while the last integral in Eq. (7.22) becomes

$$
\begin{aligned}
\oiiint_{\mathcal{V}^\circ - \mathcal{V}^{\rm INC}} \sigma_{ij}^{A'} \frac{\partial u_i^{\rm INC,tot}}{\partial x_j} dV &= \oiiint_{\mathcal{V}^\circ - \mathcal{V}^{\rm INC}} C_{ijkl}^{M} \frac{\partial u_k^{A'}}{\partial x_l} \frac{\partial u_i^{\rm INC}}{\partial x_j} dV \\
&= \oiiint_{\mathcal{V}^\circ - \mathcal{V}^{\rm INC}} C_{klij}^{M} \frac{\partial u_i^{\rm INC}}{\partial x_j} \frac{\partial u_k^{A'}}{\partial x_l} dV \\
&= \oiiint_{\mathcal{V}^\circ - \mathcal{V}^{\rm INC}} \sigma_{kl}^{\rm INC} \frac{\partial u_k^{A'}}{\partial x_l} dV \\
&= \oiint_{\mathcal{S}^\circ} \sigma_{kl}^{\rm INC} u_k^{A'} \hat{n}_l dS - \oiint_{\mathcal{S}^{\rm INC}} \sigma_{kl}^{\rm INC} u_k^{A'} \hat{n}_l dS.
\end{aligned}
\tag{7.24}
$$

Then, substituting Eqs. (7.23) and (7.24) into Eq. (7.22),

$$\oiint_{\mathcal{S}^\circ} \sigma_{ij}^{A'} u_i^{INC,tot} \hat{n}_j dS = \oiiint_{\mathcal{V}^{INC}} \sigma_{ij}^{A'} \varepsilon_{ij}^{T} dV + \oiint_{\mathcal{S}^\circ} \sigma_{kl}^{INC} u_k^{A'} \hat{n}_l dS$$
$$= \oiiint_{\mathcal{V}^{INC}} \sigma_{ij}^{A'} \varepsilon_{ij}^{T} dV \tag{7.25}$$

since $\sigma_{kl}^{INC} \hat{n}_l = 0$ on $\mathcal{S}^\circ$. Finally, substituting (7.25) into Eq. (7.21),

$$E_{int}^{INC/A} = E_{int}^{INH/A} - \oiiint_{\mathcal{V}^{INC}} \sigma_{ij}^{A'} \varepsilon_{ij}^{T} dV. \tag{7.26}$$

The interaction energy in this form is therefore the sum of two identifiable terms, i.e., the interaction energy between the inhomogeneity associated with the inhomogeneous inclusion and the imposed A field as given by Eq. (9.26), or, alternatively, Eq. (9.22) and the interaction energy between the transformation strain of the inclusion with the perturbed A field which is of the same form as the interaction energy between the transformation strain of a homogeneous inclusion with an imposed stress as given by Eq. (7.1). Also, see Exercise 7.6 for further discussion.

In Exercise 7.3, it is shown by a detailed calculation that Eqs. (7.21) and (7.26) yield identical results for the spherical inclusion treated in the following section.

7.2.2.3 *Some results for spherical inclusion in isotropic system*

Results for the case of a spherical inhomogeneous inclusion with transformation strain $\varepsilon_{ij}^{T} = \varepsilon^{T} \delta_{ij}$ interacting with the imposed A field, $\varepsilon_{ij}^{A} = \varepsilon^{A} \delta_{ij}$, in an isotropic system, can be readily determined, since all the necessary ingredients, obtained by means of the equivalent homogeneous inclusion method, are available in Chs. 6 and 9.

The transformation strain for the equivalent homogeneous inclusion corresponding to the inhomogeneous inclusion, $[\varepsilon^{T^*}]^{INC}$, is given by Eq. (6.122), and the INC field in the inclusion, $\varepsilon_{\alpha\alpha}^{INC,INC}$, needed for Eq. (7.15), is given by Eq. (6.123). The transformation strain for the equivalent homogeneous inclusion corresponding to the inhomogeneity associated with the inhomogeneous inclusion, $[\varepsilon^{T^*}]^{A'}$, is given by Eq. (9.42) with INH → INC, and the A′ field in the inclusion, $\varepsilon_{\alpha\alpha}^{A',INC}$, is given by Eq. (9.59) with INH → INC.

Interaction energy

The interaction energy, $E_{int}^{INC/A}$, between the above inclusion and the A field, given by Eq. (7.26), is determined by using Eqs. (9.43) and (7.44) to represent the first and second terms, respectively, and, therefore,

$$E_{int}^{INC/A} = -\frac{9(3K^{M}+4\mu^{M})\mathcal{V}^{INC}}{(3K^{INC}+4\mu^{M})}\left\{\frac{(K^{M}-K^{INC})}{2}(\varepsilon^{A})^{2}+K^{INC}\varepsilon^{T}\varepsilon^{A}\right\}. \tag{7.27}$$

Interaction force

If there is a gradient in the A field, the resulting force, obtained by employing Eq. (5.40), is then given, to first order, by

$$\begin{aligned} F_{l}^{INC/A} &= -\frac{\partial E_{int}^{INC/A}}{\partial \xi_{l}} \\ &= \frac{9(3K^{M}+4\mu^{M})\mathcal{V}^{INC}}{(3K^{INC}+4\mu^{M})}\frac{\partial}{\partial \xi_{l}}\left\{\frac{(K^{M}-K^{INC})}{2}(\varepsilon^{A})^{2}+K^{INC}\varepsilon^{T}\varepsilon^{A}\right\} \\ &= \frac{9(3K^{M}+4\mu^{M})\mathcal{V}^{INC}}{(3K^{INC}+4\mu^{M})}\left[(K^{M}-K^{INC})\varepsilon^{A}\frac{\partial \varepsilon^{A}}{\partial x_{l}}+K^{INC}\varepsilon^{T}\frac{\partial \varepsilon^{A}}{\partial x_{l}}\right] \end{aligned} \tag{7.28}$$

after using $\partial/\partial\xi_l = \partial/\partial x_l$. The direction of this force depends in a complex manner on the signs of the various quantities. For example, when $(K^M - K^{INC})$ is positive, and the inclusion acts as a "soft" region, the first term will urge the inclusion in the direction of increasing ε^A strain, regardless of whether it is extensive or compressive. On the other hand, if ε^T is also positive, the second term will urge it in the direction of higher extensive strain (or, equivalently, of lower compressive strain). Note that when $K^M = K^{INC} = K$, and the inclusion is homogeneous, Eq. (7.28) reduces to

$$F_{l}^{INC/A} = 9V^{INC}K\varepsilon^{T}\frac{\partial \varepsilon^{A}}{\partial x_{l}} = e^{T}V^{INC}K\frac{\partial e^{A}}{\partial x_{l}} = \Delta V^{INC}\frac{\partial}{\partial x_{l}}\left(\frac{\sigma_{mm}^{A}}{3}\right) \tag{7.29}$$

in agreement with Eq. (7.13).

Exercises

7.1. Derive Eq. (7.13) by an alternative approach that again starts with Eq. (7.3), but with $\mathcal{S}$ taken infinitesimally inside the inclusion rather than outside as in the text.

Solution. By shrinking the surface, $\mathcal{S}$, so that it is infinitesimally inside the inclusion as in the procedure used in developing Eq. (7.13), Eq. (7.3), takes the form

$$\begin{aligned}
F_l^{INC/Q} &= \oiint_{\mathcal{S}^{INC}} \left[\frac{\partial \sigma_{ij}^{Q}}{\partial x_l} (u_i^{INC^\infty,INC} + u_i^T) - \sigma_{ij}^{INC^\infty,INC} \frac{\partial u_i^Q}{\partial x_l} \right] \hat{n}_j dS \\
&= \oiint_{\mathcal{S}^{INC}} \frac{\partial \sigma_{ij}^{Q}}{\partial x_l} u_i^T \hat{n}_j dS \\
&\quad + \oiint_{\mathcal{S}^{INC}} \left[\frac{\partial \sigma_{ij}^{Q}}{\partial x_l} u_i^{INC^\infty,INC} - \sigma_{ij}^{INC^\infty,INC} \frac{\partial u_i^Q}{\partial x_l} \right] \hat{n}_j d\mathbf{S}.
\end{aligned} \tag{7.30}$$

However, the last integral in Eq. (7.30) vanishes by virtue of Eq. (2.112), and, therefore, by converting the remainder to a volume integral,

$$F_l^{INC/Q} = \oiint_{\mathcal{S}^{INC}} \frac{\partial \sigma_{ij}^{Q}}{\partial x_l} u_i^T \hat{n}_j d\mathbf{S} = \oiiint_{\mathcal{V}^{INC}} \frac{\partial}{\partial x_j} \left(\frac{\partial \sigma_{ij}^{Q}}{\partial x_l} u_i^T \right) dV. \tag{7.31}$$

Then, applying Eqs. (2.65) and (2.5),

$$\begin{aligned}
F_l^{INC/Q} &= \oiiint_{\mathcal{V}^{INC}} \frac{\partial \sigma_{ij}^{Q}}{\partial x_l} \frac{\partial u_i^T}{\partial x_j} dV \\
&= \oiiint_{\mathcal{V}^{INC}} \frac{\partial \sigma_{ij}^{Q}}{\partial x_l} \varepsilon_{ij}^T dV
\end{aligned} \tag{7.32}$$

which is identical to Eq. (7.63) in Exercise 7.7. Therefore, following the procedure following Eq. (7.63), we have, to first order

$$F_l^{INC/Q} = V^{INC} \frac{\partial}{\partial x_l} \left(\frac{\sigma_{mm}^Q}{3} \right)_{0,0,0} \tag{7.33}$$

which agrees with Eq. (7.13).

7.2. Consider a spherical homogeneous inclusion with the transformation strain, $\varepsilon_{ij}^T = \varepsilon^T \delta_{ij}$, in an infinite isotropic body and its interaction energy with a general uniform stress field, Q. Instead of mimicking the inclusion by its transformation strain and using Eq. (7.1) to obtain the interaction energy, as in Sec. 7.2.1, mimic the inclusion by a distribution of body forces and employ Eq. (5.28). Using the known solution for the elastic field of the inclusion (Sec. 6.4.3.1), find the necessary expression for the force density distribution required for use in the latter approach, and then go on to determine the interaction energy.

Solution. The displacement in the matrix, $\mathbf{u}^{INC^\infty,M}$, due to the inclusion must, under all circumstances, satisfy the Navier equation, Eq. (3.3), i.e.,

$$(\lambda^M + 2\mu^M)\nabla[\nabla \cdot \mathbf{u}^{INC^\infty,M}(\mathbf{x})] = -\mathbf{f}(\mathbf{x}). \tag{7.34}$$

The displacement appearing in Eq. (7.34) when the inclusion is mimicked by a transformation strain is given by Eq. (6.127) with $K^{INC} = K^M$, i.e.,

$$\mathbf{u}^{INC^\infty,M}(\mathbf{x}) = c\frac{\mathbf{x}}{x^3} = \frac{9K^M V^{INC} \varepsilon^T}{4\pi(3K^M + 4\mu^M)} \frac{\mathbf{x}}{x^3}. \tag{7.35}$$

Substitution of Eq. (7.35) into Eq. (7.34) then yields the required force density distribution

$$\mathbf{f}(\mathbf{x}) = -(\lambda^M + 2\mu^M)\nabla(\nabla \cdot \mathbf{u}^{INC^\infty,M}) = -(\lambda^M + 2\mu^M)4\pi c\nabla\delta(\mathbf{x}). \tag{7.36}$$

The interaction energy is then obtained by substituting Eq. (7.36) into Eq. (5.28) and using Eq. (D.4) to evaluate the resulting integral involving the derivative of the delta function, i.e.,

$$E_{int}^{INC/Q} = -\oiiint_{\mathcal{V}} u_i^Q f_i dV = (\lambda^M + 2\mu^M)4\pi c \oiiint_{\mathcal{V}} u_i^Q \frac{\partial \delta(\mathbf{x})}{\partial x_i} dV$$

$$= -(\lambda^M + 2\mu^M)4\pi c \frac{\partial u_i^Q}{\partial \mathbf{x}_i} = -(\lambda^M + 2\mu^M)4\pi c e^Q. \quad (7.37)$$

Finally, substituting the expression for c obtained from Eq. (7.35), and relationships for the elastic constants in Sec. 2.4.4, into Eq. (7.37),

$$E_{int}^{INC/Q} = -\frac{2\mu^M(1+\nu^M)V^{INC}}{1-2\nu^M}\varepsilon^T e^Q = -K^M V^{INC} e^T e^Q$$

$$= -K^M e^Q \Delta V^{INC} = -\frac{1}{3}\sigma_{mm}^Q \Delta V^{INC} \quad (7.38)$$

in agreement with Eq. (7.2).

7.3. Verify that Eqs. (7.21) and (7.26) yield identical expressions for the interaction energy, $E_{int}^{INC/A}$, between the inhomogeneous inclusion and imposed A field treated in Sec. 7.2.2.3.

Solution. The quantity $E_{int}^{INH/A}$, which appears in both equations, is given by Eq. (9.43), and substituting this into Eqs. (7.21) and (7.26), the two expressions for $E_{int}^{INC/A}$ are, respectively,

$$E_{int}^{INC/A} = \frac{9}{2}\frac{(K^{INH} - K^M)(3K^M + 4\mu^M)V^{INH}}{(3K^{INH} + 4\mu^M)}(\varepsilon^A)^2$$

$$- \oiint_{\mathcal{S}^\circ} \sigma_{ij}^{A'} u_i^{INC,tot} \hat{n}_j dS \quad (7.39)$$

and

$$E_{int}^{INC/A} = \frac{9}{2}\frac{(K^{INH} - K^M)(3K^M + 4\mu^M)V^{INH}}{(3K^{INH} + 4\mu^M)}(\varepsilon^A)^2$$

$$- \oiiint_{\mathcal{V}^{INC}} \sigma_{ij}^{A'} \varepsilon_{ij}^T dV. \quad (7.40)$$

It therefore remains to show that the surface integral in Eq. (7.39) is equal to the volume integral in Eq. (7.40).

Evaluating the surface integral by switching to spherical coordinates,

$$\oiint_{\mathcal{S}^\circ} \sigma_{ij}^{A'} u_i^{INC,tot} \hat{n}_j dS = 4\pi a^2 \sigma_{rr}^A(a) u_r^{INC}(a) \quad (7.41)$$

after using the surface condition $\sigma_{ij}^{A'}\hat{n}_j = \sigma_{ij}^{A}\hat{n}_j$ on $\mathcal{S}^{\circ}$. Then, using Eqs. (G.10) and (G.11), $\sigma_{rr}^{A}(a) = 3K^{M}\varepsilon^{A}$, and substituting this result and the expression for $u_r^{INC}(a)$ given by Eq. (8.41) into Eq. (7.41),

$$\oiint_{\mathcal{S}^{\circ}} \sigma_{ij}^{A'} u_i^{INC,tot}\hat{n}_j dS = \frac{9K^{INC}(3K^{M}+4\mu^{M})V^{INC}}{(3K^{INC}+4\mu^{M})}\varepsilon^{A}\varepsilon^{T}. \quad (7.42)$$

Next, the volume integral takes the form

$$\oiiint_{\mathcal{V}^{INC}} \sigma_{ij}^{A'}\varepsilon_{ij}^{T} dV = \sigma_{ij}^{A',INC}\varepsilon_{ij}^{T}V^{INC} = 3K^{INC}e^{A',INC}\varepsilon^{T}V^{INC}. \quad (7.43)$$

Again switching to spherical coordinates, $e^{A',INC} = \varepsilon_{rr}^{A',INC} + \varepsilon_{\theta\theta}^{A',INC} + \varepsilon_{\phi\phi}^{A',INC} = 3\varepsilon_{rr}^{A',INC}$, and using this result and Eq. (9.59) to evaluate Eq. (7.43),

$$\oiiint_{\mathcal{V}^{INC}} \sigma_{ij}^{A'}\varepsilon_{ij}^{T} dV = \frac{9K^{INC}(3K^{M}+4\mu^{M})V^{INC}}{(3K^{INC}+4\mu^{M})}\varepsilon^{A}\varepsilon^{T} \quad (7.44)$$

in agreement with Eq. (7.42).

7.4. Construct the equivalent homogeneous inclusion corresponding to the inhomogeneous inclusion of Sec. 7.2.2 while the body containing the inclusion is being subjected to applied forces that would produce a uniform A field in the absence of the inclusion. Then use this result to determine directly the strain $\varepsilon_{ij}^{(INC+A'),INC}$ given by Eq. (7.16).

Solution. The equivalent homogeneous inclusion is constructed by employing the same procedure that produced the equivalent homogeneous inclusion corresponding to the inhomogeneous inclusion in the absence of the A field described in Sec. 6.3.2.1 except that the applied forces must be applied in a sixth and final step. The applied forces must also be added to the procedure used to produce the inhomogeneous inclusion in the absence of these forces, as given in Sec. 6.3.2.1. The equal stress condition, corresponding to Eq. (6.45), then takes the form

$$\begin{gathered}\sigma_{ij}^{C,INC^*} - \sigma_{ij}^{T^*} + \sigma_{ij}^{A} = \sigma_{ij}^{C,INC} - \sigma_{ij}^{T} + \sigma_{ij}^{A',INC} \\ C_{ijkl}^{M}\left(\varepsilon_{kl}^{C,INC^*} - \varepsilon_{kl}^{T^*} + \varepsilon_{kl}^{A}\right) = C_{ijkl}^{INC}\left(\varepsilon_{kl}^{C,INC} - \varepsilon_{kl}^{T} + \varepsilon_{kl}^{A',INC}\right)\end{gathered} \quad (7.45)$$

and the equal size and shape condition, corresponding Eq. (6.44), takes the form

$$\varepsilon_{\rm ij}^{\rm C,INC^*} + \varepsilon_{\rm ij}^{\rm A} = \sigma_{\rm ij}^{\rm C,INC} + \varepsilon_{\rm ij}^{\rm A',INC}. \tag{7.46}$$

By substituting this latter equality into Eq. (7.45) and using Eqs. (6.29) and (6.48),

$$\begin{aligned}
{\rm C}_{{\rm ijk}l}^{\rm M}\left(\varepsilon_{{\rm k}l}^{\rm C,INC^*} - \varepsilon_{{\rm k}l}^{\rm T^*} + \varepsilon_{{\rm k}l}^{\rm A}\right) &= {\rm C}_{{\rm ijk}l}^{\rm INC}\left(\varepsilon_{{\rm k}l}^{\rm C,INC^*} - \varepsilon_{{\rm k}l}^{\rm T} + \varepsilon_{{\rm k}l}^{\rm A}\right),\\
{\rm C}_{{\rm ijk}l}^{\rm M}\left({\rm S}_{{\rm k}l{\rm mn}}^{\rm E}\varepsilon_{\rm mn}^{\rm T^*} - \varepsilon_{{\rm k}l}^{\rm T^*} + \varepsilon_{{\rm k}l}^{\rm A}\right) &= {\rm C}_{{\rm ijk}l}^{\rm INC}\left({\rm S}_{{\rm k}l{\rm mn}}^{\rm E}\varepsilon_{\rm mn}^{\rm T^*} - \varepsilon_{{\rm k}l}^{\rm T} + \varepsilon_{{\rm k}l}^{\rm A}\right),\\
{\rm Y}_{\rm ijmn}^{\rm INC}\varepsilon_{\rm mn}^{\rm T^*} &= {\rm C}_{\rm ijmn}^{\rm INC}\varepsilon_{\rm mn}^{\rm T} + \left({\rm C}_{\rm ijmn}^{\rm M} - {\rm C}_{\rm ijmn}^{\rm INC}\right)\varepsilon_{\rm mn}^{\rm A}.
\end{aligned} \tag{7.47}$$

Then, solving the above equation for $\varepsilon_{\rm mn}^{\rm T^*}$,

$$[\varepsilon^{\rm T^*}] = [{\rm Y}^{\rm INC}]^{-1}[{\rm C}^{\rm INC}][\varepsilon^{\rm T}] + [{\rm Y}^{\rm INC}]^{-1}([{\rm C}^{\rm M}] - [{\rm C}^{\rm INC}])[\varepsilon^{\rm A}]. \tag{7.48}$$

Equation (7.48) shows that the transformation strain of the equivalent homogeneous inclusion corresponding to the inhomogeneous inclusion subjected to the imposed stress field A, $\varepsilon_{\rm mn}^{\rm T^*}$, is just the sum of the transformation strain of the equivalent inclusion corresponding to the inhomogeneous inclusion in the absence of the A field, given by Eq. (6.49), and the transformation strain of the equivalent inclusion corresponding to the inhomogeneity associated with the inhomogeneous inclusion subjected to the A field, given by Eq. (9.10).

Now, the elastic strain in the inhomogeneous inclusion subjected to the A field is given by

$$\begin{aligned}
\varepsilon_{\rm ij}^{\rm INC} &= \varepsilon_{\rm ij}^{\rm C,INC} - \varepsilon_{\rm ij}^{\rm T} + \varepsilon_{\rm ij}^{\rm A',INC} = \varepsilon_{\rm ij}^{\rm C,INC^*} - \varepsilon_{\rm ij}^{\rm T} + \varepsilon_{\rm ij}^{\rm A}\\
&= {\rm S}_{\rm ijmn}^{\rm E}\varepsilon_{\rm mn}^{\rm T^*} - \varepsilon_{\rm ij}^{\rm T} + \varepsilon_{\rm ij}^{\rm A}
\end{aligned} \tag{7.49}$$

after employing Eqs. (7.46) and (6.29). Therefore, by substituting Eq. (7.48) into Eq. (7.49),

$$\begin{aligned}
[\varepsilon^{\rm INC}] = {}& [{\rm S}^{\rm E}][{\rm Y}^{\rm INC}]^{-1}[{\rm C}^{\rm INC}][\varepsilon^{\rm T}] + [{\rm S}^{\rm E}][{\rm Y}^{\rm INC}]^{-1}([{\rm C}^{\rm M}]\\
&-[{\rm C}^{\rm INC}])[\varepsilon^{\rm A}] - [\varepsilon^{\rm T}] + [\varepsilon^{\rm A}]
\end{aligned} \tag{7.50}$$

in agreement with Eq. (7.16) as might have been anticipated immediately from the form of Eq. (7.48).

7.5. As demonstrated in Exercise 5.6, the interaction energy between a defect, D, and a general stress field, Q, can be determined by finding the difference between the energy change when D is added to the body in the presence of Q, i.e., $E^{D/Q}$, and the energy change when it is added by itself, i.e., E^{Q}, so that

$$E_{int}^{D/Q} = E^{D/Q} - E^{D}. \tag{7.51}$$

Derive Eq. (5.13), in Sec. 5.2.1.2, for the interaction energy between a homogeneous inclusion and an internal stress, I, by starting with Eq. (7.51) instead of Eq. (5.10) as in Sec. 5.2.1.2.

Solution For this case, Eq. (7.51) takes the form

$$E_{int}^{INC/I} = E^{INC/I} - E^{INC}, \tag{7.52}$$

where E^{INC} has already been determined in the form of Eq. (6.57). The quantity $E^{INC/I}$ can be obtained by imagining that the inclusion is introduced by the process described in Sec. 6.2 except that I is present in the body from the start. At the end of step 4 the total energy in the body is given by

$$E(4) = \frac{1}{2} \iiint_{\mathcal{V}^{INC}} (\sigma_{ij}^{I,INC} - \sigma_{ij}^{T})(\varepsilon_{ij}^{I,INC} - \varepsilon_{ij}^{T}) dV + W^{I,M}, \tag{7.53}$$

where the first term is the strain energy in the inclusion and the second is the strain energy of the I field in the matrix. Then, using the same reasoning that led to Eq. (6.50) in Sec. 6.3.3, the strain energy released in the canceling operation carried out in step 5 is

$$\begin{aligned} \Delta W(4 \to 5) &= \frac{1}{2} \iint_{\mathcal{S}^{INC}} (-\sigma_{ij}^{T})(-u_i^{C,INC}) \hat{n}_j dS \\ &= \frac{1}{2} \iiint_{\mathcal{V}^{INC}} \sigma_{ij}^{T} \varepsilon_{ij}^{C,INC} dV \end{aligned} \tag{7.54}$$

and the final total energy in the body (at the end of step 5) is

$$\begin{aligned} E(5) = &\frac{1}{2} \iiint_{\mathcal{V}^{INC}} (\sigma_{ij}^{I,INC} - \sigma_{ij}^{T})(\varepsilon_{ij}^{I,INC} - \varepsilon_{ij}^{T}) dV \\ &+ W^{I,M} - \frac{1}{2} \iiint_{\mathcal{V}^{INC}} \sigma_{ij}^{T} \varepsilon_{ij}^{C,INC} dV. \end{aligned} \tag{7.55}$$

The energy of the I field is

$$\mathrm{E}^{\mathrm{I}} = \frac{1}{2}\oiiint_{\mathcal{V}^{\mathrm{INC}}} \sigma_{\mathrm{ij}}^{\mathrm{I,INC}} \varepsilon_{\mathrm{ij}}^{\mathrm{I,INC}} \mathrm{dV} + \mathrm{W}^{\mathrm{I,M}} \tag{7.56}$$

and, therefore,

$$\mathrm{E}^{\mathrm{INC/I}} = \mathrm{E}(5) - \mathrm{E}^{\mathrm{I}} = \frac{1}{2}\oiiint_{\mathcal{V}^{\mathrm{INC}}} \times(\sigma_{\mathrm{ij}}^{\mathrm{T}}\varepsilon_{\mathrm{ij}}^{\mathrm{T}} - \sigma_{\mathrm{ij}}^{\mathrm{I,INC}}\varepsilon_{\mathrm{ij}}^{\mathrm{T}} - \sigma_{\mathrm{ij}}^{\mathrm{T}}\varepsilon_{\mathrm{ij}}^{\mathrm{I,INC}} - \sigma_{\mathrm{ij}}^{\mathrm{T}}\varepsilon_{\mathrm{ij}}^{\mathrm{C,INC}})\mathrm{dV}. \tag{7.57}$$

Next, using Eqs. (6.57) and (6.1), the energy to introduce the inclusion by itself is

$$\mathrm{E}^{\mathrm{INC}} = \frac{1}{2}\oiiint_{\mathcal{V}^{\mathrm{INC}}} (\sigma_{\mathrm{ij}}^{\mathrm{T}} - \sigma_{\mathrm{ij}}^{\mathrm{C,INC}})\varepsilon_{\mathrm{ij}}^{\mathrm{T}}\mathrm{dV} \tag{7.58}$$

and, by substituting Eqs. (7.57) and (7.58) into Eq. (7.52), and using Eq. (2.102), we obtain the desired result

$$\begin{aligned}\mathrm{E}_{\mathrm{int}}^{\mathrm{INC/I}} &= -\frac{1}{2}\oiiint_{\mathcal{V}^{\mathrm{INC}}} (\sigma_{\mathrm{ij}}^{\mathrm{I,INC}}\varepsilon_{\mathrm{ij}}^{\mathrm{T}} + \sigma_{\mathrm{ij}}^{\mathrm{T}}\varepsilon_{\mathrm{ij}}^{\mathrm{I,INC}})\mathrm{dV} \\ &= -\oiiint_{\mathcal{V}^{\mathrm{INC}}} \sigma_{\mathrm{ij}}^{\mathrm{I,INC}}\varepsilon_{\mathrm{ij}}^{\mathrm{T}}\mathrm{dV}.\end{aligned} \tag{7.59}$$

7.6. Explain Eq. (7.26) for the interaction energy between an inhomogeneous inclusion and an imposed stress field, A, by imagining that the inclusion is introduced into the body containing the A field in two steps where the inhomogeneity associated with the inclusion is inserted first followed by the insertion of its transformation strain. Use only established results in the text.

Solution. Initially, when the inhomogeneity is inserted it will interact with the existing A field thereby producing a perturbed A field, indicated by A', and an interaction energy given by Eq. (9.26). Next, when the transformation strain is imposed it will interact with the existing A' field and produce an interaction energy given by Eq. (7.1) with $Q \to A'$. The total interaction energy will therefore be the sum of these two interaction energies which corresponds to Eq. (7.26).

7.7. Derive Eq. (7.13) for the force exerted on a homogeneous inclusion by an imposed stress, Q, by starting with Eq. (7.4) rather than Eq. (7.3) as in the text.

Solution. Equation (7.4) now assumes the form

$$F_{\alpha}^{INC/Q} = -\lim_{\delta x_{\alpha}\to 0}\frac{1}{\delta x_{\alpha}}(E_{int}^{\prime INC/Q} - E_{int}^{INC/Q}). \tag{7.60}$$

The required increment of interaction energy can be obtained by using Eq. (5.13) along with the relationship

$$\sigma_{ij}^{Q'} = \sigma_{ij}^{Q} + \frac{\partial \sigma_{ij}^{Q'}}{\partial x_{\alpha}}\delta\xi_{\alpha} \tag{7.61}$$

which links, to first order, the Q stresses in the inclusion before, and after, its displacement. Then, using (5.13) and (7.61),

$$\begin{aligned} E_{int}^{\prime INC/Q} - E_{int}^{INC/Q} &= -\varepsilon_{ij}^{T}\oiiint_{\mathcal{V}^{INC}}(\sigma_{ij}^{Q'} - \sigma_{ij}^{Q})dV \\ &= -\delta\xi_{\alpha}\varepsilon_{ij}^{T}\oiiint_{\mathcal{V}^{INC}}\frac{\partial\sigma_{ij}^{Q}}{\partial x_{\alpha}}dV \end{aligned} \tag{7.62}$$

since ε_{ij}^{T} remains constant during the displacement. Substitution of Eq. (7.62) into Eq. (7.60) then yields

$$F_{\alpha}^{INC/Q} = \varepsilon_{ij}^{T}\oiiint_{\mathcal{V}^{INC}}\frac{\partial\sigma_{ij}^{Q}}{\partial x_{\alpha}}dV. \tag{7.63}$$

Equation (7.63) can be integrated by placing the inclusion at the origin and expanding $\partial\sigma_{ij}^{Q}/\partial x_{\alpha}$ around (0,0,0) to first order, as in Eq. (7.5), so that Eq. (7.63) becomes

$$\begin{aligned} F_{\alpha}^{INC/Q} &= \varepsilon_{ij}^{T}\oiiint_{\mathcal{V}^{INC}}\left[\left(\frac{\partial\sigma_{ij}^{Q}}{\partial x_{\alpha}}\right)_{0,0,0} + \left(\frac{\partial^{2}\sigma_{ij}^{Q}}{\partial x_{m}\partial x_{\alpha}}\right)_{0,0,0}x_{m}\right]dV \\ &= \varepsilon_{ij}^{T}V^{INC}\left(\frac{\partial\sigma_{ij}^{Q}}{\partial x_{\alpha}}\right)_{0,0,0} + \varepsilon_{ij}^{T}\left(\frac{\partial^{2}\sigma_{ij}^{Q}}{\partial x_{m}\partial x_{\alpha}}\right)_{0,0,0}\oiiint_{\mathcal{V}^{INC}}x_{m}dV \\ &= \varepsilon_{ij}^{T}V^{INC}\left(\frac{\partial\sigma_{ij}^{Q}}{\partial x_{\alpha}}\right)_{0,0,0} \end{aligned} \tag{7.64}$$

since the integral $\iiint_{\mathcal{V}^{\mathrm{INC}}} \mathrm{x}_{\mathrm{m}} \mathrm{dV}$ vanishes. Then, since $\varepsilon_{\mathrm{ij}}^{\mathrm{T}} = \varepsilon^{\mathrm{T}} \delta_{\mathrm{ij}}$,

$$\mathrm{F}_{\alpha}^{\mathrm{INC/Q}} = \varepsilon^{\mathrm{T}} \mathrm{V}^{\mathrm{INC}} \left(\frac{\partial \sigma_{\mathrm{ii}}^{\mathrm{Q}}}{\partial \mathrm{x}_{\alpha}} \right)_{0,0,0} = \Delta \mathrm{V}^{\mathrm{INC}} \left[\frac{\partial}{\partial \mathrm{x}_{\alpha}} \left(\frac{\sigma_{\mathrm{mm}}^{\mathrm{Q}}}{3} \right) \right]_{0,0,0} \tag{7.65}$$

in agreement with Eq. (7.13).

Chapter 8

Homogeneous Inclusions in Finite and Semi-infinite Regions: Image Effects

8.1 Introduction

Having treated inclusions in an infinite matrix in Ch. 6, we now consider inclusions in finite and semi-infinite regions with interfaces therefore present. In such cases, as described in Sec. 3.8, image stresses are generated, and the inclusion stress field can be taken as the sum of the field that it would have in an infinite region plus its image stress.

Image stresses for inclusions relatively far from interfaces in large regions are considered first and found to be generally unimportant in the vicinity of the inclusion. However, it is demonstrated that the overall volume change produced by the image stress can be significant. Then, attention is focused on various types of inclusions near interfaces where image stresses must be taken into account. Solutions for their elastic fields are found using the Green's functions for point forces in regions adjoining interfaces derived earlier in Ch. 4. However, results can generally be expressed only in the form of lengthy integrals. The tractable problem of a spherical homogeneous inclusion with a uniform transformation strain near a planar free surface in a semi-infinite isotropic system is therefore solved analytically to provide physical insight. Finally, the strain energies of inclusions in finite bodies are considered.

The following notation is employed for this chapter:

$\varepsilon_{\mathrm{ij}}^{\mathrm{INC}^{\infty}}$, $\varepsilon_{\mathrm{ij}}^{\mathrm{M}^{\infty}}$ = strain in inclusion and matrix, respectively, associated with inclusion in infinite body
$\varepsilon_{\mathrm{ij}}^{\mathrm{IM}}$ = image strain associated with inclusion in finite body
$\varepsilon_{\mathrm{ij}}^{\mathrm{INC}} = \varepsilon_{\mathrm{ij}}^{\mathrm{INC}^{\infty}} + \varepsilon_{\mathrm{ij}}^{\mathrm{IM}}$ = strain in inclusion in finite body

$\varepsilon_{ij}^{M} = \varepsilon_{ij}^{M\infty} + \varepsilon_{ij}^{IM}$ = strain in matrix around inclusion in finite body
$\varepsilon_{ij}^{C,M}$ = canceling strain in matrix

8.2 Homogeneous Inclusions Far From Interfaces in Large Finite Bodies in Isotropic Systems

8.2.1 *Image stress*

Consider the simple tractable case of a spherical homogeneous inclusion of radius R, with a uniform transformation strain $\varepsilon_{ij}^{T} = \varepsilon^{T}\delta_{ij}$ at the center of a much larger spherical body of radius, a, so that $(R/a)^3 \ll 1$. The surface of the body is traction-free and, following the method for formulating the image stress in Sec. 3.8, we first write the stress in the matrix for the case where the same inclusion is in an infinite matrix, which, according to Eq. (6.187), is given by

$$\sigma_{rr}^{M\infty}(r) = -4\mu^{M}c_1^{M}\frac{1}{r^3}. \tag{8.1}$$

Next, all matrix material beyond r = a is removed while applying tractions to the new spherical surface so that the stress field in the body remains unchanged. Then, equal and opposite tractions are applied to make the surface traction-free. According to Eq. (8.1) this produces an image hydrostatic stress field throughout the matrix,

$$\sigma_{rr}^{IM} = -\sigma_{rr}^{M\infty}(a) = 4\mu^{M}c_1^{M}\frac{1}{a^3} \tag{8.2}$$

and, using Eq. (3.175), the total stress field in the matrix is then

$$\sigma_{rr}^{M}(r) = \sigma_{rr}^{M\infty}(r) + \sigma_{rr}^{IM} = -4\mu^{M}c_1^{M}\frac{1}{r^3}\left[1 - \left(\frac{r}{a}\right)^3\right]. \tag{8.3}$$

As shown in Exercise 8.4, Eq. (8.3) can also be obtained directly from Eq. (8.41). The image stress is therefore seen to be negligible compared to $\sigma_{rr}^{M\infty}(r)$ in regions of the body around the inclusion at distances where $(r/a)^3 \ll 1$. This result will be generally true for inclusions in relatively large bodies of other shapes where r is significantly smaller than the linear dimensions of the body. This is clearly because the stress fields of inclusions fall off relatively rapidly with distance (i.e., as r^{-3}) and is of prime importance, since it means that image stresses can generally be neglected in regions around inclusions that are in large bodies, but not unusually close to interfaces. This result, obtained using the tractable isotropic system above, should also apply to corresponding anisotropic systems. However, the situation is obviously quite different when the inclusion is near a surface at distances of the order of a few multiples of the inclusion size as considered below in Secs. 8.3–8.5.

8.2.2 *Volume change of body due to inclusion — effect of image stress*

Despite the above conclusion that the image stress in the vicinity of an inclusion lying well within a large homogeneous finite body with a traction-free surface can generally be neglected relative to other more significant stresses, the image stress can play a significant role in determining the volume change of the body caused by the presence of the inclusion.

Consider again the case of a spherical homogeneous inclusion with the transformation strain $\varepsilon_{ij}^{T} = \varepsilon^{T}\delta_{ij}$ embedded in a finite traction-free body, which is now taken to be of arbitrary rather than spherical shape, in an isotropic system. The volume change, ΔV^{∞}, that would occur when such an inclusion is created in an infinite homogeneous matrix is determined first. According to Eq. (6.187), the displacement field would have spherical symmetry with radial displacements given by $u^{M^{\infty}}(r) = c_1^{M} r^{-2}$. Using Cartesian coordinates so that $u_i^{M^{\infty}} = c_1^{M} x_i / x^3$, the volume change, ΔV, of any finite region, $\mathcal{V}$, bounded by $\mathcal{S}$, that lies in the infinite matrix and contains the inclusion, is given by

$$\Delta V = \oiint_{\mathcal{S}} \mathbf{u}^{M^{\infty}} \cdot \hat{\mathbf{n}} dS = c_1^{M} \oiint_{\mathcal{S}} \frac{\mathbf{x} \cdot \hat{\mathbf{n}}}{x^3} dS = 4\pi c_1^{M} = \Delta V^{\infty}. \qquad (8.4)$$

Since ΔV, in Eq. (8.4), is seen to be a constant, independent of the size and shape of $\mathcal{V}$, we can expand $\mathcal{S}$ to infinity without any change in volume and, therefore, equate ΔV^{∞} to ΔV, where ΔV^{∞} is the volume change if the body were infinite.

This result may be further understood by examining the divergence of $\mathbf{u}^{M^{\infty}}$, which corresponds to the local volume change in the matrix. If we imagine that the matrix with the displacement field $u_i^{M^{\infty}} = c_1^{M} x_i / x^3$ extends all the way to the origin, the divergence is given by

$$\nabla \cdot \mathbf{u}^{M^{\infty}} = \nabla \cdot \left(c_1^{M} \frac{\mathbf{x}}{x^3} \right) = -c_1^{M} \nabla^2 \left(\frac{1}{x} \right) = 4\pi c_1^{M} \delta(\mathbf{x}) \qquad (8.5)$$

after use of Eq. (D.6). All local volume change is therefore concentrated at the origin in the form of a delta function and vanishes everywhere else. The volume change of any finite region in the infinite matrix $\mathcal{V}'$ with boundary $\mathcal{S}'$ that encloses the inclusion (which is located the origin) is given by

$$\Delta V' = \oiiint_{\mathcal{V}'} \nabla \cdot \mathbf{u}^{M\infty} dV = 4\pi c_1^M \oiiint_{\mathcal{V}'} \delta(\mathbf{x}) dV = 4\pi c_1^M \tag{8.6}$$

which is identical to the result given by Eq. (8.4). In essence, the volume change is propagated outwards through the matrix in divergenceless fashion so that the volume change of $\mathcal{V}'$ remains constant at $4\pi c_1^M$.

If the region $\mathcal{V}$ is now cut out of the infinite matrix and converted into a free body $\mathcal{V}^\circ$, with a traction-free surface $\mathcal{S}^\circ$, as described above, it will undergo a further volume change, ΔV^{IM}, caused by the image stress and given by

$$\Delta V^{IM} = \oiiint_{\mathcal{V}^\circ} \varepsilon_{ii}^{IM} dV = \frac{1}{3K^M} \oiiint_{\mathcal{V}^\circ} \sigma_{ii}^{IM} dV. \tag{8.7}$$

Equation (8.7) can be developed further (Eshelby 1954) by introducing the vector A_j defined by $A_j = \sigma_{ij}^{IM} x_i$. With the help of Eq. (2.65), the divergence of $\mathbf{A}$ is

$$\nabla \cdot \mathbf{A} = \sigma_{ii}^{IM} \tag{8.8}$$

and, after substituting this expression into Eq. (8.7), and converting the result to a surface integral,

$$\begin{aligned} \Delta V^{IM} &= \frac{1}{3K^M} \oiiint_{\mathcal{V}^\circ} \nabla \cdot \mathbf{A} dV = \frac{1}{3K^M} \oiint_{\mathcal{S}^\circ} \mathbf{A} \cdot \hat{\mathbf{n}} dS \\ &= \frac{1}{3K^M} \oiint_{\mathcal{S}^\circ} \sigma_{ij}^{IM} x_i \hat{n}_j dS. \end{aligned} \tag{8.9}$$

Next, by substituting the relationship $\sigma_{ij}^{IM} \hat{n}_j = -\sigma_{ij}^{M\infty} \hat{n}_j$, which is valid on $\mathcal{S}^\circ$, and the further relationship $\sigma_{ij}^{M\infty} = 2\mu^M (\partial u_i^{M\infty} / \partial x_j)$, which is valid in the matrix, Eq. (8.9), with the help of $u_i^{M\infty} = c_1^M x_i / x^3$, becomes

$$\Delta V^{IM} = -\frac{2\mu^M}{3K^M} \oiint_{\mathcal{S}^\circ} x_i \frac{\partial u_i^{M\infty}}{\partial x_j} \hat{n}_j dS = -\frac{2\mu^M}{3K^M} \left(-2c_1^M \oiint_{\mathcal{S}^\circ} \frac{\mathbf{x} \cdot \hat{\mathbf{n}}}{x^3} dS \right). \tag{8.10}$$

Then, substituting Eq. (8.4) into Eq. (8.10),

$$\Delta V^{IM} = \frac{4\mu^M}{3K^M} \Delta V^\infty. \tag{8.11}$$

With the use of Eq. (6.128), the final total volume change of the traction-free finite body, ΔV, is then

$$\Delta V = \Delta V^{\infty} + \Delta V^{IM} = \Delta V^{\infty}\left(1 + \frac{4\mu^M}{3K^M}\right) = 4\pi c_1^M\left(1 + \frac{4\mu^M}{3K^M}\right)$$

$$= \frac{3K^M}{3K^M + 4\mu^M} V^{INC} e^T \left(1 + \frac{4\mu^M}{3K^M}\right) = V^{INC} e^T, \tag{8.12}$$

which is seen to be independent of the body size and shape. The image stress in this particular case is therefore responsible for an increase in the volume change due to the inclusion amounting to the factor $(1 + 4\mu^M/3K^M) = 3(1-\nu^M)/(1+\nu^M) \approx 3/2$ (when $\nu^M \approx 1/3$). This is a significant effect as first pointed out by Eshelby (1954).

8.3 Homogeneous Inclusion Near Interface in Large Semi-infinite Region

8.3.1 *Elastic field*

Consider now the elastic field of a homogeneous inclusion of arbitrary shape near the traction-free planar surface of either a half-space or a planar interface between joined half-spaces. The Green's functions for a point force in these regions have been obtained in Section 4.2: we therefore employ the Green's function method of Section 3.6.3 to obtain the all-important constraining displacement $u_k^C(\mathbf{x})$ by substituting these Green's functions into Eq. (3.168) which applies to defects represented by transformation strains such as inclusions.

Assuming a uniform transformation strain, Eq. (3.168) takes the form[1]

$$u_k^C(\mathbf{x}) = C_{imnl}\varepsilon_{nl}^T \oiiint_{\mathcal{V}^{INC}} \frac{\partial G_{km}(\mathbf{x},\mathbf{x}')}{\partial x_i'} dV', \tag{8.13}$$

where the integral is taken over the region containing the transformation strain, i.e., the inclusion. Alternatively, Eq. (8.13) can be transformed to a surface integral by applying the divergence theorem so that

$$u_k^C(\mathbf{x}) = C_{imnl}\varepsilon_{nl}^T \oiint_{\mathcal{S}^{INC}} G_{km}(\mathbf{x},\mathbf{x}')\hat{n}_i' dS'. \tag{8.14}$$

[1]Note that here $G_{km} = G_{km}(\mathbf{x},\mathbf{x}')$ rather than $G_{km} = G_{km}(\mathbf{x}-\mathbf{x}')$ because of the presence of the interface.

If Eq. (8.13) is used, the required derivative of the Green's function for the case of a half-space, obtained with the help of Eqs. (4.43), (4.58) and (4.40), is

$$\frac{\partial G_{km}(\mathbf{x},\mathbf{x}')}{\partial x'_i} = -\frac{1}{8\pi^2|\mathbf{x}-\mathbf{x}'|^2}\oint_{\hat{\mathcal{L}}}\left\{\hat{w}_i(\hat{k}\hat{k})^{-1}_{km} - \hat{k}_i(\hat{k}\hat{k})^{-1}_{kp}[(\hat{k}\hat{w})_{pj} + (\hat{w}\hat{k})_{pj}](\hat{k}\hat{k})^{-1}_{jm}\right\} ds$$

$$(\zeta > 0)$$

$$+\frac{1}{4\pi^2}\mathcal{R}_e\int_0^{2\pi}\sum_{\beta=1}^{3}\sum_{\alpha=1}^{3}\frac{A^*_{k\alpha}M^*_{\alpha j}L_{j\beta}A_{m\beta}(\hat{k}_i + p_\beta\hat{w}_i)}{[\hat{\mathbf{k}}\cdot(\mathbf{x}-\mathbf{x}') + p^*_\alpha\mathbf{x}\cdot\hat{\mathbf{w}} - p_\beta\mathbf{x}'\cdot\hat{\mathbf{w}}]^2}d\phi. \tag{8.15}$$

Evaluation of the expression for the canceling displacement field, $u^C_k(\mathbf{x})$, obtained by substituting Eq. (8.15) into (8.13), will then require: (1) line integration along s around the unit circle, $\hat{\mathcal{L}}$, traversed by $\hat{\mathbf{k}}$ in the plane perpendicular to $\hat{\mathbf{w}}$ called for by the first term in Eq. (8.15); (2) integration over ϕ as the unit vector $\hat{\mathbf{k}}$ rotates over the range $0 \le \phi \le 2\pi$ in the surface plane (which is perpendicular to $\hat{\mathbf{w}}$) called for by the second term in Eq. (8.15); and (3) final integration over the inclusion volume, as it is traversed by the source vector, $\mathbf{x}'$, called for by Eq. (8.13). Additional details regarding the integration with respect to ϕ are discussed in the text following Eq. (4.58). Having obtained both $u_i^{C,INC}$ and $u_i^{C,M}$ by this method, the elastic displacements in both the inclusion and matrix can then be obtained using Eq. (6.1).

When the interface is a planar interface between joined half-spaces, the required Green's function derivatives can be obtained with the use of Eqs. (4.59), (4.79) and (4.40), and the resulting expressions for the displacements can be integrated in a similar manner.

8.4 Homogeneous Spherical Inclusion Near Surface of Half-space in Isotropic System

8.4.1 *Elastic field*

Following Mura (1987), we now treat the tractable case of a homogeneous spherical inclusion, with a transformation strain $\varepsilon^T_{ij} = \varepsilon^T\delta_{ij}$, located a distance q from a traction-free planar surface of a half-space in an isotropic

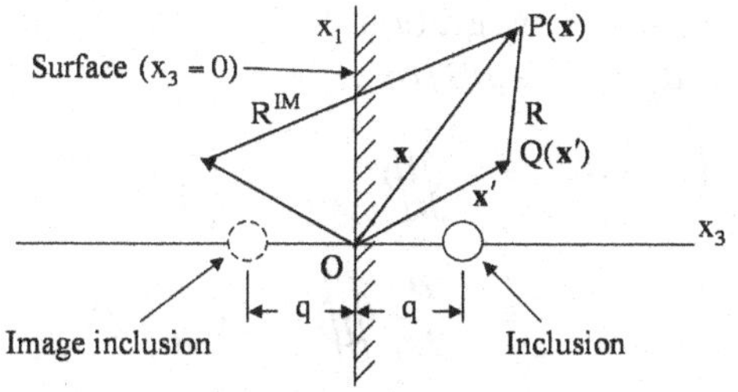

Fig. 8.1 Spherical inclusion at distance q from surface of a half-space.

system as illustrated in Fig. 8.1. Its displacement field in the matrix can be found by starting with Eq. (8.13) and modifying it so that it applies to an isotropic system by assuming the form

$$u_i^C(\mathbf{x}) = \left(\lambda^M \varepsilon_{mm}^T \delta_{jl} + 2\mu^M \varepsilon_{jl}^T\right) \oiiint_{\mathcal{V}^{INC}} \frac{\partial}{\partial x_l'} G_{ij}(\mathbf{x}, \mathbf{x}')dV'. \tag{8.16}$$

After substituting the transformation strain, $\varepsilon_{ij}^T = \varepsilon^T \delta_{ij}$, and the Green's functions, $G_{ij}(\mathbf{x}, \mathbf{x}')$, obtained from Eqs. (4.110), (4.15) and (4.116), Eq. (8.16) takes the form

$$u_i^C = \frac{(1+\nu^M)\varepsilon^T}{4\pi(1-\nu^M)} \oiiint_{\mathcal{V}^{INC}} \left\{ \frac{\partial}{\partial x_i'}\left(\frac{1}{R}\right) - (3-4\nu^M)(2\delta_{3i}-1)\frac{\partial}{\partial x_i'}\left(\frac{1}{R^{IM}}\right) \right.$$

$$\left. -2x_3 \frac{\partial^2}{\partial x_3' \partial x_i'}\left(\frac{1}{R^{IM}}\right) \right\} dV'. \tag{8.17}$$

The transformation

$$\begin{aligned}
&y_1 = x_1 \qquad y_1' = x_1' \qquad R = [(y_1 - y_1')^2 + (y_2 - y_2')^2 + (y_3 - y_3')^2]^{1/2} \\
&\qquad\qquad\qquad\qquad\qquad = |\mathbf{y} - \mathbf{y}'| \\
&y_2 = x_2 \qquad y_2' = x_2' \qquad \Lambda = y = (y_1^2 + y_2^2 + y_3^2)^{1/2} \\
&\qquad\qquad\qquad\qquad\qquad = [x_1^2 + x_2^2 + (x_3 - q)^2]^{1/2} \\
&y_3 = x_3 - q \quad y_3' = x_3' - q \quad y' = (y_1'^2 + y_2'^2 + y_3'^2)^{1/2}
\end{aligned} \tag{8.18}$$

is now made to aid in the integration of the first term in the integrand of Eq. (8.17). Using $\partial R/\partial x'_i = -\partial R/\partial x_i$,

$$\oiiint_{\mathcal{V}^{INC}} \frac{\partial}{\partial x'_i}\left(\frac{1}{R}\right) dV' = -\oiiint_{\mathcal{V}^{INC}} \frac{\partial}{\partial x_i}\left(\frac{1}{R}\right) dV'$$
$$= -\frac{\partial}{\partial x_i}\oiiint_{\mathcal{V}^{INC}} \left(\frac{1}{R}\right) dV'$$
$$= -\frac{\partial}{\partial y_i}\oiiint_{\mathcal{V}^{INC}} \frac{dV'}{|\mathbf{y}-\mathbf{y}'|} = -\frac{\partial \phi(y)}{\partial y_i}, \tag{8.19}$$

where $\phi(y)$ is the Newtonian potential of the inclusion given by Eq. (6.61). The integrations of the remaining two terms, involving $1/R^{IM}$, are carried out in similar fashion after making the transformation

$$\begin{aligned} z_1 &= x_1 \quad & z'_1 &= x'_1 \quad & R^{IM} &= [(z_1 - z'_1)^2 + (z_2 - z'_2)^2 + (z_3 - z'_3)^2]^{1/2} \\ &&&&&= |\mathbf{z} - \mathbf{z}'| \\ z_2 &= x_2 & z'_2 &= x'_2 & \Lambda^{IM} &= z = (z_1^2 + z_2^2 + z_3^2)^{1/2} \\ &&&&&= [x_1^2 + x_2^2 + (x_3 + q)^2]^{1/2} \\ z_3 &= x_3 + q & z'_3 &= -x'_3 + q & z' &= (z_1'^2 + z_{2'}^2 + z_3'^2)^{1/2} \end{aligned} \tag{8.20}$$

with the result

$$\oiiint_{\mathcal{V}^{INC}} \frac{\partial}{\partial x'_i}\left(\frac{1}{R^{IM}}\right) dV' = -\frac{\partial \phi(z)}{\partial z_i}$$
$$\oiiint_{\mathcal{V}^{INC}} \frac{\partial^2}{\partial x'_3 \partial x'_i}\left(\frac{1}{R^{IM}}\right) dV' = \frac{\partial^2 \phi(z)}{\partial z_3 \partial z_i}. \tag{8.21}$$

Substituting these expressions into Eq. (8.17), and using Eq. (6.125) for the Newtonian potential of a spherical inclusion at an exterior point in the matrix,

$$u_1^{C,M} = u_1^M = c\left\{\frac{x_1}{\Lambda^3} + \frac{(3-4\nu^M)x_1}{(\Lambda^{IM})^3} - \frac{6x_1x_3(x_3+q)}{(\Lambda^{IM})^5}\right\}$$
$$u_2^{C,M} = u_2^M = c\left\{\frac{x_2}{\Lambda^3} + \frac{(3-4\nu^M)x_2}{(\Lambda^{IM})^3} - \frac{6x_2x_3(x_3+q)}{(\Lambda^{IM})^5}\right\}$$
$$u_3^{C,M} = u_3^M = c\left\{\frac{x_3-q}{\Lambda^3} - \frac{[(3-4\nu^M)(x_3+q) - 2x_3]}{(\Lambda^{IM})^3} - \frac{6x_3(x_3+q)^2}{(\Lambda^{IM})^5}\right\} \tag{8.22}$$

in agreement with results obtained by Mindlin and Cheng (1950). The quantities Λ and Λ^{IM} are given by Eqs. (8.18) and (8.20), and $c = (1+\nu^M)\varepsilon^T V^{INC}/[4\pi(1-\nu^M)]$ is a measure of the "strength" of the inclusion acting as a center of dilatation as described in Section 6.4.3.1 [see Eqs. (6.127) and (6.128) with $\mu^{INC} = \mu^M$ and $\nu^{INC} = \nu^M$]. Comparison of Eq. (8.22) with Eq. (6.127) shows that the terms in Eq. (8.22), involving Λ, represent the displacement field that would be produced by the inclusion at the position $(0,0,q)$ in an infinite medium. The remaining terms, involving the quantity, Λ^{IM}, therefore represent the image displacement field.

The displacements at interior points of the inclusion can be obtained in the same manner by employing the Newtonian potential at interior points in a sphere given by MacMillan (1930), i.e.,

$$\phi = 2\pi\left(a^2 - \frac{x^2}{3}\right). \tag{8.23}$$

8.4.2 *Force imposed by image stress*

Using our present notation, the force imposed on a homogeneous inclusion by its image stress is given by Eq. (5.57) written as

$$\begin{aligned} F_\alpha^{INC/INC^{IM}} &= \oiint_{\mathcal{S}} \left(\frac{\partial \sigma_{ij}^{M^\infty}}{\partial x_\alpha} u_i^{IM} - \sigma_{ij}^{IM} \frac{\partial u_i^{M^\infty}}{\partial x_\alpha} \right) \hat{n}_j dS \\ &= \oiint_{\mathcal{S}} \left(\frac{\partial \sigma_{ij}^{IM}}{\partial x_\alpha} u_i^{M^\infty} - \sigma_{ij}^{M^\infty} \frac{\partial u_i^{IM}}{\partial x_\alpha} \right) \hat{n}_j dS. \end{aligned} \tag{8.24}$$

This relationship can be used to find the force tending to pull the homogeneous inclusion in Fig. 8.1 out of the isotropic half-space after putting it into the form

$$F_3^{INC^\infty/INC^{IM}} = \oiint_{\mathcal{S}} \left(\frac{\partial \sigma_{ij}^{IM}}{\partial x_3} u_i^{M^\infty} - \sigma_{ij}^{M^\infty} \frac{\partial u_i^{IM}}{\partial x_3} \right) \hat{n}_j dS. \tag{8.25}$$

The integral over $\mathcal{S}$ in Eq. (8.25) can be treated in the same manner as the surface integral in Eq. (7.3) in Section 7.2.1 by replacing σ_{ij}^{Q} with σ_{ij}^{IM} and expanding around $(0,0,q)$ rather than $(0,0,0)$. Therefore,

$$F_3^{INC^\infty/INC^{IM}} = \Delta V^{INC} \left[\frac{\partial}{\partial x_3} \left(\frac{\sigma_{mm}^{IM}}{3} \right) \right]_{0,0,q} = \Delta V^{INC} K \left[\frac{\partial e^{IM}}{\partial x_3} \right]_{0,0,q}. \tag{8.26}$$

The dilatation of the image field, e^{IM}, along $(0, 0, x_3)$ is determined by utilizing the image part of the total displacement field, $u_i^{C,M}$, given by Eq. (8.22), with the result

$$e^{IM}(0,0,x_3) = \varepsilon^{IM}_{mm}(0,0,x_3) = \left(\frac{\partial u_m^{C,M}}{\partial x_m}\right)_{0,0,x_3} = \frac{8c(1-2\nu)}{(q+x_3)^3}, \tag{8.27}$$

where $c = (1+\nu)\varepsilon^T V^{INC}/[4\pi(1-\nu)]$ is the strength of the inclusion given by Eq. (6.128) with $K^{INC} = K^M = K$. Then, substituting Eq. (8.27) into Eq. (8.26),

$$F_3^{INC^\infty/INC^{IM}} = \Delta V^{INC} K \left(\frac{\partial e^{IM}}{\partial x_3}\right)_{0,0,q} = -\frac{12\pi\mu(1-\nu)c^2}{q^4} \tag{8.28}$$

in agreement with Mura (1987) who determined the force by an alternative method. Note that the image force is relatively short-ranged and, as expected, tends to pull the inclusion out of the half-space regardless of whether the inclusion is a positive or negative center of dilatation.

8.5 Strain Energy of Inclusion in Finite Region

The strain energy of an inclusion in a presumably infinite matrix has been obtained in Sec. 6.3.3 in the form of Eq. (6.57), i.e.,

$$W = -\frac{1}{2} \oiiint_{\mathcal{V}^{INC}} \sigma_{ij}^{INC} \varepsilon_{ij}^T dV. \tag{8.29}$$

However, there is nothing in the derivation of Eq. (6.57) that precludes a finite matrix rather than an infinite one, and Eq. (8.29) therefore applies to inclusions in both finite and infinite matrices. Again, as in Sec. 6.3.3, the strain energy depends only upon the stress in the inclusion and its transformation strain.

Exercises

8.1. The dilatation associated with the image field of the homogeneous spherical inclusion, located a distance q from the surface of the isotropic half-space in Fig. 8.1, has been derived in the form of Eq. (8.27). This was accomplished by making use of the image terms in Eq. (8.22) obtained by means of the Green's function method.

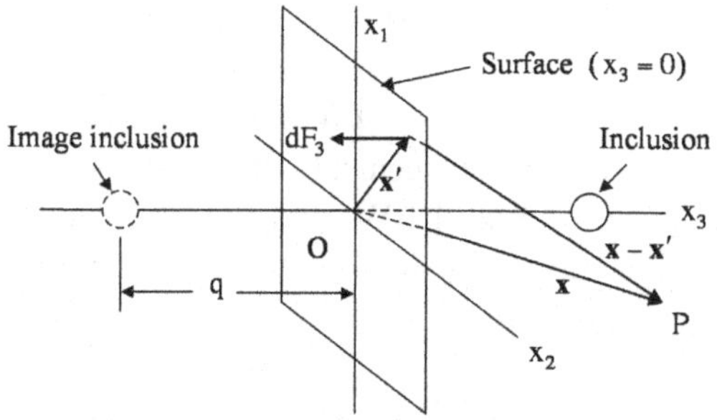

Fig. 8.2 Geometry for determining image stress of inclusion near surface ($\mathbf{x}_3 = 0$) of a half-space: x' lies in surface: dF_3 is an increment of normal force acting on the surface.

Derive Eq. (8.27) using a different approach in which an image inclusion is placed opposite the actual inclusion to partially cancel the tractions at the surface caused by the actual inclusion and then making the surface traction-free by applying an appropriate distribution of force to it.

Solution. The image inclusion is located at $(0, 0, -q)$ as in Fig. 8.2, and the combined stresses at $x_3 = 0$ due to the image and actual inclusions, obtained by use of Eq. (6.127), are then

$$\sigma_{13}^{M}(x_1, x_2, 0) = -\frac{6\mu c x_1 q}{x^3} + \frac{6\mu c x_1 q}{x^3} = 0,$$

$$\sigma_{23}^{M}(x_1, x_2, 0) = -\frac{6\mu c x_2 q}{x^3} + \frac{6\mu c x_2 q}{x^3} = 0,$$

$$\sigma_{33}^{M}(x_1, x_2, 0) = 4\mu c\left(\frac{1}{x^3} - 3\frac{q^2}{x^5}\right), \quad x = (x_1^2 + x_2^2 + q^2)^{1/2}. \tag{8.30}$$

The image inclusion therefore cancels the σ_{13}^{M} and σ_{23}^{M} tractions but not the σ_{33}^{M} tractions. The distribution of normal force that must be applied to the surface to generate the stress, $-\sigma_{33}^{M}(x_1, x_2, 0)$, and thereby produce a traction-free surface, is then

$$\frac{dF_3(x_1, x_2, 0)}{dS} = -4\mu c\left[\frac{1}{(x_1^2 + x_2^2 + q^2)^{3/2}} - 3\frac{q^2}{(x_1^2 + x_2^2 + q^2)^{5/2}}\right]. \tag{8.31}$$

The displacements due to this distribution of force density can be determined by first finding the displacements due to a single unit point force and then using this as a Green's function to integrate over the distribution. The displacements due to a unit point force ($F_3 = 1$) applied at the surface at (0,0,0) are given by $u_i^F = u_i^{F\infty} + u_i^{F^{IM}}$, where the $u_i^{F\infty}$ are the displacements due to the force if the body were infinite and given by Eq. (4.105), and the $u_i^{F^{IM}}$ are the corresponding image displacements given by Eq. (4.115) after setting $q = 0$ and $R = R^{IM} = x$. Therefore,

$$u_1^F(x_1, x_2, x_3) = \frac{1}{4\pi\mu}\left[-(1-2\nu)\frac{x_1}{x(x+x_3)} + \frac{x_1 x_3}{x^3}\right],$$

$$u_2^F(x_1, x_2, x_3) = \frac{1}{4\pi\mu}\left[-(1-2\nu)\frac{x_2}{x(x+x_3)} + \frac{x_2 x_3}{x^3}\right],$$

$$u_3^F(x_1, x_2, x_3) = \frac{1}{4\pi\mu}\left[2(1-\nu)\frac{1}{x} + \frac{x_3 x_3}{x^3}\right]. \tag{8.32}$$

Using Eq. (8.32), the dilatation at $P(x_1, x_2, x_3)$ due to a unit point force at the surface is

$$\varepsilon_{ii}^F(x_1, x_2, x_3) = \frac{\partial u_i^F}{\partial x_i} = -\frac{(1-2\nu)}{2\pi\mu}\frac{x_3}{x^3}. \tag{8.33}$$

With P on the x_3 axis, the problem is cylindrically symmetric around the axis, and by integrating over the force distribution on the surface using Eqs. (8.31) and (8.33), the total dilatation at $(0, 0, x_3)$ caused by the forces, after adjusting for sign conventions, is

$$\varepsilon_{ii}^{F(tot)}(0, 0, x_3) = -4c(1-2\nu)x_3 \int_0^\infty \frac{1}{(r'^2 + x_3^2)^{3/2}} \times \left[\frac{1}{(r'^2 + q^2)^{3/2}} - 3\frac{q^2}{(r'^2 + q^2)^{5/2}}\right] r' dr', \tag{8.34}$$

where $r'^2 = x_1'^2 + x_2'^2$. Carrying out the integration, and realizing that the dilatation due to the real inclusion and the image inclusion vanish in an infinite matrix, the final result is

$$\varepsilon_{ii}^{IM}(0, 0, x_3) = \varepsilon_{ii}^{F(tot)}(0, 0, x_3) = \frac{8c(1-2\nu)}{(q + x_3)^3} \tag{8.35}$$

in agreement with Eq. (8.27).

8.2. Show that the result

$$\Delta V = \Delta V^{\infty} + \Delta V^{IM} = \left(1 + \frac{4\mu^M}{3K^M}\right)\Delta V^{\infty} \tag{8.36}$$

given by Eq. (8.12) for the volume change experienced by a finite body containing a homogeneous inclusion in an isotropic system also applies to the case of a spherical inhomogeneous inclusion of radius R with $\varepsilon_{ij}^T = \varepsilon^T \delta_{ij}$ at the center of a spherical body of radius a possessing a free surface. Assume that $(R/a)^3 \ll 1$. Hint: determine the elastic field using the methods of Exercise 6.8.

Solution. We shall obtain Eq. (8.36) by solving for the elastic field due to the inclusion in the spherical body with a free surface to obtain ΔV and then using Eq. (6.129) to determine ΔV^{∞} due to the inclusion in an infinite body.

The elastic field in the free surface case is obtained by solving the Navier equation using the method employed in Exercise 6.8. Using Eq. (6.186), the general solution for the displacements are

$$u_r^{tot,INC}(r) = c_2^{INC} r \quad u_r^{tot,M}(r) = u_r^M(r) = \frac{c_1^M}{r^2} + c_2^M r \tag{8.37}$$

after setting $c_1^{INC} = 0$ to avoid the singularity at $r = 0$. The three remaining constants are found by satisfying the boundary conditions at the interfaces. The radial stress $\sigma_{rr}^M(r)$ must vanish at the free surface at $r = a$ and, therefore, with the use of Eqs. (G.10) and (G.11),

$$3K^M c_2^M - \frac{4\mu^M}{a^3} c_1^M = 0. \tag{8.38}$$

The radial stresses must match across the inclusion/matrix interface, and therefore,

$$3K^{INC}(c_2^{INC} - \varepsilon^T) = 3K^M c_2^M - \frac{4\mu^M}{R^3} c_1^M. \tag{8.39}$$

The total displacements must also match there so that

$$c_2^{INC} = \frac{c_1^M}{R^3} + c_2^M. \tag{8.40}$$

Then, by solving for the three constants and substituting them into Eq. (8.37), the elastic displacements in the inclusion and matrix are

$$u_r^{INC}(r) = u_r^{tot,INC}(r) - \varepsilon^T r = (c_2^{INC} - \varepsilon^T) r = -\frac{4\mu^M}{(3K^{INC} + 4\mu^M)} \varepsilon^T r,$$

$$u_r^M(r) = \frac{9K^{INC} V^{INC}}{4\pi(3K^{INC} + 4\mu^M)} \frac{\varepsilon^T}{r^2} \left[1 + \frac{4\mu^M}{3K^M}\left(\frac{r}{a}\right)^3\right]. \tag{8.41}$$

The volume change of the body due to this displacement field is then

$$\Delta V = 4\pi a^2 u_r^M(a) = \frac{9K^{INC}V^{INC}}{(3K^{INC} + 4\mu^M)}\varepsilon^T \left(1 + \frac{4\mu^M}{3K^M}\right). \quad (8.42)$$

On the other hand, the volume change in the case of an infinite body, ΔV^∞, obtained using Eq. (6.129), is

$$\Delta V^\infty = 4\pi r^2 u_r^M(r) = \frac{9K^{INC}V^{INC}}{(3K^{INC} + 4\mu^M)}\varepsilon^T. \quad (8.43)$$

Substitution of Eq. (8.43) into (8.42) then produces Eq. (8.36).

8.3. It is proven in Exercise 2.7 that the introduction of a defect of a type that is a source of internal stress, and is not mimicked by body forces, into a traction-free body causes no average elastic dilatation and, hence, no body volume change. Verify this for: (a) the edge dislocation, and (b) the screw dislocation that lie along the axes of cylindrical bodies in Section 13.3.2.2. Hint: the stresses for the edge and screw dislocations are given by Eqs. (13.57) and (13.47), respectively.

Solution. (a) Using Eq. (2.122), it is readily shown that the dilatation is proportional to $\theta = \sigma_{rr} + \sigma_{\theta\theta} + \sigma_{zz}$ and, therefore

$$e \propto (\sigma_{rr} + \sigma_{\theta\theta} + \sigma_{zz}) \quad (8.44)$$

Then, substituting the stresses given by Eq. (13.57) into Eq. (8.44) we have

$$e = A\left(\frac{2}{r} - \frac{r}{R^2}\right)\sin\theta \quad (8.45)$$

where A = constant. The dilatation, averaged over a volume $\mathcal{V}$ of the cylinder, can then be written as

$$\langle e\rangle = \frac{A}{V}\oiiint_{\mathcal{V}}\left(\frac{2}{r} - \frac{r}{R^2}\right)\sin\theta dV \quad (8.46)$$

where $dV = r dr d\theta dz$. Then, taking $\mathcal{V}$ to be a cylindrical shell of unit length lying between r_o and R, we obtain the result

$$\langle e\rangle = \frac{A}{V}\int_{r_o}^{R}\left(2 - \frac{r^2}{R^2}\right)dr\int_0^{2\pi}\sin\theta d\theta\int_0^1 dz = 0 \quad (8.47)$$

(b) As indicated by Eq. (13.47), the only stress in the cylindrical body generated by the screw dislocation is the shear stress $\sigma_{\theta z}$. There is then no elastic dilatation, and hence, no body volume change.

8.4. Use Eq. (8.41) for $u_r^M(r)$ to obtain Eq. (8.3)

Solution Using Eq. (6.192) to obtain the expression for c_1^M, we write Eq. (8.41) as

$$u_r^M(r) = c_1^M\left(\frac{1}{r^2} + \frac{4\mu^M}{3K^M a^3} r\right) \tag{8.48}$$

Then, according to Eq. G.10),

$$\varepsilon_{rr}^M = c_1^M\left(\frac{-2}{r^3} + \frac{4\mu^M}{3K^M a^3}\right)$$

$$\varepsilon_{\theta\theta}^M = \varepsilon_{\phi\phi}^M = c_1^M\left(\frac{1}{r^3} + \frac{4\mu^M}{3K^M a^3}\right) \tag{8.49}$$

and, therefore

$$e^M = c_1^M \frac{4\mu^M}{K^M a^3} \tag{8.50}$$

Then, with the help of Eq. (2.129),

$$\sigma_{rr}^M = \lambda^M e^M + 2\mu^M \varepsilon_{rr}^M = -4\mu^M c_1^M \frac{1}{r^3}\left[1 - \left(\frac{r}{a}\right)^3\right] \tag{8.51}$$

in agreement with Eq. (8.3).

Chapter 9

Inhomogeneities

9.1 Introduction

Even though an inhomogeneity does not generate its own stress field, it perturbs an imposed stress and therefore interacts with it as described in Sec. 5.4. This interaction is of importance in many phenomena (e.g., stress-induced cavity migration) and it also plays an essential role in the interaction between an inhomogeneous inclusion and an imposed stress because of the inhomogeneity associated with such inclusions as described in Sec. 7.2.2.

This chapter begins with a consideration of the perturbation of the imposed stress field and the accompanying interaction energy when the inhomogeneity is ellipsoidal, has uniform elastic properties, is in a large body, and the imposed stress field would be uniform in the absence of the inhomogeneity. The analysis is based on the equivalent homogeneous inclusion method of Eshelby described in Sec. 6.3.2. Following this, the interaction energy between the inhomogeneity and the imposed stress field is analyzed.

Finally, the interaction energy and corresponding force are considered for the more general case when both the inhomogeneity and the imposed stress field are non-uniform.

The following notation is employed for this chapter:

ε_{ij}^{A} = applied strain field, A, that would be uniform throughout the body in the absence of the inhomogeneity

$\Delta\varepsilon_{ij}^{A,INH}, \Delta\varepsilon_{ij}^{A,M}$ = perturbation of the A strain field in the inhomogeneity and matrix, respectively

$\varepsilon_{ij}^{A',INH}, \varepsilon_{ij}^{A',M}$ = perturbed A strain field in the inhomogeneity and matrix, respectively

$\varepsilon_{ij}^{INC^*}$ = elastic strain in equivalent homogeneous inclusion
$\varepsilon_{ij}^{M^*}$ = elastic strain in matrix around equivalent homogeneous inclusion
$\varepsilon_{ij}^{C,INC^*}, \varepsilon_{ij}^{C,M^*}$ = canceling strain in inclusion and matrix, respectively, in equivalent homogeneous inclusion system

9.2 Interaction Between a Uniform Ellipsoidal Inhomogeneity and Imposed Stress

Consider the relatively simple case of an ellipsoidal inhomogeneity with uniform elastic properties embedded in a large body subjected to an applied A field that would be uniform in the absence of the inhomogeneity. The resulting $\varepsilon_{ij}^{A',INH}$ and $\varepsilon_{ij}^{A',M}$ fields are determined first, by use of Eshelby's equivalent homogeneous inclusion method. Following this, the interaction energy between the inhomogeneity and the imposed A field is formulated.

9.2.1 *Elastic field in body containing inhomogeneity and imposed stress*

To apply the equivalent homogeneous inclusion method, we first construct the system, consisting of the inhomogeneity and the perturbed imposed field (i.e., the A' field) as follows:

(1) Start with the homogeneous stress-free body and introduce the inhomogeneity.
(2) Apply forces to the body that would produce a uniform applied strain field, ε_{ij}^{A}, throughout the body in the absence of the inhomogeneity.

The final elastic strains in the inhomogeneity and matrix are then

$$\varepsilon_{ij}^{A',INH} = \Delta\varepsilon_{ij}^{A,INH} + \varepsilon_{ij}^{A}, \tag{9.1}$$

where $\Delta\varepsilon_{ij}^{A,INH}$ is the perturbation of the A field, and

$$\varepsilon_{ij}^{A',M} = \Delta\varepsilon_{ij}^{A,M} + \varepsilon_{ij}^{A}. \tag{9.2}$$

It is assumed, as previously in the treatment of an inhomogeneous inclusion in the presence of an imposed stress in Sec. 7.2.2.1, that the body is relatively large and the inhomogeneity is sufficiently small and distant from the body surface so that any image strains can be neglected in its vicinity as discussed in Sec. 8.2.1 [see Eq. (8.3)].

Next, we attempt the construction of an equivalent homogeneous inclusion with a uniform transformation strain, as follows:

(1) Cut the ellipsoidal region destined to be the inclusion out of the unstressed homogeneous body, and subject it to the uniform transformation strain $\varepsilon_{ij}^{T^*}$ and then to forces producing the elastic strain $-\varepsilon_{ij}^{T^*}$.
(2) While holding these forces constant, insert the inclusion back into its cavity (where it fits exactly) and apply forces that cancel the previous forces, thereby producing the strains $\varepsilon_{ij}^{C,INC^*}$ and ε_{ij}^{C,M^*}.
(3) Apply the same forces to the body surface that were applied previously to the body containing the inhomogeneity. This produces the uniform strain field, ε_{ij}^{A}, throughout the system.

The final elastic strains in the equivalent homogeneous inclusion and matrix are then, respectively,

$$\varepsilon_{ij}^{INC^*} = \varepsilon_{ij}^{C,INC^*} - \varepsilon_{ij}^{T^*} + \varepsilon_{ij}^{A} \tag{9.3}$$

and

$$\varepsilon_{ij}^{M^*} = \varepsilon_{ij}^{C,M^*} + \varepsilon_{ij}^{A}. \tag{9.4}$$

The requirements that the equivalent homogeneous inclusion and inhomogeneity be under the same stress and have the same size and shape are, respectively,

$$\sigma_{ij}^{C,INC^*} - \sigma_{ij}^{T^*} + \sigma_{ij}^{A} = \Delta\sigma_{ij}^{A,INH} + \sigma_{ij}^{A} \tag{9.5}$$

and

$$\varepsilon_{ij}^{C,INC^*} = \Delta\varepsilon_{ij}^{A,INH}. \tag{9.6}$$

Then, by applying Hooke's law to the equal stress condition, and substituting the equal strain condition into the result,

$$C_{ijkl}^{M}(\varepsilon_{kl}^{C,INC^*} - \varepsilon_{kl}^{T^*} + \varepsilon_{kl}^{A}) = C_{ijkl}^{INH}(\varepsilon_{kl}^{C,INC^*} + \varepsilon_{kl}^{A}). \tag{9.7}$$

However, since the transformation strain, $\varepsilon_{kl}^{T^*}$, is uniform, Eq. (6.29) is valid, and substituting it into Eq. (9.7),

$$\left[\left(C_{ijkl}^{INH} - C_{ijkl}^{M}\right) S_{klmn}^{E} + C_{ijmn}^{M}\right] \varepsilon_{mn}^{T^*} = (C_{ijmn}^{M} - C_{ijmn}^{INH})\varepsilon_{mn}^{A}. \tag{9.8}$$

Finally, by introducing the tensor Y_{ijmn}^{INH} defined by

$$Y_{ijmn}^{INH} = \left(C_{ijkl}^{INH} - C_{ijkl}^{M}\right) S_{klmn}^{E} + C_{ijmn}^{M} \tag{9.9}$$

and substituting it into Eq. (9.8), the transformation strain of the equivalent homogeneous inclusion is obtained in the form

$$[\varepsilon^{T^*}] = [Y^{INH}]^{-1}\left([C^{M}] - [C^{INH}]\right)[\varepsilon^{A}]. \tag{9.10}$$

Finally, the strain in the inhomogeneity can now be determined by substituting Eqs. (9.6), (6.29), and (9.10) into Eq. (9.1), i.e.,

$$[\varepsilon^{A',INH}] = [S^{E}][\varepsilon^{T^*}] + [\varepsilon^{A}] = \left\{[S^{E}][Y^{INH}]^{-1}\left([C^{M}] - [C^{INH}]\right) + [I]\right\}[\varepsilon^{A}]. \tag{9.11}$$

The elastic field in the matrix, which is the same as the field in the matrix surrounding the inhomogeneous inclusion, can then be found by the methods described in Sec. 6.3.1 for homogeneous inclusions.

Since ε_{ij}^{A} is uniform in Eq. (9.11), $\varepsilon_{ij}^{A',INH}$ is also uniform. We have therefore obtained a valid solution for the inhomogeneity problem which demonstrates that the elastic field in an ellipsoidal inhomogeneity with uniform elastic properties, subjected to a uniform imposed stress field, is itself uniform, and that such an inhomogeneity can, indeed, be represented by a equivalent homogeneous inclusion with a uniform transformation strain.

In Exercise 9.2 this method is used to obtain the strain fields, $\varepsilon_{ij}^{A',INH}$ and $\varepsilon_{ij}^{A',M}$, for the case of a spherical inhomogeneity in the presence of the imposed A field, $\varepsilon_{ij}^{A} = \varepsilon^{A}\delta_{ij}$, in an isotropic system.

9.2.2 *Interaction energy between inhomogeneity and imposed stress*

For this analysis it is assumed that the initial homogeneous body is first subjected to the uniform applied field, *A*, and the inhomogeneity of the previous section (possessing uniform elastic properties) is then created while the *A* field is maintained constant at the body surface. The interaction energy is then found using Eq. (5.80), which requires expressions for the total energies of the stressed body with and without the inhomogeneity, i.e., $E^{A'} = W^{A'} + \Phi^{A'}$ and $E^{A} = W^{A} + \Phi^{A}$, respectively. Using Eqs. (2.133), (2.106) and the divergence theorem, and denoting the body and its surface by $\mathcal{V}^{\circ}$ and $\mathcal{S}^{\circ}$, the strain energies, W^{A} and $W^{A'}$ are

$$W^{A} = \frac{1}{2}\oiiint_{\mathcal{V}^{\circ}} \sigma_{ij}^{A}\varepsilon_{ij}^{A}dV = \frac{1}{2}\oiiint_{\mathcal{V}^{\circ}} \frac{\partial}{\partial x_j}(\sigma_{ij}^{A}u_i^{A})dV = \frac{1}{2}\oiint_{\mathcal{S}^{\circ}} \sigma_{ij}^{A}u_i^{A}\hat{n}_j dS \tag{9.12}$$

and

$$W^{A'} = \frac{1}{2}\left(\oiiint_{\mathcal{V}^{\circ}-\mathcal{V}^{INH}} \sigma_{ij}^{A',M}\varepsilon_{ij}^{A',M} + \oiiint_{\mathcal{V}^{INH}} \sigma_{ij}^{A',INH}\varepsilon_{ij}^{A',INH}\right) dV$$

$$= \frac{1}{2}\left[\oiint_{\mathcal{S}^{\circ}} \sigma_{ij}^{A',M}u_i^{A',M} - \oiint_{\mathcal{S}^{INH}} (\sigma_{ij}^{A',M}u_i^{A',M} - \sigma_{ij}^{A',INH}u_i^{A',INH})\right] \hat{n}_j dS$$

$$= \frac{1}{2}\oiint_{\mathcal{S}^{\circ}} \sigma_{ij}^{A',M}u_i^{A',M}\hat{n}_j dS \tag{9.13}$$

since the integral over $\mathcal{S}^{INH}$ vanishes because of the matching tractions and displacements at the inhomogeneity/matrix interface. Therefore, the difference, $W^{A'} - W^{A}$, is given by

$$W^{A'} - W^{A} = \frac{1}{2}\oiint_{\mathcal{S}^{\circ}} (u_i^{A',M}\sigma_{ij}^{A',M} - u_i^{A}\sigma_{ij}^{A})\hat{n}_j dS. \tag{9.14}$$

Similarly, the corresponding difference in the potential energy of the surface tractions is

$$\Phi^{A'} - \Phi^{A} = -\oiint_{\mathcal{S}^{\circ}} (u_i^{A',M}\sigma_{ij}^{A',M} - u_i^{A}\sigma_{ij}^{A})\hat{n}_j dS = 2(W^{A} - W^{A'}). \tag{9.15}$$

Therefore, according to Eq. (5.80), the interaction energy is simply

$$E_{int}^{INH/A} = W^{A} - W^{A'} \tag{9.16}$$

or, upon substituting Eqs. (9.12) and (9.13),

$$E_{int}^{INH/A} = \frac{1}{2}\left[\oiiint_{\mathcal{V}^{\circ}-\mathcal{V}^{INH}} (\sigma_{ij}^{A}\varepsilon_{ij}^{A} - \sigma_{ij}^{A',M}\varepsilon_{ij}^{A',M})\right.$$

$$\left. + \oiiint_{\mathcal{V}^{INH}} (\sigma_{ij}^{A}\varepsilon_{ij}^{A} - \sigma_{ij}^{A',INH}\varepsilon_{ij}^{A',INH})\right] dV. \tag{9.17}$$

Equation (9.17) can be put into a more useful form by first writing it as

$$E_{int}^{INH/A} = \frac{1}{2}\left[\oiiint_{\mathcal{V}^{\circ}-\mathcal{V}^{INH}} (\sigma_{ij}^{A',M}\varepsilon_{ij}^{A} - \sigma_{ij}^{A}\varepsilon_{ij}^{A',M})\right.$$

$$\left. + \oiiint_{\mathcal{V}^{INH}} (\sigma_{ij}^{A',INH}\varepsilon_{ij}^{A} - \sigma_{ij}^{A}\varepsilon_{ij}^{A',INH})\right] dV$$

$$+\frac{1}{2}\Bigg[\oiiint_{\mathcal{V}^\circ-\mathcal{V}^{\rm INH}}(\sigma_{\rm ij}^{\rm A}\varepsilon_{\rm ij}^{\rm A}+\sigma_{\rm ij}^{\rm A}\varepsilon_{\rm ij}^{\rm A',M}-\sigma_{\rm ij}^{\rm A',M}\varepsilon_{\rm ij}^{\rm A',M}-\sigma_{\rm ij}^{\rm A',M}\varepsilon_{\rm ij}^{\rm A})$$
$$+\oiiint_{\mathcal{V}^{\rm INH}}(\sigma_{\rm ij}^{\rm A}\varepsilon_{\rm ij}^{\rm A}+\sigma_{\rm ij}^{\rm A}\varepsilon_{\rm ij}^{\rm A',INH}-\sigma_{\rm ij}^{\rm A',INH}\varepsilon_{\rm ij}^{\rm A',INH}-\sigma_{\rm ij}^{\rm A',INH}\varepsilon_{\rm ij}^{\rm A})\Bigg]{\rm dV}. \tag{9.18}$$

Then, using the same method that produced Eq. (9.13), the second bracketed term can be converted to the surface integral, $\oiint_{\mathcal{S}^\circ}(\sigma_{\rm ij}^{\rm A',M}-\sigma_{\rm ij}^{\rm A})({\rm u}_{\rm i}^{\rm A',M}+{\rm u}_{\rm i}^{\rm A})\hat{\rm n}_{\rm j}{\rm dS}$, which vanishes because of the constant surface traction condition

$$\sigma_{\rm ij}^{\rm A',M}\hat{\rm n}_{\rm j}=\sigma_{\rm ij}^{\rm A}\hat{\rm n}_{\rm j}\quad(\text{on }\mathcal{S}^\circ). \tag{9.19}$$

The first term in the remaining bracketed expression also vanishes because, according to Eq. (2.102), $\sigma_{\rm ij}^{\rm A',M}\varepsilon_{\rm ij}^{\rm A}=\sigma_{\rm ij}^{\rm A}\varepsilon_{\rm ij}^{\rm A',M}$ in the region $\mathcal{V}^\circ-\mathcal{V}^{\rm INH}$. Therefore, Eq. (9.18) takes the reduced form

$${\rm E}_{\rm int}^{\rm INH/A}=\frac{1}{2}\oiiint_{\mathcal{V}^{\rm INH}}(\sigma_{\rm ij}^{\rm A',INH}\varepsilon_{\rm ij}^{\rm A}-\sigma_{\rm ij}^{\rm A}\varepsilon_{\rm ij}^{\rm A',INH}){\rm dV}. \tag{9.20}$$

Then, by using

$$\begin{aligned}\sigma_{\rm ij}^{\rm A',INH}&={\rm C}_{{\rm ijk}l}^{\rm INH}\varepsilon_{{\rm k}l}^{\rm A',INH}\\ \sigma_{\rm ij}^{\rm A}&={\rm C}_{{\rm ijk}l}^{\rm M}\varepsilon_{{\rm k}l}^{\rm A}\\ {\rm C}_{{\rm ijk}l}^{\rm M}&={\rm C}_{{\rm k}l{\rm ij}}^{\rm M}\end{aligned} \tag{9.21}$$

Eq. (9.20) can be put into the form

$${\rm E}_{\rm int}^{\rm INH/A}=\frac{1}{2}\oiiint_{\mathcal{V}^{\rm INH}}({\rm C}_{{\rm ijk}l}^{\rm INH}-{\rm C}_{{\rm ijk}l}^{\rm M})\varepsilon_{\rm ij}^{\rm A}\varepsilon_{{\rm k}l}^{\rm A',INH}{\rm dV}. \tag{9.22}$$

An alternative expression for the interaction energy can be obtained from Eq. (9.22) by introducing the transformation strain, $\varepsilon_{\rm ij}^{\rm T^*}$, of the equivalent homogeneous inclusion. Using Eqs. (9.1) and (9.6), the integrand in Eq. (9.22) can be written as

$$({\rm C}_{{\rm ijk}l}^{\rm INH}-{\rm C}_{{\rm ijk}l}^{\rm M})\varepsilon_{\rm ij}^{\rm A}\varepsilon_{{\rm k}l}^{\rm A',INH}=({\rm C}_{{\rm ijk}l}^{\rm INH}-{\rm C}_{{\rm ijk}l}^{\rm M})(\varepsilon_{{\rm k}l}^{\rm C,INC^*}+\varepsilon_{{\rm k}l}^{\rm A})\varepsilon_{\rm ij}^{\rm A} \tag{9.23}$$

and then, by substituting Eq. (9.7) in the further form

$$({\rm C}_{{\rm ijk}l}^{\rm INH}-{\rm C}_{{\rm ijk}l}^{\rm M})\varepsilon_{\rm ij}^{\rm A}\varepsilon_{{\rm k}l}^{\rm A',INH}=-{\rm C}_{{\rm ijk}l}^{\rm M}\varepsilon_{\rm ij}^{\rm A}\varepsilon_{{\rm k}l}^{\rm T^*}. \tag{9.24}$$

However,

$$C^{M}_{ijkl}\varepsilon^{A}_{ij}\varepsilon^{T^*}_{kl} = C^{M}_{klij}\varepsilon^{A}_{ij}\varepsilon^{T^*}_{kl} = \sigma^{A}_{kl}\varepsilon^{T^*}_{kl} \qquad (9.25)$$

and, finally, upon substituting Eqs. (9.24) and (9.25) into Eq. (9.22),

$$E^{INH/A}_{int} = -\frac{1}{2}\iiint_{\mathcal{V}^{INH}} \sigma^{A}_{kl}\varepsilon^{T^*}_{kl}\, dV. \qquad (9.26)$$

In Exercise 9.3, it is verified that Eqs. (9.22) and (9.26) yield the same result for the interaction energy between a spherical inhomogeneity with uniform elastic properties and an imposed stress field possessing the strains $\varepsilon^{A}_{ij} = \varepsilon^{A}\delta_{ij}$ in an isotropic system.

9.2.3 *Some results for isotropic systems*

9.2.3.1 *Elastic field in body containing inhomogeneity and imposed stress*

The system analyzed in the previous sections is now assumed to be isotropic. The equivalent homogeneous inclusion method is again employed, and the equation for the transformation strain of the equivalent homogeneous inclusion appropriate for an isotropic system is obtained by converting Eq. (9.8), by use of Eq. (2.120), to the form

$$(\lambda^{INH} - \lambda^{M})S^{E}_{mmkl}\varepsilon^{T^*}_{kl}\delta_{ij} + 2(\mu^{INH} - \mu^{M})S^{E}_{ijkl}\varepsilon^{T^*}_{kl} + \lambda^{M}e^{T^*}\delta_{ij} + 2\mu^{M}\varepsilon^{T^*}_{ij}$$
$$= (\lambda^{M} - \lambda^{INH})e^{A}\delta_{ij} + 2(\mu^{M} - \mu^{INH})\varepsilon^{A}_{ij} \qquad (9.27)$$

When $i \neq j$, the shear transformation strains for the equivalent homogeneous inclusion can be obtained directly from Eq. (9.27) in the form

$$\varepsilon^{T^*}_{\alpha\beta} = \frac{\mu^{M} - \mu^{INH}}{2(\mu^{INH} - \mu^{M})S^{E}_{\alpha\beta\alpha\beta} + \mu^{M}}\varepsilon^{A}_{\alpha\beta} \qquad (\alpha \neq \beta) \qquad (9.28)$$

by invoking the properties of the Eshelby tensor, S^{E}_{ijkl}, described in the text following Eq. (6.92), and using Greek indices to avoid the index summation convention.

When $i = j$ the corresponding normal transformation strains can be determined by solving the three simultaneous linear equations obtained from Eq. (9.27), i.e.,

$$(\lambda^{INH} - \lambda^{M})S^{E}_{mmkl}\varepsilon^{T^*}_{kl} + 2(\mu^{INH} - \mu^{M})S^{E}_{ijkl}\varepsilon^{T^*}_{kl} + \lambda^{M}e^{T^*} + 2\mu^{M}\varepsilon^{T^*}_{ij}$$
$$= (\lambda^{M} - \lambda^{INH})e^{A} + 2(\mu^{M} - \mu^{INH})\varepsilon^{A}_{ij} \qquad (i = j) \qquad (9.29)$$

As in the previous development leading to Eq. (6.120), the solution is expedited by using the contracted index notation given by Eqs. (2.88) and (2.89). Using this, Eq. (9.29) can then be expressed in the matrix form

$$[\mathrm{V}][\varepsilon^{\mathrm{T}^*}] = [\mathrm{W}][\varepsilon^{\mathrm{A}}], \tag{9.30}$$

where

$$\begin{aligned} \mathrm{V_{ij}} &= (\lambda^{\mathrm{INH}} - \lambda^{\mathrm{M}})(\mathrm{S^E_{1j}} + \mathrm{S^E_{2j}} + \mathrm{S^E_{3j}}) + 2(\mu^{\mathrm{INH}} - \mu^{\mathrm{M}})\mathrm{S^E_{ij}} + \lambda^{\mathrm{M}} + 2\mu^{\mathrm{M}}\delta_{\mathrm{ij}}, \\ \mathrm{W_{ij}} &= \lambda^{\mathrm{M}} - \lambda^{\mathrm{INH}} + 2(\mu^{\mathrm{M}} - \mu^{\mathrm{INH}})\delta_{\mathrm{ij}}. \end{aligned} \tag{9.31}$$

Then, solving Eq. (9.30) for the $\varepsilon^{\mathrm{T}^*}_{\mathrm{ij}}$,

$$[\varepsilon^{\mathrm{T}^*}] = [\mathrm{V}]^{-1}[\mathrm{W}][\varepsilon^{\mathrm{A}}] \qquad (\mathrm{i} = \mathrm{j}). \tag{9.32}$$

The elastic strains in the inhomogeneity, $\varepsilon^{\mathrm{A}',\mathrm{INH}}_{\mathrm{ij}}$, can now be found by using Eq. (9.11). The transformation strains, $\varepsilon^{\mathrm{T}^*}_{\mathrm{ij}}$, for the equivalent homogeneous inclusion are given as functions of the known strains, $\varepsilon^{\mathrm{A}}_{\mathrm{ij}}$, by Eqs. (9.28) and (9.32), and the required values of $\mathrm{S^E_{ijk\mathit{l}}}$ are available in Appendix H.

The strain in the matrix, $\varepsilon^{\mathrm{A}',\mathrm{M}}_{\mathrm{ij}}$, is the same as that around the equivalent homogeneous inclusion given by Eq. (9.4), where the required strains $\varepsilon^{\mathrm{C},\mathrm{M}^*}_{\mathrm{ij}}$ can be found using the methods described in Ch. 6 for treating homogeneous inclusions.

In Exercise 9.2 detailed expressions are found for $\varepsilon^{\mathrm{A}',\mathrm{INH}}_{\mathrm{ij}}$ and $\varepsilon^{\mathrm{A}',\mathrm{M}}_{\mathrm{ij}}$ when an initial strain field, $\varepsilon^{\mathrm{A}}_{\mathrm{ij}} = \varepsilon^{\mathrm{A}}\delta_{\mathrm{ij}}$, is imposed on a spherical inhomogeneity.

9.2.3.2 *Interaction energy between inhomogeneity and imposed stress*

The interaction energy between the ellipsoidal inhomogeneity of the previous section and the imposed strain, $\varepsilon^{\mathrm{A}}_{\mathrm{ij}}$, given by Eq. (9.22), can be converted, with the help of Eq. (2.120), to the form

$$\mathrm{E^{INH/A}_{int}} = \frac{1}{2}\oiiint_{\mathcal{V}^{\mathrm{INH}}} [(\lambda^{\mathrm{INH}} - \lambda^{\mathrm{M}})\mathrm{e^A e^{A',INH}} + 2(\mu^{\mathrm{INH}} - \mu^{\mathrm{M}})\varepsilon^{\mathrm{A}}_{\mathrm{ij}}\varepsilon^{\mathrm{A}',\mathrm{INH}}_{\mathrm{ij}}]\mathrm{dV} \tag{9.33}$$

appropriate for an isotropic system. The $\mathrm{e^{A',INH}}$ and $\varepsilon^{\mathrm{A}}_{\mathrm{ij}}\varepsilon^{\mathrm{A}',\mathrm{INH}}_{\mathrm{ij}}$ quantities required by Eq. (9.33) are obtained by first writing Eq. (9.11) in the contracted index form specified by Eqs. (2.88) and (2.89), i.e.,

$$\varepsilon^{\mathrm{A}',\mathrm{INH}}_{\mathrm{i}} = \mathrm{S^E_{ij}}\varepsilon^{\mathrm{T}^*}_{\mathrm{j}} + \varepsilon^{\mathrm{A}}_{\mathrm{i}}. \tag{9.34}$$

Then, substituting the S^E_{ij} tensor from Eq. (6.96) into Eq. (9.34),

$$[\varepsilon^{A',INH}] = [S^E][\varepsilon^{T^*}] + [\varepsilon^A] \quad \text{or} \quad \begin{bmatrix} \varepsilon_1^{A',INH} \\ \varepsilon_2^{A',INH} \\ \varepsilon_3^{A',INH} \\ \varepsilon_4^{A',INH} \\ \varepsilon_5^{A',INH} \\ \varepsilon_6^{A',INH} \end{bmatrix}$$

$$= \begin{bmatrix} S^E_{11} & S^E_{12} & S^E_{13} & 0 & 0 & 0 \\ S^E_{21} & S^E_{22} & S^E_{23} & 0 & 0 & 0 \\ S^E_{31} & S^E_{32} & S_{33} & 0 & 0 & 0 \\ 0 & 0 & 0 & 2S^E_{44} & 0 & 0 \\ 0 & 0 & 0 & 0 & 2S^E_{55} & 0 \\ 0 & 0 & 0 & 0 & 0 & 2S^E_{66} \end{bmatrix} \begin{bmatrix} \varepsilon_1^{T^*} \\ \varepsilon_2^{T^*} \\ \varepsilon_3^{T^*} \\ \varepsilon_4^{T^*} \\ \varepsilon_5^{T^*} \\ \varepsilon_6^{T^*} \end{bmatrix} + \begin{bmatrix} \varepsilon_1^A \\ \varepsilon_2^A \\ \varepsilon_3^A \\ \varepsilon_4^A \\ \varepsilon_5^A \\ \varepsilon_6^A \end{bmatrix}, \tag{9.35}$$

where the $\varepsilon_i^{T^*}$ strains are provided by Eqs. (9.28) and (9.32). Then, from Eq. (9.35), $e^{A',INH}$ and $\varepsilon^A_{ij}\varepsilon_{ij}^{A',INH}$ are obtained in the forms

$$e^{A',INH} = \varepsilon_{ii}^{A',INH} = (S^E_{1i} + S^E_{2i} + S^E_{3i})\varepsilon_i^{T^*} + e^A \tag{9.36}$$

and

$$\varepsilon^A_{ij}\varepsilon_{ij}^{A',INH} = \varepsilon^A_i\varepsilon_i^{A',INH} + 2(\varepsilon^A_4\varepsilon_4^{A',INH} + \varepsilon^A_5\varepsilon_5^{A',INH} + \varepsilon^A_6\varepsilon_6^{A',INH}). \tag{9.37}$$

9.2.3.3 *Interaction between spherical inhomogeneity and hydrostatic stress*

Interaction energy

Equation (9.33) is now used to determine the interaction energy between a spherical inhomogeneity with uniform elastic properties and the imposed strain field $\varepsilon^A_{ij} = \varepsilon^A\delta_{ij}$ in an isotropic system. It is seen immediately that the problem has spherical symmetry, and that the inhomogeneity is under hydrostatic pressure. Therefore, using Eqs. (9.36), with

$\varepsilon_1^{T^*} = \varepsilon_2^{T^*} = \varepsilon_3^{T^*} = \varepsilon^{T^*}$, (H.4), (9.37), and (2.129),

$$e^{A',INH} = \frac{1+\nu^M}{1-\nu^M}\varepsilon^{T^*} + 3\varepsilon^A = \frac{9K^M}{3K^M + 4\mu^M}\varepsilon^{T^*} + 3\varepsilon^A \tag{9.38}$$

and

$$\varepsilon_{ij}^A \varepsilon_{ij}^{A',INH} = 3\varepsilon^A \varepsilon^{A',INH} = \frac{e^A e^{A',INH}}{3}. \tag{9.39}$$

The equivalent homogeneous inclusion transformation strain ε^{T^*}, required in Eq. (9.38), is provided by Eq. (9.32) with matrix elements given by Eq. (9.31) in the forms

$$V_{11} = V_{22} = V_{33} = V = (\lambda^{INH} - \lambda^M)\frac{1+\nu^M}{3(1-\nu^M)}$$
$$+\frac{2(\mu^{INH} - \mu^M)(7-5\nu^M)}{15(1-\nu^M)} + \lambda^M + 2\mu^M,$$

$$V_{12} = V_{21} = V_{13} = V_{31} = V_{23} = V_{32} = V' = (\lambda^{INH} - \lambda^M)\frac{1+\nu^M}{3(1-\nu^M)}$$
$$+\frac{2(\mu^{INH} - \mu^M)(5\nu^M - 1)}{15(1-\nu^M)} + \lambda^M,$$

$$W_{11} = W_{22} = W_{33} = W = (\lambda^{INH} - \lambda^M) + 2(\mu^{INH} - \mu^M),$$

$$W_{12} = W_{21} = W_{13} = W_{31} = W_{23} = W_{32} = W' = \lambda^M - \lambda^{INH}. \tag{9.40}$$

Then, substituting Eq. (9.40) into Eq. (9.32),

$$\begin{bmatrix} \varepsilon^{T^*} \\ \varepsilon^{T^*} \\ \varepsilon^{T^*} \end{bmatrix} = \begin{bmatrix} V & V' & V' \\ V' & V & V' \\ V' & V' & V \end{bmatrix}^{-1} \begin{bmatrix} W & W' & W' \\ W' & W & W' \\ W' & W' & W \end{bmatrix} \begin{bmatrix} \varepsilon^A \\ \varepsilon^A \\ \varepsilon^A \end{bmatrix}. \tag{9.41}$$

Solving Eq. (9.41) for $[\varepsilon^{T^*}]$ and using Eq. (2.129),

$$\varepsilon^{T^*} = \frac{(K^M - K^{INH})(3K^M + 4\mu^M)}{K^M(3K^{INH} + 4\mu^M)}\varepsilon^A. \tag{9.42}$$

Then, substituting Eqs. (9.38), (9.39) and (9.42) into Eq. (9.33),

$$E_{\text{int}}^{\text{INH/A}} = \frac{9}{2}\frac{(K^{\text{INH}} - K^{\text{M}})(3K^{\text{M}} + 4\mu^{\text{M}})V^{\text{INH}}}{(3K^{\text{INH}} + 4\mu^{\text{M}})}(\varepsilon^{\text{A}})^2. \tag{9.43}$$

The interaction energy is therefore independent of whether ε^{A} is extensive or compressive and depends only on the square of its magnitude.

Interaction force

If the A field is no longer uniform, Eq. (5.82) can be employed to obtain an expression for the force exerted on the inhomogeneity by the hydrostatic part of the field i.e., $\varepsilon_{\text{mm}}^{\text{A}}/3$ (see Eq. (J.4)). Replacing ε^{A} by $\varepsilon_{\text{mm}}^{\text{A}}/3$ in Eq. (9.43), and then substituting the result into Eq. (5.82), this force, is then[1]

$$F_l^{\text{INH/A}} = -\frac{\partial E_{\text{int}}^{\text{INH/A}}}{\partial \xi_l} = \frac{(K^{\text{M}} - K^{\text{INH}})(3K^{\text{M}} + 4\mu^{\text{M}})V^{\text{INH}}}{(3K^{\text{INH}} + 4\mu^{\text{M}})}\varepsilon_{\text{mm}}^{\text{A}}\frac{\partial \varepsilon_{\text{mm}}^{\text{A}}}{\partial x_l}. \tag{9.44}$$

The direction of the force depends in a complex manner upon the sign of $(K^{\text{M}} - K^{\text{INH}})$ and the behavior of $\varepsilon_{\text{mm}}^{\text{A}}$. For example, if $K^{\text{INH}} < K^{\text{M}}$, and the inhomogeneity acts as a "soft" region, Eq. (9.44) indicates that it will be urged towards regions of high strain regardless of whether the strain is extensive or compressive, as might be expected. A cavity will therefore behave in this manner, whereas the force exerted on a "hard" region, such as the inhomogeneity that is associated with a hard precipitate, will act in the opposite direction.

9.3 Interaction Between an Elastically Non-uniform Inhomogeneity and a Non-uniform Imposed Stress

Finally, we consider the interaction between an inhomogeneity that is elastically non-uniform and subjected to an imposed stress field, A, that is also non-uniform. The determination of the perturbed stress field, A', in this general case is daunting,[2] and we shall therefore content

[1]This procedure may be compared with the first order procedure used to obtain Eqs. (7.13) and (7.65) for the force on a homogeneous inclusion due to a stress gradient.

[2]See, for example, the discussion in Sec. 16.4.2, of the difficulties with treating interactions involving inhomogeneous inclusions, non-uniform stresses and conditions of low symmetry.

ourselves with formulating general expressions for the interaction energy and corresponding force on the inhomogeneity assuming that A' is known.

An expression for the interaction energy, $E_{int}^{INH/A}$, given by

$$E_{int}^{INH/A} = -\frac{1}{2}\oiint_{\mathcal{S}^\circ} (\sigma_{ij}^{A'} u_i^{A'} - \sigma_{ij}^{A} u_i^{A})\hat{n}_j dS \tag{9.45}$$

is therefore obtained by employing Eqs. (9.16) and (9.14). Inspection of its formulation, which involves Eqs. (9.12), (9.13) and (9.15), shows that it is valid when the inhomogeneity and imposed stress are non-uniform as in the present case.

The corresponding force imposed upon the inhomogeneity by the A field is now shown to be given by the energy-momentum tensor. We begin by employing Eqs. (5.82), (9.14) and (9.16) to write

$$F_\alpha^{INH/A} = -\frac{\partial E^{A'}}{\partial \xi_\alpha} = -\frac{\delta(W^A - W^{A'})}{\delta \xi_\alpha} = \frac{\delta W^{A'}}{\delta \xi_\alpha}. \tag{9.46}$$

If $W^{A'}$ and $W^{A''}$ are the strain energies before and after the displacement, respectively, then, with the use of Eqs. (9.12) and (9.13),

$$\delta W^{A'} = \frac{1}{2}\oiint_{\mathcal{S}^\circ} (\sigma_{ij}^{A''} u_i^{A''} - \sigma_{ij}^{A'} u_i^{A'})\hat{n}_j dS. \tag{9.47}$$

However, since $\sigma_{ij}^{A''}\hat{n}_j = \sigma_{ij}^{A'}\hat{n}_j$ on $\mathcal{S}^\circ$ we can write Eq. (9.47) as

$$\begin{aligned}\delta W^{A'} &= \frac{1}{2}\oiint_{\mathcal{S}^\circ} (\sigma_{ij}^{A'} u_i^{A''} - \sigma_{ij}^{A''} u_i^{A'})\hat{n}_j dS = \frac{1}{2}\oiiint_{\mathcal{V}^\circ} (\sigma_{ij}^{A'} \varepsilon_{ij}^{A''} - \sigma_{ij}^{A''} \varepsilon_{ij}^{A'})dV \\ &= \frac{1}{2}\oiiint_{\mathcal{V}^\circ} (C'_{ijkm}\varepsilon_{km}^{A'}\varepsilon_{ij}^{A''} - C''_{ijkm}\varepsilon_{km}^{A''}\varepsilon_{ij}^{A'})dV\end{aligned} \tag{9.48}$$

with the help of the divergence theorem and Eqs. (2.65), (2.5) and (2.75). But,

$$C''_{ijkm} = C'_{ijkm} - \frac{\partial C'_{ijkm}}{\partial x_\alpha}\delta\xi_\alpha \tag{9.49}$$

and substituting this into Eq. (9.48) and using the symmetry properties of the C'_{ijkm} tensor,

$$\delta W^{A'} = \frac{1}{2}\oiiint_{\mathcal{V}^\circ} \frac{\partial C'_{ijkm}}{\partial x_\alpha}\varepsilon_{ij}^{A'}\varepsilon_{km}^{A'}\delta\xi_\alpha dV. \tag{9.50}$$

Therefore, with the use of Eqs. (9.46) and (2.5), the force is,

$$F_\alpha^{INH/A} = \frac{\delta W^{A'}}{\delta \xi_\alpha} = \frac{1}{2} \oiiint_{\mathcal{V}^\circ} \frac{\partial C'_{ijkm}}{\partial x_\alpha} \varepsilon_{ij}^{A'} \varepsilon_{km}^{A'} dV$$

$$= \frac{1}{2} \oiiint_{\mathcal{V}^\circ} \frac{\partial C'_{ijkm}}{\partial x_\alpha} \frac{\partial u_i^{A'}}{\partial x_j} \frac{\partial u_k^{A'}}{\partial x_m} dV. \tag{9.51}$$

To obtain the force in a form involving the surface integral of the energy-momentum tensor, as in Eq. (5.51), we can express the integrand in Eq. (9.51) as

$$\frac{\partial C'_{ijkm}}{\partial x_\alpha} \frac{\partial u_i^{A'}}{\partial x_j} \frac{\partial u_k^{A'}}{\partial x_m} = \frac{\partial}{\partial x_\alpha} (C'_{ijkm} \frac{\partial u_i^{A'}}{\partial x_j} \frac{\partial u_k^{A'}}{\partial x_m}) - 2C'_{ijkm} \frac{\partial u_i^{A'}}{\partial x_j} \frac{\partial^2 u_k^{A'}}{\partial x_\alpha \partial x_m}$$

$$= \frac{\partial}{\partial x_\alpha} \left(C'_{ijkm} \frac{\partial u_i^{A'}}{\partial x_j} \frac{\partial u_k^{A'}}{\partial x_m} \right) - 2 \frac{\partial}{\partial x_m} \left(\sigma_{km}^{A'} \frac{\partial u_k^{A'}}{\partial x_\alpha} \right) \tag{9.52}$$

with the help of the standard relationships from Ch. 2. Then, by substituting Eq. (9.52) into (9.51), and employing the divergence theorem,

$$F_\alpha^{INH/A} = \oiint_{\mathcal{S}^\circ} \frac{1}{2} \sigma_{km}^{A'} \frac{\partial u_k^{A'}}{\partial x_m} \hat{n}_\alpha dS - \oiint_{\mathcal{S}^\circ} \sigma_{km}^{A'} \frac{\partial u_k^{A'}}{\partial x_\alpha} \hat{n}_m dS$$

$$= \oiint_{\mathcal{S}^\circ} \left(w^{A'} \delta_{j\alpha} - \sigma_{ij}^{A'} \frac{\partial u_i^{A'}}{\partial x_\alpha} \right) \hat{n}_j dS \tag{9.53}$$

and the force is therefore given by the energy-momentum tensor, expressed in terms of the perturbed imposed field, A′.

Exercises

9.1. In an isotropic system find the perturbed displacement field, i.e., $u_r^{A',INH}(r)$ and $u_r^{A',M}(r)$, generated by introducing a spherical inhomogeneity of radius R and uniform elastic properties into a large but finite body initially containing the uniform applied strain field $\varepsilon_{ij}^A = \varepsilon^A \delta_{ij}$. Assume that the *A* field is maintained constant at the body surface. As in the text, focus on the common situation where the body is much larger than the inclusion, and the inhomogeneity is sufficiently distant from the body surface so that the $\varepsilon_{rr}^{A',INH}$ field is essentially uniform and independent of the location of the inhomogeneity.

Solution. For this problem a simple model system can be employed in which the inhomogeneity is at the center of a spherical body of radius a, where $(a/R)^3 \gg 1$, and the surface is subjected to tractions which maintain the strains there at $\varepsilon_{ij}^{A} = \varepsilon^{A}\delta_{ij}$. As in Exercise 6.8, where the elastic field due to an inhomogeneous inclusion in an infinite body is determined, the problem can be treated as a boundary value problem requiring the direct solution of the Navier equation in both the matrix and inhomogeneity. The problem again has spherical symmetry, and since there is no transformation strain in the system, Eq. (6.186) can be taken as the general solution of the Navier equation for the present problem when written in the form

$$u_r^{A',INH}(r) = \frac{c_1^{INH}}{r^2} + c_2^{INH} r, \quad u_r^{A',M}(r) = \frac{c_1^{M}}{r^2} + c_2^{M} r. \tag{9.54}$$

Then, after setting $c_1^{INH} = 0$ to avoid a singularity at the origin, the remaining three constants are determined by invoking the following three boundary conditions that must be satisfied at the interfaces. Using Eqs. (G.10) and (G.11), the radial stress at the body surface must be

$$\sigma_{rr}^{A',M}(a) = 3K^{M}\varepsilon^{A} \tag{9.55}$$

and, according to Eq. (9.54), the relationship

$$3K^{M}\varepsilon^{A} = 3K^{M}c_2^{M} - \frac{4\mu^{M}}{a^3}c_1^{M} \tag{9.56}$$

must therefore be satisfied. The displacements must match across the inhomogeneity/matrix interface, and therefore,

$$c_2^{INH}R = \frac{c_1^{M}}{R^2} + c_2^{M}R \tag{9.57}$$

and the corresponding normal stresses must also match so that

$$3K^{M}c_2^{M} - \frac{4\mu^{M}c_1^{M}}{R^3} = 3K^{INH}c_2^{INH}. \tag{9.58}$$

Then, solving Eqs. (9.56–9.58) for the three constants, dropping terms of magnitude $(R/a)^3$ relative to unity, and substituting the results into Eq. (9.54),

$$\begin{aligned} u_r^{A',INH}(r) &= c_2^{INH}r = \frac{3K^{M} + 4\mu^{M}}{3K^{INH} + 4\mu^{M}}\varepsilon^{A}r, \\ u_r^{A',M}(r) &= \frac{c_1^{M}}{r^2} + c_2^{M}r = \left[1 + \frac{3(K^{M} - K^{INH})}{(3K^{INH} + 4\mu^{M})}\left(\frac{R}{r}\right)^3\right]\varepsilon^{A}r. \end{aligned} \tag{9.59}$$

The strain in the inhomogeneity is therefore uniform, while the strain in the matrix varies rapidly with distance from the inhomogeneity until, at a distance of only a few multiples of R, it becomes essentially equal to the strain, ε^{A}, maintained elsewhere throughout the body. This will also be the case for an inhomogeneity that is anywhere in the body not unusually close to the body surface.

In the following Exercise 9.2 these same results are obtained by employing the equivalent homogeneous inclusion method.

9.2. Solve the problem posed in Exercise 9.1 by using the equivalent inclusion method instead of the Navier equation method.

Solution. To obtain $\varepsilon_{ij}^{A',INH}$ it is convenient to write Eq. (9.11) in the form

$$\varepsilon_{i}^{A',INH} = S_{ij}^{E}\varepsilon_{j}^{T^*} + \varepsilon_{i}^{A} \tag{9.60}$$

using contracted index notation. The required transformation strain of the equivalent homogeneous inclusion, $\varepsilon_{j}^{T^*}$, has been obtained in Sec. 9.2.3.3 in the form of Eq. (9.42), and substituting this into Eq. (9.60),

$$\begin{aligned}\varepsilon_{1}^{A',INH} = \varepsilon_{2}^{A',INH} = \varepsilon_{3}^{A',INH} = \varepsilon^{A',INH} &= \left(\frac{1+\nu^{M}}{3(1-\nu^{M})}+1\right)\varepsilon^{A}\\ &= \frac{3K^{M}+4\mu^{M}}{3K^{INH}+4\mu^{M}}\varepsilon^{A}.\end{aligned} \tag{9.61}$$

The corresponding radial displacement field in spherical coordinates is then

$$u_{r}^{A',INH} = \frac{3K^{M}+4\mu^{M}}{3K^{INH}+4\mu^{M}}\varepsilon^{A}r. \tag{9.62}$$

The strain field in the matrix is given by Eq. (9.4), which is expressed here as

$$\varepsilon_{rr}^{A',M}(r) = \varepsilon_{rr}^{C,M^*}(r) + \varepsilon^{A}. \tag{9.63}$$

The quantity $\varepsilon_{rr}^{C,M^*}(r)$ corresponds to the strain field in the matrix produced by a homogeneous inclusion possessing the transformation strain $\varepsilon_{j}^{T^*}$. The displacement field corresponding to this strain field is given by Eq. (8.41) which takes the form

$$u_{r}^{C,M^*}(r) = \frac{9K^{M}V^{INC}}{4\pi(3K^{M}+4\mu^{M})}\frac{\varepsilon^{T^*}}{r^{2}} \tag{9.64}$$

after setting $K^{INC} = K^{M}$ and $\varepsilon^{T} = \varepsilon^{T^*}$ and dropping the image term, since it is being assumed throughout that image effects can be neglected. Then, substituting Eq. (9.42) into Eq. (9.64) and substituting the result into the displacement field corresponding to Eq. (9.63),

$$u_r^{A',M}(r) = u_r^{C,M^*}(r) + \varepsilon^A r = \left[1 + \frac{3(K^M - K^{INH})}{(3K^{INH} + 4\mu^M)}\left(\frac{R}{r}\right)^3\right]\varepsilon^A r. \tag{9.65}$$

Finally, Eqs. (9.65) and (9.62) are seen to agree with Eq. (9.59).

9.3. Verify that both Eq. (9.26) and (9.22) yield the same result for the interaction energy, $E_{int}^{INH/A}$, between a spherical inhomogeneity and the uniform hydrostatic strain field $\varepsilon_{ij}^A = \varepsilon^A \delta_{ij}$ in an isotropic system.

Solution. Starting with Eq. (9.26), and using contracted indices and Eqs. (2.122) and (2.129), the integrand takes the form

$$\sigma_{ij}^A \varepsilon_{ij}^{T^*} = \sigma_i^A \varepsilon_i^{T^*} = 9K^M \varepsilon^A \varepsilon^{T^*}. \tag{9.66}$$

The transformation strain, $\varepsilon_{ij}^{T^*}$, is given by Eq. (9.42), and substituting this and Eq. (9.66) into Eq. (9.26),

$$\begin{aligned} E_{int}^{INH/A} &= -\frac{1}{2}\iiint_{\mathcal{V}^{INH}} \sigma_{ij}^A \varepsilon_{ij}^{T^*} dV = -\frac{9}{2}K^M V^{INH} \varepsilon^A \varepsilon^{T^*} \\ &= \frac{9}{2}\frac{(K^{INH} - K^M)(3K^M + 4\mu^M)V^{INH}}{(3K^{INH} + 4\mu^M)}(\varepsilon^A)^2. \end{aligned} \tag{9.67}$$

On the other hand, Eq. (9.22) takes the form of Eq. (9.33) in an isotropic system, and it is shown in the text that use of Eq. (9.33) leads to Eq. (9.43) which is identical to Eq. (9.67).

9.4. The interaction energy between a uniform spherical inhomogeneity and a uniform imposed elastic field, $\varepsilon_{ij}^A = \varepsilon^A \delta_{ij}$, as given by Eq. (9.43), can be written as proportional to the product of three bracketed terms, i.e.,

$$E_{int}^{INH/A} \propto \left[V^{INH}\right]\left[\varepsilon^A\right]\left[(K^M - K^{INH})\varepsilon^A\right]. \tag{9.68}$$

Similarly, the interaction energy between a homogeneous spherical inclusion possessing a uniform transformation strain, $\varepsilon_{ij}^T = \varepsilon^T \delta_{ij}$, and a uniform imposed elastic field, A, as given by Eq. (7.2), can be written in the form

$$E_{int}^{INC/Q} = -\sigma_{mm}^Q \varepsilon^T V^{INC} = -3Ke^Q \varepsilon^T V^{INC} \tag{9.69}$$

and, is therefore proportional to three bracketed quantities, i.e.,

$$E_{int}^{INC/Q} \propto \left[V^{INC}\right]\left[e^{Q}\right]\left[\varepsilon^{T}\right]. \tag{9.70}$$

Compare the bracketed terms in the two expressions and explain their forms in terms of simple physical concepts.

Solution The first bracketed term in both cases is simply the volume (or "size") of the defect. In each case the interaction energy is obtained by integrating over the defect volume under conditions where the integrand is constant throughout the volume. For the inhomogeneity, the integrand in Eq. (9.33), which is integrated to obtain Eq. (9.43), is constant because of the uniformity of both the imposed field and the transformation strain of the equivalent homogeneous inclusion that is employed in the treatment. For the inclusion, the integrand in Eq. (7.1), which is integrated to obtain Eq. (7.2), is constant because of the uniformity of both the imposed field and the transformation strain.

The second term in both cases is proportional to the strain in the imposed field which serves as a measure of the "strength" of the imposed field.

The third term, in the case of the inhomogeneity, can be taken as a measure of its "strength" as a defect capable of interacting with an applied stress by considering Sec. 5.4. Here, it is concluded that when applied forces are imposed on a body containing an inhomogeneity, the inhomogeneity perturbs the applied field, and the resulting perturbed field then exerts a force on the inhomogeneity. The "strength" of the inhomogeneity in playing this role should then depend upon the extent to which the bulk modulus of the inhomogeneity differs from that of the matrix and also the magnitude of the imposed strain as is indeed the case in the third bracket in Eq. (9.68). On the other hand, the corresponding "strength" of the homogeneous inclusion is simply its transformation strain.

In summary, both interaction energies are seen to be proportional to the product of three factors, i.e., the "size" of the defect (first term), the "strength" of the imposed field (second term), and the "strength" of the defect (third term).

Chapter 10

Point Defects in Infinite Homogeneous Regions

10.1 Introduction

Point defects in crystals can exist in many configurations. *Substitutional point defects* occupy substitutional lattice sites and include single *vacancies* (unoccupied substitutional sites) and single foreign solute atoms occupying substitutional sites in dilute solution. *Interstitial point defects* correspond to atoms occupying interstitial sites, i.e., sites in the interstices between substitutional sites. The interstitial atoms may be either foreign solute atoms, or the host atoms themselves: defects of the latter type are often referred to as *self-interstitial defects.* Small clusters at the atomic scale of any of the above defect types also qualify as point defects. These clusters may consist entirely of substitutional atoms, entirely of interstitial atoms, or may be of mixed character. For example, an undersized solute atom of one type may occupy a substitutional site and be bound to an undersized atom of another type occupying an adjacent interstitial site.

A common feature of all these defects is that they generally distort the host lattice and generate corresponding long-range stress and strain fields around them. If, for example, an embedded solute atom has a larger ion core radius than its host atoms it will, on average, push outwards against its near-neighbors causing a net expansion of the surrounding crystal, i.e., it will act as a *positive center of dilatation.* Conversely, a smaller solute atom will behave as a *negative center.* An interstitial atom is usually larger than the interstice in the lattice that it occupies and therefore acts as a positive center. On the other hand, the atoms around a vacancy often tend to relax inwards towards the vacant site causing the vacancy to act as a negative center. The symmetry of the surrounding stress field depends upon the symmetry of the point defect as discussed below.

Another feature of point defects, already discussed in Sec. 1.4, is the highly disturbed atomic structure of their cores. Atoms in the core have different environments than in the matrix, and the effective elastic constants in this small region therefore differ from those in the matrix, causing it to act effectively as an inhomogeneity. A point defect therefore acts approximately as a tiny inhomogeneous inclusion.

This chapter begins with a description of the symmetry of point defects in Sec. 10.2. The force multipole model for determining the elastic field is then taken up in Sec. 10.3. Here, the forces that the defect exerts on its near-neighbor atoms are replaced by an array of point forces applied to a homogeneous medium representing the crystal. The elastic displacement field produced in infinite homogeneous regions by these point forces is then found by employing point force Green's functions. The assumption of a homogeneous medium therefore neglects the inhomogeneous nature of the defect cores and any effects due to the difference between the effective elastic constants of the tiny core region and those of the lattice. Finally, in Sec. 10.4, the small inclusion model for point defects is discussed briefly. This model mimics both the forces exerted on surrounding atoms and the effect of the core inhomogeneity and can be analyzed using the methods of Chs. 6–9.

10.2 Symmetry of Point Defects

An important property of any point defect is its symmetry. When an otherwise perfect crystal contains a single point defect, the symmetry of the overall structure (which includes the strained crystal) is known as the *defect symmetry*. Because of the point nature of the defect, the defect symmetry lacks translational symmetry. Of interest is then its rotational point group symmetry and to which of the seven crystal classes it belongs, i.e., cubic, tetragonal, hexagonal, trigonal, orthorhombic, monoclinic or triclinic. The defect symmetry can be either the same as the symmetry of the host crystal or lower: in the latter case, the defect can generally assume more than one *structurally equivalent variant*[1] in the host lattice (Nowick and Berry 1972). Three examples are shown in Fig. 10.1. Figure 10.1a shows an interstitial atom defect in a FCC (i.e., face-centered-cubic) lattice where the

[1]Structurally equivalent variants are defects that have identical atomic structures but are simply rotated with respect to one another in the host lattice.

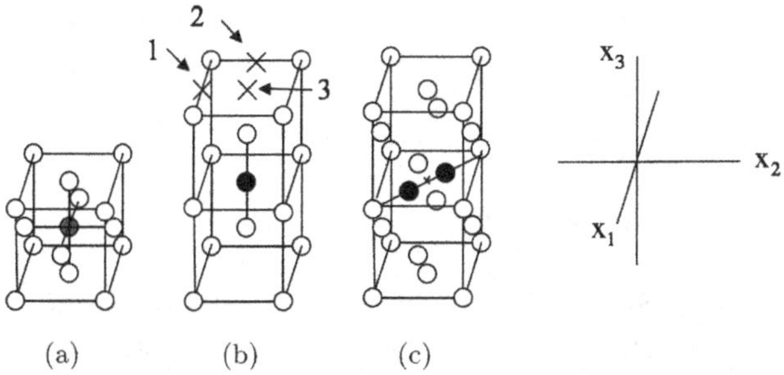

Fig. 10.1 (a) Interstitial atom (black sphere) in octahedral site in FCC lattice. (b) Interstitial atom (black sphere) in octahedral site in BCC lattice. Three structurally equivalent variants of this defect can be obtained by locating the interstitial atom on the three distinguishable 1, 2 and 3 octahedral sites. (c) Self-interstitial defect in FCC lattice where two host atoms share a lattice site in a "split" configuration extended along ⟨110⟩. Here, an extra host atom has been inserted interstitially, and relaxation has occurred so that the atom that originally occupied the face-centered site (indicated by the ×) and the inserted atom (i.e., the two black atoms) are symmetrically disposed along ⟨110⟩ around the × site which constitutes the center of the defect.

interstitial atom (black sphere) occupies an octahedral site.[2] The defect possesses the same cubic symmetry as the host lattice, and, therefore, no variants are possible. Figure 10.1b shows an interstitial atom in an octahedral site involving two nearest-neighbor and four next-nearest neighbor lattice sites in a BCC (i.e., body-centered-cubic) lattice. The defect symmetry is therefore tetragonal with the major tetragonal symmetry axis along the x_3 axis. Since this symmetry is lower than the cubic symmetry of the host lattice, this defect type can exist in the form of three variants which can be produced by locating the interstitial atom in the three structurally equivalent sites labeled 1, 2 and 3 in Fig. 10.1b where the major tetragonal axis is directed along ⟨100⟩, ⟨010⟩, and ⟨001⟩, respectively. Lastly, Fig. 10.1c shows a self-interstitial defect in a FCC lattice where two host atoms share a single face-centered lattice site in a "split" configuration extended along a ⟨110⟩ direction. This defect possesses rhombohedral symmetry which is lower than the cubic symmetry of the host lattice. Additional structurally equivalent variants can therefore exist with their extended axes along other ⟨110⟩ directions.

[2]The site is considered an octahedral site since it is located at the center of an imaginary octahedron whose vertices are located at the six nearest-neighbor lattice sites.

Further discussion of defect symmetry is given by Nowick and Berry (1972), Nowick and Heller (1963), Teodosiu (1982) and below in Sec. 10.3.4.

10.3 Force Multipole Model

As pointed out in Sec. 10.1, a point defect generally exerts forces on its neighboring atoms that are absent in the perfect crystal. These forces, in turn, produce a displacement field around the defect that extends throughout the surrounding matrix. An elastic model for the defect can therefore be constructed by replacing the defect with a homogeneous elastic continuum subjected to a localized array of point forces, i.e., a *force multipole*, that mimics the forces that are imposed on the atoms surrounding the defect. These forces fall off rapidly with distance from the defect, and it often suffices to include only those acting on nearest and next-nearest neighbors. The displacement field throughout the remainder of the body due to these forces is then found by using continuum elasticity incorporating the appropriate point force Green's function. The model is therefore expected to be applicable outside the defect core at distances beyond at least next-nearest neighbor distances. Methods of determining the forces that mimic the effect of the defect require atomistic calculations employing approaches that are beyond the scope of the present book. The continuum force multipole model is therefore developed below, assuming that the forces are known.

10.3.1 *Basic model*

The treatment mainly follows those given by Siems (1968), Leibfried and Breuer (1978) Teodosiu (1982) and Bacon, Barnett and Scattergood (1979b). As shown in Fig. 10.2, the field point is at $\mathbf{x}$, the center of the defect is at $\mathbf{x}'$, and the qth point force due to the defect is at a vector displacement $\mathbf{s}^{(q)}$ from the defect center. Assuming N point forces, and, using Eq. (3.15), the displacement at $\mathbf{x}$ is then

$$u_i(\mathbf{x}) = \sum_{q=1}^{N} G_{ij}(\mathbf{x} - \mathbf{x}' - \mathbf{s}^{(q)})F_j^{(q)}. \tag{10.1}$$

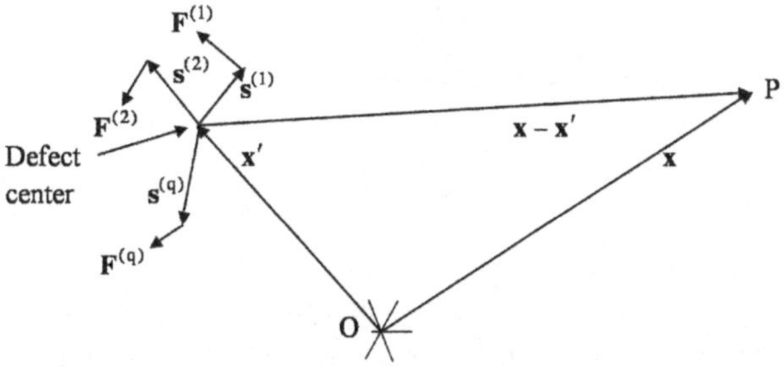

Fig. 10.2 Geometry for Eq. (10.1).

Since the distances $|\mathbf{s}^{(q)}|$ are relatively small, $G_{ij}(\mathbf{x} - \mathbf{x}' - \mathbf{s}^{(q)})$ can be expanded in a Taylor's series around $(\mathbf{x} - \mathbf{x}')$ of the form

$$\begin{aligned} G_{ij}(\mathbf{x} - \mathbf{x}' - \mathbf{s}^{(q)}) = G_{ij}(\mathbf{x} - \mathbf{x}') &+ \frac{\partial G_{ij}(\mathbf{x} - \mathbf{x}')}{\partial x'_m} s_m^{(q)} \\ &+ \frac{1}{2!} \frac{\partial^2 G_{ij}(\mathbf{x} - \mathbf{x}')}{\partial x'_m \partial x'_n} s_m^{(q)} s_n^{(q)} \\ &+ \frac{1}{3!} \frac{\partial^3 G_{ij}(\mathbf{x} - \mathbf{x}')}{\partial x'_m \partial x'_n \partial x'_r} s_m^{(q)} s_n^{(q)} s_r^{(q)} + \cdots \end{aligned} \quad (10.2)$$

Then, substituting Eq. (10.2) into Eq. (10.1),

$$\begin{aligned} u_i(\mathbf{x}) = G_{ij}(\mathbf{x} - \mathbf{x}')P_j &+ \frac{\partial G_{ij}(\mathbf{x} - \mathbf{x}')}{\partial x'_m} P_{mj} + \frac{1}{2!} \frac{\partial^2 G_{ij}(\mathbf{x} - \mathbf{x}')}{\partial x'_m \partial x'_n} P_{mnj} \\ &+ \frac{1}{3!} \frac{\partial^3 G_{ij}(\mathbf{x} - \mathbf{x}')}{\partial x'_m \partial x'_n \partial x'_r} P_{mnrj} + \cdots \end{aligned} \quad (10.3)$$

where

$$P_j = \sum_{q=1}^{N} F_j^{(q)} = 0, \quad (10.4)$$

$$P_{mj} = \sum_{q=1}^{N} s_m^{(q)} F_j^{(q)}, \quad (10.5)$$

$$P_{mnj} = \sum_{q=1}^{N} s_m^{(q)} s_n^{(q)} F_j^{(q)}, \tag{10.6}$$

$$P_{mnrj} = \sum_{q=1}^{N} s_m^{(q)} s_n^{(q)} s_r^{(q)} F_j^{(q)}. \tag{10.7}$$

The zeroth-order quantity, P_j, is just component j of the total force multipole and therefore must vanish to satisfy mechanical equilibrium. The remaining quantities P_{mj}, P_{mnj} and P_{mnrj} are second, third and fourth rank tensors and are termed the *force dipole*,[3] *quadrupole* and *octopole moment* tensors, respectively. The total torque exerted by the force distribution must vanish, so that

$$\begin{aligned}\sum_{q=1}^{N} \mathbf{s}^{(q)} \times \mathbf{F}^{(q)} &= \sum_{q=1}^{N} \left[\hat{\mathbf{e}}_1 \left(s_2^{(q)} F_3^{(q)} - s_3^{(q)} F_2^{(q)} \right) - \hat{\mathbf{e}}_2 \left(s_1^{(q)} F_3^{(q)} - s_3^{(q)} F_1^{(q)} \right) \right. \\ &\quad \left. + \hat{\mathbf{e}}_3 \left(s_1^{(q)} F_2^{(q)} - s_2^{(q)} F_1^{(q)} \right) \right] \\ &= \hat{\mathbf{e}}_1 (P_{23} - P_{32}) - \hat{\mathbf{e}}_2 (P_{13} - P_{31}) + \hat{\mathbf{e}}_3 (P_{12} - P_{21}) = 0.\end{aligned} \tag{10.8}$$

The P_{mj} force dipole tensor must therefore be symmetric, i.e.,

$$P_{mj} = P_{jm}. \tag{10.9}$$

Finally, using these results, and $\partial G_{ij}(\mathbf{x} - \mathbf{x}')/\partial x'_i = -\partial G_{ij}(\mathbf{x} - \mathbf{x}')/\partial x_i$, Eq. (10.3) takes the form (to third order),

$$\begin{aligned}u_i(\mathbf{x}) &= -\frac{\partial G_{ij}(\mathbf{x} - \mathbf{x}')}{\partial x_m} P_{mj} + \frac{1}{2} \frac{\partial^2 G_{ij}(\mathbf{x} - \mathbf{x}')}{\partial x_m \partial x_n} P_{mnj} \\ &\quad - \frac{1}{6} \frac{\partial^3 G_{ij}(\mathbf{x} - \mathbf{x}')}{\partial x_m \partial x_n \partial x_r} P_{mnrj}.\end{aligned} \tag{10.10}$$

Next, we find the dependence of each term in Eq. (10.10) on the distance of the field point from the center of the multipole. By setting $\mathbf{x}' = 0$ in Eq. (10.10), so that x is that distance, the kth term in Eq. (10.10) can be

[3] A force dipole moment is seen to be analogous to an electric dipole moment, $p = qs$, consisting of two point charges of equal but opposite sign, q and −q, separated by the distance s.

expressed as

$$[u_i(\mathbf{x})]_k = \frac{(-1)^k}{k!} \frac{\partial^k G_{ij}(\mathbf{x})}{\partial x_{q_1} \partial x_{q_2} \cdots \partial x_{q_k}} P_{q_1 q_2 \cdots q_k j} \quad (k = 1, 2, \ldots, \infty) \tag{10.11}$$

As already shown in Exercise 4.2, $G_{ij}(\mathbf{x})$ is a homogeneous function of degree −1 in the variable x. Therefore, applying the scaling constant, λ (see Exercise 4.2), to x in Eq. (10.33),

$$[u_i(\lambda\mathbf{x})]_k = \frac{(-1)^k}{k!} \frac{\partial^k}{\lambda^k \partial x_{q_1} \partial x_{q_2} \cdots \partial x_{q_k}} \left(\frac{G_{ij}(\mathbf{x})}{\lambda}\right) P_{q_1 q_2 \cdots q_k j}. \tag{10.12}$$

Then, by comparing Eqs. (10.33) and 10.34),

$$[u_i(\lambda\mathbf{x})]_k = \frac{1}{\lambda^{(k+1)}} [u_i(\mathbf{x})]_k. \tag{10.13}$$

Thus, the kth term is homogeneous of degree $-(k+1)$ in x and so must be proportional to $x^{-(k+1)}$. Therefore the displacement fields due to the force dipoles, quadrupoles and octopoles in Eq. (10.10) fall off as x^{-2}, x^{-3} and x^{-4}, respectively. The displacement field of a multipole therefore becomes increasingly short-ranged as its order increases.

Finally, it is important to emphasize that the force multipole of a defect is an intrinsic property of the defect and is independent of its position in the body.

10.3.2 *Force multipoles*

10.3.2.1 *Elementary double force*

The simplest force multipole is a *double force* consisting of two symmetrically opposed point forces, as illustrated in Fig. 10.3. Such a double force

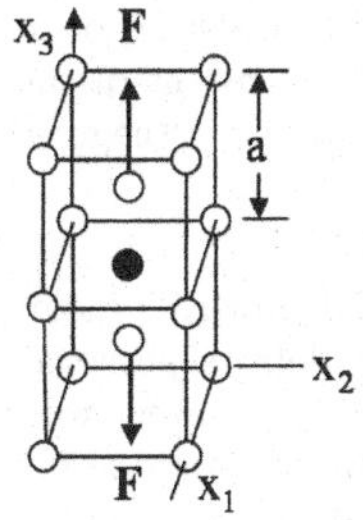

Fig. 10.3 Assumed simple nearest-neighbor double force model for interstitial defect (black sphere) in octahedral site in BCC structure. F represents forces applied to defect nearest-neighbor host atoms.

constitutes a simple nearest-neighbor model of a defect such as the octahedral interstitial in Fig. 10.1b where the main lattice forces exerted by the defect are assumed to be on its two nearest-neighbors. With the origin at the defect center and applying Eqs. (10.5)–(10.7), the quadrupole tensor vanishes since the double force defect is centrosymmetric, i.e., it possesses inversion symmetry.[4] The remaining non-zero components of the dipole and octopole tensors for the double force in Fig. 10.3 are then

$$\begin{aligned} P_{33} &= \frac{a}{2}F + \frac{a}{2}F = aF, \\ P_{3333} &= \frac{a^3}{8}F + \frac{a^3}{8}F = \frac{a^3}{4}F. \end{aligned} \tag{10.14}$$

10.3.2.2 *Combinations of double forces*

Consider next force multipoles composed of combinations of double forces. A simple example is the multipole for the defect shown in Fig. 10.4 consisting of three equal and orthogonal double forces applied to the defect

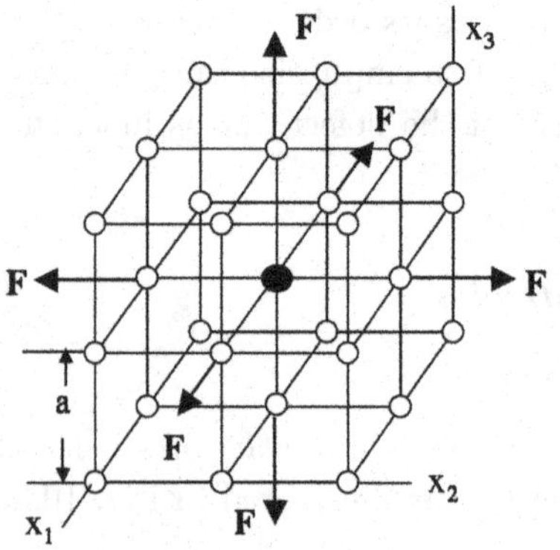

Fig. 10.4 Assumed force multipole consisting of forces, **F**, acting on the six nearest-neighbor atoms of defect (black sphere) at the distance a in simple cubic structure. The multipole is equivalent to three equal and orthogonal double forces.

[4]The inversion symmetry operation involves moving a point initially at **x** to −**x**. In the force multipole model of point defects possessing inversion symmetry such as, for example, those in Figs. 10.3 and 10.6, for every force, **F**, acting at **s**, there is a reverse force, −**F**, acting at −**s**. The contributions of each **F** and its −**F** counterpart then exactly cancel in the sum given by Eq. (10.6) for P_{mnj}, and it therefore vanishes. In fact, all force moment tensors with an odd number of indices must vanish for a point defect possessing inversion symmetry.

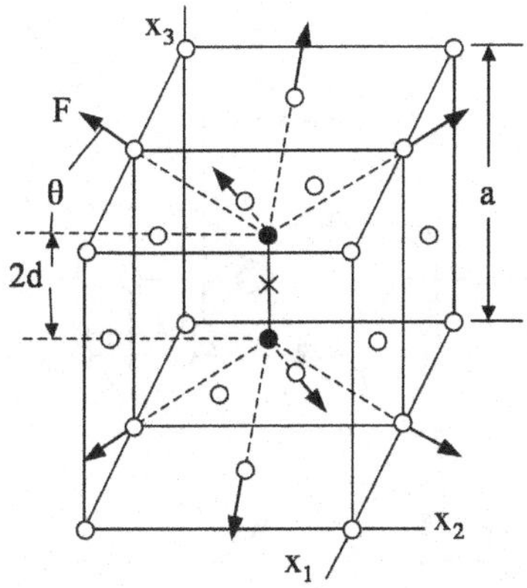

Fig. 10.5 Assumed "split" dumbbell self-interstitial defect (black spheres) in FCC structure with lattice forces acting on nearest-neighbor atoms. Here, an extra host atom has been inserted interstitially at the face-centered site, ×, and relaxation has occurred so that the original atom that occupied the × site and the inserted atom are symmetrically disposed around the × site which constitutes the defect center. All eight lattice forces are of magnitude F.

nearest-neighbors. With the defect center at the origin of the coordinate system, and employing Eqs. (10.5–10.7), the only non-zero multipole moment tensor components to third order are

$$
\begin{aligned}
P_{11} &= P_{22} = P_{33} = aF + aF = 2aF, \\
P_{1111} &= P_{2222} = P_{3333} = a^3F + a^3F = 2a^3F.
\end{aligned}
\tag{10.15}
$$

10.3.2.3 *Further force multipoles*

In many cases the force distribution representing the defect cannot be constructed using combinations of double forces. An example is the assumed "split" dumbbell self-interstitial defect in the FCC structure shown in Fig. 10.5. Here, the force multipole consists of eight non-collinear forces imposed on the eight nearest-neighbors of the defect. Using Eq. (10.5) and (10.6), the only non-zero multipole moment tensor components to second order are then

$$
\begin{aligned}
P_{11} &= P_{22} = 2aF\cos\theta, \\
P_{33} &= 4aF\sin\theta.
\end{aligned}
\tag{10.16}
$$

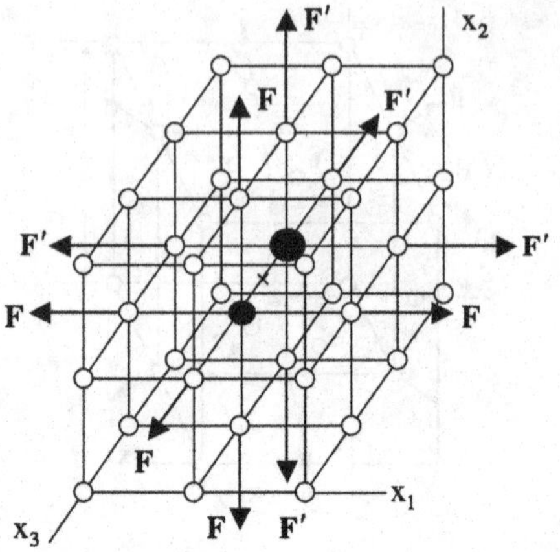

Fig. 10.6 Assumed non-centrosymmetric defect composed of two dissimilar nearest-neighbor solute atoms (black spheres) located on lattice sites in simple cubic structure with forces acting on nearest-neighbor host atoms.

All of the defects considered so far have possessed a center of inversion, and all elements of the second order quadrupole tensor have therefore vanished. Therefore, consider the defect in Fig. 10.6 which lacks a center of inversion. Here, the non-vanishing elements of the dipole and quadrupole to second order are

$$P_{11} = P_{22} = 2a(F + F'), \quad P_{131} = P_{311} = P_{232} = P_{322} = a^2(F - F'), \tag{10.17}$$
$$P_{33} = \frac{3a}{2}(F + F'), \quad P_{333} = \frac{9a^2}{4}(F - F').$$

The second order quadrupole moment components that are now present depend upon the degree to which the defect lacks a center of inversion, since they are proportional to the difference $(F - F')$ which would vanish if the defect were centrosymmetric.

10.3.3 *Elastic fields of multipoles in isotropic systems*

10.3.3.1 *Elementary double force*

The solution for the elastic field of a double force is useful, since it can be employed as a basic element in the construction of solutions for more complex force multipoles that can be made up of combinations of double forces. Using $\partial G_{ij}(\mathbf{x}-\mathbf{x}')/\partial x'_m = -\partial G_{ij}(\mathbf{x}-\mathbf{x}')/\partial x_m$, and taking the origin at the defect center so that $\mathbf{x}'$ vanishes, Eq. (10.10) for the double force in Fig. 10.3 takes the form

$$u_i(\mathbf{x}) = -aF\left(\frac{\partial G_{i3}(\mathbf{x})}{\partial x_3} + \frac{a^2}{24}\frac{\partial^3 G_{i3}(\mathbf{x})}{\partial x_3^3}\right). \tag{10.18}$$

Then, using the Green's function for infinite isotropic bodies given by Eq. (4.110),

$$u_i(\mathbf{x}) = \frac{aF}{16\pi\mu(1-\nu)}\left\{\frac{2(1-2\nu)l_3\delta_{i3} - l_i(1-3l_3^2)}{x^2} + \frac{a^2}{8}\frac{[\nu(12l_3 - 20l_3^3)\delta_{i3} + l_i(3 - 30l_3^2 + 35l_3^4)]}{x^4}\right\}, \tag{10.19}$$

where the $l_i = x_i/x$ are the direction cosines of $\mathbf{x}$. The displacement field has a complicated angular dependence. The first term, which is entirely due to the first order force dipole, falls off with distance from the center as x^{-2} as expected from the analysis following Eq. (10.10). The second term, which is due to the third order force octopole, falls off much more rapidly, i.e., as x^{-4}, and becomes relatively unimportant at a distance from the double force center only several times larger than the extension of the double force in conformity with St. Vinant's principle.

10.3.3.2 *Combinations of double forces*

For the combination of the three equal and orthogonal double forces in Fig. 10.4, Eq. (10.10) takes the form

$$u_i(\mathbf{x}) = -2aF\left(\frac{\partial G_{ij}(\mathbf{x})}{\partial x_j} + \frac{a^2}{6}\frac{\partial^3 G_{ij}(\mathbf{x})}{\partial x_j^3}\right). \tag{10.20}$$

Substitution of the Green's function given by Eq. (4.110) into Eq. (10.20) then yields the displacement field[5]

$$u_i(\mathbf{x}) = \frac{aF}{4\pi\mu(1-\nu)}\left\{\frac{(1-2\nu)l_i}{x^2} + \frac{a^2}{4}\frac{[\nu(12l_i - 20l_i^3) + 35l_i(-3/5 + l_1^4 + l_2^4 + l_3^4)]}{x^4}\right\}. \tag{10.21}$$

[5]Note that Eq. (10.21), for three orthogonal double forces, can be readily obtained from Eq. (10.19) for the single double force in Fig. 10.3 by simple summation using cyclic interchange of i = 1,2,3 and accounting for the difference in the labeling of the nearest-neighbor distance in Figs. 10.3 and 10.4.

As in the case of the previous simple double force, the first order dipole term and third order octopole term fall off as x^{-2} and x^{-4}, respectively, and the second order quadrupole term is absent because the multipole possesses a center of inversion. The first order dipole term can be written in the vector form

$$\mathbf{u}(\mathbf{x}) = u_i \hat{\mathbf{e}}_i = \frac{aF(1-2\nu)}{4\pi\mu(1-\nu)} \frac{x_i}{x^3} \hat{\mathbf{e}}_i = \frac{aF(1-2\nu)}{4\pi\mu(1-\nu)} \frac{1}{x^3} \mathbf{x}, \tag{10.22}$$

which is seen to be spherically symmetric. However, the octopole term contains the angular factor $(l_1^4 + l_2^4 + l_3^4 - 3/5)$ and odd powers of l_i and is therefore cubically symmetric.[6] The octopole term therefore contains detailed information about the local distribution of force at the defect center and decays rapidly compared to the dipole term as x increases. The far-field due to the dipole term is spherically symmetric and independent of the details of the distribution of local force as expected on the basis of St. Venant's principle. The far-field is seen to be of the same form found in Ch. 6 for the displacement field around a homogeneous spherical inclusion acting as a center of dilatation with the pure dilatational transformation strains $\varepsilon_{ij}^T = \varepsilon^T \delta_{ij}$. In fact, by employing Eqs. (10.22), (6.127) and (6.128), the fields are found to be identical when the dipole moment, 2Fa, and the strength of the spherical inclusion, c, are related by

$$\frac{(1-2\nu)}{8\pi\mu(1-\nu)} 2aF = c = \frac{(1+\nu)V^{INC}\varepsilon^T}{4\pi(1-\nu)}. \tag{10.23}$$

10.3.3.3 *Further force multipoles*

Further assumed examples of force multipoles are illustrated in Figs. 10.5 and 10.6. The forces in the multipole in Fig. 10.5, which mimic a *split dumbbell self-interstitial* in the FCC structure, are non-radial, and, using Eq. (10.9), the first order dipole component of the displacement field is[7]

$$u_i(\mathbf{x}) = -2aF \left[\frac{\partial G_{i1}(\mathbf{x})}{\partial x_1} \cos\theta + \frac{\partial G_{i2}(\mathbf{x})}{\partial x_2} \cos\theta + \frac{\partial G_{i3}(\mathbf{x})}{\partial x_3} 2\sin\theta \right] \tag{10.24}$$

with G_{ij} given for isotropic systems by Eq. (4.110).

[6] As pointed out by Eshelby (1977), the quantity $(-3/5 + l_1^4 + l_2^4 + l_3^4)$ is a surface harmonic of order four, and the only such harmonic possessing cubic symmetry.

[7] Note that this component of the displacement field becomes spherically symmetric when $\cos\theta = 2\sin\theta$, or, equivalently, when the distance 2d into which the defect is "split" in Fig. 10.5 corresponds to $2d = a/2$

For the more complex, and less symmetric, force multipole in Fig. 10.6, the displacement obtained by use of Eq. (10.10) consists of dipole and quadrupole terms given by

$$u_1(\mathbf{x}) = \frac{a l_1}{32\pi\mu(1-\nu)}\left\{(F+F')(9-16\nu-3l_3^2)\frac{1}{x^2} + \frac{3a(F-F')l_3}{4}(13-32\nu+5l_3^2)\frac{1}{x^3}\right\}. \tag{10.25}$$

Similar results are found for u_2^M and u_3^M.

10.3.4 *Elastic fields of multipoles in general anisotropic systems*

The elastic fields of force multipoles in general anisotropic systems can be found by employing the previous formalism, but with the derivatives of the Green's functions replaced by the corresponding derivatives valid for anisotropic systems given in Ch. 4 by Eqs. (4.40) and (4.42). As expected, the displacement fields due to force dipoles, quadrupoles and octopoles again fall off as x^{-2}, x^{-3} and x^{-4}, respectively. For example, in an anisotropic system the force dipole contribution to the displacement produced by the elementary double force in Fig. 10.3 is given by

$$u_i(\mathbf{x}) = -aF\frac{\partial G_{i3}(\mathbf{x})}{\partial x_3} = \frac{aF}{8\pi^2 x^2} \times \oint_{\hat{\mathcal{L}}}\left\{\hat{l}_3(\hat{k}\hat{k})^{-1}_{i3} - \hat{k}_3(\hat{k}\hat{k})^{-1}_{ip}[(\hat{k}\hat{l})_{pj} + (\hat{l}\hat{k})_{pj}](\hat{k}\hat{k})^{-1}_{j3}\right\} ds, \tag{10.26}$$

where the derivative, $\partial G_{i3}(\mathbf{x})/\partial x_3$, has been obtained from Eq. (4.40) after replacing $\hat{\mathbf{w}}$ with $\hat{\boldsymbol{l}}$.[8] The displacement falls off as x^{-2}, as expected, and has an angular dependence expressed as a function of $\hat{\boldsymbol{l}}$, just as in Eq. (10.19) for the corresponding isotropic system.

10.3.5 *The force dipole moment approximation*

The first order dipole term in Eq. (10.10) carries information about the long-range displacement field and is insensitive to the details of the local

[8]This follows from the fact that in formulating the Green's function for an infinite body the unit vector $\hat{\mathbf{w}}$, has been taken parallel to the field vector $\mathbf{x}$, which, in turn, is parallel to $\hat{\boldsymbol{l}}$ [see Fig. 4.2 and the formulation of Eq. (4.22)].

distribution of forces which mimic the defect as expected on the basis of St. Venant's principle. Conversely, the higher order shorter-ranged terms carry information about the near-field in the direct vicinity of the force multipole and are highly sensitive to the form of the force distribution. This is clearly seen, for example, in Eq. (10.21) for the displacement field due to the multipole in Fig. 10.4, consisting of three equal orthogonal double forces. Here, the dipole term is spherically symmetric, while the octopole term possesses the cubic symmetry of the force distribution. The higher order terms, obtained by the use of linear elasticity in the near vicinity of the core structures of point effects mimicked by multipoles, are generally expected to be relatively unreliable for quantitative purposes,[9] and, for many applications that depend primarily upon the longer range elastic field, it is convenient to simply omit them and adopt the *force dipole approximation* in which only the force dipole moment tensor and accompanying first derivatives of the Green's function are retained.

In this first order approximation all defects having the same dipole moment tensor therefore have the same elastic field even though their local force distributions may differ. The forces, $F_j^{(k)}$, and distances, $s_m^{(k)}$, appear linearly in the dipole moment tensor, and an infinite number of geometrically different force distributions can readily be constructed to produce the same tensor and therefore the same elastic field. For example, in the case of the double force in Fig. 10.3, whose dipole moment tensor components are given by Eq. (10.14), the distance a may be decreased and the force F increased in a manner to keep the aF product constant and the tensor unchanged.

The form of the force dipole moment tensor will, of course, depend upon the choice of coordinate system. However, it must always be symmetric because of Eq. (10.9). Its form will also be affected by the symmetry elements of the defect, since, according to Neumann's Principle:

> *"The symmetry elements of any physical property of a defect must include the symmetry elements of the defect's point group".*[10]

---

[9]Nevertheless, it should be noted that lattice models have shown that the higher order multipole terms reproduce, at least qualitatively, some of the main features of the local displacement field very near the defect core, e.g., Siems (1968).

[10]This must be true since any symmetry operation on a defect that produces an indistinguishable defect must also produce a corresponding indistinguishable physical property of the defect.

Table 10.1 Form of dipole moment tensor, P_{mj}, in conventional coordinate systems for defects belonging to different symmetry systems (Nye 1957).

Symmetry system	Characteristic symmetry	Conventional orthogonal coordinate system	Form of P_{mj}
Cubic	Four 3-fold axes	Arbitrary	$\begin{bmatrix} P & 0 & 0 \\ 0 & P & 0 \\ 0 & 0 & P \end{bmatrix}$
Tetragonal Hexagonal Trigonal	One 4-fold axis One 6-fold axis One 3-fold axis	x_3 along symmetry axis: x_1 and x_2 arbitrary	$\begin{bmatrix} P & 0 & 0 \\ 0 & P & 0 \\ 0 & 0 & P_{33} \end{bmatrix}$
Orthorhombic	Three orthogonal 2-fold axes	x_1, x_2 and x_3 along symmetry axes	$\begin{bmatrix} P_{11} & 0 & 0 \\ 0 & P_{22} & 0 \\ 0 & 0 & P_{33} \end{bmatrix}$
Monoclinic	One 2-fold axis	x_2 along symmetry axis	$\begin{bmatrix} P_{11} & 0 & P_{13} \\ 0 & P_{22} & 0 \\ P_{13} & 0 & P_{33} \end{bmatrix}$
Triclinic	Center of symmetry, or no center of symmetry	Arbitrary	$\begin{bmatrix} P_{11} & P_{12} & P_{13} \\ P_{12} & P_{22} & P_{23} \\ P_{13} & P_{23} & P_{33} \end{bmatrix}$

Nye (1957) has described how the form of a symmetrical second rank tensor representing a physical property depends on the symmetry elements when conventional choices for the orthogonal coordinate system are made. The results expected for the dipole moment tensor are shown in Table 10.1. Since P_{mj} is a second rank tensor, it can always be diagonalized regardless of the defect symmetry. As seen in Table 10.1, the conventional coordinate systems adopted are also principal coordinate systems except for those employed for the monoclinic and triclinic systems.

When the dipole moment tensor is expressed in its principal coordinate system, the displacement field given by Eq. (10.10) appears simply as

$$u_i(\mathbf{x}) = -\left[\frac{\partial G_{i1}(\mathbf{x})}{\partial x_1}P_{11} + \frac{\partial G_{i2}(\mathbf{x})}{\partial x_2}P_{22} + \frac{\partial G_{i3}(\mathbf{x})}{\partial x_3}P_{33}\right], \qquad (10.27)$$

and the force multipole is represented by three orthogonal double forces directed along the principal directions.

For certain problems it is useful to have an expression for the force density distribution that can be associated with a given dipole moment. This can be obtained by starting with a double force consisting of the forces $\mathbf{F}$ and $-\mathbf{F}$ acting at the vector positions $\mathbf{s}/2$ and $-\mathbf{s}/2$, respectively, with $\mathbf{F}$ parallel to $\mathbf{s}$. Its force density distribution can then be described by using delta functions in the usual way so that

$$\mathbf{f}(\mathbf{x}) = \mathbf{F}\left[\delta\left(\mathbf{x} - \frac{\mathbf{s}}{2}\right) - \delta\left(\mathbf{x} + \frac{\mathbf{s}}{2}\right)\right]. \tag{10.28}$$

The force dipole moment tensor corresponding to such a double force is given by

$$P^{(1)}_{mj} = s_m F_j \tag{10.29}$$

and therefore remains unchanged when the force components, F_j, are increased and the extension components, s_m, decreased while their products are maintained constant, as pointed out previously. Therefore, by making the s_m arbitrarily small, while keeping the dipole force tensor constant, and Taylor expanding the delta functions in Eq. (10.28) around $\mathbf{s} = 0$ to first order, the desired corresponding force density distribution is obtained in the form

$$\mathbf{f}(\mathbf{x}) = -s_m \mathbf{F} \frac{\partial \delta(\mathbf{x})}{\partial x_m} \quad f_j(\mathbf{x}) = -s_m F_j \frac{\partial \delta(\mathbf{x})}{\partial x_m} = -P_{mj} \frac{\partial \delta(\mathbf{x})}{\partial x_m}. \tag{10.30}$$

This expression is used in Ch. 11 in an analysis of point defects in finite bodies. Also, it is generalized in Exercise 10.3 to include higher order force moments.

10.4 Small Inclusion Model for Point Defect

As already mentioned in Sec. 10.1, point defects can also be modeled elastically as small inclusions with misfits and inhomogeneities that possess the same features as the inclusions treated in Ch. 6. In such cases the core region is replaced by a small fictitious continuum inclusion that possesses effective elastic constants that may differ from those of the matrix, and transformation strains that produce displacements that mimic those due to the point forces of the force multi-pole model. An example is the classic "ball-in-hole" model for a solute atom consisting of a sphere forced into an undersized (or oversized) spherical cavity in the matrix and then bonded (Christian 1975). This model corresponds to the spherical inhomogenous inclusion with the transformation strain $\varepsilon^T_{ij} = \varepsilon^T \delta_{ij}$ treated in Sec. 6.4.3.1. Such an inclusion model is, of course, useless for describing the detailed aspects of the point defect core, but it is useful in determining the elastic field in the surrounding matrix and its interactions with the stress fields of other defects. By adjusting the transformation strain and effective elastic constants in the fictitious inclusion, the elastic fields for corresponding point defects with various symmetries can be modeled. The small inclusion model and the force multipole model employing the force dipole approximation (Sec. 10.3.5) both yield stress fields that fall off as x^{-3}. As shown in Sec. 10.3.3.2, the elastic fields in the matrix caused by a spherically symmetric point defect simulated by a small inclusion, and by a force multipole consisting of three equal orthogonal double forces of appropriately chosen "strength" [see Eq. 10.23)], are identical when using the force dipole approximation.

It is worth noting that the first order approximations for the force exerted on an inclusion by an imposed elastic field gradient given, for example, by Eq. (7.13), should be especially acceptable in cases where a point defect is mimicked by a small inhomogeneous inclusion. Here, the accuracy of the required first order expansions benefits from the small size of the fictitious inclusion.

Exercises

10.1. Show that the expression for the force distribution due to a double force given by Eq. (10.30) can be used to obtain, directly, the first order term in Eq. (10.3) (which involves the dipole moment, P_{mj}).

Solution. Substituting Eq. (10.30) into Eq. (3.16),

$$u_i(\mathbf{x}) = -\oiiint_{\infty} G_{ij}(\mathbf{x}-\mathbf{x}')P_{mj}\frac{\partial\delta(\mathbf{x}')}{\partial x'_m}dV'. \tag{10.31}$$

To evaluate the derivative of the delta function in Eq. (10.31), the expression

$$\oiiint_{\infty} g(\mathbf{x})\frac{\partial\delta(\mathbf{x})}{\partial x_m}dV = -\left.\frac{\partial g(\mathbf{x})}{\partial x_m}\right|_{x_1=x_2=x_3=0} \tag{10.32}$$

is obtained using Eq. (D.4) (in three dimensions). Then, using this in (Eq. 10.31),

$$u_i(\mathbf{x}) = -\oiiint_{\infty} G_{ij}(\mathbf{x}-\mathbf{x}')P_{mj}\frac{\partial\delta(\mathbf{x}')}{\partial x'_m}dV' = \frac{\partial G_{ij}(\mathbf{x}-\mathbf{x}')}{\partial x'_m}P_{mj}. \tag{10.33}$$

10.2. Equation (10.21) gives the displacement field due to the force multipole illustrated in Fig. 10.4 in an infinite isotropic matrix. The first term, as expressed by Eq. (10.22), is known from Eq. (8.5) not to produce any local volume change (dilatation) in the medium around the multipole. Prove that the second term in Eq. (10.21), associated with the octopole moment tensor, behaves in the same manner.

Solution. Using Eq. (2.42), the dilatation due to the second term is

$$\begin{aligned} e(\mathbf{x}) &= -\frac{a^3F}{16\pi\mu(1-\nu)}\nabla\cdot\mathbf{e}_i\left[l_i(21-12\nu)\right. \\ &\quad \left.+20\nu l_i^3 - 35l_i(l_1^4+l_2^4+l_3^4)\right]\frac{1}{x^4} \\ &= -\frac{35a^3(1-2\nu)F}{8\pi\mu(1-\nu)}\left[(l_1^4+l_2^4+l_3^4)-\frac{3}{5}\right]\frac{1}{x^5}. \end{aligned} \qquad (10.34)$$

Then, integrating the dilatation over the region around the force multipole, the volume change is

$$\begin{aligned} \Delta V &= \oiiint_{\mathcal{V}} e dV = -\frac{35a^3(1-2\nu)F}{8\pi\mu(1-\nu)}\int_{r_\circ}^{\infty}\frac{dr}{r^3} \\ &\times \int_0^{2\pi}\int_0^{\pi}\left[\cos^4\theta\sin^4\phi + \cos^4\theta\sin^4\phi + \cos^4\phi - \frac{3}{5}\right] \\ &\times \sin\phi d\phi d\theta = 0, \end{aligned} \qquad (10.35)$$

where $r_\circ$ is a small radius around the force multipole. The integration has been carried out by switching to spherical coordinates and using spherical trigonometry to express the l_i direction cosines as functions of θ and ϕ in the forms $l_1 = \cos\theta\sin\phi$, $l_2 = \sin\theta\sin\phi$ and $l_3 = \cos\phi$.

10.3 We have derived an expression for the force density distribution associated with a force dipole moment in the form of Eq. (10.30). Now, find a general expression for the force density distribution associated with any higher order force moment. Hint: Rewrite Eq. (10.10) in the series form

$$\begin{aligned} u_i(\mathbf{x}) &= \sum_{n=1}^{\infty} u_i^{(n)}(\mathbf{x}) = \sum_{n=1}^{\infty}\frac{(-1)^n}{n!}P^{(n)}_{q_1\cdots q_n j}\frac{\partial^n G_{ij}(\mathbf{x})}{\partial x_{q_1\cdots q_n}} \\ &= \sum_{n=1}^{\infty}\phi_j^{(n)}G_{ij}(\mathbf{x}), \end{aligned} \qquad (10.36)$$

where $\phi_j^{(n)}$ is the operator

$$\phi_j^{(n)} = \frac{(-1)^n}{n!}P^{(n)}_{q_1\cdots q_n j}\frac{\partial^n}{\partial x_{q_1\cdots q_n}} \qquad (10.37)$$

and then apply $\phi_j^{(n)}$ to Eq. (3.21).

Solution Applying $\phi_j^{(n)}$ to Eq. (3.21),

$$\frac{(-1)^n}{n!} P^{(n)}_{q_1 \cdots q_n j} \frac{\partial^n}{\partial x_{q_1 \cdots q_n}} \left[C_{kpim} \frac{\partial^2 G_{ij}(\mathbf{x})}{\partial x_m \partial x_p} + \delta_{kj} \delta(\mathbf{x}) \right] = 0,$$

$$C_{kpim} \frac{\partial^2}{\partial x_m \partial x_p} \left[\frac{(-1)^n}{n!} P^{(n)}_{q_1 \cdots q_n j} \frac{\partial^n G_{ij}(\mathbf{x})}{\partial x_{q_1 \cdots q_n}} \right] + \frac{(-1)^n}{n!} P^{(n)}_{q_1 \cdots q_n k} \frac{\partial^n \delta(\mathbf{x})}{\partial x_{q_1 \cdots q_n}} = 0, \tag{10.38}$$

$$C_{kpim} \frac{\partial^2 u_i^{(n)}(\mathbf{x})}{\partial x_m \partial x_p} + \frac{(-1)^n}{n!} P^{(n)}_{q_1 \cdots q_n k} \frac{\partial^n \delta(\mathbf{x})}{\partial x_{q_1 \cdots q_n}} = 0.$$

Then, upon comparing Eq. (10.38) with Eq. (3.2), we must conclude that

$$\frac{(-1)^n}{n!} P^{(n)}_{q_1 \cdots q_n k} \frac{\partial^n \delta(\mathbf{x})}{\partial x_{q_1 \cdots q_n}} = f_k^{(n)}(\mathbf{x}). \tag{10.39}$$

Note that for a force dipole moment $n = 1$, and Eq. (10.39) agrees with Eq. (10.30).

Chapter 11

Interactions between Point Defects and Stress: Point Defects in Finite Regions

11.1 Introduction

The force multipole and small inclusion models for point defects in infinite homogeneous regions have been described in Ch. 10. The interactions of inclusions with various types of stresses have been treated in Ch. 7, and their behavior in finite and semi-infinite regions has been analyzed in Ch. 8. Consequently, there is no need to devote further attention to interactions between point defects and stress in terms of the small inclusion model. Therefore, in this chapter we employ the force multipole model to analyze the interaction between point defects and stress, and also the behavior of both single point defects and distributions of point defects in finite regions.

Section 11.2 includes a treatment of the interaction between a single point defect (represented by a force multipole) and a general internal or applied stress. In Sec. 11.3 the force multipole model is used to investigate the volume change due to a single point defect in a finite body possessing a traction-free surface where the defect image stress can play an important role. Then, with Sec. 11.3 in hand, Sec. 11.4 takes up the particularly interesting problem of the behavior of a finite traction-free body filled with a statistically uniform distribution of point defects which, for example, may be vacancies in thermal equilibrium or solute atoms dispersed throughout a solid solution. Analyses are given of the volume changes, macroscopic shape changes and lattice parameter changes (as measured by x-ray diffraction) produced by the defects. A demonstration is given of the intuitive result that a uniform concentration of point defects in a finite body with a traction-free surface produces uniform average strains throughout the body. If the defects act as spherically symmetric centers of pure dilatation, the macroscopic body either expands or contracts uniformly throughout the body (depending upon whether the centers possess positive or negative strengths) with no change in body shape. If the centers possess lower symmetry, the body again expands or contracts uniformly but undergoes a macroscopic shape change reflecting the symmetry of the defects.

11.2 Interaction Between a Point Defect (Multipole) and Stress

Equations (5.16) and (5.28) give the interaction energy between a defect, D, represented by a force multipole, and internal and applied elastic fields, respectively. For a defect force multipole in a general stress field, Q, the corresponding equation is then

$$E_{int}^{D/Q} = -\sum_q F_j^{D(q)} u_j^Q(\mathbf{s}^{D(q)}), \tag{11.1}$$

where the center of the defect is at the origin, and $\mathbf{s}^{D(q)}$ is the vector to the force $\mathbf{F}^{D(q)}$ associated with the defect force multipole. Then, expanding $u_i^Q(\mathbf{s}^{D(q)})$ around the origin to third order,

$$E_{int}^{D/Q} = -\sum_q \left[u_j^Q(0) + \frac{\partial u_j^Q}{\partial x_m} s_m^{D(q)} + \frac{1}{2!}\frac{\partial^2 u_j^Q}{\partial x_m \partial x_n} s_m^{D(q)} s_n^{D(q)} + \frac{1}{3!}\frac{\partial^3 u_j^Q}{\partial x_m \partial x_n \partial x_r} s_m^{D(q)} s_n^{D(q)} s_r^{D(q)} + \cdots \right] F_j^{D(q)} \tag{11.2}$$

and substituting the multipole quantities from Eqs. (10.4)–(10.7) into Eq. (11.2),

$$E_{int}^{D/Q} = -\left(P_{mj}^D \frac{\partial u_j^Q}{\partial x_m} + \frac{1}{2!} P_{mnj}^D \frac{\partial^2 u_j^Q}{\partial x_m \partial x_n} + \frac{1}{3!} P_{mnrj}^D \frac{\partial^3 u_j^Q}{\partial x_m \partial x_n \partial x_r} + \cdots \right). \tag{11.3}$$

If the elastic field is uniform, only the leading force dipole term contributes, and using Eq. (2.5) and the symmetry property, $P_{mj} = P_{jm}$, given by Eq. (10.9),

$$E_{int}^{D/Q} = -\varepsilon_{mj}^Q P_{mj}^D = -P_{mj}^D S_{mjik} \sigma_{ik}^Q. \tag{11.4}$$

If a field gradient is present, the corresponding force on the defect given by Eqs. (5.40), (11.3) and (2.5) takes the form

$$\mathrm{F}_{l}^{\mathrm{D/Q}} = -\frac{\partial \mathrm{E}_{\mathrm{int}}^{\mathrm{D/Q}}}{\partial \xi_{l}} = -\frac{\partial \mathrm{E}_{\mathrm{int}}^{\mathrm{D/Q}}}{\partial \mathrm{x}_{l}} = \mathrm{P}_{\mathrm{mj}}^{\mathrm{D}} \frac{\partial \varepsilon_{\mathrm{mj}}^{\mathrm{Q}}}{\partial \mathrm{x}_{l}} + \frac{1}{2!} \mathrm{P}_{\mathrm{mnj}}^{\mathrm{D}} \frac{\partial^{3} \mathrm{u}_{\mathrm{j}}^{\mathrm{Q}}}{\partial \mathrm{x}_{l} \partial \mathrm{x}_{\mathrm{m}} \partial \mathrm{x}_{\mathrm{n}}}$$

$$+ \frac{1}{3!} \mathrm{P}_{\mathrm{mnrj}}^{\mathrm{D}} \frac{\partial^{4} \mathrm{u}_{\mathrm{j}}^{\mathrm{Q}}}{\partial \mathrm{x}_{l} \partial \mathrm{x}_{\mathrm{m}} \partial \mathrm{x}_{\mathrm{n}} \partial \mathrm{x}_{\mathrm{r}}} + \cdots . \tag{11.5}$$

For the particular case where the defect is represented by the force multipole illustrated in Fig. 10.4 (corresponding to a center of dilatation) the only non-vanishing force dipoles are $\mathrm{P}_{11}^{\mathrm{D}} = \mathrm{P}_{22}^{\mathrm{D}} = \mathrm{P}_{33}^{\mathrm{D}} = 2\mathrm{aF}$, since the quadrupole force moments vanish because of the defect inversion symmetry. The force imposed on the defect in the presence of a stress gradient, given by Eq. (11.5) after dropping higher order terms, is then

$$\mathrm{F}_{l}^{\mathrm{D/Q}} = \mathrm{P}_{\mathrm{mj}}^{\mathrm{D}} \frac{\partial \varepsilon_{\mathrm{mj}}^{\mathrm{Q}}}{\partial \mathrm{x}_{l}} = 2\mathrm{aFS}_{\mathrm{mmik}} \frac{\partial \sigma_{\mathrm{ik}}^{\mathrm{Q}}}{\partial \mathrm{x}_{l}}. \tag{11.6}$$

If the stress is hydrostatic so that $\sigma_{\mathrm{ik}}^{\mathrm{Q}} = -\mathrm{P}^{\mathrm{Q}} \delta_{\mathrm{ik}}$, the above force reduces to

$$\mathrm{F}_{l}^{\mathrm{D/Q}} = -2\mathrm{aFS}_{\mathrm{mmkk}} \frac{\partial \mathrm{P}^{\mathrm{Q}}}{\partial \mathrm{x}_{l}}. \tag{11.7}$$

However, as shown below in Sec. 11.3 by Eq. (11.16), the quantity $2\mathrm{aFS}_{\mathrm{iikk}}$ in Eq. (11.7) is equal to the volume change caused by the force multipole acting as a center of dilatation, and, therefore, substituting this into Eq. (11.7),

$$\mathrm{F}_{l}^{\mathrm{D/Q}} = -\Delta \mathrm{V}^{\mathrm{D}} \frac{\partial \mathrm{P}^{\mathrm{Q}}}{\partial \mathrm{x}_{l}}. \tag{11.8}$$

This result is seen to be is similar in form to Eq. (7.13) obtained previously for the force exerted by the hydrostatic part of the imposed stress field on a spherical homogeneous inclusion acting as a center of dilatation with the transformation strain $\varepsilon_{\mathrm{ij}}^{\mathrm{T}} = \varepsilon^{\mathrm{T}} \delta_{\mathrm{ij}}$.

11.3 Volume Change of Finite Body Due to Single Point Defect

The point defect is again mimicked by an array of point forces, and we wish to find the volume change $\Delta \mathrm{V}^{\mathrm{D}}$ that it produces in a finite body $\mathcal{V}^{\circ}$

possessing a traction-free surface. The volume change ΔV^F of the body due to a single point force $\mathbf{F}$ located at the vector position $\mathbf{s}$ is therefore found first, and the total volume change is then obtained by summing the contributions made by all the point forces.

If $\mathbf{u}$ is the displacement field in the body due to the point force, $\mathbf{F}$, the volume change that it produces is

$$\Delta V^F = \oiiint_{\mathcal{V}^\circ} \nabla \cdot \mathbf{u} dV = \oiiint_{\mathcal{V}^\circ} e dV = S_{iikl} \oiiint_{\mathcal{V}^\circ} \sigma_{kl} dV \tag{11.9}$$

after using Eq. (2.93). The last integral in Eq. (11.9) can be evaluated by starting with the equation of equilibrium, Eq. (2.65),

$$\frac{\partial \sigma_{ik}(\mathbf{x})}{\partial x_k} + F_i \delta(\mathbf{x} - \mathbf{s}) = 0, \tag{11.10}$$

where $F_i \delta(\mathbf{x} - \mathbf{s})$ is the force density distribution representing the force, and $\sigma_{ik}(\mathbf{x})$ is the stress that it produces. Then, multiplying Eq. (11.10) throughout by the distance x_m and integrating it by parts over $\mathcal{V}^\circ$,

$$\oiiint_{\mathcal{V}^\circ} \left[x_m \frac{\partial \sigma_{ik}}{\partial x_k} + x_m F_i \delta(\mathbf{x} - \mathbf{s}) \right] dV = \oiiint_{\mathcal{V}^\circ} \frac{\partial}{\partial x_k}(x_m \sigma_{ik}) dV - \oiiint_{\mathcal{V}^\circ} \sigma_{im} dV + \oiiint_{\mathcal{V}^\circ} x_m F_i \delta(\mathbf{x} - \mathbf{s}) dV = 0. \tag{11.11}$$

However,

$$\oiiint_{\mathcal{V}^\circ} \frac{\partial}{\partial x_k}(x_m \sigma_{ik}) dV = \oiint_{\mathcal{S}^\circ} x_m \sigma_{ik} \hat{n}_k dS = 0 \tag{11.12}$$

as a result of the divergence theorem and the traction-free surface condition, $\sigma_{ik} \hat{n}_k = 0$. Therefore, substituting Eq. (11.12) into Eq. (11.11) and employing the properties of the delta function,

$$\oiiint_{\mathcal{V}^\circ} \sigma_{im}(\mathbf{x}) dV = \oiiint_{\mathcal{V}^\circ} x_m F_i \delta(\mathbf{x} - \mathbf{s}) dV = s_m F_i. \tag{11.13}$$

The volume change produced by the point force is then obtained by substituting Eq. (11.13) into Eq. (11.9) so that

$$\Delta V^F = S_{iikl} \oiiint_{\mathcal{V}^\circ} \sigma_{kl} dV = S_{iikl} s_l F_k. \tag{11.14}$$

Then, if N point forces are used to mimic the defect, the total volume change due to the defect is

$$\Delta V^{D} = S_{iikl} \sum_{q=1}^{N} s_{l}^{(q)} F_{k}^{(q)} = S_{iikl} P_{kl}, \tag{11.15}$$

where use has been made of Eq. 10.5. This remarkably simple result, proving that the volume change depends only upon the compliance tensor, S_{ijkl}, and the P_{ij} tensor of the defect, shows that the volume change is independent of the position of the defect and the size or shape of the body. Since the volume change due to the defect in an infinite body is a constant, the volume change due to the defect image stress must also be independent of the position of the defect and the body size or shape. This result may be compared with the closely related finding in Ch. 8 (see Eq. (8.12)) that the volume change of a finite body with a traction-free surface due to an inclusion, modeled as a center of dilatation in an isotropic system, is also independent of body size or shape.

In the case of the defect represented by the force multipole center of dilatation in Fig. 10.4, where the only non-zero components of P_{kl} are $P_{11} = P_{22} = P_{33} = 2aF$, the volume change is

$$\Delta V^{D} = S_{iikl} P_{kl} = S_{iikk} 2aF. \tag{11.16}$$

In Exercise 11.1 it is shown that Eq. (11.16), applied to an isotropic system, yields the same volume change for the defect as that produced by a small spherical inclusion of appropriately chosen 'strength', acting as a model for the defect instead of the force multipole. In Exercise 11.2 it is demonstrated that when tractions are applied to the surface of a finite body they constitute a force dipole and produce a volume change of the body given by an expression of the same form as Eq. (11.15). Also, see Exercise 11.3.

11.4 Statistically Uniform Distributions of Point Defects

11.4.1 *Defect-induced stress and volume change of finite body*

Having the above results for a single point defect, we can now find the change of volume, ΔV, of a finite homogeneous body, $\mathcal{V}^{\circ}$, of volume V, with a traction-free surface, produced by a statistically uniform distribution of N identical point defects. First, using the coordinates illustrated in

Fig. 10.2, the force density is determined at the field point $\mathbf{x}$ due to the defects distributed throughout $\mathcal{V}^\circ$ at the source points $\mathbf{x}'$. According to Eq. (10.30), there will be a force density at the field point $\mathbf{x}$ due to a defect located at $\mathbf{x}'$ possessing a dipole moment, P_{mj}, given by

$$f_j(\mathbf{x}-\mathbf{x}') = -P_{mj}\frac{\partial\delta(\mathbf{x}-\mathbf{x}')}{\partial(x_m - x'_m)}. \tag{11.17}$$

The average number of defects in any volume element dV' is $(N/V)dV'$, and the average increment of force density present at $\mathbf{x}$ due to these defects is therefore

$$d\langle f_j\rangle = \frac{N}{V}f_j(\mathbf{x}-\mathbf{x}')dV'. \tag{11.18}$$

Combining Eqs. (11.17) and (11.18), the average force density at $\mathbf{x}$ is then

$$\langle f_j(\mathbf{x})\rangle = -\frac{N}{V}P_{mj}\oiiint_{\mathcal{V}^\circ}\frac{\partial\delta(\mathbf{x}-\mathbf{x}')}{\partial(x_m - x'_m)}dx'_1dx'_2dx'_3. \tag{11.19}$$

Assume, for simplicity, that the body is a large rectangular prism of dimensions $2L_i$ with its edges aligned along the coordinate axes corresponding to the principal axes of the dipole tensor with the origin taken at the body center as in Fig. 11.1a. Then, using $dx'_i = -d(x_i - x'_i)$ and $\delta(\mathbf{x}-\mathbf{x}') = \delta(x_1 - x'_1)\delta(x_2 - x'_2)\delta(x_3 - x'_3)$, Eq. (11.19) assumes the form

$$\langle f_j(\mathbf{x})\rangle = \frac{N}{V}P_{1j}\int_{x_1+L_1}^{x_1-L_1}\frac{\partial\delta(x_1 - x'_1)}{\partial(x_1 - x'_1)}d(x_1 - x'_1)\int_{-L_2}^{L_2}\delta(x_2 - x'_2)dx'_2$$

$$\times\int_{-L_3}^{L_3}\delta(x_3 - x'_3)dx'_3 + \cdots. \tag{11.20}$$

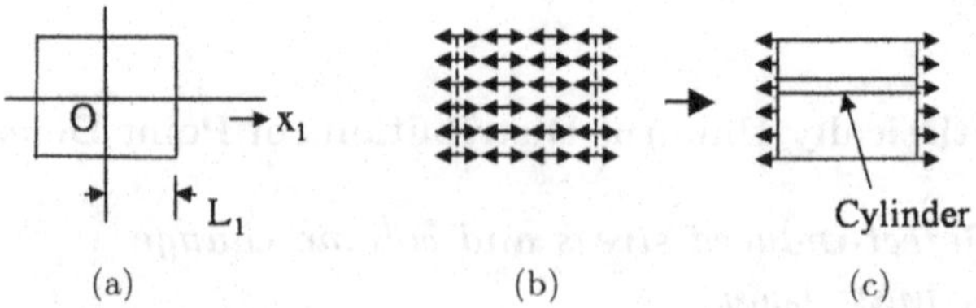

Fig. 11.1 (a) Cross-section of finite body in form of rectangular prism with dimensions $2L_i$. Coordinate axes correspond to principal axes of force dipole tensor. (b) Same body containing a uniform distribution of P_{11} type force dipoles. (c) Same body after mutual annihilation of all equal and opposite adjacent forces. The result is a residual distribution of force density on the $\pm L_1$ surfaces.

Therefore, using the properties of the delta functions,

$$\langle f_j(\mathbf{x})\rangle = \frac{N}{V}\left[P_{mj}\,|\delta(x_m - x'_m)|_{x_m+L_1}^{x_m-L_1}\right] = \frac{N}{V}P_{mj}[\delta(x_m - L_m) - \delta(x_m + L_m)] \tag{11.21}$$

which takes the form

$$\langle f_\alpha(x_\alpha)\rangle = \frac{N}{V}P_{\alpha\alpha}[\delta(x_\alpha - L_\alpha) - \delta(x_\alpha + L_\alpha)]. \quad (\alpha = 1,2,3) \tag{11.22}$$

The average force density therefore vanishes everywhere except at the body surfaces where it is present in the form of delta functions.[1] The physical interpretation of this result is that the force density associated with the force dipole moments distributed throughout the volume cancels out on average in the interior, but a residual force density is left at each surface in the form of a delta function owing to the asymmetry of the local environment at each surface as illustrated in Figs. 11.1b,c. It can now be shown that this residual force density produces an effective traction acting on each surface. The residual force at the $x_1 = \pm L_1$ surfaces is found by integrating the average force density given by Eq. (11.22) over the volume of the cylinder shown in Fig. 11.1c with the result

$$\begin{aligned} dF_1 &= dS\int_{cyl}\langle f_1\rangle dx_1 = \frac{N}{V}P_{11}dS\int_{cyl}[\delta(x_1 - L_1) - \delta(x_1 + L_1)]dx_1 \\ &= \left[\frac{N}{V}P_{11}dS\right]_{x_1=L_1} - \left[\frac{N}{V}P_{11}dS\right]_{x_1=-L_1}, \end{aligned} \tag{11.23}$$

where dS is the cross sectional area of the cylinder. The quantity NP_{11}/V can therefore be identified as an effective traction, dF_1/dS, acting at the surface in the direction x_1. Similar exercises show similar virtual tractions at the surfaces perpendicular to x_2 and x_3 due to the P_{22} and P_{33} components. These results are consistent with a picture in which the effective tractions at the surfaces cause the interior of the body to experience the average

[1] Of course, the actual force density does not vanish everywhere, since our model consists of a distribution of discrete force multipoles which do not cancel locally. Also, the spacings of the force multipoles will fluctuate locally.

uniform stresses

$$\langle\sigma_{11}\rangle = \frac{N}{V}P_{11}, \quad \langle\sigma_{22}\rangle = \frac{N}{V}P_{22}, \quad \langle\sigma_{33}\rangle = \frac{N}{V}P_{33}. \tag{11.24}$$

Using Eq. (2.93), the resulting average uniform normal strains throughout the body[2] are then

$$\begin{aligned}
\langle\varepsilon_{11}\rangle &= S_{1111}\langle\sigma_{11}\rangle + S_{1122}\langle\sigma_{22}\rangle + S_{1133}\langle\sigma_{33}\rangle \\
\langle\varepsilon_{22}\rangle &= S_{1122}\langle\sigma_{11}\rangle + S_{2222}\langle\sigma_{22}\rangle + S_{2233}\langle\sigma_{33}\rangle \\
\langle\varepsilon_{33}\rangle &= S_{1133}\langle\sigma_{11}\rangle + S_{2233}\langle\sigma_{22}\rangle + S_{3333}\langle\sigma_{33}\rangle
\end{aligned} \tag{11.25}$$

and the corresponding fractional volume change of the body is

$$\frac{\Delta V}{V} = \langle e\rangle = \langle\varepsilon_{ii}\rangle = \frac{N}{V}(S_{11ii}P_{11} + S_{22ii}P_{22} + S_{33ii}P_{33}). \tag{11.26}$$

The average strains given by Eq. (11.25) will produce a corresponding change in the macroscopic shape and size of the body. The macroscopic fractional changes in the dimensions of the body in the x_1, x_2 and x_3 directions are simply $\langle\varepsilon_{11}\rangle$, $\langle\varepsilon_{22}\rangle$ and $\langle\varepsilon_{33}\rangle$, respectively.

According to Eq. (11.26), the volume change per defect, $\Delta V^D = \Delta V/N$, is just $(S_{11ii}P_{11} + S_{22ii}P_{22} + S_{33ii}P_{33})$ which is consistent with the volume change per defect predicted earlier by Eq. (11.15). Also, using Eq. (11.13), the stress σ_{11} produced by a single defect and averaged over the volume is $(1/V)\int_{\mathcal{V}^\circ}\sigma_{11}dV = P_{11}/V$. Then, multiplying this by N to find the average stress produced by N defects, $\langle\sigma_{11}\rangle = NP_{11}/V$, in agreement with Eq. (11.24).

In summary, a finite homogeneous body with traction-free surfaces and filled with a statistically uniform distribution of point defects, acting as force multipoles, experiences a uniform average internal stress and strain throughout its volume given by Eqs. (11.24) and (11.25), respectively, and a volume change given by Eq. (11.26). The average strains throughout the body cause a change in its macroscopic shape that depends upon the symmetry of the defects and can be readily determined using the average strains.[3]

[2]The actual stress and strain fields, of course, are not uniform on a local scale, since, over most of the body, they consist of the overlapping far-fields of the distributed discrete force-multipoles, and each far-field varies rapidly with distance from its force multipole.

[3]This problem has had a tangled early history as discussed by Eshelby (1954).

It should be realized that the quantity ΔV^D is the volume change of the crystal due to the elastic displacement field produced by the defect, and that this quantity generally differs from the total volume change of the crystal if the defect is introduced into the crystal at a source in the crystal such as a dislocation or interface. If, for example, a vacancy is thermally generated in a crystal during heating in order to maintain thermodynamic equilibrium, the number of atoms in the system remains constant, and an unoccupied substitutional site (i.e., a vacancy) must be created at a vacancy source such as a dislocation (Balluffi, Allen and Carter 2005). This causes a crystal expansion of one atomic volume, Ω, if no elastic displacement field around the vacancy is allowed. However, the elastic displacement described previously causes the additional volume change ΔV^D, so that the total volume change is

$$\Delta V^{tot} = \Omega + \Delta V^D. \tag{11.27}$$

On the other hand, if the defect is a self-interstitial created at an internal source, a substitutional site is destroyed, and the total volume change is

$$\Delta V^{tot} = -\Omega + \Delta V^D. \tag{11.28}$$

Similar considerations hold for the addition of a substitutional solute atom from an external source, and Eq. (11.27) therefore applies. However, if an interstitial atom is added from an external source, the number of substitutional sites remains constant, and the total crystal volume change is simply due to the elastic displacement field so that

$$\Delta V^{tot} = \Delta V^D. \tag{11.29}$$

11.4.2 *The* $\underline{\lambda}^{(p)}$ *tensor*

To describe conveniently the change of shape of a finite traction-free body filled with a statistically uniform distribution of point defects, Nowick and Heller (1963), Nowick and Berry (1972) and Schilling (1978) have employed the second rank tensor, $\underline{\boldsymbol{\lambda}}^{(p)}$, defined by

$$\lambda_{ij}^{(p)} = \frac{1}{X^{(p)}}\langle \varepsilon_{ij}^{(p)} \rangle = \frac{1}{X^{(p)}} S_{ijkl} \langle \sigma_{kl}^{(p)} \rangle, \tag{11.30}$$

where $X^{(p)}$ is the mole fraction of type-p defects, and is therefore the average homogeneous strain produced by such defects present at a density of unit mole fraction. The defects are assumed to be present in dilute solution, and, therefore, $X^{(p)} = N^{(p)}\Omega/V$. Then, for the case analyzed previously (Fig. 11.1),

$$\lambda_{ij}^{(p)} = \frac{1}{\Omega}\left(S_{ij11}P_{11}^{(p)} + S_{ij22}P_{22}^{(p)} + S_{ij33}P_{33}^{(p)}\right) \tag{11.31}$$

and Eqs. (11.24), (11.25) and (11.26) can then be written in the respective forms

$$\begin{bmatrix} \langle\sigma_{11}^{(p)}\rangle & 0 & 0 \\ 0 & \langle\sigma_{22}^{(p)}\rangle & 0 \\ 0 & 0 & \langle\sigma_{33}^{(p)}\rangle \end{bmatrix} = \frac{X^{(p)}}{\Omega}\begin{bmatrix} P_{11}^{(p)} & 0 & 0 \\ 0 & P_{22}^{(p)} & 0 \\ 0 & 0 & P_{33}^{(p)} \end{bmatrix}$$
$$\begin{bmatrix} \langle\varepsilon_{11}^{(p)}\rangle & 0 & 0 \\ 0 & \langle\varepsilon_{22}^{(p)}\rangle & 0 \\ 0 & 0 & \langle\varepsilon_{33}^{(p)}\rangle \end{bmatrix} = \begin{bmatrix} \lambda_{11}^{(p)} & 0 & 0 \\ 0 & \lambda_{22}^{(p)} & 0 \\ 0 & 0 & \lambda_{33}^{(p)} \end{bmatrix} \tag{11.32}$$

and

$$\frac{\Delta V^{(p)}}{V} = \langle\varepsilon_{ii}^{(p)}\rangle = X^{(p)}\lambda_{ii}^{(p)}. \tag{11.33}$$

If q distinguishable types of defect are present in the body, the total average strain that they will produce can be expressed simply by the sum

$$\langle\varepsilon_{ij}\rangle = \sum_{p=1}^{q} X^{(p)}\lambda_{ij}^{(p)}, \tag{11.34}$$

where all $\boldsymbol{\lambda}^{(p)}$ tensors must be referred to a common coordinate system.

11.4.3 *Defect-induced changes in x-ray lattice parameter*

As described in the previous section, a finite traction-free crystal containing a statistically uniform distribution of point defects is strained non-uniformly on a scale on the order of the defect spacing. The strain varies between adjacent defects and becomes larger as each defect is approached. However, the average macroscopic strain experienced by the entire crystal is uniform and described by the $\boldsymbol{\lambda}$ tensor. An argument is now given that the fractional change in lattice parameter caused by the defects, as measured by x-ray diffraction, must be the same as the corresponding fractional change in body dimensions, as measured by the $\boldsymbol{\lambda}$ tensor.

Consider the crystal before the introduction of the point defects as consisting of a perfect three-dimensional network where each network node corresponds to the position of a substitutional atom and whose macroscopic shape is therefore identical to that of the crystal. The number of nodes is equal to the initial number of substitutional atoms. Now introduce the defects while keeping the network and its number of nodes intact. If a vacancy is added, a substitutional atom is removed from its node and discarded: the node remains but is now occupied by a vacancy instead of the discarded atom. If a substitutional solute is added a host atom is removed from its node and discarded and is replaced with a solute atom taken from an external reservoir. If an interstitial atom is added it is taken from the reservoir and added at an interstice between nodes. The defects will cause local non-uniform strains throughout the network, but the average macroscopic strain will be described by the $\underline{\boldsymbol{\lambda}}$ tensor. The positions of the Bragg diffraction peaks produced by x-ray diffraction will be determined by the spacings of the average planes of the network which are determined by the average macroscopic strain as measured by the $\underline{\boldsymbol{\lambda}}$ tensor. On the other hand, the local non-uniform strains in the actual crystal will produce diffuse x-ray scattering away from the Bragg diffraction peaks and will not influence the positions of the Bragg peaks. The effects of adding the point defects are analogous to the effects that would be produced by heating the crystal. When a crystal is heated, the amplitudes of the atomic vibrations increase and local displacements from the perfect crystal lattice positions occur everywhere. However, the average macroscopic strain, i.e., the macroscopic thermal expansion, is uniform, and produces shifts of the Bragg diffraction peaks, while the local atomic displacements produce thermal diffuse scattering (James 1954).

The rate of change of the macroscopic lattice parameters of a crystal of volume V due to the addition of defects may be calculated when the $\underline{\boldsymbol{\lambda}}$ tensor is known. For example, for a cubic crystal containing one type of defect with cubic symmetry the principal values of the $\underline{\boldsymbol{\lambda}}$ tensor will all be equal and the lattice parameter changes will be isotropic. Setting $\lambda = \lambda_{11} = \lambda_{22} = \lambda_{33}$, the rate of change will then be

$$\frac{1}{a}\frac{da}{dX} = \frac{1}{3V}\frac{dV}{dX} = \frac{1}{3}\frac{d(\langle\varepsilon_{11}\rangle + \langle\varepsilon_{22}\rangle + \langle\varepsilon_{33}\rangle)}{dX} = \frac{1}{3}(3\lambda) = \lambda. \quad (11.35)$$

If the defects have tetragonal symmetry and they are all oriented with their main symmetry axis aligned along the x_3 axis of the original cubic crystal, the principal values of the $\underline{\boldsymbol{\lambda}}$ tensor will be related by $\lambda_{11} = \lambda_{22} \neq \lambda_{33}$. The rate of lattice parameter change measured along either x_1 or x_2 and along

x_3 will then be, respectively,

$$\frac{1}{a_1}\frac{da_1}{dX}=\frac{1}{a_2}\frac{da_2}{dX}=\frac{d\langle\varepsilon_{11}\rangle}{dX}=\frac{d\langle\varepsilon_{22}\rangle}{dx}=\lambda_{11}=\lambda_{22},\quad \frac{1}{a_3}\frac{da_3}{dX}=\frac{d\langle\varepsilon_{33}\rangle}{dX}=\lambda_{33}. \tag{11.36}$$

The above equations are all forms of *Vegard's law* which states that the lattice parameters of dilute solid solutions vary linearly with the concentrations of solute atoms and other point defects.

The above phenomena provide a basis for determining the concentrations of point defects in thermal equilibrium in crystals. Consider a crystal with cubic crystal symmetry in the form of a macroscopic cube of edge length L whose equilibrium point defects are vacancies. If such a specimen, initially at a relatively low temperature where its equilibrium concentration of vacancies is negligible, is heated to an elevated temperature and held there, its equilibrium population of vacancies at the elevated temperature will be generated spontaneously at vacancy sources such as climbing dislocations (Seidman and Balluffi 1965). The principal values of the vacancy $\underset{\sim}{\lambda}$ tensor will all be equal, and setting them equal to λ, the change in the x-ray lattice parameter of the crystal due to the displacement fields around the vacancies that form and the thermal expansion of the lattice due to the heating will be, according to Eq. (11.35),

$$\frac{\Delta a}{a}=\lambda X^{eq}+\left(\frac{\Delta a}{a}\right)^{therm}, \tag{11.37}$$

where X^{eq} is the equilibrium mole fraction of vacancies, and $(\Delta a/a)^{therm}$ is the contribution of the thermal expansion. The corresponding fractional change in the macroscopic dimensions of the crystal, $\Delta L/L$, will be the sum of three terms, i.e.,

$$\frac{\Delta L}{L}=\frac{1}{3}\frac{N^{eq}\Omega}{V}+\lambda X^{eq}+\left(\frac{\Delta L}{L}\right)^{therm}=\frac{X^{eq}}{3}+\lambda X^{eq}+\left(\frac{\Delta L}{L}\right)^{therm}, \tag{11.38}$$

where N^{eq} is the number of vacancies created. The first term is due to the addition of the N^{eq} substitutional sites required to accommodate the N^{eq} newly created vacancies. Note that this expansion, in which N^{eq} atomic volumes are simply added to the volume, is not detected by means of x-ray diffraction, since no changes in lattice parameter are involved. In writing this term it is assumed that the creation of the N^{eq} sites at the climbing dislocation sources produces an isotropic expansion. The second term is the change in macroscopic dimension, L, due to the displacement fields of the vacancies, which, as already discussed, must be identical to the corresponding change in the x-ray lattice parameter given by Eq. (11.35). The third term is due to thermal expansion. Taking the difference, $(\Delta L/L - \Delta a/a)$, and noting that thermal expansions of the lattice parameter and macroscopic dimensions are equal, the simple result

$$X^{eq} = 3\left(\frac{\Delta L}{L} - \frac{\Delta a}{a}\right) \tag{11.39}$$

is obtained. Measurement of the difference $(\Delta L/L - \Delta a/a)$ therefore allows the determination of X^{eq} (Simmons and Balluffi 1960; Hehenkamp 1994).

Exercises

11.1. The volume change, ΔV^D, produced by the force dipole moments, 2aF, of the multipole center of dilatation in Fig. 10.4 in a homogeneous anisotropic body with a traction-free surface is given by Eq. (11.16). Suppose that the body is now isotropic. Show that Eq. (11.16) then predicts a volume change due to the multipole that agrees with the volume change predicted by Eq. (8.12) for a spherical homogeneous inclusion in the body when the strength of the inclusion (as measured by $c = c_1^M$) is related to the strength of the multipole (as measured by 2aF) by the relationship given by Eq. (10.23) which ensures that the far-displacement fields of the two defects match.

Solution. Using the relationships in Sec. 2.4.4, convert Eq. (11.16) to its form for an isotropic system to obtain

$$\Delta V^D = S_{iikk}2aF = 3(S_{11} + 2S_{12})2aF = \frac{3(1-2\nu)}{2\mu(1+\nu)}2aF \tag{11.40}$$

for the volume change due to the multipole in the isotropic body.

The volume change due to the spherical inclusion, given by Eq. (8.12), is

$$\Delta V^D = 4\pi c\frac{(3K + 4\mu)}{3K} = 4\pi c\frac{3(1-\nu)}{1+\nu}. \tag{11.41}$$

Then, substituting for c from Eq. (10.23),

$$\Delta V^{D} = 4\pi\left[\frac{aF(1-2\nu)}{4\pi\mu(1-\nu)}\right]\frac{3(1-\nu)}{(1+\nu)} = \frac{3(1-2\nu)}{2\mu(1+\nu)}2aF \quad (11.42)$$

in agreement with Eq. (11.40).

11.2. Equation (11.15) gives the volume change of a finite traction-free body caused by a defect in the body mimicked by the force dipole moment tensor P_{si}. Show that an equation of the form of Eq. (11.15) also gives the volume change of a body in cases where no defects are present but, instead, tractions due to applied forces are acting on the body surface, and therefore producing a force dipole.

Solution. The tractions will produce stresses throughout the body, σ'_{ik}, which, in the absence of applied forces within the body, must obey the equilibrium equation

$$\frac{\partial\sigma'_{ik}}{\partial x_k} = 0. \quad (11.43)$$

Following the same procedure used to obtain Eqs. (11.11),

$$\oiiint_{\mathcal{V}^\circ} x_m\frac{\partial\sigma'_{ik}}{\partial x_k}dV = \oiiint_{\mathcal{V}^\circ}\frac{\partial}{\partial x_k}(x_m\sigma'_{ik})dV - \oiiint_{\mathcal{V}^\circ}\sigma'_{im}dV = 0. \quad (11.44)$$

Then, by applying the divergence theorem,

$$\oiiint_{\mathcal{V}^\circ}\sigma'_{im}dV = \oiint_{\mathcal{S}^\circ} x_m\sigma'_{ik}\hat{n}_k dS = \oiint_{\mathcal{S}^\circ} x_m T'_i dS$$
$$= \oiint_{\mathcal{S}^\circ} x_m df'_i = P'_{mi}. \quad (11.45)$$

Here, the forces acting at the surface df'_i, due to the tractions T'_i, integrated over the surface, generate a force dipole, i.e., P'_{mi}. Then, following the same procedure that led to Eq. (11.15), the corresponding volume change is found to be

$$\Delta V' = S_{iikl}P'_{kl}. \quad (11.46)$$

See Exercise 11.6 for an application of this method.

11.3. A point defect, D, mimicked by a force dipole moment, is located in an arbitrary position in a finite body of arbitrary shape with a traction-free surface, $\mathcal{S}$. Formulate a set of equations sufficient to determine the volume change of the body, ΔV, caused by the defect.

Solution. Start by writing the volume change as $\Delta V = \Delta V^{\infty} + \Delta V^{IM}$, as in Eq. (8.12), and consider ΔV^{∞} first. Using Eq. (10.10), the displacement field throughout an infinite body containing the defect is

$$u_i^{\infty}(\mathbf{x}) = -\frac{\partial G_{ij}^{\infty}(\mathbf{x})}{\partial x_k} P_{kj}^{D}, \tag{11.47}$$

where the Green's function derivative is given by Eq. (4.40), and P_{kj}^{D} is the dipole force moment of the defect. The volume change of the region enclosed by $\mathcal{S}$, ΔV^{∞}, is then

$$\Delta V^{\infty} = \oiint_{\mathcal{S}} \mathbf{u}^{\infty} \cdot \hat{\mathbf{n}} dS = \oiint_{\mathcal{S}} u_i^{\infty} \hat{n}_i dS = -\oiint_{\mathcal{S}} \frac{\partial G_{ij}^{\infty}(\mathbf{x})}{\partial x_k} P_{kj}^{D} \hat{n}_i dS. \tag{11.48}$$

Next, ΔV^{IM} can be obtained by using Eq. (11.46), i.e.,

$$\Delta V^{IM} = S_{jjkl} P_{kl}^{IM}, \tag{11.49}$$

where P_{kl}^{IM} is the force dipole moment produced by the traction forces that must be applied to $\mathcal{S}$ to make it traction-free. Using Eq. (11.45), this moment is given by

$$P_{kl}^{IM} = \oiint_{\mathcal{S}} x_k \sigma_{li}^{IM} \hat{n}_i dS. \tag{11.50}$$

The condition for $\mathcal{S}$ to be traction-free is

$$\sigma_{li}^{IM} \hat{n}_i = -\sigma_{li}^{\infty} \hat{n}_i \tag{11.51}$$

and, therefore, substituting this and Eq. (11.50) into Eq. (11.49),[4]

$$\Delta V^{IM} = -S_{jjkl} \oiint_{\mathcal{S}} x_k \sigma_{li}^{\infty} \hat{n}_i dS. \tag{11.52}$$

Finally, using Eqs. (3.1) and (11.47),

$$\sigma_{li}^{\infty} = C_{limn} \frac{\partial u_m^{\infty}}{\partial x_n} = -C_{limn} \frac{\partial^2 G_{mr}^{\infty}(\mathbf{x})}{\partial x_n \partial x_s} P_{sr}^{D} \tag{11.53}$$

[4]Note the similarity of Eq. (11.52) with our earlier Eq. (8.9) where it is seen that the introduction of the vector A_j is essentially equivalent to the present use of the force dipole moment produced by the image tractions.

and, substituting this into Eq. (11.52),

$$\Delta V^{IM} = S_{jjkl}C_{limn} \oiint_S x_k \frac{\partial^2 G^{\infty}_{mr}(\mathbf{x})}{\partial x_n \partial x_s} P^D_{sr}\hat{n}_i dS. \tag{11.54}$$

The second derivative of the Green's function is available via Eq. (4.42), and ΔV is therefore given by the sum of Eqs. (11.48) and (11.54) with all quantities known.

11.4. A finite traction-free body contains a distribution of interstitial tetragonal defects of the type shown in Fig. 10.3 in a BCC crystal which is mimicked by a single double force. As indicated in Fig. 10.1b, such defects can exist in three crystallographically equivalent sites, labeled 1, 2 and 3, in which the double force lies along x_1, x_2 and x_3, respectively. Find an expression for the change in shape of the body due to the introduction of the defects, i.e., the average defect-induced strains, when the fraction of the total number of defects that is of type p is $\xi^{(p)}$. Note that, according to Eq. (2.96), $S_{12\alpha\alpha} = S_{13\alpha\alpha} = S_{23\alpha\alpha} = 0$ for a cubic crystal when $\alpha = 1, 2, 3$.

Solution. Using Eq. (11.31), the $\underline{\lambda}^{(p)}$ tensors for the three types of defect, referred to the coordinate system of Fig. 10.3, are

$$[\lambda^{(1)}] = \frac{aF}{\Omega}\begin{bmatrix} S_{1111} & 0 & 0 \\ 0 & S_{1122} & 0 \\ 0 & 0 & S_{1133} \end{bmatrix},$$

$$[\lambda^{(2)}] = \frac{aF}{\Omega}\begin{bmatrix} S_{2211} & 0 & 0 \\ 0 & S_{2222} & 0 \\ 0 & 0 & S_{2233} \end{bmatrix},$$

$$[\lambda^{(3)}] = \frac{aF}{\Omega}\begin{bmatrix} S_{3311} & 0 & 0 \\ 0 & S_{3322} & 0 \\ 0 & 0 & S_{3333} \end{bmatrix}. \tag{11.55}$$

From Eq. (11.34), the average defect-induced strain tensor is then

$$[\langle\varepsilon\rangle] = X^{(1)}[\lambda^{(1)}] + X^{(2)}[\lambda^{(2)}] + X^{(3)}[\lambda^{(3)}]$$

$$= \frac{aF}{\Omega}X^{tot}\left\{\xi^{(1)}\begin{bmatrix} S_{1111} & 0 & 0 \\ 0 & S_{1122} & 0 \\ 0 & 0 & S_{1133} \end{bmatrix}\right.$$

$$+\,\xi^{(2)}\begin{bmatrix} S_{2211} & 0 & 0 \\ 0 & S_{2222} & 0 \\ 0 & 0 & S_{2233} \end{bmatrix}$$

$$\left.+\,\xi^{(3)}\begin{bmatrix} S_{3311} & 0 & 0 \\ 0 & S_{3322} & 0 \\ 0 & 0 & S_{3333} \end{bmatrix}\right\}. \tag{11.56}$$

Note that when the three populations are equal, $\xi_1^{(1)} = \xi_2^{(2)} = \xi_3^{(3)} = 1/3$, and

$$[\langle\varepsilon\rangle] = \frac{aFX^{tot}}{3\Omega}\begin{bmatrix} S_{ii11} & 0 & 0 \\ 0 & S_{ii22} & 0 \\ 0 & 0 & S_{ii33} \end{bmatrix} \tag{11.57}$$

which is just the result expected for defects that are mimicked by three orthogonal double forces, such as in Fig. 10.4, and present at a concentration $X^{tot}/3$.

11.5. Imagine a finite traction-free crystal filled with a distribution of "split dumbbell" interstitial defects of the type shown in Fig. 10.5 with all dumbbell axes aligned along [001]. Find an expression for the macroscopic fractional elongation produced by these defects along the unit vector direction $\hat{\mathbf{q}}$.

Solution. Using Eqs. (10.16), (11.31) and (11.34), the average strain tensor in the coordinate system of Fig. 10.5 is

$$[\langle\varepsilon\rangle] = \frac{2aFX}{\Omega}\begin{bmatrix} \begin{matrix} S_{1111}\cos\theta \\ +S_{1122}\cos\theta \\ +2S_{1133}\sin\theta \end{matrix} & 0 & 0 \\ 0 & \begin{matrix} S_{2211}\cos\theta \\ +S_{2222}\cos\theta \\ +2S_{2233}\sin\theta \end{matrix} & 0 \\ 0 & 0 & \begin{matrix} S_{3311}\cos\theta \\ +S_{3322}\cos\theta \\ +2S_{3333}\sin\theta \end{matrix} \end{bmatrix}. \tag{11.58}$$

If a new (primed) coordinate system is chosen, with the x_3' axis parallel to $\hat{\mathbf{q}}$, the desired strain will be $\langle\varepsilon_{33}'\rangle$. Using Eq. (2.24) to transform to the new system, and employing the direction cosines,

$$[l] = \begin{bmatrix} l_{11} & l_{12} & l_{13} \\ l_{21} & l_{22} & l_{23} \\ l_{31} & l_{32} & l_{33} \end{bmatrix} = \begin{bmatrix} l_{13} & l_{12} & l_{13} \\ l_{13} & l_{13} & l_{13} \\ \hat{n}_1 & \hat{n}_2 & \hat{n}_3 \end{bmatrix} \tag{11.59}$$

the normal strain along $\hat{\mathbf{q}}$ is

$$\langle \varepsilon'_{33} \rangle = l_{3m} l_{3n} \langle \varepsilon_{mn} \rangle = \hat{q}_1^2 \langle \varepsilon_{11} \rangle + \hat{q}_2^2 \langle \varepsilon_{22} \rangle + \hat{q}_3^2 \langle \varepsilon_{33} \rangle, \tag{11.60}$$

where the $\langle \varepsilon_{\alpha\alpha} \rangle$ strains are given by Eq. (11.58).

11.6. Consider a cube with edge length L subjected to applied hydrostatic pressure, P. The resulting change in volume, using Eqs. (2.41) and (2.93) and $\sigma_{jk} = -P\delta_{jk}$, is then

$$\delta V = L^3 \varepsilon_{ii} = L^3 S_{iijk} \sigma_{jk} = -L^3 S_{iijk} P \delta_{jk} = -L^3 S_{iijj} P. \tag{11.61}$$

Obtain the same result by using the force dipole tensor approach developed in Exercise 11.2 and involving Eq. (11.46), i.e.,

$$\delta V = S_{iijk} P_{jk}, \tag{11.62}$$

where P_{jk} is the dipole moment tensor of the forces acting on the cube surface.

Solution Evaluate P_{jk} by using Eq. (10.5) and employing a $(\hat{\mathbf{e}}_1, \hat{\mathbf{e}}_2, \hat{\mathbf{e}}_3)$ coordinate system with its origin at the cube center, and its axes parallel to the cube edges. The P_{11} component (due to the forces along $\hat{\mathbf{e}}_1$ acting on the $\hat{\mathbf{n}} = (1, 0, 0)$ and $(-1, 0, 0)$ surfaces) is then obtained by integrating the differential moments due to the tractions on these surfaces, i.e.,

$$P_{11} = -LP \int_{-L/2}^{L/2} dx_2 \int_{-L/2}^{L/2} dx_3 = -L^3 P. \tag{11.63}$$

Similarly, the P_{21} component is given by

$$P_{21} = (P - P) \int_{-L/2}^{L/2} x_2 dx_2 \int_{-L/2}^{L/2} dx_3 = 0. \tag{11.64}$$

Continuing in this fashion, it is seen that $P_{jk} = -L^3 P \delta_{jk}$. Substitution of this result into Eq. (11.62) then yields Eq. (11.61).

Chapter 12

Dislocations in Infinite Homogeneous Regions

12.1 Introduction

The chapter begins with a description of the geometrical features of dislocations, including the Burgers vector and tangent vector, and the conventions that must be followed to solve dislocation problems systematically. Then, the elastic fields and strain energies of a wide variety of dislocation configurations in infinite homogeneous regions are determined using a range of approaches.

The simplest cases, involving infinitely long straight dislocations ranging in character from pure edge to pure screw, are considered first. Next, smoothly curved loops are analyzed. Finally, the elastic fields and strain energies of segmented dislocation structures involving short straight segments are derived, and it is shown how these results can be employed to find the elastic fields and strain energies of a wide range of relatively complex dislocation configurations in two and three dimensions.

12.2 Geometrical Features

A differential dislocation segment is characterized by two features, i.e., its *tangent vector*, $\hat{\mathbf{t}}$, and its *Burgers vector*, $\mathbf{b}$. To establish this, consider Fig. 12.1 which shows the geometry for the construction of a general dislocation loop, C, in an initially perfect infinite crystal. The loop is constructed via the following steps:

(1) The locus of the loop is first marked out along the closed curve, C. A cut is then made along an arbitrary surface, Σ, terminating on C as indicated by the cap-like surface, Σ.

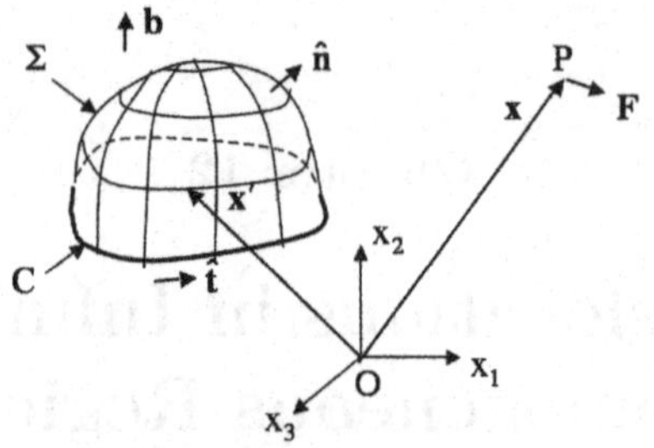

Fig. 12.1 Geometry for creation of dislocation loop, C, by cut and displacement at arbitrary surface, Σ. $\hat{\mathbf{t}}$ is the unit tangent vector to C, $\mathbf{b}$ is the Burgers vector, and $\hat{\mathbf{n}}$ is the positive unit vector normal to Σ. The conventions that determine the directions of these vectors are described in the text. Also, $\mathbf{x}$ is the field vector to the point P, and $\mathbf{x}'$ is the source vector impinging on the cut surface, Σ. $\mathbf{F}$ is a point force applied at P.

(2) The two sides of the cut are then displaced relative to one another everywhere by a constant displacement vector, i.e., the Burgers vector, $\mathbf{b}$. To accomplish this it will generally be necessary to eliminate overlaps and fill gaps by removing, or adding, material.
(3) The two surfaces are then bonded together to produce the final dislocation.

To reveal the effect of the changing direction of the dislocation line around the loop relative to the constant $\mathbf{b}$ vector, consider short segments of the loop, of length ds, at various points along C. A segment where $\hat{\mathbf{t}}$ is perpendicular to $\mathbf{b}$, illustrated in Fig. 12.2a, is termed an *edge dislocation* segment, since the edge of a sheet of extra material of thickness b, jammed into the body by the cut and displacement process, lies in the dislocation core. A segment where $\hat{\mathbf{t}}$ is parallel to $\mathbf{b}$, as in Fig. 12.2c, is termed a *right handed screw dislocation* segment, since an observer traversing circuits around the segment on planes perpendicular to the segment in a clockwise direction when looking along $\hat{\mathbf{t}}$ advances along $\hat{\mathbf{t}}$ in the manner of a right-handed screw.[1] Segments with $\hat{\mathbf{t}}$ vectors between these extremes possess intermediate structures and are termed *mixed dislocation* segments. By denoting the angle between $\mathbf{b}$ and $\hat{\mathbf{t}}$ by β, a mixed dislocation segment can be regarded as the superposition of an edge dislocation segment having a Burgers vector $\mathrm{b} \sin \beta$ (its Burgers vector component

[1]If the direction of the displacement in Fig. 12.2c were reversed, we would have a *left-handed screw dislocation* segment since the observer would advance along $-\hat{\mathbf{t}}$ in the manner of a left-hand screw.

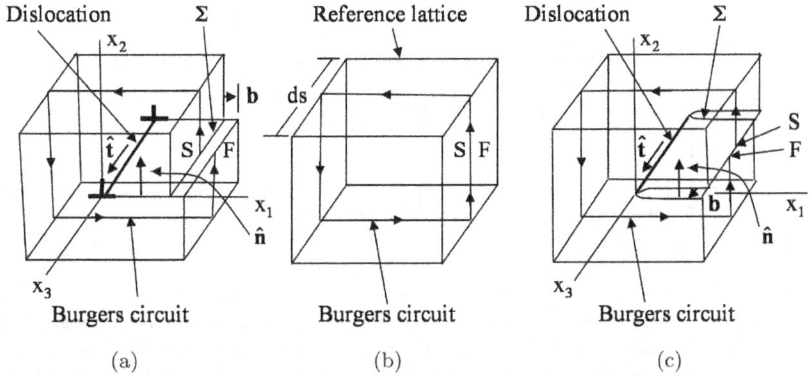

Fig. 12.2 (a) Edge dislocation segment of length ds with SF/RH Burgers circuit indicated. Also shown are the dislocation tangent vector, $\hat{\mathbf{t}}$, Burgers vector, $\mathbf{b}$, and positive unit normal vector, $\hat{\mathbf{n}}$, to Σ surface. (b) Reference lattice with same Burgers circuit as in (a). (c) Right-handed screw dislocation segment in a similar representation.

perpendicular to $\hat{\mathbf{t}}$) and a screw dislocation segment with Burgers vector $\mathrm{b}\cos\beta$ (its Burgers vector component parallel to $\hat{\mathbf{t}}$). This follows from the fact that such a segment can be produced by the cut and displacement method by imposing the displacements $\mathrm{b}\sin\beta$ and $\mathrm{b}\cos\beta$ in two successive operations.

To specify rigorously the vectors $\hat{\mathbf{t}}$, $\mathbf{b}$ and $\hat{\mathbf{n}}$, it is necessary to adopt a set of conventions. First, the positive direction along the dislocation, which is initially arbitrary, must be chosen. This establishes $\hat{\mathbf{t}}$, which is then defined as the tangent vector pointing in the positive direction. Next, to determine $\mathbf{b}$, a *Burgers circuit* is constructed using either of two alternative methods, i.e., the SF/RH or FS/RH methods (Hirth and Lothe 1982). In the SF/RH method a closed loop consisting of a series of discrete jumps between lattice points is constructed in a Reference lattice corresponding to the perfect crystal before the introduction of the dislocation as illustrated in Fig. 12.3a. This circuit is then mapped onto the crystal containing the dislocation so that it encircles the dislocation as in Fig. 12.3b. The circuit around the dislocation fails to close due to the presence of the dislocation, and if the circuit is traversed in a clockwise direction when looking along $\hat{\mathbf{t}}$, the following rule applies:

*SF/RH rule: the Burgers vector, **b**, is defined as the closure failure of a Burgers circuit of the type illustrated in Fig. 12.3b as measured by the vector from S to F.*

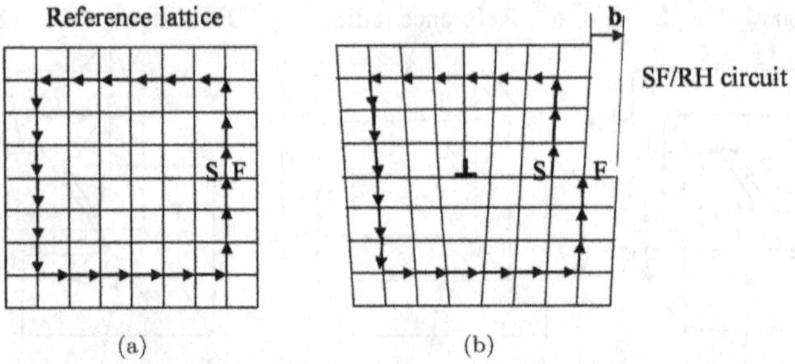

Fig. 12.3 (a) SF/RH Burgers circuit that is closed in the Reference lattice and starts at S and finishes at F. (b) Same circuit mapped on to real crystal to enclose an edge dislocation possessing tangent vector, $\hat{\mathbf{t}}$, pointing out of paper as in Fig. 12.2a. The circuit is clockwise when looking along $\hat{\mathbf{t}}$: the Burgers vector, $\mathbf{b}$, is given by closure failure of circuit using the SF/RH rule.

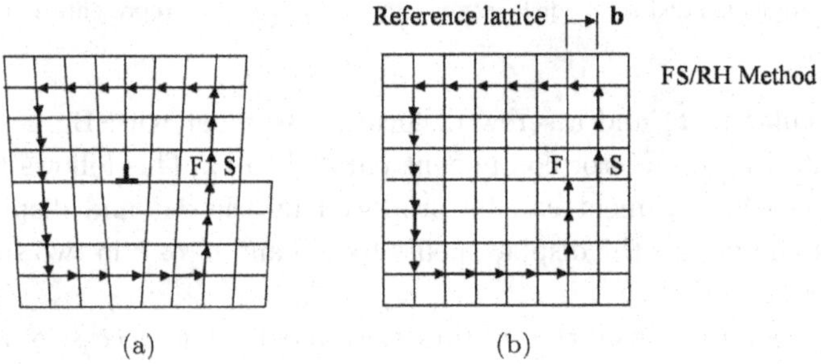

Fig. 12.4 (a) FS/RH Burgers circuit, that is closed in the crystal, and encloses an edge dislocation possessing tangent vector, $\hat{\mathbf{t}}$, pointing out of paper as in Fig. 12.2a: circuit is clockwise when looking along $\hat{\mathbf{t}}$. (b) Same circuit mapped onto Reference lattice: the Burgers vector, $\mathbf{b}$, is given by closure failure of circuit using the FS/RH rule.

Alternatively, in the FS/RH method, a closed circuit is first constructed around the dislocation in the crystal, as in Fig. 12.4a, and then mapped onto the Reference lattice where it will fail to close as seen in Fig. 12.4b. With this construction, the following rule applies:

> *FS/RH rule: the Burgers vector, **b**, is defined as the closure failure of a Burgers circuit of the type illustrated in Fig. 12.4b as measured by the vector from F to S.*

Both methods produce the same Burgers vector (except for small elastic distortions in the SF/RH case).[2] The FS/RH rule is generally preferred

[2]Note that reversing $\hat{\mathbf{t}}$ reverses the sign of $\mathbf{b}$.

when a dislocation is in a highly distorted region, such as a dislocation array, since the Burgers vector is then displayed as an unstrained lattice vector in the Reference lattice.

It now remains to state the rule determining the direction of $\hat{\mathbf{n}}$:

> *Direction of* $\hat{\mathbf{n}}$ *rule: first, if necessary, shrink the* Σ *surface down until it is completely bounded by the dislocation in the form of a closed loop. Then, if the loop is traversed in the direction of* $\hat{\mathbf{t}}$, $\hat{\mathbf{n}}$ *must point in the direction normal to the* Σ *surface which makes the traversal clockwise when sighting along* $\hat{\mathbf{n}}$.

Having defined $\hat{\mathbf{t}}$, $\mathbf{b}$ and $\hat{\mathbf{n}}$, the manner in which the two sides of the cut surface, Σ, are displaced when a dislocation is produced by the cut and displacement method is specified as follows *(Hirth and Lothe 1982)*:

> Σ *cut/displacement rule: to make the displacement at the cut, and be consistent with the SF/RH rule for* $\mathbf{b}$*, the medium on the negative side of* Σ *must be displaced with respect to the medium on the positive side by* $\mathbf{b}$*. The positive side is the side towards which* $\hat{\mathbf{n}}$ *points.*

Examples of this rule are evident in Figs. 12.2–12.4.

Invoking the above rules, the following expressions can now be written for the edge and screw Burgers vector components of a general dislocation in the respective vector forms

$$\mathbf{b}_\mathrm{e} = \hat{\mathbf{t}} \times (\mathbf{b} \times \hat{\mathbf{t}}), \tag{12.1}$$

and

$$\mathbf{b}_\mathrm{s} = (\mathbf{b} \cdot \hat{\mathbf{t}})\hat{\mathbf{t}}, \tag{12.2}$$

so that, as already discussed above,

$$\begin{aligned} \mathrm{b_e} &= \mathrm{b} \sin \beta, \\ \mathrm{b_s} &= \mathrm{b} \cos \beta. \end{aligned} \tag{12.3}$$

12.3 Infinitely Long Straight Dislocations and Lines of Force

The problem of finding the elastic field of an infinitely long straight dislocation that ranges in character from pure edge to pure screw is now taken up. The problem of deriving the field of an infinitely long straight *line of force* (which will be required later) is similar, and the dislocation and line of force problems are therefore treated in tandem.

12.3.1 *Elastic fields*

The elastic fields of both long straight dislocations and lines of force are invariant with respect to distance traveled along them, and so are two-dimensional. They can therefore be found by employing the integral formalism for two-dimensional problems introduced in Sec. 3.5.3. The source of most of the following material in Secs. 12.3.1 and 12.3.2 is the treatise of Bacon, Barnett and Scattergood (1979b).

With each entity taken parallel to the $\hat{\tau}$ axis of the $(\hat{m}, \hat{n}, \hat{\tau})$ coordinate system shown in Fig. 3.2,[3] Eq. (3.34) can be adopted as a general solution for the displacements and Eq. (3.42) taken as a general solution for the corresponding stress functions. Assuming that the function, $f(\hat{m}\cdot x+p_\alpha\hat{n}\cdot x)$, has a logarithmic form, these expressions take the forms

$$u_i = \frac{1}{2\pi i}\sum_{\alpha=1}^{6} A_{i\alpha}D_\alpha \ln(\hat{m}\cdot x + p_\alpha\hat{n}\cdot x) \tag{12.4}$$

$$\psi_j = \sum_{\alpha=1}^{6} L_{j\alpha}D_\alpha \ln(\hat{m}\cdot x + p_\alpha\hat{n}\cdot x) \tag{12.5}$$

which, however, are problematic since p_α is complex, and the logarithm is therefore multi-valued. However, this can be remedied by introducing the branch cut shown in Fig. 12.5a. With this in place, and with the complex

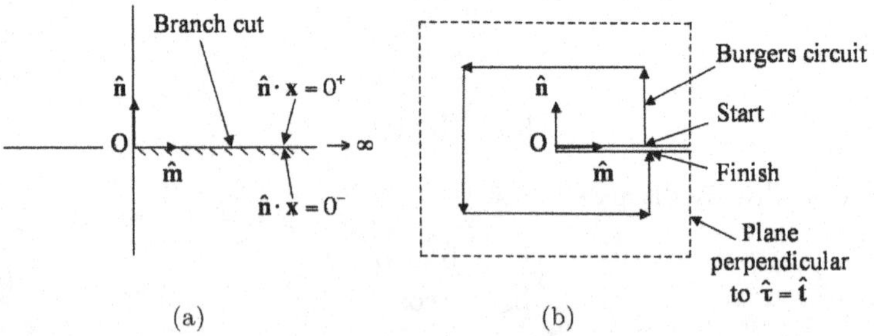

Fig. 12.5 (a) Branch cut used to make $\ln(\hat{m}\cdot x + p_\alpha\hat{n}\cdot x)$ single-valued. (b) Burgers circuit around infinitely long straight dislocation lying along $\hat{t} = \hat{\tau} = \hat{m}\times\hat{n}$ created by a cut and displacement in the plane $\hat{n}\cdot x = 0$. The Burgers circuit, in the plane perpendicular to $\hat{\tau}$ (dashed) must exhibit a closure failure equal to the Burgers vector.

[3]Since the $(\hat{m}, \hat{n}, \hat{\tau})$ base vectors are referred to the crystal coordinate system, the orientation of the dislocation in the crystal can be varied by simply rotating the $(\hat{m}, \hat{n}, \hat{\tau})$ system.

eigenvalues written as

$$\left.\begin{aligned} p_\alpha &= a_\alpha + i b_\alpha \\ p_{\alpha+3} &= a_\alpha - i b_\alpha \end{aligned}\right\} \quad (\alpha = 1, 2, 3) \tag{12.6}$$

where the a_α and b_α are real, the corresponding logarithms can be written in the single-valued form

$$\ln(\hat{\mathbf{m}} \cdot \mathbf{x} + p_\alpha \hat{\mathbf{n}} \cdot \mathbf{x}) = \ln|\hat{\mathbf{m}} \cdot \mathbf{x} + p_\alpha \hat{\mathbf{n}} \cdot \mathbf{x}| \pm i\theta,$$
$$\theta = \pm \tan^{-1} \frac{b_\alpha \hat{\mathbf{n}} \cdot \mathbf{x}}{\hat{\mathbf{m}} \cdot \mathbf{x} + a_\alpha \hat{\mathbf{n}} \cdot \mathbf{x}}, \tag{12.7}$$

where the upper sign in the $\pm$ notation applies when $\alpha = 1, 2, 3$ and the lower when $\alpha = 4, 5, 6$,[4] and θ is restricted to the range $0 \leq \theta < 2\pi$.

For a dislocation with $\hat{\mathbf{t}}$ parallel to $\hat{\boldsymbol{\tau}}$ as in Fig. 12.5b, the determination of the unknown constant in Eq. (12.4), D_α^{DIS}, is aided by noting that u_i must be of a form that produces a discontinuity in the displacement across the plane used in the cut and displacement on the surface Σ that produced the dislocation (the branch cut in this case) equal to the Burgers vector as illustrated in Fig. 12.5b. Using Eq. (12.7), it is seen that, when $\hat{\mathbf{m}} \cdot \mathbf{x} > 0$,

$$\begin{aligned} \ln(\hat{\mathbf{m}} \cdot \mathbf{x} + p_\alpha \hat{\mathbf{n}} \cdot \mathbf{x}) &\to \ln|\hat{\mathbf{m}} \cdot \mathbf{x}| \pm i0 \quad &\text{as } \hat{\mathbf{n}} \cdot \mathbf{x} \to 0^+, \\ \ln(\hat{\mathbf{m}} \cdot \mathbf{x} + p_\alpha \hat{\mathbf{n}} \cdot \mathbf{x}) &\to \ln|\hat{\mathbf{m}} \cdot \mathbf{x}| \pm i2\pi \quad &\text{as } \hat{\mathbf{n}} \cdot \mathbf{x} \to 0^-. \end{aligned} \tag{12.8}$$

According to the convention given by the Σ cut/displacement rule on p. 325, the negative side of the cut along the $\hat{\mathbf{n}} \cdot \mathbf{x} = 0$ plane is the 0^- side indicated in Fig. 12.5a. Using Eqs. (12.4) and (12.8), the discontinuity of displacement across the cut, which must equal the Burgers vector, is then

$$\begin{aligned} \Delta u_i^{DIS} &= \frac{1}{2\pi i} \left\{ \sum_{\alpha=1}^{3} A_{i\alpha} D_\alpha^{DIS} (\ln|\hat{\mathbf{m}} \cdot \mathbf{x}| + i2\pi) \right. \\ &\quad \left. + \sum_{\alpha=4}^{6} A_{i\alpha} D_\alpha^{DIS} (\ln|\hat{\mathbf{m}} \cdot \mathbf{x}| - i2\pi) - \sum_{\alpha=1}^{6} A_{i\alpha} D_\alpha^{DIS} \ln|\hat{\mathbf{m}} \cdot \mathbf{x}| \right\} \\ &= \sum_{\alpha=1}^{6} \pm A_{i\alpha} D_\alpha^{DIS} = b_i. \end{aligned} \tag{12.9}$$

For a line of force lying along $\hat{\boldsymbol{\tau}}$, and of force density f_i, a discontinuity in the function ψ_j, given by Eq. (3.44), must exist at the branch cut which is

[4] Note that this conforms to the notation introduced in Eq. (3.133).

of magnitude $\Delta\psi_j = -f_j$ according to Eq. (3.52) of Sec. 3.5.1.1. Therefore, using the same procedure as used above for the dislocation,

$$\sum_{\alpha=1}^{6} \pm L_{i\alpha} D_\alpha^{LF} = -f_i. \tag{12.10}$$

For the dislocation, D_α^{DIS} in Eq. (12.9) can now be found by noting that Eq. (12.9) has the same form as Eq. (3.72) when the arbitrary vector, $\mathbf{h}$, vanishes. Therefore, by replacing g_i with b_i and using Eq. (3.74),

$$D_\alpha^{DIS} = \pm b_s L_{s\alpha}. \tag{12.11}$$

Then, by substituting this result into Eq. (12.4), the displacement field of the dislocation is given (Bacon, Barnett and Scattergood 1979b) by

$$u_i^{DIS} = \frac{1}{2\pi i} \sum_{\alpha=1}^{6} \pm A_{i\alpha} L_{s\alpha} b_s \ln(\hat{\mathbf{m}} \cdot \mathbf{x} + p_\alpha \hat{\mathbf{n}} \cdot \mathbf{x}), \tag{12.12}$$

which is single-valued in the entire region bounding the cut, and suffers a discontinuity equal to b_i across the cut as required.

For the line force, for which Eq. (12.10) applies, the corresponding unknown quantity D_α^{LF} can be found by a similar method with the result

$$D_\alpha^{LF} = \mp f_s A_{s\alpha}. \tag{12.13}$$

Then, by substituting this into Eq. (12.4), the displacement field of the line force is

$$u_i^{LF} = -\frac{1}{2\pi i} \sum_{\alpha=1}^{6} \pm A_{i\alpha} A_{s\alpha} f_s \ln(\hat{\mathbf{m}} \cdot \mathbf{x} + p_\alpha \hat{\mathbf{n}} \cdot \mathbf{x}). \tag{12.14}$$

The elastic field of the dislocation, given by Eq. (12.12), can now be put into a form that does not require solving the Stroh eigenvalue problem. Differentiating Eq. (12.12),

$$\frac{\partial u_i^{DIS}(\mathbf{x})}{\partial x_p} = \frac{1}{2\pi i} \sum_{\alpha=1}^{6} \pm A_{i\alpha} L_{s\alpha} b_s \left(\frac{\hat{m}_p + p_\alpha \hat{n}_p}{\hat{\mathbf{m}} \cdot \mathbf{x} + p_\alpha \hat{\mathbf{n}} \cdot \mathbf{x}} \right). \tag{12.15}$$

Since $A_{i\alpha}$ and $L_{s\alpha}$ are independent of the choice of the angle ω in Fig. 3.2, $\hat{\mathbf{m}}$ can be aligned along $\mathbf{x}$ so that Eq. (12.15) becomes

$$\frac{\partial u_i^{DIS}(|\mathbf{x}|, \omega)}{\partial x_p} = \frac{b_s}{2\pi i |\mathbf{x}|} \left(\hat{m}_p \sum_{\alpha=1}^{6} \pm A_{i\alpha} L_{s\alpha} + \hat{n}_p \sum_{\alpha=1}^{6} \pm p_\alpha A_{i\alpha} L_{s\alpha} \right), \tag{12.16}$$

where it is recognized that u_i^{DIS} is a function of both the distance of the field point from the infinitely long dislocation, $|\mathbf{x}|$, and the angle, ω (since $\mathbf{x}$ is now aligned with $\hat{\mathbf{m}}$ whose direction is determined by ω). Next, an expression for $\sum_{\alpha=1}^{6} \pm p_\alpha A_{i\alpha} L_{s\alpha}$ is required. Multiplying Eq. (3.37) by $\pm L_{s\alpha}$, summing over α, and substituting Eqs. (3.134) and (3.135),

$$(\hat{n}\hat{n})_{jk} \sum_{\alpha=1}^{6} \pm p_\alpha A_{k\alpha} L_{s\alpha} = -\frac{1}{i}[4\pi B_{js} + (\hat{n}\hat{m})_{jk} S_{ks}] \tag{12.17}$$

so that

$$\sum_{\alpha=1}^{6} \pm p_\alpha A_{r\alpha} L_{s\alpha} = i(\hat{n}\hat{n})_{rj}^{-1}[4\pi B_{js} + (\hat{n}\hat{m})_{jk} S_{ks}]. \tag{12.18}$$

Then, substituting Eqs. (12.18) and (3.134) into Eq. (12.16),

$$\frac{\partial u_i^{DIS}(|\mathbf{x}|, \omega)}{\partial x_p} = \frac{b_s}{2\pi|\mathbf{x}|}\left\{-\hat{m}_p S_{is} + \hat{n}_p (\hat{n}\hat{n})_{ik}^{-1}[4\pi B_{ks} + (\hat{n}\hat{m})_{kr} S_{rs}]\right\}. \tag{12.19}$$

To obtain the displacement, u_i^{DIS}, by integrating Eq. (12.19), the general expression

$$du_i^{DIS}(|\mathbf{x}|, \omega) = \frac{\partial u_i^{DIS}}{\partial |\mathbf{x}|} d|\mathbf{x}| + \frac{\partial u_i^{DIS}}{\partial \omega} d\omega \tag{12.20}$$

is introduced. Then, using the following expressions

$$\frac{\partial u_i^{DIS}}{\partial |\mathbf{x}|} = \nabla u_i^{DIS} \cdot \hat{\mathbf{m}} = \hat{m}_p \frac{\partial u_i^{DIS}}{\partial x_p}$$

$$\frac{\partial u_i^{DIS}}{\partial \omega} = \nabla u_i^{DIS} \cdot \hat{\mathbf{n}}|\mathbf{x}| = |\mathbf{x}|\hat{n}_p \frac{\partial u_i^{DIS}}{\partial x_p} \tag{12.21}$$

for the partial derivatives in Eq. (12.20) and substituting Eq. (12.19) and (12.21) into Eq. (12.20),

$$du_i^{DIS} = \frac{b_s}{2\pi}\left\{-S_{is} d\ln|\mathbf{x}| + (\hat{n}\hat{n})_{ik}^{-1}[4\pi B_{ks} + (\hat{n}\hat{m})_{kr} S_{rs}] d\omega\right\}, \tag{12.22}$$

which can then be integrated to obtain (Bacon, Barnett and Scattergood 1979b)

$$u_i^{DIS}(|\mathbf{x}|, \omega) = \frac{b_s}{2\pi}\left[- S_{is} \ln|\mathbf{x}| + 4\pi B_{ks} \int (\hat{n}\hat{n})_{ik}^{-1} d\omega \right.$$

$$\left. + S_{rs} \int (\hat{n}\hat{n})_{ik}^{-1} (\hat{n}\hat{m})_{kr} d\omega \right], \tag{12.23}$$

where any constant of integration can be dropped, since it would represent a superfluous rigid body translation.

In Exercise 12.1 it is confirmed that if Eq. (12.23) is applied to a Burgers circuit around the dislocation, the resulting closure failure, Δu_i^{DIS}, is equal to the Burgers vector, as anticipated.

The dislocation stress field at the field vector $\mathbf{x} = |\mathbf{x}|\hat{\mathbf{m}}$, which makes the angle ω with the reference line, is readily obtained by use of Eqs. (3.1) and (12.19) in the form

$$\begin{aligned}\sigma_{mn}^{DIS}(|\mathbf{x}|, \omega) &= C_{mnip}\frac{\partial u_i^{DIS}}{\partial x_p} \\ &= \frac{b_s C_{mnip}}{2\pi|\mathbf{x}|}\left\{-\hat{m}_p S_{is} + \hat{n}_p(\hat{n}\hat{n})_{ik}^{-1}[4\pi B_{ks} + (\hat{n}\hat{m})_{kr}S_{rs}]\right\}.\end{aligned} \tag{12.24}$$

The traction caused by the dislocation stress field at the field point $\mathbf{x}$ on the plane containing both the dislocation and the field point, i.e., the plane perpendicular to the base vector $\hat{\mathbf{n}}$, is obtained using Eq. (12.24), i.e.,

$$\begin{aligned}T_n^{DIS}(|\mathbf{x}|, \omega) &= \sigma_{mn}^{DIS}\hat{n}_m = \hat{n}_m C_{mnip}\frac{\partial u_i^{DIS}}{\partial x_p} \\ &= \frac{b_s C_{mnip}}{2\pi|\mathbf{x}|}\{-\hat{n}_m\hat{m}_p S_{is} + \hat{n}_m\hat{n}_p(\hat{n}\hat{n})_{ik}^{-1}[4\pi B_{ks} + (\hat{n}\hat{m})_{kr}S_{rs}]\}.\end{aligned} \tag{12.25}$$

However, the following relationships exist:

$$\begin{aligned}C_{mnip}\hat{n}_m\hat{m}_p S_{is} &= (\hat{n}\hat{m})_{ni}S_{is}, \\ C_{mnip}\hat{n}_m\hat{n}_p(\hat{n}\hat{n})_{ik}^{-1}B_{ks} &= (\hat{n}\hat{n})_{ni}(\hat{n}\hat{n})_{ik}^{-1}B_{ks} = \delta_{nk}B_{ks} = B_{ns}, \\ C_{mnip}\hat{n}_m\hat{n}_p(\hat{n}\hat{n})_{ik}^{-1}(\hat{n}\hat{m})_{kr}S_{rs} &= (\hat{n}\hat{n})_{ni}(\hat{n}\hat{n})_{ik}^{-1}(\hat{n}\hat{m})_{kr}S_{rs} \\ &= \delta_{nk}(\hat{n}\hat{m})_{kr}S_{rs} = (\hat{n}\hat{m})_{nr}S_{rs}.\end{aligned} \tag{12.26}$$

Substitution of these results into Eq. (12.25) then produces the remarkably simple result,[5]

$$T_n^{DIS}(|\mathbf{x}|, \omega) = \frac{2b_s B_{ns}}{|\mathbf{x}|}. \tag{12.27}$$

The stresses and tractions given by Eqs. (12.24) and (12.27) fall off with the perpendicular distance from the dislocation, $|\mathbf{x}|$, as x^{-1}, which

[5] This result will be of use in Sec. 13.3.4.1.

is considerably slower than the x^{-3} decrease characteristic of inclusions and point defects. Also, singularities are present at the origin as $\rho \to 0$. These singularities are physically unrealistic, since linear elasticity cannot be applied in the bad material in the core region where realistic interatomic force laws preclude infinite stresses. However, it is demonstrated explicitly in Sec. 12.4.1.1, using results for an isotropic system because of its tractability, that even though these singularities exist, the elastic solutions given by the above results are expected to hold with acceptable accuracy up to relatively small distances from the core.

Finally, the distortion field of the line force can also be put into a form that does not require solving the Stroh eigenvalue problem by using essentially the same procedure just employed for the dislocation (Bacon, Barnett and Scattergood 1979b). First, the distortion field is obtained by differentiating Eq. (12.14) so that

$$\frac{\partial u_i^{LF}}{\partial x_l} = -\frac{1}{2\pi i}\sum_{\alpha=1}^{6} \pm A_{i\alpha}A_{s\alpha}f_s\frac{\hat{m}_l + p_\alpha \hat{n}_l}{(\hat{\mathbf{m}}\cdot\mathbf{x} + p_\alpha\hat{\mathbf{n}}\cdot\mathbf{x})}. \tag{12.28}$$

Then, aligning $\hat{\mathbf{m}}$ along $\mathbf{x}$, as in the derivation of Eq. (12.16),

$$\frac{\partial u_i^{LF}}{\partial x_l} = -\frac{1}{2\pi i|\mathbf{x}|}f_s\left(\hat{m}_l\sum_{\alpha=1}^{6}\pm A_{i\alpha}A_{s\alpha} + \hat{n}_l\sum_{\alpha=1}^{6}\pm p_\alpha A_{i\alpha}A_{s\alpha}\right). \tag{12.29}$$

The sum $\sum_{\alpha=1}^{6} \pm p_\alpha A_{i\alpha}A_{s\alpha}$ in Eq. (12.29) can be expressed in terms of the Q_{ij} and S_{ij} tensors of Sec. 3.5.2.1 by first multiplying Eq. (3.37) throughout by $\pm A_{s\alpha}$, summing over α, and using Eqs. (3.133) and (3.134) to obtain

$$\sum_{\alpha=1}^{6} \pm p_\alpha A_{r\alpha}A_{s\alpha} = i(\hat{n}\hat{n})_{rj}^{-1}\left[(\hat{n}\hat{m})_{jk}Q_{sk} + S_{sj}\right]. \tag{12.30}$$

Then, substituting Eqs. (3.133) and (12.30) into Eq. (12.29),

$$\frac{\partial u_i^{LF}}{\partial x_l} = \frac{1}{2\pi|\mathbf{x}|}f_s\left\{\hat{m}_l Q_{is} - \hat{n}_l(\hat{n}\hat{n})_{ij}^{-1}\left[(\hat{n}\hat{m})_{jk}Q_{sk} + S_{sj}\right]\right\}. \tag{12.31}$$

Equation (12.31) can be integrated to obtain the line force displacements by employing the same method used to integrate Eq. (12.19) to obtain Eq. (12.23) for the dislocation displacements, with the result

$$u_i^{LF}(|\mathbf{x}|,\omega) = \frac{f_s}{2\pi}\left[Q_{is}\ln|\mathbf{x}| - Q_{sk}\int(\hat{n}\hat{n})_{ij}^{-1}(\hat{n}\hat{m})_{jk}d\omega - S_{sj}\int(\hat{n}\hat{n})_{ij}^{-1}d\omega\right]. \tag{12.32}$$

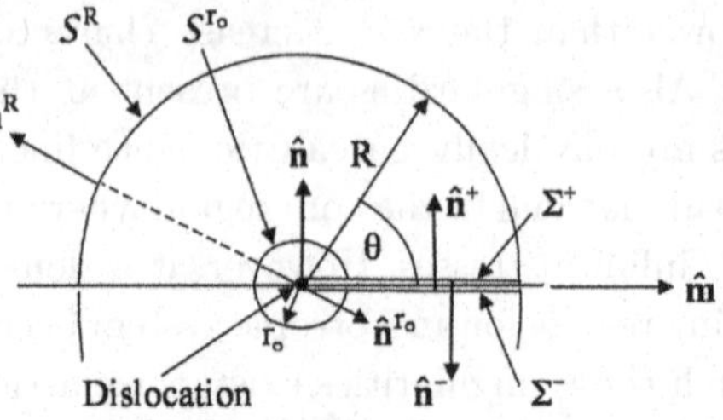

Fig. 12.6 Geometry for determining the strain energy in a cylindrical shell of outer radius R and inner radius r_o due to dislocation lying along shell axis: $\hat{\mathbf{n}}$ is positive unit normal vector to Σ cut surface used to produce dislocation: $\hat{\mathbf{n}}^+$ and $\hat{\mathbf{n}}^-$ are unit normal vectors to positive and negative sides of cut, respectively.

12.3.2 *Strain energies*

Now that the elastic field of a long straight dislocation has been obtained, its strain energy can be determined. The procedure is to avoid the elastic singularity in the core and first find the strain energy in a cylindrical shell, $\mathcal{V}$, of large outer radius R and small inner radius r_o (corresponding to the radius of the dislocation core), centered on the dislocation as in Fig. 12.6. Then, an increment of energy is added to account for the contribution made by the material in the core. The dislocation, located at the origin, lies along $\hat{\mathbf{t}} = \hat{\boldsymbol{\tau}} = \hat{\mathbf{m}} \times \hat{\mathbf{n}}$ and has been produced by a cut and displacement that displaced the Σ^+ and Σ^- surfaces relative to one another by $\mathbf{b}$. The strain energy per unit length of the shell region, i.e., $\mathcal{V}$, is then

$$w = \frac{1}{2}\oiiint_{\mathcal{V}} \sigma_{ij}\varepsilon_{ij}dV = \frac{1}{2}\oiiint_{\mathcal{V}} \frac{\partial(\sigma_{ij}u_j)}{\partial x_i}dV$$

$$= \frac{1}{2}\oiiint_{\mathcal{V}} \frac{\partial(\sigma_{ij}u_j)}{\partial x_i}dV = \frac{1}{2}\oiint_{\mathcal{S}} \sigma_{ij}u_j\hat{n}_i dS \qquad (12.33)$$

after applying the divergence theorem. The surface, $\mathcal{S}$, corresponds to $\mathcal{S} = \mathcal{S}^R + \mathcal{S}^{r_o} + \Sigma^+ + \Sigma^-$, and Eq. (12.33) may therefore be written as

$$w = \frac{1}{2}\oiint_{\Sigma^+} \sigma_{ij}^+ u_j^+ \hat{n}_i^+ dS + \frac{1}{2}\oiint_{\Sigma^-} \sigma_{ij}^- u_j^- \hat{n}_i^- dS$$

$$+\frac{1}{2}\oiint_{S^R} \sigma_{ij}^R u_j^R \hat{n}_i^R dS + \frac{1}{2}\oiint_{S^{r_o}} \sigma_{ij}^{r_o} u_j^{r_o} \hat{n}_i^{r_o} dS. \qquad (12.34)$$

However, on the surfaces of the cut, $\hat{n}_i^+ = -\hat{n}_i^-$, $u_j^- - u_j^+ = b_j$, and, $\sigma_{ij}^+\hat{n}_i^+ + \sigma_{ij}^-\hat{n}_i^- = 0$. Therefore, the strain energy per unit length of dislocation is

$$w = \frac{1}{2}\int_{r_\circ}^{R} \sigma_{ij} b_j \hat{n}_i d|\mathbf{x}| + \frac{1}{2}\oiint_{|\mathbf{x}|=R} \sigma_{ij}^R u_j^R \hat{n}_i^R dS + \frac{1}{2}\oiint_{|\mathbf{x}|=r_\circ} \sigma_{ij}^{r_\circ} u_j^{r_\circ} \hat{n}_i^{r_\circ} dS, \tag{12.35}$$

where the quantities $\sigma_{ij} = \sigma_{ij}^+ = \sigma_{ij}^-$, and $\hat{\mathbf{n}} = -\hat{\mathbf{n}}^+$ have been introduced. Now, Eq. (12.23) shows that the displacements on cylindrical surfaces of radii R and $r_\circ$ vary as

$$\begin{aligned} u_i^R &= A_i \ln R + g_i(\omega), \\ u_i^{r_\circ} &= A_i \ln r_\circ + g_i(\omega), \end{aligned} \tag{12.36}$$

where A_i = constant. Therefore, the sum of the last two integrals in Eq. (12.35), represented by I, takes the form

$$\begin{aligned} I = \frac{1}{2}\Bigg[&\oiint_{|\mathbf{x}|=R} A_j \ln R \sigma_{ij}^R \hat{n}_i^R dS + \oiint_{|\mathbf{x}|=r_\circ} A_j \ln r_\circ \sigma_{ij}^{r_\circ} \hat{n}_i^{r_\circ} dS \\ &+ \oiint_{|\mathbf{x}|=R} g_j(\omega)\sigma_{ij}^R \hat{n}_i^R dS + \oiint_{|\mathbf{x}|=r_\circ} g_j(\omega)\sigma_{ij}^{r_\circ} \hat{n}_i^{r_\circ} dS\Bigg]. \end{aligned} \tag{12.37}$$

Also, the form of Eq. (12.24) indicates that when $|\mathbf{x}|$ = constant,

$$\sigma_{ij} dS = \sigma_{ij}|\mathbf{x}| d\omega = h_{ij}(\omega) d\omega. \tag{12.38}$$

Then, substituting Eq. (12.38) into Eq. (12.37),

$$\begin{aligned} I = \frac{1}{2}\Bigg(&\oiint_{|\mathbf{x}|=R} A_j \ln R \sigma_{ij}^R \hat{n}_i^R dS + \oiint_{|\mathbf{x}|=r_\circ} A_j \ln r_\circ \sigma_{ij}^{r_\circ} \hat{n}_i^{r_\circ} dS \\ &+ \oiint_{|\mathbf{x}|=R} g_j(\omega)\hat{n}_i^R h_{ij} d\omega + \oiint_{|\mathbf{x}|=r_\circ} g_j(\omega)\hat{n}_i^{r_\circ} h_{ij} d\omega\Bigg). \end{aligned} \tag{12.39}$$

The first two integrals in Eq. (12.39) vanish, since mechanical equilibrium requires that the net forces on the cylindrical surfaces vanish. Also, the two remaining integrals cancel each other, since $\hat{n}_i^R = -\hat{n}_i^{r_\circ}$. Therefore, I = 0, and, after substituting Eq. (12.27) into Eq. (12.35) for the traction $\sigma_{ij}\hat{n}_i$,

$$w = \frac{1}{2}\int_{r_\circ}^{R} \sigma_{ij} b_j \hat{n}_i d|\mathbf{x}| = b_j b_s B_{js} \int_{r_\circ}^{R} \frac{d|\mathbf{x}|}{|\mathbf{x}|} = w_\circ \ln \frac{R}{r_\circ}, \tag{12.40}$$

where $w_\circ$, the *dislocation strain energy pre-logarithmic factor*, is given by

$$w_\circ = b_j b_s B_{js}. \tag{12.41}$$

This relatively simple result shows that the strain energy is just the work done in displacing the Σ^+ and Σ^- sides of the cut during the creation of the dislocation by the cut and displacement process.

The energy increment associated with the core is now added in a formal manner by introducing the parameter, α, defined by

$$\alpha \equiv b/r_\circ \tag{12.42}$$

into Eq. (12.40) so that the total energy per unit length assumes the form

$$w = b_j b_s B_{js} \ln \frac{\alpha R}{b}, \tag{12.43}$$

which, because of its logarithmic form, is only weakly dependent upon the magnitude of α. The determination of the correct magnitude of α requires an atomistic calculation, and typical values are found to be of order $\alpha \approx 1$ (Hirth and Lothe 1982).

Equation (12.43) shows that the energy diverges logarithmically with increasing R and becomes infinite as $R \to \infty$. However, this physically unrealistic result does not apply to a dislocation in a real finite body where image stresses (Ch. 13) and stresses due to other defects are present so that stress cancellation occurs over large distances, and finite energies result.

12.4 Infinitely Long Straight Dislocations in Isotropic Systems

12.4.1 *Elastic fields*

The elastic fields of infinitely long straight dislocations in isotropic systems are now obtained by employing Eq. (12.24) after rewriting it to apply to an isotropic system. These fields can also be obtained by alternative methods, and several of these are described.

12.4.1.1 *Edge dislocation*

By use of integral formalism

Consider an edge dislocation with $\mathbf{b} = (b, 0, 0)$ and $\hat{\mathbf{t}} = (0, 0, 1)$ corresponding to Fig. 12.2a. Taking $\hat{\mathbf{e}}_3$ and $\hat{\mathbf{e}}_1$ parallel to $\hat{\tau}$ and the reference line, respectively, its stress field can be determined by using Eq. (12.24), with the help of Eqs. (2.120), (3.141) and (3.147) to determine the necessary elastic constants and B_{ij} and S_{ij} matrices. Then, since $\hat{\mathbf{m}}$ and $\mathbf{x}$ were taken to be parallel in deriving Eq. (12.24), $\hat{\mathbf{m}}$ and $\hat{\mathbf{n}}$ can be expressed as

$$\begin{aligned} \hat{m}_1 &= \frac{x_1}{(x_1^2+x_2^2)^{1/2}}, \quad \hat{m}_2 = \frac{x_2}{(x_1^2+x_2^2)^{1/2}}, \quad \hat{m}_3 = 0, \\ \hat{n}_1 &= \frac{-x_2}{(x_1^2+x_2^2)^{1/2}}, \quad \hat{n}_2 = \frac{x_1}{(x_1^2+x_2^2)^{1/2}}, \quad \hat{n}_3 = 0. \end{aligned} \tag{12.44}$$

Using the above relationships, the edge dislocation stress field is found to be

$$\begin{aligned} \sigma_{11} &= -\frac{\mu b}{2\pi(1-\nu)}\frac{x_2(3x_1^2+x_2^2)}{(x_1^2+x_2^2)^2}, \quad \sigma_{22} = \frac{\mu b}{2\pi(1-\nu)}\frac{x_2(x_1^2-x_2^2)}{(x_1^2+x_2^2)^2}, \\ \sigma_{12} &= \frac{\mu b}{2\pi(1-\nu)}\frac{x_1(x_1^2-x_2^2)}{(x_1^2+x_2^2)^2}, \quad \sigma_{33} = \nu(\sigma_{11}+\sigma_{22}), \quad \sigma_{13} = \sigma_{23} = 0. \end{aligned} \tag{12.45}$$

Adopting polar coordinates, the stresses in Eq. (12.45) are of the form

$$\begin{aligned} \sigma_{11} &= -\frac{\mu b}{2\pi(1-\nu)}\frac{\sin\theta(2\cos^2\theta+1)}{r}, \\ \sigma_{22} &= \frac{\mu b}{2\pi(1-\nu)}\frac{\sin\theta(2\cos^2\theta-1)}{r}, \\ \sigma_{12} &= \frac{\mu b}{2\pi(1-\nu)}\frac{\cos\theta(2\cos^2\theta-1)}{r}, \\ \sigma_{33} &= \nu(\sigma_{11}+\sigma_{22}), \quad \sigma_{13} = \sigma_{23} = 0, \end{aligned} \tag{12.46}$$

and, more simply, they appear in cylindrical coordinates, after using Eq. (G.7), as

$$\begin{aligned} \sigma_{rr} &= \sigma_{\theta\theta} = -\frac{\mu b}{2\pi(1-\nu)}\frac{\sin\theta}{r}, \quad \sigma_{r\theta} = \frac{\mu b}{2\pi(1-\nu)}\frac{\cos\theta}{r}, \\ \sigma_{zz} &= \nu(\sigma_{rr}+\sigma_{\theta\theta}), \quad \sigma_{rz} = \sigma_{\theta z} = 0. \end{aligned} \tag{12.47}$$

Singularities, such as those appearing in the above equations at the origin as $|\mathbf{x}| \to 0$, are physically unrealistic. However, as is now shown, the above solutions are expected to hold with acceptable accuracy up to relatively small distances from the dislocation core.

First, hollow out the dislocation core by removing all material within the relatively small radius, $r_\circ \approx b$, while simultaneously applying surface tractions to keep the stresses in the remaining material unchanged. Next, remove these tractions by applying a stress field that cancels them, thus producing a hollow dislocation with a traction-free interior surface. This requires the cancellation of the $\sigma_{rr}(r_\circ, \theta)$ and $\sigma_{r\theta}(r_\circ, \theta)$ stresses, and can be accomplished by using the Airy stress function from Table I.1,

$$\psi = \frac{A \sin\theta}{x}, \tag{12.48}$$

where A = constant. Using the stresses associated with this stress function (given in Table I.1), and converting them to stresses in cylindrical coordinates using Eq. (G.7), $\sigma'_{rr} = -2A\sin\theta/r^3$ and $\sigma'_{r\theta} = 2A\cos\theta/r^3$. Cancellation is then achieved when $A = -\mu b r_\circ^2/[4\pi(1-\nu)]$, and by adding the Airy function stresses to the stresses given by Eq. (12.47), the total stresses due to the hollow dislocation are

$$\begin{aligned}
\sigma_{rr} &= -\frac{\mu b}{2\pi(1-\nu)}\frac{\sin\theta}{r}\left[1-\left(\frac{r_\circ}{r}\right)^2\right],\\
\sigma_{\theta\theta} &= -\frac{\mu b}{2\pi(1-\nu)}\frac{\sin\theta}{r}\left[1+\left(\frac{r_\circ}{r}\right)^2\right], \quad \sigma_{zz} = \nu(\sigma_{rr}+\sigma_{\theta\theta}),\\
\sigma_{r\theta} &= \frac{\mu b}{2\pi(1-\nu)}\frac{\cos\theta}{r}\left[1-\left(\frac{r_\circ}{r}\right)^2\right], \quad \sigma_{rz} = \sigma_{\theta z} = 0.
\end{aligned} \tag{12.49}$$

These stresses rapidly approach the stresses given by Eq. (12.47) with increasing distance from the core, showing that the stresses in the matrix, even relatively close to the core, are insensitive to the boundary conditions at $r = r_\circ$. The stress field given by Eqs. (12.45)–(12.47) is therefore acceptable with the understanding that it is reliable only at distances from the core greater that a few multiples of $r_\circ \sim b$. A more accurate analysis near the core requires an atomistic calculation to establish the correct tractions that the core exerts on the matrix. However, this is beyond our present scope.

The stress field for a straight edge dislocation given by Eq. (12.45) can be obtained by multiple methods. In Exercise 12.2 it is obtained by using the Volterra equation, and in Exercise 12.6 it is obtained by employing the Airy stress function approach, described in Sec. 3.7, since it is a case of plane strain. In addition, the displacement field corresponding to the stress field given by Eq. (12.45) is obtained below in the form of Eq. (12.54) by use of the transformation strain formalism.

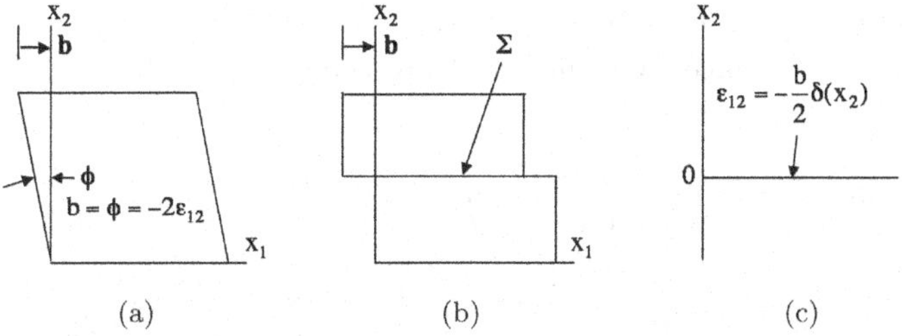

Fig. 12.7 (a) A uniform distribution of shear strain producing a shear displacement, b. (b) Localization of the shear strain on Σ surface. (c) Representation of the localized shear strain in the form of a delta function.

By use of transformation strain formalism

The elastic field, expressed in terms of displacements, can be determined by employing the transformation strain formalism of Sec. 3.6. In this method (Mura 1987) the localized displacement across the cut, used to create the edge dislocation segment in Fig. 12.2a, is represented by a highly localized transformation strain in the form of a delta function. The displacement field is then found by integrating Eq. (3.168). The delta function representation of the localized shear displacement at the Σ surface, which is of magnitude b in the $-x_1$ direction on the $x_2 = 0$ plane, is indicated in Fig. 12.7. The transformation strain is then first written as

$$\varepsilon_{12}^{T}(x_1, x_2) = -\frac{b}{2}\delta(x_2)H(x_1), \tag{12.50}$$

where $H(x_1)$ is the Heaviside step function. The transformation stress is therefore

$$\sigma_{12}^{T}(x_1, x_2) = 2\mu\varepsilon_{12}^{T}(x_1, x_2) = -\mu b\delta(x_2)H(x_1) \tag{12.51}$$

and substituting this into Eq. (3.168), and realizing that the displacements are restricted to the x_1 and x_2 directions, the $u_1(\mathbf{x})$ displacement is

$$u_1(\mathbf{x}) = -\mu b \int_{-\infty}^{\infty} H(x_1')dx_1' \int_{-\infty}^{\infty}\int_{-\infty}^{\infty} \delta(x_2') \times \left[\frac{\partial G_{12}^{\infty}(\mathbf{x}-\mathbf{x}')}{\partial x_1'} + \frac{\partial G_{11}^{\infty}(\mathbf{x}-\mathbf{x}')}{\partial x_2'}\right] dx_2' dx_3'. \tag{12.52}$$

Then, substituting Eq. (4.110), setting the field point in the $x_3 = 0$ plane (allowable since the solution is invariant along x_3), and integrating over dx'_2,

$$u_1(x_1, x_2) = -\frac{b}{8\pi(1-\nu)} \int_{-\infty}^{\infty} H(x'_1)dx'_1$$

$$\times \int_{-\infty}^{\infty} \left[\frac{(1-2\nu)x_2}{R^3} + \frac{3(x_1 - x'_1)^2 x_2}{R^5}\right] dx'_3, \quad (12.53)$$

where $R = [(x_1 - x'_1)^2 + x_2^2 + (x'_3)^2]^{1/2}$. The integration of Eq. (12.53) over dx'_3, and dx'_1 involves only elementary functions (Gradshteyn and Ryzhik 1980), and the final expressions for u_1, and also u_2, obtained by similar means, are

$$u_1(x_1, x_2) = \frac{b}{2\pi}\left[\tan^{-1}\frac{x_2}{x_1} + \frac{1}{2(1-\nu)}\frac{x_1 x_2}{(x_1^2 + x_2^2)}\right],$$
$$u_2(x_1, x_2) = -\frac{b}{8\pi(1-\nu)}\left[(1-2\nu)\ln(x_1^2 + x_2^2) + \frac{x_1^2 - x_2^2}{x_1^2 + x_2^2}\right]. \quad (12.54)$$

In Exercise 12.5 the same method is used to obtain the displacement field of a screw dislocation.

12.4.1.2 *Screw dislocation*

Consider next a screw dislocation with $\mathbf{b} = (0, 0, b)$ and $\hat{\mathbf{t}} = (0, 0, 1)$ corresponding to Fig. 12.2c. Starting with Eq. (12.24), and using the same procedure as employed above to obtain Eq. (12.45) for the edge dislocation, all non-vanishing stresses for the screw dislocation are found to be the shear stresses

$$\sigma_{13} = -\frac{\mu b}{2\pi}\frac{x_2}{(x_1^2 + x_2^2)}, \quad \sigma_{23} = \frac{\mu b}{2\pi}\frac{x_1}{(x_1^2 + x_2^2)},$$
$$\sigma_{11} = \sigma_{22} = \sigma_{33} = \sigma_{12} = 0, \quad (12.55)$$

which, in polar coordinates, appear as

$$\sigma_{13} = -\frac{\mu b}{2\pi}\frac{\sin\theta}{r}, \quad \sigma_{23} = \frac{\mu b}{2\pi}\frac{\cos\theta}{r}, \quad \sigma_{11} = \sigma_{22} = \sigma_{33} = \sigma_{12} = 0, \quad (12.56)$$

and, more simply, in cylindrical coordinates as

$$\sigma_{\theta z} = \frac{\mu b}{2\pi r}, \qquad \sigma_{rr} = \sigma_{\theta\theta} = \sigma_{zz} = \sigma_{r\theta} = \sigma_{rz} = 0. \quad (12.57)$$

The displacement field corresponding to the above stress field is derived in Exercise 12.5 by means of the transformation strain formalism (see Eq. (12.274)).

12.4.1.3 *Mixed dislocation*

As discussed in Sec. 12.2, a mixed dislocation with a general Burgers vector, **b**, may be regarded as the superposition of an edge dislocation component with Burgers vector, $b \sin\beta$, and a screw dislocation component with Burgers vector, $b \cos\beta$. Its stress field is therefore readily calculated by use of the previous results.

Finally, examination of Eqs. (12.46) and (12.56) reveals the asymmetric property of the edge and screw dislocation stress fields,

$$\sigma_{ij}(\theta + \pi) = -\sigma_{ij}(\theta). \tag{12.58}$$

12.4.2 *Strain energies*

The strain energy, per unit length, of a straight dislocation lying along x_3 is obtained by employing Eq. (12.40) with B_{js} given, for the present isotropic system, by Eq. (3.147) with the result

$$w = \frac{\mu}{4\pi}\left[\frac{1}{1-\nu}(b_1^2 + b_2^2) + b_3^2\right] \ln \frac{R}{r_\circ}. \tag{12.59}$$

The energy increment associated with the material in the core is then added, as previously in the anisotropic case (see Eqs. (12.42) and (12.43)), by introducing the parameter $\alpha = b/r_\circ$ into Eq. (12.59) so that

$$w = \frac{\mu}{4\pi}\left[\frac{1}{1-\nu}(b_1^2 + b_2^2) + b_3^2\right] \ln \frac{\alpha R}{b}. \tag{12.60}$$

Therefore, for pure edge and screw dislocations,

$$\begin{aligned} w &= \frac{\mu b^2}{4\pi(1-\nu)} \ln \frac{\alpha R}{b} \qquad \text{(edge)} \\ w &= \frac{\mu b^2}{4\pi} \ln \frac{\alpha R}{b} \qquad \text{(screw)} \end{aligned} \tag{12.61}$$

and for mixed dislocations,

$$w = \frac{\mu b^2}{4\pi}\left[\frac{\sin^2\beta}{(1-\nu)} + \cos^2\beta\right] \ln \frac{\alpha R}{b}. \tag{12.62}$$

It is demonstrated in Exercise 12.8 that Eq. (12.61) for the edge dislocation can be obtained alternatively by simply integrating the strain energy density over the volume of the cylindrical shell enclosed by $\mathcal{S}^{R}$ and $\mathcal{S}^{r_\circ}$ in Fig. 12.6.

12.5 Smoothly Curved Dislocation Loops

12.5.1 *Elastic fields*

The elastic field of a smoothly curved closed loop, such as illustrated in Fig. 12.1, can be determined in a number of alternative ways as described in the following sections.

12.5.1.1 *By use of Volterra equation*

The Volterra equation yields the displacement field produced by a general dislocation loop in the form of a surface integral taken over the cut surface, Σ, in Fig. 12.1. The equation can be obtained by employing an argument given by Hirth and Lothe (1982) where it is imagined that a constant applied point force, $\mathbf{F}$, is present at the field point, $\mathbf{x}$ (see Fig. 12.1), during the creation of the loop.[6] If the elastic displacement at $\mathbf{x}$ due to the loop creation is $\mathbf{u}(\mathbf{x})$, the change in potential energy of the force is then

$$\Delta\Phi = -\mathbf{F}\cdot\mathbf{u}(\mathbf{x}) = -\mathrm{F_i u_i}(\mathbf{x}). \tag{12.63}$$

Since the elastic field of the force is applied, and the field of the loop is internal, the interaction strain energy between them vanishes (see Eq. (5.25)). The change in potential energy of the force, $\Delta\Phi$, must, therefore, appear in the form of work, $\Delta\mathcal{W}$, done by its stress field during the cut and displacement at the Σ surface that produced the loop as demonstrated, for example, in Exercise 16.8. Therefore,

$$\Delta\mathcal{W} = \Delta\Phi. \tag{12.64}$$

The stress at $\mathbf{x}'$ at the Σ surface due to the force acting at $\mathbf{x}$, i.e., $\sigma'_{\mathrm{mn}}(\mathbf{x}-\mathbf{x}')$, generates a traction acting on the surface given by $\mathrm{T'_m}(\mathbf{x}-\mathbf{x}') = \sigma'_{\mathrm{mn}}\hat{\mathrm{n}}'_{\mathrm{n}}$.

---

[6]The Volterra equation can also be obtained by an alternative procedure (Mura 1987) in which the cut and displacement at the surface, Σ, is mimicked by a localized transformation strain in the form of a delta function as in the derivation of Eq. (12.143) in Sec. 12.5.1.6.

The displacement everywhere on the face of the cut on the negative side of Σ relative to the positive side is **b**, because of the Σ cut/displacement rule given on p. 325. Therefore, the work done by the stress field of the force on an area dS' on Σ is $d\mathcal{W}(\mathbf{x}-\mathbf{x}') = T'_m b_m dS' = \sigma'_{mn} b_m \hat{n}'_n dS'$. Then, substituting this result and Eq. (12.63) into Eq. (12.64),

$$F_i u_i(\mathbf{x}) = -b_m \iint_{\Sigma} \sigma'_{mn}(\mathbf{x}-\mathbf{x}')\hat{n}'_n dS' \tag{12.65}$$

The stress at $\mathbf{x}'$ due to the force at $\mathbf{x}$ is given by

$$\begin{aligned}
\sigma'_{mn}(\mathbf{x}-\mathbf{x}') &= C_{mnjl}\varepsilon_{jl}(\mathbf{x}-\mathbf{x}') \\
&= C_{mnjl}\frac{1}{2}\left[\frac{\partial u_j(\mathbf{x}-\mathbf{x}')}{\partial x'_l} + \frac{\partial u_l(\mathbf{x}-\mathbf{x}')}{\partial x'_j}\right] \\
&= C_{mnjl}\frac{1}{2}\left[\frac{\partial G_{ji}(\mathbf{x}-\mathbf{x}')}{\partial x'_l} + \frac{\partial G_{li}(\mathbf{x}-\mathbf{x}')}{\partial x'_j}\right] F_i \\
&= C_{mnjl}\frac{\partial G_{ji}(\mathbf{x}-\mathbf{x}')}{x'_l} F_i,
\end{aligned} \tag{12.66}$$

where $G_{ij}(\mathbf{x}-\mathbf{x}')$ is the Green's function given by Eq. (4.25), and use has been made of Eq. (3.15) and the symmetry properties of the C_{mnjl} tensor. Then, substituting Eqs. (12.66) into Eq. (12.65) and canceling out the common factor, F_i, for each component taken separately, the *Volterra equation* is obtained in the form

$$u_i(\mathbf{x}) = -C_{mnjl} b_m \iint_{\Sigma} \frac{\partial G_{ji}(\mathbf{x}-\mathbf{x}')}{\partial x'_l} \hat{n}'_n dS'. \tag{12.67}$$

Substitution of Eq. (4.40), for the derivatives of the Green's function, into Eq. (12.67) then yields the further expression

$$\begin{aligned}
u_i(\mathbf{x}) = &-\frac{C_{mnjl} b_m}{8\pi^2} \iint_{\Sigma} \frac{1}{|\mathbf{x}-\mathbf{x}'|^2} \hat{n}'_n dS' \\
&\times \oint_{\hat{\mathcal{L}}} \left\{ \hat{w}_l(\hat{k}\hat{k})^{-1}_{ji} - \hat{k}_l(\hat{k}\hat{k})^{-1}_{jp}[(\hat{k}\hat{w})_{pr} + (\hat{w}\hat{k})_{pr}](\hat{k}\hat{k})^{-1}_{ri} \right\} ds,
\end{aligned} \tag{12.68}$$

which involves a line integral along s around the unit circle, $\hat{\mathcal{L}}$ (see text following Eq. (4.40)), and a surface integral over S' corresponding to the Σ cut surface.

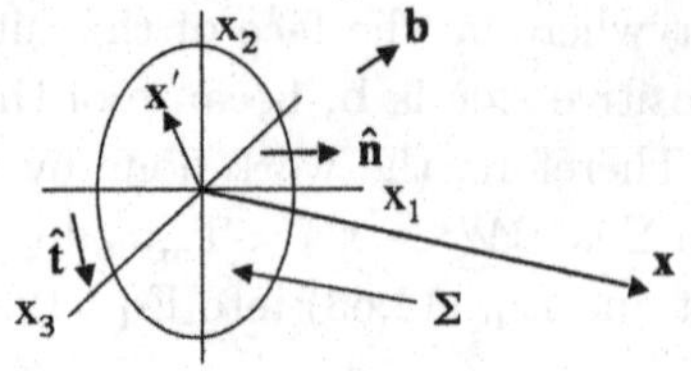

Fig. 12.8 Planar dislocation loop. Loop plane, Σ cut surface, and source vector, $\mathbf{x}'$, lie in $x_1 = 0$ plane.

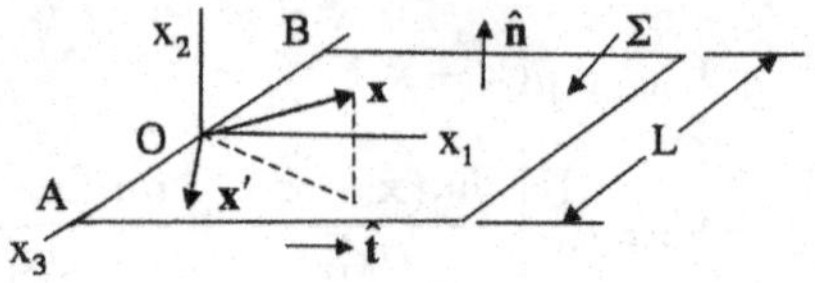

Fig. 12.9 Square planar dislocation loop. Loop plane, cut surface Σ, and source vector, x', lie in $x_2 = 0$ plane.

The evaluation of Eq. (12.68) is simplified when the loop is planar. Consider, for example, the planar loop in Fig. 12.8. Taking the cut surface, Σ, and the origin in the plane of the loop so that $x_1' = 0$, Eq. (12.68) is reduced to

$$u_i(\mathbf{x}) = -\frac{C_{m1jl}b_m}{8\pi^2}\iint_\Sigma \frac{1}{|\mathbf{x}-\mathbf{x}'|^2}dx_2'dx_3' \times \oint_{\hat{\mathcal{L}}}\left\{\hat{w}_l(\hat{k}\hat{k})^{-1}_{ji} - \hat{k}_l(\hat{k}\hat{k})^{-1}_{jp}[(\hat{k}\hat{w})_{pr} + (\hat{w}\hat{k})_{pr}](\hat{k}\hat{k})^{-1}_{ri}\right\}ds. \tag{12.69}$$

Even though the Volterra equation has been derived to apply to a closed dislocation loop it can be readily used to find the displacement field of an infinitely long straight dislocation. Consider the square planar dislocation loop of edge-length L in Fig. 12.9. Using Eq. (12.67), the loop displacement field is

$$\begin{aligned} u_i(\mathbf{x}) &= -C_{mnjl}b_m \iint_\Sigma \frac{\partial G_{ji}(\mathbf{x}-\mathbf{x}')}{\partial x_l'}\hat{n}_n' dS' \\ &= -C_{m2jl}b_m \int_0^L dx_1' \int_{-L/2}^{L/2}\left[\frac{\partial G_{ji}(\mathbf{x}-\mathbf{x}')}{\partial x_l'}\right]_{x_2'=0} dx_3'. \end{aligned} \tag{12.70}$$

If, as in Eq. (12.68), Eq. (4.40) is then substituted for the Green's function derivatives, and L is increased without limit,

$$\mathrm{u_i}(\mathbf{x}) = -\frac{\mathrm{C_{m2j}}_l \mathrm{b_m}}{8\pi^2}\int_0^\infty \mathrm{dx}_1' \int_{-\infty}^{\infty} \frac{1}{|\mathbf{x}-\mathbf{x}'|^2}$$
$$\times \oint_{\hat{\mathcal{L}}} \left\{ \hat{\mathrm{w}}_l (\hat{\mathrm{k}}\hat{\mathrm{k}})^{-1}_{\mathrm{ji}} - \hat{\mathrm{k}}_l (\hat{\mathrm{k}}\hat{\mathrm{k}})^{-1}_{\mathrm{jp}} \left[(\hat{\mathrm{k}}\hat{\mathrm{w}})_{\mathrm{ps}} + (\hat{\mathrm{w}}\hat{\mathrm{k}})_{\mathrm{ps}} \right] (\hat{\mathrm{k}}\hat{\mathrm{k}})^{-1}_{\mathrm{si}} \right\} \mathrm{ds dx}_3', \tag{12.71}$$

where the source vector, $\mathbf{x}'$, lies in the loop plane so that $|\mathbf{x}-\mathbf{x}'| = [(\mathrm{x}_1 - \mathrm{x}_1')^2 + \mathrm{x}_2^2 + (\mathrm{x}_3 - \mathrm{x}_3')^2]^{1/2}$. Equation (12.71) yields the displacement field produced by the long straight dislocation segment AB in the region around the origin in the infinite crystal, since the remaining segments of the loop, which are at infinite distances, do not contribute significantly.

In Exercise 12.2 this method is used to obtain the displacement field of a long straight edge dislocation corresponding to Fig. 12.2a in an isotropic system.

12.5.1.2 *By use of the Mura equation*

The Mura equation yields the distortions produced by a dislocation loop in terms of a line integral along the loop and is obtained by applying Stokes' theorem to the Volterra equation. First, the Volterra equation is differentiated so that

$$\frac{\partial \mathrm{u_i}(\mathbf{x})}{\partial \mathrm{x_k}} = \mathrm{C_{mnj}}_l \mathrm{b_m} \iint_{\Sigma} \frac{\partial^2 \mathrm{G_{ji}}(\mathbf{x}-\mathbf{x}')}{\partial \mathrm{x}_k' \partial \mathrm{x}_l'} \hat{\mathrm{n}}_\mathrm{n}' \mathrm{dS}'. \tag{12.72}$$

An additional expression is now needed that can be used in conjunction with Eq. (12.72) to obtain an equation amenable to Stokes' theorem. This is obtained by first substituting Eq. (3.15), i.e.,

$$\mathrm{u_j}(\mathbf{x}') = \mathrm{G_{ji}}(\mathbf{x}-\mathbf{x}')\mathrm{F_i} \tag{12.73}$$

into Eq. (3.154) in the absence of transformation strains, i.e.,

$$\sigma_{\mathrm{mn}}(\mathbf{x}') = \mathrm{C_{mnj}}_l \frac{\partial \mathrm{u_j}(\mathbf{x}')}{\partial \mathrm{x}_l'} \tag{12.74}$$

to obtain

$$\sigma_{\mathrm{mn}}(\mathbf{x}') = \mathrm{C_{mnj}}_l \frac{\partial \mathrm{G_{ji}}(\mathbf{x}-\mathbf{x}')\mathrm{F_i}}{\partial \mathrm{x}_l'}. \tag{12.75}$$

However, the stress $\sigma_{mn}(\mathbf{x}')$ must obey the equilibrium condition, Eq. (2.65), and, since $\mathbf{x}'$ is in a region free of body force, substitution of Eq. (12.75) into Eq. (2.65), with $f_i = 0$, yields

$$\frac{\partial \sigma_{mn}(\mathbf{x}')}{\partial x'_n} = C_{mnjl}\frac{\partial^2 G_{ji}(\mathbf{x}-\mathbf{x}')F_i}{\partial x'_n \partial x'_l} = 0. \qquad (12.76)$$

Then, to satisfy Eq. (12.76) for arbitrary F_i, the condition

$$C_{mnjl}\frac{\partial^2 G_{ji}(\mathbf{x}-\mathbf{x}')}{\partial x'_n \partial x'_l} = 0 \qquad (12.77)$$

must be valid. The desired expression is then obtained by using Eq. (12.77) to write

$$C_{mnjl}\frac{\partial^2 G_{ji}(\mathbf{x}-\mathbf{x}')}{\partial x'_n \partial x'_l} b_m \hat{n}_k = 0, \qquad (12.78)$$

which must also be valid. Then, by adding Eqs. (12.78) and (12.72),

$$\frac{\partial u_i(\mathbf{x})}{\partial x_k} = b_m C_{mnjl} \iint_{\Sigma} \left[\frac{\partial G^2_{ji}(\mathbf{x}-\mathbf{x}')}{\partial x'_k \partial x'_l} \hat{n}'_n - \frac{\partial^2 G_{ji}(\mathbf{x}-\mathbf{x}')}{\partial x'_n \partial x'_l} \hat{n}'_k \right] dS'. \qquad (12.79)$$

Next, Eq. (12.79) is converted to a line integral around the dislocation loop, C, illustrated in Fig. 12.10 by employing Stokes' theorem in the form of Eq. (B.4). The final result is the *Mura equation* expressed as

$$\begin{aligned}\frac{\partial u_i(\mathbf{x})}{\partial x_k} &= b_m C_{mnjl} e_{nkp} \oint_C \frac{\partial G_{ji}(\mathbf{x}-\mathbf{x}')}{\partial x'_l} dx'_p \\ &= b_m C_{mnjl} e_{nkp} \oint_C \frac{\partial G_{ji}(\mathbf{x}-\mathbf{x}')}{\partial x'_l} \hat{t}_p ds', \end{aligned} \qquad (12.80)$$

where use has been made of $d\mathbf{x}' = d\mathbf{s}' = \hat{\mathbf{t}}ds'$.

Substitution of Eq. (4.40) into Eq. (12.80), yields the further expression

$$\begin{aligned}\frac{\partial u_i(\mathbf{x})}{\partial x_k} = \frac{b_m C_{mnjl} e_{nkp}}{8\pi^2} \oint_C \frac{1}{|\mathbf{x}-\mathbf{x}'|^2} \hat{t}_p ds' \\ \times \oint_{\hat{\mathcal{L}}} \left\{ \hat{w}_l (\hat{k}\hat{k})^{-1}_{ji} - \hat{k}_l (\hat{k}\hat{k})^{-1}_{js} [(\hat{k}\hat{w})_{sr} + (\hat{w}\hat{k})_{sr}] (\hat{k}\hat{k})^{-1}_{ri} \right\} ds, \end{aligned} \qquad (12.81)$$

which involves a line integral along s around the unit circle, $\hat{\mathcal{L}}$ [see text following Eq. (4.40)], and another along s′ around the loop, C, as illustrated

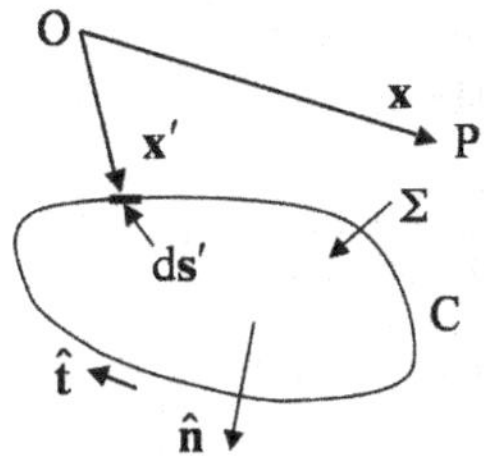

Fig. 12.10 Geometry for performing line integral around dislocation loop C.

in Fig. 12.10. Having this result, corresponding strains and stresses are readily determined. Note that Eqs. (12.80) and (12.81) are completely free of quantities associated with the cut surface in contrast to the form of the Volterra equation. This is consistent with the fact that the choice of the surface employed in the cut and displacement method for producing a given loop is arbitrary as is also demonstrated explicitly in Exercise 12.3.

As shown in Exercise 12.14, the Mura equation given by Eq. (12.80) can be reformulated to yield the stress field produced by the loop C. Also, the Mura equation, in combination with Eq. (4.40) for derivatives of the Green's function, can be used to find the elastic field of infinitely long straight dislocations in a manner similar to the method used above in Sec. 2.5.1.1 which employed the Volterra equation to obtain Eq. (12.71). This method is used in Exercise 12.4 to find the strains produced by an infinitely long edge dislocation, corresponding to Fig.12.2a, in an isotropic system.

12.5.1.3 *By use of modified Burgers equation*

The Burgers equation for the displacement field produced by a loop in an isotropic body is derived below in Sec. 12.6.1.1 (see Eq. (12.152)). More recently, (Leibfried 1953, Indenbom and Orlov 1968 and Lothe 1992b) have obtained a comparable expression for the displacement field in a general anisotropic body which can be found by assuming a solution of a form that is similar in many respects to the isotropic solution. Using Eqs. (12.151) and (12.152), the isotropic solution can be written as

$$u_i(\mathbf{x}) = -\frac{b_i}{4\pi}\Omega + \frac{1}{8\pi}\oint_C \left[-\frac{2e_{ijk}}{|\mathbf{x}-\mathbf{x}'|} - \frac{e_{jmk}}{(1-\nu)}\frac{\partial^2|\mathbf{x}-\mathbf{x}'|}{\partial x_i' \partial x_m'} \right] b_j dx_k'. \tag{12.82}$$

To obtain a corresponding solution for an anisotropic system, a solution is assumed (Lothe 1992b) of the form

$$u_i(\mathbf{x}) = -\frac{b_i}{4\pi}\Omega + \frac{1}{8\pi}\oint_C U_{ijk}(\mathbf{x}-\mathbf{x}')b_j dx'_k, \tag{12.83}$$

where $U_{ijk}(\mathbf{x}-\mathbf{x}')$ is expressed in the Fourier form, see Eq. (F.4), as

$$U_{ijk}(\mathbf{x}-\mathbf{x}') = \iiint_{-\infty}^{\infty} \overline{U}_{ijk}(\mathbf{k})e^{i\mathbf{k}\cdot(\mathbf{x}-\mathbf{x}')}dk_1dk_2dk_3. \tag{12.84}$$

The stresses are then found by substituting Eq. (12.83) into Eq. (3.1) to obtain

$$\sigma_{ij}(\mathbf{x}) = -\frac{b_i}{4\pi}C_{ijkl}b_k\frac{\partial\Omega}{\partial x_l} + \frac{1}{8\pi}C_{ijkl}\frac{\partial}{\partial x_l}\oint_C U_{kpm}(\mathbf{x}-\mathbf{x}')b_p dx'_m. \tag{12.85}$$

The quantity $\partial\Omega/\partial x_l$ is given by Eq. (12.159) which, with the help of the standard Fourier expression,

$$\frac{1}{|\mathbf{x}-\mathbf{x}'|} = \frac{1}{2\pi^2}\iiint_{-\infty}^{\infty}\frac{1}{k^2}e^{i\mathbf{k}\cdot(\mathbf{x}-\mathbf{x}')}dk_1dk_2dk_3 \tag{12.86}$$

can be written as

$$\frac{\partial\Omega}{\partial x_l} = \frac{i}{2\pi^2}\oint_C e_{lms}dx'_m\iiint_{-\infty}^{\infty}\frac{k_s}{k^2}e^{i\mathbf{k}\cdot(\mathbf{x}-\mathbf{x}')}dk_1dk_2dk_3 \tag{12.87}$$

Then, substituting Eq. (12.87) into Eq. (12.85),

$$\sigma_{ij}(\mathbf{x}) = \frac{i}{8\pi}\oint_C dx'_m\iiint_{-\infty}^{\infty} \times\left(-\frac{k_s}{\pi^2k^2}C_{ijkl}e_{lms}b_k + C_{ijkl}b_pk_l\overline{U}_{kpm}(\mathbf{k})\right)e^{i\mathbf{k}\cdot(\mathbf{x}-\mathbf{x}')}dk_1dk_2dk_3. \tag{12.88}$$

Since no force density is present in the elastic field of the dislocation, and equilibrium must be maintained, Eq. (2.65) in the form

$$\frac{\partial\sigma_{ij}}{\partial x_j} = 0 \qquad f_i = 0 \tag{12.89}$$

must be satisfied. Substituting Eq. (12.88) into Eq. (12.89) then yields

$$\frac{\partial\sigma_{ij}(\mathbf{x})}{\partial x_j} = -\frac{1}{8\pi}\oint_C dx'_m \iiint_{-\infty}^{\infty} \times\left(-\frac{k_s k_j}{\pi^2 k^2} C_{ijkl} e_{lms} b_k + C_{ijkl} b_p k_l k_j \bar{U}_{kpm}(\mathbf{k})\right) \times e^{i\mathbf{k}\cdot(\mathbf{x}-\mathbf{x}')} dk_1 dk_2 dk_3 = 0, \tag{12.90}$$

which is satisfied for all k_i if

$$-\frac{k_s k_j}{\pi^2 k^2} C_{ijkl} e_{lms} b_k + C_{ijkl} b_p k_l k_j \overline{U}_{kpm}(k) = 0. \tag{12.91}$$

Furthermore, Eq. (12.91) must be valid for all b_i, and, putting it into the form

$$\left(-\frac{k_s k_j}{\pi^2 k^2} \mathbf{C}_{ijpl} e_{lms} + \mathbf{C}_{ijkl} k_l k_j \overline{U}_{kpm}(\mathbf{k})\right) b_p = 0 \tag{12.92}$$

it is evident that the relationship

$$-\frac{k_s k_j}{\pi^2 k^2} C_{ijpl} e_{lms} + C_{ijkl} k_l k_j \overline{U}_{kpm}(\mathbf{k}) = 0 \tag{12.93}$$

must be valid as well. Equation (12.93) can now be used to solve for $\overline{U}_{kpm}(\mathbf{k})$. It is readily confirmed that

$$k_s \mathbf{e}_{lms} = (\hat{\mathbf{e}}_m \times \mathbf{k})_l, \tag{12.94}$$

where the $\hat{\mathbf{e}}_i$ are the usual coordinate base vectors. Substituting this equality into Eq. (12.93), and using the usual contracted Christoffel tensor notation,[7]

$$(kk)_{ik} \overline{U}_{kpm}(\mathbf{k}) = \frac{(k, \hat{\mathbf{e}}_m \times \mathbf{k})_{ip}}{\pi^2 k^2}. \tag{12.95}$$

Therefore, $\overline{U}_{kpm}(\mathbf{k})$ is given by

$$\overline{U}_{kpm}(\mathbf{k}) = \frac{1}{\pi^2 k^2} (kk)^{-1}_{ki} (k, \hat{\mathbf{e}}_m \times \mathbf{k})_{ip}. \tag{12.96}$$

[7]For clarity in Eqs. (12.95)–(12.99) and (12.101) we insert a comma between the two quantities present in the parenthesis of the Christoffel symbols.

Next, substituting Eq. (12.96) into Eq. (12.84),

$$\mathrm{U_{ijk}}(\mathbf{x}-\mathbf{x}') = \frac{1}{\pi^2}\iiint\limits_{-\infty}^{\infty} \frac{1}{\mathrm{k}^2}(\mathrm{kk})^{-1}_{\mathrm{is}}(\mathrm{k}, \hat{\mathbf{e}}_\mathrm{k}\times\mathbf{k})_{\mathrm{sj}} e^{i\mathbf{k}\cdot(\mathbf{x}-\mathbf{x}')}\mathrm{dk_1dk_2dk_3} \tag{12.97}$$

and substituting this result into Eq. (12.83),

$$\begin{aligned}\mathrm{u_i}(\mathbf{x}) &= -\frac{\mathrm{b_i}}{4\pi}\Omega \\ &+\frac{1}{8\pi^3}\oint_C \mathrm{dx}'_\mathrm{k}\iiint\limits_{-\infty}^{\infty} \frac{1}{\mathrm{k}^2}(\mathrm{kk})^{-1}_{\mathrm{is}}(\mathrm{k}, \hat{\mathbf{e}}_\mathrm{k}\times\mathbf{k})_{\mathrm{sj}}\mathrm{b_j} e^{i\mathbf{k}\cdot(\mathbf{x}-\mathbf{x}')}\mathrm{dk_1dk_2dk_3}.\end{aligned} \tag{12.98}$$

But, $\mathrm{d}\mathbf{x}' = \hat{\mathbf{e}}'_\mathrm{i}\mathrm{dx}'_\mathrm{i} = \mathrm{d}\mathbf{s}' = \hat{\mathbf{t}}\mathrm{ds}'$, and Eq. (12.98) can therefore be written as

$$\begin{aligned}\mathrm{u_i}(\mathbf{x}) &= -\frac{\mathrm{b_i}}{4\pi}\Omega \\ &+\frac{1}{8\pi^3}\oint_C \mathrm{ds}'\iiint\limits_{-\infty}^{\infty} \frac{1}{\mathrm{k}^2}(\mathrm{kk})^{-1}_{\mathrm{is}}(\mathrm{k}, \hat{\mathbf{t}}\times\mathbf{k})_{\mathrm{sj}}\mathrm{b_j} e^{i\mathbf{k}\cdot(\mathbf{x}-\mathbf{x}')}\mathrm{dk_1dk_2dk_3},\end{aligned} \tag{12.99}$$

where the line integral involving ds′ is illustrated in Fig. 12.10. The amplitude, $\mathrm{A}(\mathbf{k})$, of the Fourier integral embedded in Eq. (12.99) is a homogeneous function of degree −2 in the variable k, and, as pointed out by Lothe (1992b), in such a case it can be reduced to a line integral around a unit circle with its plane normal to the unit vector $\hat{\mathbf{w}} = (\mathbf{x}-\mathbf{x}')/|\mathbf{x}-\mathbf{x}'|$ by following the theorem for A given by the equation

$$\iiint\limits_{-\infty}^{\infty} \mathrm{A}^{(-2)}(\mathbf{k})e^{i\mathbf{k}\cdot(\mathbf{x}-\mathbf{x}')}\mathrm{dk_1dk_2dk_3} = \frac{\pi}{|\mathbf{x}-\mathbf{x}'|}\int_0^{2\pi}\mathrm{A}^{(-2)}(\hat{\mathbf{m}})\mathrm{d}\theta, \tag{12.100}$$

where $\hat{\mathbf{m}}$, is a unit vector lying in the plane normal to $\hat{\mathbf{w}}$ as illustrated in Fig. (12.11).

Using this theorem, Eq. (12.99) finally reduces to the expression (Lothe 1992b)

$$\mathrm{u_i}(\mathbf{x}) = -\frac{\mathrm{b_i}}{4\pi}\Omega + \frac{1}{8\pi^2}\oint_C \mathrm{ds}'\frac{1}{|\mathbf{x}-\mathbf{x}'|}\int_0^{2\pi}(\hat{\mathbf{m}}\hat{\mathbf{m}})^{-1}_{\mathrm{is}}(\hat{\mathbf{m}}, \hat{\mathbf{t}}\times\hat{\mathbf{m}})_{\mathrm{sj}}\mathrm{b_j}\mathrm{d}\theta \tag{12.101}$$

for the displacement field of the loop, C.

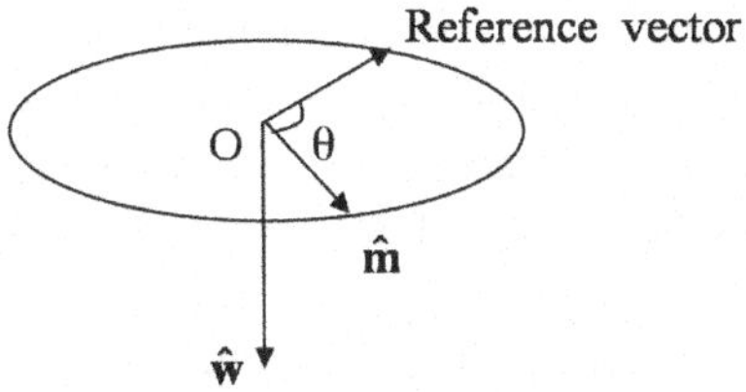

Fig. 12.11 Geometry for line integration of Eq. (12.101) around unit circle with its plane normal to $\hat{\mathbf{w}}$.

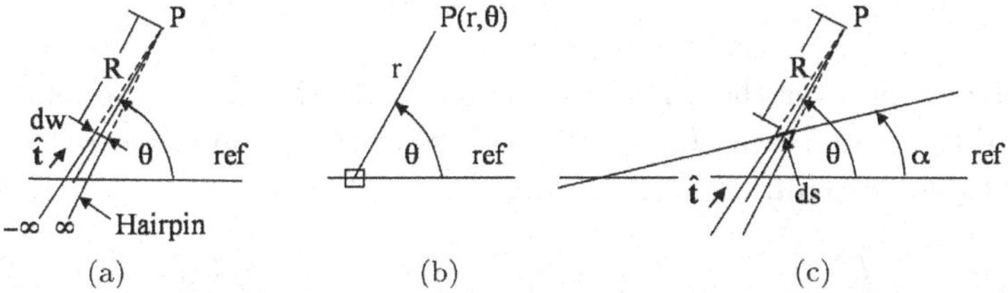

Fig. 12.12 (a) Dislocation "hairpin" configuration extending to $\pm\infty$: θ is measured from reference line. (b) Infinitesimal dislocation loop (small square). Field point P is at coordinates (r, θ). (c) Hairpin, same as in (a), but with base at oblique angle α measured from reference line. All elements are in plane of paper.

12.5.1.4 *By use of Brown's formula*

Planar loop

Brown's formula yields the stress field due to a planar dislocation loop at a field point in the plane of the loop. It has the notable feature that it is expressed in terms of parameters that are characteristic of infinitely long straight dislocations that can be determined using the methods described earlier in Sec. 12.3. Even though the formula provides only the in-plane stress, it can also be used to obtain the solutions of a wide range of problems involving three-dimensional dislocation configurations as demonstrated below, for example, in Sec. 12.7.1.3. Following Lothe (1992a,b), all dislocation configurations that can be analyzed by the application of Brown's formula, and additional related formulae that are derived below, can be produced by combining basic dislocation "hairpins" having the configuration illustrated in Fig. 12.12a.

The first task is therefore to find the stress field of such a "hairpin" at a field point P as indicated in Fig. 12.12a. For this purpose the hairpin may be regarded as an array of abutting infinitesimal loops as is done, for example, in Fig. 12.19 to find the stress due to a finite dislocation loop.

According to Eq. (12.169), the stress field of an infinitesimal loop, $d\sigma_{ij}(\mathbf{x})$, falls off as x^{-3}, possesses a complicated angular dependence, and is an even function of x.[8] It is therefore assumed that in the planar r, θ coordinate system of Fig. 12.12b the stress follows the general form

$$d\sigma_{ij}(r, \theta) = \frac{1}{r^3}\alpha_{ij}(\theta)dS, \tag{12.102}$$

where dS is its area, and the function $\alpha_{ij}(\theta)$ obeys

$$\alpha_{ij}(\theta + \pi) = \alpha_{ij}(\theta). \tag{12.103}$$

By integrating over the stresses contributed by the array of infinitesimal loops that comprise the hairpin, the stress at P at a distance R from the base of the hairpin in Fig. 12.12a is

$$d\sigma_{ij}(P) = d\theta \int_R^\infty \frac{1}{r^3}\alpha_{ij}(\theta)rdr = \alpha_{ij}(\theta)d\theta \int_R^\infty \frac{1}{r^2}dr = \frac{\alpha_{ij}(\theta)d\theta}{R} = \frac{\alpha_{ij}(\theta)dw}{R^2} \tag{12.104}$$

after using $dS = rd\theta dr$ and $dw = Rd\theta$. If the base of the hairpin is oblique at the angle α, as illustrated in Fig. 12.12c, $d\sigma_{ij}(P)$ remains unchanged to first order, and *dw* becomes $dw = \sin(\theta - \alpha)ds$. Therefore, Eq. (12.104) takes the form

$$d\sigma_{ij}(P) = \frac{\alpha_{ij}(\theta)}{R^2}\sin(\theta - \alpha)ds. \tag{12.105}$$

Next, consider Fig. 12.13 which shows how an infinitely long straight dislocation located at a distance ρ from the field point P can be constructed by employing an array of abutting hairpins. The dashed line along $\hat{\mathbf{t}}$ in Fig. 12.13a represents the desired position of the dislocation, and a segment of the dislocation, *ds*, due to the presence of a single hairpin is shown in place. In Fig. 12.13b, the dislocation is built up by adding further hairpins. All abutting hairpin elements cancel, and only the *ds* base segments are left on the dislocation line along with two semi-infinite outer segments. However, for an infinite dislocation length these latter segments can be neglected when considering the stress at P. The stress there is then just the

[8]This can be readily shown by taking derivatives of Eq. (12.169) to obtain distortions and then employing Hooke's law.

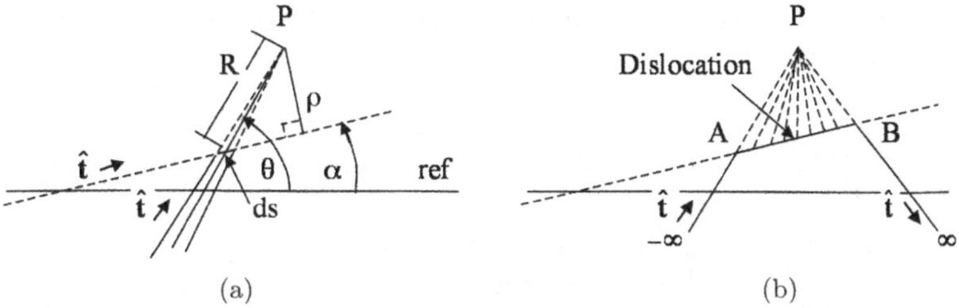

Fig. 12.13 Construction of long straight dislocation by using hairpin dislocations. (a) Dashed line along $\hat{\mathbf{t}}$ is intended position of dislocation: segment ds on the line is contributed by one hairpin. (b) The dislocation is built up by adding further hairpins. All elements in plane of paper.

sum of the stresses contributed by all hairpins, and using Eq. (12.105),

$$\sigma_{ij}(P) = \int d\sigma_{ij}(P) = \int_{-\infty}^{+\infty} \frac{\alpha_{ij}(\theta)}{R^2} \sin(\theta - \alpha) ds. \tag{12.106}$$

The function $\alpha_{ij}(\theta)$ can now be expressed in terms of parameters associated with long straight dislocations by noting that, since the stress field at P is the field of a long straight dislocation, it must fall off as ρ^{-1} (at constant α) and also be dependent on α. It can therefore be written in the general form

$$\sigma_{ij}(P) = \frac{1}{\rho} \Sigma_{ij}(\alpha). \tag{12.107}$$

Then, equating Eqs. (12.106) and (12.107),

$$\frac{1}{\rho} \Sigma_{ij}(\alpha) = \int_{-\infty}^{+\infty} \frac{\alpha_{ij}(\theta)}{R^2} \sin(\theta - \alpha) ds. \tag{12.108}$$

But, from the geometry of Fig. 12.13,

$$\rho = R \sin(\theta - \alpha), \qquad \frac{ds}{d\theta} = \frac{R}{\sin(\theta - \alpha)}. \tag{12.109}$$

Also, since ρ is constant, the relationship

$$\Sigma_{ij}(\alpha) = \int_{\alpha}^{\alpha+\pi} \alpha_{ij}(\theta) \sin(\theta - \alpha) d\theta \tag{12.110}$$

is obtained which establishes a link between the quantities $\Sigma_{ij}(\alpha)$ and $\alpha_{ij}(\theta)$. Differentiating Eq. (12.110), while invoking Leibniz's Rule (since the limits of integration are functions of α) and making use of the form of Eq. (12.103),

$$\frac{d\Sigma_{ij}(\alpha)}{d\alpha} = -\int_{\alpha}^{\alpha+\pi} \alpha_{ij}(\theta)\cos(\theta-\alpha)d\theta,$$

$$\begin{aligned}\frac{d^2\Sigma_{ij}(\alpha)}{d\alpha^2} &= -\int_{\alpha}^{\alpha+\pi} \alpha_{ij}(\theta)\sin(\theta-\alpha)d\theta + \alpha_{ij}(\alpha+\pi) + \alpha_{ij}(\alpha)\\ &= -\int_{\alpha}^{\alpha+\pi} \alpha_{ij}(\theta)\sin(\theta-\alpha)d\theta + 2\alpha_{ij}(\alpha).\end{aligned} \tag{12.111}$$

Then, using Eqs. (12.110) and (12.111),

$$\alpha_{ij}(\alpha) = \frac{1}{2}\left[\Sigma_{ij}(\alpha) + \frac{d^2\Sigma_{ij}(\alpha)}{d\alpha^2}\right]. \tag{12.112}$$

Finally, substituting Eq. (12.112) into Eq. (12.105),

$$d\sigma_{ij}(P) = \frac{1}{2R^2}\sin(\theta-\alpha)\left[\Sigma_{ij}(\theta) + \frac{d^2\Sigma_{ij}(\theta)}{d\theta^2}\right]ds \tag{12.113}$$

is obtained for the stress increment at P due to the hairpin in Fig. 12.12c.

Having this result, an expression for the stress produced by a closed planar loop at a field point P coplanar with the loop can be determined by integrating Eq. (12.113) around the loop. Figure 12.14a shows the construction of such a loop by using hairpins in the same manner as in Fig. 12.13b. All dislocation segments except those remaining on the loop cancel in this case, and by integrating Eq. (12.113) with the help of Fig. 12.14b, we obtain Brown's formula,

$$\sigma_{ij}(\mathbf{x}) = \frac{1}{2}\oint_C \frac{\sin(\theta-\alpha)}{R^2}\left[\Sigma_{ij}(\theta) + \frac{d^2\Sigma_{ij}(\theta)}{d\theta^2}\right]ds, \tag{12.114}$$

where $R = |\mathbf{x} - \mathbf{x}'|$.

The second derivative in Eq. (12.114) can be eliminated by first substituting Eq. (12.109) so that

$$\sigma_{ij}(\mathbf{x}) = \frac{1}{2}\oint_C \frac{1}{R}\left[\Sigma_{ij}(\theta) + \frac{d^2\Sigma_{ij}(\theta)}{d\theta^2}\right]d\theta \tag{12.115}$$

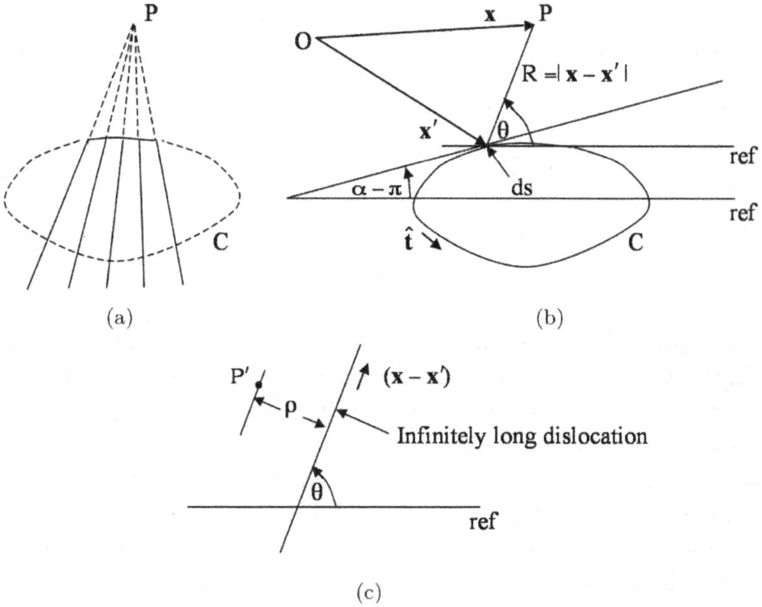

Fig. 12.14 (a) Construction of planar loop C using hairpins. (b) Geometry for finding stress due to loop at field point P by integrating around the loop. (c) Diagram for finding $\Sigma_{ij}(\theta)$ for segment *ds* in (b) (see text). All elements in plane of the paper. The Σ surface for the loop is taken in the plane of the paper, and the positive unit normal vector, $\hat{n}$, for this surface is, therefore, directed towards the reader (see rule, p. 325).

and then integrating the second derivative term by parts to obtain

$$\begin{aligned}\oint_C \frac{1}{R}\frac{d^2\Sigma_{ij}(\theta)}{d\theta^2}d\theta &= \left|\frac{1}{R}\frac{d\Sigma_{ij}(\theta)}{d\theta}\right| + \oint_C \frac{1}{R^2}\frac{d\Sigma_{ij}}{d\theta}d\mathbf{R} \\ &= -\oint_C \frac{1}{R}\frac{d\Sigma_{ij}}{d\theta}\cot(\theta-\alpha)d\theta \end{aligned} \tag{12.116}$$

after using the relation

$$dR = -R\cot(\theta-\alpha)d\theta \tag{12.117}$$

obtained from the geometry in Fig. 12.14. Then, substituting Eq. (12.116) into Eq. (12.115),

$$\sigma_{ij}(\mathbf{x}) = \frac{1}{2}\oint_C \frac{1}{R^2}\left[\sin(\theta-\alpha)\Sigma_{ij}(\theta) - \cos(\theta-\alpha)\frac{d\Sigma_{ij}(\theta)}{d\theta}\right]ds. \tag{12.118}$$

It is often useful to know the in-plane traction in the direction of **b** due to the stress field of the loop. With the use of Eq. (12.115), this is given by

$$T_b = \frac{\mathbf{T}\cdot\mathbf{b}}{b} = \frac{T_i b_i}{b} = \frac{\sigma_{ij}(P)\hat{n}_j b_i}{b} = \frac{1}{b}\oint_C \frac{\sin(\theta-\alpha)}{R^2}\left[F(\theta) + \frac{\partial^2}{\partial\theta^2}F(\theta)\right] ds, \tag{12.119}$$

where

$$F(\theta) = \frac{1}{2}\Sigma_{ij}(\theta) b_i \hat{n}_j. \tag{12.120}$$

When employing these expressions it must be emphasized that the quantity $\Sigma_{ij}(\theta)/\rho$, by virtue of Eq. (12.207), is the in-plane stress at a point P′ at a distance ρ from a fictitious infinitely long straight dislocation lying at the angle θ to the reference line as illustrated in Fig. 12.14c. The distance ρ from the straight dislocation line is in the direction of the unit vector $\hat{\mathbf{n}} \times (\mathbf{x}-\mathbf{x}')/|\mathbf{x}-\mathbf{x}'|$ (Bacon, Barnett and Scattergood (1979b). Also, of course, the straight dislocation must have the same Burgers vector as the loop, and the reference line must always have the same orientation in the crystal.

The required values of $\Sigma_{ij}(\theta)$, and its derivatives, can then be obtained from calculations of the stress fields of straight dislocations using the methods of Sec. 12.3. Bacon, Barnett and Scattergood (1979b) give detailed descriptions of optimum methods.

Non-planar loop

As is evident from the geometry of Fig. 12.15, Brown's formula can also be used to determine the 3-dimensional stress field of a closed non-planar

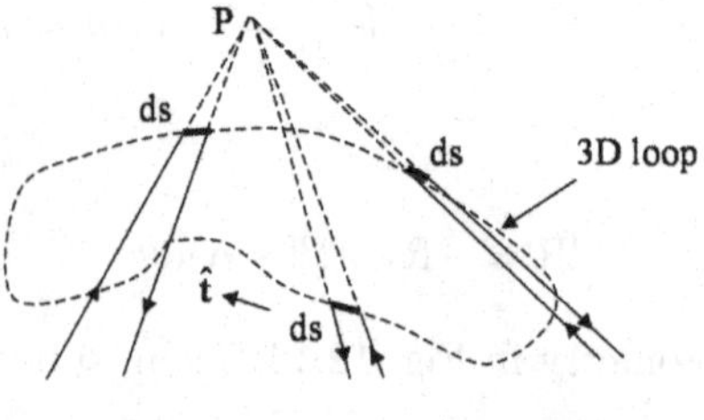

Fig. 12.15 Construction of a non-planar three-dimensional loop (dashed) using hairpins (three of which are shown) which all terminate at the field point P located at an arbitrary point in three-dimensional space. Upon completion of the loop all hairpin segments will have mutually annihilated except for the ds segments comprising the loop. The stress at P will then be the sum of the contributions of all the hairpins (see text).

loop (see caption). Since the stress at the field point P in Fig. 12.15 is just the sum of the stresses contributed by all segments, it will be given by an equation of the type of Eq. (12.118) which was derived previously for two-dimensional loops. However, in the present case the line integral is three-dimensional, and its evaluation will be more tedious, since the stress increment contributed by each particular differential segment, *ds*, must be calculated using parameters that refer to the plane containing P and that particular segment.

12.5.1.5 *By use of rational differential segments*

The elastic field of a general smoothly curved loop can also be found by the use of *rational differential segments* (Eshelby and Laub 1967; Indenbom and Dubnova 1967; Lothe 1992b) one of which is illustrated in Fig. 12.16. Such a segment consists of a differential length of dislocation, $\mathbf{ds} = \hat{\mathbf{t}}\,\mathrm{ds}$, with Burgers vector, $\mathbf{b}$, lying between an initial node, where it is joined to a very large number ($N \rightarrow \infty$) of half-infinite straight dislocations that converge radially on the node, and a terminating node, where it is joined to a similar distribution of N emanating dislocations. The incoming and outgoing dislocations are each of infinitesimal Burgers vector strength and are uniformly distributed in all radial directions. Also, the sum of the Burgers vectors of all dislocations entering the initial node and leaving the terminating node must be equal. A rational differential segment is therefore completely characterized by its Burgers vector and tangent vector.

Such a segment has the unique property that, when it is joined end-to-end to another rational segment with the same Burgers vector and positive sense, the overlapping semi-infinite radial dislocations at the junction exactly cancel regardless of the angle between the two segments since

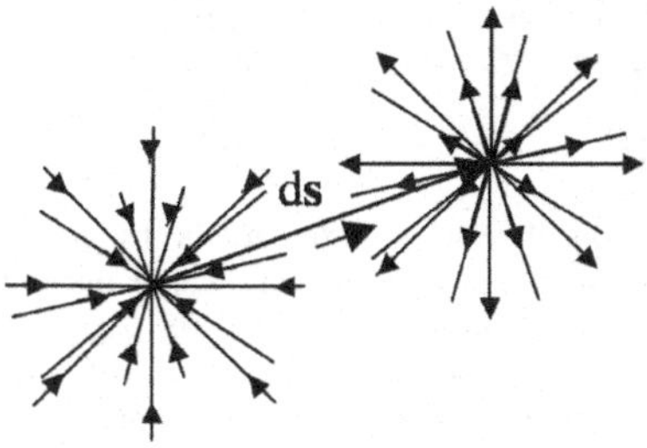

Fig. 12.16 Rational differential segment, $(\mathbf{ds}, \mathbf{b})$. Arrows indicate positive direction along dislocations.

they have opposite Burgers vectors and are uniformly distributed in all directions. A dislocation loop of any shape can therefore be produced by joining rational segments together end-to-end in the form of a closed chain, and the resulting elastic field will simply be the sum of the fields contributed by the individual segments. Then, if the elastic field of an individual segment is known, the field of the loop can be determined by means of a line integral around the loop. We therefore proceed to find the elastic field of a single rational segment following the derivation of Lothe (1992b) which employs the transformation strain formalism of Sec. 3.6.

The transformation strain required to produce a rational segment is obtained by imagining that it is constructed by introducing $N \to \infty$ semi-infinite dipole loops with infinitesimal Burgers vectors which converge radially so that their short transverse end segments superimpose to form the segment $\mathbf{ds}$ having a finite Burgers vector, $\mathbf{b}$, as illustrated in Fig. 12.17a where only a few of the $N \to \infty$ dipoles are shown. Note that the incoming and outgoing dislocations contributed by the dipoles are in the desired radial configuration at each end of the segment. Since the directions of the incoming dipoles are uniformly distributed over all radial directions, the various cuts and displacements required to create the dipole loops produce a plastic (transformation) displacement field, $u_i^T(\mathbf{x})$, throughout the surrounding crystal.

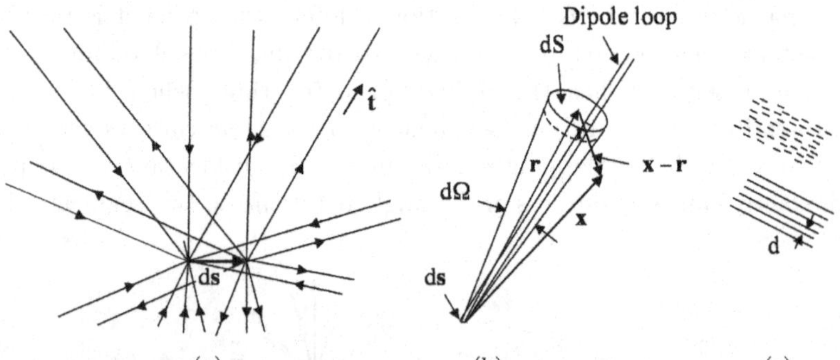

Fig. 12.17 (a) Construction of rational segment, ds, by cuts and displacements corresponding to dipole loops with their end segments superimposed on ds. (b) Dipole loop lying in solid angle $d\Omega$ centered on vector $\mathbf{r}$. Area dS is perpendicular to $\mathbf{r}$. A trace is produced on dS where the dipole loop passes through it. (c) Upper diagram; oblique view of the traces of some of the dipole loops emanating from ds and lying nearly parallel to $\mathbf{r}$ and $\mathbf{x}$ (not shown in figure) passing through an extension of the plane of dS. Lower diagram; traces of the sheets of displacement that would be obtained by consolidating all of the dipole traces shown in the upper diagram into sheets.

To determine this field, consider the plastic displacement produced at the vector position **x** relative to the displacement at the nearby vector position **r** by the radial distribution of dipole loops lying in the region between **x** and **r** in Fig. 12.17b. One of these loops is shown in Fig. 12.17b where it produces a trace when passing through the area dS. The upper diagram in Fig. 12.17c is an oblique view of the traces produced by some of the radial dipole loops lying between **x** and **r** passing through an extension of the plane of dS. The length of each trace is equal to the width of the dipole loop which is given by

$$\mathrm{w}^{\mathrm{dip}} = |\mathbf{ds} \times \hat{\mathbf{t}}| = \frac{|\mathbf{ds} \times \mathbf{r}|}{\mathrm{r}} \tag{12.121}$$

and the total number of dipole loops that pass through the plane, per unit area, is given to a good approximation (since **x** is in close proximity to **r**) by

$$n^{\mathrm{dip}} = \frac{1}{\mathrm{dS}}\mathrm{N}\frac{\mathrm{d}\Omega}{4\pi} = \frac{1}{\mathrm{dS}}\mathrm{N}\frac{\mathrm{dS}}{4\pi\mathrm{r}^2} = \frac{\mathrm{N}}{4\pi\mathrm{r}^2}. \tag{12.122}$$

Now, by means of small insignificant shifts of the inclinations of these loops they can be aligned (for purposes of calculation) so that the individual loop planes merge and form continuous and uniformly spaced sheets of displacement in the vicinity of **x** and **r** which produce the traces seen in the lower diagram of Fig. 12.17c. The spacing between sheets is then

$$\mathrm{d} = \frac{1}{\mathrm{w}^{\mathrm{dip}}n^{\mathrm{dip}}} = \frac{4\pi\mathrm{r}^3}{|\mathrm{ds} \times \mathbf{r}|\mathrm{N}} \tag{12.123}$$

and there will be a displacement across each sheet corresponding to the dipole Burgers vector given by

$$\mathbf{b}^{\mathrm{dip}} = \frac{\mathbf{b}}{\mathrm{N}}. \tag{12.124}$$

Having the above quantities, the plastic displacement at **x** relative to the displacement at **r** due to the presence of the intervening sheets is

$$\mathbf{du}^{\mathrm{T}}(\mathbf{x}) - \mathrm{du}^{\mathrm{T}}(\mathbf{r}) = \left[(\mathbf{x}-\mathbf{r})\cdot\hat{\mathbf{n}}^{\mathrm{dip}}\right]\left[\frac{1}{\mathrm{d}}\right]\left[\mathbf{b}^{\mathrm{dip}}\right] = \frac{[(\mathbf{x}-\mathbf{r})\cdot(\mathrm{ds}\times\mathbf{r})]\mathbf{b}}{4\pi\mathrm{r}^3}, \tag{12.125}$$

where $\hat{\mathbf{n}}^{\mathrm{dip}}$ is the unit normal vector to the plane of each dipole given by

$$\hat{\mathbf{n}}^{\mathrm{dip}} = \frac{\mathrm{ds}\times\mathbf{r}}{|\mathrm{ds}\times\mathbf{r}|}. \tag{12.126}$$

The first quantity in square brackets in Eq. (12.125) is the distance between $\mathbf{x}$ and $\mathbf{r}$ projected along the unit normal to the sheets, the second quantity is the number of sheets per unit distance normal to the sheets, and the third quantity is the displacement due to each sheet.

Having the displacements given by Eq. (12.125), the transformation distortion $\partial u_i^T/\partial x_j$ is given by

$$\frac{\partial u_i^T}{\partial x_j} = \frac{b_i}{4\pi r^3}\frac{\partial}{\partial x_j}(\mathbf{x}\cdot \mathbf{ds}\times \mathbf{r}) = \frac{b_i(\mathbf{ds}\times \mathbf{r})_j}{4\pi r^3} = \frac{b_i ds_m r_n e_{jmn}}{4\pi r^3} \qquad (12.127)$$

with the help of Eq. (E.5). A similar expression is obtained for $\partial u_j^T/\partial x_i$, and using Eq. (2.5), the transformation strain is then

$$d\varepsilon_{ij}^T(\mathbf{r}) = \frac{(b_i e_{jmn} + b_j e_{imn}) ds_m r_n}{8\pi r^3} = \frac{b_i e_{jmn} ds_m r_n}{4\pi r^3}. \qquad (12.128)$$

The vector $\mathbf{r}$ can now be replaced by $\mathbf{x}$ (the usual field vector). Then, by use of the equality,

$$\frac{\partial}{\partial x_n}\left(\frac{1}{x}\right) = -\frac{x_n}{x^3} \qquad (12.129)$$

Eq. (12.128) can be written as

$$d\varepsilon_{ij}^T(\mathbf{x}) = -\frac{(b_i e_{jmn} + b_j e_{imn}) ds_m}{8\pi}\frac{\partial}{\partial x_n}\left(\frac{1}{x}\right). \qquad (12.130)$$

Finally, by substituting Eq. (12.86) into the above expression, the transformation strain associated with the rational segment is obtained in the form

$$d\varepsilon_{ij}^T(\mathbf{x}) = -\frac{i(b_i e_{jmn} + b_j e_{imn}) ds_m}{16\pi^3}\iiint_{-\infty}^{\infty} \hat{k}_n k^{-1} e^{i\mathbf{k}\cdot\mathbf{x}} dk_1 dk_2 dk_3. \qquad (12.131)$$

Having Eq. (12.131), an expression for the stress due to a smooth dislocation loop can now be obtained. A comparison of Eq. (12.131) with Eq. (F.4) shows that the Fourier transform of $d\varepsilon_{ij}^T(\mathbf{x})$ is

$$d\bar{\varepsilon}_{ij}^T(\mathbf{k}) = -\frac{i(b_i e_{jmn} + b_j e_{imn}) ds_m}{16\pi^3}\hat{k}_n k^{-1} = -\frac{ik^{-2}}{16\pi^3}\left[b_i(\mathbf{ds}\times\mathbf{k})_j + b_j(\mathbf{ds}\times\mathbf{k})_i\right] \qquad (12.132)$$

and, having this, the transform of the stress due to the segment can be obtained by employing the Hooke's law-type expression between transforms given by Eq. (3.164), i.e.,

$$
\begin{aligned}
d\bar{\sigma}_{ij}(\mathbf{k}) &= C^*_{ijkl}(\mathbf{k})d\bar{\varepsilon}^T_{kl}(\mathbf{k}) = -\frac{ik^{-2}}{8\pi^3}C^*_{ijkl}(\mathbf{k})b_k(d\mathbf{s}\times\mathbf{k})_l \\
&= -\frac{ik^{-2}}{8\pi^3}C^*_{ijkl}(\mathbf{k})b_k(d\hat{\mathbf{t}}\times\mathbf{k})_l ds. \qquad (12.133)
\end{aligned}
$$

Then, by inverting Eq. (12.133) using Eq. (F.4), the increment of stress at $\mathbf{x}$ contributed by a rational differential segment at $\mathbf{x}'$ is

$$
\begin{aligned}
&d\sigma_{ij}(\mathbf{x}-\mathbf{x}') \\
&= -\frac{1}{8\pi^3}\int_{-\infty}^{\infty}\int_{-\infty}^{\infty}\int_{-\infty}^{\infty}C^*_{ijkl}(\mathbf{k})b_k(d\hat{\mathbf{t}}\times\mathbf{k})_l e^{i\mathbf{k}\cdot(\mathbf{x}-\mathbf{x}')}ik^{-2}dk_1dk_2dk_3ds'.
\end{aligned}
\qquad (12.134)
$$

Finally, by use of Eq. (12.134), the stress due to a loop, C, at a field point $\mathbf{x}$ is obtained by performing a line integral around the loop as in Fig. 12.10, i.e.,

$$
\begin{aligned}
\sigma_{ij}(\mathbf{x}) &= -\frac{1}{8\pi^3}\oint_C ds' \\
&\quad\times\int_{-\infty}^{\infty}\int_{-\infty}^{\infty}\int_{-\infty}^{\infty}C^*_{ijkl}(\mathbf{k})b_k(d\hat{\mathbf{t}}\times\mathbf{k})_l e^{i\mathbf{k}\cdot(\mathbf{x}-\mathbf{x}')}ik^{-2}dk_1dk_2dk_3.
\end{aligned}
\qquad (12.135)
$$

12.5.1.6 *By use of infinitesimal loops*

We now describe the final method that we shall consider for obtaining the stress field of a loop, i.e., the method of using infinitesimal loops. The concept of an infinitesimal dislocation loop (Kroupa 1962; 1966 and Hirth and Lothe 1982) is introduced, and its use in the anisotropic elasticity theory of finite loops is described.

A general infinitesimal dislocation loop of area dS, shown in Fig. 12.18a, is characterized by its Burgers vector and its positive unit normal vector, $\hat{\mathbf{n}}$, as defined by the Σ cut/displacement rule given on p. 325. The displacement field produced by the loop can be readily obtained by the method employed in Sec. 12.4.1.1 and Exercise 12.5 where the loop is created by a cut and displacement on the surface Σ which is mimicked by a concentrated transformation strain in the form of a delta function. The displacement field is then obtained by determining the corresponding transformation stress using

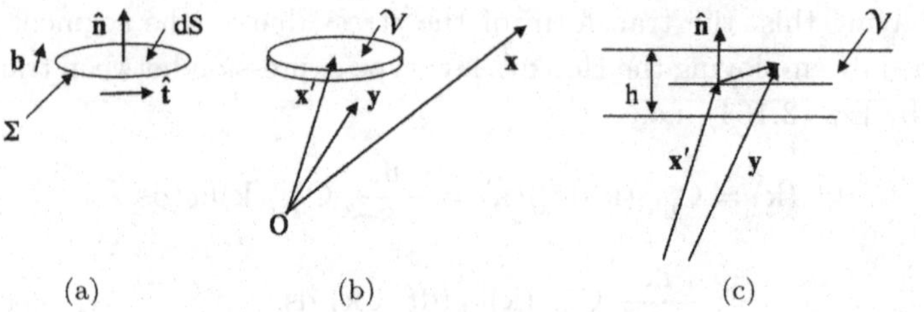

Fig. 12.18 (a) Infinitesimal dislocation loop of area dS. (b) Region $\mathcal{V}$ containing transformation strain that produces the loop in (a). (c) Enlarged view of the region $\mathcal{V}$ in the vicinity of its center located at $\mathbf{x}'$. The viewing direction is parallel to the two broad faces of $\mathcal{V}$.

Hooke's law and substituting the stress into Eq. (3.168). The transformation strain is initially taken to be a homogeneous strain distributed throughout the region $\mathcal{V}$ in Fig. 12.18b in which the upper surface is displaced relative to the lower surface by $-\mathbf{b}$ in accordance with the Σ cut/displacement rule given on p. 325. If the thickness of $\mathcal{V}$ is h, and the center of $\mathcal{V}$ is at $\mathbf{x}'$ as in Fig. 12.18c, the transformation displacement at $\mathbf{y}$ is

$$\mathbf{u}^{\mathrm{T}}(\mathbf{y}) = -\mathbf{b}\frac{(\mathbf{y}-\mathbf{x}')\cdot\hat{\mathbf{n}}}{h} \quad \text{or} \quad u_k^{\mathrm{T}}(\mathbf{y}) = -\frac{b_k(y_i - x_i')\hat{n}_i}{h} \qquad (12.136)$$

and the corresponding strain is

$$\varepsilon_{kj}^{\mathrm{T}} = \frac{1}{2}\left[\frac{\partial u_k^{\mathrm{T}}}{\partial y_j} + \frac{\partial u_j^{\mathrm{T}}}{\partial y_k}\right] = -\frac{1}{2h}(b_k\hat{n}_j + b_j\hat{n}_k). \qquad (12.137)$$

Having this, the transformation strain can be localized in the plane of the cut by writing it in the delta function form[9]

$$\varepsilon_{kj}^{\mathrm{T}} = -\frac{1}{2}(b_k\hat{n}_j' + b_j\hat{n}_k)\delta(\xi), \qquad (12.138)$$

where $\delta(\xi)$ is a one-dimensional delta function and ξ measures distance from dS along $\hat{\mathbf{n}}$. Then, substituting Eq. (12.138) into Eq. (3.168), and using Hooke's law, and the symmetry properties of the C_{jkmn} tensor, the differential displacement at $\mathbf{x}$ due to an infinitesimal loop of area dS

[9]See, as an example, Fig. 12.7.

located at $\mathbf{x}'$ is

$$du_i = -C_{jkmn} b_m \int_{-\infty}^{\infty} \frac{\partial G_{ik}^{\infty}(\mathbf{x}-\mathbf{x}')}{\partial x_j'} \hat{n}_n' \delta(\xi') d\xi' dS' \qquad (12.139)$$

or

$$du_i = -C_{jkmn} b_m \frac{\partial G_{ik}^{\infty}(\mathbf{x}-\mathbf{x}')}{\partial x_j'} \hat{n}_n' dS'. \qquad (12.140)$$

Then, substituting Eq. (4.40) for the Green's function derivatives,

$$du_i = -\frac{C_{jkmn} b_m \hat{n}_n'}{8\pi^2} \frac{1}{|\mathbf{x}-\mathbf{x}'|^2} \times \oint_{\hat{\mathcal{L}}} \left\{ \hat{w}_j (\hat{k}\hat{k})_{ik}^{-1} - \hat{k}_j (\hat{k}\hat{k})_{is}^{-1} \left[(\hat{k}\hat{w})_{sr} + (\hat{w}\hat{k})_{sr} \right] (\hat{k}\hat{k})_{rk}^{-1} \right\} ds dS'. \qquad (12.141)$$

According to St. Venant's principle, the displacement field of a dislocation loop should be independent of the detailed shape of the loop at distances from the loop larger than approximately its largest dimension. The differential displacement, du_i, given by Eq. (12.141) is seen to be proportional to the differential area of the loop, dS. Therefore Eq. (12.141) is expected to give the finite displacement field, $u_i(\mathbf{x})$, of a finite loop of area S (obtained by increasing dS to S) at distances that are large compared to the loop dimensions. Also, as shown in Fig. 12.19, a finite loop is equivalent to an array of abutting infinitesimal loops, and, therefore, the field of a finite loop at all distances from the loop outside of its core can be obtained (Kroupa 1966) by summing the contributions made at the field point by the infinitesimal loops comprising the array according to

$$u_i = \iint_{\Sigma} du_i, \qquad (12.142)$$

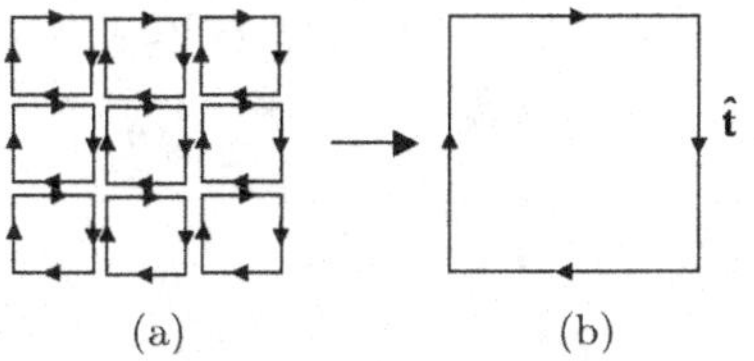

Fig. 12.19 Equivalence between a finite dislocation loop in (b) and an array of abutting infinitesimal loops in (a). In (b) all adjacent interior segments present in (a) have mutually annihilated one another.

where Σ is any surface region bounded by the loop. Therefore, substituting Eq. (12.140) into Eq. (12.142),

$$u_i(\mathbf{x}) = -C_{jkmn} b_m \iint_{\Sigma} \frac{\partial G^{\infty}_{ik}(\mathbf{x}-\mathbf{x}')}{\partial x'_j} \hat{n}'_n dS'. \tag{12.143}$$

Equation (12.143) is seen to be identical to the Volterra equation, i.e., Eq. (12.67), as might have been anticipated. The preceding development leading to Eq. (12.143) is, therefore, an alternative derivation of this basic equation in essence.

12.5.2 *Strain energies*

The strain energy of a dislocation loop can be readily obtained from the interaction energy between two dislocation loops as expressed by Eq. (16.22) in Sec. 16.3.1.2. To demonstrate this, consider a general dislocation loop labeled $C^{(1)}$. Now imagine superposing on this loop a second loop $C^{(2)}$, that is an exact duplicate, to produce the arrangement shown in Fig. 12.20 which is elastically equivalent to a dislocation with twice the Burgers vector of either loop. The elastic strain energy of any loop must vary as b^2, since both its stress field and strain field are proportional to b. Therefore, the strain energy of the superposed loops, $W^{C^{(1)}+C^{(2)}}$, will be four times as large as the strain energy of loop $C^{(1)}$, i.e.,$W^{C^{(1)}}$. This establishes the strain energy relationship,

$$W^{C^{(1)}+C^{(2)}} = 4W^{C^{(1)}} = W^{C^{(1)}} + W^{C^{(2)}} + W_{\text{int}}^{C^{(1)}/C^{(2)}}, \tag{12.144}$$

where $W_{\text{int}}^{C^{(1)}/C^{(2)}}$ is the interaction energy between the two superposed loops, and $W^{C^{(1)}} = W^{C^{(2)}}$. Thus,

$$W^{C^{(1)}} = \frac{1}{2} W_{\text{int}}^{C^{(1)}/C^{(2)}}, \tag{12.145}$$

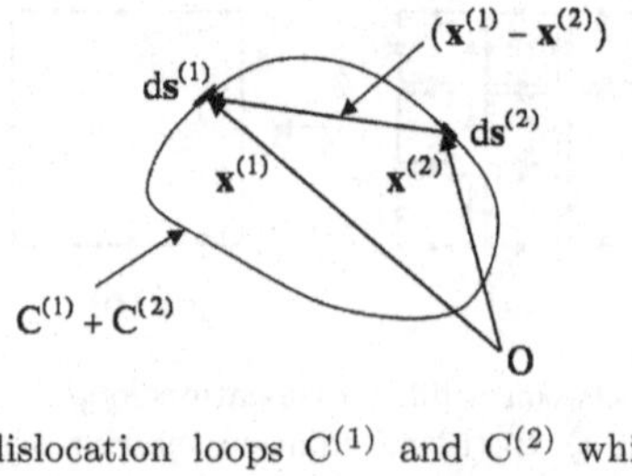

Fig. 12.20 Superposed dislocation loops $C^{(1)}$ and $C^{(2)}$ which are duplicates of one another.

proving that the loop strain energy is just half of the interaction energy between the loop and its duplicate. This latter quantity is given by Eq. (16.22) with $C^{(1)} = C$, $C^{(2)} = C$, $\mathbf{b}^{(1)} = \mathbf{b}^{(2)} = \mathbf{b}$, and therefore, the strain energy of a loop C is given by

$$W^C = \frac{1}{16\pi^2} \oint_{C^{(1)}=C} \oint_{C^{(2)}=C} \frac{1}{|\mathbf{x}^{(1)} - \mathbf{x}^{(2)}|} \times \int_0^{2\pi} C^*_{ijkl}(\hat{\mathbf{m}}) b_k (d\mathbf{s}^{(1)} \times \hat{\mathbf{m}})_l b_i (d\mathbf{s}^{(2)} \times \hat{\mathbf{m}})_j d\theta. \quad (12.146)$$

The energy given by Eq. (12.146) will diverge in an unrealistic manner when segments $d\mathbf{s}^{(1)}$ and $d\mathbf{s}^{(2)}$ become close together during the double line integration, and $1/|\mathbf{x}^{(1)} - \mathbf{x}^{(2)}| \to \infty$. This singularity can be avoided by assuming that the segments do not interact when they are closer than a *cutoff distance* denoted by ρ. Note that this is just another example of dealing with the breakdown of linear elasticity at the dislocation core. This cutoff approach is also used in Secs. 12.6.2 and 12.8.2 where isotropic systems are considered, and an example of its application appears in Exercise 12.11.

12.6 Smoothly Curved Dislocation Loops in Isotropic Systems

12.6.1 *Elastic fields*

The elastic field of a smoothly curved loop in an isotropic system can be obtained by using the methods described above in Sec. 12.5.1 modified for an isotropic system. However, it is often more convenient to employ a modified Volterra equation in the form of either the *Burgers equation* or the *Peach–Koehler equation*, and these are therefore derived in the following two sections. Both involve line integrals around the loop, but the Burgers equation yields the displacement field, while the Peach–Koehler equation yields the stress field. However, as pointed out by Khraishi, Hirth and Zbib (2000), the task of obtaining analytical closed-form solutions of these equations is non-trivial. The integrations, when possible, are generally complicated and tedious and will not be pursued here in any detail. Examples of successful integrations include the integration of the Peach–Koehler equation to obtain the stress field of a circular loop with a general Burgers vector by Khraishi, Hirth, and Zbib (2000), and the integration of the

Burgers equation to obtain its displacement field by Khraishi, Hirth, Zbib and Khaleel (2000).

Finally, the method of using infinitesimal loops to determine the elastic field of a smoothly curved loop is discussed in Sec. 12.6.1.3.

12.6.1.1 *By use of Burgers equation*

The Burgers equation (Burgers 1939a,b; Hirth and Lothe 1982) for the displacement field due to a loop in an isotropic system is derived from the Volterra equation by converting the elastic constants to isotropic constants, introducing the isotropic Green's function, and then recasting the equation as a line integral with the help of Stokes' theorem. Substitution of Eq. (2.120) into Eq. (12.67) gives

$$u_i(\mathbf{x}) = \iint_{\Sigma} \left[-\lambda \frac{\partial G_{ij}(\mathbf{x}-\mathbf{x}')}{\partial x'_j} b_m \hat{n}'_m - \mu \frac{\partial G_{im}(\mathbf{x}-\mathbf{x}')}{\partial x'_n} b_m \hat{n}'_n - \mu \frac{\partial G_{in}(\mathbf{x}-\mathbf{x}')}{\partial x'_m} b_m \hat{n}'_n \right] dS', \tag{12.147}$$

where the geometry for the surface integration is shown in Fig. 12.1. Next, Eq. (4.110) for the Green's function is substituted to obtain

$$u_i(\mathbf{x}) = -\frac{1}{4\pi} \iint_{\Sigma} \left[b_i \hat{n}'_n \frac{\partial}{\partial x'_n} \left(\frac{1}{|\mathbf{x}-\mathbf{x}'|} \right) + b_m \hat{n}'_i \frac{\partial}{\partial x'_m} \left(\frac{1}{|\mathbf{x}-\mathbf{x}'|} \right) + \frac{\lambda}{\mu} b_m \hat{n}'_m \frac{\partial}{\partial x'_i} \left(\frac{1}{|\mathbf{x}-\mathbf{x}'|} \right) - \frac{1}{4(1-\nu)} \frac{\lambda}{\mu} b_m \hat{n}'_m \frac{\partial^3 |\mathbf{x}-\mathbf{x}'|}{\partial x'_i \partial x'_j \partial x'_j} - \frac{1}{2(1-\nu)} b_m \hat{n}'_n \frac{\partial^3 |\mathbf{x}-\mathbf{x}'|}{\partial x'_i \partial x'_n \partial x'_m} \right] dS'. \tag{12.148}$$

Equation (12.148) is next put into a form suitable for the application of Stokes' theorem by substituting the equality

$$\frac{\lambda}{\mu} b_m \hat{n}'_m \frac{\partial}{\partial x'_i} \left(\frac{1}{|\mathbf{x}-\mathbf{x}'|} \right) - \frac{1}{4(1-\nu)} \frac{\lambda}{\mu} b_m \hat{n}'_m \frac{\partial^3 |\mathbf{x}-\mathbf{x}'|}{\partial x'_i \partial x'_j \partial x'_j} = \frac{1}{2(1-\nu)} b_m \hat{n}'_m \frac{\partial^3 |\mathbf{x}-\mathbf{x}'|}{\partial x'_i \partial x'_j \partial x'_j} - b_m \hat{n}'_m \frac{\partial}{\partial x'_i} \left(\frac{1}{|\mathbf{x}-\mathbf{x}'|} \right) \tag{12.149}$$

so that

$$u_i(\mathbf{x}) = -\frac{1}{4\pi}\iint_{\Sigma} b_i \hat{n}'_n \frac{\partial}{\partial x'_n}\left(\frac{1}{|\mathbf{x}-\mathbf{x}'|}\right) dS'$$
$$-\frac{1}{4\pi}\iint_{\Sigma}\left[b_m \hat{n}'_i \frac{\partial}{\partial x'_m}\left(\frac{1}{|\mathbf{x}-\mathbf{x}'|}\right) - b_m \hat{n}'_m \frac{\partial}{\partial x'_i}\left(\frac{1}{|\mathbf{x}-\mathbf{x}'|}\right)\right] dS'$$
$$+\frac{1}{8\pi(1-\nu)}\iint_{\Sigma}\left[b_m \hat{n}'_n \frac{\partial^3|\mathbf{x}-\mathbf{x}'|}{\partial x'_i \partial x'_m \partial x'_n} - b_m \hat{n}'_m \frac{\partial^3|\mathbf{x}-\mathbf{x}'|}{\partial x'_i \partial x'_j \partial x'_j}\right] dS'. \tag{12.150}$$

Applying Stokes' theorem given by Eq. (B.4), the second and third surface integrals in Eq. 12.105 are converted into line integrals so that

$$u_i(\mathbf{x}) = -\frac{1}{4\pi}\iint_{\Sigma} b_i \hat{n}'_n \frac{\partial}{\partial x'_n}\left(\frac{1}{|\mathbf{x}-\mathbf{x}'|}\right) dS' - \frac{1}{4\pi}\oint_C e_{imk} b_m \frac{1}{|\mathbf{x}-\mathbf{x}'|} dx'_k$$
$$-\frac{1}{8\pi(1-\nu)}\oint_C e_{mjk} b_m \frac{\partial^2|\mathbf{x}-\mathbf{x}'|}{\partial x'_i \partial x'_j} dx'_k. \tag{12.151}$$

The geometry for the line integration is illustrated in Fig. 12.10. Alternatively, Eq. (12.151) may be converted to a vector form by expressing its three terms as

$$\mathbf{u}(\mathbf{x}) = -\frac{\mathbf{b}}{4\pi}\Omega - \frac{1}{4\pi}\oint_C \frac{\mathbf{b}\times d\mathbf{s}'}{|\mathbf{x}-\mathbf{x}'|} - \frac{1}{8\pi(1-\nu)}\nabla\oint_C \frac{[\mathbf{b}\times(\mathbf{x}-\mathbf{x}')]\cdot d\mathbf{s}'}{|\mathbf{x}-\mathbf{x}'|}, \tag{12.152}$$

where

$$\Omega = \iint_{\Sigma} \frac{(\mathbf{x}-\mathbf{x}')\cdot\hat{\mathbf{n}}'}{|\mathbf{x}-\mathbf{x}'|^3} dS'. \tag{12.153}$$

The quantity Ω is the solid angle subtended by the cut surface Σ, when viewed from the field point $\mathbf{x}$, and is positive when its positive side is in view. Also, the magnitude of the first term in Eq. (12.152) undergoes a discontinuity of 4π when $\mathbf{x}$ passes through Σ. This term therefore produces a discontinuity in $\mathbf{u}$ equal to $\Delta\mathbf{u} = \mathbf{b}$ (see Exercise 12.9) which is consistent with the cut and displacement procedure for producing the dislocation.

12.6.1.2 *By use of Peach–Koehler equation*

The Peach–Koehler equation for the stress field due to a loop in an isotropic system (Peach and Koehler 1950) is obtained by taking derivatives of the displacements given by the Burgers equation. First, following deWit (1960) and Hirth and Lothe (1982), Eqs. (3.1) and (2.120) are used to write the stress in the form

$$\sigma_{\alpha\beta}(\mathbf{x}) = [\lambda\delta_{\alpha\beta}\delta_{ml} + \mu(\delta_{\alpha l}\delta_{\beta m} + \delta_{\alpha m}\delta_{\beta l})]\frac{\partial u_m(\mathbf{x})}{\partial x_l}. \tag{12.154}$$

The displacement in Eq. (12.154), is now replaced by the first term from Eq. (12.152) and the second and third terms from Eq. (12.151) so that

$$\begin{aligned}\sigma_{\alpha\beta}(\mathbf{x}) = &-\frac{b_m}{4\pi}[\lambda\delta_{\alpha\beta}\delta_{lm} + \mu(\delta_{\alpha l}\delta_{\beta m} + \delta_{\alpha m}\delta_{\beta l})]\frac{\partial\Omega}{\partial x_l}\\ &-\frac{1}{4\pi}[\lambda\delta_{\alpha\beta}\delta_{lm} + \mu(\delta_{\alpha l}\delta_{\beta m} + \delta_{\alpha m}\delta_{\beta l})]\oint_C e_{mik}b_i\frac{\partial}{\partial x_l}\left(\frac{1}{|\mathbf{x}-\mathbf{x}'|}\right)dx'_k\\ &-\frac{1}{8\pi(1-\nu)}[\lambda\delta_{\alpha\beta}\delta_{lm} + \mu(\delta_{\alpha l}\delta_{\beta m} + \delta_{\alpha m}\delta_{\beta l})]\\ &\times\oint_C e_{ijk}b_i\frac{\partial^3|\mathbf{x}-\mathbf{x}'|}{\partial x_l\partial x'_m\partial x'_j}dx'_k\end{aligned} \tag{12.155}$$

after changing some of the dummy indices. An expression for $\partial\Omega/\partial x_l$ in the first term can be obtained in the form of a line integral by considering Fig. 12.21. The quantity Ω has been identified in Sec. 12.6.1.1 as the solid angle subtended by the surface, Σ, when viewed from the field point at $\mathbf{x}$. If the point at $\mathbf{x}$ is shifted by $\delta\mathbf{x}$, then the surface will appear from the viewing point to have shifted by $-\delta\mathbf{x}$. The resulting change in Ω, $\delta\Omega$, can then be determined from the geometry of Fig. 12.21. The incremental change in area is $\delta\mathbf{\Sigma} = \delta\mathbf{x}' \times (-\delta\mathbf{x}) = \delta x_m(\hat{\mathbf{e}}_m \times \delta\mathbf{x}')$, and, therefore,

$$\delta\Omega = -\oint_C \frac{(\mathbf{x}-\mathbf{x}')\cdot[\delta x_m(\hat{\mathbf{e}}_m \times \delta\mathbf{x}')]}{|\mathbf{x}-\mathbf{x}'|^3} \tag{12.156}$$

and

$$\frac{\delta\Omega}{\delta x_j} = -\oint_C \frac{(\mathbf{x}-\mathbf{x}')\cdot(\hat{\mathbf{e}}_j \times \delta\mathbf{x}')}{|\mathbf{x}-\mathbf{x}'|^3}. \tag{12.157}$$

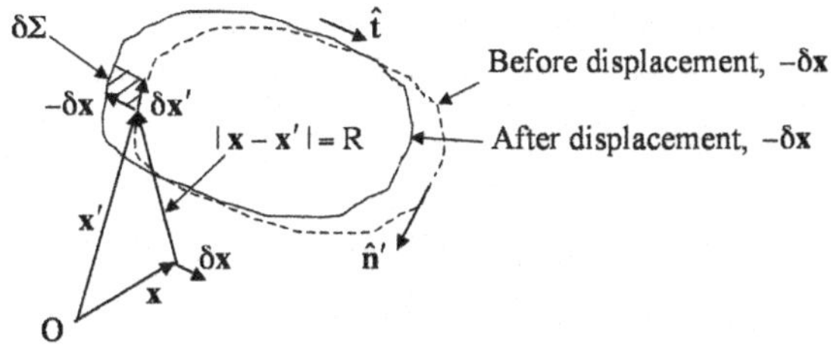

Fig. 12.21 Geometry for the determination of $\partial\Omega/\partial \mathbf{x}_l$.

Since

$$\begin{aligned}\frac{\delta\Omega}{\delta x_1} &= -\oint_C \frac{(\mathbf{x}-\mathbf{x}')\cdot(\hat{\mathbf{e}}_1\times\delta\mathbf{x}')}{|\mathbf{x}-\mathbf{x}'|^3}\\ &= -\oint_C \frac{(x_3-x_3')\delta x_2' - (x_2-x_2')\delta x_3'}{|\mathbf{x}-\mathbf{x}'|^3}\\ &= -\oint_C e_{i1k}\frac{\partial}{\partial x_i'}\left(\frac{1}{|\mathbf{x}-\mathbf{x}'|}\right)dx_k',\end{aligned} \tag{12.158}$$

Eq. (12.157) can be written as

$$\frac{\delta\Omega}{\delta x_j} = -\oint_C e_{ijk}\frac{\partial}{\partial x_i'}\left(\frac{1}{|\mathbf{x}-\mathbf{x}'|}\right)dx_k'. \tag{12.159}$$

Therefore, substituting Eq. (12.159) into Eq. (12.155), and changing some of the dummy indices, using the properties of the alternator (Appendix E), and employing the relationship $\partial/\partial x_l = -\partial/\partial \mathbf{x}_l'$,

$$\begin{aligned}\sigma_{\alpha\beta} = {} & \frac{\mu}{4\pi}\left[(\delta_{\alpha l}\delta_{\beta m}+\delta_{\alpha m}\delta_{\beta l})e_{ilk} + (\delta_{\alpha i}\delta_{\beta l}+\delta_{\alpha l}\delta_{\beta i})e_{lmk}\right]b_m\\ & \times\oint_C \frac{\partial}{\partial x_i'}\left(\frac{1}{|\mathbf{x}-\mathbf{x}'|}\right)dx_k'\\ & - \frac{\mu}{4\pi(1-\nu)}e_{imk}b_m\oint_C \frac{\partial^3|\mathbf{x}-\mathbf{x}'|}{\partial x_i'\partial x_\alpha'\partial x_\beta'}dx_k'\\ & + \frac{\mu\nu}{2\pi(1-\nu)}\delta_{\alpha\beta}e_{imk}b_m\oint_C \frac{\partial}{\partial x_i'}\left(\frac{1}{|\mathbf{x}-\mathbf{x}'|}\right)dx_k'.\end{aligned} \tag{12.160}$$

Equation (12.160) can then put into a more symmetrical form by employing the identity $e_{ijk}e_{klm} = \delta_{il}\delta_{jm} - \delta_{im}\delta_{jl}$ from Eq. (E.4) to obtain[10] the Peach–Koehler equation in the form

$$\begin{aligned}\sigma_{\alpha\beta} = &-\frac{\mu}{4\pi}e_{im\alpha}b_m \oint_C \frac{\partial}{\partial x'_i}\left(\frac{1}{|\mathbf{x}-\mathbf{x}'|}\right)dx'_\beta \\ &-\frac{\mu}{4\pi}e_{im\beta}b_m \oint_C \frac{\partial}{\partial x'_i}\left(\frac{1}{|\mathbf{x}-\mathbf{x}'|}\right)dx'_\alpha \\ &-\frac{\mu}{4\pi(1-\nu)}e_{imk}b_m \oint_C \frac{\partial^3|\mathbf{x}-\mathbf{x}'|}{\partial x'_i\partial x'_\alpha\partial x'_\beta}dx'_k \\ &+\frac{\mu}{2\pi(1-\nu)}\delta_{\alpha\beta}e_{imk}b_m \oint_C \frac{\partial}{\partial x'_i}\left(\frac{1}{|\mathbf{x}-\mathbf{x}'|}\right)dx'_k \end{aligned} \quad (12.161)$$

or, alternatively, with the use of $\partial/\partial x_i = -\partial/\partial x'_i$, $R = |\mathbf{x}-\mathbf{x}'|$, and $\nabla^2 R = \nabla'^2 R = 2/R$,

$$\begin{aligned}\sigma_{\alpha\beta}(\mathbf{x}) = &\frac{\mu}{8\pi}e_{im\alpha}b_m \oint_C \frac{\partial}{\partial x_i}(\nabla^2 R)dx'_\beta + \frac{\mu}{8\pi}e_{im\beta}b_m \oint_C \frac{\partial}{\partial x_i}(\nabla^2 R)dx'_\alpha \\ &+\frac{\mu}{4\pi(1-\nu)}e_{imk}b_m\left[\oint_C \frac{\partial^3 R}{\partial x_i \partial x_\alpha \partial x_\beta} - \delta_{\alpha\beta}\oint_C \frac{\partial}{\partial x_i}(\nabla^2 R)\right]dx'_k, \end{aligned} \quad (12.162)$$

where the geometry for the line integration is shown in Fig. 12.10. This equation can be written in a more concise vector form by employing the tensor product of two vectors (see Appendix C). Consider, for example, the first term, denoted by $\sigma_{\alpha\beta}(1)$, which may be written as

$$\begin{aligned}\sigma_{\alpha\beta}(1) = -\frac{\mu}{4\pi}\oint_C \Big\{ &e_{12\alpha}b_2\frac{\partial}{\partial x'_1} + e_{13\alpha}b_3\frac{\partial}{\partial x'_1} + e_{21\alpha}b_1\frac{\partial}{\partial x'_2} \\ &+e_{23\alpha}b_3\frac{\partial}{\partial x'_2} + e_{31\alpha}b_1\frac{\partial}{\partial x'_3} + e_{32\alpha}b_2\frac{\partial}{\partial x'_3}\Big\}\frac{1}{R}ds'_\beta \end{aligned} \quad (12.163)$$

[10]The procedure, described in detail by Hirth and Lothe (1982), is tedious, but straightforward.

or in matrix form as

$$[\sigma(1)] = \frac{\mu}{4\pi}\oint_C \frac{1}{R}$$

$$\times \begin{vmatrix} \left(b_2\frac{\partial}{\partial x_3'} - b_3\frac{\partial}{\partial x_2'}\right) ds_1' & \left(b_2\frac{\partial}{\partial x_3'} - b_3\frac{\partial}{\partial x_2'}\right) ds_2' & \left(b_2\frac{\partial}{\partial x_3'} - b_3\frac{\partial}{\partial x_2'}\right) ds_3' \\ \left(b_3\frac{\partial}{\partial x_1'} - b_1\frac{\partial}{\partial x_3'}\right) ds_1' & \left(b_3\frac{\partial}{\partial x_1'} - b_1\frac{\partial}{\partial x_3'}\right) ds_2' & \left(b_3\frac{\partial}{\partial x_1'} - b_1\frac{\partial}{\partial x_3'}\right) ds_3' \\ \left(b_1\frac{\partial}{\partial x_2'} - b_2\frac{\partial}{\partial x_1'}\right) ds_1' & \left(b_1\frac{\partial}{\partial x_2'} - b_2\frac{\partial}{\partial x_1'}\right) ds_2' & \left(b_1\frac{\partial}{\partial x_2'} - b_2\frac{\partial}{\partial x_1'}\right) ds_3' \end{vmatrix}. \tag{12.164}$$

A comparison of Eq. (12.164) with Eqs. (C.1) and (C.2) shows that the tensor $\underline{\boldsymbol{\sigma}}(1)$ can be written as the tensor product of the two vectors $(\mathbf{b} \times \nabla')\frac{1}{R}$ and $d\mathbf{s}'$, i.e.,

$$\underline{\boldsymbol{\sigma}}(1) = \frac{\mu}{4\pi}\oint_C (\mathbf{b} \times \nabla')\frac{1}{R} \otimes d\mathbf{s}'. \tag{12.165}$$

The remaining terms in Eq. (12.162) can be rewritten using the same method to produce the relatively concise result

$$\underline{\boldsymbol{\sigma}}(\mathbf{x}) = \frac{\mu}{4\pi}\oint_C (\mathbf{b} \times \nabla')\frac{1}{R} \otimes d\mathbf{s}' + \frac{\mu}{4\pi}\oint_C d\mathbf{s}' \otimes (\mathbf{b} \times \nabla')\frac{1}{R}$$
$$- \frac{\mu}{4\pi(1-\nu)}\oint_C [\nabla' \cdot (\mathbf{b} \times d\mathbf{s}')][\nabla \otimes \nabla - \underline{\mathbf{I}}\nabla^2]R \tag{12.166}$$

12.6.1.3 *By use of infinitesimal dislocation loops*

The use of infinitesimal dislocation loops for finding the elastic fields of finite loops has been described in Sec. 12.5.1.6. Equation (12.140) of that section can be modified to apply to the present case of an isotropic system by using Eq. (2.120) to obtain the differential displacement at $\mathbf{x}$ due to a differential dislocation loop located at $\mathbf{x}'$ in the form

$$du_i = -[\lambda b_l \hat{n}_l' \delta_{kj} + \mu(b_k \hat{n}_j' + b_j \hat{n}_k')]\frac{\partial G_{ik}^{\infty}(\mathbf{x} - \mathbf{x}')}{\partial x_j'} dS' \tag{12.167}$$

or, equivalently,

$$du_i = -\mu b_j \left[\frac{\partial G_{ij}^{\infty}(\mathbf{x} - \mathbf{x}')}{\partial x_k'} + \frac{\partial G_{ik}^{\infty}(\mathbf{x} - \mathbf{x}')}{\partial x_j'}\right.$$
$$\left. + \frac{2\nu}{1-2\nu}\delta_{jk}\frac{\partial G_{im}^{\infty}(\mathbf{x} - \mathbf{x}')}{\partial x_m'}\right] \hat{n}_k' dS'. \tag{12.168}$$

Then, substituting the Green's function given by Eq. (4.110), the expression

$$du_i(\mathbf{x}) = -\frac{1}{8\pi(1-\nu)} \times \left[\frac{(1-2\nu)(\hat{n}_i b_k x_k + b_i \hat{n}_k x_k - b_k \hat{n}_k x_i)}{x^3} + \frac{3 b_k \hat{n}_l x_i x_k x_l}{x^5}\right] dS \tag{12.169}$$

is obtained for the infinitesimal displacement at $\mathbf{x}$ produced by an infinitesimal loop characterized by its Burgers vector $\mathbf{b}$, positive unit normal $\hat{\mathbf{n}}$, and area dS, located at the origin ($\mathbf{x}' = 0$).

In Exercise 12.12, Eq. (12.169) is employed to formulate an integral expression for the displacement field produced by a finite prismatic dislocation loop.

As demonstrated by Groves and Bacon (1969) and Bacon and Groves (1970), the displacement fields of a number of infinitesimal loops are elastically equivalent to the fields produced by classical *nuclei of strain* (Love 1944). For example, for an infinitesimal prismatic edge-type loop with $\mathbf{b} = (0, 0, b)$ and $\hat{\mathbf{n}} = (0, 0, 1)$, Eq. (12.168) reduces to

$$du_i = 2\mu b \left(\frac{\partial G^{\infty}_{i3}(\mathbf{x}-\mathbf{x}')}{\partial x_3} + \frac{\nu}{1-2\nu}\frac{\partial G^{\infty}_{im}(\mathbf{x}-\mathbf{x}')}{\partial x_m}\right) dS'. \tag{12.170}$$

By comparing Eq. (12.170) with Eqs. (10.18) and (10.20), it may be seen that the first term of Eq. (12.170) produces a displacement field of the same functional form as that produced by a *double force without moment* in the force dipole moment approximation of Sec. 10.3.5, while the second term produces a field attributable to a *center of dilatation.* For a shear loop with $\mathbf{b} = (b, 0, 0)$ and $\hat{\mathbf{n}} = (0, 0, 1)$, Eq. (12.168) yields the displacements

$$du_i = \mu b \left(\frac{\partial G^{\infty}_{i1}(\mathbf{x}-\mathbf{x}')}{\partial x_3} + \frac{\partial G^{\infty}_{i3}(\mathbf{x}-\mathbf{x}')}{\partial x_1}\right) dS' \tag{12.171}$$

corresponding to the displacements produced by the *pair of double forces with equal, but opposite, moments* shown in Fig. 12.22, since use of Eqs. (10.5) and (10.10) shows that

$$u_i = 2aF \left(\frac{\partial G^{\infty}_{i1}(\mathbf{x}-\mathbf{x}')}{\partial x_3} + \frac{\partial G^{\infty}_{i3}(\mathbf{x}-\mathbf{x}')}{\partial x_1}\right). \tag{12.172}$$

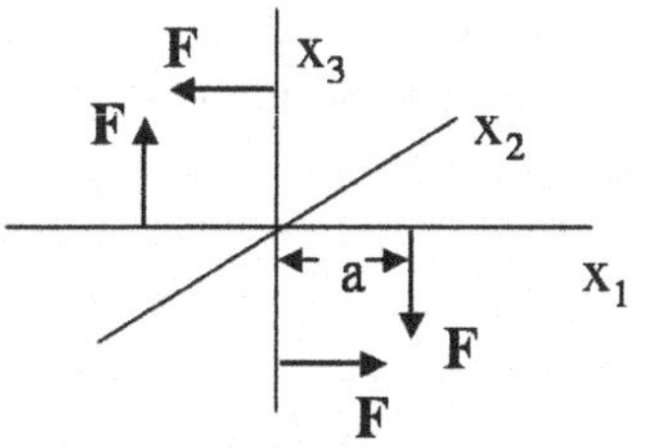

Fig. 12.22 Pair of double forces with equal, but opposite, moments around the $\mathbf{x}_2$ axis. All moment arms of length a.

Nuclei of strain such as these are useful, since they can be used as displacement functions to construct solutions for a variety of elasticity problems.

12.6.2 *Strain energies*

The strain self-energy of a dislocation loop is found in Sec. 12.5.2 to be just half the interaction energy between two superimposed loops that are duplicates of one another. The interaction energy between two loops in an isotropic system is given by Eq. (16.34), and, therefore, after setting $\mathbf{b}^{(1)} = \mathbf{b}^{(2)} = \mathbf{b}$, and $\mathrm{C}^{(1)} = \mathrm{C}^{(2)} = \mathrm{C}$ and dividing by two, the strain self-energy of a single loop C in an isotropic system is (Hirth and Lothe 1982)

$$W^{C} = \frac{\mu}{8\pi} \oint_{C^{(1)}=C} \oint_{C^{(2)}=C} \frac{\left(\mathbf{b}\cdot d\mathbf{s}^{(1)}\right)\left(\mathbf{b}\cdot d\mathbf{s}^{(2)}\right)}{R} + \frac{\mu}{8\pi(1-\nu)} \oint_{C^{(1)}=C} \oint_{C^{(2)}=C} \left(\mathbf{b}\times d\mathbf{s}^{(1)}\right)\cdot \underline{\mathbf{T}} \cdot \left(\mathbf{b}\times d\mathbf{s}^{(2)}\right). \quad (12.173)$$

As in the case of Eq. (12.146), singularities are avoided in the double integration by assuming that the segments do not interact when they are closer than a *cutoff distance*, ρ.

In Exercise 12.11, Eq. (12.173) is used to find the strain energy of a circular planar loop with $\mathbf{b}$ lying in the loop plane in an isotropic system.

12.7 Segmented Dislocation Structures

Many common dislocation structures consist of assemblies of straight finite segments joined together in various configurations. These include segmented dislocation lines, polygonal dislocation loops and networks of dislocation segments of various types including, for example, small-angle grain boundaries (Fig. 14.11). Furthermore, smoothly curved dislocation lines or loops can be well approximated by short contiguous straight segments linked together in chains. Many three-dimensional dislocation models composed of ensembles of relatively short interconnected segments have been constructed which approximate the complex dislocation structures produced by plastic deformation (Zbib, Rhee and Hirth 1998; Devincre, Kubin, Lemarchand and Madec 2001; Devincre, Kubin and Hoc 2006).

In this section we develop methods for determining the elastic stress fields produced by such segmented structures. First, the stress fields of relatively simple segmented structures, which can serve as building blocks for constructing more complex structures, are formulated. Then, a number of more complex segmented structures are produced by assembling such building blocks, and the elastic fields of these structures are found by simply superimposing the known elastic fields of their simpler components. References include Asaro and Barnett (1976), Bacon, Barnett and Scattergood (1979a), Bacon, Barnett and Scattergood (1979b), Hirth and Lothe (1982), and Lothe (1992b).

12.7.1 *Elastic fields*

12.7.1.1 *Straight segment which is part of a closed loop*

The simplest segmented dislocation structure, that can also be used to construct more complex segmented structures, is a single straight finite segment which is part of a closed loop as illustrated in Fig. 12.23. As shown by Willis (1970), Steeds and Willis (1979) and Lothe (1992b), the portion of the total elastic field of such a loop at the field point P that is

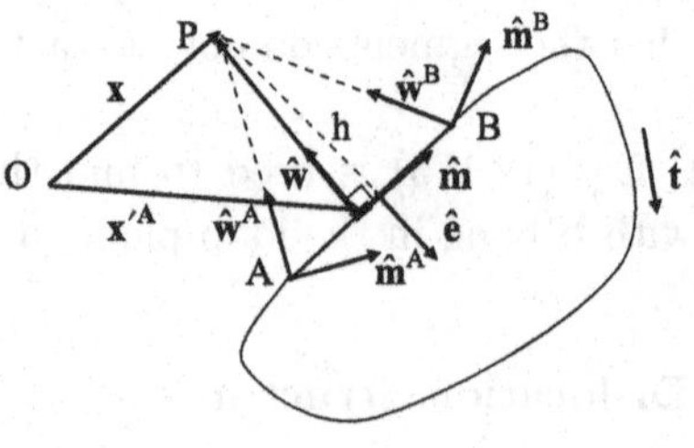

Fig. 12.23 Straight finite dislocation segment AB, which is part of a closed loop. The dislocation-based $(\hat{\mathbf{e}}, \hat{\mathbf{n}}, \hat{\mathbf{t}})$ coordinate system is used to obtain the portion of the total elastic field of the loop at the field point P that is contributed by the segment. The unit base vector $\hat{\mathbf{n}} = \hat{\mathbf{t}} \times \hat{\mathbf{e}}$ (not visible) points into the paper. Also, shown are the unit vectors $\hat{\mathbf{w}} = (\mathbf{x} - \mathbf{x}')/|\mathbf{x} - \mathbf{x}'|$ and $\hat{\mathbf{m}}$. Note that $\hat{\mathbf{n}}$ is given alternatively by $\hat{\mathbf{n}} = \hat{\mathbf{w}} \times \hat{\mathbf{m}}$.

contributed by the segment AB can be found by using an expression for the total field of the loop in the form of a line integral around the loop and then integrating the expression between the points A and B. Following Lothe (1992b), the Mura equation, given by Eq. (12.80), is appropriate for this purpose, and the distortion contributed by the AB segment at P(**x**) is then[11]

$$\frac{\partial u_m(\mathbf{x})}{\partial x_s} = b_i C_{ijkl} e_{jsn} \int_{\mathbf{x}'^A}^{\mathbf{x}'^B} \frac{\partial G_{km}(\mathbf{x}-\mathbf{x}')}{\partial x'_l} dx'_n \qquad (12.174)$$

Then, after substituting for the derivative of the Green's function by use of Eq. (4.40), Eq. (12.174) takes the form

$$\frac{\partial u_m(\mathbf{x})}{\partial x_s} = \frac{b_i C_{ijkl} e_{jsn}}{8\pi^2} \int_{\mathbf{x}'^A}^{\mathbf{x}'^B} \frac{dx'_n}{|\mathbf{x}-\mathbf{x}'|^2} \times \oint_{\hat{\mathcal{L}}} \left\{ \hat{w}_l \left(\hat{k}\hat{k}\right)^{-1}_{km} - \hat{k}_l \left(\hat{k}\hat{k}\right)^{-1}_{kp} \left[\left(\hat{k}\hat{w}\right)_{pj} + \left(\hat{w}\hat{k}\right)_{pj} \right] \left(\hat{k}\hat{k}\right)^{-1}_{jm} \right\} ds, \qquad (12.175)$$

where it is recalled (Fig. 4.2) that the line integral along s goes around the unit circle, $\hat{\mathcal{L}}$, traversed by the unit vector $\hat{\mathbf{k}}$ as it rotates in the plane perpendicular to $\hat{\mathbf{w}}$. However, Lothe (1992b) has shown that Eq. (12.174) can be integrated along AB using the geometry shown in Fig. 12.23. Here, a dislocation-based $(\hat{\mathbf{e}}, \hat{\mathbf{n}}, \hat{\mathbf{t}})$ coordinate system is employed, so that the vector $\mathbf{x} - \mathbf{x}'$ can be expressed by

$$\mathbf{x}' - \mathbf{x} = h\hat{\mathbf{e}} + w\hat{\mathbf{n}} + l\hat{\mathbf{t}}. \qquad (12.176)$$

As a consequence, $G_{km}(\mathbf{x}-\mathbf{x}') \rightarrow G_{km}(h, w, l)$ where the coordinates in the crystal $(\hat{\mathbf{e}}_1, \hat{\mathbf{e}}_2, \hat{\mathbf{e}}_3)$ system and $(\hat{\mathbf{e}}, \hat{\mathbf{n}}, \hat{\mathbf{t}})$ system are related by

$$\begin{bmatrix} (x'_1 - x_1) \\ (x'_2 - x_2) \\ (x'_3 - x_3) \end{bmatrix} = \begin{bmatrix} \hat{e}_1 & \hat{n}_1 & \hat{t}_1 \\ \hat{e}_2 & \hat{n}_2 & \hat{t}_2 \\ \hat{e}_3 & \hat{n}_3 & \hat{t}_3 \end{bmatrix} \begin{bmatrix} h \\ w \\ l \end{bmatrix},$$

$$\begin{bmatrix} h \\ w \\ l \end{bmatrix} = \begin{bmatrix} \hat{e}_1 & \hat{e}_2 & \hat{e}_3 \\ \hat{n}_1 & \hat{n}_2 & \hat{n}_3 \\ \hat{t}_1 & \hat{t}_2 & \hat{t}_3 \end{bmatrix} \begin{bmatrix} (x'_1 - x_1) \\ (x'_2 - x_2) \\ (x'_3 - x_3) \end{bmatrix}. \qquad (12.177)$$

[11]Note that, once the distortions are known, the strains are readily determined using Eq. (2.5)

The operator $\partial/\partial x'_l$ in Eq. (12.174) is therefore given by

$$\frac{\partial}{\partial x'_l} = \hat{e}_l\frac{\partial}{\partial h} + \hat{n}_l\frac{\partial}{\partial w} + \hat{t}_l\frac{\partial}{\partial l} \tag{12.178}$$

and by employing Eq. (12.178) in Eq. (12.174), and using the relation $dx'_n = \hat{t}_n ds$,

$$\frac{\partial u_m}{\partial x_s} = b_i C_{ijkl} e_{jsn} \hat{t}_n \left(\hat{e}_l \int_{l^A}^{l^B} \frac{\partial G_{km}}{\partial h} ds + \hat{n}_l \int_{l^A}^{l^B} \frac{\partial G_{km}}{\partial w} ds + \hat{t}_l \int_{l^A}^{l^B} \frac{\partial G_{km}}{\partial l} ds \right). \tag{12.179}$$

Lothe (1992b) has evaluated the three integrals by an unusually lengthy procedure, and we therefore merely present the result expressed in terms of the integral formalism in the form

$$\frac{\partial u_m}{\partial x_s} = \frac{1}{4\pi h} b_i C_{ijkl} e_{jsn} \hat{t}_n \times \left\{ -\hat{m}_l Q_{mk} + \hat{n}_l \left[(\hat{n}\hat{n})^{-1}_{mr} (\hat{n}\hat{m})_{rp} Q_{pk} + (\hat{n}\hat{n})^{-1}_{mp} S_{pk} \right] \right\}_A^B, \tag{12.180}$$

where the vector $\hat{\mathbf{m}}$ is shown in Fig. 12.23, and the vector $\hat{\mathbf{n}}$ corresponds $\hat{\mathbf{n}} = \hat{\mathbf{w}} \times \hat{\mathbf{m}}$. The vector $\hat{\mathbf{w}}$ is given by $\hat{\mathbf{w}} = (\mathbf{x} - \mathbf{x}')/|\mathbf{x} - \mathbf{x}'|$, and the matrices Q_{ij} and S_{ij} are given by Eqs. (3.133) and (3.134) where $\hat{\mathbf{m}}$ and $\hat{\mathbf{n}}$ in the integrals in these equations rotate by the angle ω in the plane perpendicular to $\hat{\mathbf{w}}$.

The method by which two dislocation loops containing straight segments of the type in Fig. 12.23 can be combined to construct a more complex segmented structure, i.e., an angular dislocation, is illustrated in Fig. 12.31.

12.7.1.2 *Bi-angular planar dislocation*

Another relatively simple segmented structure, which can also be used to construct more complex segmented structures, is shown in Fig. 12.24. Following Bacon, Barnett and Scattergood (1979b), it is termed a *bi-angular dislocation.* The bi-angular dislocation lies in one plane, and the total in-plane stress $\sigma_{ij}^{AB}(P)$ that it produces at the field point P can be obtained by simply integrating Eq. (12.113), which holds for a single hairpin, along the AB segment. After changing variables by use of Eq. (12.109), and expressing the integral in terms of the angular limits θ^A and θ^B, we

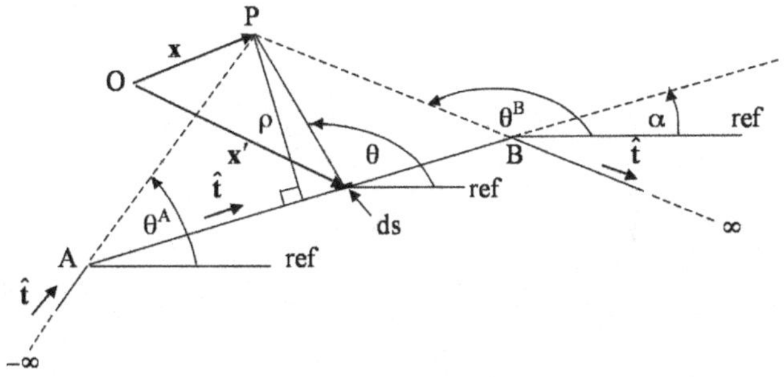

Fig. 12.24 Bi-angular dislocation showing the geometry used for integrating Eq. (12.113) along the dislocation segment AB. All elements of diagram are in plane of the paper.

have

$$\sigma_{ij}^{AB}(P) = \frac{1}{2\rho}\int_{\theta_A}^{\theta_B} \sin(\theta-\alpha)\left[\Sigma_{ij}(\theta) + \frac{d^2\Sigma_{ij}(\theta)}{d\theta^2}\right] d\theta \qquad (12.181)$$

and, after integrating by parts twice,

$$\sigma_{ij}^{AB}(P) = \frac{1}{2\rho}\left| -\cos(\theta-\alpha)\Sigma_{ij}(\theta) + \sin(\theta-\alpha)\frac{d\Sigma_{ij}(\theta)}{d\theta}\right|_{\theta^A}^{\theta^B}. \qquad (12.182)$$

According to Eq. (12.182), the stress depends only upon ρ, α, the two terminating angles θ^A and θ^B, and the data for infinitely long straight dislocations which determines $\Sigma_{ij}(\theta)$. Note that when the segment length becomes infinite, $\theta^B \rightarrow \alpha + \pi$, and $\theta^A \rightarrow \alpha$, and Eq. (12.182) takes the form $\sigma_{ij}^{-\infty,\infty}(P) = \Sigma_{ij}(\alpha)/\rho$, which corresponds to Fig. 12.14c with $\theta = \alpha$, and is characteristic of an infinitely long straight dislocation according to Eq. (12.107).

The method by which bi-angular dislocations can be combined to construct a segmented three-fold dislocation node is illustrated in Fig. 12.25.

12.7.1.3 *N-fold planar dislocation nodes*

The three-fold planar dislocation node, shown in Fig. 12.25, must obey the nodal Burgers vector condition

$$\mathbf{b}^{AB} + \mathbf{b}^{AC} + \mathbf{b}^{AD} = 0 \qquad (12.183)$$

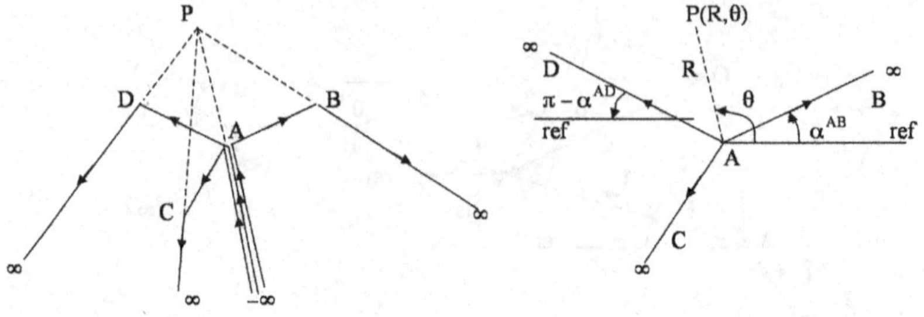

(a) (b)

Fig. 12.25 (a) Construction of a three-fold node at A by assembling three bi-angular dislocations containing finite straight segments, AB, AC and AD, respectively. The three semi-infinite end segments lying between A and $-\infty$ mutually annihilate because of Eq. (12.183). If the three segments are then lengthened by moving their end points at B, C and D, respectively, out to infinity, the final node shown in (b) will be obtained. The three segment lengths beyond B, C and D in (a) (that are now at infinity) are not shown in (b), since the stresses that they contribute to the total stress field of the node at finite distances from A are vanishingly small. All elements are in plane of paper.

since the positive directions of all three segments emanate from the node. As is demonstrated in Fig. 12.25b, the node can therefore be constructed by assembling three bi-angular dislocations. The in-plane stress contributed by the segment AB to the total stress due to the node at P in Fig. 12.25b, $\sigma_{ij}^{AB}(P)$, can then be obtained by employing Eq. (12.182). In this case, $\rho = R\sin(\theta - \alpha)$, and when B is extended to infinity, as in Fig. 12.25b, $\theta^{B} \rightarrow \pi + \alpha$. Therefore, upon setting $\theta^{A} = \theta$, Eq. (12.182) takes the form

$$\sigma_{ij}^{AB}(P) = \frac{1}{2R}\left[\csc(\theta - \alpha)\Sigma_{ij}(\alpha) + \cot(\theta - \alpha)\Sigma_{ij}(\theta) - \frac{d\Sigma_{ij}(\theta)}{d\theta}\right]. \tag{12.184}$$

Similar expressions are obtained for the remaining segments AC and AD. If additional segments are added so that N segments are present, the total stress at P can then be expressed as the sum

$$\sigma_{ij}(P) = \frac{1}{2R}\sum_{m=1}^{N}\left[\csc(\theta - \alpha^{(m)})\Sigma_{ij}^{(m)}(\alpha^{(m)}) + \cot(\theta - \alpha^{(m)})\Sigma_{ij}^{(m)}(\theta) - \frac{d\Sigma_{ij}^{(m)}(\theta)}{d\theta}\right], \tag{12.185}$$

where the segments are now labeled in the order $1, 2, \ldots N$. However, the relationship

$$\sum_{m=1}^{N} \frac{d\Sigma_{ij}^{(m)}(\theta)}{d\theta} = 0 \tag{12.186}$$

is valid, since $d\Sigma_{ij}^{(m)}(\theta)/d\theta$ for segment m in the above sum is linear and homogeneous with respect to its Burgers vector, $\mathbf{b}^{(m)}$. Therefore, applying the nodal condition given by Eq. (12.183), the sum corresponding to Eq. (12.186) vanishes, and Eq. (12.185) finally reduces to (Asaro and Barnett 1976)

$$\sigma_{ij}(P) = \frac{1}{2R} \sum_{m=1}^{N} \left[\csc(\theta - \alpha^{(m)}) \Sigma_{ij}^{(m)}(\alpha^{(m)}) + \cot(\theta - \alpha^{(m)}) \Sigma_{ij}^{(m)}(\theta) \right]. \tag{12.187}$$

12.7.1.4 *Angular dislocation*

An angular dislocation, illustrated in Fig. 12.26, consists of two joined non-collinear semi-infinite straight segments. Comparison with Fig. 12.25b shows that it is nothing more than a two-fold node, with the Burgers vector of one segment reversed. Therefore, using Eq. (12.187) with $N = 2$, and reversing the Burgers vector of segment 1, the in-plane stress at P is given by

$$\sigma_{ij}(P) = \frac{1}{2R} \left[\csc(\theta - \alpha^{(m)}) \Sigma_{ij}^{(m)}(\alpha^{(m)}) + \cot(\theta - \alpha^{(m)}) \Sigma_{ij}^{(m)}(\theta) \right]_{m=1}^{m=2}. \tag{12.188}$$

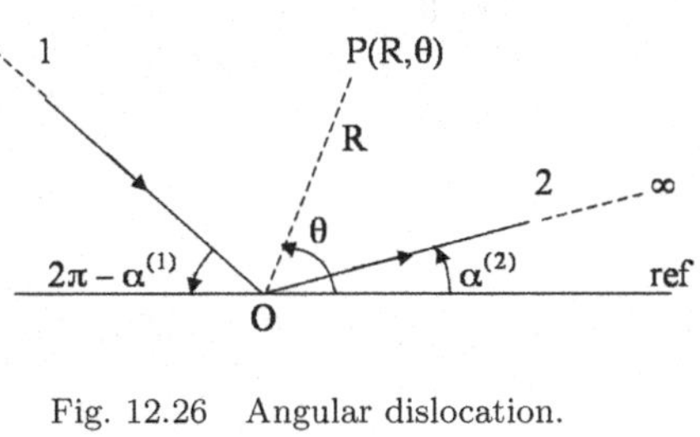

Fig. 12.26 Angular dislocation.

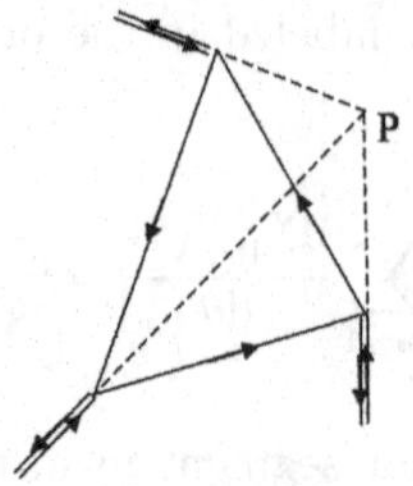

Fig. 12.27 Construction of planar triangular loop by assembling three bi-angular dislocations of the type shown in Fig. 12.24. The anti-parallel semi-infinite end segments emanating from the corners mutually annihilate.

12.7.1.5 *N-sided polygonal planar dislocation loop*

As shown in Fig. 12.27, a triangular loop can be produced by assembling three bi-angular dislocations. The in-plane stress produced at P could be determined by using Eq. (12.182), which is valid for bi-angular dislocations, to sum the stress contributions of the three segment sides. However, when the segments comprise a closed loop, as in the present case, a simpler formulation can be found based on Eq. (12.118) which is valid for a closed loop. The contribution of segment AB, obtained by integrating Eq. (12.118) along AB, is

$$\sigma_{\mathrm{ij}}^{\mathrm{AB}}(\mathrm{P}) = \frac{1}{2\rho}\int_{\theta^{\mathrm{A}}}^{\theta^{\mathrm{B}}} \left[\sin(\theta-\alpha)\Sigma_{\mathrm{ij}}(\theta) - \cos(\theta-\alpha)\frac{\mathrm{d}\Sigma_{\mathrm{ij}}(\theta)}{\mathrm{d}\theta}\right]\mathrm{d}\theta \tag{12.189}$$

after applying Eq. (12.109). The second term can be integrated by parts to obtain

$$\int_{\theta^{\mathrm{A}}}^{\theta^{\mathrm{B}}} \cos(\theta-\alpha)\frac{\mathrm{d}\Sigma_{\mathrm{ij}}(\theta)}{\mathrm{d}\theta}\mathrm{d}\theta = \int_{\theta^{\mathrm{A}}}^{\theta^{\mathrm{B}}} \sin(\theta-\alpha)\Sigma_{\mathrm{ij}}(\theta)\mathrm{d}\theta + \left[\cos(\theta-\alpha)\Sigma_{\mathrm{ij}}(\theta)\right]_{\theta^{\mathrm{A}}}^{\theta^{\mathrm{B}}} \tag{12.190}$$

and, after substituting Eq. (12.190) into Eq. (12.189),

$$\sigma_{\mathrm{ij}}^{\mathrm{AB}}(\mathrm{P}) = \frac{1}{2\rho}\left[-\cos(\theta-\alpha)\Sigma_{\mathrm{ij}}(\theta)\right]_{\theta^{\mathrm{A}}}^{\theta^{\mathrm{B}}}. \tag{12.191}$$

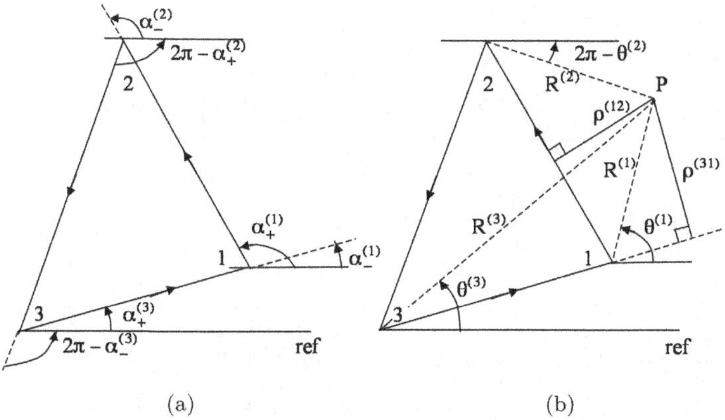

Fig. 12.28 Triangular dislocation loop. The three vertices are indicated by (1,2,3). $\alpha_+^{(i)}$ is the angle, with respect to reference line, of the segment leaving junction (i); $\alpha_-^{(i)}$ is the corresponding angle of the segment entering junction (i). (b) Same loop as in (a), but the lines from the field point P to the vertices, and the angles between these lines and the reference line, now shown.

Equation (12.191) is simpler than Eq. (12.182) and has the advantage that it does not require knowledge of the derivative of $\Sigma_{ij}(\theta)$ which can be troublesome to determine accurately as discussed by Bacon, Barnett and Scattergood (1979b). It is therefore the preferred expression for segments of closed loops.

Therefore, using Eq. (12.191) and the parameters illustrated in Fig. 12.28, the stress at P due to the three segments of the triangular loop in Fig. 12.28 is obtained in the form

$$\begin{aligned}\sigma_{ij}(P) = {} & \frac{1}{2\rho^{(12)}}\left[-\cos\left(\theta^{(2)}-\alpha_+^{(1)}\right)\Sigma_{ij}(\theta^{(2)})\right.\\ & \left.+\cos(\theta^{(1)}-\alpha_+^{(1)})\Sigma_{ij}\left(\theta^{(1)}\right)\right]+\frac{1}{2\rho^{(23)}}\\ & \times\left[-\cos(\theta^{(3)}-\alpha_+^{(2)})\Sigma_{ij}(\theta^{(3)})+\cos(\theta^{(2)}-\alpha_+^{(2)})\Sigma_{ij}(\theta^{(2)})\right]\\ & +\frac{1}{2\rho^{(31)}}\left[-\cos(\theta^{(1)}-\alpha_+^{(3)})\Sigma_{ij}(\theta^{(1)})\right.\\ & \left.+\cos(\theta^{(3)}-\alpha_+^{(3)})\Sigma_{ij}(\theta^{(3)})\right]. \qquad (12.192)\end{aligned}$$

Then, the two terms containing the common factor $\Sigma_{ij}(\theta^{(1)})$ can be combined in the form

$$\frac{\cos\left(\theta^{(1)}-\alpha_+^{(1)}\right)\Sigma_{ij}\left(\theta^{(1)}\right)}{2\rho^{(12)}}-\frac{\cos\left(\theta^{(1)}-\alpha_+^{(3)}\right)\Sigma_{ij}\left(\theta^{(1)}\right)}{2\rho^{(31)}}$$
$$=\frac{\sin\left(\alpha_+^{(1)}-\alpha_-^{(1)}\right)\csc\left(\theta^{(1)}-\alpha_+^{(1)}\right)\csc\left(\theta^{(1)}-\alpha_-^{(1)}\right)\Sigma_{ij}\left(\theta^{(1)}\right)}{2\mathrm{R}^{(1)}} \tag{12.193}$$

after using the following relationships derived from Fig. 12.28,

$$\alpha_+^{(3)}=\alpha_-^{(1)},\quad \rho^{(12)}=\mathrm{R}^{(1)}\sin(\theta^{(1)}-\alpha_+^{(1)}),$$
$$\rho^{(31)}=\mathrm{R}^{(1)}\sin(\theta^{(3)}-\alpha_-^{(1)}). \tag{12.194}$$

Similarly, two additional terms of the same general form can be obtained by combining the remaining two pairs of terms. Furthermore, additional sides can be added to the original triangle to form an N-sided planar polygonal loop. As a result, each side will contribute a term of the form of Eq. (12.193), leading to the total stress at P, due to a N-sided planar polygonal loop,

$$\sigma_{ij}(\mathrm{P})=\frac{1}{2}\sum_{m=1}^{N}$$
$$\times\frac{\sin\left(\alpha_+^{(m)}-\alpha_-^{(m)}\right)\csc\left(\theta^{(m)}-\alpha_+^{(m)}\right)\csc\left(\theta^{(m)}-\alpha_-^{(m)}\right)\Sigma_{ij}\left(\theta^{(m)}\right)}{\mathrm{R}^{(m)}}. \tag{12.195}$$

12.7.1.6 *Three-dimensional multi-segment structures*

As demonstrated in Sec. 12.5.1.4 (Fig. 12.15), the Brown's formula method can be used to find the stress produced by a smoothly curved non-planar loop in three dimensions. The method can also be applied to 3-dimensional multi-segmented structures, and several examples, which are fabricated by assembling bi-angular dislocations, are shown in Fig. 12.29. Figure 12.29a shows a non-planar polygonal loop where the semi-infinite end segments mutually annihilate, and the five planes containing the field point P and the

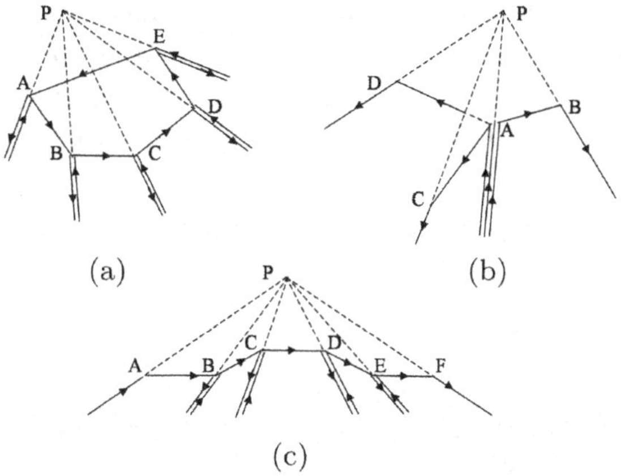

Fig. 12.29 (a) Non-planar polygonal loop ABCDE. (b) Non-planar 3-fold node with segments AB, AC and AD. (c) Non-planar segmented line.

segments, i.e., PAB, PBC, PCD, PDE and PEA, are non-coplanar.[12]. Figure 12.29b shows a non-planar 3-fold node where the three parallel semi-infinite end segments emanating from A mutually annihilate because of Eq. (12.183) The segment ends at B, C and D can be extended out to infinity, and the planes containing P and the various segments are non-coplanar. Figure 12.29c shows a segmented line where the parallel semi-infinite segments emanating from the junctions mutually annihilate, the semi-infinite segment ends at A and F can be extended out to infinity, and the five planes containing P and the various segments are non-coplanar. In all cases the stress at P contributed by each segment can be determined by use of the previous methods with the quantities involved for each segment referred to the local plane containing P and the segment.

12.7.2 *Strain energies*

12.7.2.1 *Straight segment*

Just as the stress field contributed by a straight segment that is part of a closed dislocation loop can be obtained by integrating the equation for the

[12]Note that Fig. 12.29a is just a coarsely polygonized version of Fig. 12.15.

stress field of the loop along the length of the segment, the strain energy contributed by the segment can be obtained by integrating the equation for the loop strain energy along the length of the segment. Employing Eq. (12.146), the strain energy contribution of an AB segment is therefore

$$W^{AB} = \frac{1}{16\pi^2} \int_{\mathcal{L}^{(1)}=AB} \int_{\mathcal{L}^{(2)}=AB} \frac{1}{|\mathbf{x}^{(1)} - \mathbf{x}^{(2)}|} \times \int_0^{2\pi} C^*_{ijkl}(\hat{\mathbf{m}}) b_k (d\mathbf{s}^{(1)} \times \hat{\mathbf{m}})_l b_i (d\mathbf{s}^{(2)} \times \hat{\mathbf{m}})_j d\theta, \quad (12.196)$$

where the line integrals, $\mathcal{L}^{(1)}$ and $\mathcal{L}^{(2)}$, involving $d\mathbf{s}^{(1)}$ and $d\mathbf{s}^{(2)}$, are each taken along the AB segment.

12.7.2.2 *Multi-segment structure*

Hirth and Lothe (1982) have shown that the strain energy associated with a segmented structure can be determined with acceptable accuracy as the sum of the strain energies contributed by all segments comprising the structure plus the interaction energies between all distinguishable pairs of segments in the structure as obtained by the use of our present models. The energy contributed by each segment is obtained by employing Eq. (12.196), and the interaction energy contributed by each pair of segments by employing Eq. (16.23).

The demonstration given by Hirth and Lothe (1982) employs an isotropic system where analytical results for simple tractable cases can be readily derived, and is described in detail in Sec. 12.8.2.2.

12.8 Segmented Dislocation Structures in Isotropic Systems

12.8.1 *Elastic fields*

12.8.1.1 *Straight segment*

To find the stress at the field point, P, contributed by a straight finite segment which is part of a closed loop in an isotropic system, we follow deWit (1967) and Devincre (1995) and integrate the Peach–Koehler equation along the segment using the geometry in Fig. 12.30. The origin of the coordinate system can be positioned arbitrarily, the vectors $\mathbf{R} = \mathbf{x} - \mathbf{x}'$ and

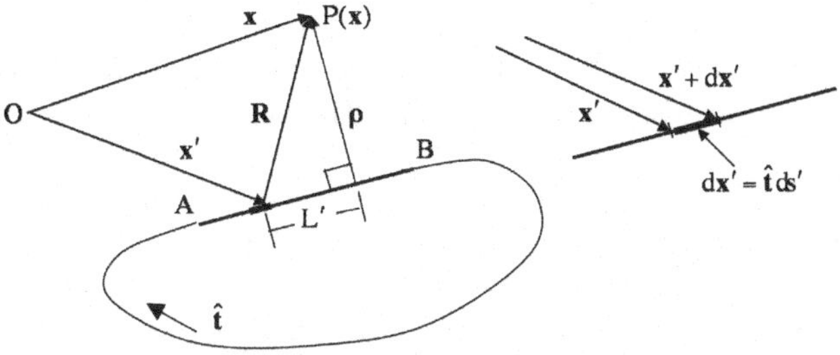

Fig. 12.30 (a) Geometry used to find stress at field point P due to straight dislocation segment AB which is part of a closed loop. (b) Detail of geometry for line integral along segment.

$\boldsymbol{\rho}$ run from the segment to P, L' is the projection of $\mathbf{R}$ on $\hat{\mathbf{t}}$, and

$$\hat{\mathbf{t}} = \frac{d\mathbf{x}'}{ds'}, \quad dx'_i = \hat{t}_i ds'. \tag{12.197}$$

The following relationships then hold for the quantities in Fig. 12.30:

$$L' = (\mathbf{x} - \mathbf{x}') \cdot \hat{\mathbf{t}} = \mathbf{R} \cdot \hat{\mathbf{t}}, \quad \boldsymbol{\rho} = \mathbf{R} - L'\hat{\mathbf{t}}, \quad \frac{dL'}{ds'} = -1,$$

$$\frac{dR}{ds'} = -\frac{L'}{R}, \quad \frac{\partial L'}{\partial x_i} = \hat{t}_i, \quad \frac{\partial R}{\partial x_i} = \frac{R_i}{R}, \tag{12.198}$$

$$\frac{\partial(\rho^2)}{\partial x_i} = 2\rho_i, \quad \frac{\partial \rho_i}{\partial x_j} = \delta_{ij} - \hat{t}_i\hat{t}_j.$$

Following Devincre (1995), and using $\partial/\partial x'_m = -\partial/\partial x_m$ and the relation

$$\frac{\partial}{\partial x_m}\left(\frac{1}{R}\right) = \frac{\partial}{\partial x_m}\left(\frac{1}{2}\nabla^2 R\right) = \frac{1}{2}\frac{\partial^3 R}{\partial x_m \partial x_p \partial x_p}, \tag{12.199}$$

the Peach-Koehler equation, i.e., Eq. (12.162), is rewritten in the form

$$\sigma_{ij}(\mathbf{x}) = \frac{\mu b_n}{8\pi}\oint_C \left[\frac{\partial^3 R}{\partial x_m \partial x_p \partial x_p}(e_{imn}dx'_j + e_{jmn}dx'_i)\right.$$
$$\left. + \frac{2}{1-\nu}e_{mnk}\left(\frac{\partial^3 R}{\partial x_m \partial x_i \partial x_j} - \delta_{ij}\frac{\partial^3 R}{\partial x_m \partial x_p \partial x_p}\right)dx'_k\right]. \tag{12.200}$$

For present purposes, this can then be put into a more usable form by introducing the variable, q, defined by the indefinite integral

$$q \equiv \frac{1}{\hat{t}_i}\oint_C R dx'_i = \oint_C R ds = -\oint_C [\rho^2 + (L')^2] dL'$$
$$= -\frac{1}{2}[\rho^2 \ln(R + L') + L'R] + f(\rho), \tag{12.201}$$

where use has been made of Eqs. (12.197) and (12.198), and $f(\rho)$ is the constant of integration. Then, substituting Eq. (12.201) into Eq. (12.200), with the constant of integration assigned the value $f(\rho) = \rho^2/4$,[13]

$$\sigma_{ij}(\mathbf{x}) = \frac{\mu b_n}{8\pi}\left[\frac{\partial^3 q}{\partial x_m \partial x_p \partial x_p}(e_{imn}\hat{t}_j + e_{jmn}t_i') + \frac{2}{1-\nu}\left(\frac{\partial^3 q}{\partial x_m \partial x_i \partial x_j} - \delta_{ij}\frac{\partial^3 q}{\partial x_m \partial x_p \partial x_p}\right)e_{mnk}\hat{t}_k\right]. \quad (12.202)$$

Useful expressions for the derivatives of q in Eq. (12.202) are now required. Introducing the vector

$$Y_i \equiv R_i + R\hat{t}_i, \quad (12.203)$$

derivatives of q are obtained of the forms

$$\frac{\partial q}{\partial x_i} = -\frac{1}{2}\left\{2\rho_i\left[\ln(R+L') - \frac{1}{2}\right] + \frac{\rho^2\hat{t}_i}{(R+L')} + Y_i\right\}, \quad (12.204)$$

$$\begin{aligned}\frac{\partial^2 q}{\partial x_i \partial x_j} &= -\frac{1}{2}\Big[2(\delta_{ij} - \hat{t}_i\hat{t}_j)\ln(R+L') \\ &\quad + \frac{2\rho_i Y_j + Y_j(R+L')\hat{t}_i + R\rho_j\hat{t}_i + (L'-R)\hat{t}_i Y_j}{R(R+L')}\Big] \\ &= -\left[(\delta_{ij} - \hat{t}_i\hat{t}_j)\ln(R+L') + \frac{\rho_i\hat{t}_j + \rho_j\hat{t}_i + L'\hat{t}_i\hat{t}_j}{(R+L')} + \frac{R_j(\rho_i + L'\hat{t}_i)}{R(R+L')}\right] \\ &= -\left[(\delta_{ij} - \hat{t}_i\hat{t}_j)\ln(R+L') + \frac{\rho_i\hat{t}_j + \rho_j\hat{t}_i + L'\hat{t}_i\hat{t}_j}{R} + \frac{(\rho_i\rho_j}{R(R+L')}\right]\end{aligned} \quad (12.205)$$

---

[13]This requirement on $f(\rho)$ is obtained by demanding that the final formalism developed with the use of $f(\rho) = \rho^2/4$, i.e., Eqs. (12.202) and (12.211), yields the known correct stresses produced by infinitely long straight dislocations. The use of this formalism in determining such stresses is illustrated in Exercise 12.7.

and

$$\frac{\partial^3 q}{\partial x_m \partial x_i \partial x_j} = -\left(\frac{\partial \rho_i}{\partial x_m}\hat{t}_j + \frac{\partial \rho_j}{\partial x_m}\hat{t}_i + \hat{t}_i\hat{t}_j\hat{t}_m\right)\frac{1}{R}$$
$$-\left(\frac{\partial \rho_i}{\partial x_j}Y_m + \frac{\partial \rho_i}{\partial x_m}\rho_j + \frac{\partial \rho_j}{\partial x_m}\rho_i\right)\frac{1}{R(R+L')}$$
$$+\left(\rho_i\hat{t}_j + \rho_j\hat{t}_i + \hat{t}_i\hat{t}_j L'\right)\frac{R_m}{R^3}$$
$$+\rho_i\rho_j\left[\frac{R_m}{R^3(R+L')} + \frac{Y_m}{R^2(R+L')^2}\right]. \tag{12.206}$$

However, when Eq. (12.206) is substituted into Eq. (12.202), factors of the form

$$e_{kmn}\hat{t}_k\hat{t}_m = 0 \tag{12.207}$$

appear. Since they vanish, Eq. (12.206) can be simplified by removing the terms leading to such factors with the result

$$\frac{\partial^3 q}{\partial x_m \partial x_i \partial x_j} = -\frac{\delta_{im}Y_j + \delta_{jm}Y_i}{R(R+L')} - \rho_m\left\{\frac{\partial \rho_i}{\partial x_j}\frac{1}{R(R+L')} - \left(\rho_i\hat{t}_j + \rho_j\hat{t}_i + \hat{t}_i\hat{t}_j L'\right)\right.$$
$$\left.\times\frac{1}{R^3} - \rho_i\rho_j\left[\frac{1}{R^3(R+L')} + \frac{1}{R^2(R+L')^2}\right]\right\}. \tag{12.208}$$

In addition, the following derivatives are obtained:

$$\frac{\partial^2 q}{\partial x_q \partial x_q} = -2\ln(R+L') - 1 \tag{12.209}$$

and

$$\frac{\partial^3 q}{\partial x_m \partial x_q \partial x_q} = -\frac{2Y_m}{R(R+L')}. \tag{12.210}$$

Finally, the stress at $P(\mathbf{x})$ contributed by the segment AB is obtained by evaluating $\sigma_{ij}(\mathbf{x})$, as given by Eq. (12.202) at the limits $\mathbf{x}' = \mathbf{x}'^B$ and $\mathbf{x}' = \mathbf{x}'^A$, i.e.,

$$\sigma_{ij}^{AB}(\mathbf{x}) = |\sigma_{ij}(\mathbf{x})|_{\mathbf{x}'=\mathbf{x}'^A}^{\mathbf{x}'=\mathbf{x}'^B}. \tag{12.211}$$

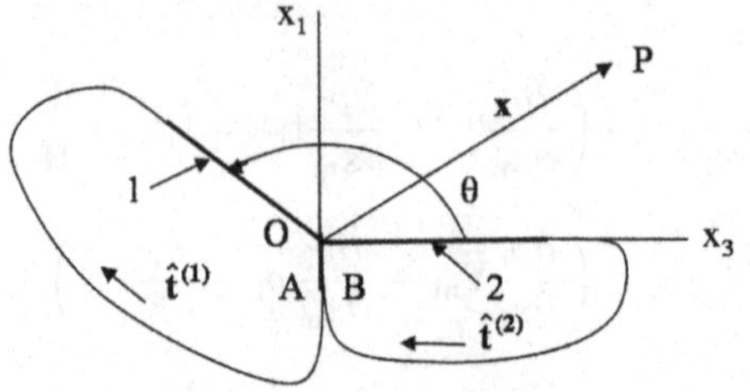

Fig. 12.31 Construction of angular dislocation by combining two straight dislocation segments, 1 and 2, which are initially parts of two closed loops. If the two segments and the two loops are configured as indicated, the loop lengths in the vicinity of A and B will mutually annihilate, and the two loops will merge to form a single closed loop containing the two abutting straight segments. Then, if the loop, along with the lengths of its two straight segments, is expanded outwards to infinity, the configuration will correspond to an angular dislocation with its two ends at infinity connected by the remainder of the closed loop. The stress at finite distances from the origin will then be due solely to the stresses contributed by the angular dislocation, since the contribution from the remainder of the closed loop at infinity will be vanishingly small.

Results for a wide range of segmented configurations can be determined using the above method by integrating along each segment referred to a common origin. Also, stresses for smoothly curved loops can be obtained by approximating the loops as N-sided polygons. For example, Khraishi, Hirth and Zbib (2000) have used this approximation to obtain the stresses due to circular dislocation loops and have shown that results closely approaching the exact analytical solution can be obtained by employing manageable values of N.

This method is employed in the next section to obtain the stress field due to the angular dislocation shown in Fig. 12.31. In Exercise 12.7 Eq. (12.202) is used to obtain the σ_{11} stress due to an infinitely long straight edge dislocation of the type illustrated in Fig. 12.2a.

12.8.1.2 *Angular dislocation*

The stress field of the angular dislocation in Fig. 12.31, in an isotropic system, can be obtained (see caption) by summing the stress fields contributed by the segments 1 and 2 which are given by expressions of the type corresponding to Eq. (12.211). The results are cumbersome, and, therefore, only the σ_{23} stress is considered for the case when $\mathbf{b} = (0, \mathrm{b}, 0)$. For segment 1, using the geometry in Fig. 12.31, $\hat{\mathbf{t}}^{(1)} = (-\sin\theta, 0, -\cos\theta)$, and

Eq. (12.202) then reduces to

$$\sigma_{23}^{(1)}(P) = \frac{\mu b}{4\pi(1-\nu)}\left(\frac{\partial^3 q}{\partial x_3 \partial x_2 \partial x_3}\sin\theta - \frac{\partial^3 q}{\partial x_1 \partial x_2 \partial x_3}\cos\theta\right). \quad (12.212)$$

Using Eq. (12.208), with m = 3, i = 2 and j = 3 and then with m = 1, i = 2 and j = 3 to evaluate the first and second terms, respectively,

$$\sigma_{23}^{(1)}(P) = \frac{\mu b}{4\pi(1-\nu)}\left\{\left[\frac{\rho_2\cos\theta}{R^3} - \frac{\rho_2\rho_3(L'+2R)}{R^3(L'+R)^2}\right]\right.$$

$$\left.\times(\rho_1\cos\theta - \rho_3\sin\theta) - \frac{R_2\sin\theta}{R(L'+R)}\right\}_{x'\to-\infty}^{x'=0}. \quad (12.213)$$

Then, after evaluating the limits,

$$\sigma_{23}^{(1)}(P) = \frac{\mu b x_2}{4\pi(1-\nu)x}\left[\frac{-\sin\theta}{(x-\xi_3)} + \frac{\xi_1\cos\theta}{x^2}\right.$$

$$\left. + \frac{2\xi_1^2\sin\theta}{x(x-\xi_3)^2} - \frac{\xi_1^2\xi_3\sin\theta}{x^2(x-\xi_3)^2}\right] \quad (12.214)$$

with

$$\xi_1 = x_1\cos\theta - x_3\sin\theta, \quad \xi_3 = x_1\sin\theta + x_3\cos\theta. \quad (12.215)$$

For segment 2, with $\hat{\mathbf{t}}^{(2)} = (0, 0, 1)$, Eq. (12.202) reduces to

$$\sigma_{23}^{(2)}(P) = \frac{\mu b}{4\pi(1-\nu)}\frac{\partial^3 q}{\partial x_1 \partial x_2 \partial x_3}. \quad (12.216)$$

Then, using Eq. (12.208), with m = 1, i = 2 and j = 3,

$$\sigma_{23}^{(2)}(P) = \frac{\mu b}{4\pi(1-\nu)}\left|\frac{\rho_1\rho_2}{R^3}\right|_{x'=O}^{x'\to\infty} = -\frac{\mu b}{4\pi(1-\nu)}\frac{x_1 x_2}{x^3}. \quad (12.217)$$

The total stress is then

$$\sigma_{23}(P) = \sigma_{23}^{(1)}(P) + \sigma_{23}^{(2)}(P)$$

$$= \frac{\mu b x_2}{4\pi(1-\nu)x}\left[\frac{-\sin\theta}{(x-\xi_3)} - \frac{x_1}{x^2} + \frac{\xi_1\cos\theta}{x^2} + \frac{\xi_1^2\sin\theta}{x(x-\xi_3)^2} + \frac{\xi_1^2\sin\theta}{x^2(x-\xi_3)}\right] \quad (12.218)$$

in agreement with results obtained by Yoffe (1960, 1961), who first determined the stress field by integrating the Burgers equation, and Hirth and

Lothe (1982), who summed the stresses of the segments after finding each in its local coordinate system.

12.8.1.3 *Three-dimensional multi-segment structures using triangular loops*

More complex three-dimensional multi-segment structures in isotropic systems can obviously be found by employing the basic elements and methods described previously. However, as pointed out by Barnett (1985), the displacement field of a triangular dislocation loop can often be used as a basic element for the efficient determination of the displacement fields produced by a wide range of segmented structures. For example, as shown in Fig. 12.32, a non-planar dislocation loop and a segmented dislocation line, respectively, can be constructed by assembling abutting triangular loops. If the displacement field of each triangular loop is known at the field point P in a common coordinate system, the total field can be determined by simple summation.

We therefore follow Barnett (1985) and obtain the displacement field of a single triangular loop by integrating the general Burgers displacement equation, i.e., Eq. (12.152), around the loop. The result is obtained in a coordinate-free vectorial form so that the field due to any ensemble of abutting loops is expressed directly in a common coordinate system. The geometry is shown in Fig. 12.33, and Eq. (12.152) therefore takes

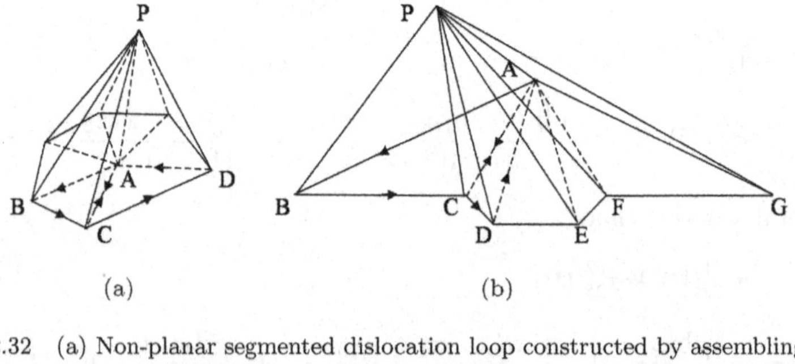

Fig. 12.32 (a) Non-planar segmented dislocation loop constructed by assembling abutting triangular loops such as ABC and ACD. Planes of the triangles such as ABC and ACD are not necessarily coplanar. Abutting segments, such as along AC, mutually annihilate leaving only the peripheral loop. (b) Non-planar segmented dislocation BCDEFG constructed, as in (a), by assembling abutting triangular loops such as ABC and ACD. If the points A, B and G are moved out to infinity, the result will be an infinite line containing the two offsets CD and EF.

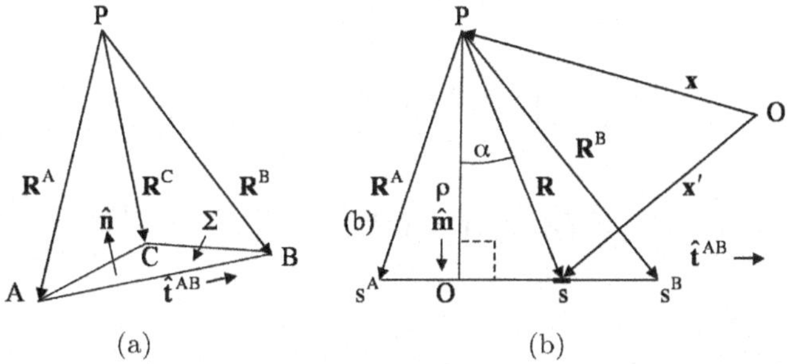

Fig. 12.33 (a) Triangular dislocation loop ABC with vectors $\mathbf{R}^A$, $\mathbf{R}^B$ and $\mathbf{R}^C$ emanating from field point P. (b) Geometry for integrating Burgers equation along the loop segment AB.

the form

$$\mathbf{u}(P) = -\frac{\mathbf{b}}{4\pi}\Omega + \mathbf{F}^{AB} + \mathbf{F}^{BC} + \mathbf{F}^{CA}, \tag{12.219}$$

where Ω is the total solid angle subtended by the loop, and the contribution of segment AB is of the form

$$\mathbf{F}^{AB} = -\frac{1}{4\pi}(\mathbf{b} \times \hat{\mathbf{t}}^{AB}) \int_{s^A}^{s^B} \frac{ds}{R} - \nabla \int_{s^A}^{s^B} \frac{(\mathbf{b} \times \hat{\mathbf{t}}^{AB}) \cdot \mathbf{R}}{R} ds \tag{12.220}$$

after using $d\mathbf{x}' = \hat{\mathbf{t}}^{AB} ds$. $\mathbf{F}^{BC}$ and $\mathbf{F}^{CA}$ are obtained by the cyclic interchange of A, B and C.

The $\mathbf{F}^{AB}$ term is evaluated first using the geometry of Fig. 12.33b. For the first integral,

$$(\mathbf{b} \times \hat{\mathbf{t}}^{AB}) \int_{s^A}^{s^B} \frac{ds}{R} = (\mathbf{b} \times \hat{\mathbf{t}}^{AB}) \int_{s^A}^{s^B} \frac{ds}{[s^2 + \rho^2]^{1/2}}$$
$$= (\mathbf{b} \times \hat{\mathbf{t}}^{AB}) \ln\left(\frac{R^B + \mathbf{R}^B \cdot \hat{\mathbf{t}}^{AB}}{R^A + \mathbf{R}^A \cdot \hat{\mathbf{t}}^{AB}}\right). \tag{12.221}$$

For the second integral,

$$-\nabla \int_{s^A}^{s^B} \frac{(\mathbf{b} \times \hat{\mathbf{t}}^{AB}) \cdot \mathbf{R}}{R} ds = -e_{ijk} b_j \hat{t}_k^{AB} \nabla \int_{s^A}^{s^B} \frac{(x_i' - x_i)}{R} ds. \tag{12.222}$$

However, since

$$\nabla\left(\frac{x_i' - x_i}{R}\right) = -\frac{\hat{e}_i}{R} + \frac{(x_i' - x_i)\mathbf{R}}{R^3}, \tag{12.223}$$

$$-e_{ijk}b_j t_k^{AB}\nabla\int_{s^A}^{s^B}\frac{(x_i' - x_i)}{R}ds = (\mathbf{b}\times\mathbf{t}^{AB})\int_{s^A}^{s^B}\frac{ds}{R} - \int_{s^A}^{s^B}\frac{[(\mathbf{b}\times\mathbf{t}^{AB})\cdot\mathbf{R}]\mathbf{R}}{R^3}ds. \tag{12.224}$$

To evaluate the latter integral, $\mathbf{R}$ is needed as a function of s. Using the unit vector, $\hat{\mathbf{m}}$, in Fig. 12.33b, $\mathbf{R} = \rho\hat{\mathbf{m}} + s\hat{\mathbf{t}}^{AB}$, and

$$(\mathbf{b}\times\hat{\mathbf{t}}^{AB})\cdot\mathbf{R} = (\mathbf{b}\times\hat{\mathbf{t}}^{AB})\cdot(\rho\hat{\mathbf{m}} + s\hat{\mathbf{t}}^{AB}) = (\mathbf{b}\times\hat{\mathbf{t}}^{AB})\cdot\rho\hat{\mathbf{m}}. \tag{12.225}$$

Then substituting Eq. (12.225) into Eq. (12.224), and integrating,

$$\begin{aligned}-\int_{s^A}^{s^B}\frac{[(\mathbf{b}\times\hat{\mathbf{t}}^{AB})\cdot\mathbf{R}]\mathbf{R}}{R^3}ds &= -(\mathbf{b}\times\hat{\mathbf{t}}^{AB})\cdot\rho\hat{\mathbf{m}}\int_{s^A}^{s^B}\frac{(\rho\hat{\mathbf{m}} + s\hat{\mathbf{t}}^{AB})}{(\rho^2 + s^2)^{3/2}}ds \\ &= -(\mathbf{b}\times\hat{\mathbf{t}}^{AB})\cdot\hat{\mathbf{m}}\left|\frac{(s\hat{\mathbf{m}} - \rho\hat{\mathbf{t}}^{AB})}{(\rho^2 + s^2)^{1/2}}\right|_{s^A}^{s^B}.\end{aligned} \tag{12.226}$$

However,

$$\frac{\mathbf{R}}{R} = \frac{\rho\hat{\mathbf{m}} + s\hat{\mathbf{t}}^{AB}}{(\rho^2 + s^2)^{1/2}} = \hat{\mathbf{m}}\cos\alpha + \hat{\mathbf{t}}^{AB}\sin\alpha \tag{12.227}$$

and, therefore,

$$\frac{s\hat{\mathbf{m}} - \rho\hat{\mathbf{t}}^{AB}}{(\rho^2 + s^2)^{1/2}} = -\frac{d}{d\alpha}\left(\frac{\mathbf{R}}{R}\right). \tag{12.228}$$

This result identifies the quantity $(s\hat{\mathbf{m}} - \rho\hat{\mathbf{t}}^{AB})/(\rho^2 + s^2)^{1/2}$ as a unit vector that can be written as

$$\frac{s\hat{\mathbf{m}} - \rho\hat{\mathbf{t}}^{AB}}{(\rho^2 + s^2)^{1/2}} = \frac{\mathbf{R}\times\hat{\mathbf{n}}^{AB}}{R}, \tag{12.229}$$

where $\hat{\mathbf{n}}^{AB}$ is the unit vector

$$\hat{\mathbf{n}}^{AB} = \frac{\mathbf{R}^A\times\mathbf{R}^B}{|\mathbf{R}^A\times\mathbf{R}^B|}. \tag{12.230}$$

Then, substituting Eq. (12.229) into Eq. (12.226), and using $\hat{\mathbf{t}}^{\mathrm{AB}} \times \hat{\mathbf{m}} = -\hat{\mathbf{n}}^{\mathrm{AB}}$,

$$-(\mathbf{b} \times \hat{\mathbf{t}}^{\mathrm{AB}}) \cdot \hat{\mathbf{m}} \left| \frac{(\mathrm{s}\hat{\mathbf{m}} - \rho\hat{\mathbf{t}}^{\mathrm{AB}})}{(\rho^2 + \mathrm{s}^2)^{1/2}} \right|_{\mathrm{s}^{\mathrm{A}}}^{\mathrm{s}^{\mathrm{B}}} = -(\mathbf{b} \times \hat{\mathbf{t}}^{\mathrm{AB}}) \cdot \hat{\mathbf{m}} \left| \frac{\mathbf{R} \times \hat{\mathbf{n}}^{\mathrm{AB}}}{\mathrm{R}} \right|_{\mathrm{s}^{\mathrm{A}}}^{\mathrm{s}^{\mathrm{B}}}$$
$$= (\mathbf{b} \cdot \hat{\mathbf{n}}^{\mathrm{AB}}) \left(\frac{\mathbf{R}^{\mathrm{B}}}{\mathrm{R}^{\mathrm{B}}} - \frac{\mathbf{R}^{\mathrm{A}}}{\mathrm{R}^{\mathrm{A}}} \right) \times \hat{\mathbf{n}}^{\mathrm{AB}}. \tag{12.231}$$

Finally, after gathering all terms required in Eq. (12.220) for $\mathbf{F}^{\mathrm{AB}}$,

$$\mathbf{F}^{\mathrm{AB}} = -\frac{(1-2\nu)}{8\pi(1-\nu)}(\mathbf{b} \times \hat{\mathbf{t}}^{\mathrm{AB}}) \ln \left(\frac{\mathrm{R}^{\mathrm{B}} + \mathbf{R}^{\mathrm{B}} \cdot \hat{\mathbf{t}}^{\mathrm{AB}}}{\mathrm{R}^{\mathrm{A}} + \mathbf{R}^{\mathrm{A}} \cdot \hat{\mathbf{t}}^{\mathrm{AB}}} \right)$$
$$+ \frac{1}{8\pi(1-\nu)}(\mathbf{b} \cdot \hat{\mathbf{n}}^{\mathrm{AB}}) \left(\frac{\mathbf{R}^{\mathrm{B}}}{\mathrm{R}^{\mathrm{B}}} - \frac{\mathbf{R}^{\mathrm{A}}}{\mathrm{R}^{\mathrm{A}}} \right) \times \hat{\mathbf{n}}^{\mathrm{AB}}. \tag{12.232}$$

Using Eq. (12.219), the displacement field can then be conveniently expressed as

$$\mathbf{u}(\mathrm{P}) = -\frac{\mathbf{b}\Omega}{4\pi} - \frac{(1-2\nu)}{8\pi(1-\nu)}[\mathbf{f}^{\mathrm{AB}} + \mathbf{f}^{\mathrm{BC}} + \mathbf{f}^{\mathrm{CA}}]$$
$$+ \frac{1}{8\pi(1-\nu)}[\mathbf{g}^{\mathrm{AB}} + \mathbf{g}^{\mathrm{BC}} + \mathbf{g}^{\mathrm{CA}}], \tag{12.233}$$

where, for example, for the AB segment,

$$\mathbf{f}^{\mathrm{AB}} = (\mathbf{b} \times \hat{\mathbf{t}}^{\mathrm{AB}}) \ln \left[\frac{\mathrm{R}^{\mathrm{B}}}{\mathrm{R}^{\mathrm{A}}} \frac{(1 + \hat{\boldsymbol{\lambda}}^{\mathrm{B}} \cdot \hat{\mathbf{t}}^{\mathrm{AB}})}{(1 + \hat{\boldsymbol{\lambda}}^{\mathrm{A}} \cdot \hat{\mathbf{t}}^{\mathrm{AB}})} \right]$$
$$\mathbf{g}^{\mathrm{AB}} = \frac{[\mathbf{b} \cdot (\hat{\boldsymbol{\lambda}}^{\mathrm{A}} \times \hat{\boldsymbol{\lambda}}^{\mathrm{B}})](\hat{\boldsymbol{\lambda}}^{\mathrm{A}} + \hat{\boldsymbol{\lambda}}^{\mathrm{B}})}{(1 + \hat{\boldsymbol{\lambda}}^{\mathrm{A}} \cdot \hat{\boldsymbol{\lambda}}^{\mathrm{B}})} \tag{12.234}$$

and $\hat{\boldsymbol{\lambda}}^{\mathrm{A}}$, $\hat{\boldsymbol{\lambda}}^{\mathrm{B}}$ and $\hat{\boldsymbol{\lambda}}^{\mathrm{C}}$ are unit vectors parallel to $\mathbf{R}^{\mathrm{A}}$, $\mathbf{R}^{\mathrm{B}}$ and $\mathbf{R}^{\mathrm{C}}$, respectively.

It now remains to find an expression for the solid angle Ω, which, as shown by Barnett (1985), can be obtained by means of spherical trigonometry. If one imagines a unit sphere centered on the field point at P, as in Fig. 12.34a, the solid angle subtended by the triangular loop ABC when viewed from the field point defines a spherical triangle on the surface of the

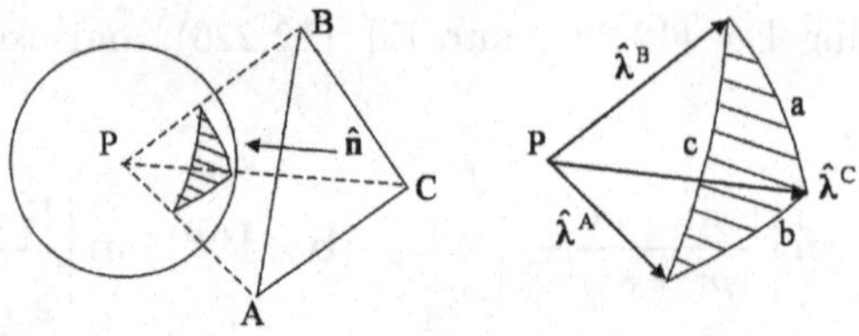

Fig. 12.34 (a) Spherical triangle (shaded) delineated on surface of unit sphere by solid angle subtended by the triangular loop ABC when viewed from the field point, P, located at sphere center: $\hat{\mathbf{n}}$ is the positive unit vector normal to the ABC plane. (b) Enlarged view of spherical triangle in (a).

unit sphere (shaded area). As seen in the enlarged view of the triangle in Fig. 12.34b, the unit vectors $\hat{\boldsymbol{\lambda}}^{\mathrm{A}}$, $\hat{\boldsymbol{\lambda}}^{\mathrm{B}}$ and $\hat{\boldsymbol{\lambda}}^{\mathrm{C}}$ connect the field point to the three corners of the triangle, and the three sides of the triangle (expressed in radians) are therefore given by

$$\mathrm{a} = \cos^{-1}(\hat{\boldsymbol{\lambda}}^{\mathrm{B}} \cdot \hat{\boldsymbol{\lambda}}^{\mathrm{C}}), \quad \mathrm{b} = \cos^{-1}(\hat{\boldsymbol{\lambda}}^{\mathrm{A}} \cdot \hat{\boldsymbol{\lambda}}^{\mathrm{C}}),$$
$$\mathrm{c} = \cos^{-1}(\hat{\boldsymbol{\lambda}}^{\mathrm{A}} \cdot \hat{\boldsymbol{\lambda}}^{\mathrm{B}}). \tag{12.235}$$

From spherical trigonometry (Reitz, Reilly and Woods 1936), the area of the triangle, $\mathbf{S}^{\mathrm{T}}$, is then given by

$$\tan^2\left(\frac{\mathrm{S}^{\mathrm{T}}}{4}\right) = \tan\left(\frac{\mathrm{s}}{2}\right)\tan\left(\frac{\mathrm{s}-\mathrm{a}}{2}\right)\tan\left(\frac{\mathrm{s}-\mathrm{b}}{2}\right)\tan\left(\frac{\mathrm{s}-\mathrm{c}}{2}\right), \tag{12.236}$$

where $\mathrm{s} = (\mathrm{a}+\mathrm{b}+\mathrm{c})/2$. Since the area of the triangle is equal to the solid angle that it defines on the unit sphere,

$$\Omega = -\mathrm{sgn}(\hat{\boldsymbol{\lambda}}^{\mathrm{A}} \cdot \hat{\mathbf{n}})\mathrm{S}^{\mathrm{T}} \tag{12.237}$$

after taking our sign conventions into account.

In Exercise 12.9, Eq. (12.237) is used to demonstrate formally that $\mathbf{u}$ undergoes a discontinuous change, $\Delta\mathbf{u} = \mathbf{b}$, when the field point just penetrates the ABC plane of the loop from its positive side. Barnett and Balluffi (2007) have shown that Eq. (12.232) reduces to an expression for the displacement given by Hirth and Lothe (1982) for the more restrictive case where a Cartesian $(\hat{\mathbf{e}}_1, \hat{\mathbf{e}}_2, \hat{\mathbf{e}}_3)$ coordinate system is employed and the AB segment lies along the $\hat{\mathbf{e}}_3$ axis, and the field point is located in the $\mathrm{x}_2 = 0$ plane. In Exercise 12.10, Eq. (12.233) is employed to derive the displacement field of an infinitely long straight screw dislocation in an isotropic system.

12.8.2 *Strain energies*

12.8.2.1 *Straight finite segment*

As discussed in Sec. 12.7.2.1, the strain energy contributed by a straight finite segment, that is part of a closed loop, can be obtained by integrating the equation for the strain energy of the loop along the length of the segment. Using Eq. (12.173), the strain energy contributed by a segment AB in an isotropic system is then

$$\mathrm{W} = \frac{\mu}{8\pi}\int_{\mathcal{L}^{(1)}=\mathrm{AB}}\int_{\mathcal{L}^{(2)}=\mathrm{AB}} \frac{(\mathbf{b}\cdot \mathrm{d}\mathbf{s}^{(1)})(\mathbf{b}\cdot \mathrm{d}\mathbf{s}^{(2)})}{\mathrm{R}} + \frac{\mu}{8\pi(1-\nu)}\int_{\mathrm{L}^{(1)}=\mathrm{AB}}\int_{\mathrm{L}^{(2)}=\mathrm{AB}} (\mathbf{b}\times \mathrm{d}\mathbf{s}^{(1)})\cdot \underline{\mathbf{T}}\cdot(\mathbf{b}\times \mathrm{d}\mathbf{s}^{(2)}), \tag{12.238}$$

where the line integrals, $\mathcal{L}^{(1)}$ and $\mathcal{L}^{(2)}$, involving $\mathrm{d}\mathbf{s}^{(1)}$ and $\mathrm{d}\mathbf{s}^{(2)}$, are each taken along the AB segment, and

$$\mathrm{R} = [(\mathrm{x}_1^{(1)} - \mathrm{x}_1^{(2)})^2 + (\mathrm{x}_2^{(1)} - \mathrm{x}_2^{(2)})^2 + (\mathrm{x}_3^{(1)} - \mathrm{x}_3^{(2)})^2]^{1/2}, \tag{12.239}$$

$$\mathrm{T}_{ij} = \frac{\partial \mathrm{R}}{\partial \mathrm{x}_i^{(1)}\partial \mathrm{x}_j^{(1)}} = \frac{1}{\mathrm{R}}\delta_{ij} - \frac{(\mathrm{x}_i^{(1)} - \mathrm{x}_i^{(2)})(\mathrm{x}_j^{(1)} - \mathrm{x}_j^{(2)})}{\mathrm{R}^3}$$

The x_3 axis of the coordinate system can be taken along the segment without loss of generality so that $\mathrm{d}\mathbf{s}^{(1)} = \hat{\mathbf{t}}\mathrm{dx}_3^{(1)}$, $\mathrm{d}\mathbf{s}^{(2)} = \hat{\mathbf{t}}\mathrm{dx}_3^{(2)}$, $\mathrm{T}_{11} = \mathrm{T}_{22} = 1/\mathrm{R}$, $\mathrm{T}_{33} = 0$, $\mathrm{T}_{ij}(i \neq j) = 0$, and $\mathrm{R} = \mathrm{x}_3^{(1)} - \mathrm{x}_3^{(2)}$. Then, substituting these quantities into Eq. (12.238), and writing $\mathrm{dx}_3^{(1)} = \mathrm{ds}^{(1)}$ and $\mathrm{dx}_3^{(2)} = \mathrm{ds}^{(2)}$,

$$\mathrm{W} = \frac{\mu}{8\pi}\left[(\mathbf{b}\cdot\hat{\mathbf{t}})^2 + \frac{|\mathbf{b}\times\hat{\mathbf{t}}|^2}{1-\nu}\right]\int_{\mathcal{L}^{(1)}=\mathrm{AB}}\int_{\mathcal{L}^{(2)}=\mathrm{AB}} \frac{1}{\mathrm{s}^{(1)} - \mathrm{s}^{(2)}}\mathrm{ds}^{(1)}\mathrm{ds}^{(2)}. \tag{12.240}$$

However, for a segment of length L, and using the cutoff parameter ρ introduced in Sec. 12.5.2, the double integral in Eq. (12.240) takes the form

$$\int_{\mathcal{L}^{(1)}=\mathrm{AB}}\int_{\mathcal{L}^{(2)}=\mathrm{AB}} \frac{1}{\mathrm{s}^{(1)} - \mathrm{s}^{(2)}}\mathrm{ds}^{(1)}\mathrm{ds}^{(2)} = \int_0^{\mathrm{L}} \mathrm{ds}^{(1)}\left[\int_0^{\mathrm{s}^{(1)}-\rho} \frac{\mathrm{ds}^{(2)}}{\mathrm{s}^{(1)} - \mathrm{s}^{(2)}} + \int_{\mathrm{s}^{(1)}+\rho}^{\mathrm{L}} \frac{\mathrm{ds}^{(2)}}{\mathrm{s}^{(1)} - \mathrm{s}^{(2)}}\right] = 2\mathrm{L}\ln\frac{\mathrm{L}}{\rho e} \tag{12.241}$$

and, therefore, the strain energy contributed by a straight segment of length L is (Hirth and Lothe 1982)

$$W = \frac{\mu}{4\pi}\left[(\mathbf{b}\cdot\hat{\mathbf{t}})^2 + \frac{|\mathbf{b}\mathrm{x}\hat{\mathbf{t}}|^2}{1-\nu}\right] L \ln \frac{L}{\rho e} \tag{12.242}$$

12.8.2.2 *Multi-segment structure*

As stated in Sec. 12.7.2.2, the total strain energy associated with a multi-segment structure is given, with acceptable accuracy, by the sum of the strain energies contributed by the various segments that comprise the structure, and the interaction energies contributed by all distinguishable pairs of segments in the structure, as obtained using our present models. Following Hirth and Lothe (1982), the validity of this conclusion is now demonstrated by means of detailed calculations for two simple cases in an isotropic system.

The energy contributed by each segment is given by Eq. (12.242), while the interaction energy contributed by each pair of segments is obtained by regarding each segment as belonging to its own closed loop and then integrating the equation for the interaction energy between the two loops along the lengths of the two segments. The interaction energy between two loops is given by Eq. (16.34), and, therefore, using this equation, the interaction energy associated with the two segments AB and CD is of the form

$$\begin{aligned} W_{\mathrm{int}}^{\mathrm{AB/CD}} &= -\frac{\mu}{2\pi}\int_{\mathcal{L}^{\mathrm{AB}}}\int_{\mathcal{L}^{\mathrm{CD}}} \frac{(\mathbf{b}^{\mathrm{AB}}\times\mathbf{b}^{\mathrm{CD}})\cdot(\mathbf{ds}^{\mathrm{AB}}\times\mathbf{ds}^{\mathrm{CD}})}{R} \\ &+\frac{\mu}{4\pi}\int_{\mathcal{L}^{\mathrm{AB}}}\int_{\mathcal{L}^{\mathrm{CD}}} \frac{(\mathbf{b}^{\mathrm{AB}}\cdot\mathbf{ds}^{\mathrm{AB}})(\mathbf{b}^{\mathrm{CD}}\cdot\mathbf{ds}^{\mathrm{CD}})}{R} \\ &+\frac{\mu}{4\pi(1-\nu)}\int_{\mathcal{L}^{\mathrm{AB}}}\int_{\mathcal{L}^{\mathrm{CD}}} (\mathbf{b}^{\mathrm{AB}}\times\mathbf{ds}^{\mathrm{AB}})\cdot\underline{\mathbf{T}}\cdot(\mathbf{b}^{\mathrm{CD}}\times\mathbf{ds}^{\mathrm{CD}}), \end{aligned} \tag{12.243}$$

where the line integrals, $\mathcal{L}^{\mathrm{AB}}$ and $\mathcal{L}^{\mathrm{CD}}$, involving $\mathrm{ds}^{\mathrm{AB}}$ and $\mathrm{ds}^{\mathrm{CD}}$, are taken along AB and CD, respectively.

Consider first the simple case of a straight segment AC with a point B lying between the end points A and C. Such a structure may be regarded as either a single segment of length $(L^{\mathrm{AB}} + L^{\mathrm{BC}})$ or a multi-segment structure consisting of two individual collinear segments of lengths L^{AB} and L^{BC}, respectively. For simplicity, the segments are assumed to be of pure screw orientation. According to the two-segment interpretation, the total strain energy should then be

$$W = W^{AB} + W^{CD} + W_{int}^{AB/BC}, \tag{12.244}$$

where, W^{AB} and W^{CD} are each given by Eq. (12.242), and $W_{int}^{AB/BC}$ by Eq. (12.243) in the form

$$\begin{aligned} W_{int}^{AB/BC} &= \frac{\mu}{4\pi}\int_{\mathcal{L}^{AB}}\int_{\mathcal{L}^{CD}} \frac{(\mathbf{b}^{AB}\cdot d\mathbf{s}^{AB})(\mathbf{b}^{BC}\cdot d\mathbf{s}^{BC})}{R} \\ &= \frac{\mu b^2}{4\pi}\int_0^{L^{AB}} ds^{AB}\int_0^{L^{BC}} \frac{ds^{BC}}{s^{AB}+s^{BC}} \\ &= \frac{\mu b^2}{4\pi}\left[L^{AB}\ln\left(\frac{L^{AB}+L^{BC}}{L^{AB}}\right) + L^{BC}\ln\left(\frac{L^{AB}+L^{BC}}{L^{BC}}\right)\right]. \end{aligned} \tag{12.245}$$

Then, substituting Eqs. (12.242) and (12.245) into (Eq. 12.244), the total strain energy contributed by the two segments is

$$W = \frac{\mu b^2}{4\pi}(L^{AB} + L^{BC})\ln\left(\frac{L^{AB}+L^{BC}}{\rho e}\right). \tag{12.246}$$

On the other hand, if the structure is regarded as a single segment of length $L^{AB} + L^{BC}$, the total strain energy, according to Eq. (12.242), is

$$W = \frac{\mu b^2}{4\pi}L\ln\frac{L}{\rho e} = \frac{\mu b^2}{4\pi}(L^{AB} + L^{BC})\ln\left(\frac{L^{AB}+L^{BC}}{\rho e}\right), \tag{12.247}$$

which is seen to be identical to the result given by Eq. (12.246).

A second demonstration makes use of the structure shown in Fig. 12.35 which consists of two long straight parallel dislocation segments, AB and

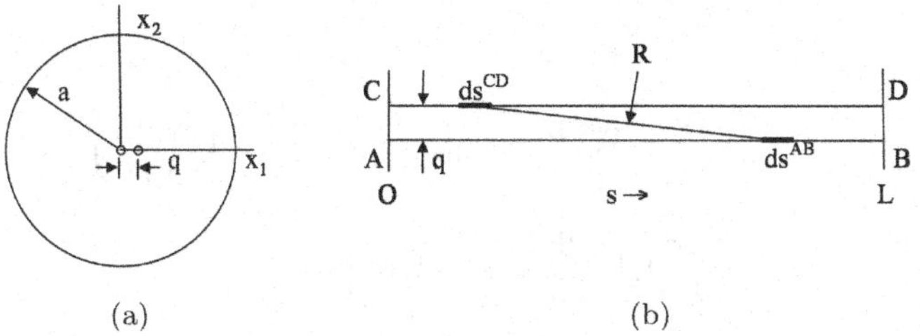

Fig. 12.35 (a) End view of dislocation segments AB and CD in cylinder of radius a. AB lies along cylinder axis. (b) View along x_2 of parallel segments AB and CD separated by distance q.

CD, of screw character with $b^{AB} = -b^{CD}$ and length L. The segment AB lies along the axis of a cylinder of radius a, and segment CD is off-axis by the distance q, where $q \ll L$. In this case Eq. (12.244) takes the form

$$W = W^{AB} + W^{CD} + W_{int}^{AB/BC} = \frac{\mu b^2}{2\pi} L \ln\left(\frac{L}{\rho e}\right)$$

$$-\frac{\mu b^2}{4\pi}\int_0^L ds^{CD}\int_0^L \frac{ds^{AB}}{[q^2 + (s^{AB} - s^{CD})^2]^{1/2}}. \qquad (12.248)$$

Performing the first integration in Eq. (12.248),

$$\int_0^L \frac{ds^{AB}}{[q^2 + (s^{AB} - s^{CD})^2]^{1/2}} = \ln\left[\frac{L - s^{CD} + \sqrt{(L - s^{CD})^2 + q^2}}{-s^{CD} + \sqrt{(s^{CD})^2 + q^2}}\right]$$

$$= \ln\left[\frac{(L - s^{CD})(1 + \sqrt{1 + \varepsilon_1^2})}{s^{CD}(\sqrt{1 + \varepsilon_2^2} - 1)}\right], \qquad (12.249)$$

where

$$\varepsilon_1 = \frac{q}{L - s^{CD}}, \quad \varepsilon_2 = \frac{q}{s^{CD}}. \qquad (12.250)$$

To perform the second integration, it is recognized that $\varepsilon_1^2 \ll 1$ and $\varepsilon_2^2 \ll 1$ over essentially the entire range of the integral, and therefore, expanding the integrand to first order,

$$\int_0^L \ln\left[\frac{(L - s^{CD})(1 + \sqrt{1 + \varepsilon_1^2})}{s^{CD}(\sqrt{1 + \varepsilon_2^2} - 1)}\right] ds^{CD} = \int_0^L \ln\left[\frac{4(L - s^{CD})s^{CD}}{q^2}\right] ds^{CD}$$

$$= 2L \ln\frac{2L}{eq}. \qquad (12.251)$$

Finally,

$$W = W^{AB} + W^{CD} + W_{int}^{AB/BC} = \frac{\mu b^2}{2\pi} L \ln\left(\frac{L}{\rho e}\right)$$

$$-\frac{\mu b^2}{2\pi} L \ln\frac{2L}{eq} = \frac{\mu b^2}{2\pi} L \ln\left(\frac{q}{2\rho}\right). \qquad (12.252)$$

This result may be compared with the total strain energy calculated by carrying out a calculation (Hirth and Lothe 1982) in which the AB segment is first introduced into the cylinder and then the CD segment is introduced by the cut and displacement method. The final strain energy is then the sum of the strain energy of the initial segment AB, plus the work required to displace the two sides of the cut during the introduction of the CD segment. The AB segment is assumed to be a screw dislocation of the type in Fig. 12.2c with the cut surface taken on the $x_2 = 0$ plane. The increment of work required to displace an area Ldx_1 on this cut surface is

$$d\mathcal{W}^{CD} = -\int_0^b db'\sigma_{23}(x_1, 0)Ldx_1, \tag{12.253}$$

where $\sigma_{23}(x_1, 0)$ is the stress acting at the cut surface given by

$$\sigma_{23}(x_1, 0) = \frac{\mu b}{2\pi}\frac{1}{x_1} - \frac{\mu b'}{2\pi}\frac{1}{(x_1 - q)} + \frac{\mu b'}{2\pi}\frac{1}{(x_1 - R^2/q)}. \tag{12.254}$$

The first term is the stress due to the presence of the AB segment (see Eq. (12.55)), the second is the stress that would be generated by the introduction of the CD segment if the cylinder were infinite, and the third is the image stress which causes the surface of the cylinder to be traction-free.[14] Substituting Eq. (12.254) into Eq. (12.253) and integrating over the cut surface, the final total strain energy is

$$\begin{aligned} W = W^{AB} + \mathcal{W}^{CD} &= \frac{\mu b^2}{4\pi}L\ln\frac{R}{r_o} - \frac{\mu b}{2\pi}L\int_q^R\int_0^b\frac{dx_1}{x_1}db' \\ &+ \frac{\mu}{2\pi}L\int_{q+r_o}^R\int_0^b\frac{dx_1}{(x_1 - q)}b'db' - \frac{\mu}{2\pi}L\int_q^R\int_0^b\frac{dx_1}{(x_1 - R^2/q)}b'db' \\ &= \frac{\mu b^2}{2\pi}L\ln\frac{q}{r_o} = \frac{\mu b^2}{2\pi}L\ln\frac{q\alpha}{b}, \end{aligned} \tag{12.255}$$

after dropping a second order term in q/R and using Eq. (12.61). Comparing this result with that given by Eq. (12.252), it is seen that they are identical if $\rho = b/(2\alpha)$. This is close to what should be expected since α is of order unity. Furthermore, the strain energy is relatively insensitive to the exact value of these factors since they appear in the argument of the logarithm. It is therefore again verified that the formalism proposed

[14] As shown in Exercise 13.4, the image stress in this case is the stress generated by a screw dislocation of opposite Burgers vector at the position $x_1 = R^2/q$.

initially for determining the total strain energy of a segmented structure is of acceptable accuracy.[15]

Exercises

12.1. Verify that, if Eq. (12.23) is applied to a Burgers circuit that encloses the dislocation, the resulting closure failure, Δu_i, is equal to the Burgers vector.

Solution. Using Eq. (12.23), and choosing a Burgers circuit of sense similar to the one in Fig. 12.5, but circular and of constant $|\mathbf{x}|$,

$$\Delta u_i = \frac{b_s}{2\pi}\left[4\pi B_{ks}\int_0^{2\pi}(\hat{n}\hat{n})^{-1}_{ik}d\omega + S_{rs}\int_0^{2\pi}(\hat{n}\hat{n})^{-1}_{ik}(\hat{n}\hat{m})_{kr}d\omega\right]. \tag{12.256}$$

Then, substituting Eqs. (3.133) and (3.134),

$$\Delta u_i = -b_s(4\pi B_{ks}Q_{ki} + S_{rs}S_{ir}) \tag{12.257}$$

and, finally, applying Eq. (3.140),

$$\Delta u_i = -b_s(4\pi B_{ks}Q_{ki} + S_{rs}S_{ir}) = b_s\delta_{si} = b_i. \tag{12.258}$$

12.2. Obtain the displacement field of a straight edge dislocation of the type in Fig. 12.2a in an isotropic body by employing the Volterra equation in the form of Eq. (12.70).

Solution. Here, $\mathbf{b} = (b, 0, 0)$, $\hat{\mathbf{t}} = (0, 0, 1)$ and $\hat{\mathbf{n}} = (0, 1, 0)$ so that Eq. (12.70) for $u_1(\mathbf{x})$ becomes

$$\begin{aligned} u_1(\mathbf{x}) &= -b\int_0^{\infty}dx'_1\int_{-\infty}^{\infty}C_{12jl}\left[\frac{\partial G_{1j}(\mathbf{x}-\mathbf{x}')}{\partial x'_l}\right]_{x'_2=0}dx'_3 \\ &= -\mu b\int_0^{\infty}dx'_1\int_{-\infty}^{\infty}\left[\frac{\partial G_{11}(\mathbf{x}-\mathbf{x}')}{\partial x'_2} + \frac{\partial G_{12}(\mathbf{x}-\mathbf{x}')}{\partial x'_1}\right]_{x'_2=0}dx'_3. \end{aligned} \tag{12.259}$$

[15] Hirth and Lothe (1982) perform a similar exercise for a pair of edge dislocation segments and reach the same conclusion.

Then, using Eq. (4.110) for G^{∞}_{ij}, and setting $x_3 = 0$ (allowable, since u_1 is constant along x_3),

$$u_1(x_1, x_2) = -\frac{b}{8\pi(1-\nu)} \int_0^{\infty} dx'_1 \times \int_{-\infty}^{\infty} \left[\frac{(1-2\nu)x_2}{R^3} + \frac{3x_2(x_1 - x'_1)^2}{R^5} \right] dx'_3, \quad (12.260)$$

with $R = [(x_1 - x'_1)^2 + x_2^2 + (x'_3)^2]^{1/2}$, and, by integrating over dx'_3 and dx'_1,

$$u_1(x_1, x_2) = \frac{b}{2\pi} \left[\tan^{-1}\left(\frac{x_2}{x_1}\right) + \frac{1}{2(1-\nu)} \frac{x_1 x_2}{(x_1^2 + x_2^2)} \right]. \quad (12.261)$$

Also, by applying the same procedure to find $u_2(x_1, x_2)$,

$$u_2(x_1, x_2) = -\frac{b}{8\pi(1-\nu)} \left[(1-2\nu)\ln(x_1^2 + x_2^2) + \frac{x_1^2 - x_2^2}{x_1^2 + x_2^2} \right]. \quad (12.262)$$

12.3. The Mura equation, i.e., Eq. (12.80), which involves only a line integral around a dislocation loop, indicates that a dislocation produced by the cut and displacement method must be independent of the surface, Σ, chosen for the cut. Verify this explicitly in an isotropic system by using the Volterra equation to show that the displacement field of the straight edge dislocation produced in Fig. 12.36 by making the cut on the $x_1 = 0$ surface is identical to the one found in Exercise 12.2 by making the cut on the $x_2 = 0$ surface.

Solution. Using the Volterra equation, i.e., Eq. (12.67), the displacement for the dislocation in Fig. 12.36 with $\mathbf{b} = (b, 0, 0)$,

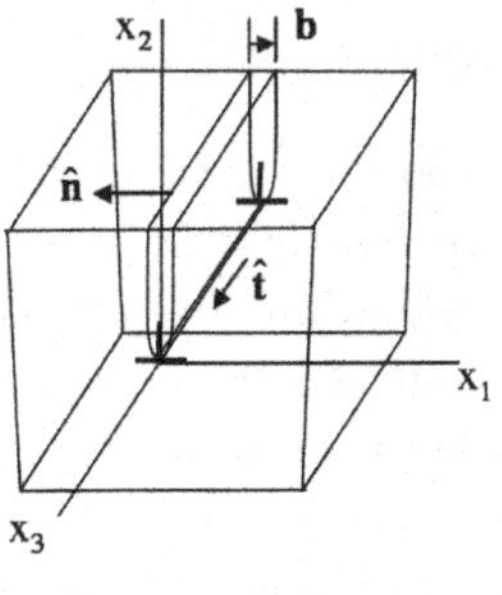

Fig. 12.36 Straight edge dislocation produced by a cut on the $x_1 = 0$ surface.

$\hat{\mathbf{t}} = (0,0,1)$ and $\hat{\mathbf{n}} = (-1,0,0)$, is given by

$$u_1(\mathbf{x}) = b\int_0^\infty dx_2' \int_{-\infty}^\infty C_{11jl}\left[\frac{\partial G_{1j}^\infty}{\partial x_l'}\right]_{x_1'=0} dx_3' = \frac{2\mu b}{1-2\nu}\int_0^\infty dx_2'$$
$$\times \int_{-\infty}^\infty \left[(1-\nu)\frac{\partial G_{11}^\infty}{\partial x_1'} + \nu\left(\frac{\partial G_{12}^\infty}{\partial x_2'} + \frac{\partial G_{13}^\infty}{\partial x_3'}\right)\right]_{x_1'=0} dx_3'. \tag{12.263}$$

Then, using Eq. (4.110) for G_{ij}^∞ and setting $x_3 = 0$ (since u_1 is constant along x_3),

$$u_1(x_1,x_2) = \frac{b}{8\pi(1-\nu)}\int_0^\infty dx_2' \int_{-\infty}^\infty \left[\frac{(1-2\nu)x_1}{R^3} + \frac{3x_1^3}{R^5}\right] dx_3' \tag{12.264}$$

with $R = [x_1^2 + (x_2 - x_2')^2 + (x_3')^2]^{1/2}$. After integrating over dx_3' and dx_1',

$$u_1(x_1,x_2) = \frac{b}{2\pi}\left[\tan^{-1}\left(\frac{x_2}{x_1}\right) + \frac{1}{2(1-\nu)}\frac{x_1x_2}{(x_1^2+x_2^2)}\right] + a_1, \tag{12.265}$$

where a_1 = constant that can be dropped, since it simply represents a rigid body translation. Therefore, Eq. (12.265) is in agreement with Eq. (12.261) for the dislocation in Fig. 12.2a. A similar result is obtained for u_2.

12.4. Use the Mura equation, i.e., Eq. (12.80), to find the strains produced by a long straight edge dislocation of the type shown in Fig. 12.2a in an isotropic system.

Solution. Employ the same general approach as that used in Sec. 12.5.1.1 to obtain the elastic field of a long straight dislocation by use of the Volterra equation. Assume the square edge dislocation loop with sides of length 2L in the coordinate system of Fig. 12.37 and focus on the dislocation segment AB. When L is increased without limit, the strains in the vicinity of the origin can be attributed to the segment AB, and it is therefore only necessary to evaluate the line integral along AB in the limit $L \to \infty$. For ε_{11}, the Mura

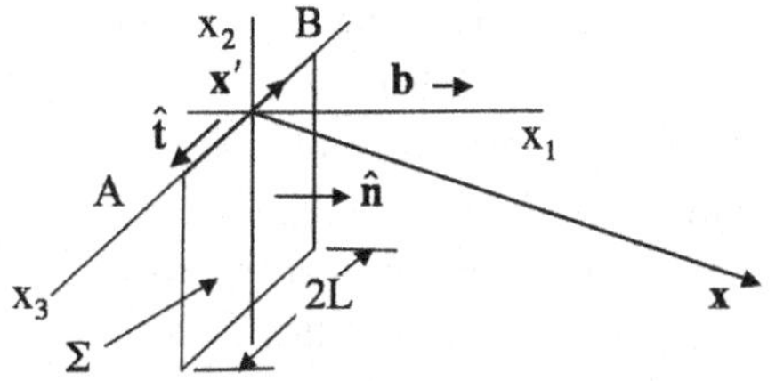

Fig. 12.37 Square edge dislocation loop. The loop plane, Σ cut surface, and $\mathbf{x}'$ lie in the $x_1 = 0$ plane.

equation, with $\mathbf{b} = (b, 0, 0)$, takes the form

$$\varepsilon_{11} = \frac{\partial u_1}{\partial x_1} = bC_{1njl}e_{n1p} \oint_C \frac{\partial G^{\infty}_{j1}}{\partial x'_l} dx'_p$$

$$= bC_{13jl} \oint_C \frac{\partial G^{\infty}_{j1}}{\partial x'_l} dx'_2 - bC_{12jl} \oint_C \frac{\partial G^{\infty}_{j1}}{\partial x'_l} dx'_3$$

$$= \mu b \oint_C \left[\left(\frac{\partial G^{\infty}_{11}}{\partial x'_3} + \frac{\partial G^{\infty}_{13}}{\partial x'_1} \right) dx'_2 - \left(\frac{\partial G^{\infty}_{11}}{\partial x'_2} + \frac{\partial G^{\infty}_{12}}{\partial x'_1} \right) dx'_3 \right] \tag{12.266}$$

with the help of Eq. (2.120), and after substituting the Green's function given by Eq. (4.110),

$$\varepsilon_{11} = \frac{b}{8\pi(1-\nu)} \oint_C \left\{ \left[\frac{(1-2\nu)(x_3 - x'_3)}{R^3} + \frac{3(x_1 - x'_1)^2(x_3 - x'_3)}{R^5} \right] dx'_2 \right.$$

$$\left. - \left[\frac{(1-2\nu)(x_2 - x'_2)}{R^3} + \frac{3(x_1 - x'_1)^2(x_2 - x'_2)}{R^5} \right] dx'_3 \right\}. \tag{12.267}$$

Then, for the integration along the $\hat{\mathbf{e}}_3$ axis in the interval $-L < x'_3 < L$, Eq. (12.267) is expressed as

$$\varepsilon_{11} = -\frac{b}{8\pi(1-\nu)} x_2 \int_{-L}^{L} \left[\frac{[1-2\nu]}{R^3} + \frac{3x_1^2}{R^5} \right] dx'_3 \tag{12.268}$$

with $R = |\mathbf{x} - \mathbf{x}'| = |\mathbf{x} - \hat{\mathbf{t}}x'_3| = (x_1^2 + x_2^2 + x_3'^2)^{1/2}$ after setting $x_3 = 0$. Finally, performing the integration, and taking the limit as $L \to \infty$,

$$\varepsilon_{11} = -\frac{b}{4\pi(1-\nu)} \left[\frac{(1-2\nu)x_2^3 + (3-2\nu)x_1^2 x_2}{(x_1^2 + x_2^2)^2} \right]. \tag{12.269}$$

It is readily confirmed that this result agrees with the ε_{11} strain obtained by using Eq. (12.261) and $\varepsilon_{11} = \partial u_1/\partial x_1$. Similar results are obtained for the remaining strains.

12.5. The displacement field for an edge dislocation of the type in Fig. 12.2a in an isotropic system was obtained in Sec. 12.4.1.1 by use of the transformation strain formalism. Use this method to find the displacement field of a screw dislocation of the type in Fig. 12.2c.

Solution. In this case the transformation strain is written as

$$\varepsilon_{23}^{T}(x_1, x_2) = -\frac{b}{2}\delta(x_2)H(x_1) \tag{12.270}$$

and the transformation stress is then

$$\sigma_{23}^{T}(x_1, x_2) = 2\mu\varepsilon_{23}^{T}(x_1, x_2) = -\mu b\delta(x_2)H(x_1). \tag{12.271}$$

Substituting Eq. (12.271) into Eq. (3.168), and realizing that the only displacements are in the x_3 direction,

$$u_3(\mathbf{x}) = -\mu b\int_{-\infty}^{\infty} H(x_1')dx_1' \int_{-\infty}^{\infty}\int_{-\infty}^{\infty} \delta(x_2') \times \left[\frac{\partial G_{32}^{\infty}(\mathbf{x}-\mathbf{x}')}{\partial x_3'} + \frac{\partial G_{33}^{\infty}(\mathbf{x}-\mathbf{x}')}{\partial x_2'}\right] dx_2' dx_3'. \tag{12.272}$$

Then, substituting Eq. (4.110) for the Green's function, setting the field point in the $x_3 = 0$ plane (since the solution is invariant along x_3), and integrating over dx_2',

$$u_3(x_1, x_2) = -\frac{b}{8\pi(1-\nu)}\int_{-\infty}^{\infty} H(x_1')dx_1' \times \int_{-\infty}^{\infty}\left[\frac{(1-2\nu)x_2}{R^3} + \frac{3x_2(x_3')^2}{R^5}\right] dx_3', \tag{12.273}$$

where $R = [(x_1 - x_1')^2 + x_2^2 + (x_3')^2]^{1/2}$. Integration of Eq. (12.273) over dx_3' and dx_1' to obtain u_3 then yields the displacement field

$$u_1(x_1, x_2) = u_2(x_1, x_2) = 0,$$
$$u_3(x_1, x_2) = -\frac{bx_2}{2\pi}\int_0^{\infty} \frac{dx_1'}{(x_1 - x_1')^2 + x_2^2} = \frac{b}{2\pi}\tan^{-1}\frac{x_2}{x_1} = \frac{b}{2\pi}\theta. \tag{12.274}$$

It is readily confirmed that the above displacement field is consistent with previous results for a screw dislocation. For example, using Eq. (12.274),

$$\sigma_{13} = 2\mu\varepsilon_{13} = \mu\left(\frac{\partial u_1}{\partial x_3} + \frac{\partial u_3}{\partial x_1}\right) = -\frac{\mu b}{2\pi}\frac{x_2}{(x_1^2 + x_2^2)} \tag{12.275}$$

in agreement with Eq. (12.55).

12.6. Show that in an isotropic system the stress field of an edge dislocation of the type in Fig. 12.2a given by Eq. (12.445) can be obtained by employing the Airy stress function method described in Sec. 3.7, since it is a case of plane strain. Hint: try the first stress function listed in Table I.1.

Solution. The displacements are obtained first by integrating the strains. For plane strain Eq. (2.122), with $\sigma_{33} = \nu(\sigma_{11} + \sigma_{22})$, yields the strains

$$\begin{aligned} \varepsilon_{11} &= \frac{\partial u_1}{\partial x_1} = \frac{1}{2\mu}[\sigma_{11} - \nu(\sigma_{11} + \sigma_{22})], \\ \varepsilon_{22} &= \frac{\partial u_2}{\partial x_2} = \frac{1}{2\mu}[\sigma_{22} - \nu(\sigma_{11} + \sigma_{22})]. \end{aligned} \tag{12.276}$$

Selecting the first stress function in Table I.1, and writing it in the form $\psi = A x \ln x \sin\theta$ where $A = \text{cons tan t}$, and substituting the corresponding stresses given in Table I.1 into Eq. (12.276),

$$\begin{aligned} \frac{\partial u_1}{\partial x_1} &= \frac{A x_2}{2\mu}\left[\frac{3x_1^2 + x_2^2 - 2\nu x_1^2 - 2\nu x_2^2}{(x_1^2 + x_2^2)^2}\right], \\ \frac{\partial u_2}{\partial x_2} &= \frac{A x_2}{2\mu}\left[\frac{-x_1^2 + x_2^2 - 2\nu x_1^2 - 2\nu x_2^2}{(x_1^2 + x_2^2)^2}\right]. \end{aligned} \tag{12.277}$$

Then, integrating Eq. (12.277),

$$\begin{aligned} u_1 &= -\frac{A(1-\nu)}{\mu}\left[\tan^{-1}\frac{x_2}{x_1} + \frac{x_1 x_2}{2(1-\nu)(x_1^2 + x_2^2)}\right], \\ u_2 &= \frac{A}{4\mu}\left[(1-2\nu)\ln(x_1^2 + x_2^2) + \frac{(x_1^2 - x_2^2)}{(x_1^2 + x_2^2)}\right]. \end{aligned} \tag{12.278}$$

The constant, A, can be determined by making a Burgers circuit enclosing the dislocation and equating the resulting closure failure and discontinuity in u_1 to the Burgers vector. Therefore, using Eq. (12.278), with $\Delta \tan^{-1}(x_2/x_1) = 2\pi$,

$$\Delta u_1 = -\frac{A(1-\nu)}{\mu} 2\pi = b \tag{12.279}$$

so that $A = -\mu b/[2\pi(1-\nu)]$. Substituting this into Eq. (12.278) then produces agreement with the expressions for u_1 and u_2 given by Eqs. (12.261) and (12.262).

12.7. Show that in an isotropic system the σ_{11} stress for an infinitely long straight edge dislocation of the type in Fig. 12.2a given by Eq. (12.45) can be obtained by employing Eq. (12.202) for the stress field contributed by a straight dislocation segment that is part of a closed loop.

Solution. Since $\mathbf{b} = (1,0,0)$ and $\hat{\mathbf{t}} = (0,0,1)$, Eq. (12.202) reduces for the case of the σ_{11} stress to

$$\sigma_{11}(\mathbf{x}) = -\frac{\mu b}{4\pi(1-\nu)} \left(\frac{\partial^3 q}{\partial x_2 \partial x_1 \partial x_1} - \frac{\partial^3 q}{\partial x_2 \partial x_p \partial x_p} \right). \tag{12.280}$$

Next, using Eqs. (12.208), with $m = 2$ and $i = j = 1$, and Eq. (12.210) to evaluate the derivatives,

$$\sigma_{11}(\mathbf{x}) = \frac{\mu b}{4\pi(1-\nu)} \left\{ \frac{\rho_2}{R(L'+R)} - \rho_2 \rho_1^2 \right.$$
$$\left. \times \left[\frac{1}{R^3(L'+R)} + \frac{1}{R^2(L'+R)^2} \right] - \frac{2R_2}{R(L'+R)} \right\}. \tag{12.281}$$

The stress due to an infinitely long segment can now be obtained by taking the origin on the segment as in Fig. 12.38 and evaluating Eq. (12.281) at the limits $L' \to \pm\infty$. Since $\boldsymbol{\rho} = (x_1, x_2, 0)$, and L'

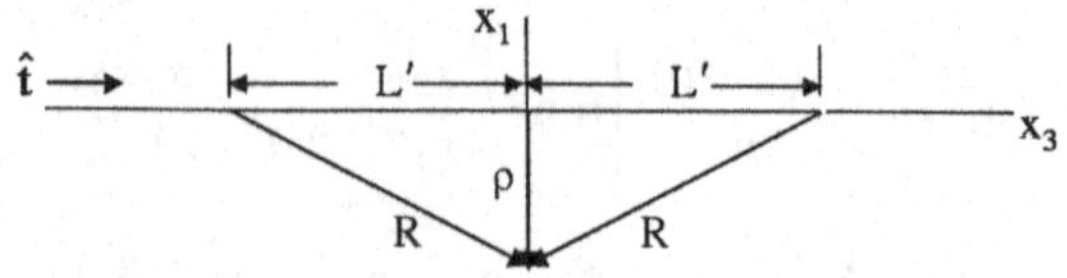

Fig. 12.38 Geometry for determining stress due to infinitely long straight dislocation along x_3.

is negative when $x_3 > 0$, use of Eq. (12.211) shows that the stress due to an infinite length is

$$\sigma_{11}(\mathbf{x}) = -\lim_{|L'|\to\infty} \frac{\mu b x_2}{4\pi(1-\nu)R}\left\{\left[\frac{1}{[-|L'|+R]} + x_1^2\left(\frac{1}{R^2(-|L'|+R)} + \frac{1}{R(-|L'|+R)^2}\right)\right] - \left[\frac{1}{(|L'|+R)} + x_1^2\left(\frac{1}{R^2(|L'|+R)} + \frac{1}{R(|L'|+R)^2}\right)\right]\right\}. \tag{12.282}$$

Since $R^2 = (L')^2 + \rho^2 = (L')^2 + x_1^2 + x_2^2$, and $R \to |L'|$ at large $|L'|$, R is given to first order by

$$R = |L'| + \frac{x_1^2 + x_2^2}{2|L'|}. \tag{12.283}$$

Then, substituting Eq. (12.283) into Eq. (12.282) and taking the limit,

$$\sigma_{11} = -\frac{\mu b}{2\pi(1-\nu)}\frac{x_2(3x_1^2 + x_2^2)}{(x_1^2 + x_2^3)^2} \tag{12.284}$$

in agreement with Eq. (12.45).

12.8. Equation (12.61) gives the strain energy produced by a straight edge dislocation in a coaxial cylindrical shell of outer radius, R, and inner radius, $r_\circ = b/\alpha$, in an isotropic system (see Fig. 12.6). The dislocation is of the type in Fig. 12.2a and possesses a stress field given by Eq. (12.47). Obtain the same expression for the strain energy by simply integrating the strain energy density throughout the volume of the shell.

Solution. The strain energy density in the shell given by Eq. (2.136), expressed in cylindrical coordinates, assumes the form

$$w = \frac{1}{2E}[(1+\nu)(\sigma_{rr}^2 + \sigma_{\theta\theta}^2 + \sigma_{zz}^2 + 2\sigma_{r\theta}^2 + 2\sigma_{rz}^2 + 2\sigma_{\theta z}^2) - \nu\Theta^2]. \tag{12.285}$$

Substituting Eq. (12.47), and integrating over the volume of the shell, the strain energy per unit shell length is then

$$\mathcal{W} = \frac{1}{2\mu}\int_0^{2\pi}\int_{r_o}^{R}\left[\sigma_{r\theta}^2 + \frac{1}{2(1+\nu)}(\sigma_{rr}^2 + \sigma_{\theta\theta}^2 - 2\nu\sigma_{rr}\sigma_{\theta\theta} - \sigma_{zz}^2)\right] r dr d\theta$$

$$= \frac{\mu b^2}{4\pi(1-\nu)}\ln\frac{R}{r_o}. \quad (12.286)$$

12.9. Assume that the cut surface, Σ, used to produce a triangular dislocation loop corresponds to the plane of the loop. Now, use Eq. (12.237) to demonstrate formally that the displacement, **u**, due to the loop undergoes a discontinuous change, $\Delta\mathbf{u} = \mathbf{b}$, when the field point just penetrates the plane of the loop from its positive side.

Solution. When the field point is on the positive side of the plane of the loop and is just about to penetrate it, $\boldsymbol{\lambda}^A \cdot \hat{\mathbf{n}} < 0$. Therefore, according to Eq. (12.237), $\Omega = S^T$. The spherical triangle is a unit circle with $a + b + c = 2\pi$, and so, according to Eq. (2.236), $\Omega = S^T = 2\pi$. Just after penetration, $\boldsymbol{\lambda}^A \cdot \hat{\mathbf{n}} > 0$, and Ω changes sign so that $\Omega = -2\pi$. The change in Ω due to the penetration is therefore, $\Delta\Omega = -2\pi - 2\pi = -4\pi$, and the change in **u**, according to Eq. (12.152), is $\Delta\mathbf{u} = \mathbf{b}$. Note that this result is consistent with the Σ cut/displacement rule given on p. 325.

12.10. Use Eq. (12.233), for the displacement field of a triangular dislocation loop, to obtain the displacement field of a straight screw dislocation in an isotropic system.

Solution. Adopt the coordinate system in Fig. 12.39 where the AB segment is a segment of screw dislocation of length 2L, and the field point, P, is located at $\mathbf{x} = (x_1, x_2, 0)$. Then, determine $\mathbf{u}(x_1, x_2)$ under conditions where L is very large compared to x_1 and x_2 so that $\mathbf{u}(x_1, x_2)$ is not significantly affected by the presence of the distant segments BC and CA. Using the vectors

$$\begin{aligned} &\mathbf{R}^A = (-x_1, -x_2, -L) && \hat{\mathbf{t}}^{AB} = (0, 0, 1) \\ &\mathbf{R}^B = (-x_1, -x_2, L) && \hat{\mathbf{t}}^{BC} = (\sqrt{3}/2, 0, -1/2) \\ &\mathbf{R}^C = ((\sqrt{3}L - x_1), -x_2, 0) && \hat{\mathbf{t}}^{CA} = (-\sqrt{3}/2, 0, -1/2) \end{aligned} \quad (12.287)$$

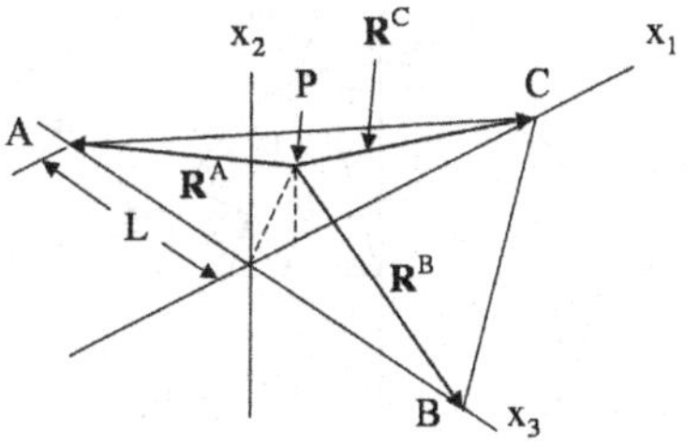

Fig. 12.39 Coordinate system for triangular equilateral dislocation loop ABC.

and $\mathbf{b} = (0, 0, b)$, and Eq. (12.234), it is found that $\mathbf{f}^{AB} = \mathbf{g}^{AB} = 0$, $\mathbf{f}^{BC} + \mathbf{f}^{CA} = 0$ and $\mathbf{g}^{BC} + \mathbf{g}^{CA} = 0$. Therefore, Eq. (12.233) simplifies to

$$\mathbf{u}(P) = -\frac{\mathbf{b}}{4\pi}\Omega. \tag{12.288}$$

In this case Ω may be calculated by making the triangle sufficiently large relative to x_1 and x_2 that the angle subtended by the triangle when viewed from P is adequately approximated by the angle subtended by the entire half plane $x_2 = 0$, $x_1 > 0$. Then, using Eq. (12.153) with $\hat{\mathbf{n}} = (0, 1, 0)$ and $(\mathbf{x}' - \mathbf{x}) = ((x_1' - x_1), x_2, x_3')$,

$$\begin{aligned}\Omega &= x_2 \int_0^\infty dx_1' \int_{-\infty}^\infty \frac{dx_3'}{[(x_1' - x_1)^2 + x_2^2 + x_3'^2]^{3/2}} \\ &= 2\pi - 2\tan^{-1}\frac{x_2}{x_1}.\end{aligned} \tag{12.289}$$

Finally, employing Eq. (12.288),

$$\mathbf{u}(P) = -\frac{\mathbf{b}}{4\pi}\Omega = \frac{\mathbf{b}}{2\pi}\tan^{-1}\frac{x_2}{x_1} - \frac{\mathbf{b}}{2} \tag{12.290}$$

in agreement with Eq. (12.274), since the term, $\mathbf{b}/2$, in Eq. (12.290) is simply a rigid body translation that may be discarded.

12.11. Determine the strain energy of a circular planar dislocation loop with its Burgers vector lying in the loop plane in an isotropic system.

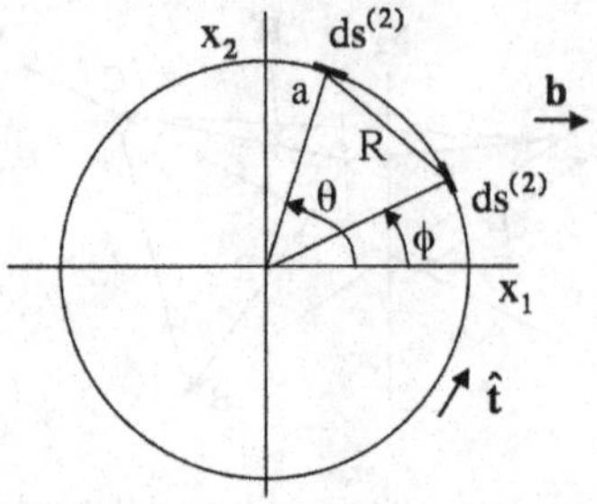

Fig. 12.40 Circular dislocation loop of radius a. The loop plane lies in the $x_3 = 0$ plane.

Solution. Employ Eq. (12.173) with **b** parallel to x_1 as in Fig. 12.40. The various quantities in Eq. (12.173) are then given by,

$$\begin{aligned}
\mathbf{b}\cdot \mathbf{ds}^{(1)} &= -\mathrm{b}\sin\phi \mathrm{ds}^{(1)},\\
\mathbf{b}\cdot \mathbf{ds}^{(2)} &= -\mathrm{b}\sin\theta \mathrm{ds}^{(1)},\\
\mathbf{b}\times \mathbf{ds}^{(1)} &= \hat{\mathbf{e}}_3\mathrm{b}\cos\phi \mathrm{ds}^{(1)},\\
\mathbf{b}\times \mathbf{ds}^{(2)} &= \hat{\mathbf{e}}_3\mathrm{b}\cos\theta \mathrm{ds}^{(2)},
\end{aligned} \tag{12.291}$$

$$(\mathbf{b}\times \mathbf{ds}^{(1)})\cdot \underline{\mathbf{T}}\cdot(\mathbf{b}\times \mathbf{ds}^{(2)}) = \mathrm{b}^2\cos\phi\cos\theta \mathrm{T}_{33}\mathrm{ds}^{(1)}\mathrm{ds}^{(2)},$$

$$\mathrm{T}_{33} = \left[\frac{\partial^2 \mathrm{R}}{\partial \mathrm{x}_3^{(1)}\partial \mathrm{x}_3^{(1)}}\right]_{\mathrm{x}_3^{(1)}=\mathrm{x}_3^{(2)}=0}$$

$$= \left[\frac{1}{\mathrm{R}}\delta_{33} - \frac{(\mathrm{x}_3^{(1)}-\mathrm{x}_3^{(2)})^2}{\mathrm{R}^3}\right]_{\mathrm{x}_3^{(1)}=\mathrm{x}_3^{(2)}=0} = \frac{1}{\mathrm{R}}.$$

Substituting these results into Eq. (12.173), while assuming a cutoff distance, ρ, within which the segments do not interact,

$$\mathrm{W} = \frac{\mu \mathrm{b}^2}{8\pi}\oint_{\mathrm{C}}\sin\phi \mathrm{ds}^{(1)}\int_{\rho}^{2\pi \mathrm{a}-\rho}\frac{\sin\theta \mathrm{ds}^{(2)}}{\mathrm{R}}$$

$$+\frac{\mu \mathrm{b}^2}{8\pi(1-\nu)}\oint_{\mathrm{C}}\cos\phi \mathrm{ds}^{(1)}\int_{\rho}^{2\pi \mathrm{a}-\rho}\frac{\cos\theta \mathrm{ds}^{(2)}}{\mathrm{R}}. \tag{12.292}$$

Next, substituting

$$\mathrm{ds}^{(1)} = \mathrm{a}d\phi \quad \mathrm{ds}^{(2)} = \mathrm{a}d\theta \quad \mathrm{R} = 2\mathrm{a}\sin\frac{\theta-\phi}{2} \tag{12.293}$$

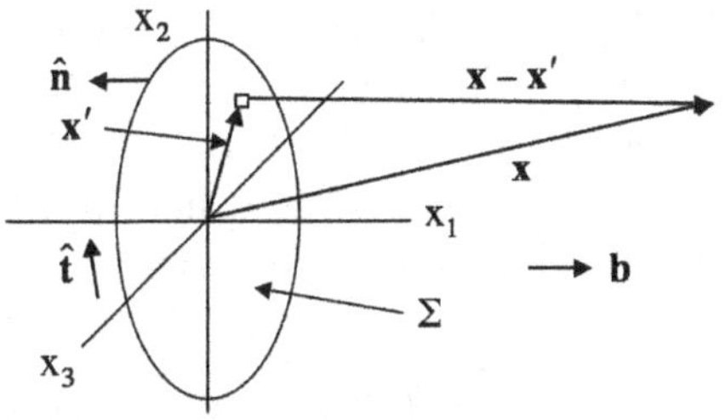

Fig. 12.41 Circular prismatic dislocation loop. The loop plane, Σ cut surface, and $\mathbf{x}'$ lie in the $x_1 = 0$ plane.

into Eq. (12.292) and integrating, the strain energy is

$$W = -\frac{\mu b^2 a}{4}\frac{(2-\nu)}{(1-\nu)}\left[\ln\left(\tan\frac{\rho}{4a}\right) + 2\cos\left(\frac{\rho}{2a}\right)\right]. \quad (12.294)$$

However, for the linear elastic model to remain valid, $a \gg \rho$. Therefore, using the approximations $\tan(\rho/4a) \approx \rho/4a$ and $\cos(\rho/2a) \approx 1$, Eq. (12.294) assumes the form

$$W = \frac{\mu b^2 a}{4}\frac{(2-\nu)}{(1-\nu)}\left[\ln\left(\frac{4a}{\rho}\right) - 2\right]. \quad (12.295)$$

12.12. Derive an integral expression for the displacement field of the circular prismatic dislocation loop in Fig. 12.41 in an isotropic system.

Solution. One method is to consider the loop as an array of infinitesimal loops as in Fig. 12.19 and integrate their displacements over the array. Using Eq. (12.169), the displacement at $\mathbf{x}$ produced by an infinitesimal loop with $\mathbf{b} = (b, 0, 0)$ and $\hat{\mathbf{n}} = (-1, 0, 0)$, and area dS located at the origin, is

$$du_1(\mathbf{x}) = \frac{b}{8\pi(1-\nu)}\left[\frac{(1-2\nu)x_1}{x^3} + \frac{3x_1^3}{x^5}\right] dS. \quad (12.296)$$

The displacement at $\mathbf{x}$ in Fig. 12.41, obtained as the sum of the displacements contributed by the infinitesimal loops distributed over the loop area Σ at source points $\mathbf{x}'$, is then

$$u_1(\mathbf{x}) = \iint_\Sigma du_1 = \frac{b}{8\pi(1-\nu)}\iint_\Sigma \left[\frac{(1-2\nu)x_1}{R^3} + \frac{3x_1^3}{R^5}\right] dx_2' dx_3', \quad (12.297)$$

where $R = [x_1^2 + (x_2 - x_2')^2 + (x_3 - x_3')^2]^{1/2}$.

This result, of course, can be obtained more directly by simply starting with the Volterra equation, i.e., Eq. (12.67), and employing the Green's function given by Eq. (4.110).

12.13. Show that the in-plane stress at a field point P lying within a smoothly curved planar loop, C, is given by

$$\sigma_{ij}(P) = \frac{1}{2}\oint_C \kappa \Sigma_{ij}(\theta)\csc^3(\theta - \alpha)d\theta, \tag{12.298}$$

where κ is the local loop curvature (Asaro and Barnett 1976). Hint: use Eqs. (12.115) and (12.116).

Solution. Substituting Eq. (12.116) into Eq. (12.115),

$$\sigma_{ij}(P) = \frac{1}{2}\oint_C \frac{1}{R}\left[\Sigma_{ij}(\theta) - \frac{d\Sigma_{ij}(\theta)}{d\theta}\cot(\theta - \alpha)\right] d\theta. \tag{12.299}$$

An expression for the second term in Eq. (12.299) can be obtained by introducing the derivative

$$d\left[\frac{1}{R}\Sigma_{ij}(\theta)\cot(\theta - \alpha)\right] = -\frac{1}{R}\Sigma_{ij}(\theta)\csc^2(\theta - \alpha)(d\theta - d\alpha)$$
$$+ \cot(\theta - \alpha)\left[\frac{d\Sigma_{ij}(\theta)}{d\theta} - \frac{\Sigma_{ij}(\theta)}{R}\frac{dR}{d\theta}\right]\frac{1}{R}d\theta. \tag{12.300}$$

Then, taking the line integral of the above derivative around the loop,

$$\oint_C d\left[\frac{1}{R}\Sigma_{ij}(\theta)\cot(\theta - \alpha)\right] = 0$$
$$= \oint_C \left\{-\frac{1}{R}\Sigma_{ij}(\theta)\csc^2(\theta - \alpha)(d\theta - d\alpha)\right.$$
$$\left. + \cot(\theta - \alpha)\left[\frac{d\Sigma_{ij}(\theta)}{d\theta} - \frac{\Sigma_{ij}(\theta)}{R}\frac{dR}{d\theta}\right]\frac{1}{R}d\theta\right\} \tag{12.301}$$

which can be rearranged with the use of Eq. (12.117) to produce the equality

$$\oint_C \frac{1}{R}\frac{d\Sigma_{ij}(\theta)}{d\theta}\cot(\theta-\alpha)d\theta = \oint_C \frac{1}{R}\Sigma_{ij}(\theta)\left\{[\csc^2(\theta-\alpha) - \cot^2(\theta-\alpha)]d\theta - \csc^2(\theta-\alpha)d\alpha\right\}. \tag{12.302}$$

Next, Eq. (12.302) is substituted into Eq. (12.299) with the result

$$\sigma_{ij}(P) = \frac{1}{2}\oint_C \frac{1}{R}\Sigma_{ij}(\theta)\csc^2(\theta-\alpha)d\alpha. \tag{12.303}$$

The curvature is defined by $\kappa = d\alpha/ds$, and, therefore, with the use of Eq. (12.109),

$$d\alpha = \frac{\kappa R}{\sin(\theta-\alpha)}d\theta. \tag{12.304}$$

Finally, substituting Eq. (12.304) into Eq. (12.303),

$$\sigma_{ij}(P) = \frac{1}{2}\oint_C \kappa\Sigma_{ij}(\theta)\csc^3(\theta-\alpha)d\theta. \tag{12.305}$$

12.14. Show that the Mura equation, i.e. Eq. (12.80), can be reformulated to yield the stress field produced by a general dislocation loop in an infinite body expressed in the form

$$\sigma_{\alpha\beta}(\mathbf{x}) = -ib_k C_{\alpha\beta js} C_{kpim} e_{rps} \oint_C e^{-i\mathbf{k}\cdot\mathbf{x}')} dx'_r \times \int_{-\infty}^{\infty}\int_{-\infty}^{\infty}\int_{-\infty}^{\infty} (kk)^{-1}_{ij} e^{i\mathbf{k}\cdot\mathbf{x}} k_m dk_1 dk_2 dk_3. \tag{12.306}$$

Solution. Using Eqs. (2.75) and (2.5), Eq. (12.80) takes the form

$$\sigma_{\alpha\beta}(\mathbf{x}) = -b_k C_{\alpha\beta js} C_{kpim} e_{rps} \oint_C \frac{\partial G_{ij}(\mathbf{x}-\mathbf{x}')}{\partial x_m} dx'_r. \tag{12.307}$$

Next, using Eqs. (F.4) and (3.23), the Green's function is given by,

$$\begin{aligned} G_{ij}(\mathbf{x}-\mathbf{x}') &= \int_{-\infty}^{\infty}\int_{-\infty}^{\infty}\int_{-\infty}^{\infty} \overline{G}_{ij}^{\infty}(\mathbf{k})e^{i\mathbf{k}\cdot(\mathbf{x}-\mathbf{x}')}dk_1dk_2dk_3 \\ &= \int_{-\infty}^{\infty}\int_{-\infty}^{\infty}\int_{-\infty}^{\infty} (kk)_{ij}^{-1}e^{i\mathbf{k}\cdot(\mathbf{x}-\mathbf{x}')}dk_1dk_2dk_3 \end{aligned} \tag{12.308}$$

and the derivative required in Eq. (12.307) is,

$$\frac{\partial G_{ij}(\mathbf{x}-\mathbf{x}')}{\partial x_m} = i\int_{-\infty}^{\infty}\int_{-\infty}^{\infty}\int_{-\infty}^{\infty} (kk)_{ij}^{-1}(\mathbf{k})e^{i\mathbf{k}\cdot(\mathbf{x}-\mathbf{x}')}k_m dk_1dk_2dk_3. \tag{12.309}$$

Substitution of this result into Eq. (12.307) then yields Eq. (12.306).

Chapter 13

Interactions between Dislocations and Stress: Image Effects

13.1 Introduction

We now take up interactions between dislocations and various types of stress fields. Forces between dislocations and imposed internal or applied fields are considered first, and are shown to be described by the Peach–Koehler force equation. Following this, the interaction between a dislocation and its image stress is analyzed in situations where the dislocation lies in a finite, or semi-infinite, homogenous region in the proximity of an interface.

With these results in hand, a number of dislocation image stress problems are treated that involve straight dislocations lying parallel to several types of planar interfaces, straight dislocations impinging on planar interfaces at various angles, and dislocation loops lying near planar interfaces.

Treatments and reviews of a wide variety of dislocation image problems have been published by Eshelby (1979), Lothe (1992c) and Belov (1992).

The following notation is employed for this chapter:

$\varepsilon_{\mathrm{ij}} = \varepsilon_{\mathrm{ij}}^{\infty} + \varepsilon_{\mathrm{ij}}^{\mathrm{IM}}$ = strain produced by dislocation in finite homogeneous region

$\varepsilon_{\mathrm{ij}}^{\infty}$ = strain produced by dislocation in infinite homogeneous region

$\varepsilon_{\mathrm{ij}}^{\mathrm{IM}}$ = image strain of dislocation in finite homogeneous body

$\varepsilon_{\mathrm{ij}}^{\mathrm{Q}}$ = imposed internal or applied strain

13.2 Interaction of Dislocation with Imposed Internal or Applied Stress: The Peach–Koehler Force Equation

We first find the interaction energy between a general dislocation loop and an imposed stress field, Q, which can be either internal or applied, by employing Eq. (5.5) in the form

$$E_{int}^{DIS/Q} = E^{DIS/Q} - E^{DIS}. \tag{13.1}$$

The dislocation is introduced, in the presence of the constant Q field by the usual cut and displacement method described in Sec. 12.2, and the extra work required to produce the loop due to the presence of the Q field is then the work required to displace the two sides of the cut by the Burgers vector against the tractions due to the Q field integrated over the surface of the cut. This is given by

$$\mathcal{W} = b_i \iint_\Sigma \sigma_{ij}^Q \hat{n}_j dS. \tag{13.2}$$

This work is just the energy difference in Eq. (13.1), i.e.,

$$\mathcal{W} = E^{DIS/Q} - E^{DIS} \tag{13.3}$$

and, substitution of Eqs. (13.3) and (13.2) into Eq. (13.1) then yields

$$E_{int}^{DIS/Q} = \mathcal{W} = b_i \iint_\Sigma \sigma_{ij}^Q \hat{n}_j dS. \tag{13.4}$$

Having the interaction energy, the corresponding force per unit length experienced by the loop at any point along its length can be found by virtually displacing the loop everywhere along its perimeter by the vector $\delta\boldsymbol{\xi}(\mathbf{x})$. Each segment of loop length ds then sweeps out a differential area $\delta\boldsymbol{\Sigma}$ of the cut surface Σ in Eq. (13.4), thus changing its area by

$$\delta\boldsymbol{\Sigma} = \delta\boldsymbol{\xi} \times \hat{\mathbf{t}} ds = e_{lsj}\delta\xi_l \hat{t}_s \hat{\mathbf{e}}_j ds. \tag{13.5}$$

The corresponding change in the interaction energy between the loop and the imposed stress, $\delta E_{int}^{DIS/Q}$, is then obtained by use of Eqs. (13.4) and (13.5) in the form

$$\delta E_{int}^{DIS/Q} = b_i \delta \iint_{\delta\Sigma} \sigma_{ij}^Q \hat{n}_j dS = e_{lsj} b_i \oint_C \sigma_{ij}^Q \delta\xi_l \hat{t}_s ds, \tag{13.6}$$

where component j of the incremental surface area vector, $\mathbf{dS} = dS\hat{\mathbf{n}}$, corresponding to $dS\hat{n}_j$ in Eq. (13.4) has been replaced by $e_{lsj}\delta\xi_l \hat{t}_s ds$ obtained from Eq. (13.5), and the surface integration over $\delta\Sigma$ has been replaced by line integration along C. Then, if $f_l^{DIS/Q}$ is the local force (per unit length) exerted on the loop by the stress Q at any point along its length, the total work performed during the loop displacement is

$$\delta \mathcal{W} = \oint_{C} f_{l}^{DIS/Q} \delta \xi_{l} ds, \tag{13.7}$$

which decreases the interaction energy between the loop and the imposed stress according to

$$\delta E_{int}^{DIS/Q} = -\delta \mathcal{W}. \tag{13.8}$$

Substitution of Eqs. (13.6) and (13.7) into Eq. (13.8) then yields the force per unit length

$$f_{l}^{DIS/Q} = -e_{lsj} \sigma_{ij}^{Q} b_{i} \hat{t}_{s} = e_{jsl} \sigma_{ij}^{Q} b_{i} \hat{t}_{s}. \tag{13.9}$$

Equation (13.9) is known as the *Peach–Koehler force equation* (Peach and Koehler 1950) and is often written in vector form by introducing the vector **d** given by

$$[d] = [b][\sigma^{Q}] \tag{13.10}$$

so that

$$\boldsymbol{f}^{DIS/Q} = \mathbf{d} \times \hat{\mathbf{t}} \tag{13.11}$$

with components now given by

$$f_{l}^{DIS/Q} = e_{jsl} d_{j} \hat{t}_{s}, \quad d_{j} = b_{i} \sigma_{ij}^{Q}. \tag{13.12}$$

The Peach–Koehler force equation is an essential relationship that yields the force, per unit length, exerted by an imposed stress field on a dislocation lying in either a finite or infinite homogeneous region. Note that the equation predicts that the force is always perpendicular to the dislocation line, i.e., perpendicular to $\hat{\mathbf{t}}$, as expected, since as shown by Eq. (13.5) the only component of the displacement vector $\delta\boldsymbol{\xi}$ that makes a non-vanishing contribution to the area swept out by motion of the dislocation is perpendicular to $\hat{\mathbf{t}}$.

Applications of the Peach–Koehler equation are carried out in Exercises 13.1 and 13.3. It is demonstrated in Exercise 13.2 that the equation can be obtained by an alternative derivation involving the direct determination of the extra work required to displace a dislocation in the presence of an imposed stress field.[1]

[1]Note that, since the displacement of the dislocation causes a change in the body shape, this force can also be considered as an example of the type 4 interaction listed in Sec. 5.1.

13.3 Interaction of Dislocation with its Image Stress

13.3.1 *General formulation*

Consider now the case where a dislocation loop is in a finite body with a traction-free surface, $\mathcal{S}^\circ$, in the presence of its image stress. The interaction energy between the loop and its image stress can be determined by employing the same basic method used to obtain the force between a loop and the imposed Q stress in the previous section. Therefore, following the same procedure,

$$\mathrm{E}_{\mathrm{int}}^{\mathrm{DIS/DIS^{IM}}} = \mathrm{E}^{\mathrm{DIS/DIS^{IM}}} - \mathrm{E}^{\mathrm{DIS}}. \tag{13.13}$$

However, the work performed at the cut is now of the form

$$\mathcal{W} = \frac{1}{2}\mathrm{b_i} \iint_{\Sigma} \sigma_{\mathrm{ij}}^{\mathrm{IM}} \hat{\mathrm{n}}_{\mathrm{j}} \mathrm{dS}, \tag{13.14}$$

where the factor of 1/2 appears because the image stress increases linearly as the displacement across the cut increases in contrast to the situation in the previous analysis where the Q stress is constant, and the work is in the form of Eq. (13.2). Finally,

$$\mathrm{E}_{\mathrm{int}}^{\mathrm{DIS/DIS^{IM}}} = \mathrm{E}^{\mathrm{DIS/DIS^{IM}}} - \mathrm{E}^{\mathrm{DIS}} = \mathcal{W} = \frac{1}{2}\mathrm{b_i} \iint_{\Sigma} \sigma_{\mathrm{ij}}^{\mathrm{IM}} \hat{\mathrm{n}}_{\mathrm{j}} \mathrm{dS}. \tag{13.15}$$

Having the interaction energy, the force per unit length exerted on the loop at any point along its perimeter can be determined, as in the previous section, by virtually displacing the loop along its perimeter by $\delta\boldsymbol{\xi}(\mathbf{x})$. Previously, the displacement was carried out in the presence of a constant imposed stress, and the only result was a change in the area of the cut surface, Σ. However, in the present case an added complication is present, since the image stress changes as the loop is displaced. Therefore, starting with Eq. (13.15) and using Eq. (13.5) to account for the change in the area of Σ, as previously, and including the effect of the change in image stress, the interaction energy can be written as the sum of two terms of the form

$$\delta\mathrm{E}_{\mathrm{int}}^{\mathrm{DIS/DIS^{IM}}} = \frac{1}{2}\mathrm{e}_{l\mathrm{sj}}\mathrm{b_i} \oint_{\mathrm{C}} \sigma_{\mathrm{ij}}^{\mathrm{IM}} \delta\xi_l \hat{\mathrm{t}}_{\mathrm{s}} \mathrm{ds} + \frac{1}{2}\mathrm{b_i} \iint_{\Sigma} \delta\sigma_{\mathrm{ij}}^{\mathrm{IM}} \hat{\mathrm{n}}_{\mathrm{j}} \mathrm{dS}. \tag{13.16}$$

It is now shown, following Gavazza and Barnett (1975), that the line and surface integrals in Eq. (13.16) are equal. For this purpose, we introduce Fig. 13.1 which shows the loop in the region, $\mathcal{V}$, enclosed by the surface, $\mathcal{S}$,

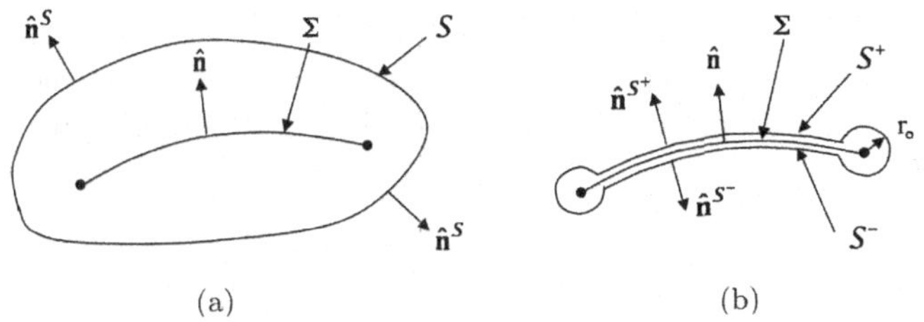

Fig. 13.1 Cross section of dislocation loop created by cut and displacement on the surface Σ. Loop enclosed by surface $\mathcal{S}$ with unit normal vector, $\hat{\mathbf{n}}^{\mathcal{S}}$. (b) View after shrinking $\mathcal{S}$ down around the dislocation loop and the Σ surface. The $\mathcal{S}$ surface now consists of a tube of radius r_o, coaxial with the loop, and two surfaces, $\mathcal{S}^+$ (with unit normal $\hat{\mathbf{n}}^{\mathcal{S}^+}$) and $\mathcal{S}^-$ (with unit normal $\hat{\mathbf{n}}^{\mathcal{S}^-}$) which are infinitesimally close to the Σ surface.

which, in turn, is assumed to be embedded in a larger region $\mathcal{V}^\circ$ enclosed by a traction-free surface $\mathcal{S}^\circ$.

We now consider the system in Fig. 13.1 before, and after, the loop displacement, $\delta\boldsymbol{\xi}(\mathbf{x})$. The elastic fields present before, and after, are denoted by

$$\begin{aligned}
\sigma_{ij} &= \sigma_{ij}^{\infty} + \sigma_{ij}^{IM}, \quad \sigma'_{ij} = \sigma_{ij}^{\prime\infty} + \sigma_{ij}^{\prime IM},\\
\varepsilon_{ij} &= \varepsilon_{ij}^{\infty} + \varepsilon_{ij}^{IM}, \quad \varepsilon'_{ij} = \varepsilon_{ij}^{\prime\infty} + \varepsilon_{ij}^{\prime IM},\\
u_i &= u_i^{\infty} + u_i^{IM}, \quad u'_i = u_i^{\prime\infty} + u_i^{\prime IM},
\end{aligned} \tag{13.17}$$

where the prime indicates a quantity after the displacement. Next, consider the integral

$$\oiiint_{\mathcal{V}^\circ-\mathcal{V}} \left[(\sigma_{ij}^{\infty} + \sigma_{ij}^{IM})(\varepsilon_{ij}^{\prime\infty} + \varepsilon_{ij}^{\prime IM}) - (\sigma_{ij}^{\prime\infty} + \sigma_{ij}^{\prime IM})(\varepsilon_{ij}^{\infty} + \varepsilon_{ij}^{IM})\right] dV. \tag{13.18}$$

Since the primed and unprimed fields are corresponding fields in $(\mathcal{V}^\circ - \mathcal{V})$ that obey Eq. (2.106), we can convert Eq. (13.18) to the surface integral

$$\begin{aligned}
&\oiint_{\mathcal{S}} \left[(\sigma_{ij}^{\infty} + \sigma_{ij}^{IM})(u_i^{\prime\infty} + u_i^{\prime IM}) - (\sigma_{ij}^{\prime\infty} + \sigma_{ij}^{\prime IM})(u_i^{\infty} + u_i^{IM})\right] \hat{n}_j dS\\
&= \oiint_{\mathcal{S}} \left[(\sigma_{ij}^{\infty} u_i^{\prime\infty} - \sigma_{ij}^{\prime\infty} u_i^{\infty}) + (\sigma_{ij}^{\infty} u_i^{\prime IM} - \sigma_{ij}^{\prime\infty} u_i^{IM})\right.\\
&\quad \left. + (\sigma_{ij}^{IM} u'_i - \sigma_{ij}^{\prime IM} u_i)\right] \hat{n}_j dS = 0
\end{aligned} \tag{13.19}$$

after recognizing that $(\sigma_{ij}^{\infty}+\sigma_{ij}^{IM})\hat{n}_j$ and $(\sigma_{ij}^{\prime\infty}+\sigma_{ij}^{\prime IM})\hat{n}_j$ both vanish on $\mathcal{S}^{\circ}$. Equation (13.19) must remain valid if the $\mathcal{S}^{\circ}$ surface is expanded out to infinity so that the image field on $\mathcal{S}$ vanishes. Therefore, it must be concluded that

$$\oiint_{\mathcal{S}} (\sigma_{ij}^{\infty}u_i^{\prime\infty} - \sigma_{ij}^{\prime\infty}u_i^{\infty})\hat{n}_j dS = 0 \tag{13.20}$$

and, consequently, Eq. (13.19) assumes the form

$$\oiint_{\mathcal{S}} (\sigma_{ij}^{\infty}u_i^{\prime IM} - \sigma_{ij}^{\prime\infty}u_i^{IM})\hat{n}_j dS + \oiint_{\mathcal{S}} (\sigma_{ij}^{IM}u_i^{\prime} - \sigma_{ij}^{\prime IM}u_i)\hat{n}_j dS = 0. \tag{13.21}$$

Next, shrink the surface $\mathcal{S}$ in Eq. (13.21) down around the initial and displaced loops as illustrated in Fig. 13.1b. The first term in Eq. (13.21) then vanishes because the image displacement field is continuous across the Σ surface, and the surface integral on the reduced surface $\mathcal{S}$ in Fig. 13.1b therefore vanishes. Since the dislocation displacement field has a discontinuity equal to the Burgers vector on the Σ and Σ' surfaces, the remaining second term can be written as

$$b_i \iint_{\Sigma'} \sigma_{ij}^{IM}\hat{n}_j dS = b_i \iint_{\Sigma} \sigma_{ij}^{\prime IM}\hat{n}_j dS. \tag{13.22}$$

However, writing $\delta\sigma_{ij}^{IM} = \sigma_{ij}^{\prime IM} - \sigma_{ij}^{IM}$ and substituting this into Eq. (13.22), and using Eq. (13.5),

$$\frac{1}{2}b_i \iint_{\Sigma} \delta\sigma_{ij}^{IM}\hat{n}_j dS = \frac{1}{2}b_i \iint_{\Sigma'-\Sigma} \sigma_{ij}^{IM}\hat{n}_j dS = \frac{1}{2}e_{lsj}b_i \oint_C \sigma_{ij}^{IM}\delta\xi_l\hat{t}_s ds \tag{13.23}$$

which proves the equality of the two integrals in Eq. (13.16). Substitution of Eq. (13.23) into Eq. (13.16) then yields

$$\delta E_{int}^{DIS/DIS^{IM}} = e_{lsj}b_i \oint_C \sigma_{ij}^{IM}\delta\xi_l\hat{t}_s ds. \tag{13.24}$$

The force experienced by the loop as a consequence of this interaction energy is then obtained from Eq. (13.24) by following the same procedure used to derive Eq. (13.9), i.e., the Peach–Koehler force equation, from Eq. (13.6). This yields

$$f_l^{DIS/DIS^{IM}} = -e_{lsj}\sigma_{ij}^{IM}b_i\hat{t}_s = e_{jsl}\sigma_{ij}^{IM}b_i\hat{t}_s \tag{13.25}$$

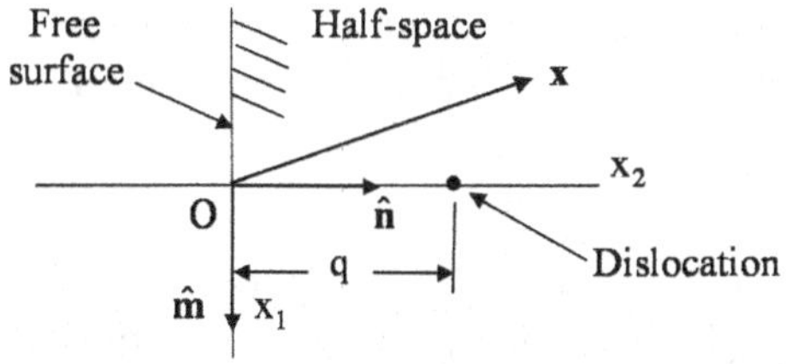

Fig. 13.2 End view of long straight dislocation lying parallel to planar free surface of half-space at distance q in $(\hat{\mathbf{m}}, \hat{\mathbf{n}}, \hat{\boldsymbol{\tau}})$ coordinate system: $\hat{\boldsymbol{\tau}} = \hat{\mathbf{m}} \times \hat{\mathbf{n}}$.

which is seen to be of the same form as Eq. (13.9) with the imposed stress σ_{ij}^{Q} and image stress σ_{ij}^{IM} playing equivalent roles. The force exerted on a dislocation by its image stress in a body with a free surface can therefore be obtained by substituting the image stress in the Peach-Koehler force equation.

13.3.2 *Straight dislocations parallel to free surfaces*

13.3.2.1 *General dislocation in half-space with planar surface*

Elastic field

Consider a long straight general dislocation with $\hat{\mathbf{t}} = \hat{\boldsymbol{\tau}}$ lying parallel to a planar traction-free surface of a half-space at a distance q in the $(\hat{\mathbf{m}}, \hat{\mathbf{n}}, \hat{\boldsymbol{\tau}})$ coordinate system illustrated in Fig. 13.2. A solution for its stress field can be obtained (Barnett and Lothe 1974) by assuming a displacement field of the form

$$u_i = u_i^{\infty} + u_i^{IM} = u_i^{\infty} + \sum_{\alpha=1}^{3} A_{i\alpha} \int_0^{\infty} D_{\alpha}(k) e^{ik[\hat{\mathbf{m}}\cdot\mathbf{x} + p_{\alpha}(\hat{\mathbf{n}}\cdot\mathbf{x} - q)]} dk$$

$$+ \sum_{\alpha=1}^{3} A_{i\alpha}^{*} \int_0^{\infty} D_{\alpha}^{*}(k) e^{-ik[\hat{\mathbf{m}}\cdot\mathbf{x} + p_{\alpha}^{*}(\hat{\mathbf{n}}\cdot\mathbf{x} - q)]} dk. \quad (13.26)$$

The first term is the displacement field that the dislocation would have in an infinite body, while the remaining terms constitute the image field. Equation (12.12) can then be used for the $u_i^{\infty}(\mathbf{x})$ displacement in the form

$$u_i^{\infty}(\mathbf{x}) = \frac{1}{2\pi i} \sum_{\alpha=1}^{6} \pm A_{i\alpha} L_{s\alpha} b_s \ln(\hat{\mathbf{m}} \cdot \mathbf{x} + p_{\alpha}(\hat{\mathbf{n}} \cdot \mathbf{x} - q)) \quad (13.27)$$

since the dislocation is displaced from the origin by $\mathbf{x} \cdot \hat{\mathbf{n}} = q$. Then, substituting Eq. (13.27) into Eq. (13.26) and employing Eq. (3.1),

$$\begin{aligned}\sigma_{mn} &= \sigma_{mn}^{\infty} + \sigma_{mn}^{IM} \\ &= C_{mnip} \sum_{\alpha=1}^{3} \left\{ \frac{b_s}{2\pi i} \left[\frac{A_{i\alpha} L_{s\alpha} (\hat{m}_p + p_\alpha \hat{n}_p)}{\hat{\mathbf{m}} \cdot \mathbf{x} + p_\alpha (\hat{\mathbf{n}} \cdot \mathbf{x} - q)} - \frac{A_{i\alpha}^* L_{s\alpha}^* (\hat{m}_p + p_\alpha^* \hat{n}_p)}{\hat{\mathbf{m}} \cdot \mathbf{x} + p_\alpha^* (\hat{\mathbf{n}} \cdot \mathbf{x} - q)} \right] \right. \\ &\quad + A_{i\alpha} \int_0^{\infty} D_\alpha(k) e^{ik[\hat{\mathbf{m}} \cdot \mathbf{x} + p_\alpha (\hat{\mathbf{n}} \cdot \mathbf{x} - q)]} (\hat{m}_p + p_\alpha \hat{n}_p) ikdk \\ &\quad \left. - A_{i\alpha}^* \int_0^{\infty} D_\alpha^*(k) e^{-ik[\hat{\mathbf{m}} \cdot \mathbf{x} + p_\alpha^* (\hat{\mathbf{n}} \cdot \mathbf{x} - q)]} (\hat{m}_p + p_\alpha^* \hat{n}_p) ikdk \right\}. \end{aligned} \tag{13.28}$$

The unknown coefficients $D_\alpha(k)$ and $E_\alpha^*(k)$ can be found by invoking the boundary condition of vanishing tractions on the surface at $\mathbf{x} \cdot \hat{\mathbf{n}} = 0$ expressed by

$$\begin{aligned} T_n &= (\sigma_{mn} n_m)_{\mathbf{x} \cdot \hat{\mathbf{n}} = 0} \\ &= \sum_{\alpha=1}^{3} \left\{ \frac{b_s}{2\pi i} \left[-\frac{L_{n\alpha} L_{s\alpha}}{(\hat{\mathbf{m}} \cdot \mathbf{x} - p_\alpha q)} + \frac{L_{n\alpha}^* L_{s\alpha}^*}{(\hat{\mathbf{m}} \cdot \mathbf{x} - p_\alpha^* q)} \right] \right. \\ &\quad - -L_{n\alpha} \int_0^{\infty} D_\alpha(k) e^{ik(\hat{\mathbf{m}} \cdot \mathbf{x} - p_\alpha q)} ikdk \\ &\quad \left. + L_{n\alpha}^* \int_0^{\infty} D_\alpha^*(k) e^{-ik(\hat{\mathbf{m}} \cdot \mathbf{x} - p_\alpha^* q)} ikdk \right\} = 0. \end{aligned} \tag{13.29}$$

Then, by substituting the two equalities

$$\begin{aligned} (\hat{\mathbf{m}} \cdot \mathbf{x} - p_\alpha q)^{-1} &= i \int_0^{\infty} e^{-ik(\hat{\mathbf{m}} \cdot \mathbf{x} - p_\alpha q)} dk \\ (\hat{\mathbf{m}} \cdot \mathbf{x} - p_\alpha^* q)^{-1} &= -i \int_0^{\infty} e^{-ik(\hat{\mathbf{m}} \cdot \mathbf{x} - p_\alpha^* q)} dk \end{aligned} \tag{13.30}$$

into Eq. (13.29), and collecting the resulting coefficients of $e^{ik(\hat{\mathbf{m}} \cdot \mathbf{x})}$ and $e^{-ik(\hat{\mathbf{m}} \cdot \mathbf{x})}$, and setting the collected coefficients in each case equal to zero to satisfy Eq. (13.29), the two expressions

$$\begin{aligned} \sum_{\alpha=1}^{3} \left(\frac{-L_{n\alpha} L_{s\alpha} b_s}{2\pi} e^{ikp_\alpha q} + ikD_\alpha^* L_{n\alpha}^* e^{ikp_\alpha^* q} \right) &= 0 \\ \sum_{\alpha=1}^{3} \left(\frac{L_{n\alpha}^* L_{s\alpha}^* b_s}{2\pi} e^{-ikp_\alpha^* q} + ikD_\alpha L_{n\alpha} e^{-ikp_\alpha q} \right) &= 0 \end{aligned} \tag{13.31}$$

are obtained. Then, multiplying the first Eq. (13.31) throughout by $M^*_{\beta n}$, which has the inverse properties with respect to $L^*_{n\alpha}$ indicated by Eq. (3.118), to solve for D^*_α and multiplying the second Eq. (13.31) by $M_{\beta n}$, with the inverse properties given by Eq. (3.117), to solve for D_α,[2]

$$\begin{aligned} D_\beta &= -\frac{M_{\beta n}}{2\pi ik} e^{ikp_\beta q} \sum_{\alpha=1}^{3} L^*_{n\alpha} L^*_{s\alpha} b_s e^{-ikp^*_\alpha q}, \\ D^*_\beta &= \frac{M^*_{\beta n}}{2\pi ik} e^{-ikp^*_\beta q} \sum_{\alpha=1}^{3} L_{n\alpha} L_{s\alpha} b_s e^{ikp_\alpha q}. \end{aligned} \tag{13.32}$$

The image stress is then obtained by substituting Eq. (13.32) into Eq. (13.28) to obtain

$$\sigma^{IM}_{mn} = \frac{C_{mnip}}{2\pi i} \sum_{\alpha=1}^{3} \sum_{\beta=1}^{3} \left\{ \frac{A_{i\alpha} M_{\alpha j} L^*_{j\beta} L^*_{s\beta} b_s (\hat{m}_p + p_\alpha \hat{n}_p)}{[\hat{\mathbf{m}} \cdot \mathbf{x} + p_\alpha (\hat{\mathbf{n}} \cdot \mathbf{x}) - p^*_\beta q]} - \frac{A^*_{i\alpha} M^*_{\alpha j} L_{j\beta} L_{s\beta} b_s (\hat{m}_p + p^*_\alpha \hat{n}_p)}{[\hat{\mathbf{m}} \cdot \mathbf{x} + p^*_\alpha (\hat{\mathbf{n}} \cdot \mathbf{x}) - p_\beta q]} \right\} \tag{13.33}$$

Image force

The force exerted on the dislocation by the above image stress can now be determined by substituting Eq. (13.33) (with $\hat{\mathbf{m}} \cdot \mathbf{x} = 0$ and $\hat{\mathbf{n}} \cdot \mathbf{x} = q$) into the Peach–Koehler force equation, i.e., Eq. (13.11). Symmetry requires that the force be normal to the surface, and, after arranging quantities so that a positive result indicates a force towards the surface, the force assumes the form

$$f^{DIS^\infty/DIS^{IM}} = \sigma^{IM}_{mn} b_n \hat{n}_m = \sigma^{IM}_{mn} b_n \hat{m}_m. \tag{13.34}$$

Making the substitution and applying Eq. (3.38), and interchanging α and β in the second term of the result,

$$f^{DIS^\infty/DIS^{IM}} = -\frac{b_n b_s}{2\pi iq} \sum_{\alpha=1}^{3} \sum_{\beta=1}^{3} \left(\frac{p_\alpha L_{n\alpha} M_{\alpha j} L^*_{j\beta} L^*_{s\beta} + p^*_\beta L^*_{n\beta} M^*_{\beta j} L_{j\alpha} L_{s\alpha}}{p^*_\beta - p_\alpha} \right) \tag{13.35}$$

and then, by interchanging n and s in the first term,

$$f^{DIS^\infty/DIS^{IM}} = -\frac{b_n b_s}{2\pi iq} \sum_{\alpha=1}^{3} \sum_{\beta=1}^{3} \left[\frac{L_{s\alpha} L^*_{n\beta} (p^*_\beta M^*_{\beta j} L_{j\alpha} + p_\alpha M_{\alpha j} L^*_{j\beta})}{p^*_\beta - p_\alpha} \right]. \tag{13.36}$$

[2] As indicated in Sec. 3.5.1.2, the [M] matrices are inverse with respect to corresponding [L] matrices, i.e., [M][L] = [I]. Note the similarity of this procedure to that used to solve Eq. (4.54) for $E^*_{k\alpha}$.

Next, by substituting Eq. (3.123) into Eq. (13.36) and using Eq. (13.137),

$$f^{\mathrm{DIS}^{\infty}/\mathrm{DIS}^{\mathrm{IM}}} = -\frac{b_n b_s}{2\pi i q}\sum_{\alpha=1}^{3} L_{j\alpha}L_{s\alpha}\sum_{\beta=1}^{3} L^{*}_{n\beta}M^{*}_{\beta j}$$

$$= -\frac{b_n b_s}{2\pi i q}\delta_{nj}\sum_{\alpha=1}^{3} L_{j\alpha}L_{s\alpha} = -\frac{b_j b_s}{2\pi i q}\sum_{\alpha=1}^{3} L_{j\alpha}L_{s\alpha} = \frac{b_j b_s B_{js}}{q}. \tag{13.37}$$

However, according to Eq. (12.41), the quantity $b_j b_s B_{js}$ is equal to the pre-logarithmic energy factor, $w_o = b_j b_s B_{js}$, for the same dislocation in an infinite body, and therefore Eq. (13.37) assumes the relatively simple form

$$f^{\mathrm{DIS}^{\infty}/\mathrm{DIS}^{\mathrm{IM}}} = \frac{w_o}{q}. \tag{13.38}$$

Since the strain energy is always positive, the image force is directed towards the surface and varies inversely with the distance q. Also, since w_o depends only upon the $\hat{\mathbf{t}}$ and $\mathbf{b}$ vectors of the dislocation and the C_{ijkl} tensor of the crystal, Eq. (13.38) establishes the noteworthy result that all free surfaces parallel to a dislocation of fixed $\hat{\mathbf{t}}$ and $\mathbf{b}$, and at a distance q, exert the same image force, despite the fact that they are not crystallographically equivalent.

13.3.2.2 *Some results for isotropic system*

Screw dislocation in half-space with planar surface

A particularly simple situation is a straight screw dislocation, with $\mathbf{b} = (0, 0, b)$ and $\hat{\mathbf{t}} = (0, 0, 1)$, lying parallel to a nearby planar free surface of a half-space as illustrated in Fig. 13.3. As now shown, the image stress

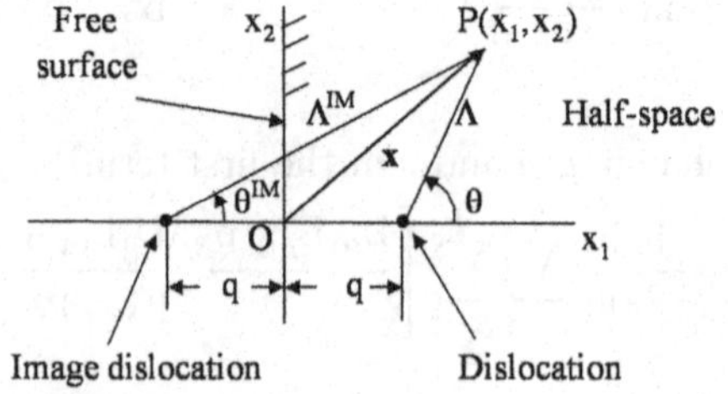

Fig. 13.3 End view of a long straight screw dislocation, and its image dislocation, lying parallel to x_3 at $x_1 = q$ and $x_1 = -q$, respectively. Dislocation is in a half-space having a free surface in the $x_1 = 0$ plane.

required to produce a traction-free surface is just the stress field produced by the image screw dislocation shown in Fig. 13.3 possessing a Burgers vector opposite to that of the actual dislocation. Using Eq. (12.274), the displacement field at P produced by the two dislocations, with their Burgers vectors related by $\mathbf{b} = -\mathbf{b}^{\mathrm{IM}}$, is

$$u_3(x_1, x_2) = \frac{b}{2\pi}\left(\theta - \theta^{\mathrm{IM}}\right) = \frac{b}{2\pi}\left(\tan^{-1}\frac{x_2}{x_1 - q} - \tan^{-1}\frac{x_2}{x_1 + q}\right) \tag{13.39}$$

and the corresponding stresses are

$$\sigma_{13}(x_1, x_2) = 2\mu\varepsilon_{13} = \mu\frac{\partial u_3}{\partial x_1} = -\frac{\mu b x_2}{2\pi}\left\{\frac{1}{\Lambda^2} - \frac{1}{(\Lambda^{\mathrm{IM}})^2}\right\},$$

$$\sigma_{23}(x_1, x_2) = \frac{\mu b}{2\pi}\left\{\frac{x_1 - q}{\Lambda^2} - \frac{x_1 + q}{(\Lambda^{\mathrm{IM}})^2}\right\}, \tag{13.40}$$

where

$$\Lambda^2 = (x_1 - q)^2 + x_2^2, \quad (\Lambda^{\mathrm{IM}})^2 = (x_1 + q)^2 + x_2^2. \tag{13.41}$$

The only stress that must be cancelled on the surface is the σ_{13} stress, and this is seen to vanish as required.

The image stresses, obtained from Eq. (13.40), are then

$$\sigma_{13}^{\mathrm{IM}} = \sigma_{13} - \sigma_{13}^{\infty} = \frac{\mu b}{2\pi}\frac{x_2}{(\Lambda^{\mathrm{IM}})^2}, \quad \sigma_{23}^{\mathrm{IM}} = \sigma_{23} - \sigma_{23}^{\infty} = -\frac{\mu b}{2\pi}\frac{(x_1 + q)}{(\Lambda^{\mathrm{IM}})^2}. \tag{13.42}$$

and the force they impose on the dislocation is readily found by inserting them into the Peach–Koehler force equation, i.e., Eq. (13.11), with $\mathbf{b} = (0, 0, b)$ and $\hat{\mathbf{t}} = (0, 0, 1)$. Therefore,

$$\boldsymbol{f}^{\mathrm{DIS}^{\infty}/\mathrm{DIS}^{\mathrm{IM}}} = \mathbf{d} \times \hat{\mathbf{t}} = b\sigma_{23}^{\mathrm{IM}}(q, 0)\hat{\mathbf{e}}_1 - b\sigma_{13}^{\mathrm{IM}}(q, 0)\hat{\mathbf{e}}_2 = -\frac{\mu b^2}{4\pi q}\hat{\mathbf{e}}_1. \tag{13.43}$$

The image stresses therefore tends to pull the dislocation out of the body with a force that varies inversely with its distance from the surface.

Screw dislocation along axis of cylindrical body

Consider a straight screw dislocation with $\mathbf{b} = (0, 0, b)$ and $\hat{\mathbf{t}} = (0, 0, 1)$ in an isotropic system lying along the z axis of a long traction-free cylinder of radius r with end faces at $z = \pm L$, where $L \gg R$. According to Eq. (12.57), the stress field in cylindrical coordinates of the same dislocation in an infinite body is given by $\sigma_{\theta z}^{\infty}(r) = \mu b/(2\pi r)$. Therefore, to obtain a traction-free surface on the cylinder, this stress must be cancelled on the end faces by an equal and opposite image stress. Note that no cancellation is required on the cylindrical surface. Using Eq. (12.57), the integrated effect of imposing this distribution of image stress on the end faces is equivalent to that produced by applying a torque around the z axis given by

$$\tau^{\mathrm{IM}} = -2\pi \int_0^{\mathrm{R}} \sigma_{\theta \mathrm{z}}^{\infty} \mathrm{r}^2 \mathrm{dr} = -\frac{\mu \mathrm{b} \mathrm{R}^2}{2}. \tag{13.44}$$

Such a torque will induce a twist angle per unit length along the z axis, θ, and a corresponding shear strain $\varepsilon_{\theta \mathrm{z}}^{\mathrm{IM}} = \theta \mathrm{r}/2$ and shear stress $\sigma_{\theta \mathrm{z}}^{\mathrm{IM}} = \mu \theta \mathrm{r}$. Since the image torque is given by

$$\tau^{\mathrm{IM}} = 2\pi \int_0^{\mathrm{R}} \sigma_{\theta \mathrm{z}}^{\mathrm{IM}} \mathrm{r}^2 \mathrm{dr} = \frac{\pi \mu \mathrm{R}^4}{2} \theta \tag{13.45}$$

θ must be of magnitude $\theta = -\mathrm{b}/(\pi \mathrm{R}^2)$. Even though application of this torque and associated image stress eliminates the net torque experienced by the cylinder it does not produce detailed cancellation of the $\sigma_{\theta \mathrm{z}}^{\infty}(\mathrm{r}) = \mu \mathrm{b}/(2\pi \mathrm{r})$ stress distribution everywhere on the end faces. However, according to St. Venant's principle, this failure to achieve detailed cancellation will be significant only in relatively small regions of the cylinder at distances from the cylinder ends less than about 2R. Neglecting these regions, a suitable image stress is then

$$\sigma_{\theta \mathrm{z}}^{\mathrm{IM}} = \mu \theta \mathrm{r} = -\frac{\mu \mathrm{b}}{\pi \mathrm{R}^2} \mathrm{r}. \tag{13.46}$$

The final total stress in essentially the entire traction-free cylinder is then

$$\sigma_{\theta \mathrm{z}} = \sigma_{\theta \mathrm{z}}^{\infty} + \sigma_{\theta \mathrm{z}}^{\mathrm{IM}} = \frac{\mu \mathrm{b}}{2\pi \mathrm{r}} - \frac{\mu \mathrm{b}}{\pi \mathrm{R}^2} \mathrm{r} = \frac{\mu \mathrm{b}}{2\pi \mathrm{r}} \left[1 - 2 \left(\frac{\mathrm{r}}{\mathrm{R}} \right)^2 \right] \tag{13.47}$$

and the contribution of the image stress is therefore negligible in regions where r is small relative to R.

The corresponding strain energy per unit length due to the screw dislocation is readily found by substituting Eq. (13.47) into Eq. (2.136) to obtain

$$\begin{aligned} w &= \frac{1}{2\mu} \int_0^{2\pi} \int_{\mathrm{r}_\circ}^{\mathrm{R}} \sigma_{\theta \mathrm{z}}^2 \mathrm{rdrd}\theta = \frac{1}{2\mu} \int_0^{2\pi} \int_{\mathrm{r}_\circ}^{\mathrm{R}} \left\{ \frac{\mu \mathrm{b}}{2\pi \mathrm{r}} \left[1 - 2 \left(\frac{\mathrm{r}}{\mathrm{R}} \right)^2 \right] \right\}^2 \mathrm{rdrd}\theta \\ &= \frac{\mu \mathrm{b}^2}{4\pi} \left[\ln \frac{\mathrm{R}}{\mathrm{r}_\circ} - 1 \right]. \end{aligned} \tag{13.48}$$

Then, accounting for the core energy by employing the α parameter [see Eq. (12.42)],

$$w = \frac{\mu b^2}{4\pi}\left[\ln\frac{\alpha R}{b} - 1\right] \tag{13.49}$$

which may be compared with Eq. (12.61). As expected, the strain energy is therefore reduced when the image stress is added to the system.

Edge dislocation in half-space with planar surface

The problem of finding the image stress for a straight edge dislocation in an isotropic system lying parallel to a nearby planar surface of a half-space is more complicated than for the above screw dislocation, since a traction-free surface is not obtained by simply introducing a negative image dislocation. Consider an edge dislocation, with $\mathbf{b} = (b, 0, 0)$ and $\hat{\mathbf{t}} = (0, 0, 1)$, lying parallel to x_3 at $x_1 = q$ along with a negative image dislocation lying along $x_1 = -q$ as in Fig. 13.3. Equation (12.45) shows that, while the resulting total σ_{11} stress vanishes at $x_1 = 0$, the total σ_{12} stress does not. Additional image stresses must therefore be added.

Following Dundurs (1969), a solution can be found in the form of a stress function that is a combination of Airy stress functions taken from Table I.1, i.e.,

$$\psi = -\frac{\mu b}{2\pi(1-\nu)}\left[\Lambda \ln \Lambda \sin\theta - \Lambda^{IM}\ln\Lambda^{IM}\sin\theta^{IM} + a_1 \sin 2\theta^{IM} + a_2\frac{\sin\theta^{IM}}{x^{IM}}\right], \tag{13.50}$$

where the quantities Λ, Λ^{IM}, θ and θ^{IM} are shown in Fig. 13.3. As seen from Exercise 12.6, the first two terms are the stress functions of the real dislocation and its negative image, respectively, while the two additional functions are added to produce a traction-free surface. Using Eq. (3.171), the stresses associated with the stress function $x^{(1)} \ln x^{(1)} \sin\theta^{(1)}$ are, for example,

$$\sigma_{11} = -Ax_2\frac{3(x_1-q)^2 + x_2^2}{\Lambda^4}, \quad \sigma_{22} = Ax_2\frac{(x_1-q)^2 - x_2^2}{\Lambda^4},$$
$$\sigma_{12} = A\frac{(x_1-q)[(x_1-q)^2 - x_2^2]}{\Lambda^4}, \tag{13.51}$$

where $A = \mu b/[2\pi(1-\nu)]$. Using Eq. (3.171) to determine the stresses contributed by the remaining terms in Eq. (13.50), a traction-free surface at $x_1 = 0$ is obtained when $a_1 = q$ and $a_2 = -2q^2$. The final total stress field is then

$$\sigma_{11} = -Ax_2\left[\frac{3(x_1-q)^2+x_2^2}{\Lambda^4} - \frac{3(x_1+q)^2+x_2^2}{(\Lambda^{IM})^4} - 4qx_1\frac{3(x_1+q)^2-x_2^2}{(\Lambda^{IM})^6}\right],$$

$$\sigma_{22} = Ax_2\left[\frac{(x_1-q)^2-x_2^2}{\Lambda^4} - \frac{(x_1+q)^2-x_2^2}{(\Lambda^{IM})^4} + 4q\frac{2(x_1+q)^3-3x_1(x_1+q)^2+2(x_1+q)x_2^2+x_1x_2^2}{(\Lambda^{IM})^6}\right],$$

$$\sigma_{12} = A\left\{\frac{(x_1-q)[(x_1-q)^2-x_2^2]}{\Lambda^4} - \frac{(x_1+q)[(x_1+q)^2-x_2^2]}{(\Lambda^{IM})^4} + 2q\frac{(x_1+q)^4-2x_1(x_1+q)^3+6x_1(x_1+q)x_2^2-x_2^4}{(\Lambda^{IM})^6}\right\}. \quad (13.52)$$

The force exerted on the dislocation by its image stress is determined in Exercise 13.6.

Edge dislocation along axis of cylindrical body

Consider a straight edge dislocation with $\mathbf{b} = (b, 0, 0)$ and $\hat{\mathbf{t}} = (0, 0, 1)$ in an isotropic system lying along the axis of a long traction-free cylinder of radius R with flat end faces at $x_3 = \pm L$, where $L \gg R$. According to Eq. (12.47), the same dislocation in an infinite body would generate stress components σ_{rr}^{∞} and $\sigma_{r\theta}^{\infty}$ that produce tractions on cylindrical surfaces and a σ_{zz}^{∞} component that produces tractions on constant z surfaces. However, integrals of the latter tractions taken over the end faces of the cylinder vanish. Assuming a solution that is the sum of an infinite body solution and an image stress, there is then no need to introduce an image stress to produce detailed cancellation of these tractions, since, according to St. Venant's principle, their effects essentially vanish at insignificant distances (approximately *2R*) from each end.

Turning to the cylindrical surface, the Airy stress function

$$\psi^{IM} = Ar^3 \sin\theta = A(x_1^2 + x_2^2)x_2 \quad (A = \text{constant}) \quad (13.53)$$

listed in Table I.1 provides stresses that can be used to cancel the infinite body tractions on this surface. Using Eq. (3.171) and the equations of elasticity in cylindrical coordinates given in Appendix G, the required stresses provided by this stress function are

$$\sigma_{rr}^{IM} = 2Ar\sin\theta \quad \text{and} \quad \sigma_{r\theta}^{IM} = -2Ar\cos\theta. \tag{13.54}$$

The conditions for a traction-free cylindrical surface are then

$$\left.\begin{aligned}\sigma_{rr}(R,\theta) &= \sigma_{rr}^{\infty}(R,\theta) + \sigma_{rr}^{IM}(R,\theta) = -\frac{\mu b\sin\theta}{2\pi(1-\nu)R} + 2AR\sin\theta = 0\\ \sigma_{r\theta}(R,\theta) &= \sigma_{r\theta}^{\infty}(R,\theta) + \sigma_{r\theta}^{IM}(R,\theta) = -\frac{\mu b\cos}{2\pi(1-\nu)R} - 2AR\cos\theta = 0\end{aligned}\right\} \tag{13.55}$$

after use of Eq. (12.47). Then, solving for A,

$$A = \frac{\mu b}{4\pi(1-\nu)R^2} \tag{13.56}$$

and the stress field in the finite cylinder with a traction-free surface is

$$\begin{aligned}\sigma_{rr} &= \sigma_{rr}^{\infty} + \sigma_{rr}^{IM} = -\frac{\mu b\sin\theta}{2\pi(1-\nu)r}\left[1-\left(\frac{r}{R}\right)^2\right],\\ \sigma_{\theta\theta} &= \sigma_{\theta\theta}^{\infty} = -\frac{\mu b\sin\theta}{2\pi(1-\nu)r}\\ \sigma_{r\theta} &= \sigma_{r\theta}^{\infty} + \sigma_{r\theta}^{IM} = \frac{\mu b\cos\theta}{2\pi(1-\nu)r}\left[1-\left(\frac{r}{R}\right)^2\right],\\ \sigma_{zz} &= \sigma_{zz}^{\infty} = \nu(\sigma_{rr}+\sigma_{\theta\theta}),\\ \sigma_{rz} &= \sigma_{rz}^{\infty} = 0, \quad \sigma_{\theta z} = \sigma_{\theta z}^{\infty} = 0.\end{aligned} \tag{13.57}$$

Comparing these stresses with the infinite body stresses given by Eq. (12.47), it is seen that the image stress can be neglected in the vicinity of the dislocation where r is small relative to R.

Following Hirth and Lothe (1982), the strain energy in the traction-free cylinder can be determined by calculating the change in strain energy due to the introduction of the image stresses and adding this to the strain energy before the introduction of the image stresses given by Eq. (12.61). Assume that the cylinder containing the dislocation is first cut out of the infinite matrix while tractions are simultaneously applied to the R = constant surface so that no stress relaxation occurs. Then, introduce the Airy stresses to cancel these tractions. This causes the displacements u_r^{IM} and u_θ^{IM}, thereby requiring that work be done against the existing tractions. Since these tractions decrease linearly as these displacements increase, the work performed per unit cylinder length, or the corresponding change in strain energy per unit length, is

$$\Delta w = \frac{1}{2}\int_0^{2\pi} [\sigma_{rr}^{\infty}(R,\theta)u_r^{IM}(R,\theta) + \sigma_{r\theta}^{\infty}(R,\theta)u_\theta^{IM}(R,\theta)]R d\theta. \quad (13.58)$$

The displacements u_r^{IM} and u_θ^{IM}, required in Eq. (13.58), are obtained by integrating the image strains according to

$$\begin{aligned} u_r^{IM} &= \int \frac{\partial u_r^{IM}}{\partial r} dr = \int \varepsilon_{rr}^{IM} dr, \\ u_\theta^{IM} &= \int \frac{\partial u_\theta^{IM}}{\partial \theta} d\theta = \int (r\varepsilon_{\theta\theta}^{IM} - u_r^{IM}) d\theta. \end{aligned} \quad (13.59)$$

The strains required in Eq. (13.59), obtained from the image stress function, Eq. (13.53), by use of Eq. (3.171) and the standard relationships in appendix G, are

$$\begin{aligned} \varepsilon_{rr}^{IM} &= \frac{A(1-4\nu)}{\mu} r \sin\theta \\ \varepsilon_{\theta\theta}^{IM} &= \frac{A(3-4\nu)}{\mu} r \sin\theta \end{aligned} \quad (13.60)$$

Then, substituting these strains into Eq. (13.59) and performing the integrations, and substituting the resulting displacements into Eq. (13.58) and performing the integration,

$$\Delta w = -\frac{\mu b^2 (3-4\nu)}{16\pi(1-\nu^2)} \quad (13.61)$$

after using Eq. (13.56). Then, finally adding this result to the strain energy per unit length given by Eq. (12.61), the expression

$$w = \frac{\mu b^2}{4\pi(1-\nu)}\left[\ln\frac{\alpha R}{b} - \frac{(3-4\nu)}{4(1+\nu)}\right] \quad (13.62)$$

is obtained for the strain energy (per unit length) of an edge dislocation along the axis of a finite traction-free cylinder. As in the case of Eq. (13.49) for the screw dislocation in a traction-free cylinder, the relaxation due to the addition of the image stress produces a relatively small decrease of the strain energy.

13.3.3 *Straight dislocation parallel to planar interface between elastically dissimilar half-spaces*

13.3.3.1 *General dislocation*

Elastic field

Consider the general straight dislocation in half-space 1 near the planar interface between the joined half-spaces 1 and 2 illustrated in Fig. 13.4. Barnett and Lothe (1974) have determined its elastic field in both half-spaces by imagining that the arrangement shown in Fig. 13.4 is reached in the two-step procedure illustrated in Fig. 13.5. In Fig. 13.5a, the two half-spaces are not yet joined, and possess traction-free planar surfaces facing each other. In step 1, the dislocation is introduced into half-space 1 by a cut and displacement along the Σ surface illustrated in Fig. 13.5b. In step 2, the half-spaces are forcibly bonded so that points that were opposite one another across the gap in Fig. 13.5a are rejoined as illustrated in Fig. 13.5c.

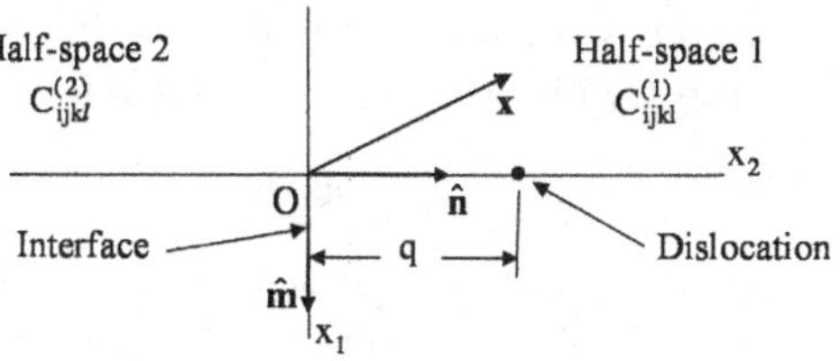

Fig. 13.4 End view of long straight dislocation in half-space 1 lying parallel to planar interface between dissimilar half-spaces 1 and 2 at a distance q in $(\hat{\mathbf{m}}, \hat{\mathbf{n}}, \hat{\boldsymbol{\tau}})$ coordinate system: $\hat{\boldsymbol{\tau}} = \hat{\mathbf{m}} \times \hat{\mathbf{n}}$.

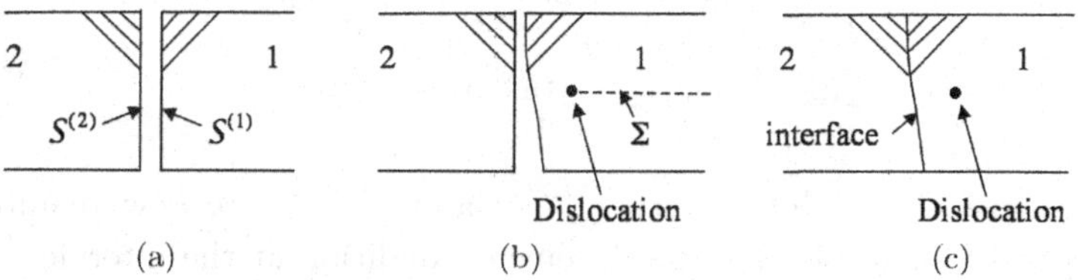

Fig. 13.5 (a) Half-spaces 1 and 2 with traction-free surfaces $\mathcal{S}^{(1)}$ and $\mathcal{S}^{(2)}$ initially facing each other. (b) Long straight dislocation (end view) introduced into half-space 1 via Σ cut surface. (c) Half-spaces 1 and 2 joined together to create interface.

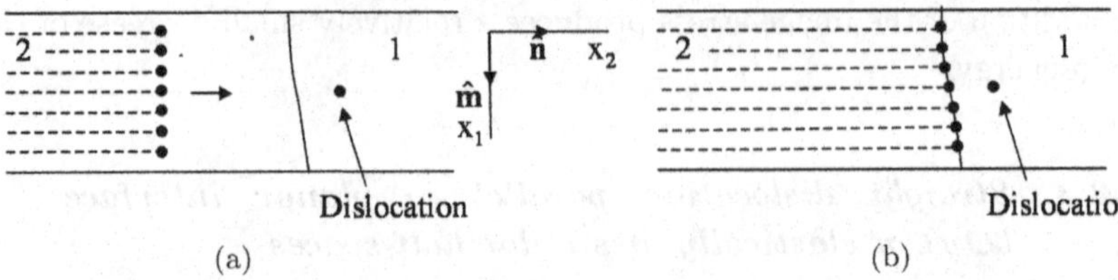

Fig. 13.6 (a) End view of the introduction of an infinite number of parallel infinitesimal dislocations into half-space 2 of the joined half-spaces 1 and 2 shown previously in Fig. 13.5c. (b) Final positions of dislocations in interface.

The stress field in Fig. 13.5b due to the dislocation, denoted by $\sigma_{ij}^{(1)'}$, is already known from the results of Sec. 13.3.2.1. The stresses induced in half-spaces 1 and 2 by step 2, denoted by $\sigma_{ij}^{(1)''}$ and $\sigma_{ij}^{(2)''}$, respectively, can be found by imagining that an infinite number of infinitesimal straight parallel dislocations is passed through half-space 2 and deposited in the interface in the manner illustrated in Fig. 13.6.

If the distribution of the Burgers vector strength of these dislocations, as a function of distance along x_1, is chosen, so that the plastic displacements at the interface produced by the passage of the dislocations are equal and opposite to the known elastic displacements at the interface in half-space 1 at the end of step 1, the stress field in half-space 2 will vanish, and the stress field in half-space 1 will return to $\sigma_{ij}^{(1)'}$. Therefore,

$$\begin{aligned} \sigma_{ij}^{(1)''} &= -\sigma_{ij}^{DIS(1)}, \\ \sigma_{ij}^{(2)''} &= -\sigma_{ij}^{DIS(2)}, \end{aligned} \tag{13.63}$$

where $\sigma_{ij}^{DIS(1)}$ and $\sigma_{ij}^{DIS(2)}$ are the stress fields in half-spaces 1 and 2 due to the dislocations in the interface. The final stresses in the half-spaces are then

$$\begin{aligned} \sigma_{ij}^{(1)} &= \sigma_{ij}^{(1)'} + \sigma_{ij}^{(1)''} = \sigma_{ij}^{(1)'} - \sigma_{ij}^{DIS(1)}, \\ \sigma_{ij}^{(2)} &= \sigma_{ij}^{(2)''} = -\sigma_{ij}^{DIS(2)}. \end{aligned} \tag{13.64}$$

Expressions for the stresses $\sigma_{ij}^{DIS(1)}$ and $\sigma_{ij}^{DIS(2)}$ are now required to complete the analysis. The displacement condition at the interface which must be satisfied is,

$$u_i^{DIS(2)}(x_1) = -u_i^{(1)'}(x_1, 0), \tag{13.65}$$

where $u_i^{DIS(2)}(x_1)$ is the plastic displacement along the interface due to the passage of the dislocations, and $u_i^{(1)'}(x_1, 0)$ is the elastic displacement at the interface due to the dislocation at $(0, q)$. If the Burgers vector strength of the infinitesimal dislocations lying in the interface between x_1 and $x_1 + dx_1$ is $B_i(x_1)dx_1$, the plastic displacement they produce in half-space 2 is related to $B_i(x_1)$ by

$$B_i(x_1) = -\frac{du_i^{DIS(2)}}{dx_1}. \tag{13.66}$$

The required distribution $B_i(x_1)$ is therefore obtained by substituting Eq. (13.65) into Eq. (13.66) so that

$$B_i(x_1) = \left[\frac{\partial u_i^{(1)'}(x_1, x_2)}{\partial x_1}\right]_{\hat{n}\cdot x=0}. \tag{13.67}$$

Then, substituting Eqs. (13.26), (13.27), and (13.32) for $u_i^{(1)'}(x_1, x_2)$, and using the variable *s* to measure distance along the interface in the direction of $\hat{\mathbf{m}}$,

$$B_i(s) = \frac{b_s}{2\pi i}\left(\sum_{\alpha=1}^{6}\frac{\pm A_{i\alpha}^{(1)}L_{s\alpha}^{(1)}}{s - p_\alpha^{(1)}q} + \sum_{\alpha=1}^{3}A_{i\alpha}^{(1)}M_{\alpha j}^{(1)}\sum_{\beta=1}^{3}\frac{L_{j\beta}^{*(1)}L_{s\beta}^{*(1)}}{s - p_\beta^{*(1)}q}\right.$$
$$\left. - \sum_{\alpha=1}^{3}A_{i\alpha}^{*(1)}M_{\alpha j}^{*(1)}\sum_{\beta=1}^{3}\frac{L_{j\beta}^{(1)}L_{s\beta}^{(1)}}{s - p_\beta^{(1)}q}\right). \tag{13.68}$$

Next, the stresses, $\sigma_{ij}^{DIS(1)}$ and $\sigma_{ij}^{DIS(2)}$, due to the distribution of Burgers vector strength in the interface, $B_i(s)$, are obtained by employing Eq. (14.76). For a single dislocation in an interface the quantities $E_\alpha^{(1)}$ and $E_\alpha^{(2)}$ in Eq. (14.76) are expected to be proportional to its Burgers vector and therefore expressible in the forms

$$E_\alpha^{(1)} = \pm J_{s\alpha}^{(1)}b_s, \quad E_\alpha^{(2)} = \pm J_{s\alpha}^{(2)}, \quad (\alpha = 1, 2 \ldots, 6) \tag{13.69}$$

where the $J_{s\alpha}^{(i)}$ are constants. For example, for the special case where the two half-spaces are identical, a comparison of Eqs. (14.76) and (12.12) shows that $E_\alpha = L_{s\alpha}b_s$. For the present dislocation distribution, the quantities $E_\alpha^{(1)}$ and $E_\alpha^{(2)}$ are again expected to be proportional to the Burgers vector strength and, therefore, expressible in the forms

$$E_\alpha^{(1)} = \pm J_{s\alpha}^{(1)} B_s(s), \quad E_\alpha^{(2)} = \pm J_{s\alpha}^{(2)} B_s(s), \quad (\alpha = 1, 2, \ldots, 6) \quad (13.70)$$

where the $J_{s\alpha}^{(i)}$ are the same constants as in Eq. (13.69) and are to be determined. Then, by using Eqs. (14.76) and (13.70) for half-space 1, and integrating over the distribution of Burgers vector strength along the interface,

$$\sigma_{mn}^{DIS(1)}(\mathbf{x}) = C_{mnip} \frac{\partial u_i^{(1)}}{\partial x_p} = \frac{1}{2\pi i} \sum_{\alpha=1}^{6} C_{mnip}^{(1)} A_{i\alpha}^{(1)} (\pm J_{s\alpha}^{(1)})(\hat{m}_p + p_\alpha^{(1)} \hat{n}_p)$$

$$\times \int_{-\infty}^{\infty} \frac{B_s(s) ds}{\hat{\mathbf{m}} \cdot \mathbf{x} - s + p_\alpha^{(1)} \hat{\mathbf{n}} \cdot \mathbf{x}}. \quad (13.71)$$

The only remaining unknown quantity is now $J_{s\alpha}^{(1)}$ in Eq. (13.71), and this can be obtained (Barnett and Lothe 1974) by the following rather lengthy but straightforward process. Consider a single dislocation in the interface and begin by multiplying Eq. (14.80) by $L_{i\beta}^{(2)}$ and Eq. (14.88) by $A_{i\beta}^{(2)}$, adding the results, and substituting Eq. (3.71) to obtain

$$\sum_{\alpha=1}^{6} (A_{i\alpha}^{(1)} L_{i\beta}^{(2)} + A_{i\beta}^{(2)} L_{i\alpha}^{(1)}) E_\alpha^{(1)} = \sum_{\alpha=1}^{6} (A_{i\beta}^{(2)} L_{i\alpha}^{(2)} + A_{i\alpha}^{(2)} L_{i\beta}^{(2)}) E_\alpha^{(2)}$$

$$= \sum_{\alpha=1}^{6} \delta_{\alpha\beta} E_\alpha^{(2)} = E_\beta^{(2)}. \quad (13.72)$$

Similarly, by using $L_{i\beta}^{(1)}$ and $A_{i\beta}^{(1)}$ as multipliers,

$$\sum_{\alpha=1}^{6} (A_{i\alpha}^{(1)} L_{i\beta}^{(1)} + A_{i\beta}^{(1)} L_{i\alpha}^{(1)}) E_\alpha^{(1)} = E_\beta^{(1)} = \sum_{\alpha=1}^{6} (A_{i\beta}^{(1)} L_{i\alpha}^{(2)} + A_{i\alpha}^{(2)} L_{i\beta}^{(1)}) E_\alpha^{(2)}. \quad (13.73)$$

Then, after setting

$$K_{\alpha\beta} = (A_{i\beta}^{(1)} L_{i\alpha}^{(2)} + A_{i\alpha}^{(2)} L_{i\beta}^{(1)}), \quad (13.74)$$

Eqs. (13.72) and (13.73) reduce to

$$E_\beta^{(1)} = \sum_{\alpha=1}^{6} K_{\alpha\beta} E_\alpha^{(2)}, \quad E_\beta^{(2)} = \sum_{\alpha=1}^{6} K_{\beta\alpha} E_\alpha^{(1)}. \quad (13.75)$$

Furthermore, it can be confirmed, by direct substitution and the use of the completeness relationships, Eqs. (3.76)–(3.78), and also Eq. (3.71), that

$$\sum_{\alpha=1}^{6} \mathrm{K}_{\alpha\beta}\mathrm{K}_{\alpha\mu} = \delta_{\beta\mu} \tag{13.76}$$

and

$$\sum_{\beta=1}^{6} \mathrm{K}_{\alpha\beta}\mathrm{L}^{(1)}_{\mathrm{s}\beta} = \mathrm{L}^{(2)}_{\mathrm{s}\alpha}, \quad \sum_{\beta=1}^{6} \mathrm{K}_{\beta\alpha}\mathrm{L}^{(2)}_{\mathrm{s}\beta} = \mathrm{L}^{(1)}_{\mathrm{s}\alpha}, \tag{13.77}$$

$$\sum_{\beta=1}^{6} \mathrm{K}_{\alpha\beta}\mathrm{A}^{(1)}_{\mathrm{s}\beta} = \mathrm{A}^{(2)}_{\mathrm{s}\alpha}, \quad \sum_{\beta=1}^{6} \mathrm{K}_{\beta\alpha}\mathrm{A}^{(2)}_{\mathrm{s}\beta} = \mathrm{A}^{(1)}_{\mathrm{s}\alpha}. \tag{13.78}$$

Another relationship is obtained by substituting Eq. (13.75) into Eqs. (14.89) and (14.81), multiplying the results by $\mathrm{A}^{(2)}_{\mathrm{i}\beta}$ and $\mathrm{L}^{(2)}_{\mathrm{i}\beta}$, respectively, and adding the results so that

$$\sum_{\alpha=1}^{6} \left\{ \left[(\pm\mathrm{L}^{(1)}_{\mathrm{i}\alpha})\mathrm{A}^{(2)}_{\mathrm{i}\beta}) + (\pm\mathrm{A}^{(1)}_{\mathrm{i}\alpha})\mathrm{L}^{(2)}_{\mathrm{i}\beta})\right] \mathrm{E}^{(1)}_{\alpha} \right.$$
$$\left. + \left[(\pm\mathrm{L}^{(2)}_{\mathrm{i}\alpha})\mathrm{A}^{(2)}_{\mathrm{i}\beta}) + (\pm\mathrm{A}^{(2)}_{\mathrm{i}\alpha})\mathrm{L}^{(2)}_{\mathrm{i}\beta})\right] \sum_{\mu=1}^{6} \mathrm{K}_{\alpha\mu}\mathrm{E}^{(1)}_{\mu} \right\} = 2\mathrm{L}^{(2)}_{\mathrm{i}\beta}\mathrm{b}_{\mathrm{i}}. \tag{13.79}$$

Then, substituting Eqs. (13.74) and (3.71) into Eq. (13.79),

$$\sum_{\alpha=1}^{6} (\pm\mathrm{K}_{\beta\alpha})\mathrm{E}^{(1)}_{\alpha} + \sum_{\alpha=1}^{6}\sum_{\mu=1}^{6} (\pm\delta_{\mu\beta})\mathrm{K}_{\mu\alpha}\mathrm{E}^{(1)}_{\alpha} = 2\mathrm{L}^{(2)}_{\mathrm{i}\beta}\mathrm{b}_{\mathrm{i}}. \tag{13.80}$$

When $\beta = 1, 2, 3$, and when $\beta = 4, 5, 6$, Eq. (13.80) reduces, respectively, to

$$\sum_{\alpha=1}^{3} \mathrm{K}_{\beta\alpha}\mathrm{E}^{(1)}_{\alpha} = \mathrm{L}^{(2)}_{\mathrm{i}\beta}\mathrm{b}_{\mathrm{i}}, \quad (\beta = 1, 2, 3) \tag{13.81}$$

$$\sum_{\alpha=4}^{6} \mathrm{K}_{\beta\alpha}\mathrm{E}^{(1)}_{\alpha} = -\mathrm{L}^{(2)}_{\mathrm{i}\beta}\mathrm{b}_{\mathrm{i}}. \quad (\beta = 4, 5, 6) \tag{13.82}$$

The solution for $\mathrm{E}^{(1)}_{\alpha} (\alpha = 1, 2, 3)$ can now be obtained. First, multiply Eq. (13.81) throughout by the matrix, $\mathrm{M}^{(2)}_{\beta\mathrm{s}}$, which is the inverse of $\mathrm{L}^{(2)}_{\mathrm{s}\alpha}$ as indicated by Eq. (3.117), sum over $\beta = 1, 2, 3$, and use Eq. (13.74) to obtain

$$\sum_{\beta=1}^{3} L_{i\beta}^{(2)} M_{\beta s}^{(2)} b_i = b_s = \sum_{\beta=1}^{3}\sum_{\alpha=1}^{3} M_{\beta s}^{(2)} K_{\beta\alpha} E_{\alpha}^{(1)}$$

$$= \sum_{\beta=1}^{3}\sum_{\alpha=1}^{3} (A_{i\alpha}^{(1)} L_{i\beta}^{(2)} M_{\beta s}^{\{2\}} + A_{i\beta}^{(2)} L_{i\alpha}^{(1)} M_{\beta s}^{(2)}) E_{\alpha}^{(1)}. \quad (13.83)$$

Then, by again using Eq. (3.117), $\sum_{\beta=1}^{3}\sum_{\alpha=1}^{3} (A_{i\alpha}^{(1)} L_{i\beta}^{(2)} M_{\beta s}^{(2)} = \sum_{\beta=1}^{3}\sum_{\alpha=1}^{3} (A_{s\beta}^{(1)} L_{i\alpha}^{(1)} M_{\beta i}^{(1)}$, and Eq. (13.83) can be expressed in the form

$$\sum_{\alpha=1}^{3} F_{si} L_{i\alpha}^{(1)} E_{\alpha}^{(1)} = b_s, \quad (13.84)$$

where

$$F_{si} = \sum_{\beta=1}^{3} (A_{s\beta}^{(1)} M_{\beta i}^{(1)} + A_{i\beta}^{(2)} M_{\beta s}^{(2)}). \quad (13.85)$$

Using matrix notation, the solution of Eq. (13.84) for $E_{\alpha}^{(1)}$ $(\alpha = 1, 2, 3)$ is

$$\begin{aligned} &[F][L^{(1)}][E^{(1)}] = [b], \\ &[L^{(1)}][E^{(1)}] = [F]^{-1}[b], \\ &[E^{(1)}] = [L^{(1)}]^{-1}[F]^{-1}[b] = [M^{(1)}][F]^{-1}[b]. \end{aligned} \quad (13.86)$$

Now, from Eq. (13.70), $E_{\alpha}^{(1)} = J_{s\alpha}^{(1)} b_s$ $(\alpha = 1, 2, 3)$. Also, from Eq. (13.86), $[E^{(1)}]$, in component form, is given by $E_{\alpha}^{(1)} = M_{\alpha k}^{(1)} F_{ks}^{-1} b_s$ $(\alpha = 1, 2, 3)$. Therefore, finally,

$$J_{s\alpha}^{(1)} = M_{\alpha k}^{(1)} F_{ks}^{-1}. \quad (\alpha = 1, 2, 3) \quad (13.87)$$

As now shown, the matrix F_{kj} has the properties

$$F_{ks} = -F_{sk}^{*} \quad \text{or} \quad [F] = -[F^{*}]^{T}, \quad (13.88)$$

$$F_{ks}^{-1} = -F_{sk}^{*-1} \quad \text{or} \quad [F]^{-1} = -\{[F^{*}]^{-1}\}^{T}. \quad (13.89)$$

Equation (13.88) can be derived by substituting Eq. (3.129) into Eq. (13.85) so that

$$F_{ks} = -\sum_{\beta=1}^{3} (A_{s\beta}^{*(1)} M_{\beta k}^{*(1)} + A_{k\beta}^{*(2)} M_{\beta s}^{*(2)}). \quad (13.90)$$

However, $F^*_{ks} = \sum_{\beta=1}^{3} (A^{*(1)}_{k\beta} M^{*(1)}_{\beta s} + A^{*(2)}_{s\beta} M^{*(2)}_{\beta k})$, and, upon comparing this with F_{ks}, expressed by Eq. (13.90), Eq. (13.88) is obtained. Equation (13.89) can be obtained by the following operations:

$$\begin{aligned} [F] &= -[F^*]^T, \\ [F][F]^{-1} &= -[F^*]^T[F]^{-1} = [I], \\ -\{[F]^{-1}\}^T[F^*][F^*]^{-1} &= -\{[F]^{-1}\}^T = [I][F^*]^{-1} = [F^*]^{-1}, \\ [F]^{-1} &= -\{[F^*]^{-1}\}^T. \end{aligned} \tag{13.91}$$

The quantity $J^{(1)}_{s\alpha}(\alpha = 4,5,6)$ can be obtained in a similar manner starting with Eq. (13.82) with the result

$$J^{(1)}_{s,\alpha+3} = J^{*(1)}_{s\alpha} = M^{*(1)}_{\alpha k} F^{*-1}_{ks}. \quad (\alpha = 1,2,3) \tag{13.92}$$

These equations therefore suffice to determine $\sigma^{DIS(1)}_{ij}(\mathbf{x})$. The corresponding field in half-space 2 can be obtained by minor extensions of the analysis (Barnett and Lothe 1974), and the final stresses in the two half-spaces are then given by Eq. (13.64).

Image force

The image force on the dislocation can be obtained by substituting the final stress fields given by Eq. (13.64) into the Peach–Koehler force equation. The force due to the $\sigma^{(1)'}_{ij}$ stress is already known, since, by use of Eq. (13.38), it takes the form $f^{DIS^\infty/DIS^{IM'}} = w'_\circ/q$ where $w'_\circ$ is the pre-logarithmic strain energy factor for the same dislocation in an infinite body which is given by Eq. (12.41). Barnett and Lothe (1974) show, by an extension of the results in Sec. 13.3.3.1, that the force due to the $\sigma^{(1)''}_{ij}$ stress is of the similar form $f^{DIS^\infty/DIS^{IM''}} = -w''_\circ/q$ where $w''_\circ$ is the pre-logarithmic strain energy factor for the dislocation when it is in the interface and is given by Eq. (14.97). The total image force is then

$$f^{DIS^\infty/DIS^{IM}} = f^{DIS^\infty/DIS^{IM'}} + f^{IDIS^\infty/DIS^{IM''}} = (w'_\circ - w''_\circ)/q. \tag{13.93}$$

Note that when the two half-spaces are elastically identical, $f^{DIS^\infty/DIS^{IM}} = 0$, as must be the case. In Sec. 13.3.2.1 it is shown that all free surfaces that are parallel to a dislocation lying in a half-space with a fixed $\hat{\mathbf{t}}$ and $\mathbf{b}$, and at a distance q from the surface, exert the same image force, $f^{DIS^\infty/DIS^{IM'}}$ (see Eq. (13.38)). Furthermore, the results of Barnett and Lothe (1974) show that $w''_\circ$ is a function only of $\hat{\mathbf{t}}$ and $\mathbf{b}$ of the dislocation and the $C^{(i)}_{ijkl}$ tensors of the half-spaces. Therefore, all interfaces between dissimilar half-spaces that are parallel to a dislocation lying in one half-space with $\hat{\mathbf{t}}$ and $\mathbf{b}$ fixed, and at a distance q from the interface, exert the same image force, $f^{DIS^\infty/DIS^{IM}}$, on the dislocation.[3]

13.3.3.2 *Some results for isotropic system*

Screw dislocation

Consider the image field associated with a straight screw dislocation with $\mathbf{b} = (0, 0, b)$ and $\hat{\mathbf{t}} = (0, 0, 1)$ in an isotropic system lying in the half-space region 1 parallel to the nearby planar interface illustrated in Fig. 13.7. This image problem can be treated by employing a somewhat more complicated image dislocation arrangement than used previously for the free surface case. With the real dislocation located at q, place an image dislocation at $x_1 = -q$ with the modified Burgers vector $(0, 0, -a_1 b)$, and calculate the elastic field in region 1 assuming that $\mu = \mu^{(1)}$ everywhere. To obtain the field in half-space region 2 place a single screw dislocation at $x_1 = q$ with the modified Burgers vector $(0, 0, a_2)$, and assume that $\mu = \mu^{(2)}$ everywhere. The quantities a_1 and a_2 are mismatch parameters introduced to account for the difference in elastic constants across the interface and will be determined by the boundary conditions at the interface. Therefore, with the use of

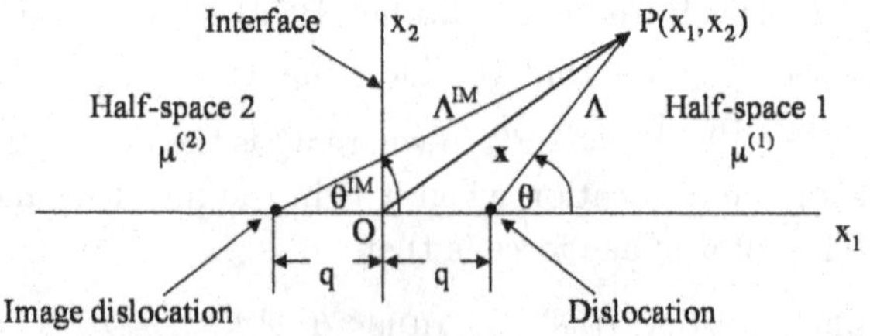

Fig. 13.7 Dissimilar half-spaces 1 and 2 joined along planar interface in $x_1 = 0$ plane. Long straight dislocation in half-space 1 lying parallel to x_3 at $x_1 = q$. Parallel image dislocation at $x_1 = -q$.

[3]This result may be compared to the similar result obtained at the end of Sec. 13.3.2.1 for a dislocation lying parallel to the free surface of a half-space.

Eq. (12.274),

$$u_3^{(1)}(x_1, x_2) = \frac{b}{2\pi}\left(\tan^{-1}\frac{x_2}{x_1 - q} - a_1 \tan^{-1}\frac{x_2}{x_1 + q}\right)$$

$$u_3^{(2)}(x_1, x_2) = \frac{b}{2\pi}\left(a_2 \tan^{-1}\frac{x_2}{x_1 - q}\right). \tag{13.94}$$

The boundary condition $u_3^{(1)}(0, x_2) = u_3^{(2)}(0, x_2)$ is then satisfied when

$$a_1 + 1 = a_2. \tag{13.95}$$

The corresponding σ_{13} stresses are then

$$\sigma_{13}^{(1)} = \mu^{(1)}\frac{\partial u_3^{(1)}}{\partial x_1} = \frac{\mu^{(1)} b}{2\pi}\left[\frac{-x_2}{(x_1 - q)^2 + x_2^2} + a_1 \frac{x_2}{(x_1 + q)^2 + x_2^2}\right]$$

$$\sigma_{13}^{(2)} = -\frac{\mu^{(2)} b}{2\pi}\left[\frac{(a_1 + 1)x_2}{(x_1 - q)^2 + x_2^2}\right] \tag{13.96}$$

and the boundary condition $\sigma_{13}^{(1)}(0, x_2) = \sigma_{13}^{(2)}(0, x_2)$ is satisfied when

$$a_1 = \frac{\mu^{(1)} - \mu^{(2)}}{\mu^{(1)} + \mu^{(2)}}. \tag{13.97}$$

The assumed arrangement of image dislocations therefore provides a solution that satisfies all conditions. By introducing the distances Λ^{IM} and Λ, defined by Eq. (13.41) (also Fig. 13.7), the corresponding σ_{23} stresses are readily obtained in the forms

$$\sigma_{23}^{(1)} = \frac{\mu^{(1)} b}{2\pi}\left[\frac{x_1 - q}{\Lambda^2} - \frac{a_1(x_1 + q)}{(\Lambda^{IM})^2}\right],$$

$$\sigma_{23}^{(2)} = \frac{\mu^{(2)} b}{2\pi}\left[\frac{(1 + a_1)(x_1 - q)}{\Lambda^2}\right]. \tag{13.98}$$

The final image stresses, from the above results, are therefore

$$\sigma_{13}^{(1),IM} = \frac{\mu^{(1)} b a_1}{2\pi}\frac{x_2}{(\Lambda^{IM})^2}, \quad \sigma_{13}^{(2),IM} = -\frac{\mu^{(2)} b a_1}{2\pi}\frac{x_2}{\Lambda^2},$$

$$\sigma_{23}^{(1),IM} = -\frac{\mu^{(1)} b a_1}{2\pi}\frac{(x_1 + q)}{(\Lambda^{IM})^2}, \quad \sigma_{23}^{(2),IM} = \frac{\mu^{(2)} b a_1}{2\pi}\frac{(x_1 - q)}{\Lambda^2}. \tag{13.99}$$

Note that when $\mu^{(1)} = \mu^{(2)}$, and $a_1 = q = 0$, the above solution reduces to Eq. (12.55) for the dislocation in an infinite homogeneous body. When $\mu^{(2)} = 0$, and $a_1 = 1$, it reduces for region 1 to Eq. (13.40).

The corresponding force exerted on the dislocation is found by inserting the image stress into the Peach–Koehler force equation with the result

$$\begin{aligned} \mathbf{f}^{\mathrm{DIS}^{\infty}/\mathrm{DIS}^{\mathrm{IM}}} &= \mathbf{d} \times \hat{\mathbf{t}} = b\sigma_{23}^{(1),\mathrm{IM}}(q,0)\hat{\mathbf{e}}_1 - b\sigma_{13}^{(1),\mathrm{IM}}(q,0)\hat{\mathbf{e}}_2 \\ &= b\left[\frac{-\mu^{(1)}ba_1}{2\pi(x_1+q)}\right]_{x_1=q}\hat{\mathbf{e}}_1 = -\left(\frac{\mu^{(1)}-\mu^{(2)}}{\mu^{(1)}+\mu^{(2)}}\right)\frac{\mu^{(1)}b^2}{4\pi q}\hat{\mathbf{e}}_1. \end{aligned} \tag{13.100}$$

When $\mu^{(1)} > \mu^{(2)}$, and half-space 1 is stiffer than half-space 2, a_1 is positive. The dislocation is then urged along $-\hat{\mathbf{e}}_1$, i.e., towards the softer region as expected.

Edge dislocation

Consider next the image field associated with a straight edge dislocation with $\mathbf{b} = (b, 0, 0)$ and $\hat{\mathbf{t}} = (0, 0, 1)$ in an isotropic system lying in the half-space region 1 parallel to a nearby planar interface, as illustrated in Fig. 13.7. This problem cannot be solved by simply introducing image dislocations as in the case of the above screw dislocation. However, Dundurs (1969) has constructed a solution for the image field using a stress function approach employing several of the stress functions listed in Table I.1. The results are lengthy and tedious and have been written out in full for a variety of situations by Asaro and Lubarda (2006), and, therefore, will not be reproduced here. As in the case of the screw dislocation, the edge dislocation is generally repelled by the stiffer half-space and attracted to the softer one.

13.3.4 *Straight dislocation impinging on planar free surface of half-space*

13.3.4.1 *Elastic field of general dislocation*

Consider the long straight general dislocation impinging on the free surface of the half-space in Fig. 13.8a. As shown by Lothe, Indenbom and Chamrov (1982) and Lothe (1992c), its elastic field in the half-space can be obtained as the sum of two fields, i.e., the elastic field of a corresponding dislocation of infinite length in an infinite body (running parallel to the semi-infinite impinged dislocation) and the field of a *dislocation fan* in an

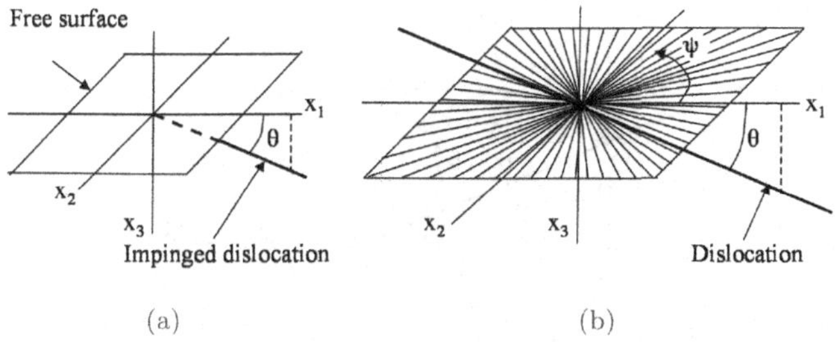

Fig. 13.8 (a) Long straight dislocation impinged on free surface of half-space at angle θ. (b) Corresponding infinitely long straight dislocation in infinite body piercing $x_3 = 0$ plane at angle θ. Also present is a dislocation fan lying in the $x_3 = 0$ plane and centered on the point of its intersection with the long straight dislocation.

infinite body lying in the $x_3 = 0$ plane and centered on the point at which the dislocation pierces the plane as shown in Fig. 13.8b. The fan consists of a planar array of an infinite number of infinitely long straight dislocations of infinitesimal Burgers vector strength intersecting at a point in a fan-like radial configuration (Lothe 1992b). The Burgers vector strength of the dislocations comprising the fan has an angular distribution, $\mathbf{b}(\psi)$, where $d\mathbf{b} = \mathbf{b}(\psi)d\psi$ is the total Burgers vector strength of the dislocations lying in the fan at angles between ψ and $\psi + d\psi$. The distribution, $\mathbf{b}(\psi)$, is a variable of the model that, when properly chosen, causes the tractions on the $x_3 = 0$ plane due to the elastic field of the fan to cancel the tractions due to the dislocation, thus producing a traction-free surface. The elastic field in the half-space in Fig. 13.8a can therefore be expressed as

$$\sigma_{ij} = \sigma_{ij}^{\infty} + \sigma_{ij}^{FAN}, \qquad (13.101)$$

where σ_{ij}^{∞} is the stress in the infinite body due to the dislocation, and σ_{ij}^{FAN} is the stress in the infinite body due to the fan serving as an image stress. We proceed by first determining the traction on the $x_3 = 0$ plane due to the dislocation and then the traction due to the fan in order to find the condition on $\mathbf{b}(\psi)$ for vanishibg net traction. Having this, σ_{ij}^{FAN} and σ_{ij}^{∞} are determined and then substituted into Eq. (13.101).

Traction on $x_3 = 0$ *plane at P due to dislocation*

The coordinate systems and geometry in Fig. 13.9 are employed. Here, the field point P lies in the plane O′APB, which is perpendicular to the dislocation, and is located at the polar coordinates (ρ, ω) with the origin

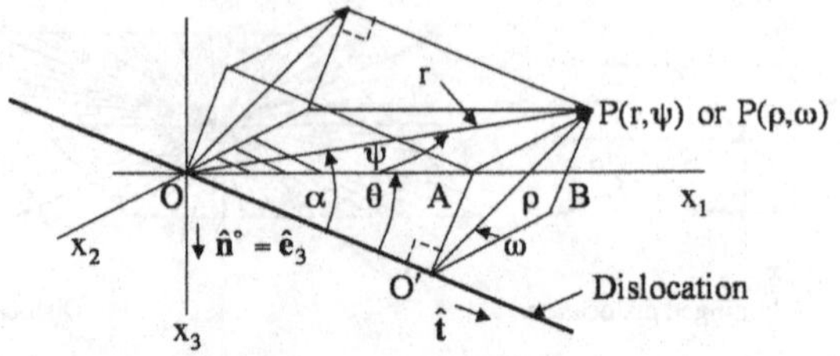

Fig. 13.9 Coordinate systems and geometry for analyzing the traction at field point $P(\rho, \omega)$ on the $x_3 = 0$ plane due to an infinitely long straight dislocation intersecting it at the angle θ. Also shown is an element of a dislocation fan, of the type illustrated in Fig. 13.8b, lying on the $x_3 = 0$ plane and centered at O. Entire system is embedded in an infinite homogeneous body.

taken at O′. Alternatively, P lies in the $x_3 = 0$ plane, located at the polar coordinates (r, ψ) with the origin taken at O.

To obtain the traction acting on the $x_3 = 0$ plane at $P(\rho, \omega)$ due to the elastic field of the dislocation, the corresponding distortion is first determined by integrating the Mura equation, Eq. (12.80) using the method employed in Exercise 12.4. Here, the elastic field of an infinitely long straight dislocation is obtained by starting with a loop initially containing a long straight segment of the desired dislocation type and then extending it to the limit of infinite length in the subsequent line integration. Using Eq. (12.80), the distortion can therefore be written as

$$\frac{\partial u_j^{\infty}(\mathbf{x})}{\partial x_s} = -b_k C_{kpim} e_{qps} \int_{-\infty}^{\infty} \frac{\partial G_{ij}^{\infty}(\mathbf{x} - \hat{\mathbf{t}}s')}{\partial x_m} \hat{t}_q ds', \tag{13.102}$$

where s' measures distance along the dislocation, and $\mathbf{x}' = s'\hat{\mathbf{t}}$. The required derivative of the Green's function can be obtained by use of Eq. (4.31) which can be put into a useful form by employing the delta function relationship

$$\frac{d\delta(\hat{\mathbf{k}} \cdot \hat{\mathbf{w}})}{d(\hat{\mathbf{k}} \cdot \hat{\mathbf{w}})} = |\mathbf{x} - \mathbf{x}'|^2 \frac{d\delta([\hat{\mathbf{k}} \cdot (\mathbf{x} - \mathbf{x}')]}{d[\hat{\mathbf{k}} \cdot (\mathbf{x} - \mathbf{x}')]}, \tag{13.103}$$

which can be established using Eqs. (c) and (d) in Table D.1. Insertion of Eq. (13.103) into Eq. (4.31) then yields

$$\frac{\partial G_{km}^{\infty}(\mathbf{x} - \mathbf{x}')}{\partial x_i} = \frac{1}{8\pi^2} \oiint_{\hat{S}} (\hat{k}\hat{k})_{km}^{-1} \hat{k}_i \frac{d\delta[\hat{\mathbf{k}} \cdot (\mathbf{x} - \mathbf{x}')]}{d[\hat{\mathbf{k}} \cdot (\mathbf{x} - \mathbf{x}')]} dS, \tag{13.104}$$

where the surface integral is over the unit sphere $\hat{\mathcal{S}}$ ($\hat{k} = 1$). Insertion of Eq. (13.104) into Eq. (13.102) then yields

$$\frac{\partial u_j^\infty(\mathbf{x})}{\partial x_s} = -\frac{1}{8\pi^2} b_k C_{kpim} e_{qps} \oiint_{\hat{\mathcal{S}}} (\hat{k}\hat{k})_{ij}^{-1} \hat{k}_m \int_{-\infty}^{\infty} \frac{d\delta[\hat{\mathbf{k}} \cdot (\mathbf{x} - \hat{\mathbf{t}}s')]}{d[\hat{\mathbf{k}} \cdot (\mathbf{x} - \hat{\mathbf{t}}s')]} \hat{t}_q ds' dS. \tag{13.105}$$

However, according to Eq. (h) in Table D.1,

$$\int_{-\infty}^{\infty} \frac{d\delta[\hat{\mathbf{k}} \cdot (\mathbf{x} - \hat{\mathbf{t}}s')]}{d[\hat{\mathbf{k}} \cdot (\mathbf{x} - \hat{\mathbf{t}}s')]} ds' = \frac{-2\delta(\hat{\mathbf{k}} \cdot \hat{\mathbf{t}})}{(\hat{\mathbf{k}} \cdot \mathbf{x})} \tag{13.106}$$

and, substituting this into Eq. (13.105),

$$\frac{\partial u_j^\infty(\mathbf{x})}{\partial x_s} = \frac{1}{4\pi^2} b_k C_{kpim} e_{qps} \oiint_{\hat{\mathcal{S}}} \frac{(\hat{k}\hat{k})_{ij}^{-1} \hat{k}_m}{(\hat{\mathbf{k}} \cdot \mathbf{x})} \delta(\hat{\mathbf{k}} \cdot \hat{\mathbf{t}}) \hat{t}_q dS. \tag{13.107}$$

However, the delta function $\delta(\hat{\mathbf{k}} \cdot \hat{\mathbf{t}})$ in Eq. (13.107) reduces the surface integral over the unit sphere $\hat{\mathcal{S}}$ ($\hat{k} = 1$) to a line integral in which the unit vector, $\hat{\mathbf{k}}$, traverses the unit circle $\hat{\mathcal{L}}$ ($\hat{k} = 1$) lying in the plane $\hat{\mathbf{k}} \cdot \hat{\mathbf{t}} = 0$. Using Eq. (13.107), the distortion at $\mathbf{x}$ due to the dislocation can therefore be expressed in the alternative line integral form

$$\frac{\partial u_j^\infty(\mathbf{x})}{\partial x_s} = \frac{1}{4\pi^2} b_k C_{kpim} e_{qps} \oint_{\hat{\mathcal{L}}} \frac{(\hat{k}\hat{k})_{ij}^{-1} \hat{k}_m}{(\hat{\mathbf{k}} \cdot \mathbf{x})} \hat{t}_q d\phi, \tag{13.108}$$

where the integral is around a unit circle in a plane perpendicular to $\hat{\mathbf{t}}$ (i.e., the dislocation), and ϕ measures the rotation of $\hat{k}$ around the unit circle.

At this point the coordinates shown in Fig. 13.10a are introduced: these include the usual $(\hat{\mathbf{m}}, \hat{\mathbf{n}}, \hat{\boldsymbol{\tau}})$ coordinates of the integral formalism and the base vectors of a $(\hat{\mathbf{m}}', \hat{\mathbf{n}}', \hat{\boldsymbol{\tau}})$ system obtained by rotating the $(\hat{\mathbf{m}}, \hat{\mathbf{n}}, \hat{\boldsymbol{\tau}})$ system around its $\hat{\boldsymbol{\tau}} = \hat{\mathbf{t}}$ axis by the relative angle $(\omega' - \omega)$. The dislocation lies along $\hat{\boldsymbol{\tau}} = \hat{\mathbf{t}}$, and all remaining vectors lie in the O′BPA plane of Fig. 13.9. After putting the field point at $\mathbf{x} = \rho\,\hat{\mathbf{m}}$,[4] replacing the vector $\hat{\mathbf{k}}$ with the vector $\hat{\mathbf{n}}'$, and using $\hat{\mathbf{k}} \cdot \mathbf{x} \rightarrow \hat{\mathbf{n}}' \cdot \mathbf{x} = \rho\hat{\mathbf{n}}' \cdot \hat{\mathbf{m}} = -\rho \sin(\omega' - \omega)$ and

[4]See Eq. (12.16).

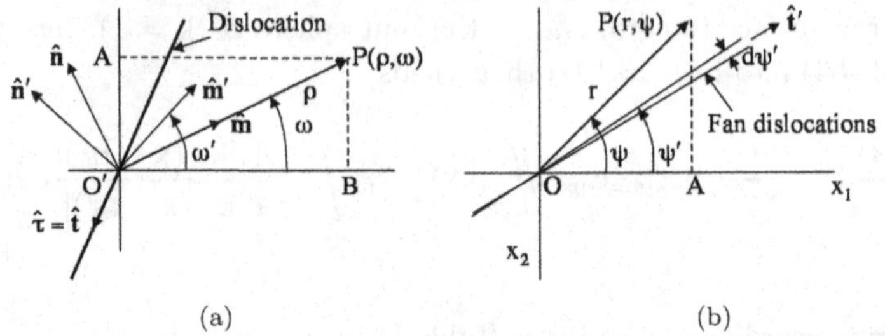

Fig. 13.10 (a) Coordinate systems used for finding the distortions at the field point $P(\rho, \omega)$ of Fig. 13.9 due to the dislocation lying along $\hat{\mathbf{t}} = \hat{\boldsymbol{\tau}} = \hat{\mathbf{m}} \times \hat{\mathbf{n}}$. All vectors except $\hat{\boldsymbol{\tau}} = \hat{\mathbf{t}}$ lie in the O′BPA plane which is perpendicular to $\hat{\boldsymbol{\tau}} = \hat{\mathbf{t}}$. (b) Coordinate system for finding the traction at $P(r, \psi)$ on the $x_3 = 0$ plane of Fig. 13.9 due to a dislocation fan lying in the $x_3 = 0$ plane. Only elements of the fan at the angles ψ' and $\psi' + d\psi'$ are shown: the positive directions of the fan dislocations are taken outwards from the origin.

$\phi = \omega' + \pi/2$, Eq. (13.108) becomes

$$\frac{\partial u_j^{\infty}(\rho,\omega)}{\partial x_s} = -\frac{b_k C_{kpim}}{2\pi^2 \rho} e_{qps} \int_0^{\pi} \frac{(\hat{n}'\hat{n}')_{ij}^{-1} \hat{n}'_m}{\sin(\omega' - \omega)} \hat{t}_q d\omega'. \tag{13.109}$$

Also, since $\hat{\mathbf{m}}' \times \hat{\mathbf{n}}' = \hat{\mathbf{t}}$, $e_{qps}\hat{t}_q = \hat{m}'_p \hat{n}'_s - \hat{m}'_s \hat{n}'_p$, and putting this into Eq. (13.109),

$$\frac{\partial u_j^{\infty}(\rho,\omega)}{\partial x_s} = -\frac{b_k C_{kpim}}{2\pi^2 \rho} \int_0^{\pi} \frac{(\hat{m}'_p \hat{n}'_s - \hat{m}'_s \hat{n}'_p)(\hat{n}'\hat{n}')_{ij}^{-1} \hat{n}'_m}{\sin(\omega' - \omega)} d\omega' \tag{13.110}$$

which, upon introducing Christoffel stiffness matrices, takes the simpler form

$$\frac{\partial u_j^{\infty}(\rho,\omega)}{\partial x_s} = -\frac{b_k}{2\pi^2 \rho} \int_0^{\pi} \frac{[\hat{n}'_s (\hat{m}'\hat{n}')_{ki} (\hat{n}'\hat{n}')_{ij}^{-1} - \hat{m}'_s \delta_{kj}]}{\sin(\omega' - \omega)} d\omega'. \tag{13.111}$$

The traction on the $x_3 = 0$ plane at $P(\rho, \omega)$ due to the dislocation is then

$$\begin{aligned} T_j^{\infty}(\rho,\omega) &= \sigma_{ij}^{\infty} \hat{n}_i^{\circ} = C_{ijrs} \frac{\partial u_r^{\infty}}{\partial x_s} \hat{n}_i^{\circ} \\ &= -\frac{b_k}{2\pi^2 \rho} \int_0^{\pi} \frac{[(\hat{n}_i^{\circ} C_{ijrs} \hat{n}'_s)(\hat{m}'\hat{n}')_{kp} (\hat{n}'\hat{n}')_{pr}^{-1} - (\hat{n}_i^{\circ} C_{ijrs} \hat{m}'_s)\delta_{kr}]}{\sin(\omega' - \omega)} d\omega' \\ &= \frac{b_k}{2\pi^2 \rho} \int_0^{\pi} \frac{[(\hat{n}^{\circ}\hat{m}')_{jk} - (\hat{n}^{\circ}\hat{n}')_{jr} (\hat{n}'\hat{n}')_{rp}^{-1} (\hat{n}'\hat{m}')_{pk}]}{\sin(\omega' - \omega)} d\omega', \end{aligned} \tag{13.112}$$

where the relation $(\hat{m}'\hat{n}')_{ki} = (\hat{n}'\hat{m}')_{ik}$ has been used.

Traction on $x_3 = 0$ *plane at P due to fan*

The traction at $P(r, \psi)$ on the $x_3 = 0$ plane due to the dislocation fan can be determined with the help of the coordinate system in Fig. 13.10b. The perpendicular distance from each source of Burgers vector strength in the fan is $|\mathbf{x}| = r\sin(\psi - \psi')$, and, therefore, using Eq. (12.27), the traction at the field point $P(r, \psi)$, obtained by integrating the contributions of all dislocations in the fan, is

$$T_j^{FAN}(r,\psi) = \frac{2}{r}\int_0^{\pi} \frac{B_{jk}(\psi')b_k^{FAN}(\psi')}{\sin(\psi - \psi')}d\psi', \tag{13.113}$$

where $b_k^{FAN}(\psi')$ is the angular distribution of Burgers vector strength in the fan. The condition for a traction-free surface at P in Fig. 13.9 is then

$$\begin{aligned} T_j^{FAN}(r,\psi) &= -T_j^{\infty}(\rho,\omega) = \frac{2}{r}\int_0^{\pi} \frac{B_{jk}(\psi')b_k^{FAN}(\psi')}{\sin(\psi - \psi')}d\psi' \\ &= \frac{b_k}{2\pi^2\rho}\int_0^{\pi} \frac{[(\hat{n}^{\circ}\hat{m}')_{jk} - (\hat{n}^{\circ}\hat{n}')_{jr}(\hat{n}'\hat{n}')^{-1}_{rp}(\hat{n}'\hat{m}')_{pk}]}{\sin(\omega - \omega')}d\omega'. \end{aligned} \tag{13.114}$$

Both integrals in Eq. (13.114) are principal value type integrals because of the singularities that occur as $\omega' \to \omega$ and $\psi' \to \psi$. However, the integral equation can be solved without evaluating either integral. Recognizing that the angle θ in Fig. 13.9 is constant, the following relationships hold between the angles, θ, and α, ω, and ψ, and their primed counterparts:

$$\begin{aligned} \sin\alpha &= \rho/r, & \sin\alpha' &= \rho'/r', \\ \sin\omega &= \sin\psi/\sin\alpha, & \sin\omega' &= \sin\psi'/\sin\alpha', \\ \cos\omega &= \sin\theta\cos\psi/\sin\alpha, & \cos\omega' &= \sin\theta\cos\psi'/\sin\alpha', \\ \cot\omega &= \sin\theta\cot\psi, & \cot\omega' &= \sin\theta\cot\psi'. \end{aligned} \tag{13.115}$$

Then, by employing these expressions, the two further relationships,

$$\left.\begin{aligned} \sin(\omega - \omega') &= \frac{\sin\theta}{\sin\alpha\sin\alpha'}\sin(\psi - \psi') \\ d\omega' &= \frac{\sin\theta}{\sin^2\alpha'}d\psi' \end{aligned}\right\} \tag{13.116}$$

are obtained, and upon substituting them into Eq. (3.114), we obtain

$$\int_0^\pi \frac{B_{jk}(\psi')b_k^{FAN}(\psi')}{\sin(\psi-\psi')}d\psi'$$

$$= \frac{b_k}{4\pi^2}\int_0^\pi \frac{[(\hat{n}^\circ\hat{m}')_{jk} - (\hat{n}^\circ\hat{n}')_{jr}(\hat{n}'\hat{n}')^{-1}_{rp}(\hat{n}'\hat{m}')_{pk}]}{\sin\alpha'\sin(\psi-\psi')}d\psi'. \quad (13.117)$$

Equation (13.117) is therefore satisfied if

$$B_{jk}(\psi')b_k^{FAN}(\psi') = \frac{b_k}{4\pi^2\sin\alpha'}[(\hat{n}^\circ\hat{m}')_{jk} - (\hat{n}^\circ\hat{n}')_{jr}(\hat{n}'\hat{n}')^{-1}_{rp}(\hat{n}'\hat{m}')_{pk}]. \quad (13.118)$$

The angular distribution for the fan, $b_k^{FAN}(\psi')$, required to produce a traction-free surface, is then

$$b_r^{FAN}(\psi') = \frac{B_{rj}^{-1}(\psi')b_k}{4\pi^2\sin\alpha'}[(\hat{n}^\circ\hat{m}')_{jk} - (\hat{n}^\circ\hat{n}')_{jq}(\hat{n}'\hat{n}')^{-1}_{qp}(\hat{n}'\hat{m}')_{pk}]. \quad (13.119)$$

Stress field in half-space containing impinged dislocation

Having the above results, the stress field in the half-space containing the impinged dislocation is obtained by use of Eq. (13.101). To determine the dislocation contribution, σ_{ij}^{∞}, we introduce the coordinate system and geometry in Fig. 13.11a where the dislocation lies along OO′, and the field point,

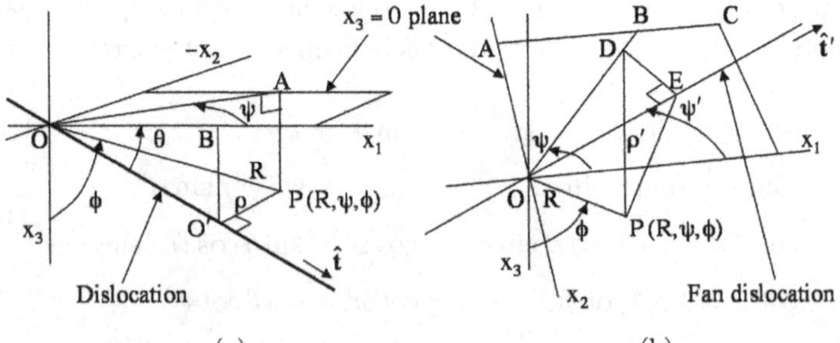

(a) (b)

Fig. 13.11 (a) Coordinate system and geometry for determining stress at field point $P(R,\psi,\phi)$ due to dislocation lying along OO′ in infinite body. Planes OAP and OBO′are perpendicular to $x_3 = 0$ plane. (b) Geometry for determining stress at P due to dislocation fan on $x_3 = 0$ plane in infinite body. A dislocation which is part of the fan is shown along OE. Plane PDE is perpendicular to OE. Plane POD is perpendicular to $x_3 = 0$ plane. Points A, B, C, D and E lie in $x_3 = 0$ plane.

P, is at the spherical coordinates ($r = R, \psi, \phi$) at a perpendicular distance from the dislocation, ρ. Using Eq. (12.24), the stress at $P(R, \psi, \phi)$ due to the dislocation is then

$$\begin{aligned}\sigma_{mn}^{\infty}(R, \psi, \phi) &= C_{mnip}\frac{\partial u_i^{\infty}}{\partial x_p} \\ &= \frac{b_s C_{mnip}}{2\pi\rho}\left\{-\hat{m}_p S_{is} + \hat{n}_p(\hat{n}\hat{n})_{ik}^{-1}[4\pi B_{ks} + (\hat{n}\hat{m})_{kr}S_{rs}]\right\},\end{aligned} \tag{13.120}$$

where ρ, $\hat{\mathbf{m}}$ and $\hat{\mathbf{n}}$ are given below by Eq. (13.122). To obtain these quantities it is noted from Fig. 13.11a that the vectors from O to P and from O to O′ are given by

$$\begin{aligned}\mathbf{OP} &= R(\sin\phi\cos\psi\hat{\mathbf{e}}_1 - \sin\phi\sin\psi\hat{\mathbf{e}}_2 + \sin\phi\cos\phi\hat{\mathbf{e}}_3) \\ \mathbf{OO'} &= (\mathbf{OP}\cdot\hat{\mathbf{t}})\hat{\mathbf{t}} = R(\sin\phi\cos\psi\hat{t}_1 - \sin\phi\sin\psi\hat{t}_2 + \sin\phi\cos\phi\hat{t}_3)\hat{\mathbf{t}}.\end{aligned} \tag{13.121}$$

Then, since $\mathbf{O'P} = \mathbf{OP} - (\mathbf{OP}\cdot\hat{\mathbf{t}})\hat{\mathbf{t}}$, and $\rho = |\mathbf{O'P}|$, and $\hat{\mathbf{m}} = \mathbf{O'P}/\rho$,

$$\begin{cases}\rho = R\left\{[\sin\phi\cos\psi - f(\phi,\psi)\hat{t}_1]^2 + [\sin\phi\sin\psi - f(\phi,\psi)\hat{t}_2]^2 \right. \\ \qquad \left. + [\sin\phi\cos\phi - f(\phi,\psi)\hat{t}_3]^2\right\}^{1/2}, \\ \hat{\mathbf{m}} = \dfrac{\begin{array}{c}\{[\sin\phi\cos\psi - f(\phi,\psi)\hat{t}_1]\hat{\mathbf{e}}_1 + [\sin\phi\sin\psi - f(\phi,\psi)\hat{t}_2]\hat{\mathbf{e}}_2 \\ +[\sin\phi\cos\phi - f(\phi,\psi)\hat{t}_3]\hat{\mathbf{e}}_3\}\end{array}}{\begin{array}{c}\{[\sin\phi\cos\psi - f(\phi,\psi)\hat{t}_1]^2 + [\sin\phi\sin\psi - f(\phi,\psi)\hat{t}_2]^2 \\ + [\sin\phi\cos\phi - f(\phi,\psi)\hat{t}_3]^2\}^{1/2}\end{array}}, \\ \hat{\mathbf{n}} = \hat{\mathbf{t}}\times\hat{\mathbf{m}},\end{cases} \tag{13.122}$$

where $f(\phi, \psi) = (\sin\phi\cos\psi\hat{t}_1 - \sin\phi\sin\psi\hat{t}_2 + \sin\phi\cos\phi\hat{t}_3)$.

The stress at P contributed by the fan, determined by use of the coordinate system and geometry of Fig. 13.11b and Eq. (12.24), is

$$\begin{aligned}\sigma_{mn}^{FAN}(R, \psi, \phi) &= C_{mnip}\frac{\partial u_i^{FAN}}{\partial x_p} = \frac{1}{2\pi}\int_0^{\pi}\frac{b_s^{FAN}(\psi')C_{mnip}}{\rho'} \\ &\quad \times\left\{-\hat{m}'_p S'_{is} + \hat{n}'_p(\hat{n}'\hat{n}')_{ik}^{-1}[4\pi B'_{ks} + (\hat{n}'\hat{m}')_{kr}S'_{rs}]\right\}d\psi',\end{aligned} \tag{13.123}$$

where $b_s^{FAN}(\psi')$ is obtained from Eq. (3.119), and ρ', $\hat{\mathbf{m}}'$ and $\hat{\mathbf{n}}'$ are given below by Eq. (13.125). To obtain these quantities refer to Fig. 13.11b where

$$\begin{cases} \mathbf{DP} = \mathrm{R}\cos\phi\hat{\mathbf{e}}_3, \\ \mathbf{ED} = -\mathrm{R}\sin\phi\sin(\psi-\psi')[\sin\psi'\hat{\mathbf{e}}_1 + \cos\psi'\hat{\mathbf{e}}_2], \end{cases} \tag{13.124}$$

Then, using $\boldsymbol{\rho}' = (\mathbf{DE} + \mathbf{DP})$,

$$\begin{cases} \rho' = \mathrm{R}[\cos^2\phi + \sin^2\phi\sin^2(\psi-\psi')]^{1/2} \\ \hat{\mathbf{m}}' = \dfrac{\mathrm{R}}{\rho'}[-\sin\phi\sin(\psi-\psi')\sin\psi'\hat{\mathbf{e}}_1 \\ \qquad -\sin\phi\sin(\psi-\psi')\cos\psi'\hat{\mathbf{e}}_2 + \cos\phi\hat{\mathbf{e}}_3], \\ \hat{\mathbf{n}}' = \hat{\mathbf{t}}' \times \hat{\mathbf{m}}'. \end{cases} \tag{13.125}$$

Finally, the stress in the half-space is obtained by substituting Eqs. (13.120) and (13.123) into Eq. (13.101).

13.3.4.2 *Some results for isotropic system*

A great many solutions have been found for the image stress fields associated with dislocations impinging upon free surfaces in isotropic systems using a variety of methods including applications of potential theory (see reviews by Eshelby 1979; Lothe 1992c).[5]

Screw dislocation at normal incidence

A relatively simple case is that of a screw dislocation with $\mathbf{b} = (0, 0, \mathrm{b})$ normal to a planar free surface as in Fig. 13.12 (Eshelby and

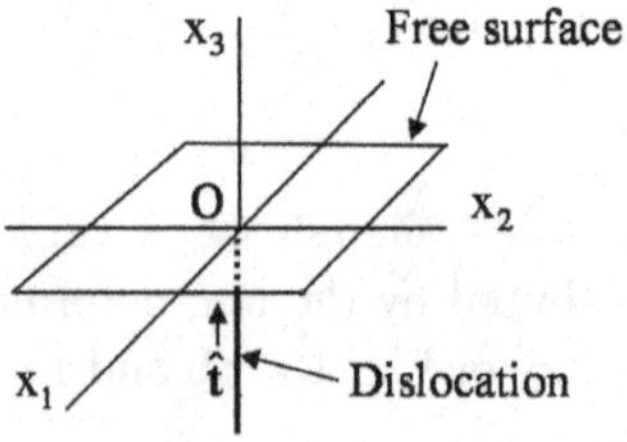

Fig. 13.12 Dislocation impinging at normal incidence on planar free surface of half-space.

[5]Eshelby (1979) states that "by ransacking textbooks of electrostatics and hydrodynamics we could write down the solutions for an indefinite number of special cases."

Stroh 1951; Yoffe 1961). According to Eq. 12.55, if the dislocation were of infinite length in an infinite body, it would produce the tractions

$$T_1 = \sigma_{13}^{\infty} = -\frac{\mu b x_2}{2\pi(x_1^2 + x_2^2)} \quad T_2 = \sigma_{23}^{\infty} = \frac{\mu b x_1}{2\pi(x_1^2 + x_2^2)} \tag{13.126}$$

on the $x_3 = 0$ plane. An image field is therefore sought to cancel these tractions. As shown previously by Eq. (13.44), the above stresses produce a twisting torque around the x_3 axis on the $x_3 = 0$ plane. As discussed by Eshelby (1979), this suggests an image displacement field of the form

$$\mathbf{u}^{IM} = \nabla \times [0, 0, a_1\phi(x_1, x_2, x_3)] = a_1\frac{\partial\phi}{\partial x_2}\hat{\mathbf{e}}_1 - a_1\frac{\partial\phi}{\partial x_1}\hat{\mathbf{e}}_2, \tag{13.127}$$

where $\phi(x_1, x_2, x_3)$ is a harmonic function and a_1 is a constant. Such a field corresponds to rotation around the x_3 axis which presumably could cancel the torque produced by the infinite dislocation. Also, since ϕ is harmonic, it satisfies the Navier equation, as may be verified by direct substitution into Eq. (3.3), and so would be an acceptable solution. The image stresses due to this displacement field are

$$\sigma_{13}^{IM} = 2\mu\varepsilon_{13}^{IM} = \mu a_1\frac{\partial^2\phi}{\partial x_3\partial x_2}, \quad \sigma_{23}^{IM} = 2\mu\varepsilon_{23}^{IM} = -\mu a_1\frac{\partial^2\phi}{\partial x_3\partial x_1}. \tag{13.128}$$

In a first try at finding the unknown ϕ, the harmonic function $\phi = a_1/x$ corresponding to the potential for a point electrical charge, as given by Eq. (3.10), is attempted. However, this produces the stresses

$$\sigma_{13}^{IM}(x_1, x_2, x_3) = \frac{3a_1\mu x_2 x_3}{x^5} \quad \sigma_{23}^{IM}(x_1, x_2, x_3) = -\frac{3a_1\mu x_1 x_3}{x^5} \tag{13.129}$$

which fall off with distance from the dislocation as x^{-3} and vanish on the $x_3 = 0$ plane and thus are unsuitable. In a second try, a candidate potential of the form $\phi = a_1 \ln(x_3 + x)$ is obtained by uniformly spacing point charges along the dislocation (i.e., along x_3), and integrating their contributions (i.e., the a_1/x quantities) along x_3. However, use of this potential results in image stresses on the $x_3 = 0$ plane of the form

$$\sigma_{13}^{IM}(x_1, x_2, 0) = -\frac{a_1\mu x_2}{(x_1^2 + x_2^2)^{3/2}} \quad \sigma_{23}^{IM}(x_1, x_2, 0) = \frac{a_1\mu x_1}{(x_1^2 + x_2^2)^{3/2}} \tag{13.130}$$

which are finite but fall off as x^{-2} and are therefore also unsuitable. However, in a third try, use of the potential $\phi = a_1[x_3 \ln(x_3 + x) - x]$, obtained by integrating the above result a second time along x_3, produces the image stresses

$$\sigma_{13}^{IM}(x_1, x_2, x_3) = \frac{a_1 \mu x_2}{x(x + x_3)} \quad \sigma_{23}^{IM}(x_1, x_2, x_3) = -\frac{a_1 \mu x_1}{x(x + x_3)} \tag{13.131}$$

which satisfactorily fall off as x^{-1} and cancel the surface tractions of the infinite dislocation when $x_3 = 0$ and $a_1 = b/(2\pi)$. Using these image stresses, the final stress field for the impinging screw dislocation is then

$$\left.\begin{aligned} \sigma_{13}(x_1, x_2, x_3) &= \sigma_{13}^{\infty} + \sigma_{13}^{IM} = -\frac{\mu b x_2}{2\pi(x_1^2 + x_2^2)} + \frac{\mu b x_2}{2\pi x(x + x_3)} \\ \sigma_{23}(x_1, x_2, x_3) &= \sigma_{23}^{\infty} + \sigma_{23}^{IM} = \frac{\mu b x_1}{2\pi(x_1^2 + x_2^2)} - \frac{\mu b x_1}{2\pi x(x + x_3)} \end{aligned}\right\} \tag{13.132}$$

Edge dislocation at normal incidence

When the dislocation in Fig. 13.12 is an edge dislocation with $\mathbf{b} = (b, 0, 0)$, in an isotropic system, use of Eq. (12.45) shows that the only surface traction requiring cancellation is the normal traction

$$T_3 = \sigma_{33} = -\frac{\mu b \nu}{\pi(1 - \nu)} \frac{x_2}{(x_1^2 + x_2^2)}. \tag{13.133}$$

Yoffe (1961) and Eshelby (1979) have shown that a harmonic function can be found in this case which can be used to construct a suitable image stress.

Inclined dislocations

As demonstrated by Yoffe (1961), solutions for the image fields of inclined dislocations impinged on free surfaces in isotropic systems, such as illustrated in Fig. 13.13, can be constructed with the help of angular dislocations whose stress fields have been analyzed in Sec. 12.8.1.2. The three cases of interest occur when the Burgers vector is parallel to x_1, x_2, and x_3, respectively. In the first case when $\mathbf{b} = (b, 0, 0)$, corresponding to an edge dislocation, Yoffe (1961) finds that if the impinged dislocation in Fig. 13.13a is replaced by an angular dislocation as in Fig. 13.13b with segment 1 identical to the impinged dislocation and segment 2 in the image position shown, all tractions on the surface vanish except a normal $T_3 = \sigma_{33}$ traction.

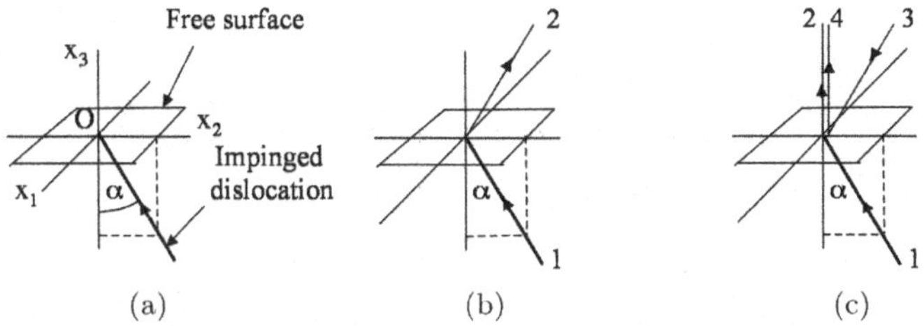

Fig. 13.13 (a) Dislocation impinged on the free surface of half-space at angle α. (b) Angular dislocation with segment 1 identical to impinged dislocation in (a) and segment 2 in its image position. (c) Angular dislocations 1–2 and 3–4 present with segment 1 identical to the impinged dislocation in (a) and segment 3 in its image position.

A harmonic stress function can then be found which yields an image stress that cancels this traction.

A similar result is found when $\mathbf{b} = (0, \mathrm{b}, 0)$. However, when $\mathbf{b} = (0, 0, \mathrm{b})$, the angular dislocation produces shear tractions on the surface. This situation can be partially remedied by aligning segment 2 with x_3 and adding a second angular dislocation, with acute segments 3 and 4, in the configuration shown in Fig. 13.13c. This arrangement results in shear stresses that are the same as the shear stresses produced by a screw dislocation normal to the surface plus a normal $T_3 = \sigma_{33}$ traction. The final total stress field resulting in a traction-free surface then consists of the fields due to the two angular dislocations (known from the results of Sec. 12.8.1.2), the image field for the normal screw dislocation, known from Eq. (13.132), and an image field that cancels the $T_3 = \sigma_{33}$ traction.

Further aspects of the above problems are considered by Shaibani and Hazzledine (1981) who also point out several minor misprints in Yoffe (1961).

13.3.5 *Dislocation loop near planar free surface of half-space*

13.3.5.1 *Image field*

To find the image displacement field for a general dislocation loop near a planar free surface of a half-space, the geometrical arrangement in Fig. 13.14 is adopted, and the Volterra equation, i.e., Eq. (12.67), is employed. This will yield the image displacement field of the above loop if the derivative of the Green's function required by the Volterra equation simply corresponds to the derivative of the image Green's function for a point force in

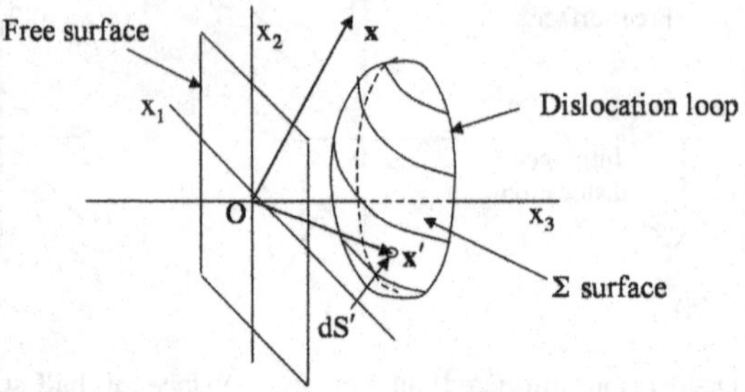

Fig. 13.14 General dislocation loop in half-space near planar free surface. The loop is threaded by the x_3 axis, and the Σ surface is arbitrary. The field point is at $\mathbf{x}$, and the source vector, $\mathbf{x}'$, goes to source points on the Σ surface.

a half-space given by the second term in Eq. (8.15), i.e.,

$$\frac{\partial G_{ik}^{IM}(\mathbf{x},\mathbf{x}')}{\partial x_j'} = \frac{1}{4\pi^2}\mathcal{R}_e\int_0^{2\pi}\sum_{\beta=1}^{3}\sum_{\alpha=1}^{3}\frac{A_{i\alpha}^* M_{\alpha r}^* L_{r\beta}A_{k\beta}(\hat{k}_j + p_\beta\hat{w}_j)}{[\hat{\mathbf{k}}\cdot(\mathbf{x}-\mathbf{x}') + p_\alpha^*\mathbf{x}\cdot\hat{\mathbf{w}} - p_\beta\mathbf{x}'\cdot\hat{\mathbf{w}}]^2}d\phi. \tag{13.134}$$

Therefore, after substituting this expression into Eq. (12.67), the image displacement field due to the loop in Fig. 13.14 is given by

$$u_i^{IM}(\mathbf{x},\mathbf{x}) = -\frac{C_{jkmn}b_m}{4\pi^2}\mathcal{R}_e\iint_\Sigma\int_0^{2\pi}\sum_{\beta=1}^{3}\sum_{\alpha=1}^{3}$$

$$\times\frac{A_{i\alpha}^* M_{\alpha r}^* L_{r\beta}A_{k\beta}(\hat{k}_j + p_\beta\hat{w}_j)}{[\hat{\mathbf{k}}\cdot(\mathbf{x}-\mathbf{x}') + p_\alpha^*\mathbf{x}\cdot\hat{\mathbf{w}} - p_\beta\mathbf{x}'\cdot\hat{\mathbf{w}}]^2}d\phi\hat{n}_n' dS'. \tag{13.135}$$

13.3.5.2 *Some results for isotropic systems*

The image displacement field for a finite loop near a planar free surface of an isotropic half-space can be found, as in the previous section, by employing the Volterra equation, i.e., Eq. (12.67), with the required Green's function given in this case by the image Green's function for isotropic half-spaces given by Eq. (4.116). Even though Eq. (4.116) is relatively simple compared with its anisotropic counterpart, the results are lengthy and therefore will not be written out here.

Using the same approach, the image displacement field, du_i^{IM} of an infinitesimal loop located at $\mathbf{x}'$ in a half-space can be obtained by employing the Volterra equation in the form of Eq. (12.168) after modifying it so that its Green's function is the image Green's function for a half-space so that it assumes the form

$$du_i^{IM}(\mathbf{x},\mathbf{x}') = -\mu b_j \hat{n}_k \times \left[\frac{\partial G_{ij}^{IM}(\mathbf{x},\mathbf{x}')}{\partial x'_k} + \frac{\partial G_{ik}^{IM}(\mathbf{x},\mathbf{x}')}{\partial x'_j} + \frac{2\nu}{1-2\nu}\delta_{jk}\frac{\partial G_{im}^{IM}(\mathbf{x},\mathbf{x}')}{\partial x'_m} \right] dS \tag{13.136}$$

with $G_{ij}^{IM}(\mathbf{x},\mathbf{x}')$ given by Eq. (4.116). Again, writing out the final results for du_i^{IM} will be lengthy, but, fortunately, only differentiation is required.

Bacon and Groves (1970) have obtained solutions for a variety of different types of infinitesimal loops located at $\mathbf{x}' = (0, 0, q)$ by a somewhat different, but essentially equivalent, route that again requires only differentiation. Here, an image dislocation loop is introduced at $(0, 0, -q)$, and the solution is expressed as the sum of the displacement fields that the actual loop and its image would produce in an infinite body plus an additional term that is needed to produce a traction-free surface. The additional term is found by using Mindlin's (1936) solution for the displacements due to a point force in a half-space which serves as the basis for the Green's function, given by Eq. (4.116), and which is called for in Eq. (13.136).

13.3.6 *Dislocation loop near planar interface between elastically dissimilar half-spaces*

13.3.6.1 *Image field*

The image displacement field for a finite loop in half-space 1 near a planar interface between joined dissimilar half-spaces 1 and 2 can be obtained by the same approach used above for a loop in a half-space with a free surface. For the present problem, the Volterra equation, Eq. (12.67), can again be used but with its Green's function now given by the Green's functions for joined dissimilar half-spaces given by Eq. (4.79) for anisotropic systems and by Eqs. (4.105) and (4.106) for isotropic systems.

For the anisotropic case, the derivative of the image Green's function for half-space 1 required for the Volterra equation, obtained with the use of Eq. (4.79), is

$$\frac{\partial g_{ik}^{IM(1)}(\mathbf{x},\mathbf{x}')}{\partial x_j'} = \frac{1}{4\pi^2}\mathcal{R}_e \int_0^{2\pi} \sum_{\alpha=1}^{3}\sum_{\gamma=1}^{3} \times \frac{A_{i\gamma}^{*(1)} \mathcal{W}_{\gamma r} U_{r\alpha} A_{k\alpha}^{(1)} (\hat{k}_j + p_\alpha^{(1)}\hat{w}_j)}{[\hat{\mathbf{k}}\cdot(\mathbf{x}-\mathbf{x}') + p_\gamma^{*(1)}\mathbf{x}\cdot\hat{\mathbf{w}} - p_\alpha^{(1)}\mathbf{x}'\cdot\hat{\mathbf{w}}]^2} d\phi. \quad (13.137)$$

Substituting this into the Volterra equation, the image displacement field in half-space 1 (containing the dislocation loop) is then

$$u_i^{IM(1)}(\mathbf{x}) = -\frac{C_{jkmn}b_m}{4\pi^2}\mathcal{R}_e \iint_0^{2\pi} \sum_{\alpha=1}^{3}\sum_{\gamma=1}^{3} \times \frac{A_{i\gamma}^{*(1)} \mathcal{W}_{\gamma r} U_{r\alpha} A_{k\alpha}^{(1)} (\hat{k}_j + p_\alpha^{(1)}\hat{w}_j)}{[\hat{\mathbf{k}}\cdot(\mathbf{x}-\mathbf{x}') + p_\gamma^{*(1)}\mathbf{x}\cdot\hat{\mathbf{w}} - p_\alpha^{(1)}\mathbf{x}'\cdot\hat{\mathbf{w}}]^2} d\phi \hat{n}_n' dS'. \quad (13.138)$$

The corresponding displacement field induced in the adjoining half-space 2 can be obtained by similar means using the Green's function for half-space 2 given by Eq. (4.79).

Exercises

13.1. Find the forces exerted on long straight edge and screw dislocations of the types illustrated in Fig. 12.2 by a general stress tensor, σ_{ij}^Q.

Solution. For the edge dislocation in Fig. 12.2a, $\mathbf{b} = (b,0,0)$ and $\hat{\mathbf{t}} = (0,0,1)$, and Eq. (13.10) yields

$$[d] = [b][\sigma^Q] = [b \quad 0 \quad 0]\begin{bmatrix} \sigma_{11}^Q & \sigma_{12}^Q & \sigma_{13}^Q \\ \sigma_{12}^Q & \sigma_{22}^Q & \sigma_{23}^Q \\ \sigma_{13}^Q & \sigma_{23}^Q & \sigma_{33}^Q \end{bmatrix} = b[\sigma_{11}^Q \quad \sigma_{12}^Q \quad \sigma_{13}^Q]. \quad (13.139)$$

Then, using Eq. (13.11),

$$\boldsymbol{f}^{DIS/\sigma}(\text{edge}) = b\det\begin{vmatrix} \hat{\mathbf{e}}_1 & \hat{\mathbf{e}}_2 & \hat{\mathbf{e}}_3 \\ \sigma_{11}^Q & \sigma_{12}^Q & \sigma_{13}^Q \\ 0 & 0 & 1 \end{vmatrix} = b(\sigma_{12}^Q\hat{\mathbf{e}}_1 - \sigma_{11}^Q\hat{\mathbf{e}}_2). \quad (13.140)$$

For the screw dislocation in Fig. 12.2c, $\mathbf{b} = (0, 0, \mathrm{b})$ and $\hat{\mathbf{t}} = (0, 0, 1)$. Then, by similar calculations,

$$[\mathrm{d}] = [\mathrm{b}][\sigma^{\mathrm{Q}}] = \mathrm{b}[\sigma^{\mathrm{Q}}_{13} \quad \sigma^{\mathrm{Q}}_{23} \quad \sigma^{\mathrm{Q}}_{33}],$$
$$\boldsymbol{f}^{\mathrm{DIS}/\sigma}(\text{screw}) = \mathrm{b}(\sigma^{\mathrm{Q}}_{23}\hat{\mathbf{e}}_1 - \sigma^{\mathrm{Q}}_{13}\hat{\mathbf{e}}_2). \tag{13.141}$$

Having these results, the force on any long straight dislocation is easily determined. Note that the σ^{Q}_{22} and σ^{Q}_{33} stress components are not involved as might have been anticipated, since neither dislocation can move in a direction that allows these imposed stresses to perform work.

13.2. Instead of starting with Eq. (13.1), as in the text, derive the Peach–Koehler equation given by Eq. (13.09) by simply displacing a differential segment of a general dislocation line a differential distance in the presence of an imposed stress field and determining directly the extra work required due to the presence of the stress field.

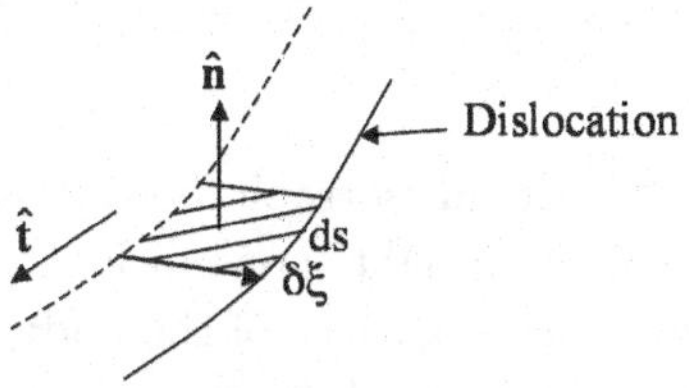

Fig. 13.15 Differential segment of dislocation line, *ds*, displaced by $\delta\boldsymbol{\xi}$.

Solution. When a dislocation segment of length *ds* is displaced by $\delta\boldsymbol{\xi}$ as in Fig. 13.15 in the presence of a stress system Q, the force on the dislocation due to the stress does the work $\delta\mathcal{W} = (\boldsymbol{f}^{\mathrm{DIS/Q}} \cdot \delta\boldsymbol{\xi})\mathrm{ds}$. This must be equal to the work done by the stress when the two sides of the cut associated with the dislocation are displaced with respect to each other by the Burgers vector during the displacement. Taking the Σ cut/displacement rule on p. 325 into account, this latter work is given by $\delta\mathcal{W} = -\mathbf{b} \cdot \mathbf{T}^{\mathrm{Q}}|\delta\boldsymbol{\xi}|\mathrm{ds}$ where $\mathbf{T}^{\mathrm{Q}}$ is the traction on the side of the cut having the normal vector $-\hat{\mathbf{n}}$. Then, equating these expressions, and using Eq. (E.5),

$$f^{\mathrm{DIS/Q}}_{\mathrm{n}}\delta\xi_{\mathrm{n}} = \mathrm{b_i}\sigma^{\mathrm{Q}}_{\mathrm{ij}}\hat{\mathrm{n}}_{\mathrm{j}}|\delta\boldsymbol{\xi}| = \mathrm{b_i}\sigma^{\mathrm{Q}}_{\mathrm{ij}}(\hat{\mathbf{t}} \times \delta\boldsymbol{\xi})_{\mathrm{j}} = \mathrm{e_{mnj}}\mathrm{b_i}\sigma^{\mathrm{Q}}_{\mathrm{ij}}\hat{\mathrm{t}}_{\mathrm{m}}\delta\xi_{\mathrm{n}}. \tag{13.142}$$

Finally, since $\delta\xi_n$ can be varied independently,

$$f_n^{DIS/Q} = e_{jmn} b_i \sigma_{ij}^Q \hat{t}_m \qquad (13.143)$$

in agreement with Eq. (13.9).

13.3. Show that the net force on a dislocation loop lying in a uniform imposed stress field, Q, vanishes.

Solution. Using Eq.(13.9), the total force is given by the line integral

$$F_l^{LOOP/Q} = \oint_C f_l^{LOOP/Q} ds = e_{jkl} \sigma_{ij}^Q b_i \oint_C \hat{t}_k ds \qquad (13.144)$$

since σ_{ij}^Q and b_i are constant around the loop. Then, substituting $\hat{t}_k ds = dx_k$,

$$F_l^{LOOP/Q} = e_{jkl} \sigma_{ij}^Q b_i \oint_C \hat{t}_k ds = e_{jkl} \sigma_{ij}^Q b_i \oint_C dx_k = 0. \qquad (13.145)$$

13.4. Suppose a long cylindrical isotropic body of radius R co-axial with the x_3 axis contains a straight screw dislocation lying parallel to x_3 at $x_1 = q$. The cylindrical surface of the body can be made traction-free by adding the elastic field of a parallel image screw dislocation of opposite Burgers vector lying along $x_1 = h$. Find the required distance, *h*, in terms of *q* and *R*.

Solution. Using Eq. (12.55), the stress produced by the real dislocation in an infinite body is

$$\sigma_{13}^{\infty} = \frac{\mu b}{2\pi} \left[\frac{-x_2}{(x_1 - q)^2 + x_2^2} + \frac{x_2}{(x_1 - h)^2 + x_2^2} \right],$$

$$\sigma_{23}^{\infty} = \frac{\mu b}{2\pi} \left[\frac{(x_1 - s)}{(x_1 - q)^2 + x_2^2} - \frac{(x_1 - h)}{(x_1 - h)^2 + x_2^2} \right]. \qquad (13.146)$$

Using Eq. (2.59), this field produces a traction component at the point $P(x_1, x_2)$ on the cylindrical surface of radius R given by

$$T_3 = \sigma_{13}^{\infty} \frac{x_1}{R} + \sigma_{23}^{\infty} \frac{x_2}{R}. \qquad (13.147)$$

Combining Eqs. (3.146), (13.147) and $x_1^2 + x_2^2 = R^2$, the condition for the cylindrical surface to be traction-free is

$$-\frac{x_2x_1}{R[R^2 - 2qx_1 + q^2]} + \frac{x_2x_1}{R[R^2 - 2hx_1 + h^2]} + \frac{(x_1 - q)x_2}{R[R^2 - 2qx_1 + q^2]} - \frac{(x_1 - h)x_2}{R[R^2 - 2hx_1 + h^2]} = 0. \qquad (13.148)$$

Then, solving Eq. (13.148) for h, $h = R^2/q$.

13.5. Figure 13.16 shows a positive straight screw dislocation, +D, parallel to the $\mathcal{S}^{(1)}$ and $\mathcal{S}^{(2)}$ free surfaces of a large flat isotropic plate of thickness *d*. If +D is at the distance *q* from surface, $\mathcal{S}^{(1)}$, find an expression for the image stress at the dislocation.

Solution. Following Hirth and Lothe (1982), first put a negative screw dislocation (i.e., −D) at $x_1 = 2q$ to cancel σ_{13} on $\mathcal{S}^{(1)}$ due to +D at $x_1 = 0$. Put a +D at $x_1 = -2d$ to cancel σ_{13} on $\mathcal{S}^{(2)}$ due to −D at $x_1 = 2q$. Put a −D at $x_1 = 2(d+q)$ to cancel σ_{13} on $\mathcal{S}^{(1)}$ due to +D at $x_1 = -2d$. Put a +D at $x_1 = -4d$ to cancel σ_{13} on $\mathcal{S}^{(2)}$ due to −D at $x_1 = 2(d+q)$, etc. Put a −D at $x_1 = 2(-d+q)$ to cancel σ_{13} on $\mathcal{S}^{(2)}$ due to +D at $x_1 = 0$. Put a +D at $x_1 = 2d$ to cancel σ_{13} on $\mathcal{S}^{(1)}$ due to −D at $x_1 = 2(-d+q)$. Put a −D at $x_1 = 2(-2d+q)$ to cancel σ_{13} on $\mathcal{S}^{(2)}$ due to +D at $x_1 = 2d$. Put a +D at $x_1 = 4d$ to cancel σ_{13} on $\mathcal{S}^{(1)}$ due to −D at $x_1 = 2(-2d+q)$, etc. An infinite number of image dislocations is therefore required, and the final image stresses at the dislocation are found by summing their contributions at $x_1 = 0$. Using Eq. 12.55, all σ_{13}^{IM} contributions vanish, while the σ_{23}^{IM} contributions can be written as the infinite series

$$\sigma_{23}^{IM} = \frac{\mu b}{2\pi} \sum_{n=-\infty}^{\infty} \frac{1}{2(nd + q)} \qquad (13.149)$$

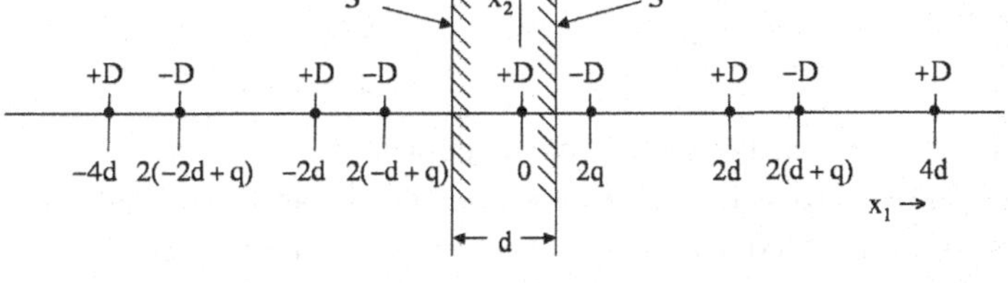

Fig. 13.16 End view of positive straight screw dislocation, +D, lying parallel to x_3 at $x_1 = x_2 = 0$ at a distance q from the surface, $\mathcal{S}^{(1)}$, of an infinite thin plate. The remaining ±Ds represent image dislocations.

which, in turn, corresponds to the function (Morse and Feshbach 1953)

$$\sigma_{23}^{\mathrm{IM}} = \frac{\mu b}{4d} \cot \frac{\pi q}{d}. \tag{13.150}$$

13.6. Determine the image force exerted on a long straight edge dislocation parallel to a nearby flat surface of an isotropic region (Fig. 13.3) whose total stress field is given by Eq. (13.52).

Solution. The image force can be found by substituting the image stress at the dislocation obtained from Eq. (13.52) into the Peach–Koehler force equation given by Eq. (13.11). Then, since $\mathbf{b} = (b, 0, 0)$ and $\hat{\mathbf{t}} = (0, 0, 1)$,

$$\begin{aligned}[d] &= [b \quad 0 \quad 0] \begin{bmatrix} \sigma_{11}^{\mathrm{IM}} & \sigma_{12}^{\mathrm{IM}} & 0 \\ \sigma_{12}^{\mathrm{IM}} & \sigma_{22}^{\mathrm{IM}} & 0 \\ 0 & 0 & \sigma_{33}^{\mathrm{IM}} \end{bmatrix} \\ &= b[\sigma_{11}^{\mathrm{IM}} \quad \sigma_{12}^{\mathrm{IM}} \quad 0] \end{aligned} \tag{13.151}$$

so that

$$\begin{aligned} \boldsymbol{f}^{\mathrm{DIS}^\infty/\mathrm{DIS}^{\mathrm{IM}}} &= \mathbf{d} \times \hat{\mathbf{t}} = \det \begin{vmatrix} \hat{\mathbf{e}}_1 & \hat{\mathbf{e}}_2 & \hat{\mathbf{e}}_3 \\ b\sigma_{11}^{\mathrm{IM}} & b\sigma_{12}^{\mathrm{IM}} & 0 \\ 0 & 0 & 1 \end{vmatrix} \\ &= b\sigma_{11}^{\mathrm{IM}}(q,0)\hat{\mathbf{e}}_1 - b\sigma_{12}^{\mathrm{IM}}(q,0)\hat{\mathbf{e}}_2. \end{aligned} \tag{13.152}$$

The image stresses according to Eq. (13.52) are

$$\begin{aligned} \sigma_{12}^{\mathrm{IM}}(q,0) &= -A/(2q) \\ \sigma_{11}^{\mathrm{IM}} &= 0 \end{aligned} \tag{13.153}$$

and substituting them into Eq. (13.152), the image force is

$$\boldsymbol{f}^{\mathrm{DIS}^\infty/\mathrm{DIS}^{\mathrm{IM}}} = -\frac{\mu b^2}{4\pi(1-\nu)q}\hat{\mathbf{e}}_1 \tag{13.154}$$

which is towards the surface, as expected.

13.7. Equation (13.4) for the interaction energy between a dislocation loop and an imposed stress field, Q, has been obtained by employing the basic formulation for defect interaction energies given by Eq. (5.5). Show that Eq. (13.4) can just as well be derived by starting with

Eq. (5.10), which, in turn, has been obtained by employing the basic formulation for defect interaction energies given by Eq. (5.4).

Solution. Equation (5.10), for the present case takes the form

$$E_{\text{int}}^{\text{DIS/Q}} = \oiint_{\mathcal{S}} (\sigma_{ij} u_i^Q - \sigma_{ij}^Q u_i) \hat{n}_j dS. \tag{13.155}$$

Assume that the surface $\mathcal{S}$ is in the configuration illustrated in Fig. 13.1b. Now, if $\mathcal{S}$ is shrunk to the point where $\mathcal{S}^+ \rightarrow \Sigma^+$, $\mathcal{S}^- \rightarrow \Sigma^-$, $r_\circ \rightarrow 0$, $\hat{\mathbf{n}}^{\mathcal{S}^+} \rightarrow \hat{\mathbf{n}}$ and $\hat{\mathbf{n}}^{\mathcal{S}^-} \rightarrow -\hat{\mathbf{n}}$, Eq. (13.155) can be written as the sum of two integrals over the Σ surface, given by[6]

$$E_{\text{int}}^{\text{DIS/Q}} = \iint_{\Sigma^+} (\sigma_{ij} u_i^Q - \sigma_{ij}^Q u_i) \hat{n}_j dS - \iint_{\Sigma^-} (\sigma_{ij} u_i^Q - \sigma_{ij}^Q u_i) \hat{n}_j dS \tag{13.156}$$

with boundary conditions at these surfaces expressed by

$$\begin{aligned} \sigma_{ij}(\Sigma^+)\,\hat{n}_j &= \sigma_{ij}(\Sigma^-)\,\hat{n}_j, \\ u_i(\Sigma^-) - u_i(\Sigma^+) &= b_i, \\ \sigma_{ij}^Q(\Sigma^+)\,\hat{n}_j &= \sigma_{ij}^Q(\Sigma^-)\,\hat{n}_j, \\ u_i^Q(\Sigma^+) &= u_i^Q(\Sigma^-). \end{aligned} \tag{13.157}$$

Substitution of these conditions into Eq. (13.156) then yields the interaction energy in the form

$$E_{\text{int}}^{\text{DIS/Q}} = b_i \iint_{\Sigma} \sigma_{ij}^Q \hat{n}_j dS \tag{13.158}$$

in agreement with Eq. (13.4).

[6]See Eshelby (1956) and Bacon, Barnett and Scattergood (1979b) for further details regarding this procedure.

Chapter 14

Interfaces

14.1 Introduction

A variety of distinguishable types of internal interfaces exist in crystalline materials. Many contain isolated dislocations, or arrays of regularly spaced discrete dislocations with localized cores (Sutton and Balluffi 2006), and therefore possess stress fields that extend significant distances into the two adjoining crystals. In this chapter we consider some of the major types of such interfaces and focus on their structures, elastic fields and strain energies.

In order to treat various cases it is convenient to divide the interfaces into two major classes, i.e., interfaces where the elastic properties of the two crystals adjoining the interface are effectively the same (i.e., *iso-elastic interfaces*), and those where they differ significantly (i.e., *hetero-elastic interfaces*).[1]

The chapter begins with a description of the essential geometrical features of interfaces containing dislocations in crystalline bodies and then goes on to analyze the elastic fields of iso-elastic and hetero-elastic interfaces possessing arrays of discrete dislocations. The elastic fields of single isolated dislocations in hetero-elastic interfaces are also determined as an essential part of the overall treatment.

[1]Note, for example, that in the case of a large-angle grain boundary, the two crystals adjoining the boundary are of the same phase and so have identical elastic constants when referred to their own crystal axes. However, since they have significantly different crystal orientations, their elastic constants will effectively differ across the boundary. Such an interface is therefore classified as hetero-elastic.

The following notation is employed for this chapter:

$\varepsilon_{ij}^{\infty}$ = strain field due to single dislocation in either infinite homogeneous body or iso-elastic interface in infinite body

$\varepsilon_{ij}^{(1)}$ = strain field in half-space 1 due to single dislocation in hetero-elastic interface between half-spaces 1 and 2

$\varepsilon_{ij}^{\text{array}^{\infty}}$ = strain field due to dislocation array in iso-elastic interface in infinite body

$\varepsilon_{ij}^{\text{array}^{(1)}}$ = strain field in half-space 1 due to dislocation array in heterophase interface

$\varepsilon_{ij}^{\text{LF}^{\infty}}$ = strain field due to single line force in infinite homogeneous body

$\varepsilon_{ij}^{\text{LFarray}^{\infty}}$ = strain field due to line force array in infinite homogeneous body

14.2 Geometrical Features of Interfaces — Degrees of Freedom

Interfaces are most easily studied in the form of planar configurations in bicrystals (Sutton and Balluffi 2006). An infinite number of distinguishable structures can then be produced by varying the crystal misorientation of the two crystals adjoining the interface and the inclination of the interface plane with respect to the crystal axes of the adjoining crystals. There are then five macroscopic *degrees of freedom* that can be independently varied to produce the full range of possible *homophase* and *heterophase* interfaces [2] These can be chosen in several ways. Consider homophase interfaces first. A useful set for present purposes is obtained by constructing a planar interface in a bicrystal as follows:

(1) Start with a single crystal labeled as crystal 1 and establish the desired interface plane by specifying its unit normal vector, $\hat{\mathbf{n}}$, in the coordinate system of crystal 1. Since $\hat{\mathbf{n}}$ is a unit vector, this involves two degrees of freedom, i.e., two direction cosines.

(2) Rotate the crystal region on the side of the interface towards which $\hat{\mathbf{n}}$ is pointing around an axis, $\hat{\mathbf{a}}$, lying in the interface plane, thus converting this region to crystal 2. The $\hat{\mathbf{a}}$ axis lies at an angle α with respect to a reference line inscribed in the interface plane, and the angle of rotation

[2] For a homophase interface the two crystals adjoining the interface are of the same phase: for a heterophase interface the phases of the two crystals differ.

around $\hat{\mathbf{a}}$ is θ. This introduces the two additional degrees of freedom, α and θ.

(3) Finally, rotate crystal 2 around $\hat{\mathbf{n}}$ by the angle ϕ which is then the fifth degree of freedom.

For heterophase interfaces we need only add a further step in which crystal 2 is transformed into the required new phase. However, this does not involve a continuous variable and does not contribute a further degree of freedom.

14.3 Iso-elastic Interfaces

An important type of iso-elastic interface is the *small-angle homophase interface* where the crystals adjoining the interface are of the same phase and only slightly misoriented. For such interfaces the effect of the relatively small difference in crystal orientation across the interface on the elastic constants can be neglected, and the elastic fields of dislocations in the interface can be determined to a good approximation as if they were embedded in a single crystal. As discussed below, certain *large-angle homophase interfaces* also support discrete dislocations. Interfaces of this type are iso-elastic in isotropic systems but not in anisotropic systems, since the difference in the effective elastic constants caused by the relatively large misorientation across the interface is then significant and must be taken into account.

14.3.1 *Geometrical features*

14.3.1.1 *Small-angle tilt, twist and mixed homophase interfaces*

A *small-angle tilt homophase interface*, consisting of an array of discrete, straight, parallel, edge dislocations, is shown in Fig. 14.1c. The bicrystal containing this interface can be constructed by the following steps:

(1) Start with a single crystal and bend it elastically until its outer extremes are tilted with respect to each other by the desired angle θ as in Fig. 14.1a. Long-range applied bending stress is therefore present throughout the crystal.
(2) Introduce edge dislocations throughout the crystal to accommodate the bending and eliminate the long-range bending stress as shown in Fig. 14.1b. Note that local stress due to the individual dislocations is present, but the long-range bending stress has been eliminated.

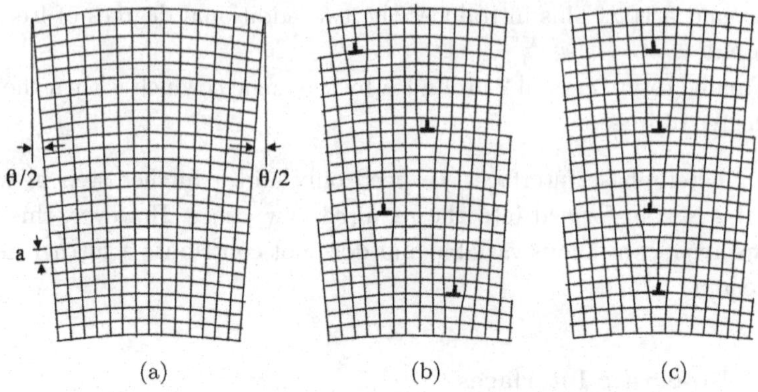

Fig. 14.1 Construction of small-angle tilt interface. Crystal structure is simple cubic with lattice parameter a. (a) Single crystal is elastically bent around axis normal to plane of paper to produce a tilt angle θ between left and right free surfaces. (b) Distribution of edge dislocations is introduced to accommodate the bending and eliminate imposed long-range bending stress in (a). (c) Dislocations introduced in (b) are now gathered in planar array to decrease the elastic strain energy further and form the small-angle tilt interface.

(3) Gather the dislocations into a planar interface consisting of an array of uniformly spaced dislocations as in Fig. 14.1c. In this arrangement the strain fields of the individual dislocations tend to partially cancel, their total elastic strain energy is thereby reduced, and the two crystalline regions adjoining the inerface are tilted with respect to each other by the angle θ and are free of long-range stress.

A *small-angle twist interface* consisting of a square array of discrete screw dislocations can be produced by essentially the same method used above. In Fig. 14.2a the initial single crystal is first twisted elastically around the intended interface normal by the angle ϕ. A distribution of screw dislocations is then introduced (not shown) to eliminate the long-range stress that was required to twist the crystal initially. Then, these dislocations are arranged in a square planar grid of screw dislocations to produce the interface shown in Fig. 14.2b. Figure 14.2c shows a section of the grid in more detail.

More complex small-angle interfaces can be constructed by combining tilt and twist rotations (i.e., θ and ϕ rotations) to produce *small-angle mixed interfaces.* In such cases arrays of several types of dislocations must be introduced to accommodate the crystal misorientation.

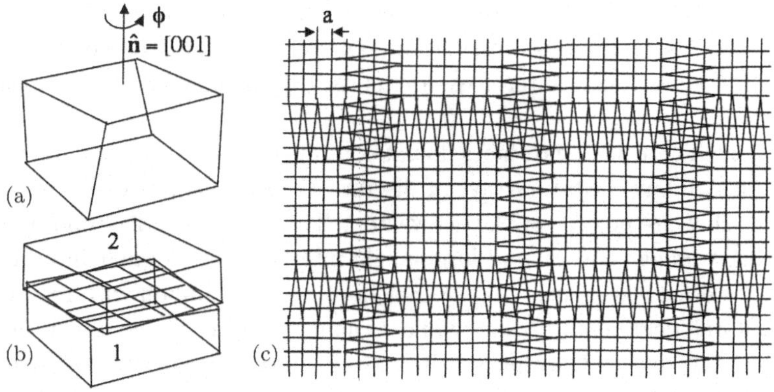

Fig. 14.2 Construction of small-angle twist interface. Crystal structure is simple cubic. (a) Single crystal initially twisted elastically around $\hat{\mathbf{n}}$ by angle ϕ. (b) Final interface consisting of square grid of screw dislocations which accommodates the twist misorientation between the upper and lower regions of the crystal in (a) in the absence of long-range torsional stress. (c) Detailed view along $-\hat{\mathbf{n}}$ of dislocation grid showing the first lattice planes of the upper and lower crystals adjoining the interface.

In Exercise 14.3 the above procedure, in conjunction with Eq. (14.11), is employed to find the dislocation structures (including the dislocation types and spacings) of the interfaces in Figs. 14.1 and 14.2. In Exercise 14.1 the dislocation structure of the interface in Fig. 14.2 is found by the alternative method of applying the Frank–Bilby equation [derived below in the form of Eq. (14.24)].

14.3.1.2 *Large-angle homophase vicinal interfaces*

The dislocation spacings in the small-angle homophase interfaces described above decrease as the crystal misorientation increases, since larger numbers of dislocations are required to accommodate the increased misorientation. For example, for the tilt interface in Fig. 14.1c each vertical plane that terminates in the interface injects an edge dislocation into the interface, and it is readily seen that the dislocation spacing in the interface plane, *d*, is therefore given by

$$\mathrm{d} = \frac{\mathrm{a}}{2\sin(\theta/2)} \cong \frac{\mathrm{a}}{\theta}. \tag{14.1}$$

At angles exceeding about 15 degrees or so, the dislocation spacing becomes sufficiently small to cause the dislocation cores to overlap, and the interfaces therefore no longer consist of arrays of discretely separated dislocations possessing Burgers vectors that are vectors of the crystal lattice. Such interfaces are classified as *large-angle homophase interfaces.*

Even though such interfaces do not contain arrays of discrete crystal dislocations[3] many contain arrays of discrete dislocations possessing Burgers vectors that are not crystal lattice vectors. This may be understood by considering *singular* and *vicinal* large-angle interfaces. A singular interface is defined as one possessing a free energy that is at a minimum with respect to at least one degree of freedom, while a vicinal interface is one that is near a singular interface with respect to its degrees of freedom. It is therefore energetically favorable for a vicinal interface to adopt a structure consisting of the structure of the nearby singular interface plus a superimposed array of discrete line defects (which may be dislocations, dislocations with step character, or steps) whose function is to accommodate the difference between the degrees of freedom of the two interfaces (Sutton and Balluffi 2006). An example where the line defects are dislocations is illustrated in Fig. 14.3a which shows a singular large-angle symmetrical tilt interface possessing a structure of short periodicity and relatively low energy. Then Fig. 14.3b

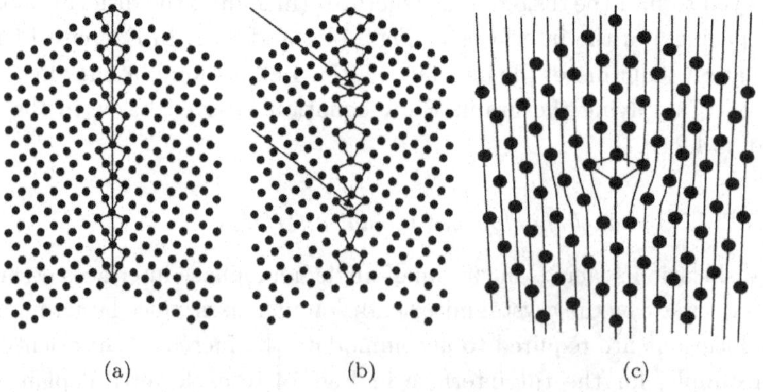

Fig. 14.3 (a) Large-angle singular interface. (b) Large-angle interface with slightly larger tilt angle than the singular interface in (a) and therefore vicinal to it. The interface consists of array of edge dislocations embedded in the interface structure of the singular interface. (c) Detailed view of the core structure of the dislocations indicated by arrows in (b).

[3]A *crystal dislocation* is a dislocation having a Burgers vector that is a translation vector of the crystal lattice.

shows an interface with a tilt angle that is only slightly larger than that of the singular interface and is therefore vicinal with respect to it. The vicinal interface possesses a structure corresponding to that of the nearby singular interface plus a superimposed array of discrete edge dislocations (at arrows) that accommodates the difference between the tilt angles of the two interfaces. Furthermore, the Burgers vector of these dislocations is not a crystal lattice vector as seen in Fig. 14.3c which presents an enlarged view of their core structure.[4] The vicinal interface in Fig. 14.3b can be produced by the same general method employed previously to produce the small-angle tilt interface of Fig. 14.1c by starting with a bicrystal containing the singular interface instead of a single crystal. The bicrystal is then elastically bent to produce the added tilt angle, and the accommodation dislocations are introduced to eliminate the resulting long-range bending stress.

14.3.2 *The Frank–Bilby equation*

The above pure tilt and twist interfaces possessed relatively simple structures, and the dislocation structures required to accommodate their tilts and twists in the absence of long-range stress were readily found. A general method for accomplishing this for any interface whose five degrees of freedom are specified is now developed based on the Frank–Bilby equation. But, first it is necessary to introduce two essential tensors, i.e., the α_{ij} tensor which specifies the dislocation content of a dislocated crystal, and the κ_{ij} tensor which describes the lattice rotation which is present locally within a crystal when it is bent and/or twisted. Then, a formalism linking these tensors under the condition of vanishing long-range stress is developed. Finally, having these results, the Frank–Bilby equation is formulated.

14.3.2.1 *The "state of dislocation" tensor,* α_{ij}

To specify the dislocation content of a dislocated crystal we follow Nye (1953) and introduce the *"state of dislocation tensor"*, α_{ij}, by starting with a body containing a distribution of discrete crystal dislocations. The dislocations are first smeared out into an infinite number of dislocations having

[4]Further description of vicinal large-angle interfaces, along with methods for determining the Burgers vectors of the accommodating dislocations, is given by Sutton and Balluffi (2006).

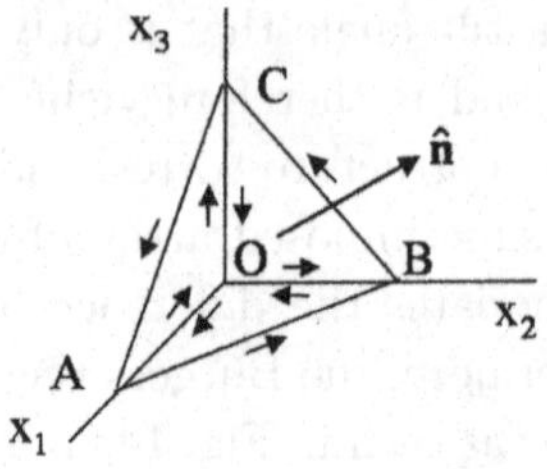

Fig. 14.4 Small tetrahedron embedded in dislocated crystal: $\hat{\mathbf{n}}$ is the unit normal vector to ABC face. Clockwise Burgers circuits around each face are indicated.

infinitesimal Burgers vectors so that the dislocation content is smoothly distributed on a local scale at a density that varies continuously with the field vector, $\mathbf{x}$. A small tetrahedron is then constructed in the crystal, and Burgers circuits are performed around each face as illustrated in Fig. 14.4. Since the circuit ABC on the front face is equivalent to the sum of the three circuits around the back faces,

$$\mathbf{B}^{\mathrm{ABC}} = \mathbf{B}^{\mathrm{OBC}} + \mathbf{B}^{\mathrm{OCA}} + \mathbf{B}^{\mathrm{OAB}}, \tag{14.2}$$

where, for example, $\mathbf{B}^{\mathrm{OBC}}$ is the sum of the Burgers vectors of all dislocations threading the area OBC which is perpendicular to $\hat{\mathbf{e}}_1$. The $\underline{\boldsymbol{\alpha}}$ tensor is now defined as the tensor whose component, α_{ij}, is equal to the sum of the i components of the Burgers vectors of all dislocations that thread a unit area in the body perpendicular to the j axis. The quantities in Eq. (14.2) can then be expressed in the form

$$\begin{aligned} \mathbf{B}^{\mathrm{OBC}} &= (\alpha_{11}\hat{\mathbf{e}}_1 + \alpha_{21}\hat{\mathbf{e}}_2 + \alpha_{31}\hat{\mathbf{e}}_3)\mathrm{A}^{\mathrm{OBC}}, \\ \mathbf{B}^{\mathrm{OCA}} &= (\alpha_{12}\hat{\mathbf{e}}_1 + \alpha_{22}\hat{\mathbf{e}}_2 + \alpha_{32}\hat{\mathbf{e}}_3)\mathrm{A}^{\mathrm{OCA}}, \\ \mathbf{B}^{\mathrm{OAB}} &= (\alpha_{13}\hat{\mathbf{e}}_1 + \alpha_{23}\hat{\mathbf{e}}_2 + \alpha_{33}\hat{\mathbf{e}}_3)\mathrm{A}^{\mathrm{OAB}}, \end{aligned} \tag{14.3}$$

where, for example, $\mathrm{A}^{\mathrm{OBC}}$ is the area of face OBC. Next, setting $\boldsymbol{\mathcal{B}}$ equal to the sum of the Burgers vectors of all dislocations threading unit area perpendicular to $\hat{\mathbf{n}}$,

$$\boldsymbol{\mathcal{B}} = \frac{\mathbf{B}^{\mathrm{ABC}}}{\mathrm{A}^{\mathrm{ABC}}} = \frac{\mathbf{B}^{\mathrm{OBC}} + \mathbf{B}^{\mathrm{OCA}} + \mathbf{B}^{\mathrm{OAB}}}{\mathrm{A}^{\mathrm{ABC}}}. \tag{14.4}$$

Then, substituting Eq. (14.3) into Eq. (14.4),

$$\mathcal{B}_{\mathrm{i}} = \alpha_{\mathrm{ij}}\hat{n}_{\mathrm{j}}. \tag{14.5}$$

The second rank tensor α_{ij} therefore couples the vectors $\boldsymbol{\mathcal{B}}$ and $\hat{\mathbf{n}}$. (Note that $\mathcal{B}_i$ and α_{ij} both have the dimensions m^{-1}).

If the dislocations throughout the body can be effectively grouped into families of parallel dislocations, where family k consists of $N^{(k)}$ dislocations per unit area perpendicular to $\hat{\mathbf{t}}^{(k)}$ with Burgers vector $\mathbf{b}^{(k)}$, an expression for the corresponding α_{ij} tensor can be formulated, since

$$\mathcal{B}_i = \sum_k N^{(k)} b_i^{(k)} (\hat{\mathbf{t}}^{(k)} \cdot \hat{\mathbf{n}}) = \sum_k N^{(k)} b_i^{(k)} \hat{t}_j^{(k)} \hat{n}_j \tag{14.6}$$

and, therefore,

$$\alpha_{ij} = \sum_k N^{(k)} b_i^{(k)} \hat{t}_j^{(k)}, \tag{14.7}$$

where the diagonal and off-diagonal components of α_{ij} correspond to screw and edge components, respectively.

14.3.2.2 *The curvature tensor, κ_{ij}*

If a crystal is filled with a distribution of dislocations as specified by the tensor, α_{ij}, the dislocations will generally induce lattice rotations which cause the crystal to be macroscopically bent and/or twisted as, for example, in Fig. 14.1b. In the case where the crystal is bent around the i axis, as in Fig. 14.5a, the crystal lattice rotates by the angle $\delta\phi_i$ around the i axis when an advance is made in the crystal along the j axis of the distance δx_j. For small advances $\delta x_j = R\delta\phi_i$ as in Fig. 14.5b, and the crystal curvature associated with the bending can be expressed as

$$\kappa_{ij} = \lim_{\delta x_j \to 0} \frac{\delta\phi_i}{\delta x_j} = \frac{1}{R} = \frac{\partial\phi_i}{\partial x_j}. \tag{14.8}$$

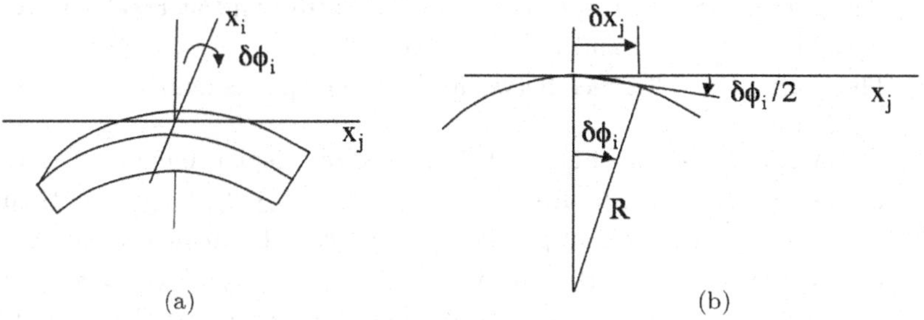

Fig. 14.5 (a) Crystal bent around $\hat{\mathbf{e}}_i$ axis. (b) Rotation, $\delta\phi_i$, around $\hat{\mathbf{e}}_i$ axis produced by traversing bent crystal by distance, δx_j, along $\hat{\mathbf{e}}_j$ axis. View is along the $\hat{\mathbf{e}}_i$ axis.

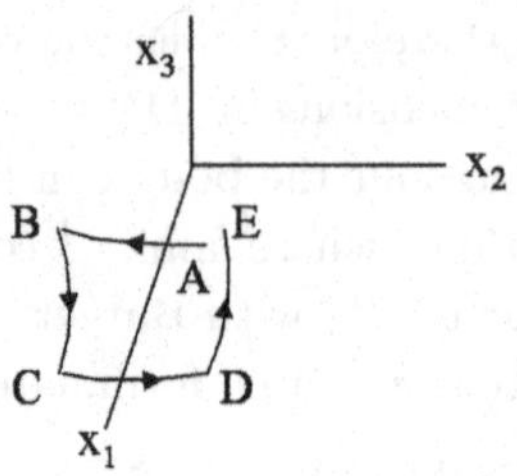

Fig. 14.6 Unit square Burgers circuit in dislocated crystal with its plane normal to the x_1 axis.

The quantity κ_{ij} therefore serves as a general *curvature tensor* which yields the crystal rotation that occurs when an advance is made in a bent and/or twisted crystal. Its diagonal components represent pure twists while its off-diagonal components represent pure bends.

14.3.2.3 *Relationship between the α_{ij} and κ_{ij} tensors*

The relationship that must exist between the α_{ij} and κ_{ij} tensors for a bent and/or twisted dislocated crystal under the condition of vanishing long-range stress is now obtained by examining the closure failures of unit square Burgers circuits embedded in the crystal with their planes normal to the coordinate axes as illustrated, for example, in Fig. 14.6.

First, the closure failure of the circuit in Fig. 14.6 is obtained in terms of the curvature tensor by the following procedure:

(1) Take point C as a fixed reference point and determine the displacement of the starting point at A relative to C due to crystal rotation as the unit segments CB and BA are traversed.
(2) Next, determine the displacement of the finish point at E relative to C as the segments CD and DE are traversed.
(3) Then, take the difference between the two displacements.

Using diagrams such as Fig. 14.6, the vector displacements at E due to traversing CD and DE are $\mathbf{d}^{\mathrm{CD}} = ((\kappa_{22} - \kappa_{32}/2), -\kappa_{12}, \kappa_{12}/2)$ and $\mathbf{d}^{\mathrm{DE}} = (\kappa_{23}/2, -\kappa_{13}/2, 0)$, respectively. Similarly, the displacements at A due to traversing CB and BA are $\mathbf{d}^{\mathrm{CD}} = ((\kappa_{23}/2 - \kappa_{33}), -\kappa_{13}/2, \kappa_{13})$ and $\mathbf{d}^{\mathrm{BA}} = (-\kappa_{32}/2, 0, \kappa_{12}/2)$, respectively. The closure failure of the circuit, measured from start to finish (i.e., from A to E), is then $\mathbf{d}^{\mathrm{CD}} + \mathbf{d}^{\mathrm{DE}} - \mathbf{d}^{\mathrm{CD}} - \mathbf{d}^{\mathrm{BA}}$.

Now, according to the SF/RH rule (p. 323), if no long-range stress is present, this closure failure must be equal to the sum of the Burgers vectors of all dislocations threading the circuit, which, in turn, is equal to the vector $\boldsymbol{\mathcal{B}}$. Therefore,

$$\mathbf{d}^{\mathrm{CD}} + \mathbf{d}^{\mathrm{DE}} - \mathbf{d}^{\mathrm{CD}} - \mathbf{d}^{\mathrm{BA}} = ((\kappa_{22} + \kappa_{33}), -\kappa_{12}, -\kappa_{13}) = \boldsymbol{\mathcal{B}} \quad (14.9)$$

and, with the help of Eq. (14.5),

$$\mathcal{B}_1 = \alpha_{11} = (\kappa_{22} + \kappa_{33}), \qquad \mathcal{B}_2 = \alpha_{21} = -\kappa_{12}, \qquad \mathcal{B}_3 = \alpha_{31} = -\kappa_{13}. \quad (14.10)$$

By combining these results with results obtained for similar circuits normal to axes 2 and 3, the relationship[5]

$$\alpha_{\mathrm{ij}} = \delta_{\mathrm{ij}}\kappa_{\mathrm{kk}} - \kappa_{\mathrm{ji}} \quad (14.11)$$

is finally obtained connecting the state of the dislocations in a dislocated crystal to the corresponding curvature of the crystal when it is free of long-range stress.

14.3.2.4 *Dislocation content of an interface — the Frank–Bilby equation*

The relationship that must exist between the dislocation content of an interface and its five degrees of freedom in the absence of long-range stress, i.e., the Frank–Bilby equation, is now found by producing the interface by the method illustrated in Fig. 14.1. For simplicity the crystals adjoining the interface are assumed to possess the simple cubic structure present in Figs. 14.1–14.3.[6] Starting with a dislocated crystal that is generally bent or twisted (or both) and free of long-range stress, as illustrated, for example, for simple bending in Fig. 14.1b, construct a small local Burgers circuit in the crystal by employing the FS/RH method illustrated in Fig. 12.4. Now, let

$$\mathrm{d}\mathbf{v} = \mathrm{d}\xi_{\mathrm{i}}\mathbf{a}_{\mathrm{i}}^{\mathrm{C}}(\mathbf{x}) \quad (14.12)$$

be a small vector in the dislocated crystal where the $\mathbf{a}_{\mathrm{i}}^{\mathrm{C}}(\mathbf{x})$ are the base vectors of the crystal (which are functions of position because of the

[5]The sign of Eq. (14.11) is opposite to that given originally by Nye (1953) because of a difference in dislocation sign conventions.

[6]The analysis can be readily generalized to include general crystal structures (Sutton and Balluffi 2006).

continuously varying rotations in the crystal), and the ξ_i are dimensionless coefficients. Also, express the base vectors of the crystal as a linear combination of the base vectors of the Reference lattice so that

$$\mathbf{a}_i^C = D_{ij}\mathbf{a}_j^R, \qquad \mathbf{a}_i^R = D_{ij}^{-1}\mathbf{a}_j^C. \tag{14.13}$$

The Burgers circuit (which is closed in the dislocated crystal because of the use of the FS/RH method) is then represented by the line integral,

$$\oint_C d\mathbf{v} = \oint_C d\xi_i \mathbf{a}_i^C(\mathbf{x}) = 0. \tag{14.14}$$

Then map this circuit onto the Reference lattice by the procedure illustrated in Fig. 12.4. In the mapping process each vector displacement increment, $d\mathbf{v} = d\xi_i \mathbf{a}_i^C = d\xi_i D_{ij}\mathbf{a}_j^R$, in the crystal circuit is transformed into a corresponding increment $d\mathbf{v}' = d\xi_i \mathbf{a}_i^R$ in the reference circuit. The mapped circuit fails to close, and using the FS/RH rule (p. 324), the sum of the Burgers vectors of all dislocations threading the circuit in the dislocated crystal, $\mathbf{b}^{tot}$, is given by

$$-\mathbf{b}^{tot} = \oint_C d\mathbf{v}' - d\mathbf{v} = d\xi_i \left(\mathbf{a}_i^R - D_{ij}\mathbf{a}_j^R\right). \tag{14.15}$$

Upon transforming the circuit to a Cartesian coordinate by means of the coordinate transformation $\mathbf{a}_i^R = a\hat{\mathbf{e}}_i$, and writing coordinate displacements in the Cartesian system as $dx_i = a d\xi_i$, $\mathbf{b}^{tot}$ assumes the form

$$-\mathbf{b}^{tot} = \oint_C d\xi_i \left(\mathbf{a}_i^R - D_{ij}\mathbf{a}_j^R\right) = \oint_C (dx_i - D_{ji}dx_j)\hat{\mathbf{e}}_i \tag{14.16}$$

or, equivalently,

$$-b_i^{tot} = \oint_C (dx_i - D_{ji}dx_j) = \oint_C (\delta_{ij} - D_{ji})dx_j. \tag{14.17}$$

Equation (14.16) can now be converted to a surface integral by using Stokes' theorem, as given by Eq. (B.3); i.e.,

$$\begin{aligned} -\mathbf{b}^{tot} &= \oint_C (dx_i - D_{ji}dx_j)\hat{\mathbf{e}}_i = e_{jkn}\iint_{\mathcal{S}} \frac{\partial D_{ni}}{\partial x_k}\hat{\mathbf{e}}_i\hat{n}_j dS, \\ -b_i^{tot} &= e_{jkn}\iint_{\mathcal{S}} \frac{\partial D_{ni}}{\partial x_k}\hat{n}_j dS, \end{aligned} \tag{14.18}$$

where $\mathcal{S}$ is any surface terminating on C. Now, using Eq. (14.5), the sum of the b_i of all dislocations threading the surface $\iint_{\mathcal{S}} \hat{n}_j dS$ can also be expressed by

$$b_i^{tot} = \iint_{\mathcal{S}} \alpha_{ij}\hat{n}_j dS. \tag{14.19}$$

Therefore, upon comparing Eqs. (14.18) and (14.19),

$$\alpha_{ij} = -e_{jkn}\frac{\partial D_{ni}}{\partial x_k}. \tag{14.20}$$

Next, to produce the interface, as, for example, in Fig. 14.1c, the distribution of dislocations in the dislocated crystal is gathered into a thin slab of thickness q as illustrated in Fig. 14.7. The two regions adjoining the slab then become crystals 1 and 2 misoriented with respect to one another. Also, since α_{ij} has vanished in both crystals outside the slab, the deformation tensor, D_{ni}, according to Eq. (14.20), is constant in each crystal and changes discontinuously upon crossing the interface slab.

The slab thickness, q, can now be made arbitrarily small while α_{ij} is correspondingly increased to maintain the dislocation content in the slab constant. A closed Burgers circuit enclosing a representative sample of the dislocations in the slab can then be constructed following the path $C \Rightarrow \mathbf{p}^{(2)} - \mathbf{p}^{(1)}$ where $\mathbf{p}^{(2)}$ and $\mathbf{p}^{(1)}$ lie in crystals 2 and 1 infinitesimally outside the slab as shown in Fig. 14.7. To apply Eq. (14.18) to this circuit, the derivative in the integrand is written as

$$\frac{\partial D_{ni}}{\partial x_k} = \frac{\left(D_{ni}^{(2)} - D_{ni}^{(1)}\right)}{q}\hat{\mathbf{n}} \cdot \hat{\mathbf{e}}_k = \frac{\left(D_{ni}^{(2)} - D_{ni}^{(1)}\right)}{q}\hat{n}_k, \tag{14.21}$$

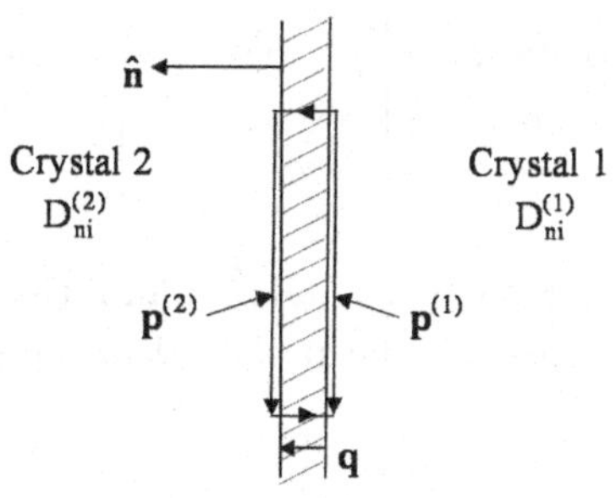

Fig. 14.7 Thin slab of thickness q containing dislocations which cause misorientation of dislocation-free crystals 1 and 2.

where $\hat{\mathbf{n}}$ is the interface normal. Then, substituting Eq. (14.21) into Eq. (14.18),

$$-\mathrm{b}_i^{tot} = \mathrm{e}_{jkn}\frac{\left(\mathrm{D}_{ni}^{(2)} - \mathrm{D}_{ni}^{(1)}\right)}{\mathrm{q}}\hat{\mathrm{n}}_k \iint_{\mathcal{S}} \hat{\mathrm{n}}_j^S dS, \tag{14.22}$$

where $\hat{\mathrm{n}}_j^S$ is the normal to the circuit in Fig. 14.7. But, with the use of Eq. (E.5),

$$\iint_{\mathcal{S}} \hat{\mathrm{n}}_j^S dS = \mathrm{q}(\hat{\mathbf{n}} \times \mathbf{p})_j = -\mathrm{q}\mathrm{e}_{jsr}\hat{\mathrm{n}}_r \mathrm{p}_s \tag{14.23}$$

and substituting this into Eq. (14.22), and using Eq. (E.4), we obtain the Frank–Bilby equation in the form

$$\mathrm{b}_i^{tot} = \mathbf{e}_{jkn}\mathbf{e}_{jsr}\left(\mathrm{D}_{ni}^{(2)} - \mathrm{D}_{ni}^{(1)}\right)\hat{\mathrm{n}}_k\hat{\mathrm{n}}_r\mathrm{p}_s = \left(\mathrm{D}_{ni}^{(2)} - \mathrm{D}_{ni}^{(1)}\right)\mathrm{p}_n = \mathrm{p}_i^{(2)} - \mathrm{p}_i^{(1)}. \tag{14.24}$$

Here, information about the five degrees of interface freedom is embedded in the D_{ij} tensors, which, in turn, are coupled to the Burgers vector content of the interface represented by $\mathbf{b}^{tot}$.[7]

As an example of its use, consider the small-angle symmetrical tilt interface in Fig. 14.1c and assume that its degrees of freedom are specified but that its dislocation content is unknown and to be determined. Crystal 1, Crystal 2, the Reference lattice and the Cartesian axes are then as shown in Fig. 14.8. Crystals 2 and 1 are symmetrically tilted around the $\hat{\mathbf{e}}_3$ axis (corresponding to the $\mathbf{a}_3^C$ crystal axis) by the angles $\theta/2$ and $-\theta/2$, respectively, and, therefore, according to Eq. (14.13),

$$\mathrm{D}_{ji}^{(2)} = \begin{bmatrix} \cos(\theta/2) & -\sin(\theta/2) & 1 \\ \sin(\theta/2) & \cos(\theta/2) & 1 \\ 0 & 0 & 1 \end{bmatrix}, \qquad \mathrm{D}_{ji}^{(1)} = \begin{bmatrix} \cos(\theta/2) & \sin(\theta/2) & 1 \\ -\sin(\theta/2) & \cos(\theta/2) & 1 \\ 0 & 0 & 1 \end{bmatrix}. \tag{14.25}$$

The interface plane corresponds to $(001)^R$, and if the vector $\mathbf{p}$, which must lie in the interface, is chosen to be $\mathbf{p} = (0, 0, \mathrm{p})$, and therefore parallel to

[7]The Frank–Bilby equation, which is based purely on geometrical considerations, also applies to interfaces where the adjoining crystals possess different lattice structures, as can be shown (Sutton and Balluffi 2006) by an easy generalization of the D_{ij} tensors.

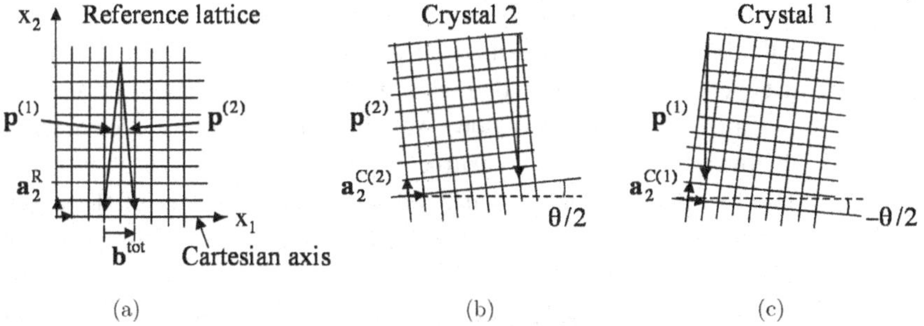

Fig. 14.8 Arrangement of Crystal 1, Crystal 2, the Reference lattice and the Cartesian axes for determining the dislocation content of the small-angle symmetric tilt interface in Fig. 14.1c. $|\mathbf{a}_i^R| = |\mathbf{a}_i^{C(1)}| = |\mathbf{a}_i^{C(2)}| = \mathbf{a}$.

the tilt axis, Eq. (14.24) yields

$$\begin{bmatrix} b_1^{tot} \\ b_2^{tot} \\ b_3^{tot} \end{bmatrix} = \begin{bmatrix} 0 & -2\sin(\theta/2) & 0 \\ 2\sin(\theta/2) & 0 & 0 \\ 0 & 0 & 0 \end{bmatrix} \begin{bmatrix} 0 \\ 0 \\ p \end{bmatrix} = \begin{bmatrix} 0 \\ 0 \\ 0 \end{bmatrix}. \tag{14.26}$$

On the other hand, if $\mathbf{p} = (0, -p, 0)$, and is therefore normal to the tilt axis,

$$\begin{bmatrix} b_1^{tot} \\ b_2^{tot} \\ b_3^{tot} \end{bmatrix} = \begin{bmatrix} 0 & -2\sin(\theta/2) & 0 \\ 2\sin(\theta/2) & 0 & 0 \\ 0 & 0 & 0 \end{bmatrix} \begin{bmatrix} 0 \\ -p \\ 0 \end{bmatrix} = \begin{bmatrix} 2p\sin(\theta/2) \\ 0 \\ 0 \end{bmatrix}. \tag{14.27}$$

Equation (14.26) rules out all families of straight parallel dislocations except those parallel to the tilt axis, while Eq. (14.27) shows that a family running parallel to the tilt axis, with Burgers vectors whose sum satisfies the condition $\mathbf{b}^{tot} = (2p\sin(\theta/2), 0, 0)$ constitutes a satisfactory dislocation structure. The vector $\mathbf{b}^{tot}$ is shown in Fig. 14.8a, and the sum condition is therefore satisfied if $N = (2p/a)\sin(\theta/2)$ dislocations with Burgers vectors $\mathbf{b} = \mathbf{a}_1^R$ intersect $\mathbf{p}$. The necessary dislocation spacing is then $d = p/N = a/[2\sin(\theta/2)]$ in agreement with Eq. (14.1).

It is evident from this exercise that the vector $\mathbf{p}$ can be used as a general probe to find interface dislocation structures composed of families of straight parallel dislocations that are consistent with the Frank–Bilby equation. Obviously, an infinite number of different arrays satisfying the equation can be constructed for any given interface. The physically preferred array is the one of minimum energy, and this can be identified only by means of energy calculations (as, for example, described in Sec. 14.3.5). Arrays of the present type have been termed *impotent arrays* (Mura 1987) or *surface dislocations* (Sutton and Balluffi 2006). When such an array moves normal to itself without changing its structure, it produces a macroscopic shape change of the bicrystal consistent with the dislocation content of the array and the corresponding change in crystal orientation across it as discussed in Secs. 15.3.1 and 15.3.2.

In Exercise 14.1, Eq. (14.24) is employed to find the screw dislocation structure of the twist interface in Fig. 14.2. Further applications of the Frank–Bilby equation to more complex mixed interfaces possessing both tilt and twist components are described by Hirth and Lothe (1982) and Sutton and Balluffi (2006).

14.3.3 *Elastic fields of interfaces consisting of arrays of parallel dislocations*

Even though interfaces that consist of arrays of parallel dislocations consistent with the Frank–Bilby equation are free of long-range stress, a near-stress field exists which extends into each adjoining crystal a distance approximately equal to the dislocation spacing in the interface in conformity with St. Venant's principle. Therefore, consider the elastic fields produced by arrays of straight parallel dislocations in iso-elastic interfaces using the coordinate system and geometry in Fig. 14.9a where the elastic constants are assumed to be effectively the same in each half-space. The contribution of each dislocation in the array to the total displacement field of the array then corresponds to the field produced by a single straight dislocation in a

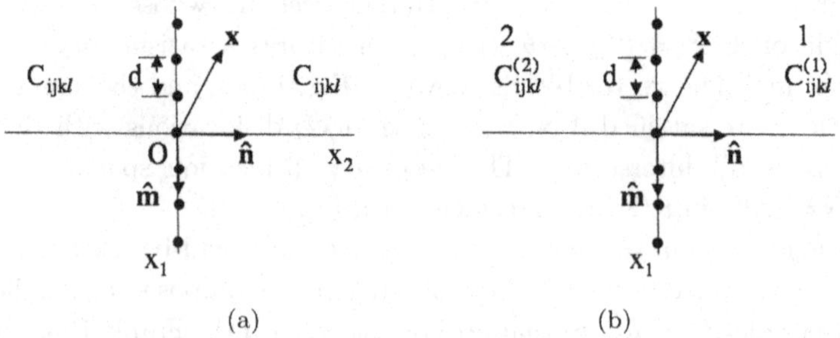

Fig. 14.9 End view of array of straight parallel dislocations at the spacing d. (a) In iso-elastic interface. (b) In hetero-elastic interface.

single crystal given by Eq. (12.12). Using this, the total displacement field of the array takes the form of the sum

$$u_j^{\mathrm{array}^\infty}(\mathbf{x}) = \frac{1}{2\pi i}\sum_{N=-\infty}^{\infty}\sum_{\alpha=1}^{6} \pm A_{j\alpha}L_{k\alpha}b_k \ln(\hat{\mathbf{m}}\cdot\mathbf{x} - Nd + p_\alpha\hat{\mathbf{n}}\cdot\mathbf{x}), \tag{14.28}$$

where the N are integers.

If the system containing the array is consistent with the Frank–Bilby equation, Eq. (14.28) should yield a displacement field corresponding to a near-field possessing significant strains and a far-field that is stress-free but causes the regions on opposite sides of the array to be rotated with respect to each other. This can be verified for the case of the symmetric tilt boundary in Fig. 14.1c by considering the distortions associated with the displacement field which, with the use of Eq. (14.28), are given by

$$\frac{\partial u_j^{\mathrm{array}^\infty}(\mathbf{x})}{\partial x_l} = \frac{1}{2\pi i}\sum_{N=-\infty}^{\infty}\sum_{\alpha=1}^{6} \pm A_{j\alpha}L_{k\alpha}b_k \frac{(\hat{m}_l + p_\alpha\hat{n}_l)}{(\hat{\mathbf{m}}\cdot\mathbf{x} - Nd + p_\alpha\hat{\mathbf{n}}\cdot\mathbf{x})}. \tag{14.29}$$

Equation (14.29) can be developed further by employing the following procedure (Hirth, Barnett and Lothe 1979) in which the sum over N in Eq. (14.29) can be expressed (Morse and Feshbach 1953) as

$$\sum_{N=-\infty}^{\infty} \frac{1}{(\hat{\mathbf{m}}\cdot\mathbf{x} - Nd + p_\alpha\hat{\mathbf{n}}\cdot\mathbf{x})} = \frac{\pi}{d}\cot\frac{\pi}{d}(\hat{\mathbf{m}}\cdot\mathbf{x} + p_\alpha\hat{\mathbf{n}}\cdot\mathbf{x}). \tag{14.30}$$

For the far-field region where $\hat{\mathbf{n}}\cdot\mathbf{x} \gg d$, Eq. (14.30) is then reduced by first setting $(\pi/d)(\hat{\mathbf{m}}\cdot\mathbf{x} + p_\alpha\hat{\mathbf{n}}\cdot\mathbf{x}) = z$ so that

$$\frac{\pi}{d}\cot\frac{\pi}{d}(\hat{\mathbf{m}}\cdot\mathbf{x} + p_\alpha\hat{\mathbf{n}}\cdot\mathbf{x}) = \frac{\pi i}{d}\frac{(e^{iz} + e^{-iz})}{(e^{iz} - e^{-iz})}. \tag{14.31}$$

By substituting Eq. (12.6) for p_α, it is seen that when $\hat{\mathbf{n}}\cdot\mathbf{x} \gg d$

$$\frac{\pi}{d}\cot\frac{\pi}{d}(\hat{\mathbf{m}}\cdot\mathbf{x} + p_\alpha\hat{\mathbf{n}}\cdot\mathbf{x}) = -\frac{\pi i}{d}\mathrm{sgn}(b_\alpha\hat{\mathbf{n}}\cdot\mathbf{x}) = -\frac{\pi i}{d}[\pm\,\mathrm{sgn}(\hat{\mathbf{n}}\cdot\mathbf{x})], \tag{14.32}$$

where b_α is the imaginary part of p_α, and the $\pm$ symbol has been employed. Then, substituting Eq. (14.32) into Eq. (14.29), the far-field distortions are

$$\frac{\partial u_j^{\mathrm{array}^{(1)}}(\mathrm{far-field})}{\partial x_l} = -\frac{1}{2d}\mathrm{sgn}(\hat{\mathbf{n}}\cdot\mathbf{x})\sum_{\alpha=1}^{6} A_{j\alpha}L_{k\alpha}b_k(\hat{m}_l + p_\alpha\hat{n}_l), \tag{14.33}$$

and, upon substituting Eq. (3.78) into Eq. (14.33),

$$\frac{\partial u_j^{\text{array}^{(1)}}(\text{far-field})}{\partial x_l} = -\frac{1}{2d}\text{sgn}(\hat{\mathbf{n}}\cdot\mathbf{x})\left(b_j\hat{m}_l + \sum_{\alpha=1}^{6} A_{j\alpha}L_{k\alpha}b_k p_\alpha \hat{n}_l\right). \tag{14.34}$$

The sum in Eq. (14.34) can be evaluated by substituting Eq. (3.113) to obtain

$$\sum_{\alpha=1}^{6} A_{j\alpha}L_{k\alpha}b_k p_\alpha \hat{n}_l = b_k\hat{n}_l \sum_{\alpha=1}^{6} A_{j\alpha}L_{k\alpha}p_\alpha = -(\hat{n}\hat{n})^{-1}_{jr}(\hat{n}\hat{m})_{rk}b_k\hat{n}_l. \tag{14.35}$$

For the present edge dislocations $\mathbf{b} = (0, b, 0)$ so that $\hat{n}_i b = b_i$, and the quantity $(\hat{n}\hat{m})_{rk}b_k$ in Eq. (14.35) takes the form

$$\begin{aligned}(\hat{n}\hat{m})_{rk}b_k &= \hat{n}_i C_{irkj}\hat{m}_j b_k = \hat{n}_i C_{irkj}\hat{m}_j \hat{n}_k b = \hat{n}_i C_{irkj}\hat{n}_k \hat{m}_j b \\ &= \hat{n}_i C_{irjk}\hat{n}_j \hat{m}_k b = \hat{n}_i C_{irkj}\hat{n}_j \hat{m}_k b = (\hat{n}\hat{n})_{rk}\hat{m}_k b.\end{aligned} \tag{14.36}$$

Then, substituting Eqs. (14.35) and (14.36) into Eq. (14.34), the far-field distortion tensor due to the array is given by (Hirth, Barnett and Lothe 1979)

$$\begin{aligned}\frac{\partial u_j^{\text{array}^{(1)}}(\text{far} - \text{field})}{\partial x_l} &= -\frac{1}{2d}\text{sgn}(\hat{\mathbf{n}}\cdot\mathbf{x})\left(b_j\hat{m}_l - b_l\hat{m}_j\right) \\ &= \frac{1}{2d}\text{sgn}(\hat{\mathbf{n}}\cdot\mathbf{x})\begin{bmatrix} 0 & b & 0 \\ -b & 0 & 0 \\ 0 & 0 & 0\end{bmatrix}.\end{aligned} \tag{14.37}$$

For the far-field region where $\hat{\mathbf{n}}\cdot\mathbf{x} > 0$, substitution of Eq. (14.37) into Eq. (2.14) shows that the only non-vanishing rotation is

$$\omega_3^{\text{array}^{(1)}}(\text{far-field}) = -\frac{b}{2d} \tag{14.38}$$

corresponding to a rigid body rotation of this region around the x_3 axis given by $\delta\theta = -b/2d$. For the opposite far-field region where $\hat{\mathbf{n}}\cdot\mathbf{x} < 0$, $\delta\theta = b/2d$, and it is therefore concluded that, as anticipated for the present array, the two regions indeed rotate with respect to one another according to the Frank–Bilby equation. Also, as indicated by Eqs. (14.37) and (2.5), the two far-fields are free of elastic strain.

However, for arrays which do not satisfy the Frank–Bilby equation (see Exercise 14.2), non-vanishing far-field distortions that produce elastic strains will be present. This is demonstrated in the following section for the arrays in Figs. 14.10b and 14.10c in isotropic systems.

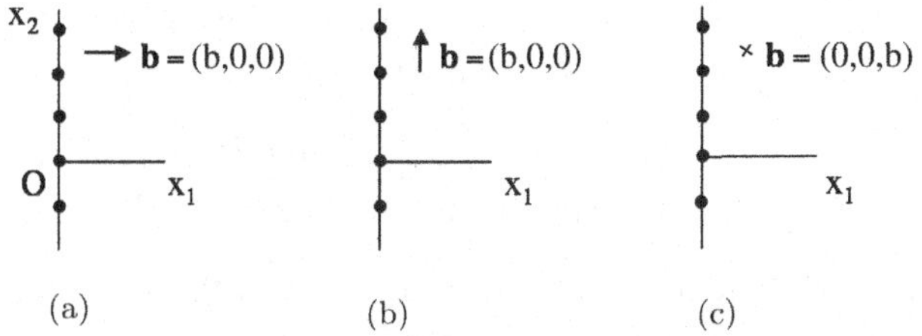

Fig. 14.10 End views of arrays of straight parallel dislocations with Burgers vectors $\mathbf{b} = (b, 0, 0)$, $\mathbf{b} = (0, b, 0)$ and $\mathbf{b} = (0, 0, b)$.

14.3.4 *Elastic fields of arrays of parallel dislocations in isotropic systems*

It is helpful to have the elastic fields for the three arrays in Fig. 14.10 with the three Burgers vectors shown, since, with their use, the field for an array with any Burgers vector can be obtained by simple superposition. To demonstrate the procedure for finding these fields in isotropic systems (see Hirth and Lothe 1982), the σ_{12} field due to the array of edge dislocations in Fig. 14.10a is determined. The σ_{12} field due to a single dislocation in this array is given by Eq. (12.45), and, therefore, summing over the array,

$$\sigma_{12}^{\text{array}^\infty} = \frac{\mu b}{2\pi(1-\nu)} \sum_{N=-\infty}^{\infty} \frac{x_1[x_1^2 - (x_2 - Nd)^2]}{[x_1^2 + (x_2 - Nd)^2]^2}. \tag{14.39}$$

The sum in Eq. (14.39) can be evaluated by starting with the identity (Morse and Feshbach 1953)

$$\sum_{N=-\infty}^{\infty} \frac{1}{N+a} = \pi \cot \pi a. \tag{14.40}$$

Then, by substituting the complex numbers $a = p + iq$ and then $p = a - iq$ into Eq. (14.40), adding the results, and using the standard relationships

$$\begin{aligned} \sin(x + iy) &= \sin(x)\cosh(y) + i[\cos(x)\sinh(y)] \\ \cos(x + iy) &= \cos(x)\cosh(y) - i[\sin(x)\sinh(y)] \end{aligned} \tag{14.41}$$

the further identity

$$\begin{aligned} \sum_{N=-\infty}^{\infty} \frac{N+p}{q^2 + (N+p)^2} &= \frac{\pi}{2}[\cot \pi(p + iq) + \cot \pi(p - iq)] \\ &= \frac{\pi \sin(2\pi p)}{\cosh(2\pi q) - \cos(2\pi p)} \end{aligned} \tag{14.42}$$

is obtained. (It is noted here that, if the results obtained with Eq. (14.40) above are subtracted instead of added, the equality

$$\sum_{N=-\infty}^{\infty} \frac{1}{q^2 + (N+p)^2} = \frac{\pi}{q} \frac{\sinh(2\pi q)}{\cosh(2\pi q) - \cos(2\pi p)} \qquad (14.43)$$

is obtained which will be of later use in Sec. 14.4.2.2.). Differentiation of Eq. (14.42) with respect to p then produces

$$\sum_{N=-\infty}^{\infty} \frac{q^2 - (N+p)^2}{[q^2 + (N+p)^2]^2} = 2\pi^2 \frac{\cosh(2\pi q)\cos(2\pi p) - 1}{[\cosh(2\pi q) - \cos(2\pi p)]^2}. \qquad (14.44)$$

Having this result, the sum in Eq. (14.39) can be evaluated, and $\sigma_{12}^{\mathrm{array}^\infty}$ is finally obtained in the form

$$\sigma_{12}^{\mathrm{array}^\infty} = \frac{\mu b \pi x_1 [\cosh(2\pi x_1/d)\cos(2\pi x_2/d) - 1]}{d^2(1-\nu)[\cosh(2\pi x_1/d) - \cos(2\pi x_2/d)]^2}. \qquad (14.45)$$

The remaining stresses and also all stresses for the other two families in Fig. 14.10 are obtained in a similar manner and are presented and discussed by Hirth and Lothe (1982) and Sutton and Balluffi (2006).

When x_1 is greater than about d, the $\sigma_{12}^{\mathrm{array}^\infty}$ stress given by Eq. (14.45) is well represented by

$$\sigma_{12}^{\mathrm{array}^\infty} = \frac{\mu b 2\pi x_1}{d^2(1-\nu)} \exp\left(-\frac{2\pi x_1}{d}\right) \cos\frac{2\pi x_2}{d} \qquad (14.46)$$

and is seen to be relatively small, as expected, and to essentially vanish at distances that are a few multiples of d.[8] The absence of a long-range stress field for this family is as expected, since it corresponds to the simple symmetric tilt interface in Fig. 14.1c which obeys the Frank–Bilby equation.

In contrast to the above results for the family in Fig. 14.10a, the families in Figs. 14.10b and 14.10c possess non-vanishing long-range elastic fields. As demonstrated in Exercise 14.2, this is consistent with the fact that the Frank–Bilby equation cannot be satisfied for these two arrays.

When more than one family of dislocations is present in an interface and intersect, they may be able to reduce the total strain energy by interacting to form a cellular network consisting of straight segments as illustrated, for example, in Fig. 14.11. Here, a new network (dashed) is formed as a

[8]Relative to, for example, the σ_{12} stress at the same distance from a single edge dislocation as given by Eq. (12.45).

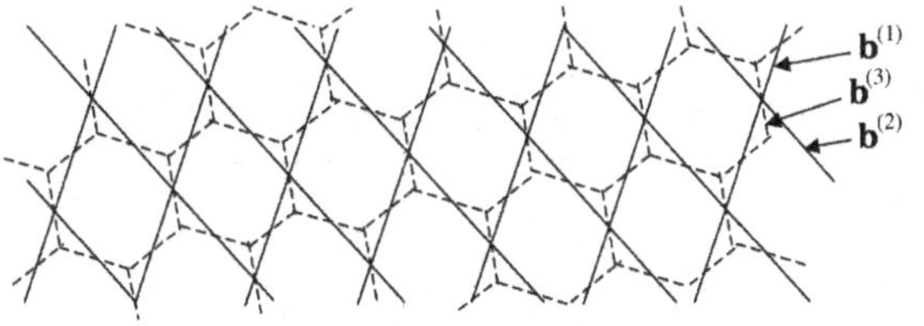

Fig. 14.11 Two arrays of intersecting straight parallel dislocations (solid lines) interacting to form cellular dislocation network (dashed lines).

result of local dislocation interactions at the intersections of the two original arrays where lengths of dislocations with Burgers vectors $\mathbf{b}_1$ and $\mathbf{b}_2$, respectively, form segments of dislocations with Burgers vector $\mathbf{b}_3$ according to the reaction

$$\mathbf{b}^{(1)} + \mathbf{b}^{(2)} = \mathbf{b}^{(3)}. \tag{14.47}$$

The total Burgers vector strength of the interface is conserved, as far as the Frank–Bilby equation is concerned, and the degrees of freedom of the interface remain unchanged. However, the determination of the elastic field becomes more complicated requiring summations of the elastic fields contributed by individual segments by methods described in Sec. 12.7.

14.3.5 *Interfacial strain energies in isotropic systems*

The energies of small-angle interfaces can be determined by a conceptually simple method (Hirth and Lothe 1982) in which two interfaces of the desired type, but with opposite misorientions, are generated along the intended interface plane and then moved apart against the attractive force that tends to pull them back together. The process is illustrated in Fig. 14.12 for the simple case of a symmetrical tilt interface. Here, two interfaces with opposite tilt angles are created by generating an array consisting of pairs of long straight edge dislocations having opposite Burgers vectors and then moving them infinitely far apart on their slip planes. The energy of each interface is then half of the total work done in separating the two arrays against the attractive force between them. This force, in an isotropic system, can be determined by noting that the $\sigma_{12}^{\text{array}^\infty}$ stress on each slip plane can be obtained from Eq. (14.45) in the form

$$\sigma_{12}^{\text{array}^\infty} = \frac{\mu b \pi x_1}{2d^2(1-\nu)\sinh^2(\pi x_1/d)}. \tag{14.48}$$

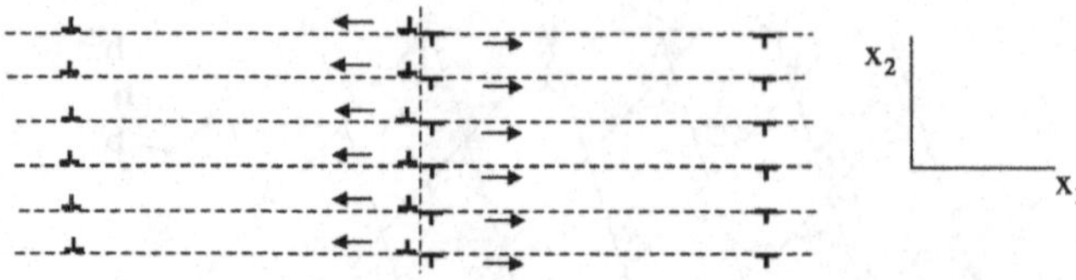

Fig. 14.12 Creation of two small-angle symmetric tilt interfaces with opposite tilt angles.

Then, using the Peach–Koehler force equation, Eq. (13.11), the restraining force per unit length experienced by each dislocation is $f = \mathrm{b}\sigma_{12}^{\mathrm{array}^{\infty}}$, and, therefore, the corresponding work (per unit area) to create the two interfaces is

$$\mathcal{W}' = \frac{1}{\mathrm{d}}\int_{\mathrm{b}/\alpha}^{\infty} \mathrm{b}\sigma_{12}^{\mathrm{array}^{\infty}}\mathrm{dx}_1 = \frac{\mu \mathrm{b}^2\pi}{2\mathrm{d}^3(1-\nu)}\int_{\mathrm{b}/\alpha}^{\infty}\frac{\mathrm{x}_1\mathrm{dx}_1}{\sinh^2(\pi \mathrm{x}_1/\mathrm{d})}, \tag{14.49}$$

where Eq. (12.42) has been used to obtain the core cutoff radius. Then, performing the integration,

$$\mathcal{W}' = \frac{\mu \mathrm{b}^2}{2\pi \mathrm{d}(1-\nu)}\left\{\frac{\pi \mathrm{b}}{\alpha \mathrm{d}}\coth\left(\frac{\pi \mathrm{b}}{\alpha \mathrm{d}}\right) - \ln\left[2\sinh\left(\frac{\pi \mathrm{b}}{\alpha \mathrm{d}}\right)\right]\right\} \tag{14.50}$$

since $\mathrm{b}/\mathrm{d} = \theta$, and $\theta \ll 1$, $\pi\mathrm{b}/\alpha\mathrm{d} = \pi\theta/\alpha \ll 1$. Equation (14.50) can then be expanded to first order to obtain the energy (per unit area) of a single interface in the form

$$\gamma = \frac{\mathcal{W}'}{2} = \frac{\mu \mathrm{b}^2}{4\pi(1-\nu)\mathrm{d}}\ln\left(\frac{\alpha \mathrm{d} e}{2\pi \mathrm{b}}\right). \tag{14.51}$$

This expression for the energy may be compared with an approximate, but simple, determination of the interface energy based on the previous results. The previous solution for the elastic near-field indicates that the field within a cylinder of radius $\approx\mathrm{d}/2$ around each dislocation is dominated by the field due to that dislocation. Therefore, using Eq. (12.61) for the energy associated with the edge dislocation in each cylinder, the interface energy is

$$\gamma \approx \frac{\mu \mathrm{b}^2}{4\pi(1-\nu)\mathrm{d}}\ln\left(\frac{\alpha \mathrm{d}}{2\mathrm{b}}\right), \tag{14.52}$$

which is essentially identical to Eq. (14.51). By substituting $\mathrm{b}/\mathrm{d} = \theta$, the interface energy given by Eq. (14.51) can be expressed, as first shown by

Read and Shockley (1950), in the relatively simple functional form

$$\gamma = \gamma_\circ\theta(\mathrm{A} - \ln\theta), \tag{14.53}$$

where $\gamma_\circ = \mu b/[4\pi(1-\nu)]$ and $\mathrm{A} = \ln(\alpha e/2\pi)$.

The energies of other types of small-angle interfaces can also be determined by the above method. In most cases the two interfaces created will not be glissile, and dislocation climb will be required to move them apart while not altering their structure as discussed in Sec. 15.3. However, this does not affect the energy calculation. Arguments given by Read (1953), and extended by Hirth and Lothe (1982), indicate that the general functional form of Eq. (14.53) is quite general and should be valid for all small-angle interfaces.

14.4 Hetero-Elastic Interfaces

14.4.1 *Geometrical features*

As is the case for the previous iso-elastic interfaces, a variety of distinguishable types of hetero-elastic interfaces containing arrays of discrete dislocations exist (Sutton and Balluffi 2006). A common type is the *epitaxial interface* which occurs when two dissimilar phases possess atomic planes that are almost commensurate. A stable interface can then be produced where these planes are parallel to the interface, and the small mismatch between them is accommodated by a dislocation array. A simple example is the interface in Fig. 14.13 where it is assumed that phases 1 and 2 are exactly commensurate in the x_3 direction and nearly commensurate along x_1. The dislocation spacing in this interface, consistent with a vanishing elastic far-field, is readily found by means of the Frank–Bilby equation. Choosing the phase 1 lattice as the Reference lattice, and expressing the

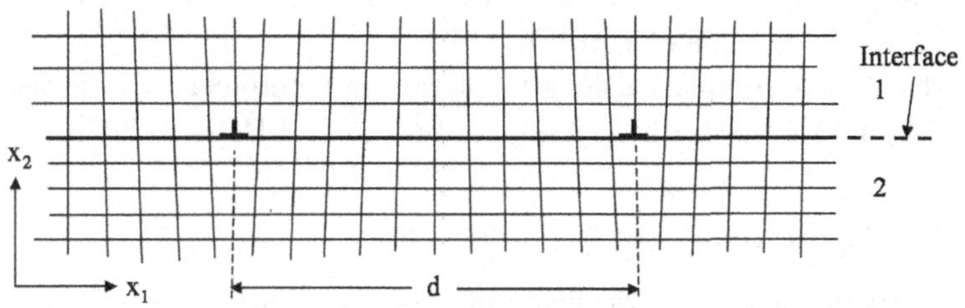

Fig. 14.13 Cross-section of edge dislocation array in epitaxial hetero-elastic interface between dissimilar phases occupying half-spaces 1 and 2.

base vectors of the relevant lattices as

$$\begin{aligned} \mathbf{a}_1^{(2)} &= (1+\varepsilon)\mathbf{a}_1^{\mathrm{R}}, \quad \mathbf{a}_1^{(1)} = \mathbf{a}_1^{\mathrm{R}}, \\ \mathbf{a}_2^{(2)} &= (\mathrm{a}_2^{(2)}/\mathrm{a}_2^{(1)})\mathbf{a}_2^{\mathrm{R}}, \quad \mathbf{a}_2^{(1)} = \mathbf{a}_2^{\mathrm{R}}, \\ \mathbf{a}_3^{(2)} &= \mathbf{a}_3^{\mathrm{R}}, \quad \mathbf{a}_3^{(1)} = \mathbf{a}_3^{\mathrm{R}}, \end{aligned} \tag{14.54}$$

the D_{ij} tensors of Eq. (14.13) are

$$\mathrm{D}_{ji}^{(2)} = \begin{bmatrix} 1+\varepsilon & 0 & 0 \\ 0 & \mathrm{a}_2^{(2)}/\mathrm{a}_2^{(1)} & 0 \\ 0 & 0 & 1 \end{bmatrix}, \quad \mathrm{D}_{ji}^{(1)} = \delta_{ji}. \tag{14.55}$$

Then, by first taking $\mathbf{p}$ parallel to axis 3, and then parallel to axis 1, Eq. (14.24) yields, respectively,

$$\begin{bmatrix} \mathrm{b}_1^{\mathrm{tot}} \\ \mathrm{b}_2^{\mathrm{tot}} \\ \mathrm{b}_3^{\mathrm{tot}} \end{bmatrix} = \begin{bmatrix} \varepsilon & 0 & 0 \\ 0 & [\mathrm{a}_2^{(2)}/\mathrm{a}_2^{(1)}]-1 & 0 \\ 0 & 0 & 0 \end{bmatrix} \begin{bmatrix} 0 \\ 0 \\ \mathrm{p} \end{bmatrix} = \begin{bmatrix} 0 \\ 0 \\ 0 \end{bmatrix},$$

$$\begin{bmatrix} \mathrm{b}_1^{\mathrm{tot}} \\ \mathrm{b}_2^{\mathrm{tot}} \\ \mathrm{b}_3^{\mathrm{tot}} \end{bmatrix} = \begin{bmatrix} \varepsilon & 0 & 0 \\ 0 & [\mathrm{a}_2^{(2)}/\mathrm{a}_2^{(1)}]-1 & 0 \\ 0 & 0 & 0 \end{bmatrix} \begin{bmatrix} \mathrm{p} \\ 0 \\ 0 \end{bmatrix} = \begin{bmatrix} \varepsilon\mathrm{p} \\ 0 \\ 0 \end{bmatrix}. \tag{14.56}$$

The first relationship is satisfied by a family of dislocations running parallel to axis 3. The second, which requires that $\mathrm{b}_1^{\mathrm{tot}} = \varepsilon\mathrm{p}$, can be satisfied by assigning the dislocations a Burgers vector $\mathbf{b} = (\mathrm{a}_1^{(1)}, 0, 0)$ and a spacing

$$\mathrm{d} = \frac{\mathrm{a}_1^{(1)}}{\varepsilon} \tag{14.57}$$

since, if it is assumed that when $\mathbf{p}$ is parallel to axis 1 it intersects N dislocations at the spacing *d*, the above requirement becomes $\mathrm{b}_1^{\mathrm{tot}} = \varepsilon\mathrm{p} = \varepsilon\mathrm{Nd} = \mathrm{Na}_1^{(1)}$ in agreement with Eq. (14.57).

Further more complex dislocation arrays in hetero-elastic interfaces are treated by Hirth and Lothe (1982) and Sutton and Balluffi (2006).

14.4.2 *Elastic fields*

The treatment of the elastic fields of dislocation arrays in hetero-elastic interfaces is considerably more complicated than in iso-elastic interfaces because of the difference in the effective elastic constants that now exists across the interface. It is therefore helpful to consider first this problem in the case of an isotropic system before taking up the more difficult anisotropic case.

14.4.2.1 *Single dislocation in planar interface in isotropic system*

Let us first find the elastic field of a single straight dislocation lying in a planar hetero-elastic interface in an isotropic system following the analysis of Nakahara, Wu and Li (1972). This treatment makes use of Eq. (12.65) which expresses the relationship between the displacement $u_i(\mathbf{x})$ at a field point $\mathbf{x}$ due to a dislocation loop and the stress $\sigma'_{mn}(\mathbf{x}-\mathbf{x}')$ at a point $\mathbf{x}'$ on the cut surface Σ used to create the loop caused by a point force F_i applied at $\mathbf{x}$ (see Fig. 12.1). The results are lengthy and cumbersome, and the analysis will therefore be restricted to the case of the edge dislocation with Burgers vector $\mathbf{b} = (0, b, 0)$ and $\hat{\mathbf{t}} = (0, 0, 1)$ in the interface shown in Fig. 14.14. This will lay the groundwork for going on to the analysis of a simple tilt boundary consisting of an array of parallel edge dislocations.

Start with a planar dislocation loop lying in the interface of Fig. 14.14 with $\mathbf{b} = (0, b, 0)$ produced by a cut along a planar Σ surface in the $x_2 = 0$ plane with positive unit normal vector $\hat{\mathbf{n}} = (0, -1, 0)$. For such a loop, Eq. (12.65) reduces to

$$\begin{aligned} F_i u_i(\mathbf{x}) &= b \iint_{\Sigma} \sigma'_{22}(\mathbf{x}-\mathbf{x}')dx'_1 dx'_3 \\ &= b \iint_{\Sigma} [\sigma'^{F_1}_{22}(\mathbf{x}-\mathbf{x}') + \sigma'^{F_2}_{22}(\mathbf{x}-\mathbf{x}') + \sigma'^{F_3}_{22}(\mathbf{x}-\mathbf{x}')]dx'_1 dx'_3, \end{aligned} \tag{14.58}$$

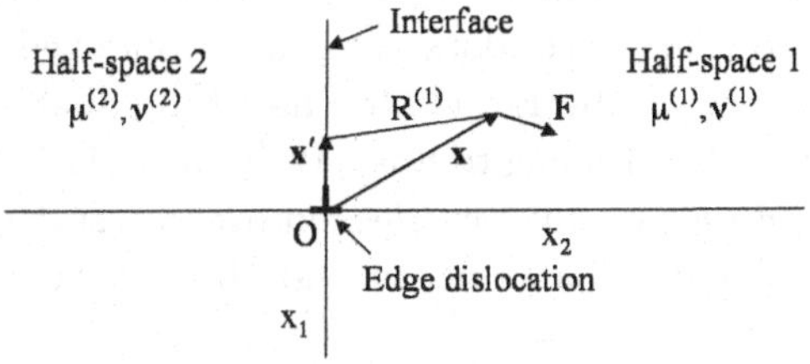

Fig. 14.14 Geometry for determining displacement field of single edge dislocation in hetero-elastic interface in isotropic system.

where $\sigma_{22}^{\prime F_i}(\mathbf{x}-\mathbf{x}')$ is the stress produced at a source point $\mathbf{x}'$ in the plane of the loop by component i of the force, $\mathbf{F}$, applied at the field point $\mathbf{x}$. The $\sigma_{22}^{\prime F_i}$ stresses in half-space 1 can be obtained by substituting the Papkovitch functions given by Eqs. (4.101) and (4.102) into Eq. (4.86) to obtain the displacements and then using Eq. (2.5) and Hooke's law. This yields (Nakahara, Wu and Li 1972)

$$\begin{aligned}
\sigma_{22}^{\prime F_1} &= F_1 D\frac{x_1 - x_1'}{(R^{(1)})^3}\left(3\frac{x_2^2}{(R^{(1)})^2} - \frac{3 - A_{12}}{2} + 2\nu^{(1)}\right), \\
\sigma_{22}^{\prime F_2} &= F_2 D\frac{x_2}{(R^{(1)})^3}\left(3\frac{x_2^2}{(R^{(1)})^2} + \frac{1 + A_{12}}{2} - 2\nu^{(1)}\right), \\
\sigma_{22}^{\prime F_3} &= F_3 D\frac{x_3 - x_3'}{(R^{(1)})^3}\left(3\frac{x_2^2}{(R^{(1)})^2} - \frac{3 - A_{12}}{2} + 2\nu^{(1)}\right),
\end{aligned} \tag{14.59}$$

where

$$\begin{aligned}
A_{12} &= \frac{\mu^{(1)}K_{12}}{\mu^{(2)}K_{21}} = \frac{\mu^{(1)} + (3 - 4\nu^{(1)})\mu^{(2)}}{\mu^{(2)} + (3 - 4\nu^{(2)})\mu^{(1)}}, \\
D &= \frac{\mu^{(2)}K_{21}}{2\pi\mu^{(1)}}, \\
K_{12} &= \frac{\mu^{(2)}}{[\mu^{(2)} + (3 - 4\nu^{(2)})\mu^{(1)}]}, \quad K_{21} = \frac{\mu^{(1)}}{[\mu^{(1)} + (3 - 4\nu^{(1)})\mu^{(2)}]}, \\
R^{(1)} &= [(x_1 - x_1')^2 + x_2^2 + (x_3 - x_3')^2]^{1/2}.
\end{aligned} \tag{14.60}$$

Having this, the displacements in half-space 1 in Fig. 14.14 due to the long straight edge dislocation with Burgers vector $\mathbf{b} = (0, b, 0)$ and $\hat{\mathbf{t}} = (0, 0, 1)$ lying in the interface along x_3 can be obtained by substituting each of the stresses given by Eq. (14.59) into Eq. (14.58) and integrating according to the method leading to Eqs. (12.70) and (12.71) where the loop is expanded so that the integral is carried out over a Σ surface corresponding to an entire half-plane bounded on one side by the straight dislocation.[9]

[9] Also, see Exercise 12.2.

This yields

$$\begin{aligned}
u_1^{(1)} &= \frac{b(1 - K_{12} - K_{21})}{4\pi} \ln(x_1^2 + x_2^2) - \frac{\mu^{(2)} K_{21} b}{2\pi\mu^{(1)}} \frac{(x_1^2 - x_2^2)}{(x_1^2 + x_2^2)}, \\
u_2^{(1)} &= \frac{b(1 + K_{12} - K_{21})}{2\pi} \tan^{-1}\left(\frac{x_2}{x_1}\right) - \frac{\mu^{(2)} K_{21} b}{\pi\mu^{(1)}} \frac{x_1 x_2}{(x_1^2 + x_2^2)}, \\
u_3^{(1)} &= 0.
\end{aligned} \tag{14.61}$$

Then, applying Eq. (2.5) and Hooke's law, the corresponding stresses are

$$\begin{aligned}
\sigma_{11}^{(1)} &= \frac{b x_1}{\pi(x_1^2 + x_2^2)} \left(-\mu^{(1)} K_{12} + \mu^{(2)} K_{21} \frac{3x_1^2 - x_2^2}{x_1^2 + x_2^2} \right), \\
\sigma_{33}^{(1)} &= \frac{b x_1}{\pi(x_1^2 + x_2^2)} 4\nu^{(1)} \mu^{(2)} K_{21}, \\
\sigma_{22}^{(1)} &= \frac{b x_1}{\pi(x_1^2 + x_2^2)} \left(\mu^{(1)} K_{12} + \mu^{(2)} K_{21} \frac{x_1^2 + 5x_2^2}{x_1^2 + x_2^2} \right), \\
\sigma_{12}^{(1)} &= \frac{x_2}{x_1} \sigma_{11}^{(1)}.
\end{aligned} \tag{14.62}$$

Corresponding results for half-space 2 can be obtained by simply interchanging indices.

14.4.2.2 *Planar array of parallel dislocations in isotropic system*

Consider now a simple dislocation array in an isotropic system consisting of uniformly spaced parallel edge dislocations of the type analyzed previously in the interface of Fig. 14.14. This array corresponds to a small-angle tilt interface whose elastic field is expected to produce far-field rotations in the adjoining crystals that conform to the Frank–Bilby equation and far-field stresses that vanish in each half-space. We now determine the elastic field and verify this expectation. However, the analysis will prove to be more complicated than for the previous case of the iso-elastic interface in Sec. 14.3.3, since the anticipated result cannot be obtained by simply summing the fields of the individual dislocations.

The procedure is initially similar to that employed previously in Sec. 14.3.3 for an iso-elastic interface. The only relevant distortions in half-space 1 due to a single dislocation in the array are $\partial u_1^{(1)}/\partial x_2$ and $\partial u_2^{(1)}/\partial x_1$, which, by use of Eq. (14.61), are given by

$$\frac{\partial u_1^{(1)}}{\partial x_2} = \frac{b}{2\pi}\frac{x_2^2}{(x_1^2+x_2^2)}\left[(1-K_{12}-K_{21}) + \frac{4\mu^{(2)}K_{21}}{\mu^{(1)}}\frac{x_1^2}{(x_1^2+x_2^2)}\right],$$
$$\frac{\partial u_2^{(1)}}{\partial x_1} = \frac{b}{2\pi}\frac{x_2^2}{(x_1^2+x_2^2)}\left[-(1+K_{12}-K_{21}) + \frac{2\mu^{(2)}K_{21}}{\mu^{(1)}}\frac{(x_1^2-x_2^2)}{(x_1^2+x_2^2)}\right]. \tag{14.63}$$

Next, by substituting these results into Eq. (2.14), and summing over all the dislocations, the only non-vanishing rotation in half-space 1 due to the array is

$$\delta\omega_3^{\mathrm{array}^{(1)}} = -\frac{b}{2\pi}\left[1+K_{21}\left(\frac{\mu^{(2)}}{\mu^{(1)}}-1\right)\right]x_2\sum_{N=-\infty}^{\infty}\frac{1}{(x_1+Nd)^2+x_2^2}. \tag{14.64}$$

Then, using Eq. (14.43), the rotation of the far-field region of half-space 1, where $x_2 \gg d$, is

$$\delta\omega_3^{\mathrm{array}^{(1)}}(\text{far-field}) = -\frac{b}{2d}\left[1+K_{21}\left(\frac{\mu^{(2)}}{\mu^{(1)}}-1\right)\right]. \tag{14.65}$$

This corresponds to a left-handed rotation of the far-field region of half-space 1 around the x_3 axis equal to $\delta\theta^{(1)} = \delta\omega_3^{\mathrm{array}^{(1)}}$ (far-field). The rotation of half-space 2 can be obtained by a similar procedure, and by combining the results, the far-field rotation of half-space 1 relative to that of half-space 2 is

$$\delta\theta = \delta\theta^{(1)} - \delta\theta^{(2)} = -\frac{b}{d} - \frac{b}{2d}\left[\left(\frac{\mu^{(2)}K_{21}}{\mu^{(1)}} - K_{21}\right) + \left(\frac{\mu^{(1)}K_{12}}{\mu^{(2)}} - K_{12}\right)\right]. \tag{14.66}$$

This result is seen to be inconsistent with the Frank–Bilby equation, since then the equation would predict $\delta\theta = -b/d$. Note, however, that if $\mu^{(1)} = \mu^{(2)}$, as it would be if the interface were iso-elastic, Eq. (14.66) reduces to $\delta\theta = -b/d$, and the Frank–Bilby result is obtained.

Consider next the stress field of the array in half-space 1, $\sigma_{12}^{\mathrm{array}^{(1)}}(x_1, x_2)$, obtained by summing the contributions of the individual dislocations. Using Eq. (14.62), and taking the origin at one of the dislocations,

$$\sigma_{12}^{\mathrm{array}^{(1)}}(x_1,x_2) = \frac{bx_2}{\pi}\sum_{N=-\infty}^{\infty}\frac{1}{(x_1+Nd)^2+x_2^2}$$

$$\times\left(-\mu^{(1)}K_{12}+\mu^{(2)}K_{21}\frac{3(x_1+Nd)^2-x_2^2}{(x_1+Nd)^2+x_2^2}\right). \quad (14.67)$$

To obtain the far-field at large x_2, x_1 can be set equal to zero, and the sums in the above expression replaced by integrals after smearing out the array of discrete dislocations along x_1 into a uniform distribution of an infinite number of dislocations with infinitesimal Burgers vectors with a Burgers vector line density b/d. The expression obtained can then be integrated to obtain the far-field stress

$$\sigma_{12}^{\mathrm{array}^{(1)}}(\mathrm{far-field}) = \frac{bx_2}{\pi d}\left[-\mu^{(1)}K_{12}\int_{-\infty}^{\infty}\frac{ds}{s^2+x_2^2}\right.$$

$$\left.+\,\mu^{(2)}K_{21}\int_{-\infty}^{\infty}\frac{(3s^2-x_2^2)}{(s^2+x_2^2)^2}ds\right]$$

$$= -\frac{b}{d}(\mu^{(1)}K_{12}-\mu^{(2)}K_{21}). \quad (14.68)$$

Similar calculations show that the other far-field stress components vanish. Nevertheless, Eq. (14.68) establishes the existence of a non-vanishing far-field stress, which is not surprising, since the far-field rotation field does not conform to the Frank–Bilby result. The $\sigma_{12}^{\mathrm{array}^{(1)}}$ stress given by Eq. (14.68) has the notable feature that it is independent of x_2, thereby producing the same shearing traction on all surfaces parallel to the interface. Also, it vanishes for an iso-elastic interface where $\mu^{(1)} = \mu^{(2)}$.

For a bicrystal with surfaces parallel to the interface, the far-field $\sigma_{12}^{\mathrm{array}^{(1)}}$ stress given by Eq. (14.68) can now be eliminated in half-space 1 by applying a canceling uniform image stress, $\sigma_{12}^{\mathrm{IM}^{(1)}} = -\sigma_{12}^{\mathrm{array}^{(1)}}$, thus producing a uniform shear stress and strain

$$\sigma_{12}^{\mathrm{IM}^{(1)}}(\mathrm{far-field}) = \frac{b}{d}(\mu^{(1)}K_{12}-\mu^{(2)}K_{21})$$

$$\varepsilon_{12}^{\mathrm{IM}^{(1)}}(\mathrm{far-field}) = \frac{\sigma_{12}^{\mathrm{IM}^{(1)}}(\mathrm{far-field})}{2\mu^{(1)}} = \frac{b}{2d}\left(K_{12}-\frac{\mu^{(2)}K_{21}}{\mu^{(1)}}\right) \quad (14.69)$$

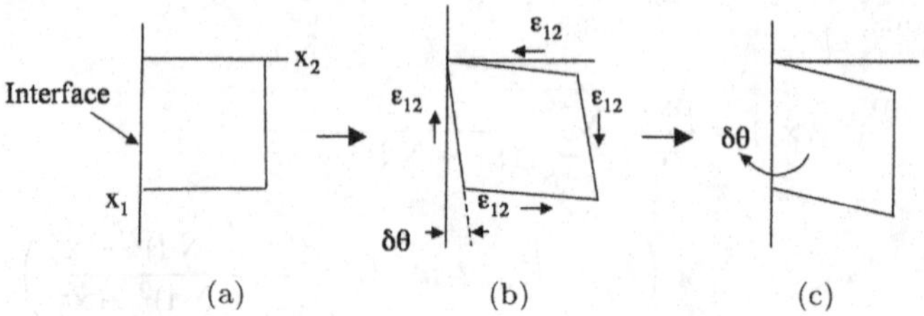

Fig. 14.15 Rotation, $\delta\theta$, of half-space caused by ε_{12} shear strain.

without producing any incompatibility at the interface as may be seen in the two-step process illustrated in Fig. 14.15. In the first step, (a) → (b), a uniform ε_{12} shear strain is applied causing a shape change described by the two-dimensional displacement field $\mathbf{u}(\mathbf{x}) = \varepsilon_{12}(x_2\hat{\mathbf{e}}_1 + x_1\hat{\mathbf{e}}_2)$ which, according to Eqs. (2.13) and (2.14), does not produce any rigid body rotation. In the second step, (b) → (c), compatibility is restored by means of the rigid body rotation $\delta\theta = -\varepsilon_{12}$ around the x_3 axis. Using Eq. (14.69), and the results of a similar analysis of the effect of applying a corresponding image stress to half-space 2, the final change in tilt angle between half-spaces 1 and 2 produced by the application of the image stresses to the two half-spaces is

$$\delta\theta^{IM} = \delta\theta^{IM^{(1)}} - \delta\theta^{IM^{(2)}} = -\varepsilon_{12}^{IM^{(1)}} + \varepsilon_{12}^{IM^{(2)}}$$
$$= \frac{b}{2d}\left[\left(\frac{\mu^{(2)}K_{21}}{\mu^{(1)}} - K_{12}\right) + \left(\frac{\mu^{(1)}K_{12}}{\mu^{(2)}} - K_{21}\right)\right]. \quad (14.70)$$

The rotation given by Eq.(14.70) is seen to cancel exactly the portion of the rotation in Eq. (14.66) that deviates from the prediction of the Frank–Bilby equation. The final result is therefore a bicrystal free of far-field stress and in conformity with the Frank–Bilby equation as originally expected.

Insight into the source of the far-field $\sigma_{12}^{array^{(1)}}$ stress produced above by the array of edge dislocations, given by Eq. (14.68), is gained by first realizing that the elastic field in half-space 1 due to each edge dislocation in the interface, given by Eq. (14.61), is equal to the sum of the fields produced in an infinite body, with the same elastic properties as half-space 1, by a fictitious edge dislocation with a Burgers vector, $b^{fict^{(1)}}$, and a fictitious line force of strength, $f_1^{fict^{(1)}}$, given by

$$b^{fict^{(1)}} = b(1 + K_{12} - K_{21}), \quad f_1^{fict^{(1)}} = -2b(\mu^{(2)}K_{21} - \mu^{(1)}K_{12}), \quad (14.71)$$

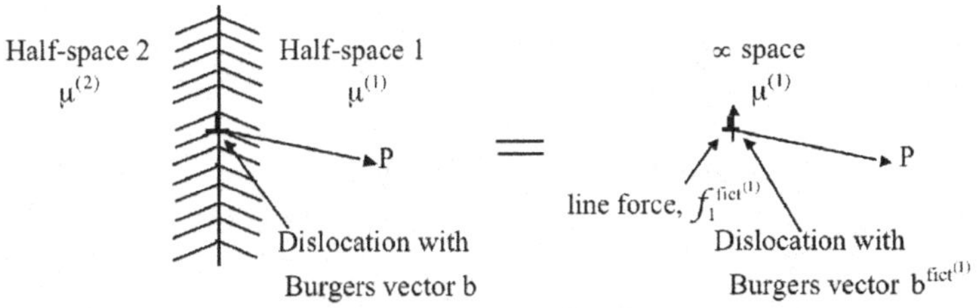

Fig. 14.16 Equivalence between the elastic field in the left diagram produced at a field point P in half-space 1 by an interfacial dislocation with Burgers vector b, and the field produced in the right diagram at a field point P (at the same vector position as on the left) in an infinite space having the same elastic properties as half-space 1, by a superimposed dislocation with Burgers vector $\mathrm{b}^{\mathrm{fict}^{(1)}}$ and a line force of strength $f_1^{\mathrm{fict}^{(1)}}$.

where b is the Burgers vector of the real dislocations in the interface (Dundurs and Sendeckyj 1965; Sutton and Balluffi 2006). This result, which is illustrated in Fig. 14.16, also holds for half-space 2, when the Burgers vector and line force strength of the fictitious dislocation and line force are, respectively, $\mathrm{b}^{\mathrm{fict}^{(1)}} = \mathrm{b}(1 + \mathrm{K}_{21} - \mathrm{K}_{12})$ and $f_1^{\mathrm{fict}^{(2)}} = -f_1^{\mathrm{fict}^{(1)}}$. In this arrangement the directions of the two line forces are therefore opposed, and the net line force acting on the system vanishes, as must be the case.

This equivalence can be validated by first expressing it in terms of the condition

$$\mathrm{u}_\mathrm{i}^{(1)} = \mathrm{u}_\mathrm{i}^{\mathrm{fict}^\infty} + \mathrm{u}_\mathrm{i}^{\mathrm{LF}^\infty}, \tag{14.72}$$

where $\mathrm{u}_\mathrm{i}^{\mathrm{fict}^\infty}$ and $\mathrm{u}_\mathrm{i}^{\mathrm{LF}^\infty}$ are the displacements, respectively, due to a single fictitious dislocation and a line force in the infinite body. According to Eqs. (14.61), (12.54) and (14.125), and use of Eq. (14.71), these displacements are given by

$$\begin{aligned}
\mathrm{u}_1^{(1)} &= \frac{\mathrm{b}}{2\pi}\left[\frac{(1-\mathrm{K}_{12}-\mathrm{K}_{21})}{2}\ln(\mathrm{x}_1^2+\mathrm{x}_2^2) - \frac{\mu^{(2)}\mathrm{K}_{21}}{\mu^{(1)}}\frac{(\mathrm{x}_1^2-\mathrm{x}_2^2)}{(\mathrm{x}_1^2+\mathrm{x}_2^2)}\right],\\
\mathrm{u}_1^{\mathrm{fict}^\infty} &= \frac{\mathrm{b}(1+\mathrm{K}_{12}-\mathrm{K}_{21})}{8\pi(1-\nu^{(1)})}\left[(1-2\nu^{(1)})\ln(\mathrm{x}_1^2+\mathrm{x}_2^2) - \frac{(\mathrm{x}_1^2-\mathrm{x}_2^2)}{(\mathrm{x}_1^2+\mathrm{x}_2^2)}\right],\\
\mathrm{u}_1^{\mathrm{LF}^\infty} &= \frac{\mathrm{b}(\mu^{(2)}\mathrm{K}_{21}-\mu^{(1)}\mathrm{K}_{12})}{8\pi\mu^{(1)}(1-\nu^{(1)})}\\
&\quad\times\left[(3-4\nu^{(1)})\ln(\mathrm{x}_1^2+\mathrm{x}_2^2) - \frac{2\mathrm{x}_1^2}{(\mathrm{x}_1^2+\mathrm{x}_2^2)}\right],
\end{aligned} \tag{14.73}$$

$$u_2^{(1)} = \frac{b}{2\pi}\left[(1+K_{12}-K_{21})\tan^{-1}\left(\frac{x_2}{x_1}\right) - \frac{2\mu^{(2)}K_{21}}{\mu^{(1)}}\frac{x_1x_2}{(x_1^2+x_2^2)}\right],$$

$$u_2^{\text{fict}\infty} = \frac{b(1+K_{12}-K_{21})}{2\pi}\left[\tan^{-1}\left(\frac{x_2}{x_1}\right) - \frac{1}{2(1-\nu^{(1)})}\frac{x_1x_2}{(x_1^2+x_2^2)}\right],$$

$$u_2^{\text{LF}\infty} = -\frac{b(\mu^{(2)}K_{21}-\mu^{(1)}K_{12})}{4\pi\mu^{(1)}(1-\nu^{(1)})}\left[\frac{x_1x_2}{(x_1^2+x_2^2)}\right].$$

Substitution of Eq. (14.73) into Eq. (14.72), and use of Eq. (14.60), then shows that the condition given by Eq. (14.72) is indeed satisfied.

It is now shown that the far-field stress in half-space 1 obtained by summing the fields of the dislocations in the actual array [Eq. (14.68)] is equal to the far-field stress due to a corresponding array of the fictitious line forces in an infinite body with $\nu = \nu^{(1)}$. Using Eqs. (14.125), (2.5), Hooke's law, and Eq. (14.71) for the fictitious line force strength, the stress due to a single line force in the infinite body is

$$\sigma_{12}^{\text{LF}\infty}(x_1,x_2) = \frac{b(\mu^{(2)}K_{21}-\mu^{(1)}K_{12})}{2\pi(1-\nu)}\left[\frac{(1-2\nu)x_2}{x_1^2+x_2^2} + \frac{2x_2x_1^2}{(x_1^2+x_2^2)^2}\right]. \tag{14.74}$$

Then, assembling the array and smearing out the line forces into a uniform distribution of an infinite number of infinitesimal lines forces along x_1, the far-field stress due to the array is obtained by integration in the form

$$\begin{aligned}\sigma_{12}^{\text{LFarray}\infty}(\text{far-field}) &= \frac{b(\mu^{(2)}K_{21}-\mu^{(1)}K_{12})}{2\pi d(1-\nu)} \\ &\quad\times\int_{-\infty}^{\infty}\left[\frac{(1-2\nu)x_2}{s^2+x_2^2} + \frac{2x_2}{(s^2+x_2^2)^2}s^2\right]ds \\ &= -\frac{b}{d}(\mu^{(1)}K_{12}-\mu^{(2)}K_{21})\end{aligned} \tag{14.75}$$

in agreement with Eq. (14.68).

These results show that, in the fictitious dislocation/line force representation, the array of line forces is the source of the non-vanishing far-field stress obtained [Eq. (14.68)] when an effort is made to obtain the stress field of the real dislocation array by summing the stress fields of the individual dislocations. The final stress field, with vanishing far-field stresses in each half-space, can therefore be regarded as the sum of three fields, i.e., (1) the field of the fictitious array of dislocations, (2) the field of the fictitious array of line forces, and (3) the image stress whose role is to cancel the far-field tractions produced by the array of line forces.

14.4.2.3 *Single dislocation in planar interface*

Displacement field

With the above results for hetero-elastic interfaces in isotropic systems in hand, we now turn to hetero-elastic interfaces in general anisotropic systems. The first step is to determine the elastic field of a single dislocation in such an interface. This problem has been treated by Barnett and Lothe (1974) and Nakahara and Willis (1973), and we shall follow Barnett and Lothe by considering the arrangement in Fig. 14.17, where an infinitely long straight dislocation lies in a planar interface between two dissimilar half-spaces. Using the integral formalism, a solution for the displacement field in the half-spaces 1 and 2 is assumed of the form

$$\left.\begin{aligned} u_i^{(1)}(\mathbf{x}) &= \frac{1}{2\pi i}\sum_{\alpha=1}^{6} A_{i\alpha}^{(1)}E_\alpha^{(1)}\ln(\hat{\mathbf{m}}\cdot\mathbf{x}+p_\alpha^{(1)}\hat{\mathbf{n}}\cdot\mathbf{x}) \quad (\mathbf{x}\cdot\mathbf{n}>0) \\ u_i^{(2)}(\mathbf{x}) &= \frac{1}{2\pi i}\sum_{\alpha=1}^{6} A_{i\alpha}^{(2)}E_\alpha^{(2)}\ln(\hat{\mathbf{m}}\cdot\mathbf{x}+p_\alpha^{(2)}\hat{\mathbf{n}}\cdot\mathbf{x}) \quad (\mathbf{x}\cdot\mathbf{n}<0) \end{aligned}\right\} \tag{14.76}$$

where the twelve unknown quantities $E_\alpha^{(1)}$ and $E_\alpha^{(2)}$ are to be determined by the boundary conditions at the interface. A branch cut is introduced along

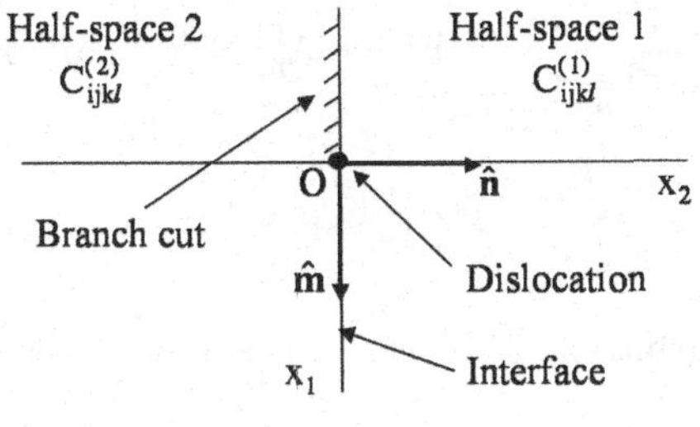

Fig. 14.17 End view of infinitely long straight dislocation lying along $\hat{\mathbf{t}} = \hat{\boldsymbol{\tau}} = \hat{\mathbf{m}} \times \hat{\mathbf{n}}$ in planar interface between dissimilar half-spaces 1 and 2.

$\hat{\mathbf{m}}\cdot\mathbf{x} < 0$ in the interface plane to make the complex logarithm in Eq. (14.76) single-valued [see Sec. 12.3.1 and Fig. (12.5)]. It is assumed that the surface for the cut and displacement that created the dislocation is the same surface used for the branch cut, and a discontinuity in the displacement across this plane equal to the Burgers vector is therefore required. Using the same procedure as in the previous development of Eq. (12.8), it is found that when $\hat{\mathbf{m}} \cdot \mathbf{x} < 0$,

$$\left.\begin{aligned} &\ln(\hat{\mathbf{m}} \cdot \mathbf{x} + p_\alpha^{(1)} \hat{\mathbf{n}} \cdot \mathbf{x}) \to \ln|\hat{\mathbf{m}} \cdot \mathbf{x}| \pm i\pi \quad \text{as } \hat{\mathbf{n}} \cdot \mathbf{x} \to 0^- \\ &\ln(\hat{\mathbf{m}} \cdot \mathbf{x} + p_\alpha^{(2)} \hat{\mathbf{n}} \cdot \mathbf{x}) \to \ln|\hat{\mathbf{m}} \cdot \mathbf{x}| \mp i\pi \quad \text{as } \hat{\mathbf{n}} \cdot \mathbf{x} \to 0^+ \end{aligned}\right\} \tag{14.77}$$

where the upper and lower signs in the $\pm$ notation correspond to our usual convention for the summation index α. According to the convention given by the Σ cut/displacement rule on p. 325, the negative side of the cut is the 0^- side in the above formulation, and, therefore, the displacement across the cut is related to the Burgers vector by

$$\begin{aligned} \Delta u_i = \frac{1}{2\pi i}\Bigg[&\sum_{\alpha=1}^{3} A_{i\alpha}^{(1)} E_\alpha^{(1)} (\ln|\hat{\mathbf{m}} \cdot \mathbf{x}| + i\pi) \\ &+ \sum_{\alpha=4}^{6} A_{i\alpha}^{(1)} E_\alpha^{(1)} (\ln|\hat{\mathbf{m}} \cdot \mathbf{x}| - i\pi) \\ &- \sum_{\alpha=1}^{3} A_{i\alpha}^{(2)} E_\alpha^{(2)} (\ln|\hat{\mathbf{m}} \cdot \mathbf{x}| - i\pi) \\ &- \sum_{\alpha=4}^{6} A_{i\alpha}^{(2)} E_\alpha^{(2)} (\ln|\hat{\mathbf{m}} \cdot \mathbf{x}| + i\pi)\Bigg] = b_i \end{aligned} \tag{14.78}$$

or,

$$\begin{aligned} &\frac{1}{2\pi i} \sum_{\alpha=1}^{6} (A_{i\alpha}^{(1)} E_\alpha^{(1)} - A_{i\alpha}^{(2)} E_\alpha^{(2)}) \ln|\hat{\mathbf{m}} \cdot \mathbf{x}| \\ &+ \frac{1}{2} \sum_{\alpha=1}^{6} (\pm A_{i\alpha}^{(1)} E_\alpha^{(1)} \pm A_{i\alpha}^{(2)} E_\alpha^{(2)}) = b_i. \end{aligned} \tag{14.79}$$

Since $\ln|\hat{\mathbf{m}} \cdot \mathbf{x}|$ is arbitrary, Eq. (14.79) is satisfied when

$$\sum_{\alpha=1}^{6} (A_{i\alpha}^{(1)} E_\alpha^{(1)} - A_{i\alpha}^{(2)} E_\alpha^{(2)}) = 0 \tag{14.80}$$

and

$$\sum_{\alpha=1}^{6}(\pm \mathrm{A}_{\mathrm{i}\alpha}^{(1)}\mathrm{E}_{\alpha}^{(1)} \pm \mathrm{A}_{\mathrm{i}\alpha}^{(2)}\mathrm{E}_{\alpha}^{(2)}) = 2\mathrm{b}_{\mathrm{i}}. \tag{14.81}$$

No net tractions can exist along the interface, and the solution must therefore also satisfy the boundary condition

$$\sigma_{\mathrm{mn}}^{(1)}\hat{\mathrm{n}}_{\mathrm{n}} = \sigma_{\mathrm{mn}}^{(2)}\hat{\mathrm{n}}_{\mathrm{n}}. \quad (\mathrm{x}_2 = 0) \tag{14.82}$$

Therefore, by employing Eqs. (3.1), (14.76) and (3.37), Eq. (14.82) takes the successive forms

$$\frac{\sum_{\alpha=1}^{6}\mathrm{A}_{\mathrm{i}\alpha}^{(1)}\mathrm{E}_{\alpha}^{(1)}\hat{\mathrm{n}}_{\mathrm{n}}\mathrm{C}_{\mathrm{nmip}}^{(1)}(\hat{\mathrm{m}}_{\mathrm{p}} + \mathrm{p}_{\alpha}^{(1)}\hat{\mathrm{n}}_{\mathrm{p}})}{(\hat{\mathbf{m}}\cdot\mathbf{x} + \mathrm{p}_{\alpha}^{(1)}\hat{\mathbf{n}}\cdot\mathbf{x})}$$

$$= \frac{\sum_{\alpha=1}^{6}\mathrm{A}_{\mathrm{i}\alpha}^{(2)}\mathrm{E}_{\alpha}^{(2)}\hat{\mathrm{n}}_{\mathrm{n}}\mathrm{C}_{\mathrm{nmip}}^{(2)}(\hat{\mathrm{m}}_{\mathrm{p}} + \mathrm{p}_{\alpha}^{(2)}\hat{\mathrm{n}}_{\mathrm{p}})}{(\hat{\mathbf{m}}\cdot\mathbf{x} + \mathrm{p}_{\alpha}^{(2)}\hat{\mathbf{n}}\cdot\mathbf{x})}$$

$$\frac{\sum_{\alpha=1}^{6}\mathrm{L}_{\mathrm{i}\alpha}^{(1)}\mathrm{E}_{\alpha}^{(1)}}{(\hat{\mathbf{m}}\cdot\mathbf{x} + \mathrm{p}_{\alpha}^{(1)}\hat{\mathbf{n}}\cdot\mathbf{x})} = \frac{\sum_{\alpha=1}^{6}\mathrm{L}_{\mathrm{i}\alpha}^{(2)}\mathrm{E}_{\alpha}^{(2)}}{(\hat{\mathbf{m}}\cdot\mathbf{x} + \mathrm{p}_{\alpha}^{(2)}\hat{\mathbf{n}}\cdot\mathbf{x})}. \tag{14.83}$$

However, since $\mathrm{p}_{\alpha}^{(1)}$ and $\mathrm{p}_{\alpha}^{(2)}$ are complex, the terms in the denominators of Eq. (14.83) take the limiting forms

$$\begin{aligned}
(\hat{\mathbf{m}}\cdot\mathbf{x} + \mathrm{p}_{\alpha}^{(1)}\hat{\mathbf{n}}\cdot\mathbf{x}) &\to \hat{\mathbf{m}}\cdot\mathbf{x} \pm \mathrm{i}0^{-} \quad \text{as } \hat{\mathbf{n}}\cdot\mathbf{x} \to 0^{-} \\
(\hat{\mathbf{m}}\cdot\mathbf{x} + \mathrm{p}_{\alpha}^{(2)}\hat{\mathbf{n}}\cdot\mathbf{x}) &\to \hat{\mathbf{m}}\cdot\mathbf{x} \mp \mathrm{i}0^{+} \quad \text{as } \hat{\mathbf{n}}\cdot\mathbf{x} \to 0^{+}
\end{aligned} \tag{14.84}$$

when $\mathbf{x}$ is at the negative and positive sides of the branch cut. Then, substituting this result into Eq. (14.83),

$$\frac{\sum_{\alpha=1}^{6}\mathrm{L}_{\mathrm{i}\alpha}^{(1)}\mathrm{E}_{\alpha}^{(1)}}{\hat{\mathbf{m}}\cdot\mathbf{x} \pm \mathrm{i}0^{-}} = \frac{\sum_{\alpha=1}^{6}\mathrm{L}_{\mathrm{i}\alpha}^{(2)}\mathrm{E}_{\alpha}^{(2)}}{\hat{\mathbf{m}}\cdot\mathbf{x} \mp \mathrm{i}0^{+}}. \tag{14.85}$$

However, Eq. (14.85) can be put into another form by employing the identity (Gelfand and Shilov 1964)

$$\frac{1}{\hat{\mathbf{m}}\cdot\mathbf{x} \pm \mathrm{i}0} = \frac{1}{\hat{\mathbf{m}}\cdot\mathbf{x}} \pm \mathrm{i}\pi\delta(\hat{\mathbf{m}}\cdot\mathbf{x}). \tag{14.86}$$

Therefore, by substituting Eq. (14.86) into Eq. (14.85), and rearranging,

$$\sum_{\alpha=1}^{6}(L_{i\alpha}^{(1)}E_{\alpha}^{(1)} - L_{i\alpha}^{(2)}E_{\alpha}^{(2)})\frac{1}{\hat{\mathbf{m}}\cdot\mathbf{x}}$$

$$+\sum_{\alpha=1}^{6}(\pm L_{i\alpha}^{(1)}E_{\alpha}^{(1)} \pm L_{i\alpha}^{(2)}E_{\alpha}^{(2)})i\pi\delta(\hat{\mathbf{m}}\cdot\mathbf{x}) = 0. \quad (14.87)$$

The condition given by Eq. (14.87) is then satisfied if

$$\sum_{\alpha=1}^{6}(L_{i\alpha}^{(1)}E_{\alpha}^{(1)} - L_{i\alpha}^{(2)}E_{\alpha}^{(2)}) = 0 \quad (14.88)$$

and

$$\sum_{\alpha=1}^{6}(\pm L_{i\alpha}^{(1)}E_{\alpha}^{(1)} \pm L_{i\alpha}^{(2)}E_{\alpha}^{(2)}) = 0. \quad (14.89)$$

Equations (14.80), (14.81), (14.88) and (14.89) constitute a set of twelve equations that suffice to determine the twelve $E_{\alpha}^{(1)}$ and $E_{\alpha}^{(2)}$ quantities appearing in Eq. (14.76), and the solution is therefore complete.

Strain energy

To obtain the strain energy of the interfacial dislocation shown in Fig. 14.18 we continue to follow the treatment of Barnett and Lothe (1974) which employs essentially the same method used previously to obtain the strain energy associated with a crystal dislocation (see Sec. 12.3.2 and Fig. 12.6). As in Eq. (12.33), the strain energy in the cylindrical shell of radii R and $r_{\circ}$ centered on the dislocation is given by

$$W = \frac{1}{2}\oiiint_{\mathcal{V}}\sigma_{ij}\varepsilon_{ij}dV = \frac{1}{2}\oiint_{\mathcal{S}}\sigma_{ij}u_{j}\hat{n}_{i}dS, \quad (14.90)$$

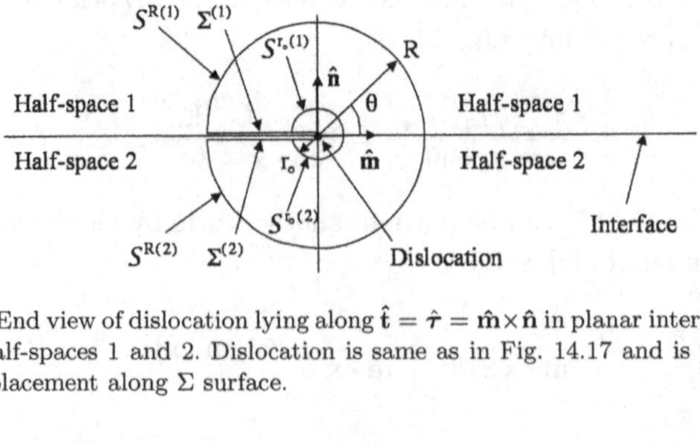

Fig. 14.18 End view of dislocation lying along $\hat{\mathbf{t}} = \hat{\tau} = \hat{\mathbf{m}}\times\hat{\mathbf{n}}$ in planar interface between dissimilar half-spaces 1 and 2. Dislocation is same as in Fig. 14.17 and is produced by cut and displacement along Σ surface.

where the surface $\mathcal{S}$ is now the total surface corresponding to $\mathcal{S} = \mathcal{S}^{R(1)} + \mathcal{S}^{R(2)} + \mathcal{S}^{r_\circ(1)} + \mathcal{S}^{r_\circ(2)} + \Sigma^{(1)} + \Sigma^{(2)}$. The displacement given by Eq. (14.76) is of the same general logarithmic form as the displacement for the crystal dislocation given by Eq. (12.12). As in the case of the crystal dislocation (and as shown next) the contributions of the cylindrical surfaces in Fig. 14.18 to the surface integral in Eq. (14.90) vanish. The strain energy (per unit dislocation length) is then just the work done in displacing the two sides of the cut by the Burgers vector during the introduction of the dislocation and, in a manner similar to Eq. (12.40), is given by

$$w = \frac{1}{2}\int_{R}^{r_\circ} \sigma_{ij} b_j \hat{n}_i d|\mathbf{x}|. \quad (\hat{\mathbf{n}} \cdot \mathbf{x} = 0) \tag{14.91}$$

To show that the contributions of the cylindrical surfaces vanish, $\mathbf{x}$ is written as $\mathbf{x} = |\mathbf{x}|(\cos\theta\hat{\mathbf{m}} + \sin\theta\hat{\mathbf{n}})$ so that Eq. (14.76) for half-space 1 takes the form

$$u_i^{(1)}(\mathbf{x}) = \frac{1}{2\pi i}\sum_{\alpha=1}^{6} A_{i\alpha}^{(1)} E_\alpha^{(1)}[\ln|\mathbf{x}| + \ln(\cos\theta + p_\alpha^{(1)}\sin\theta)] \tag{14.92}$$

which is of the same general form as Eq. (12.36) for a cylindrical surface when $|\mathbf{x}|$ is equal to $R^{(1)}$ or $r_\circ^{(1)}$. The corresponding stress is

$$\sigma_{mn}^{(1)} = C_{mnip}^{(1)}\frac{\partial u_i^{(1)}}{\partial x_p} = \frac{1}{2\pi i|\mathbf{x}|}\sum_{\alpha=1}^{6} C_{mnip}^{(1)} A_{i\alpha}^{(1)} E_\alpha^{(1)}\left(\frac{\hat{m}_p + p_\alpha^{(1)}\hat{n}_p}{\cos\theta + p_\alpha^{(1)}\sin\theta}\right) \tag{14.93}$$

so that, for an element of area, dS, on any cylindrical surface,

$$\sigma_{mn}^{(1)} dS = \sigma_{mn}^{(1)}|\mathbf{x}|d\theta = f_{mn}^{(1)}(\theta)d\theta \tag{14.94}$$

which is of the same general form as Eq. (12.38). Similar results are obtained for half-space 2. The argument in Sec. 12.3.2 that the integrals over the cylindrical surfaces do not contribute to the dislocation strain energy is therefore valid in this case, and the strain energy is given by Eq. (14.91).

To evaluate Eq. (14.91), Eq. (14.93) is employed with $\cos\theta = -1$ and $\sin\theta = 0$, along with Eqs. (3.37) and (13.69), to obtain

$$\begin{aligned} \sigma_{mn}^{(1)}\hat{n}_m &= \frac{-1}{2\pi i|\mathbf{x}|}\sum_{\alpha=1}^{6} A_{i\alpha}^{(1)} E_\alpha^{(1)}[(\hat{n}\hat{m})_{ni} + p_\alpha^{(1)}(\hat{n}\hat{n})_{ni}] \\ &= \frac{1}{2\pi i|\mathbf{x}|}\sum_{\alpha=1}^{6} E_\alpha^{(1)} L_{n\alpha}^{(1)} = \frac{1}{2\pi i|\mathbf{x}|}\sum_{\alpha=1}^{6} \pm J_{s\alpha}^{(1)} b_s L_{n\alpha}^{(1)}. \end{aligned} \tag{14.95}$$

Then, substitution of Eq. (14.95) into Eq. (14.91) yields

$$\begin{aligned} w &= -\frac{1}{4\pi i}\sum_{\alpha=1}^{6} \pm J^{(1)}_{s\alpha} L^{(1)}_{j\alpha} b_j b_s \int_{r_o}^{R} \frac{d|\mathbf{x}|}{|\mathbf{x}|} \\ &= w_o \ln \frac{r_o}{R}, \end{aligned} \tag{14.96}$$

where w_o, the interfacial dislocation pre-logarithmic strain energy factor, (see Eq. (12.40)) is given by (Barnett and Lothe (1974)

$$w_o = -\frac{1}{4\pi i}\sum_{\alpha=1}^{6} \pm J^{(1)}_{s\alpha} L^{(1)}_{j\alpha} b_j b_s. \tag{14.97}$$

14.4.2.4 *Planar array of parallel dislocations*

Having the above results for a single interfacial dislocation, the elastic field of the array of edge dislocations illustrated in Fig. 14.9b is now determined following the work of Barnett and Lothe (1974) and Hirth, Barnett and Lothe (1979).

Using Eq. (14.76), the displacement fields in the half-spaces 1 and 2, obtained by summing the contributions of the dislocations in the array, are given by

$$\left.\begin{aligned} u_j^{\text{array}^{(1)}}(\mathbf{x}) &= \frac{1}{2\pi i}\sum_{N=-\infty}^{\infty}\sum_{\alpha=1}^{6} A^{(1)}_{j\alpha} E^{(1)}_{\alpha} \ln(\hat{\mathbf{m}}\cdot\mathbf{x} - Nd + p^{(1)}_{\alpha}\hat{\mathbf{n}}\cdot\mathbf{x}) \\ &\qquad (x_2 > 0) \\ u_j^{\text{array}^{(2)}}(\mathbf{x}) &= \frac{1}{2\pi i}\sum_{N=-\infty}^{\infty}\sum_{\alpha=1}^{6} A^{(2)}_{j\alpha} E^{(2)}_{\alpha} \ln(\hat{\mathbf{m}}\cdot\mathbf{x} - Nd + p^{(2)}_{\alpha}\hat{\mathbf{n}}\cdot\mathbf{x}) \\ &\qquad (x_2 < 0) \end{aligned}\right\} \tag{14.98}$$

and the distortions are then

$$\left.\begin{aligned} \frac{\partial u_j^{\text{array}^{(1)}}(\mathbf{x})}{\partial x_l} &= \frac{1}{2\pi i}\sum_{N=-\infty}^{\infty}\sum_{\alpha=1}^{6} A^{(1)}_{j\alpha} E^{(1)}_{\alpha} \frac{\hat{m}_l + p^{(1)}_{\alpha}\hat{n}_l}{(\hat{\mathbf{m}}\cdot\mathbf{x} - Nd + p^{(1)}_{\alpha}\hat{\mathbf{n}}\cdot\mathbf{x})} \\ &\qquad (x_2 > 0) \\ \frac{\partial u_j^{\text{array}^{(2)}}(\mathbf{x})}{\partial x_l} &= \frac{1}{2\pi i}\sum_{N=-\infty}^{\infty}\sum_{\alpha=1}^{6} A^{(2)}_{j\alpha} E^{(2)}_{\alpha} \frac{\hat{m}_l + p^{(2)}_{\alpha}\hat{n}_l}{(\hat{\mathbf{m}}\cdot\mathbf{x} - Nd + p^{(2)}_{\alpha}\hat{\mathbf{n}}\cdot\mathbf{x})} \\ &\qquad (x_2 < 0) \end{aligned}\right\} \tag{14.99}$$

Next, using Eqs. (14.30) and (14.31) to evaluate the sums over N, and employing the ± symbol for summing over α, the far-field distortion tensors are

$$\left.\begin{aligned} \frac{\partial u_j^{array^{(1)}}(\mathrm{far-field})}{\partial x_l} &= -\frac{1}{2d}\sum_{\alpha=1}^{6} \pm A_{j\alpha}^{(1)} E_\alpha^{(1)}(\hat{m}_l + p_\alpha^{(1)}\hat{n}_l) \quad (x_2 > 0) \\ \frac{\partial u_j^{array^{(2)}}(\mathrm{far-field})}{\partial x_l} &= \frac{1}{2d}\sum_{\alpha=1}^{6} \pm A_{j\alpha}^{(2)} E_\alpha^{(2)}(\hat{m}_l + p_\alpha^{(2)}\hat{n}_l) \quad (x_2 < 0) \end{aligned}\right\} \tag{14.100}$$

Now, subtracting the second equation from the first and substituting the condition given by Eq. (14.81),

$$\frac{\partial[u_j^{array^{(1)}}(\mathrm{far-field}) - u_j^{array^{(2)}}(\mathrm{far-field})]}{\partial x_l} = -\frac{b_j\hat{m}_l}{d} - \frac{1}{2d}\sum_{\alpha=1}^{6}(\pm A_{j\alpha}^{(1)} E_\alpha^{(1)} p_\alpha^{(1)} \pm A_{j\alpha}^{(2)} E_\alpha^{(2)} p_\alpha^{(2)})\hat{n}_l. \tag{14.101}$$

Then, applying the deformation tensor given by Eq. (14.101) to a vector in the far-field corresponding to the unit vector $\hat{\mathbf{m}}$ (Fig. 14.17),

$$\delta\hat{m}_j = \frac{\partial[u_j^{array^{(1)}}(\mathrm{far-field}) - u_j^{array^{(2)}}(\mathrm{far-field})]}{\partial \mathbf{x}_l}\hat{m}_l = -\frac{b_j}{d} \tag{14.102}$$

which, for an array of edge dislocations with $\mathbf{b} = (0, b, 0)$ and $\hat{\mathbf{t}} = (0, 0, 1)$ of the type illustrated in Fig. 14.14, yields

$$\delta\hat{m}_1 = \delta\hat{m}_3 = 0, \quad \delta\hat{m}_2 = -b/d = \theta. \tag{14.103}$$

This result indicates a far-field rotation of half-space 1 relative to half-space 2 around the x_3 axis consistent with the Frank–Bilby equation (see discussion following Eq. (14.66) for this same interface in an isotropic system).

Despite this result that the Frank–Bilby equation is obeyed, the far-field distortions given by Eq. (14.100) are non-vanishing and predict far-field stresses, which Barnett and Lothe (1974) have shown produce tractions on all surfaces parallel to the interface in a manner analogous to the tractions caused by the stress given by Eq. (14.68) for the same array in an isotropic system.[10] However, as in the isotropic case, and now shown, these tractions can be eliminated by adding image fields that are uniform and also obey the expression

$$\delta \hat{m}_j = \left(\frac{\partial u_j^{IM^{(1)}}}{\partial x_l} - \frac{\partial u_j^{IM^{(2)}}}{\partial x_l} \right) \hat{m}_l = 0 \tag{14.104}$$

so that the compliance with the Frank–Bilby equation found above is preserved.

Following Hirth, Barnett and Lothe (1979), candidate image displacements for the above purpose are

$$\left.\begin{aligned} u_j^{IM^{(1)}}(\mathbf{x}) &= \sum_{\alpha=1}^{6} A_{j\alpha}^{IM^{(1)}} E_\alpha^{IM^{(1)}} (\hat{\mathbf{m}} \cdot \mathbf{x} + p_\alpha^{IM^{(1)}} \hat{\mathbf{n}} \cdot \mathbf{x}) \quad (x_2 > 0) \\ u_j^{IM^{(2)}}(\mathbf{x}) &= \sum_{\alpha=1}^{6} A_{j\alpha}^{IM^{(2)}} E_\alpha^{IM^{(2)}} (\hat{\mathbf{m}} \cdot \mathbf{x} + p_\alpha^{IM^{(2)}} \hat{\mathbf{n}} \cdot \mathbf{x}) \quad (x_2 < 0) \end{aligned}\right\} \tag{14.105}$$

To preserve coherence at the interface it is necessary that $u_j^{IM^{(1)}}(\hat{\mathbf{n}} \cdot \mathbf{x} = 0) = u_j^{IM^{(2)}}(\hat{\mathbf{n}} \cdot \mathbf{x} = 0)$, and this is satisfied when

$$\sum_{\alpha=1}^{6} A_{j\alpha}^{IM^{(1)}} E_\alpha^{IM^{(1)}} = \sum_{\alpha=1}^{6} A_{j\alpha}^{IM^{(2)}} E_\alpha^{IM^{(2)}}. \tag{14.106}$$

Using Eqs. (14.105), (3.1) and (3.37), the traction produced by $u_j^{IM^{(1)}}(\mathbf{x})$ on a surface parallel to the interface is then

$$\begin{aligned} T_m^{IM^{(1)}} = \sigma_{mn}^{IM^{(1)}} \hat{n}_n &= \sum_{\alpha=1}^{6} A_{j\alpha}^{IM^{(1)}} E_\alpha^{IM^{(1)}} (\hat{n}_n C_{mnjk}^{(1)} \hat{m}_k + p_\alpha^{IM^{(1)}} \hat{n}_n C_{mnjk}^{(1)} \hat{n}_k) \\ &= -\sum_{\alpha=1}^{6} L_{m\alpha}^{IM^{(1)}} E_\alpha^{IM^{(1)}} \end{aligned} \tag{14.107}$$

which is independent of the position of the interface, as must be the case. Since the tractions must match across the interface, $T_m^{IM^{(1)}} = T_m^{IM^{(2)}} = T_m^{IM}$,

[10]The determination of these tractions is lengthy and is not presented here; for details see Barnett and Lothe (1974).

and, therefore, the relationship

$$\sum_{\alpha=1}^{6} L_{m\alpha}^{IM^{(1)}} E_{\alpha}^{IM^{(1)}} = \sum_{\alpha=1}^{6} L_{m\alpha}^{IM^{(2)}} E_{\alpha}^{IM^{(2)}} \tag{14.108}$$

must be satisfied. Using Eqs. (3.76) and (3.78), the conditions given by Eqs. (14.106) and (14.108) are satisfied when

$$E_{\alpha}^{IM^{(1)}} = A_{m\alpha}^{IM^{(1)}} T_m^{IM} \quad E_{\alpha}^{IM^{(2)}} = A_{m\alpha}^{IM^{(2)}} T_m^{IM}. \tag{14.109}$$

Then, substituting Eq. (14.109) into Eq. (14.105),

$$\left.\begin{aligned} u_j^{IM^{(1)}}(\mathbf{x}) &= \sum_{\alpha=1}^{6} A_{j\alpha}^{IM^{(1)}} A_{m\alpha}^{IM^{(1)}} T_m^{IM}(\hat{\mathbf{m}}\cdot\mathbf{x} + p_{\alpha}^{IM^{(1)}}\hat{\mathbf{n}}\cdot\mathbf{x}) \quad (x_2 > 0) \\ u_j^{IM^{(2)}}(\mathbf{x}) &= \sum_{\alpha=1}^{6} A_{j\alpha}^{IM^{(2)}} A_{m\alpha}^{IM^{(2)}} T_m^{IM}(\hat{\mathbf{m}}\cdot\mathbf{x} + p_{\alpha}^{IM^{(2)}}\hat{\mathbf{n}}\cdot\mathbf{x}) \quad (x_2 < 0) \end{aligned}\right\} \tag{14.110}$$

which finally takes the form

$$\left.\begin{aligned} u_j^{IM^{(1)}}(\mathbf{x}) &= -(nn)_{jm}^{(1)^{-1}} T_m^{IM}\hat{\mathbf{n}}\cdot\mathbf{x} \quad (x_2 > 0) \\ u_j^{IM^{(2)}}(\mathbf{x}) &= -(nn)_{jm}^{(2)^{-1}} T_m^{IM}\hat{\mathbf{n}}\cdot\mathbf{x} \quad (x_2 < 0) \end{aligned}\right\} \tag{14.111}$$

upon use of Eqs. (3.76) and (3.110). The corresponding image distortions are then

$$\left.\begin{aligned} \frac{\partial u_j^{IM^{(1)}}(\mathbf{x})}{\partial x_l} &= -(\hat{n}\hat{n})_{jm}^{(1)^{-1}} T_m^{IM}\hat{n}_l \quad (x_2 > 0) \\ \frac{\partial u_j^{IM^{(2)}}(\mathbf{x})}{\partial x_l} &= -(\hat{n}\hat{n})_{jm}^{(2)^{-1}} T_m^{IM}\hat{n}_l \quad (x_2 < 0) \end{aligned}\right\} \tag{14.112}$$

and are seen to satisfy Eq. (14.104), as required.

By adding these image fields to the infinite body solution obtained previously, a solution for the array is obtained that is free of long-range stress and also consistent with the Frank–Bilby equation in a manner analogous to the solution for the isotropic case obtained in Sec. 14.4.2.2

Exercises

14.1. Use the Frank–Bilby equation, Eq. (14.24), to find the screw dislocation structure of the twist interface in Fig. 14.2.

Solution. All lattices and coordinate systems are the same as in Fig. 14.8 but with the tilt angle θ replaced with the twist angle ϕ. The matrices $D_{ji}^{(2)}$ and $D_{ji}^{(1)}$ are therefore again given by Eq. (14.25). Insertion of $\mathbf{p} = (0, -p, 0)$ into Eq. (14.24) yields $\mathbf{b}^{tot} = (2p\sin(\phi/2), 0, 0)$. This condition is satisfied by a family of screw dislocations with Burgers vector $\mathbf{a}_1^R$ running parallel to $\mathbf{a}_1^R$ at the spacing $d = a/[2\sin(\phi/2)] \cong a/\phi$. Next, insertion of $\mathbf{p} = (p, 0, 0)$ into Eq. (14.24) yields $\mathbf{b}^{tot} = (0, 2p\sin(\phi/2), 0)$ which is satisfied by a family with Burgers vector $\mathbf{a}_2^R$ running parallel to $\mathbf{a}_2^R$ at the spacing $d \cong a/\phi$.

14.2. Show that the long-range stress fields produced by the dislocation families in Figs. 14.10b and 14.10c are consistent with the fact that the Frank–Bilby equation cannot be satisfied for either family.

Solution. For this purpose the Frank-Bilby equation, Eq. (14.24), for a planar array of lattice dislocations, can be written in the relatively simple form

$$[b^{tot}] = [\omega][p] \text{ or } \begin{bmatrix} b_1^{tot} \\ b_2^{tot} \\ b_3^{tot} \end{bmatrix} = \begin{bmatrix} 0 & -\omega_3 & \omega_2 \\ \omega_3 & 0 & -\omega_1 \\ -\omega_2 & \omega_1 & 0 \end{bmatrix} \begin{bmatrix} p_1 \\ p_2 \\ p_3 \end{bmatrix} \quad (14.113)$$

where $\mathbf{p}$ is a vector lying in the plane of the array, $\mathbf{b}^{tot}$ is the sum of the Burgers vectors of the dislocations intersected by $\mathbf{p}$, and $\underline{\boldsymbol{\omega}}$ is the rotation tensor (see Eq. (2.13)) which describes the crystal misorientation which must exist across the array in order to avoid long-range stress.

For the array in Fig. 14.10a, Eq. (14.113) yields the relationships

$$\begin{bmatrix} b_1^{tot} \\ 0 \\ 0 \end{bmatrix} \begin{bmatrix} 0 & -\omega_3 & \omega_2 \\ \omega_3 & 0 & -\omega_1 \\ -\omega_2 & \omega_1 & 0 \end{bmatrix} \begin{bmatrix} 0 \\ 1 \\ 0 \end{bmatrix} = \begin{bmatrix} -\omega_3 \\ 0 \\ \omega_1 \end{bmatrix}$$

$$\begin{bmatrix} 0 \\ 0 \\ 0 \end{bmatrix} \begin{bmatrix} 0 & -\omega_3 & \omega_2 \\ \omega_3 & 0 & -\omega_1 \\ -\omega_2 & \omega_1 & 0 \end{bmatrix} \begin{bmatrix} 0 \\ 0 \\ 1 \end{bmatrix} = \begin{bmatrix} \omega_2 \\ -\omega_1 \\ 0 \end{bmatrix} \quad (14.114)$$

when $\mathbf{p} = (0, 1, 0)$ and $(0, 0, 1)$, respectively. These relationships can be satisfied by constructing a $\underline{\boldsymbol{\omega}}$ tensor with components $\omega_1 = \omega_2 = 0$ and $\omega_3 = -b_1^{tot}$ corresponding to a tilt misorientation around the $(0, 0, 1)$ axis. The Frank-Bilby equation is therefore satisfied, and, has already been established, no long-range stress is generated by this array. By carrying out the same procedure for the dislocation families in Figs. 14.10b and 14.10c it is easily verified that Eq. (14.113) cannot be satisfied by any $\underline{\omega}$ tensor for either family. Therefore, there are no rigid body rotations possible that will alleviate the long-range stresses produced by these families. (See Exercise 14.6 for further insight into this exercise).

14.3. In Figs. 14.1 and 14.2 the construction of a small-angle symmetrical tilt interface and a twist interface, respectively, has been accomplished by a process in which:

(1) A dislocation-free single crystal is first elastically bent or twisted.
(2) A suitable distribution of dislocations is introduced to eliminate the long-range stresses generated in step 1.
(3) The distributed dislocations are gathered into a planar array to form the desired interface.

Use this process, and, with the help of Eq. (14.11), find the dislocation structures of the tilt and twist interfaces including specification of the required types of dislocations and their spacings.

Solution. (a) To produce the tilt interface, start with a single crystal corresponding to a unit cube with {100} faces and elastically bend it around [001] as in Fig. (14.1). The required dislocation distribution to eliminate the resulting stresses is then obtained by use of Eq. (14.11) which takes the form

$$\begin{bmatrix} \alpha_{11} & \alpha_{12} & \alpha_{13} \\ \alpha_{21} & \alpha_{22} & \alpha_{23} \\ \alpha_{31} & \alpha_{32} & \alpha_{33} \end{bmatrix} = \begin{bmatrix} 0 & 0 & 0 \\ -\kappa_{12} & 0 & 0 \\ 0 & 0 & 0 \end{bmatrix} \tag{14.115}$$

since κ_{12} is the only non-vanishing component of $\underline{\kappa}$. The solution of Eq. (14.115) is then

$$\alpha_{21} = -\kappa_{12} = -\frac{\partial \phi_1}{\partial x_2} \tag{14.116}$$

with all other κ_{ij} equal to zero. If we insert N edge dislocations with $\mathbf{b} = (0, a, 0)$ and $\hat{\mathbf{t}} = (1, 0, 0)$,

$$\alpha_{21} = aN \tag{14.117}$$

and the long-range stress will be eliminated. Finally, put all N dislocations into a planar array as in Fig. 14.1c. Since the crystal is bent downwards (Fig.14.1a), the quantity $\partial\phi_1/\partial x_2$ is negative, and therefore, upon traversing the crystal the unit distance along x_2,

$-\partial\phi_1/\partial x_2 = \theta$. Then, putting this result and Eq. (14.117) into Eq. (14.116), the dislocation spacing in the interface is

$$d = \frac{1}{N} = \frac{a}{\alpha_{21}} = -\frac{a}{(\partial\phi_1/\partial x_2)} = \frac{a}{\theta} \tag{14.118}$$

in agreement with Eq. (14.1).

(b) For the twist interface, the single crystal is first twisted elastically around $\hat{\mathbf{n}} = (0, 0, 1)$ as in Fig. 14.2a. Equation (14.11) is then given by

$$\begin{bmatrix} \alpha_{11} & \alpha_{12} & \alpha_{13} \\ \alpha_{21} & \alpha_{22} & \alpha_{23} \\ \alpha_{31} & \alpha_{32} & \alpha_{33} \end{bmatrix} = \begin{bmatrix} \kappa_{33} & 0 & 0 \\ 0 & \kappa_{33} & 0 \\ 0 & 0 & 0 \end{bmatrix} \tag{14.119}$$

with a solution

$$\alpha_{11} = \alpha_{22} = \kappa_{33} = \frac{\partial\phi_3}{\partial x_3}. \tag{14.120}$$

In this case, to eliminate the long range stress, N screw dislocations with $\mathbf{b}^{(1)} = (a, 0, 0)$ and $\hat{\mathbf{t}}^{(1)} = (1, 0, 0)$ are inserted along with N orthogonal screw dislocations with $\mathbf{b}^{(2)} = (0, a, 0)$ and $\hat{\mathbf{t}}^{(2)} = (0, 1, 0)$, so that

$$\alpha_{11} = \alpha_{22} = aN. \tag{14.121}$$

Then, following the same procedure as in (a), $\partial\phi_3/\partial x_3 = \phi$, where ϕ is the twist angle for the interface shown in Fig. 14.2, and

$$\begin{aligned} d^{(1)} &= a/\phi \\ d^{(2)} &= a/\phi \end{aligned} \tag{14.122}$$

is obtained for the spacings of the two families of screw dislocations in agreement with the results in Exercise 14.1.

14.4. Determine the displacement field due to component 1 of an infinitely long straight line of force in an infinite isotropic body. Assume that it lies along the x_3 axis of the coordinate system in Fig. 14.14.

Solution: Start with Eq. (12.32), which reduces immediately for the present case by use of Eq. (3.147) to

$$\begin{aligned}
u_1^{LF\infty} &= \frac{f_1}{2\pi}[Q_{11}\ln|\mathbf{x}| - Q_{11}\int[(\hat{n}\hat{n})_{11}^{-1}(\hat{n}\hat{m})_{11} + (\hat{n}\hat{n})_{12}^{-1}(\hat{n}\hat{m})_{21}]d\omega] \\
&\quad - S_{12}\int(\hat{n}\hat{n})_{12}^{-1}d\omega, \\
u_2^{LF\infty} &= \frac{f_1}{2\pi}[-Q_{11}\int[(\hat{n}\hat{n})_{21}^{-1}(\hat{n}\hat{m})_{11} + (\hat{n}\hat{n})_{22}^{-1}(\hat{n}\hat{m})_{21}]d\omega \\
&\quad - S_{12}\int(\hat{n}\hat{n})_{22}^{-1}d\omega], \qquad (14.123) \\
u_3^{LF\infty} &= 0.
\end{aligned}$$

Then, by use of Eq. (3.141), with $\hat{n}_1 = -\sin\omega$, $\hat{n}_2 = \cos\omega$, $\hat{m}_1 = \cos\omega$, and $\hat{m}_2 = \sin\omega$,

$$\begin{aligned}
u_1^{LF\infty} &= \frac{f_1}{2\pi}\left[-\frac{3-4\nu}{4\mu(1-\nu)}\ln(x_1^2+x_2^2)^{1/2} - \frac{1}{2\mu(1-\nu)}\int\sin\omega\cos\omega d\omega\right] \\
u_2^{LF\infty} &= \frac{f_1}{8\pi\mu(1-\nu)}\int(1-2\sin^2\omega)d\omega \qquad (14.124)
\end{aligned}$$

and then, upon integration,

$$\begin{aligned}
u_1^{LF\infty} &= -\frac{f_1}{16\pi\mu(1-\nu)}\left[(3-4\nu)\ln(x_1^2+x_2^2) - \frac{2x_1^2}{(x_1^2+x_2^2)}\right], \\
& \qquad (14.125) \\
u_2^{LF\infty} &= \frac{f_1}{8\pi\mu(1-\nu)}\frac{x_1x_2}{(x_1^2+x_2^2)},
\end{aligned}$$

See Exercise 14.5 for an alternative derivation of Eq. (14.125).

14.5. In Exercise 14.4 we found the displacement field of a line of force by starting with Eq. (12.32). However, it should be possible to solve this problem by treating a line of force as an array of point forces along a line which is then smeared out into an array consisting of an infinite number of infinitesimal point forces whose displacement field can be found by employing the point force Green's function. Use this method to obtain the solution to the problem posed in Exercise 14.4.

Solution Take $\mathbf{x}$ normal to the line so that $x = (x_1^2 + x_2^2)^{1/2}$, and $\mathbf{x}'$ along the line so that $x' = x_3'$. The point force strength of component 1 of the line of force present in distance dx_3' along the line is then

$f_1 dx'_3$, and the resulting displacement field, after use of Eq. (4.110), is given by

$$u_i(\mathbf{x}) = f_1 \int_{-\infty}^{\infty} G_{i1}(\mathbf{x} - \mathbf{x}')dx'_3 = \frac{f_1}{16\pi\mu(1-\nu)}$$

$$\times \mathrm{Lim}_{L\to\infty} \int_{-L}^{L} \left[\frac{(3-4\nu)\delta_{i1}}{|\mathbf{x} - \mathbf{x}'|} + \frac{(x_i - x'_1)x_1}{|\mathbf{x} - \mathbf{x}'|^3} \right] dx'_3 \quad (14.126)$$

with $|\mathbf{x} - \mathbf{x}'| = [x_1^2 + x_2^2 + (x'_3)^2]^{1/2}$. For $u_1(\mathbf{x})$, the two integrals in Eq. (14.126) take the forms

$$\int_{-L}^{L} \frac{dx'_3}{|\mathbf{x} - \mathbf{x}'|} = 2\int_0^L \frac{dx'_3}{[x_1^2 + x_2^2 + (x'_3)^2]^{1/2}}$$

$$= -\ln(x_1^2 + x_2^2) + 2\ln L \left[1 + \left(1 + \frac{x_1^2 + x_2^2}{L^2} \right) \right],$$

$$\int_{-L}^{L} \frac{dx'_3}{|\mathbf{x} - \mathbf{x}'|^{3/2}} = 2\int_0^L \frac{dx'_3}{[x_1^2 + x_2^2 + (x'_3)^2]^{3/2}}$$

$$= \frac{2}{(x_1^2 + x_2^2)} \left(1 + \frac{x_1^2 + x_2^2}{L^2} \right)^{-1/2}. \quad (14.127)$$

Then, substituting these results into Eq. (14.126), taking the limit, and discarding the superfluous constant,

$$u_1(\mathbf{x}) = -\frac{f_1}{16\pi\mu(1-\nu)} \left[(3-4\nu)\ln(x_1^2 + x_2^2) - \frac{2x_1^2}{x_1^2 + x_2^2} \right] \quad (14.128)$$

in agreement with Eq. (14.125). Similar agreement is obtained for $u_2(\mathbf{x})$ and $u_3(\mathbf{x})$.

14.6. Show that the long-range stress field of an infinite array of straight parallel dislocations, as in Fig. 14.19, generally vanishes when **b** is parallel to $\hat{\mathbf{n}}$. Note: an example of this result is given in Sec. 14.3.4 where it is shown that the long-range field of the array in Fig. 14.10a vanishes. Hint: assume that the array has been created by parallel cuts and displacements in Crystal 2 (towards which $\hat{\mathbf{n}}$ points), and employ the general transformation strain method in Sec. 12.5.1.5 that led to Eq. (12.134).

Solution. Smear out the array into an infinite number of infinitesimal dislocations produced by an infinite number of cuts and displacements in Crystal 2. Then, according to the method cited above,

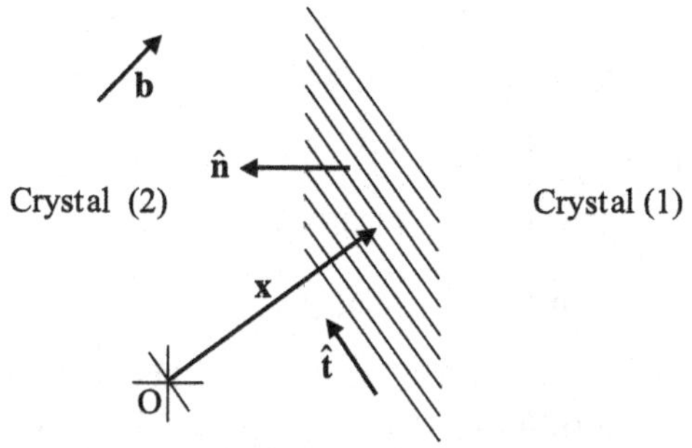

Fig. 14.19 Infinite array of straight parallel dislocations: $\hat{\mathbf{n}}$ is normal to plane of array.

the plastic transformation displacement due to these cuts and displacements which accumulates over the length of the vector $\mathbf{x}$ is

$$\mathbf{u}^{\mathrm{T}} = \frac{1}{\mathrm{d}}[\mathbf{x} \cdot (\hat{\mathbf{t}} \times \hat{\mathbf{n}})]\mathbf{b} = \frac{1}{\mathrm{d}}[\hat{\mathbf{t}} \cdot (\hat{\mathbf{n}} \times \mathbf{x})]\mathbf{b} = \frac{1}{\mathrm{d}}(\hat{\mathrm{n}}_{\mathrm{i}}\mathrm{x}_{\mathrm{j}}\mathrm{e}_{\mathrm{ijk}}\hat{\mathrm{t}}_{\mathrm{k}})\mathbf{b}. \quad (14.129)$$

The corresponding strain, which is spatially uniform, is then

$$\varepsilon^{\mathrm{T}}_{\mathrm{rs}} = \frac{1}{2}\left(\frac{\partial \mathrm{u}^{\mathrm{T}}_{\mathrm{r}}}{\partial \mathrm{x}_{\mathrm{s}}} + \frac{\partial \mathrm{u}^{\mathrm{T}}_{\mathrm{s}}}{\partial \mathrm{x}_{\mathrm{r}}}\right) = \frac{1}{2\mathrm{d}}\hat{\mathrm{n}}_{\mathrm{i}}\hat{\mathrm{t}}_{\mathrm{k}}(\mathrm{e}_{\mathrm{isk}}\mathrm{b}_{\mathrm{r}} + \mathrm{e}_{\mathrm{irk}}\mathrm{b}_{\mathrm{s}}). \quad (14.130)$$

Next, representing this strain by its Fourier transform, i.e.,

$$\bar{\varepsilon}^{\mathrm{T}}_{\mathrm{rs}}(\mathbf{k}) = \frac{1}{2\mathrm{d}}\hat{\mathrm{n}}_{\mathrm{i}}\hat{\mathrm{t}}_{\mathrm{k}}(\mathrm{e}_{\mathrm{isk}}\mathrm{b}_{\mathrm{r}} + \mathrm{e}_{\mathrm{irk}}\mathrm{b}_{\mathrm{s}}) \int_{-\infty}^{\infty}\int_{-\infty}^{\infty}\int_{-\infty}^{\infty} \mathrm{e}^{\mathrm{i}\mathbf{k}\cdot\mathbf{x}}\mathrm{dx}_1\mathrm{dx}_2\mathrm{dx}_3 \quad (14.131)$$

and substituting this into Eq. (3.164),

$$\bar{\sigma}_{\mathrm{pq}}(\mathbf{k}) = -\frac{1}{2\mathrm{d}}\mathrm{C}^{*}_{\mathrm{pqrs}}\hat{\mathrm{n}}_{\mathrm{i}}\hat{\mathrm{t}}_{\mathrm{k}}(\mathrm{e}_{\mathrm{isk}}\mathrm{b}_{\mathrm{r}} + \mathrm{e}_{\mathrm{irk}}\mathrm{b}_{\mathrm{s}}) \times \int_{-\infty}^{\infty}\int_{-\infty}^{\infty}\int_{-\infty}^{\infty} \mathrm{e}^{\mathrm{i}\mathbf{k}\cdot\mathbf{x}}\mathrm{dx}_1\mathrm{dx}_2\mathrm{dx}_3. \quad (14.132)$$

Now, in the special case when $\mathbf{b}$ is parallel to $\hat{\mathbf{n}}$, $\mathrm{b}_{\mathrm{i}} = \mathrm{b}\hat{\mathrm{n}}_{\mathrm{i}}$, and Eq. (14.132) assumes the form

$$\bar{\sigma}_{\mathrm{pq}}(\mathbf{k}) = -\frac{\mathrm{b}}{2\mathrm{d}}\mathrm{C}^{*}_{\mathrm{pqrs}}\hat{\mathrm{n}}_{\mathrm{i}}\hat{\mathrm{t}}_{\mathrm{k}}(\mathrm{e}_{\mathrm{isk}}\hat{\mathrm{n}}_{\mathrm{r}} + \mathrm{e}_{\mathrm{irk}}\hat{\mathrm{n}}_{\mathrm{s}}) \times \int_{-\infty}^{\infty}\int_{-\infty}^{\infty}\int_{-\infty}^{\infty} \mathrm{e}^{\mathrm{i}\mathbf{k}\cdot\mathbf{x}}\mathrm{dx}_1\mathrm{dx}_2\mathrm{dx}_3. \quad (\mathbf{b}||\hat{\mathbf{n}}). \quad (14.133)$$

However, as now shown, the factor $\hat{n}_r C^*_{pqrs}$ in Eq. (14.133) vanishes. Using Eq. (3.163) and the symmetry properties of C^*_{pqrs},

$$\begin{aligned}\hat{n}_r C^*_{pqrs} &= \hat{n}_r C^*_{rspq} = \hat{n}_r C_{rspq} - \hat{n}_r C_{rsij}\hat{n}_j(\hat{n}\hat{n})^{-1}_{im}\hat{n}_n C_{nmpq} \\ &= \hat{n}_r C_{rspq} - \hat{n}_n(\hat{n}\hat{n})_{si}(\hat{n}\hat{n})^{-1}_{im} C_{nmpq} \\ &= \hat{n}_r C_{rspq} - \hat{n}_n \delta_{sm} C_{nmpq} = 0. \qquad (14.134)\end{aligned}$$

The factor $\hat{n}_s C^*_{pqrs}$ in Eq. (14.133) vanishes in similar fashion, and, therefore, when $\mathbf{b}||\hat{\mathbf{n}}$, $\bar{\sigma}_{pq}(\mathbf{k})$ vanishes, and its inverse, i.e., the stress at $\mathbf{x}$, $\sigma_{pq}(\mathbf{x})$, vanishes as well.

Chapter 15

Interactions between Interfaces and Stress

15.1 Introduction

An interface can experience a number of different types of forces including chemical forces (due to chemical free energy differences across the interface), curvature forces (due to interface curvature) and mechanical forces (Sutton and Balluffi 1996; Asaro and Lumbarda 2006). In view of the focus of the present book we consider only mechanical forces due to the presence of applied forces and elastic strain fields.

In general, an interface experiences a mechanical force when it lies between two adjoining regions containing different elastic fields and therefore different strain energy densities and elastic displacement fields. In such cases movement of the interface, in which one region grows at the expense of the other, can produce a decrease in the overall energy of the body and thereby give rise to a force on the interface expressed by Eq. (5.38). Such a force can occur under a variety of circumstances. For example, during the recrystallization of a plastically deformed crystalline body relatively strain-free crystals form and then grow into the surrounding plastically deformed and dislocated matrix. Here, the reduction in energy that occurs as the strain-free crystals grow at the expense of the dislocated (and strained) matrix produces outward forces on the interfaces bounding the strain-free crystals. In other situations elastic fields that differ across interfaces, and therefore generate a mechanical interface force, often occur in polycrystalline materials in the form of compatibility stresses arising as a result of elastic anisotropy, anisotropic thermal expansion, and differing modes of plastic deformation in the crystals adjoining the interfaces (e.g., Sutton and Balluffi 2006).

An interface can also experience a mechanical force when it is present in a body subjected to applied forces, and is of a type whose displacement in the body produces a shape change of the body. This allows the applied forces to perform work, thereby lowering the potential energy of the system and giving rise to an effective force on the interface, again expressed by Eq. (5.38). As will be shown, interfaces of this type are generally able to support discrete localized interfacial dislocations and move via the movement of these dislocations. In such cases the transfer of atoms across the interface, which is necessary for the interface motion, occurs in a highly ordered and correlated "military fashion" resulting in a body shape change (Christian 1975; Sutton and Balluffi 2006).[1] Eshelby (1956) distinguished between the two types of forces described above and appropriately termed the latter force the *dislocation force.*

Both forces are treated in this chapter. In Sec. 15.2 the force arising from elastic field differences across the interface is formulated in terms of the energy-momentum tensor introduced earlier in Sec. 5.3.2.1 (see Eqs. (5.51) and (5.52)). We therefore term it the *energy-momentum tensor force.*

In Sec. 15.3 the geometrical features of interfaces whose motion causes body shape changes are described in terms of their interfacial dislocation content. Then, expressions are formulated for the forces imposed on such interfaces when forces are applied to the body. Following Eshelby, we term this force the *interfacial dislocation force.*

15.2 The Energy-Momentum Tensor Force

To find the energy-momentum tensor force on an interface separating two regions with different strain fields consider the displacement of the interface, $\mathcal{S}$, separating regions $\mathcal{V}^{(1)}$ and $\mathcal{V}^{(2)}$ shown in Fig. 15.1. The two regions possess different strain fields, and we wish to determine the resulting local force acting on the interface by use of Eq. (5.38) which requires the determination of the change in total elastic energy of the system as the interface is displaced by the vector, $\delta\boldsymbol{\xi}$. Assume that as the interface advances into region 1 the medium in the overrun region 1 is converted into the medium of

[1]Many interfaces are unable to support arrays of discrete localized dislocations and so move by the chaotic uncorrelated transfer of atoms across the interface in "civilian fashion". Hence, their motion does not produce a body shape change, and they do not experience an interfacial dislocation force.

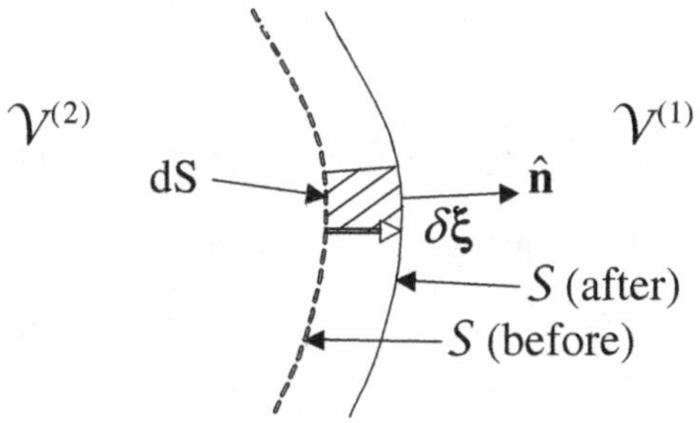

Fig. 15.1 Vector displacement, $\delta\boldsymbol{\xi}$, of interface $\mathcal{S}$ lying between regions $\mathcal{V}^{(1)}$ and $\mathcal{V}^{(2)}$. Dashed and solid lines are, respectively, the traces of the interface before and after displacement of interface towards region 1.

region 2 possessing the same elastic field that exists in the directly adjoining region 2. Following Asaro and Lubarda (2006), this can be accomplished in each local region (such as the hatched region in Fig. 15.1) by the following steps:

(1) Remove the intervening medium between the "before" and "after" boundary positions while applying tractions, $\mathbf{T}$, to the newly created surfaces to prevent elastic relaxation.
(2) Transform the intervening medium into medium 2 possessing the same elastic field as the adjoining region 2.
(3) Insert the intervening medium back into the gap from which it was removed, while maintaining its elastic field, and bond the surfaces together to produce a system containing the displaced interface.

No energy change occurs in step 1. In step 2, the strain energies corresponding to the fields in region 1 and 2 are exchanged, and the change in strain energy in the differential volume, $\mathrm{dV} = \delta\xi \mathrm{dS}$, (where dS is the interfacial area associated with dV) is therefore

$$\delta \mathrm{W} = \left(\mathrm{w}^{(2)} - \mathrm{w}^{(1)}\right) \delta \mathrm{V} = \left(\mathrm{w}^{(2)} - \mathrm{w}^{(1)}\right) \delta\xi \mathrm{dS} \tag{15.1}$$

However, in step 3, the intervening medium will generally not fit back into the gap due to a mismatch that has accumulated along the vector $\delta\boldsymbol{\xi}$, which in turn is due to the difference in the elastic displacement fields of regions 1 and 2. This mismatch is given by

$$\delta\left(\mathrm{u}_\mathrm{i}^{\mathrm{tot},2} - \mathrm{u}_\mathrm{i}^{\mathrm{tot},1}\right) = \nabla\left(\mathrm{u}_\mathrm{i}^{\mathrm{tot},2} - \mathrm{u}_\mathrm{i}^{\mathrm{tot},1}\right) \cdot \hat{\mathbf{n}}\delta\xi = \frac{\partial\left(\mathrm{u}_\mathrm{i}^{\mathrm{tot},2} - \mathrm{u}_\mathrm{i}^{\mathrm{tot},1}\right)}{\partial \mathrm{x}_l} \hat{\mathrm{n}}_l \delta\xi \tag{15.2}$$

and the work that must be performed in rebonding the surfaces in step 3 is then[2]

$$\delta\mathcal{W} = \mathrm{T_i}\delta\left(\mathrm{u_i^{tot,1}} - \mathrm{u_i^{tot,2}}\right)\mathrm{dS} = \sigma_{\mathrm{ij}}\hat{\mathrm{n}}_{\mathrm{j}}\delta\left(\mathrm{u_i^{tot,1}} - \mathrm{u_i^{tot,2}}\right)\mathrm{dS}. \quad (15.3)$$

Employing the previous equations, the change in total energy is therefore

$$\delta\mathrm{E} = \delta\mathrm{W} + \delta\mathcal{W} = \left[\left(\mathrm{w}^{(2)} - \mathrm{w}^{(1)}\right) - \sigma_{\mathrm{ij}}\hat{\mathrm{n}}_{\mathrm{j}}\left(\frac{\partial \mathrm{u_i^{tot,2}}}{\partial \mathrm{x}_l} - \frac{\partial \mathrm{u_i^{tot,1}}}{\partial \mathrm{x}_l}\right)\hat{\mathrm{n}}_l\right]\delta\xi\mathrm{dS} \quad (15.4)$$

Using Eq. (5.38), the local force per unit area is then

$$\mathcal{F} = -\frac{\delta\mathrm{E}}{\mathrm{dS}\delta\xi} = (\mathrm{w}^{(1)} - \mathrm{w}^{(2)}) - \sigma_{\mathrm{ij}}\hat{\mathbf{n}}_{\mathrm{j}}\left(\frac{\partial \mathbf{u}_{\mathrm{i}}^{\mathrm{tot,1}}}{\partial \mathrm{x}_l} - \frac{\partial \mathbf{u}_{\mathrm{i}}^{\mathrm{tot,2}}}{\partial \mathrm{x}_l}\right)\hat{\mathrm{n}}_l, \quad (15.5)$$

or, alternatively

$$\mathcal{F} = (\mathrm{w}^{(1)} - \mathrm{w}^{(2)}) - \mathrm{T_i}\left(\frac{\partial \mathrm{u_i^{tot,1}}}{\partial \mathrm{x}_l} - \frac{\partial \mathrm{u_i^{tot,2}}}{\partial \mathrm{x}_l}\right)\hat{\mathrm{n}}_l. \quad (15.6)$$

In this formulation the first term arises from the change in strain energy density that occurs when the interface is displaced, while the second arises from the work performed by the traction acting on the interface during the displacement. Finally, writing Eq. (15.6) in the further form

$$\mathcal{F} = \left[(\mathrm{w}^{(1)} - \mathrm{w}^{(2)})\delta_{\mathrm{ij}} - \sigma_{\mathrm{ij}}\left(\frac{\partial \mathbf{u}_{\mathrm{i}}^{\mathrm{tot,1}}}{\partial \mathrm{x}_l} - \frac{\partial \mathrm{u_i^{tot,2}}}{\partial \mathrm{x}_l}\right)\right]\hat{\mathrm{n}}_l\hat{\mathrm{n}}_{\mathrm{j}}, \quad (15.7)$$

and introducing Eq. (5.52), we obtain

$$\mathcal{F} = \left(\mathrm{P}_{\mathrm{j}l}^{(1)} - \mathrm{P}_{\mathrm{j}l}^{(2)}\right)\hat{\mathrm{n}}_l\hat{\mathrm{n}}_{\mathrm{j}}. \quad (15.8)$$

Equation (15.8) shows that the force depends upon the difference between the energy-momentum tensors of the two regions and is the justification for terming the force the Energy-Momentum Tensor Force.

---

[2]The total displacement may include a transformation strain (see Sec. 3.6 and Eq. (3.152)), or, in the case of a heterophase interface, a term accounting for a difference in the atomic volumes of the two adjoining media.

An analysis of this force acting on the interface between an inclusion and its matrix is given in Sec. 17.5.

15.3 The Interfacial Dislocation Force

Many interfaces move in a body as a result of the movement of discrete interfacial dislocations (Sutton and Balluffi 2008) thus changing the shape of the body. This can be seen most clearly by considering cases where such interfaces are present in bicrystals with simple shapes.

15.3.1 *Small-angle symmetric tilt interfaces*

The simplest cases involve small-angle interfaces composed of arrays of crystal dislocations as described in Sec. 14.2.2.1. Consider first a small-angle symmetric tilt interface in a bicrystal with the shape shown in Fig. 15.2. The interface is of the type illustrated in Fig. 14.1, and if it is displaced normal to itself (by the vector $\delta\boldsymbol{\xi}$) by the simultaneous glide of its edge dislocations so that its structure remains unchanged, the bicrystal undergoes the shape change illustrated in Fig. 15.2b.[3] The displacement of the interface by the distance $\delta\xi$ causes a shear displacement of the surface of crystal 1 parallel

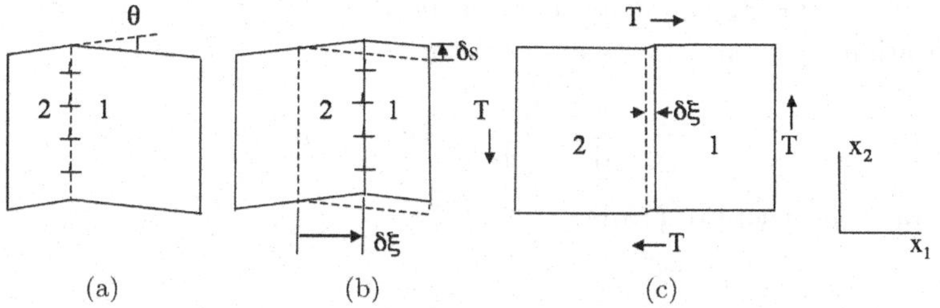

Fig. 15.2 (a,b) Change in bicrystal shape due to displacement, $\delta\xi$, of small-angle symmetric tilt interface. (b) Surface on right side is displaced by δs. (c) Shape change of bicrystal under shear tractions, T, due to the displacement of the tilt interface, $\delta\xi$.

[3]This is readily verified by simply constructing the same tilt interface at different distances from a fixed reference point located in crystal 2.

to the interface given by

$$\delta \mathrm{s} = 2\tan(\theta/2)\delta\xi \cong \delta\xi\theta. \tag{15.9}$$

If the bicrystal is subjected to an applied traction, as in Fig. 15.2c, the change in potential energy of the system resulting from the interface displacement (per unit area of interface) is then

$$\delta\Phi = -\mathrm{T}\delta\mathrm{s} = \theta\mathrm{T}\delta\xi \tag{15.10}$$

and, by using Eq. (5.38), the force per unit area on the interface is

$$\mathcal{F}^{\mathrm{INT/T}} = -\lim_{\delta\xi\to 0}\frac{\delta \mathrm{E}}{\delta\xi} = -\lim_{\delta\xi\to 0}\frac{\delta\Phi}{\delta\xi} = \theta\mathrm{T}. \tag{15.11}$$

Since the movement of the dislocations comprising the interface are responsible for the bicrystal shape change, the above force should be equal to the total force per unit area exerted on the individual dislocations in the unit area by the applied tractions. The edge dislocations are characterized by $\mathbf{b} = (\mathrm{b}, 0, 0)$ and $\hat{\mathbf{t}} = (0, 0, 1)$, and the tractions impose the shear stress, σ_{12}. Therefore, applying the Peach–Koehler force equation in the form of Eq. (13.11), the force per unit length on each dislocation is

$$\boldsymbol{f} = \mathbf{d} \times \hat{\mathbf{t}} = \mathrm{b}\sigma_{12}\hat{\mathbf{e}}_1. \tag{15.12}$$

Since the dislocation spacing is given by $\mathrm{d} = \mathrm{b}/\theta$, the total force on all dislocations per unit interface area is then

$$\mathcal{F} = \frac{f}{\mathrm{d}} = \frac{\mathrm{b}\sigma_{12}\theta}{\mathrm{b}} = \theta\sigma_{12} = \theta\mathrm{T} \tag{15.13}$$

in agreement with Eq. (15.11).

15.3.2 *Small-angle asymmetric tilt interfaces*

Consider next the small-angle asymmetric tilt interface shown in Fig. 15.3d. The first task is to determine its dislocation structure by use of the Frank–Bilby equation, with the Reference lattice, Crystals 1 and 2 and the Cartesian coordinate system all arranged, as shown in Fig. 15.3, in the same manner as in the case of the symmetric tilt interface treated previously in Ch. 14 (see Fig. 14.8 and Eqs. (14.25)–(14.27)). However, the interface is now inclined asymmetrically between the two crystals and lies at the angle α with respect to the (010) plane of the Reference lattice rather than at the

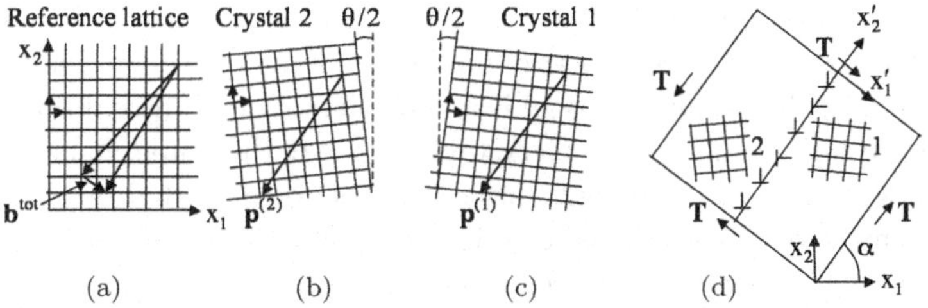

Fig. 15.3 Lattices and other quantities used for analysis of small-angle asymmetric tilt interface.

previous symmetrical (100) inclination. Equation (14.24), with $\mathbf{p}$ chosen as $\mathbf{p} = -\mathrm{p}(\cos\alpha, \sin\alpha, 0)$, then yields

$$\begin{bmatrix} b_1^{tot} \\ b_2^{tot} \\ b_3^{tot} \end{bmatrix} = \begin{bmatrix} 0 & -2\sin(\theta/2) & 0 \\ 2\sin(\theta/2) & 0 & 0 \\ 0 & 0 & 0 \end{bmatrix} \begin{bmatrix} -\mathrm{p}\cos\alpha \\ -\mathrm{p}\sin\alpha \\ 0 \end{bmatrix}$$
$$= \begin{bmatrix} 2\mathrm{p}\sin\alpha\sin(\theta/2) \\ -2\mathrm{p}\cos\alpha\sin(\theta/2) \\ 0 \end{bmatrix}, \tag{15.14}$$

which may be compared with Eq. (14.27) for the symmetrical case. Equation (15.14) is satisfied by two families of straight edge dislocations lying in the interface parallel to the x_3 axis of Fig. 15.3 with Burgers vectors $\mathbf{b}^\alpha = (\mathrm{a}, 0, 0)$ and $\mathbf{b}^\beta = (0, -\mathrm{a}, 0)$ at the spacings

$$\begin{aligned} d^\alpha &= \frac{a}{2\sin\alpha\sin(\theta/2)} \cong \frac{a}{\theta\sin\alpha}, \\ d^\beta &= \frac{a}{2\cos\alpha\sin(\theta/2)} \cong \frac{a}{\theta\cos\alpha}. \end{aligned} \tag{15.15}$$

The total Burgers vector strength of the dislocations in the boundary per unit distance normal to the tilt axis is therefore

$$\mathbf{b}^\alpha \frac{1}{d^\alpha} + \mathbf{b}^\beta \frac{1}{d^\beta} = (\sin\alpha\hat{\mathbf{e}}_1 - \cos\alpha\hat{\mathbf{e}}_2)\theta, \tag{15.16}$$

which is normal to the interface and of magnitude, θ. This is seen to be equal to the corresponding Burgers vector strength of the dislocations in the symmetrical tilt interface of Fig. 15.2 when the two interfaces have the same tilt angle.[4] The asymmetric interface can move without changing its structure, by the simultaneous glide and climb motion of these dislocations, causing a shape change of the bicrystal identical to the shape change caused by the motion of the symmetric interface. The force exerted on the asymmetric interface by the applied tractions shown in Fig. 15.3d must therefore be of the same form as the force on the symmetric boundary, i.e., the force given by Eq. (15.13). This result is verified in Exercise 15.1 by a calculation of the total force exerted on the individual interfacial dislocations. Further aspects of the force are considered in Exercise 15.2.

15.3.3 *Large-angle homophase interfaces*

Many large-angle homophase interfaces, which are either singular or vicinal and support localized interfacial dislocations, can move by the lateral motion of interfacial dislocations which are associated with steps in the interface plane and therefore possess step character (Balluffi 2006; Pond, Ma, Chai and Hirth 2007). An example is shown in Fig. 15.4 consisting of a large-angle symmetrical tilt interface possessing an interfacial edge dislocation with Burgers vector **b**, which is associated with a step in the interface of height h.[5] The change in body shape which occurs when such a dislocation moves across a bicrystal is illustrated in Fig. 15.5. As indicated by Fig. 15.5c, this consists of the displacement of Crystal 2 with respect to Crystal 1 by the Burgers vector and the simultaneous displacement of the interface by the step height. If the surface of the bicrystal parallel to the

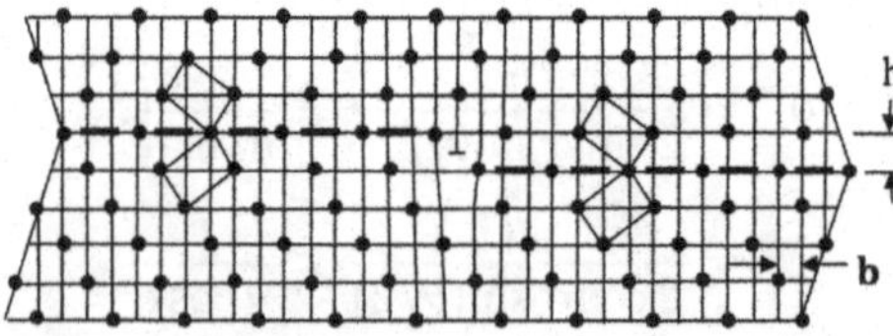

Fig. 15.4 Large-angle symmetric tilt interface containing an interfacial dislocation with Burgers vector **b**, associated with a step in the interface of height h.

[4]It is readily shown that the total Burgers vector strength in any tilt interface must be normal to the interface to avoid long-range stress (see Exercise 14.6).

[5]These types of interfacial line defects have been given many different names in the literature including *transformation dislocations*, *dislocations with step character* (Sutton and Balluffi 2006), and *disconnections* (Pond, Ma, Chai and Hirth 2007).

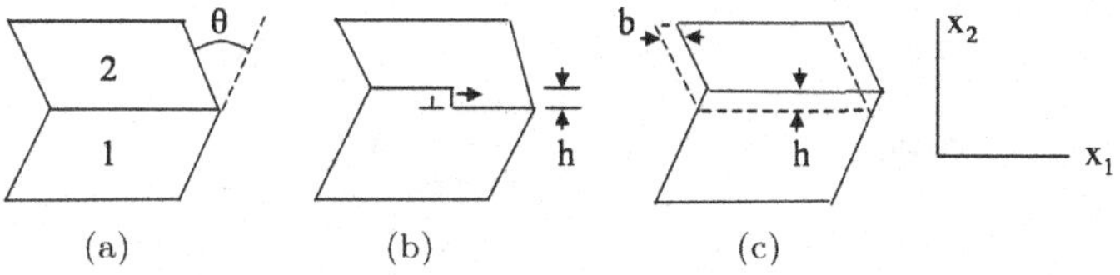

Fig. 15.5 Change in body shape of large-angle symmetric tilt interface by lateral motion of interfacial dislocation possessing step character. (a) Bicrystal containing initial interface. (b) Bicrystal after partial passage of dislocation, possessing step height h, across interface. (c) Bicrystal after complete passage.

interface is subjected to a shear traction $T_{12} = \sigma_{12}$, the change in potential energy due to the interface displacement is

$$\delta\Phi = -\sigma_{12}b \tag{15.17}$$

and the force on the interface per unit area is then

$$\mathcal{F} = -\frac{\delta E}{\delta\xi} = -\frac{\delta\Phi}{\delta\xi} = \frac{\sigma_{12}b}{h}. \tag{15.18}$$

15.3.4 *Heterophase interfaces*

A variety of heterophase interfaces, which are either singular or vicinal and can support localized interfacial dislocations, can also move by the lateral motion of interfacial dislocations possessing step character (Sutton and Balluffi 2006: Pond, Ma, Chai and Hirth 2007). In such cases the Burgers vectors of these dislocations generally possess components that are both normal and parallel to the interface, since the interplanar spacings of the relevant planes in the adjoining crystals (which are of different phase) do not match exactly. The lateral movement of such dislocations across the interface therefore causes a dilatation normal to the interface as well as a shear displacement.

In martensitic phase transformations the dislocation structure of the interface between the growing martensite phase and shrinking parent phase is relatively complex and frequently contains two arrays of interfacial dislocations where the first array possesses step character and the second acts to cancel the long-range stresses produced by the first (Pond, Ma, Chai and Hirth 2007). The forward motion of the interface is coupled to the lateral motion of the dislocations with step character and transforms

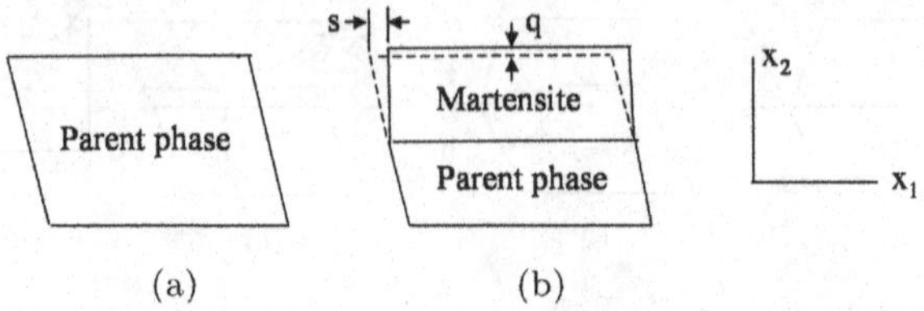

Fig. 15.6 Change of body shape due to martensitic phase transformation. (a) Before transformation. (b) Upper region transformed to martensite. Habit plane between phases is invariant.

the parent phase into the martensite phase. On a macroscopic scale, the interface (habit plane of the transformation) is an *invariant plane* of the transformation, meaning that it is neither strained nor rotated. The shape change due to the transformation, illustrated in Fig. 15.6, therefore involves a shear displacement parallel to the interface, s, and a dilatational displacement normal to the interface, q, owing to the difference in atomic density that generally exists between the martensitic and parent phases (Balluffi, Allen and Carter 2005).

Since s and q increase linearly with the displacement of the interface, a simple expression can be written for the force per unit interfacial area, $\mathcal{F}$, exerted on the interface by applied tractions in the coordinate system of Fig. 15.6. Expressing the increments $\delta \mathrm{s}$ and $\delta \mathrm{q}$ that result from an interface displacement $\delta\xi$ as

$$\begin{aligned} \delta \mathrm{s} &= \mathrm{k}^{\mathrm{s}}\delta\xi, \\ \delta \mathrm{q} &= \mathrm{k}^{\mathrm{q}}\delta\xi, \end{aligned} \tag{15.19}$$

where k^{s} and k^{q} depend upon the detailed dislocation structure of the interface, the force, per unit interfacial area, can be written in the simple form

$$\mathcal{F} = -\frac{\partial \Phi}{\partial \xi} = \frac{\sigma_{22}\delta \mathrm{q} + \sigma_{12}\delta \mathrm{s}}{\delta\xi} = \sigma_{22}\mathrm{k}^{\mathrm{q}} + \sigma_{12}\mathrm{k}^{\mathrm{s}}. \tag{15.20}$$

Exercises

15.1 Determine the force per unit area exerted on the asymmetric tilt interface of Fig. 15.3d by the tractions indicated in the figure, by summing the forces on the individual dislocations.

Solution. Using the primed coordinate system in Fig. 15.3d, the Burgers vectors of the two sets of dislocations are

$$\begin{aligned} \mathbf{b}^{\alpha} &= (\mathrm{a}\sin\alpha, \mathrm{a}\cos\alpha, 0), \\ \mathbf{b}^{\beta} &= (\mathrm{a}\cos\alpha, -\mathrm{a}\sin\alpha, 0). \end{aligned} \tag{15.21}$$

Next, by applying the Peach–Koehler equation, the forces exerted on the dislocations per unit length by the tractions $\mathrm{T}_{12} = \sigma_{12}$ are

$$\begin{aligned} \boldsymbol{f}^{\alpha} &= \mathrm{a}\sigma_{12}(\sin\alpha\hat{\mathbf{e}}'_1 - \cos\alpha\hat{\mathbf{e}}'_2), \\ \boldsymbol{f}^{\beta} &= \mathrm{a}\sigma_{12}(\cos\alpha\hat{\mathbf{e}}'_1 + \sin\alpha\hat{\mathbf{e}}'_2). \end{aligned} \tag{15.22}$$

Then, summing all the forces on the dislocations in unit interface area, and employing Eqs. (15.22) and (15.15), the total force per unit interfacial area is

$$\boldsymbol{f}^{\alpha}\frac{1}{\mathrm{d}^{\alpha}} + \boldsymbol{f}^{\beta}\frac{1}{\mathrm{d}^{\beta}} = \theta\sigma_{12}\hat{\mathbf{e}}'_1 \tag{15.23}$$

which is normal to the interface and in agreement with Eq. (15.13), as anticipated.

15.2 It has been shown in Sec. 15.3.2 that the force on the asymmetrical tilt interface of Fig. 15.3(d) is given by Eq. (15.13) if the interface moves by the forward motion of its edge dislocations by combined glide and climb without changing its structure. It might be conjectured that the interface could also move forward by the purely glissile motion of its edge dislocations on their respective intersecting slip planes. In such a case the interface structure would vary continuously. Show that for this type of motion, which requires the simultaneous gain and loss of dislocations from, and to, the bicrystal, the applied tractions in Fig. 15.3d exert no force on the interface.

Solution. We first show that if the interface moves as conjectured above, the macroscopic shape of the bicrystal will not change and the applied tractions will therefore be unable to exert any force on the interface. When the α dislocations in the interface glide through Crystal 1 (Fig. 15.3) as the interface advances they will shear the crystal. This can be described in terms of a *deformation matrix*, [D], defined as the matrix which, when applied to any vector in the non-deformed crystal, yields the corresponding deformed vector. After taking into account the Burgers vector and the spacing and inclination of the glide planes of the α dislocations, the deformation matrix describing the shear deformation due to the gliding dislocations, expressed in the $(\hat{e}_1', \hat{e}_2', \hat{e}_3')$ coordinate system, is found to be

$$[\mathrm{D}]^{\alpha} = \begin{bmatrix} 1-\theta\sin\alpha\cos\alpha & \theta\sin^2\alpha & 0 \\ -\theta\cos^2\alpha & 1+\theta\sin\alpha\cos\alpha & 0 \\ 0 & 0 & 1 \end{bmatrix} \tag{15.24}$$

and the corresponding matrix for the β dislocations is

$$[\mathrm{D}]^{\beta} = \begin{bmatrix} 1+\theta\sin\alpha\cos\alpha & \theta\cos^2\alpha & 0 \\ -\theta\sin^2\alpha & 1-\theta\sin\alpha\cos\alpha & 0 \\ 0 & 0 & 1 \end{bmatrix}. \tag{15.25}$$

When Crystal 1 is traversed by both families, the total distortion matrix (to first order in θ) is then

$$[\mathrm{D}] = [\mathrm{D}]^{\beta}[\mathrm{D}]^{\alpha} = \begin{bmatrix} 1 & \theta & 0 \\ -\theta & 1 & 0 \\ 0 & 0 & 1 \end{bmatrix} \tag{15.26}$$

which is seen to be a simple rotation of $-\theta$ around the $\hat{e}_3'$ axis. Then, to restore contact with Crystal 2, we must rotate crystal 1 back by θ so that the final distortion matrix is

$$[\mathrm{D}] = \begin{bmatrix} 1 & -\theta & 0 \\ \theta & 1 & 0 \\ 0 & 0 & 1 \end{bmatrix}\begin{bmatrix} 1 & \theta & 0 \\ -\theta & 1 & 0 \\ 0 & 0 & 1 \end{bmatrix} = \begin{bmatrix} 1 & 0 & 0 \\ 0 & 1 & 0 \\ 0 & 0 & 1 \end{bmatrix} \tag{15.27}$$

and the shape of the bicrystal is unchanged.

The same conclusion regarding the force on the interface when it moves by the above two processes can be reached by comparing the work which must be performed by the applied tractions in the two cases. When the interface moves the distance $\delta x_1'$ by the glide and climb of its dislocations without changing structure, the work done by the applied tractions must equal the work done to move the dislocations which is given by

$$\delta \mathcal{W} = \left(\frac{f_1^{\alpha}\sin\alpha}{d^{\alpha}} + \frac{f_1^{\beta}\cos\alpha}{d^{\beta}}\right)\delta x_1'. \tag{15.28}$$

This assumes the form

$$\delta \mathcal{W} = \theta\sigma_{12}\delta x_1' \tag{15.29}$$

after substituting Eqs. (15.15) and (15.22).

On the other hand, if the interface moves by the purely glissile motion of its dislocations,

$$\delta\mathcal{W} = \left(\frac{\boldsymbol{f}^{\alpha} \cdot \mathbf{b}^{\alpha}}{|\mathbf{b}^{\alpha}|\mathrm{d}^{\alpha}} \frac{1}{\sin\alpha} + \frac{\boldsymbol{f}^{\beta} \cdot \mathbf{b}^{\beta}}{|\mathbf{b}^{\beta}|\mathrm{d}^{\beta}} \frac{1}{\cos\alpha} \right) \delta \mathrm{x}_1' = 0 \qquad (15.30)$$

after use of Eqs. (15.15), (15.21) and (15.22). In this case, the applied tractions are unable to perform work and exert a force on the interface, since, in agreement with our previous conclusion, there is no change in the bicrystal shape when the interface moves.

Chapter 16

Interactions between Defects

16.1 Introduction

The general procedure for determining the interaction between two individual defects is to obtain the elastic field due to one defect at the location of the other, and then to determine the interaction of the latter defect with this field. The elastic fields produced by the defects considered in this book and the interactions between these defects and various elastic fields have been formulated in previous chapters: the fundamentals necessary to analyze the interaction energies and forces that can occur between the various defects are, therefore, now in place. Since the number of possible defect–defect interactions is far too large to consider comprehensively, we restrict this chapter to a limited number of representative interactions which serve to demonstrate basic methods.

16.2 Point Defect–Point Defect Interactions

First, a general formulation is given of the interaction energy between two point defects that are represented by force multipoles as in Chs. 10 and 11. Following this, the interaction between two point defects, each possessing cubic defect symmetry, is analyzed for the case of an isotropic system.

16.2.1 *General formulation*

The interaction energy between a single point defect and a general elastic field has been formulated in Ch. 11 (see Eqs. (11.1)–(11.5)), and the displacement field due to a force multipole is given by Eq. (10.10). The interaction energy between two point defects, i.e., D1 and D2, can therefore be obtained by imagining that D2 is created at the origin in the presence of D1 located at the position, $\mathbf{x}$, and using the previous results for the interaction energy and required elastic field.

After replacing the general displacement field Q in Eq. (11.1) by the displacement field of D2, the interaction energy assumes the form

$$E_{int}^{D1/D2}(\mathbf{x}) = -\sum_{q} u_j^{D2}\left(\mathbf{x} + \mathbf{s}^{D1(q)}\right) F_j^{D1(q)}, \tag{16.1}$$

where $u_j^{D2}(\mathbf{x} + \mathbf{s}^{D1(q)})$ is the displacement at the point $\mathbf{x} + \mathbf{s}^{D1(q)}$ due to the creation of D2 at the origin, and $F_j^{D1(q)}$ is the point force at $\mathbf{x} + \mathbf{s}^{D1(q)}$ associated with the D1 multipole whose center is at $\mathbf{x}$. Then, by expanding $u_j^{D2}(\mathbf{x} + \mathbf{s}^{D1(q)})$ around $\mathbf{x}$,

$$\begin{aligned} E_{int}^{D1/D2}(\mathbf{x}) = -\sum_{q} \Bigg[& u_j^{D2}(\mathbf{x}) + \frac{\partial u_j^{D2}}{\partial x_m} s_m^{D1(q)} + \frac{1}{2!}\frac{\partial^2 u_j^{D2}}{\partial x_m \partial x_n} s_m^{D1(q)} s_n^{D1(q)} \\ & + \frac{1}{3!}\frac{\partial^3 u_j^{D2}}{\partial x_m \partial x_n \partial x_r} s_m^{D1(q)} s_n^{D1(q)} s_r^{D1(q)} + \cdots \Bigg] F_j^{D1(q)} \end{aligned} \tag{16.2}$$

which can be expressed more compactly in the series form

$$E_{int}^{D1/D2}(\mathbf{x}) = -\sum_{s=1}^{\infty} \frac{1}{s!} P_{r_1 r_2 \ldots r_s i}^{D1} \frac{\partial^s u_i^{D2}(\mathbf{x})}{\partial x_{r_1} \partial x_{r_2} \ldots \partial x_{r_s}}. \tag{16.3}$$

Next, using Eq. (10.10) to write the displacement field of defect D2 in the series form

$$u_i^{D2}(\mathbf{x}) = \sum_{k=1}^{\infty} \frac{(-1)^k}{k!} \frac{\partial^k G_{ij}(\mathbf{x})}{\partial x_{q_1} \partial x_{q_2} \ldots \partial x_{q_k}} P_{q_1 q_2 \ldots q_k j}^{D2} \tag{16.4}$$

and substituting Eq. (16.4) into Eq. (16.3), the interaction energy between the two defects, separated by the vector $\mathbf{x}$, is given by

$$\begin{aligned} E_{int}^{D1/D2}(\mathbf{x}) = & -\sum_{s=1}^{\infty} \frac{1}{s!} P_{r_1 r_2 \ldots r_s i}^{D1} \sum_{k=1}^{\infty} \frac{(-1)^k}{k!} \\ & \times \frac{\partial^{k+s} G_{ij}(\mathbf{x})}{\partial x_{q_1} \partial x_{q_2} \ldots \partial x_{q_k} \partial x_{r_1} \partial x_{r_2} \ldots \partial x_{r_s}} P_{q_1 q_2 \ldots q_k j}^{D2}, \end{aligned} \tag{16.5}$$

where the subscripts q_k and r_s assume the usual values $q_k = (1, 2, 3)$ and $r_s = (1, 2, 3)$.

Interaction energies can then be determined by using Eq. (4.42) for the required Green's function derivatives and expressions for the force multipoles determined by the methods of Ch. 10.

16.2.2 *Between two point defects in isotropic system*

As an example of the application of Eq. (16.5), consider the case of two identical multipoles with cubic defect symmetry of the type illustrated in Fig. 10.4 in an isotropic system. Retaining only force dipole and octopole terms (quadrupole terms vanish because of the defect inversion symmetry), Eq. (16.5) takes the form

$$\begin{aligned}
E_{int}^{D1/D2}(\mathbf{x}) &= P_{r_1 i}P_{q_1 j}\frac{\partial^2 G_{ij}(\mathbf{x})}{\partial x_{q_1}\partial x_{r_1}} + \frac{1}{6}\left[P_{r_1 i}P_{q_1 q_2 q_3 j}\frac{\partial^4 G_{ij}(\mathbf{x})}{\partial x_{q_1}\partial x_{q_2}\partial x_{q_3}\partial x_{r_1}}\right. \\
&\quad \left. + P_{q_1 j}P_{r_1 r_2 r_3 i}\frac{\partial^4 G_{ij}(\mathbf{x})}{\partial x_{r_1}\partial x_{r_2}\partial x_{r_3}\partial x_{q_1}}\right] \\
&= P^{(1)}P^{(1)}\frac{\partial^2 G_{ij}(\mathbf{x})}{\partial x_i \partial x_j} + \frac{1}{3}P^{(1)}P^{(3)}\frac{\partial^4 G_{ij}(\mathbf{x})}{\partial x_i \partial x_j^3}
\end{aligned} \tag{16.6}$$

where $P^{(1)}$ and $P^{(3)}$ are the magnitudes of the force dipoles and octopoles, respectively. However, the leading dipole-dipole term vanishes since use of Eq. (4.110) shows that

$$\frac{\partial G_{ij}}{\partial x_j} = -\frac{(1-2\nu)}{8\pi\mu(1-\nu)}\frac{x_i}{x^3}, \qquad \frac{\partial^2 G_{ij}}{\partial x_i \partial x_j} = 0. \tag{16.7}$$

To evaluate the remaining dipole-octopole term, Eq. (16.6) is used to obtain

$$\frac{\partial^4 G_{ij}(\mathbf{x})}{\partial x_i \partial x_j^3} = \frac{21(1-2\nu)}{8\pi\mu(1-\nu)x^5}\left[-3 + 5\frac{(x_1^4 + x_2^4 + x_3^4)}{x^4}\right]. \tag{16.8}$$

Then, substituting this result into Eq. (16.6), and setting $l_i = x_i/x$,

$$E_{int}^{D1/D2}(\mathbf{x}) = P^{(1)}P^{(3)}\frac{35(1-2\nu)}{8\pi\mu(1-\nu)x^5}\left[-\frac{3}{5} + l_1^4 + l_2^4 + l_3^4\right], \tag{16.9}$$

where the angular factor is seen to be the same cubically symmetric factor that appeared previously in Eq. (10.21). This result shows that the first order dipole-dipole interaction energy vanishes for two cubically symmetric point defects in an isotropic material, but that a relatively short-range cubically symmetric interaction energy remains due to the dipole-octopole interaction. According to Eqs. (10.21) and (10.22) for the displacement fields of the present multipoles, each defect contributes a displacement field that is cubically symmetric, and divergenceless with e = 0 in the region around the multipole (see Eq. (8.5) and Exercise 10.2). As pointed out by Eshelby (1977), each defect looks for a hydrostatic pressure that its colleague does not provide. However, while this behavior is characteristic of isotropic systems, it is not so for anisotropic systems where the dipole-dipole term in Eq. (16.6) remains finite and falls off as x^{-3}. Further point defect interactions are considered in Exercises 16.5 and 16.6.

16.3 Dislocation–Dislocation Interactions

16.3.1 *Interaction energies*

16.3.1.1 *Between two rational infinitesimal segments*

The total interaction energy between two dislocations can be regarded as the sum of all the interaction energies between the differential segments comprising one dislocation and the differential segments of the other. Consider, therefore, the two loops in Fig. 16.1. If the interaction energy between segment $ds^{(1)}$ on dislocation $C^{(1)}$ and segment $ds^{(2)}$ on dislocation $C^{(2)}$, $W_{int}^{ds^{(1)}/ds^{(2)}}$, is known, then the total interaction energy between the loops, $W_{int}^{(1)/(2)}$, is given simply by the double line integral over both loops,

$$W_{int}^{(1)/(2)} = \oint_{C^{(2)}} \oint_{C^{(1)}} dW_{int}^{ds^{(1)}/ds^{(2)}} . \qquad (16.10)$$

A tractable formalism for obtaining the differential interaction energy, $dW_{int}^{ds^{(1)}/ds^{(2)}}$, has been obtained by Lothe (1992b), by treating the segments as rational differential segments possessing the properties described in

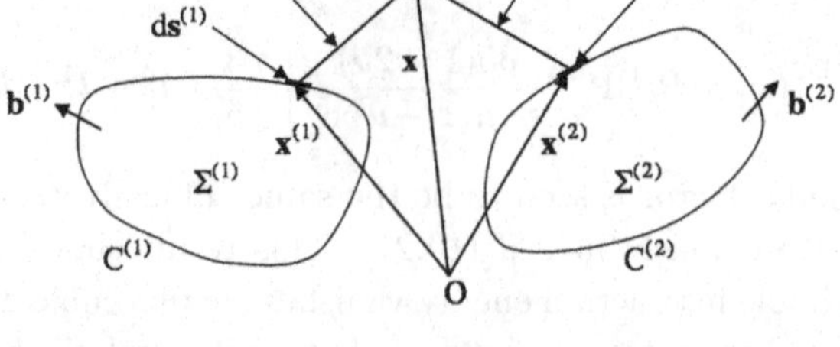

Fig. 16.1 Geometry for determining the interaction energy between dislocation loops $C^{(1)}$ and $C^{(2)}$.

Sec. 12.5.1.5. Following Lothe (1992b), the interaction energy $dW_{int}^{ds^{(1)}/ds^{(2)}}$ between two such segments, $\mathbf{ds}^{(1)}$ and $\mathbf{ds}^{(2)}$, arranged as in Fig. 16.1, can be obtained by imagining that segment $\mathbf{ds}^{(2)}$ is created in the presence of the stress field due to segment $\mathbf{ds}^{(1)}$ and finding the work, $d\mathcal{W}$, done by this stress field during the creation of segment $\mathbf{ds}^{(2)}$ by the cuts and displacements required to produce the dipoles that create segment $ds^{(2)}$ as described in Sec. 12.5.1.5 (Fig. 12.17). Using the results in Sec. 12.5.1.5, the work performed (per unit volume) in a differential disk-shaped volume of area dS and thickness dr (see Fig. 12.17b) can be expressed as

$$\begin{aligned} d\mathcal{W}' &= \left(\frac{d\sigma_{ij}^{(1)}\hat{n}_j^{dip}b_i^{dip}}{N}\right)\left(\frac{|\mathbf{ds}^{(2)}\times\mathbf{r}|}{r}dr\right)\left(\frac{N}{4\pi r^2}dS\right)\frac{1}{dSdr} \\ &= \frac{d\sigma_{ij}^{(1)}\hat{n}_j^{dip}b_i^{(2)}|\mathbf{ds}^{(2)}\times\mathbf{r}|}{4\pi r^3}, \end{aligned} \tag{16.11}$$

where the first bracketed quantity is the work performed per unit area of dipole created, the second bracketed quantity, obtained with the use of Eq. (12.121), is the area within the volume dV = dSdr per dipole created, and the third, obtained with the use of Eq. (12.122), is the number of dipoles threading the area dS. Equation (16.11) can be further developed by using the relationship corresponding to Eq. (12.126), written in the form $\hat{\mathbf{n}}^{dip} = \mathbf{ds}^{(2)}\times\mathbf{r}/|\mathbf{ds}^{(2)}\times\mathbf{r}|$, so that

$$\hat{n}_j^{dip}|\mathbf{ds}^{(2)}\times\mathbf{r}| = (\mathbf{ds}^{(2)}\times\mathbf{r})_j = ds_m^{(2)}r_n e_{jmn} \tag{16.12}$$

after using Eq. (E.5). Then, substituting Eq. (16.12) into Eq. (16.11),

$$d\mathcal{W}' = \frac{d\sigma_{ij}^{(1)}b_i^{(2)}e_{jmn}ds_m^{(2)}r_n}{4\pi r^3}. \tag{16.13}$$

Equation (16.13) can now be expressed very simply by substituting Eq. (12.128), which represents the transformation strain produced by all the dipoles, into it to obtain

$$d\mathcal{W}' = d\sigma_{ij}^{(1)}d\varepsilon_{ij}^{T(2)}. \tag{16.14}$$

However, the interaction energy between segments per unit volume of the elastic field, $dw_{int}^{dl^{(1)}/dl^{(2)}}$, must be equal to $-d\mathcal{W}'$, and, therefore,

$$dw_{int}^{dl^{(1)}/dl^{(2)}} = -d\mathcal{W}' = -d\sigma_{ij}^{(1)}d\varepsilon_{ij}^{T(2)}. \tag{16.15}$$

By employing the coordinates of Fig. 16.1, the total differential interaction energy is obtained by integrating over the entire volume so that

$$dW_{int}^{ds^{(1)}/ds^{(2)}} = -\int_{-\infty}^{\infty}\int_{-\infty}^{\infty}\int_{-\infty}^{\infty} d\sigma_{ij}^{(1)}(\mathbf{x}-\mathbf{x}^{(1)}) \times d\varepsilon_{ij}^{T(2)}(\mathbf{x}-\mathbf{x}^{(2)})dx_1dx_2dx_3. \quad (16.16)$$

However, if two functions, such as $f(\mathbf{x}-\mathbf{x}^{(1)})$ and $g(\mathbf{x}-\mathbf{x}^{(2)})$, have the respective inverse transforms,

$$\begin{aligned} f(\mathbf{x}-\mathbf{x}^{(1)}) &= \int_{-\infty}^{\infty}\int_{-\infty}^{\infty}\int_{-\infty}^{\infty} \bar{f}(\mathbf{k})e^{i\mathbf{k}\cdot\mathbf{x}}dk_1dk_2dk_3 \\ g(\mathbf{x}-\mathbf{x}^{(2)}) &= \int_{-\infty}^{\infty}\int_{-\infty}^{\infty}\int_{-\infty}^{\infty} \bar{g}(\mathbf{k})e^{i\mathbf{k}\cdot\mathbf{x}}dk_1dk_2dk_3 \end{aligned} \quad (16.17)$$

Parseval's theorem (Sneddon 1951) states that

$$\int_{-\infty}^{\infty}\int_{-\infty}^{\infty}\int_{-\infty}^{\infty} f(\mathbf{x}-\mathbf{x}^{(1)})g(\mathbf{x}-\mathbf{x}^{(2)})dx_1dx_2dx_2 = (2\pi)^3\int_{-\infty}^{\infty}\int_{-\infty}^{\infty}\int_{-\infty}^{\infty} \bar{f}(\mathbf{k})\bar{g}(-\mathbf{k})e^{-i\mathbf{k}\cdot(\mathbf{x}^{(1)}-\mathbf{x}^{(2)})}dk_1dk_2dk_3. \quad (16.18)$$

Then, by employing this theorem, Eq. (16.16) can be written as

$$dW_{int}^{ds^{(1)}/ds^{(2)}} = -(2\pi)^3\int_{-\infty}^{\infty}\int_{-\infty}^{\infty}\int_{-\infty}^{\infty} d\bar{\sigma}_{ij}^{(1)}(\mathbf{k}) \times d\bar{\varepsilon}_{ij}^{T(2)}(-\mathbf{k})e^{-i\mathbf{k}\cdot(\mathbf{x}^{(1)}-\mathbf{x}^{(2)})}dk_1dk_2dk_3 \quad (16.19)$$

The transforms $d\bar{\sigma}_{ij}^{(1)}(\mathbf{k})$ and $d\bar{\varepsilon}_{ij}^{T(2)}(-\mathbf{k})$ appearing in Eq. (16.19) can be obtained from Eqs. (12.133) and (12.132), respectively, and substituting these quantities into Eq. (16.19), and applying the symmetry properties of the C^*_{ijkl} tensor,

$$dW_{int}^{ds^{(1)}/ds^{(2)}} = \frac{1}{8\pi^3}\int_{-\infty}^{\infty}\int_{-\infty}^{\infty}\int_{-\infty}^{\infty} C^*_{ijkl}(\mathbf{k})k^{-4}b_k^{(1)}(d\mathbf{s}^{(1)}\times\mathbf{k})_l b_i^{(2)} \times (d\mathbf{s}^{(2)}\times\mathbf{k})_j e^{-i\mathbf{k}\cdot(\mathbf{x}^{(1)}-\mathbf{x}^{(2)})}dk_1dk_2dk_3. \quad (16.20)$$

The amplitude of the Fourier expression given by Eq. (16.20), with C^*_{ijkl} given by Eq. (3.163), is a homogeneous function of degree −2 in the variable k, and the theorem given by Eq. (12.100) can therefore be employed to obtain the interaction energy between two differential segments in the form

$$\mathrm{dW}_{\mathrm{int}}^{\mathrm{ds}^{(1)}/\mathrm{ds}^{(2)}} = \frac{\mathrm{b}_\mathrm{k}^{(1)}\mathrm{b}_\mathrm{i}^{(2)}}{8\pi^2|\mathbf{x}^{(1)}-\mathbf{x}^{(2)}|}\int_0^{2\pi} C^*_{\mathrm{ijk}l}(\hat{\mathbf{m}})(\mathbf{ds}^{(1)}\times\hat{\mathbf{m}})_l(\mathbf{ds}^{(2)}\times\hat{\mathbf{m}})_\mathrm{j}\mathrm{d}\theta, \tag{16.21}$$

where $\hat{\mathbf{m}}$ lies in the plane perpendicular to the vector $\hat{\mathbf{w}} = (\mathbf{x}^{(1)} - \mathbf{x}^{(2)})/|\mathbf{x}^{(1)} - \mathbf{x}^{(2)}|$, and the integration with respect to θ is around a unit circle in that plane as illustrated in Fig. 12.11.

16.3.1.2 *Between two loops*

Having obtained an expression for the interaction energy between differential segments $\mathbf{ds}^{(1)}$ and $\mathbf{ds}^{(2)}$ on the respective dislocation loops $C^{(1)}$ and $C^{(2)}$ in Fig. 16.1, the total interaction energy between the two loops can now be readily obtained by means of the double line integration expressed by Eq. (16.10). Therefore, substituting Eq. (16.21) into Eq. (16.10),

$$\begin{aligned}\mathrm{W}_{\mathrm{int}}^{(1)/(2)} &= \oint_{C^{(1)}}\oint_{C^{(2)}} \mathrm{dW}_{\mathrm{int}}^{\mathrm{ds}^{(1)}/\mathrm{ds}^{(2)}}\\ &= \frac{\mathrm{b}_\mathrm{k}^{(1)}\mathrm{b}_\mathrm{i}^{(2)}}{8\pi^2}\oint_{C^{(1)}}\oint_{C^{(2)}}\frac{1}{|\mathbf{x}^{(1)}-\mathbf{x}^{(2)}|}\\ &\quad\times\int_0^{2\pi} C^*_{\mathrm{ijk}l}(\hat{\mathbf{m}})(\mathbf{ds}^{(1)}\times\hat{\mathbf{m}})_l(\mathbf{ds}^{(2)}\times\hat{\mathbf{m}})_\mathrm{j}\mathrm{d}\theta.\end{aligned} \tag{16.22}$$

16.3.1.3 *Between two straight segments*

The interaction energy associated with two straight dislocation segments that are parts of loops $C^{(1)}$ and $C^{(2)}$, respectively, can be obtained in a similar manner as previously by integrating Eq. (16.21) along the two segments so that

$$\begin{aligned}\mathrm{W}_{\mathrm{int}}^{(1)/(2)} &= \frac{\mathrm{b}_\mathrm{k}^{(1)}\mathrm{b}_\mathrm{i}^{(2)}}{8\pi^2}\int_{\mathcal{L}^{(1)}}\int_{\mathcal{L}^{(2)}}\frac{1}{|\mathbf{x}^{(1)}-\mathbf{x}^{(2)}|}\\ &\quad\times\int_0^{2\pi} C^*_{\mathrm{ijk}l}(\hat{\mathbf{m}})(\mathbf{ds}^{(1)}\times\hat{\mathbf{m}})_l(\mathbf{ds}^{(2)}\times\hat{\mathbf{m}})_\mathrm{j}\mathrm{d}\theta.\end{aligned} \tag{16.23}$$

Parallel segments

When the segments are parallel, the coordinate system shown in Fig. 16.2 is convenient. Then,

$$ds^{(1)} = \hat{t}dx_3^{(1)} \qquad ds^{(2)} = \hat{t}dx_3^{(2)}$$

$$\left|x^{(1)} - x^{(2)}\right| = R = [p^2 + (x_3^{(1)} - x_3^{(2)})^2]^{1/2} \tag{16.24}$$

and Eq. (16.23) takes the form

$$W_{\text{int}}^{(1)/(2)} = \frac{b_k^{(1)} b_i^{(2)}}{8\pi^2} \int_{\mathcal{L}^{(1)}} \int_{\mathcal{L}^{(2)}} \frac{dx_3^{(1)} dx_3^{(2)}}{\left[p^2 + (x_3^{(1)} - x_3^{(2)})^2\right]^{1/2}}$$

$$\times \int_0^{2\pi} C_{ijkl}^*(\hat{m})(\hat{t} \times \hat{m})_l(\hat{t} \times \hat{m})_j d\theta. \tag{16.25}$$

Non-parallel segments

When the segments are non-parallel, Hirth and Lothe (1982) have shown that it is convenient to adopt the oblique coordinate system with base vectors $(\hat{t}^{(1)}, \hat{t}^{(2)}, \hat{e}_3)$ illustrated in Fig. 16.3. Here, $\hat{e}_3$ lies along the line of

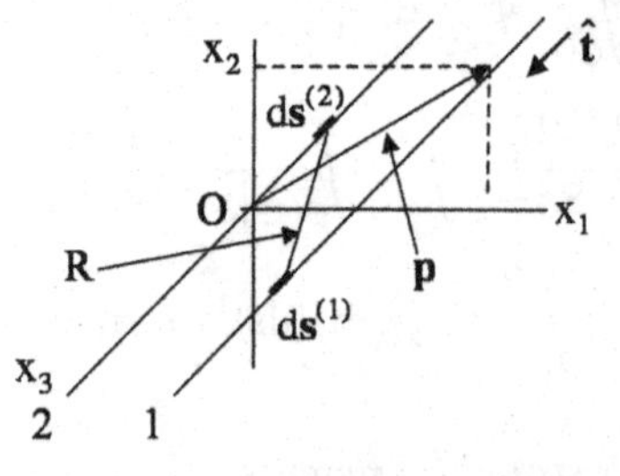

Fig. 16.2 Coordinate system for determining interaction energy between straight parallel segments 1 and 2 separated by the distance p.

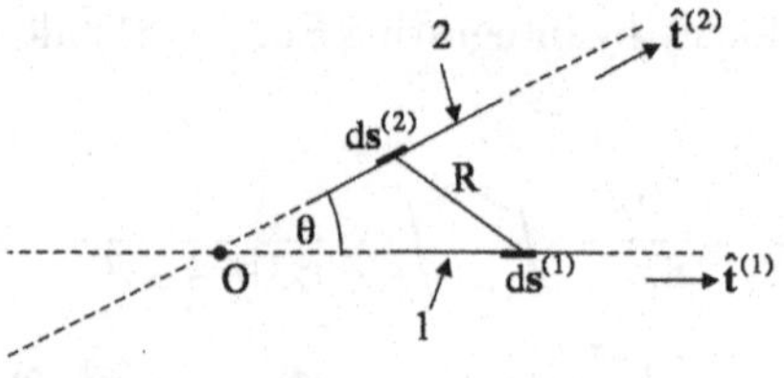

Fig. 16.3 Oblique coordinate system for determining interaction energy between non-parallel segments 1 and 2. $\hat{e}_3$ is normal to the paper pointing towards the reader.

closest approach of the two lines collinear with the two segments (shown dashed) and is given by $\hat{\mathbf{e}}_3 = (\hat{\mathbf{t}}^{(1)} \times \hat{\mathbf{t}}^{(2)})/|\hat{\mathbf{t}}^{(1)} \times \hat{\mathbf{t}}^{(2)}|$. Distances measured along the three axes are represented by $(\mathrm{s}^{(1)}, \mathrm{s}^{(2)}, \mathrm{z})$, and therefore,

$$\mathbf{ds}^{(1)} = \hat{\mathbf{t}}\mathrm{ds}^{(1)} \qquad \mathbf{ds}^{(2)} = \hat{\mathbf{t}}\mathrm{ds}^{(2)}$$

$$|\mathbf{x}^{(1)} - \mathbf{x}^{(2)}| = \mathrm{R} = [(\mathrm{s}^{(1)})^2 + (\mathrm{s}^{(2)})^2 + \mathrm{s}^{(1)}\mathrm{s}^{(2)}\cos\theta + \mathrm{z}^2]^{1/2} \qquad (16.26)$$

and Eq. (16.23) assumes the form

$$\mathrm{W}_{\mathrm{int}}^{(1)/(2)} = \frac{\mathrm{b}_\mathrm{k}^{(1)}\mathrm{b}_\mathrm{i}^{(2)}}{8\pi^2}\int_{\mathcal{L}^{(1)}}\int_{\mathcal{L}^{(2)}}\frac{\mathrm{ds}^{(1)}\mathrm{ds}^{(2)}}{[(\mathrm{s}^{(1)})^2 + (\mathrm{s}^{(2)})^2 + \mathrm{s}^{(1)}\mathrm{s}^{(2)}\cos\theta + \mathrm{z}^2]^{1/2}}$$
$$\times \int_0^{2\pi} \mathrm{C}^*_{\mathrm{ijk}l}(\hat{\mathbf{m}})(\hat{\mathbf{t}}^{(1)} \times \hat{\mathbf{m}})_l(\hat{\mathbf{t}}^{(2)} \times \hat{\mathbf{m}})_\mathrm{j}\mathrm{d}\theta. \qquad (16.27)$$

16.3.2 *Interaction energies in isotropic systems*

16.3.2.1 *Between two loops*

Instead of following the procedure employed in Sec. 16.3.1.2 to determine the interaction energy between loops $\mathrm{C}^{(1)}$ and $\mathrm{C}^{(2)}$, another approach, used by Blin (1955) and Hirth and Lothe (1982), can be employed. Here, the extra work, $\mathcal{W}$, performed by the stress field of loop $\mathrm{C}^{(2)}$ when loop $\mathrm{C}^{(1)}$ is created in its presence by the cut and displacement method is determined. The result is then transformed by use of Stokes' theorem into a double line integral involving the differential segments comprising both loops.

The two loops are shown in Fig. 16.4, where the traction on the surface of the cut, $\Sigma^{(1)}$, due to the stress field of loop $\mathrm{C}^{(2)}$ is $\sigma_{\mathrm{pq}}^{(2)}\hat{\mathrm{n}}_\mathrm{q}^{(1)}$. The interaction energy between the loops is then

$$\mathrm{W}_{\mathrm{int}}^{(1)/(2)} = -\mathcal{W} = -\iint_{\Sigma^{(1)}} \sigma_{\mathrm{pq}}^{(2)}\hat{\mathrm{n}}_\mathrm{q}^{(1)}\mathrm{b}_\mathrm{p}^{(1)}\mathrm{dS}. \qquad (16.28)$$

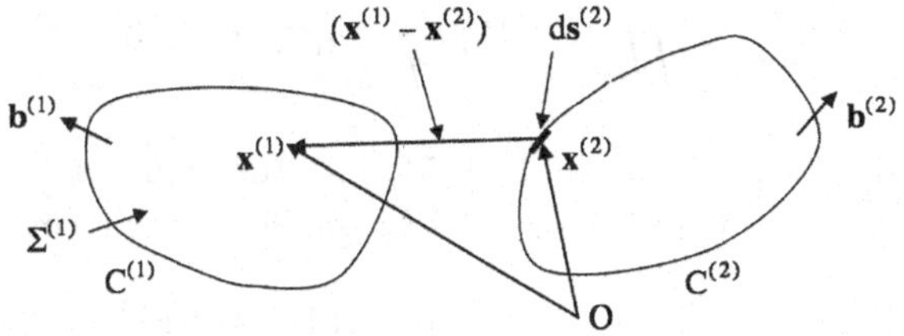

Fig. 16.4 Geometry for determining the interaction energy between dislocation loops $\mathrm{C}^{(1)}$ and $\mathrm{C}^{(2)}$.

For an isotropic system, the stress, $\sigma^{(2)}_{pq}$, can be obtained from the Peach–Koehler equation, i.e., Eq. (12.162), which consists of four terms. Starting with the first term, and substituting it into Eq. (16.28),

$$W^{(1)/(2)}_{int}(1) = -\mathcal{W}(1) = -\frac{\mu}{8\pi}e_{imp}b^{(2)}_m b^{(1)}_p \times \iint_{\Sigma^{(1)}} \frac{\partial}{\partial x^{(1)}_i}(\nabla^2 R)\hat{n}^{(1)}_q dS \oint_{C^{(2)}} dx^{(2)}_q. \quad (16.29)$$

Note that here the source vector has been changed according to $\mathbf{x}' \to \mathbf{x}^{(2)}$ and the field vector by $\mathbf{x} \to \mathbf{x}^{(1)}$. Also, $R = |\mathbf{x}^{(1)} - \mathbf{x}^{(2)}|$, and $\partial/\partial x^{(1)}_i = -\partial/\partial x^{(2)}_i$. The surface integral in Eq. (16.29) can now be converted to a line integral over $C^{(1)}$ by applying Stokes' theorem in the form given by Eq. (B.7) so that

$$\iint_{\Sigma^{(1)}} \frac{\partial}{\partial x^{(1)}_i}(\nabla^2 R)\hat{n}^{(1)}_q dS = \iint_{\Sigma^{(1)}} \frac{\partial}{\partial x^{(1)}_q}(\nabla^2 R)\hat{n}^{(1)}_i dS + e_{qik}\oint_{C^{(1)}} \nabla^2 R dx^{(1)}_k. \quad (16.30)$$

Then, substituting this into Eq. (16.29),

$$W^{(1)/(2)}_{int}(1) = -\frac{\mu}{8\pi}e_{imp}b^{(2)}_m b^{(1)}_p e_{qik}\oint_{C^{(1)}} (\nabla^2 R)dx^{(1)}_k \oint_{C^{(2)}} dx^{(2)}_q \quad (16.31)$$

since,

$$\begin{aligned}\iint_{\Sigma^{(1)}} \frac{\partial}{\partial x^{(1)}_q}(\nabla^2 R)\hat{n}^{(1)}_i dS \oint_{C^{(2)}} dx^{(2)}_q &= -\iint_{\Sigma^{(1)}} \hat{n}^{(1)}_i dS \oint_{C^{(2)}} \frac{\partial}{\partial x^{(2)}_q}(\nabla^2 R)dx^{(2)}_q \\ &= -\iint_{\Sigma^{(1)}} \hat{n}^{(1)}_i dS \oint_{C^{(2)}} d(\nabla^2 R) = 0. \quad (16.32)\end{aligned}$$

Then, by using Eqs. (E.3) and (E.5) and the equality $\nabla^2 R = 2/R$, Eq. (16.31) can be put into the relatively simple vector form

$$W^{(1)/(2)}_{int}(1) = -\frac{\mu}{4\pi}\oint_{C^{(1)}}\oint_{C^{(2)}} \frac{(\mathbf{b}^{(1)} \times \mathbf{b}^{(2)}) \cdot (\mathbf{ds}^{(1)} \times \mathbf{ds}^{(2)})}{R}. \quad (16.33)$$

The remaining three terms from the Peach–Koehler equation can be substituted and treated in a similar manner as shown in detail by Hirth and Lothe (1982). These results can then be consolidated into the final expression for the interaction energy between loops $C^{(1)}$ and $C^{(2)}$,

$$\begin{aligned} W_{\text{int}}^{(1)/(2)} &= -\frac{\mu}{2\pi}\oint_{C^{(1)}}\oint_{C^{(2)}}\frac{(\mathbf{b}^{(1)}\times\mathbf{b}^{(2)})\cdot(\mathbf{ds}^{(1)}\times\mathbf{ds}^{(2)})}{R} \\ &+\frac{\mu}{4\pi}\oint_{C^{(1)}}\oint_{C^{(2)}}\frac{(\mathbf{b}^{(1)}\cdot\mathbf{ds}^{(1)})(\mathbf{b}^{(2)}\cdot\mathbf{ds}^{(2)})}{R} \\ &+\frac{\mu}{4\pi(1-\nu)}\oint_{C^{(1)}}\oint_{C^{(2)}}(\mathbf{b}^{(1)}\times\mathbf{ds}^{(1)})\cdot\underline{\mathbf{T}}\cdot(\mathbf{b}^{(2)}\times\mathbf{ds}^{(2)}), \end{aligned} \tag{16.34}$$

where $\underline{\mathbf{T}}$ is the tensor with components

$$T_{ij} = \frac{\partial^2 R}{\partial x_i^{(1)}\partial x_j^{(1)}} = \frac{\partial^2 R}{\partial x_i^{(2)}\partial x_j^{(2)}}. \tag{16.35}$$

16.3.2.2 *Between two straight segments*

Having Eq. (16.34), the interaction energy associated with two straight dislocation segments in an isotropic system can be obtained by imagining that segments 1 and 2 are parts of loops $C^{(1)}$ and $C^{(2)}$, respectively, and then integrating Eq. (16.34) along the two segments so that

$$\begin{aligned} W_{\text{int}}^{(1)/(2)} &= -\frac{\mu}{2\pi}\int_{\mathcal{L}^{(1)}}\int_{\mathcal{L}^{(2)}}\frac{(\mathbf{b}^{(1)}\times\mathbf{b}^{(2)})\cdot(\mathbf{ds}^{(1)}\times\mathbf{ds}^{(2)})}{R} \\ &+\frac{\mu}{4\pi}\int_{\mathcal{L}^{(1)}}\int_{\mathcal{L}^{(2)}}\frac{(\mathbf{b}^{(1)}\cdot\mathbf{ds}^{(1)})(\mathbf{b}^{(2)}\cdot\mathbf{ds}^{(2)})}{R} \\ &+\frac{\mu}{4\pi(1-\nu)}\int_{\mathcal{L}^{(1)}}\int_{\mathcal{L}^{(2)}}(\mathbf{b}^{(1)}\times\mathbf{ds}^{(1)})\cdot\underline{\mathbf{T}}\cdot(\mathbf{b}^{(2)}\times\mathbf{ds}^{(2)}), \end{aligned} \tag{16.36}$$

where the line integrals, $\mathcal{L}^{(1)}$ and $\mathcal{L}^{(2)}$, involving $\mathbf{ds}^{(1)}$ and $\mathbf{ds}^{(2)}$ are taken along the segments 1 and 2, respectively.

Parallel segments

When the segments are parallel, it is again convenient to employ the coordinate system in Fig. 16.2. In Exercise 16.1 this coordinate system is used to integrate Eq. (16.36) to obtain the interaction energy (per unit length) associated with two infinitely long parallel dislocations separated by the distance p, in the form

$$w_{\text{int}}^{(1)/(2)} = -\frac{\mu}{2\pi}\left\{\left[\mathbf{b}^{(1)}\cdot\hat{\mathbf{t}})(\mathbf{b}^{(2)}\cdot\hat{\mathbf{t}}) + \frac{1}{1-\nu}(\mathbf{b}^{(1)}\times\hat{\mathbf{t}})\cdot(\mathbf{b}^{(2)}\times\hat{\mathbf{t}})\right]\ln\frac{p}{p_\circ}\right.$$
$$\left. + \frac{1}{1-\nu}\frac{[(\mathbf{b}^{(1)}\times\hat{\mathbf{t}})\cdot\mathbf{p}][(\mathbf{b}^{(2)}\times\hat{\mathbf{t}})\cdot\mathbf{p}]}{p^2}\right\}. \tag{16.37}$$

Hirth and Lothe (1982) have evaluated many of the integrals that arise in applying Eq. (16.36) to parallel coplanar segments lying in the $x_2 = 0$ plane and have shown that they can be obtained as elementary functions.

Non-parallel segments

When the segments are non-parallel, it is again convenient to employ the oblique coordinate system in Fig. 16.3 with distances along the three axes measured by $\left(s^{(1)}, s^{(2)}, z\right)$, along with Eq. (16.26). To evaluate Eq. (16.36), an expression for the tensor, T_{ij}, in the oblique system is needed. Writing the del operator as

$$\nabla = \frac{\partial}{\partial s^{(1)}}\hat{\mathbf{t}}^{(1)} + \frac{\partial}{\partial s^{(2)}}\hat{\mathbf{t}}^{(2)} + \frac{\partial}{\partial z}\hat{\mathbf{e}}_3 = \nabla_1\hat{\mathbf{t}}^{(1)} + \nabla_2\hat{\mathbf{t}}^{(2)} + \nabla_3\hat{\mathbf{e}}_3, \tag{16.38}$$

T_{ij} can be expressed as

$$T_{ij} = \nabla_i\nabla_j R. \tag{16.39}$$

The ∇_1 and ∇_2 components can then be obtained using the vector diagram in Fig. 16.5.

Since $R\left(s^{(1)}, s^{(2)}, z\right)$ lies between the points $\left(s^{(1)}, 0, z\right)$ and $\left(0, s^{(2)}, 0\right)$, the ∇ operator describes the effect of moving the end points of r along the

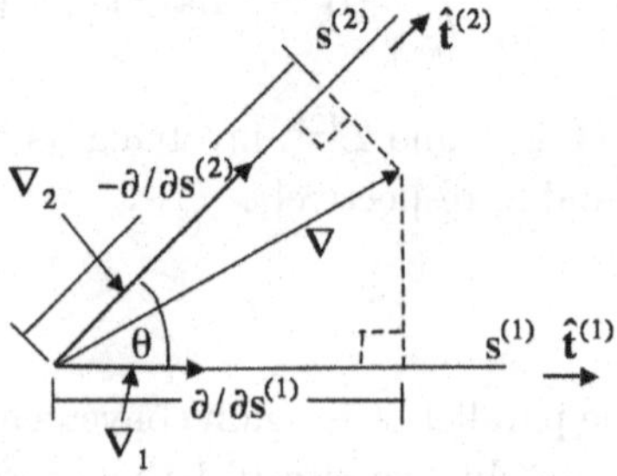

Fig. 16.5 Vector diagram, projected along $-\hat{\mathbf{e}}_3$, showing components ∇_1 and ∇_2 of the ∇ operator in the oblique $\left(\hat{\mathbf{t}}^{(1)}, \hat{\mathbf{t}}^{(2)}, \hat{\mathbf{e}}_3\right)$ coordinate system.

coordinate axes. Since the effect of moving the $(s^{(1)}, 0, z)$ end along $\mathbf{t}^{(2)}$ is equivalent to moving the $(0, s^{(2)}, 0)$ end the same distance in the opposite direction along $-\mathbf{t}^{(2)}$, the operator must have projections along the three coordinate axes corresponding to $\partial/\partial s^{(1)}$, $-\partial/\partial s^{(2)}$ and $\partial/\partial z$ as shown (in projection) in Fig. 16.5. The components of ∇ must then have the forms

$$\nabla_1 = \frac{\partial}{\partial s^{(1)}} - \nabla_2 \cos\theta, \quad \nabla_2 = -\frac{\partial}{\partial s^{(2)}} - \nabla_1 \cos\theta, \quad \nabla_3 = \frac{\partial}{\partial z}. \qquad (16.40)$$

Solving these three equations for the ∇_i quantities, and substituting the results into Eq. (16.38),

$$\begin{aligned}\nabla = {} & \frac{1}{\sin^2\theta}\left(\frac{\partial}{\partial s^{(1)}} + \cos\theta\frac{\partial}{\partial s^{(2)}}\right)\mathbf{t}^{(1)} \\ & - \frac{1}{\sin^2\theta}\left(\cos\theta\frac{\partial}{\partial s^{(1)}} + \frac{\partial}{\partial s^{(2)}}\right)\mathbf{t}^{(2)} + \frac{\partial}{\partial z}\mathbf{e}_3. \end{aligned} \qquad (16.41)$$

The components of $\underline{\mathbf{T}}$ can now be determined by use of Eq. (16.39). For example,

$$\begin{aligned} T_{12} &= \nabla_1\nabla_2 R = -\frac{1}{\sin^4\theta}\left(\cos\theta\frac{\partial^2 R}{\partial x^2} + (1+\cos^2\theta)\frac{\partial^2 R}{\partial x \partial y} + \cos\theta\frac{\partial^2 R}{\partial y^2}\right), \\ T_{22} &= \nabla_2\nabla_2 R = \frac{1}{\sin^4\theta}\left(\cos\theta\frac{\partial^2 R}{\partial x^2} + 2\cos\theta\frac{\partial^2 R}{\partial x \partial y} + \cos\theta\frac{\partial^2 R}{\partial y^2}\right). \end{aligned} \qquad (16.42)$$

Hirth and Lothe (1982) have evaluated many of the integrals that arise in applying the previous formalism to non-parallel segments in the present oblique coordinate system. As for the previous case of parallel segments, the results can be expressed as elementary functions.

16.3.3 *Interaction forces*

When the interaction energy between two dislocations is known, the corresponding force between them may always be determined by differentiating the interaction energy with respect to variations in their relative coordinates. However, it is often more efficient to determine the forces directly by use of the Peach–Koehler force equation. This approach is therefore pursued in this section, and the force between differential segments that are parts of two different loops is obtained. Having this, the forces between closed loops and straight finite segments are derived by means of line integrations.

16.3.3.1 *Between two rational infinitesimal segments*

Consider the two differential rational segments $\mathbf{ds}^{(1)}$ and $\mathbf{ds}^{(2)}$ on loops $C^{(1)}$ and $C^{(2)}$, respectively, in Fig. 16.6. Using the Peach–Koehler force equation, Eq. (13.11), segment $\mathbf{ds}^{(2)}$ exerts a force on segment $\mathbf{ds}^{(1)}$ given by

$$\mathbf{dF} = \left(\mathbf{d} \times \hat{\mathbf{t}}^{(1)}\right) \mathrm{ds}^{(1)} = b_i^{(1)} d\sigma_{ij}^{(2)} \left(\hat{\mathbf{e}}_j \times \mathbf{ds}^{(1)}\right), \tag{16.43}$$

where $d\sigma_{ij}^{(2)}$ is the stress imposed on the segment $\mathbf{ds}^{(1)}$ at the position $\mathbf{x}^{(1)}$ by the segment $\mathbf{ds}^{(2)}$ at $\mathbf{x}^{(2)}$. According to Eq. (12.134), this quantity is given by

$$d\sigma_{ij}^{(2)}(\mathbf{x}^{(1)} - \mathbf{x}^{(2)}) = -\frac{1}{8\pi^3} \int_{-\infty}^{\infty} \int_{-\infty}^{\infty} \int_{-\infty}^{\infty} C^*_{ijkl}(\mathbf{k}) b_k^{(2)} (\mathbf{ds}^{(2)} \times \mathbf{k})_l \times e^{i\mathbf{k}\cdot(\mathbf{x}^{(1)} - \mathbf{x}^{(2)})} ik^{-2} dk_1 dk_2 dk_3 \tag{16.44}$$

and substituting Eq. (16.44) into Eq. (16.43), the differential force is therefore

$$\mathbf{dF} = -\frac{1}{8\pi^3} b_i^{(1)} b_k^{(2)} \int_{-\infty}^{\infty} \int_{-\infty}^{\infty} \int_{-\infty}^{\infty} C^*_{ijkl}(\mathbf{k}) (\hat{\mathbf{e}}_j \times \mathbf{ds}^{(1)}) \times (\mathbf{ds}^{(2)} \times \mathbf{k})_l e^{i\mathbf{k}\cdot(\mathbf{x}^{(1)} - \mathbf{x}^{(2)})} ik^{-2} dk_1 dk_2 dk_3. \tag{16.45}$$

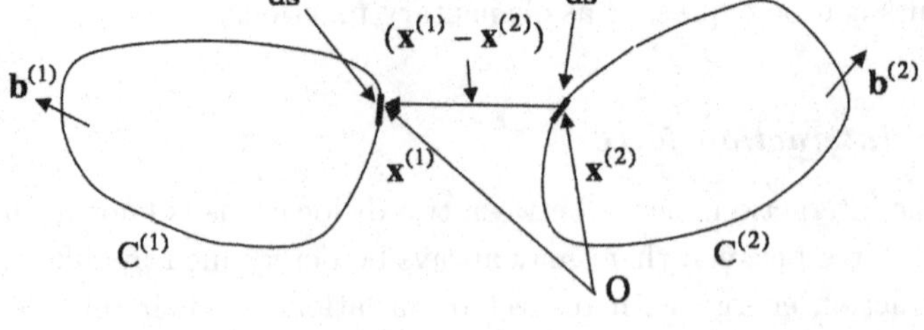

Fig. 16.6 Geometry for determining force exerted on segment $\mathbf{ds}^{(1)}$ by segment $\mathbf{ds}^{(2)}$.

16.3.3.2 *Between two loops*

The total force between two loops, such as in Fig. 16.6, is obtained by performing a line integration of Eq. (16.45) around each loop, i.e.,

$$\mathbf{F} = -\frac{1}{8\pi^3} b_i^{(1)} b_k^{(2)} \oint_{C^{(1)}} (\hat{\mathbf{e}}_j \times \mathbf{ds}^{(1)}) \times \oint_{C^{(2)}} \int_{-\infty}^{\infty}\int_{-\infty}^{\infty}\int_{-\infty}^{\infty} C^*_{ijkl}(\mathbf{k})(\mathbf{ds}^{(2)} \times \mathbf{k})_l \times e^{i\mathbf{k}\cdot(\mathbf{x}^{(1)}-\mathbf{x}^{(2)})} i k^{-2} dk_1 dk_2 dk_3. \tag{16.46}$$

16.3.3.3 *Between two straight segments*

The force between two straight segments, such as in Fig. 16.2, or 16.3, is obtained by performing a line integration of Eq. (16.45) along each segment, i.e.,

$$\mathbf{F} = -\frac{1}{8\pi^3} b_i^{(1)} b_k^{(2)} \int_{\mathcal{L}^{(1)}} (\hat{\mathbf{e}}_j \times \mathbf{ds}^{(1)}) \times \int_{\mathcal{L}^{(2)}} \int_{-\infty}^{\infty}\int_{-\infty}^{\infty}\int_{-\infty}^{\infty} C^*_{ijkl}(\mathbf{k})(\mathbf{ds}^{(2)} \times \mathbf{k})_l \times e^{i\mathbf{k}\cdot(\mathbf{x}^{(1)}-\mathbf{x}^{(2)})} i k^{-2} dk_1 dk_2 dk_3. \tag{16.47}$$

16.3.4 *Interaction forces in isotropic systems*

16.3.4.1 *Between two infinitesimal segments*

The force between two differential segments, such as illustrated in Fig. 16.6, is again given by Eq. (16.43) where $d\sigma_{ij}^{(2)}$ is the stress imposed on segment 1 by segment 2. In the present isotropic case $d\sigma_{ij}^{(2)}$ can be obtained from the Peach–Koehler equation, Eq. (12.162), in the modified form

$$d\sigma_{\alpha\beta}^{(2)}\left(\mathbf{x}^{(1)}\right) = \frac{\mu}{4\pi} e_{im\alpha} b_m^{(2)} \frac{\partial}{\partial x_i^{(1)}}\left(\frac{1}{R}\right) ds_\beta^{(2)} + \frac{\mu}{4\pi} e_{im\beta} b_m^{(2)} \frac{\partial}{\partial x_i^{(1)}}\left(\nabla\frac{1}{R}\right) ds_\alpha^{(2)} + \frac{\mu}{4\pi(1-\nu)} e_{imk} b_m^{(2)} \times \left[\frac{\partial^3 R}{\partial x_i^{(1)} \partial x_\alpha^{(1)} \partial x_\beta^{(1)}} - \frac{\delta_{\alpha\beta}}{2}\frac{\partial}{\partial x_i^{(1)}}\left(\nabla\frac{1}{R}\right)\right] ds_k^{(2)} \tag{16.48}$$

with $R = |\mathbf{x}^{(1)} - \mathbf{x}^{(2)}|$ and $\partial/\partial x_i^{(1)} = -\partial/\partial x_i^{(2)}$. Substituting Eq. (16.48) into Eq. (16.43), a result is obtained, after considerable algebra, that can be expressed in the vector form

$$\begin{aligned} d\mathbf{F} = &-\frac{\mu}{8\pi}[(\mathbf{b}^{(2)} \times \mathbf{b}^{(1)}) \cdot \nabla(\nabla^2 R)][d\mathbf{s}^{(1)} \times d\mathbf{s}^{(2)}] \\ &- \frac{\mu}{8\pi}\{[\mathbf{b}^{(2)} \times \nabla(\nabla^2 R)] \times d\mathbf{s}^{(1)}\}[\mathbf{b}^{(1)} \cdot d\mathbf{s}^{(2)}] \\ &- \frac{\mu}{4\pi(1-\nu)}[(\mathbf{b}^{(2)} \times d\mathbf{s}^{(2)}) \cdot \nabla][d\mathbf{s}^{(1)} \times (\mathbf{b}^{(1)}\underline{\mathbf{T}})] \\ &+ \frac{\mu}{4\pi(1-\nu)}[(\mathbf{b}^{(2)} \times d\mathbf{s}^{(2)}) \cdot \nabla(\nabla^2 R)][d\mathbf{s}^{(1)} \times \mathbf{b}^{(1)}], \end{aligned} \tag{16.49}$$

where $\nabla = (\partial/\partial x_i^{(1)})\hat{\mathbf{e}}_i$, $\underline{\mathbf{T}}$ is the tensor

$$T_{ij} = \frac{\partial^2 R}{\partial \mathbf{x}_i^{(1)} \partial \mathbf{x}_j^{(1)}} \tag{16.50}$$

and $\mathbf{b}^{(1)}\underline{\mathbf{T}}$ corresponds to the vector $b_j^{(1)} T_{ji} \hat{\mathbf{e}}_i$.

16.3.4.2 *Between two loops and between two straight segments*

The force between two loops in an isotropic system can be obtained by integrating Eq. (16.49) around the two loops using the same approach employed above to formulate Eq. (14.46). Similarly, the force between two straight segments can be found by integrating Eq. (16.49) along the two segments using the same approach employed previously to formulate Eq. (16.47). Hirth and Lothe (1982) treat the problem of the force between two straight segments in detail and show that the line integrals involved can be obtained as elementary functions. However, the analysis is lengthy, and the reader is referred to their work for detailed results.

In Exercise 16.3, Eq. (16.49) is employed to find the force exerted by a straight screw dislocation segment on an orthogonal differential screw segment.

16.4 Inclusion–Inclusion Interactions

16.4.1 *Between two homogeneous inclusions*

The interaction energy between homogeneous inclusions 1 and 2 in a body, $\mathcal{V}^\circ$, with a free surface, $\mathcal{S}^\circ$, is now determined. The first task is to formulate the total energy of the system, which, since transformation strains are present (Sec. 3.6), is written as

$$\begin{aligned} E = W &= \frac{1}{2} \oiiint\limits_{\mathcal{V}^\circ} \left(\sigma_{ij}^{(1)} + \sigma_{ij}^{(2)}\right)\left(\varepsilon_{ij}^{(1)} + \varepsilon_{ij}^{(2)}\right) dV \\ &= \frac{1}{2} \oiiint\limits_{\mathcal{V}^\circ} \left(\sigma_{ij}^{(1)} + \sigma_{ij}^{(2)}\right)\left[\left(\varepsilon_{ij}^{tot(1)} - \varepsilon_{ij}^{T(1)}\right) + \left(\varepsilon_{ij}^{tot(2)} - \varepsilon_{ij}^{T(2)}\right)\right] dV, \end{aligned} \tag{16.51}$$

where the sources of the stresses and strain are indicated by the superscripts. Then, using Eqs. (3.153) and (2.65),

$$\begin{aligned} W = &\frac{1}{2} \oiiint\limits_{\mathcal{V}^\circ} \frac{\partial}{\partial x_j}\left[\left(\sigma_{ij}^{(1)} + \sigma_{ij}^{(2)}\right)\left(u_i^{tot(1)} + u_i^{tot(2)}\right)\right] dV \\ &- \frac{1}{2} \oiiint\limits_{\mathcal{V}^\circ} \left(\sigma_{ij}^{(1)} + \sigma_{ij}^{(2)}\right)\left(\varepsilon_{ij}^{T(1)} + \varepsilon_{ij}^{T(2)}\right) dV \end{aligned} \tag{16.52}$$

and converting the first integral to a surface integral,

$$\begin{aligned} W = &\frac{1}{2} \oiint\limits_{\mathcal{S}^\circ} \left(\sigma_{ij}^{(1)} + \sigma_{ij}^{(2)}\right)\left(u_i^{tot(1)} + u_i^{tot(2)}\right) \hat{n}_j dS \\ &- \frac{1}{2} \oiiint\limits_{\mathcal{V}^\circ} \left(\sigma_{ij}^{(1)} + \sigma_{ij}^{(2)}\right)\left(\varepsilon_{ij}^{T(1)} + \varepsilon_{ij}^{T(2)}\right) dV. \end{aligned} \tag{16.53}$$

However, since $\sigma_{ij}^{(1)}\hat{n}_j = \sigma_{ij}^{(2)}\hat{n}_j = 0$ on $\mathcal{S}^\circ$,

$$\begin{aligned} W &= -\frac{1}{2} \oiiint\limits_{\mathcal{V}^\circ} \left(\sigma_{ij}^{(1)} + \sigma_{ij}^{(2)}\right)\left(\varepsilon_{ij}^{T(1)} + \varepsilon_{ij}^{T(2)}\right) dV \\ &= -\frac{1}{2}\left[\oiiint\limits_{\mathcal{V}^{(1)}} \left(\sigma_{ij}^{(1)} + \sigma_{ij}^{(2)}\right)\varepsilon_{ij}^{T(1)} dV + \oiiint\limits_{\mathcal{V}^{(2)}} \left(\sigma_{ij}^{(1)} + \sigma_{ij}^{(2)}\right)\varepsilon_{ij}^{T(2)} dV\right]. \end{aligned} \tag{16.54}$$

Using Eq. (6.57), the inclusion self-energies are

$$\begin{aligned} W^{(1)} &= -\frac{1}{2} \oiiint\limits_{\mathcal{V}^{(1)}} \sigma_{ij}^{(1)}\varepsilon_{ij}^{T(1)} dV \\ W^{(2)} &= -\frac{1}{2} \oiiint\limits_{\mathcal{V}^{(2)}} \sigma_{ij}^{(2)}\varepsilon_{ij}^{T(2)} dV \end{aligned} \tag{16.55}$$

and the interaction energy between inclusions, according to Eq. (5.4), is therefore

$$E_{int}^{INC(1)/INC(2)} = W - W^{(1)} - W^{(2)}$$
$$= -\frac{1}{2}\left[\oiiint_{\mathcal{V}^{(1)}} \sigma_{ij}^{(2)}\varepsilon_{ij}^{T(1)}dV + \oiiint_{\mathcal{V}^{(2)}} \sigma_{ij}^{(1)}\varepsilon_{ij}^{T(2)}dV\right]. \quad (16.56)$$

The two integrals, $\oiiint_{\mathcal{V}^{(1)}} \sigma_{ij}^{(2)}\varepsilon_{ij}^{T(1)}dV$ and $\oiiint_{\mathcal{V}^{(2)}} \sigma_{ij}^{(1)}\varepsilon_{ij}^{T(2)}dV$, in Eq. (16.56) can now be shown to be equal. Begin by writing the first integral as

$$\oiiint_{\mathcal{V}^{(1)}} \sigma_{ij}^{(2)}\varepsilon_{ij}^{T(1)}dV = \oiiint_{\mathcal{V}^{\circ}} \sigma_{ij}^{(2)}\varepsilon_{ij}^{T(1)}dV$$
$$= \oiiint_{\mathcal{V}^{\circ}} \sigma_{ij}^{(2)}\left(\varepsilon_{ij}^{T(1)} - \frac{\partial u_i^{tot(1)}}{\partial x_j} + \frac{\partial u_i^{tot(1)}}{\partial x_j}\right)dV. \quad (16.57)$$

However,

$$\sigma_{ij}^{(2)}\left(\varepsilon_{ij}^{T(1)} - \frac{\partial u_i^{tot(1)}}{\partial x_j}\right) = \sigma_{ij}^{(2)}\left(\varepsilon_{ij}^{T(1)} - \varepsilon_{ij}^{tot(1)}\right) = -\sigma_{ij}^{(2)}\varepsilon_{ij}^{(1)} \quad (16.58)$$

and, with the use of Eq. (2.65), the divergence theorem, and the condition, $\sigma_{ij}^{(2)}\hat{n}_j = 0$ on $\mathcal{S}^{\circ}$,

$$\oiiint_{\mathcal{V}^{\circ}} \sigma_{ij}^{(2)}\frac{\partial u_i^{tot(1)}}{\partial x_j}dV = \oiiint_{\mathcal{V}^{\circ}} \frac{\partial\left(\sigma_{ij}^{(2)}u_i^{tot(1)}\right)}{\partial x_j}dV$$
$$= \oiint_{\mathcal{S}^{\circ}} \sigma_{ij}^{(2)}u_i^{tot(1)}\hat{n}_j dS = 0. \quad (16.59)$$

Therefore, substituting Eqs. (16.58) and (16.59) into Eq. (16.57)

$$\oiiint_{\mathcal{V}^{(1)}} \sigma_{ij}^{(2)}\varepsilon_{ij}^{T(1)}dV = -\oiiint_{\mathcal{V}^{\circ}} \sigma_{ij}^{(2)}\varepsilon_{ij}^{(1)}dV. \quad (16.60)$$

A similar exercise shows that

$$\oiiint_{\mathcal{V}^{(2)}} \sigma_{ij}^{(1)}\varepsilon_{ij}^{T(2)}dV = -\oiiint_{\mathcal{V}^{\circ}} \sigma_{ij}^{(1)}\varepsilon_{ij}^{(2)}dV. \quad (16.61)$$

Now, using the symmetry properties of the C_{ijkl} tensor,

$$\sigma_{ij}^{(1)}\varepsilon_{ij}^{(2)} = C_{ijkl}\varepsilon_{kl}^{(1)}\varepsilon_{ij}^{(2)} = C_{klij}\varepsilon_{ij}^{(1)}\varepsilon_{kl}^{(2)}$$
$$= C_{ijkl}\varepsilon_{ij}^{(1)}\varepsilon_{kl}^{(2)} = \sigma_{ij}^{(2)}\varepsilon_{ij}^{(1)}. \quad (16.62)$$

Then, substituting Eq. (16.62) into Eq. (16.61), and comparing the result with Eq. (16.60), it is seen that the equality

$$\oiiint\limits_{\mathcal{V}^{(2)}} \sigma_{ij}^{(1)}\varepsilon_{ij}^{T(2)}dV = \oiiint\limits_{\mathcal{V}^{(1)}} \sigma_{ij}^{(2)}\varepsilon_{ij}^{T(1)}dV \quad (16.63)$$

is valid. Finally, substituting Eq. (16.63) into Eq. (16.56), the interaction energy is given by either of the two forms,

$$E_{int}^{INC(1)/INC(2)} = -\oiiint\limits_{\mathcal{V}^{(1)}} \sigma_{ij}^{(2)}\varepsilon_{ij}^{T(1)}dV = -\oiiint\limits_{\mathcal{V}^{(2)}} \sigma_{ij}^{(1)}\varepsilon_{ij}^{T(2)}dV \quad (16.64)$$

which are seen to be of the same form as, for example, Eq. (5.13).

The two forms of this result can be readily understood by imagining first that inclusion 1 is added to the system in the presence of inclusion 2. The increase in the energy of the system is then the self-energy of inclusion 1 plus the interaction energy between the two inclusions which should be equal to the interaction energy between inclusion 1 and the stress field of inclusion 2 as given by Eq. (7.1), i.e., $-\oiiint_{\mathcal{V}^{(1)}} \sigma_{ij}^{(2)}\varepsilon_{ij}^{T(1)}dV$, as is seen to be the case. On the other hand, if inclusion 2 is added in the presence of inclusion 1, the interaction energy should correspond to the interaction energy between inclusion 2 and the stress field of inclusion 1 as given by Eq. (7.1), which, again, is the case. (See the related discussion in Exercises 5.4 and 5.5.)

The interaction energy is then obtained by placing one inclusion at the origin and determining its stress field by the methods of Ch. 6 and then employing Eq. (16.64) to integrate the product of this stress with the transformation strain of the second inclusion over the volume of the second inclusion.

16.4.2 *Between two inhomogeneous inclusions*

The determination of the interaction energy between two inhomogeneous inclusions is considerably more difficult than for the previous homogeneous inclusions because the inhomogeneity associated with each inclusion perturbs the elastic field generated by the other as discussed in Sec. 7.2.2.

The strain fields within each inclusion are therefore non-uniform, and the problem generally possesses low symmetry and so cannot be solved in simple closed form using standard functions. Such problems, including the closely related problem of two inhomogeneities under an applied stress, have therefore been treated in the literature as boundary value problems or by the equivalent inclusion method where solutions are obtained by the use of polynomial expansions, series expansions and series representations. Some examples may be found in Mura (1987), Meisner and Kouris (1995) and Chalon and Montheillet (2003) and the references given there. In many cases solutions have been constructed by employing Papkovitch (Sec. 4.3.1.1), Boussinesq (1885), or Neuber (1944) displacement functions. The results are generally lengthy and require numerical methods for evaluation. We therefore will not pursue this approach further.

16.5 Point Defect–Dislocation Interactions

16.5.1 *General formulation*

The interaction between a point defect, i.e., *D*, and a dislocation can be formulated by modeling the point defect as a force multipole lying with its center at the position $\mathbf{x}$ in the elastic field of the dislocation located at the origin. The interaction energy is then given by Eq. (16.1) or (16.3) after writing them in the respective forms

$$E_{int}^{D/DIS}(\mathbf{x}) = -\sum_{q} u_j^{DIS}(\mathbf{x} + \mathbf{s}^{D(q)})F_j^{D(q)} \tag{16.65}$$

and

$$E_{int}^{D/DIS}(\mathbf{x}) = -\sum_{s=1}^{\infty} \frac{1}{s!} P^{D}_{r_1 r_2 \ldots r_s i} \frac{\partial^s u_i^{DIS}(\mathbf{x})}{\partial x_{r_1} \partial x_{r_2} \ldots \partial x_{r_s}}. \tag{16.66}$$

Solutions for various cases can then be obtained by using expressions for force multipoles from Ch. 10 and dislocation elastic fields from Ch. 6. In the following, a solution is obtained for the relatively simple case of a point defect with tetragonal symmetry in the field of an infinitely long straight screw dislocation in an isotropic system. Also, see Exercise 16.7 for an alternative derivation of Eq. (16.65).

16.5.2 *Between point defect and screw dislocation in isotropic system*

The classic approach of Cochardt, Schoeck and Wiedersich (1955) is followed to determine the interaction energy between an interstitial tetragonal point defect in the BCC structure, with its tetragonal axis along [100], as shown in Fig. 16.7a, and an infinitely long straight screw dislocation with $\mathbf{b} = (0, 0, b)$ and $\hat{\mathbf{t}} = (1, 1, 1)/\sqrt{3}$. However, the point defect is modeled here as a force multipole rather than a small misfitting region corresponding to a transformation strain as in the Cochardt, Schoeck and Wiedersich treatment.

To determine the interaction energy when the point defect is in different regions of the elastic field of the dislocation, use is made of the (x_1, x_2, x_3) and (x'_1, x'_2, x'_3) coordinate systems shown in Fig. 16.7b originally introduced by Cochardt, Schoeck and Wiedersich (1955). The geometry of this figure has the following features:

(1) The (x_1, x_2, x_3) system is aligned with the crystal axes.
(2) The (x'_1, x'_2, x'_3) system is coupled to the dislocation; x'_3 is parallel to the dislocation; x'_1 lies in the (111) plane at the angle, ϕ, with respect to the reference direction $(2, -1, -1)/\sqrt{6}$ corresponding to the projection

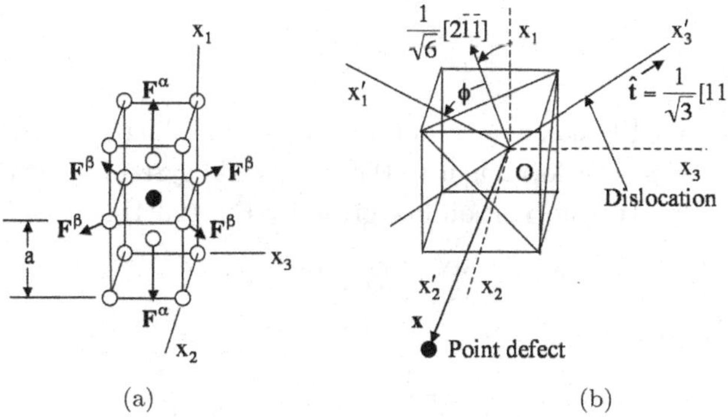

Fig. 16.7 Interstitial point defect with tetragonal symmetry in BCC structure and interacting screw dislocation. (a) Detailed view of defect and assumed forces acting on near-neighbor host atoms. (b) Straight screw dislocation with $\mathbf{b} = (0, 0, b)$ and $\hat{\mathbf{t}} = (1, 1, 1)/\sqrt{3}$ passing through origin. Point defect, shown in (a), now located at vector position $\mathbf{x}$. Coordinate systems (x_1, x_2, x_3) and (x'_1, x'_2, x'_3) are indicated.

of x_1 on the (111) plane; x'_2 lies in the (111) plane and is therefore perpendicular to the dislocation.

(3) The defect lies at the field position, **x**, which lies along x'_2. The position of the defect relative to the dislocation can then be varied by varying x, its perpendicular distance from the dislocation, and ϕ, causing it to rotate around the dislocation.

Using Eq. (16.66), and adopting the force dipole moment approximation (Sec. 10.3.5), the interaction energy, expressed in the (x'_1, x'_2, x'_3) system, takes the form

$$
\begin{aligned}
E_{int}^{D/DIS} &= -P_{ij}^{D'} \frac{\partial u_j^{DIS'}}{\partial x_i} = -P_{ij}^{D'} \varepsilon_{ij}^{DIS'} \\
&= -P_{ij}^{D'} \left(\frac{-\lambda \Theta^{DIS'}}{2\mu(3\lambda + 2\mu)} \delta_{ij} + \frac{1}{2\mu} \sigma_{ij}^{DIS'} \right)
\end{aligned} \tag{16.67}
$$

after making use of Eqs. (10.9) and (2.123). It is readily verified, by use of Eq. (12.55), that the only non-vanishing component of the dislocation stress tensor at the defect is

$$\sigma_{13}^{DIS'} = -\frac{\mu b}{2\pi x} \tag{16.68}$$

and substituting this into Eq. (16.67),

$$E_{int}^{D/DIS} = \frac{b}{2\pi x} P_{13}^{D'}. \tag{16.69}$$

The force dipole $P_{13}^{D'}$ can now be obtained by first finding it in the (x_1, x_2, x_3) system and then transforming it to the (x'_1, x'_2, x'_3) system by employing the standard tensor transformation law given by Eq. (2.24) so that,

$$P_{ij}^{D'} = l_{im} l_{jn} P_{mn}^{D}. \tag{16.70}$$

From Fig. 16.7a,

$$P_{mn}^{D} = a \begin{bmatrix} F^{\alpha} & 0 & 0 \\ 0 & \sqrt{2}F^{\beta} & 0 \\ 0 & 0 & \sqrt{2}F^{\beta} \end{bmatrix} \tag{16.71}$$

and, substituting this into Eq. (16.70),

$$P_{13}^{D'} = a \left[l_{11} l_{31} F^{\alpha} + (l_{12} l_{23} + l_{13} l_{33}) \sqrt{2} F^{\beta} \right]. \tag{16.72}$$

From Fig. 16.7b the matrix of direction cosines is deduced to be

$$l_{ij} = \begin{bmatrix} 2\cos\phi/\sqrt{6} & -\cos\phi/\sqrt{6}+\sin\phi/\sqrt{2} & -\cos\phi/\sqrt{6}-\sin\phi/\sqrt{2} \\ -2\sin\phi/\sqrt{6} & \sin\phi/\sqrt{6}-1/\sqrt{2} & \sin\phi/\sqrt{6}+1/\sqrt{2} \\ 1/\sqrt{3} & 1/\sqrt{3} & 1/\sqrt{3} \end{bmatrix} \tag{16.73}$$

and, therefore, substituting Eq. (16.73) into (16.72),

$$P_{13}^{D'} = \frac{\sqrt{2}a\cos\phi}{3}(F^{\alpha} - \sqrt{2}F^{\beta}). \tag{16.74}$$

Finally, upon substituting Eq. (16.74) into Eq. (16.69),

$$E_{int}^{D/DIS}(\mathbf{x}, \phi) = \frac{ba}{3\sqrt{2}\pi}(F^{\alpha} - \sqrt{2}F^{\beta})\frac{\cos\phi}{x}. \tag{16.75}$$

The interaction energy is seen to depend upon the degree of tetragonality of the point defect, as measured by the magnitude of the quantity $(F^{\alpha} - \sqrt{2}F^{\beta})$, and the defect position with respect to the dislocation as measured by the distance x and the angle ϕ. Note from Fig. 16.7a that for the special condition, $F^{\alpha} = \sqrt{2}F^{\beta}$, the defect is represented by three equal and orthogonal double forces, and both the tetragonality and interaction energy vanish.

16.6 Point Defect–Inclusion Interactions

16.6.1 *General formulation*

The interaction between a point defect, D, and an inclusion can be formulated by modeling the point defect as a force dipole lying with its center at the position **x** in the elastic field of the inclusion located at the origin. The interaction energy is then given by Eqs. (16.1) and (16.3) in the forms

$$E_{int}^{D/INC}(\mathbf{x}) = -\sum_{q} u_j^{INC}(\mathbf{x} + \mathbf{s}^{D(q)})F_j^{D(q)} \tag{16.76}$$

and

$$E_{int}^{D/INC}(\mathbf{x}) = -\sum_{s=1}^{\infty} \frac{1}{s!} P_{r_1 r_2 \ldots r_s i}^{D} \frac{\partial^s u_i^{INC}(\mathbf{x})}{\partial x_{r_1} \partial x_{r_2} \ldots \partial x_{r_s}}. \tag{16.77}$$

Solutions for various cases can then be obtained by using force multipoles from Ch. 10 and inclusion displacement fields from Ch. 6. In the following, a solution is obtained for the simple case of a point defect with tetragonal symmetry in the field of a spherical inhomogeneous inclusion with $\varepsilon^{\mathrm{T}}_{\mathrm{ij}} = \varepsilon^{\mathrm{T}}\delta_{\mathrm{ij}}$ in an isotropic system.

16.6.2 *Between point defect and spherical inhomogeneous inclusion with $\varepsilon^{\mathrm{T}}_{\mathrm{ij}} = \varepsilon^{\mathrm{T}}\delta_{\mathrm{ij}}$ in isotropic system*

Assuming the tetragonal point defect in Fig. 10.3, with the non-vanishing multipole tensor components $P_{33} = aF$ and $P_{3333} = a^3F/4$ given by Eq. (10.14), Eq. (16.77) takes the form

$$E^{D/INC}_{int}(\mathbf{x}) = -\left(P^{D}_{33}\frac{\partial u^{INC}_3}{\partial x_3} + \frac{1}{6}P^{D}_{3333}\frac{\partial^3 u^{INC}_3}{\partial x^3_3}\right)$$
$$= -aF\left(\frac{\partial u^{INC}_3}{\partial x_3} + \frac{a^2}{24}\frac{\partial^3 u^{INC}_3}{\partial x^3_3}\right). \quad (16.78)$$

The displacement field in the matrix due to the inclusion is given for an isotropic system by Eqs. (6.127) and (6.128), so that the inclusion distortion field is

$$\frac{\partial u^{INC}_i}{\partial x_j} = c\left(\frac{\delta_{ij}}{x^3} - \frac{3x_i x_j}{x^5}\right). \quad (16.79)$$

Then, substituting Eq. (16.79) into Eq. (16.78), the interaction energy is

$$E^{D/INC}_{int}(\mathbf{x}) = -aFc\left[\left(1 - 3l^2_3\right)\frac{1}{x^3} - \frac{3a^2}{8}\left(1 - 10l^2_3 + 15l^4_3\right)\frac{1}{x^5}\right] \quad (16.80)$$

with c given by Eq. (6.128).

16.7 Dislocation–Inclusion Interactions

16.7.1 *General formulation*

The force imposed on a dislocation by an inclusion can be formulated by use of the Peach–Koehler force equation and an expression for the elastic field of the inclusion from Ch. 6. Taking the inclusion at the origin as in Fig. 16.8, the force exerted by the inclusion on segment *ds* of the dislocation is obtained from Eq. (13.12) in the form

$$dF^{DIS/INC}_l = f^{DIS/INC}_l ds = e_{jsl} b^{DIS}_i \sigma^{INC}_{ij} \hat{t}_s ds, \quad (16.81)$$

where σ^{INC}_{ij} is the stress at the dislocation due to the inclusion which, for a variety of different types of inclusions, can be obtained from the results

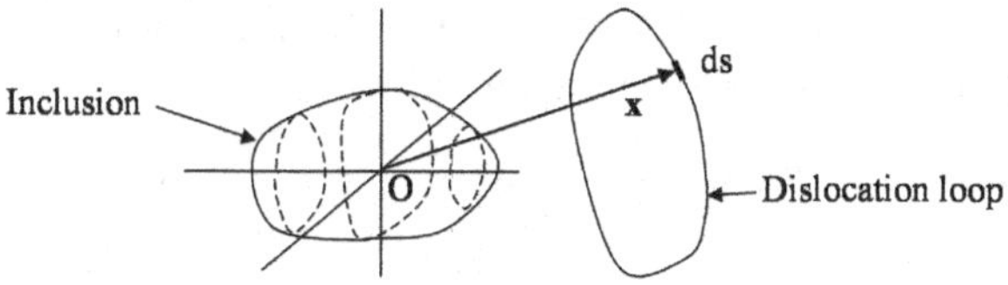

Fig. 16.8 Dislocation in vicinity of inclusion located at the origin.

in Ch. 6. In the following, an expression is obtained for this force for the tractable case of a spherical inhomogeneous inclusion with the transformation strain $\varepsilon_{ij}^{T} = \varepsilon^{T}\delta_{ij}$ and a general dislocation in an isotropic system.

16.7.2 *Between dislocation and spherical inhomogeneous inclusion with $\varepsilon_{ij}^{T} = \varepsilon^{T}\delta_{ij}$ in isotropic system*

For this case the distortion field in the matrix is given, as above, by Eq. (16.79), so that the corresponding stress field is

$$\sigma_{ij}^{INC} = \frac{2\mu c}{x^3}\left(\delta_{ij} - \frac{3x_i x_j}{x^2}\right) \tag{16.82}$$

after use of Eq. (2.123). Then, substituting this stress into Eq. (16.81), the force is

$$dF_l^{DIS/INC} = e_{jsl}\sigma_{ij}^{INC} b_i^{DIS}\hat{t}_s ds = 2\mu c e_{jsl} b_i^{DIS}\hat{t}_s\left(\frac{\delta_{ij}}{x^3} - \frac{3x_i x_j}{x^5}\right) ds \tag{16.83}$$

with c given by Eq. (6.128).

Exercises

16.1 Starting with Eq. (16.36), derive Eq. (16.37) for the interaction energy between the two infinitely long parallel dislocations in Fig. 16.2 in an isotropic system.

Solution. From the geometry of Fig. 16.2,

$$R = \left[p^2 + \left(x_3^{(1)} - x_3^{(2)}\right)^2\right]^{1/2}, \quad ds^{(i)} = \hat{t}dx_3^{(i)},$$

$$T_{ij} = \frac{\partial^2 R}{\partial x_i^{(1)}\partial x_j^{(1)}} = \frac{\delta_{ij}}{R} - \frac{\left(x_i^{(1)} - x_i^{(2)}\right)\left(x_j^{(1)} - x_j^{(2)}\right)}{R^3}$$

$$= \frac{\delta_{ij}}{R} - \frac{p_i p_j}{R^3}. \quad (i = 1, 2) \tag{16.84}$$

Substituting these relationships into Eq. (16.36), the interaction energy per unit dislocation length is reduced to

$$
\begin{aligned}
w_{\text{int}}^{(1)/(2)} = \frac{\mu}{4\pi} \Bigg\{ & \left[(\mathbf{b}^{(1)} \cdot \hat{\mathbf{t}})(\mathbf{b}^{(2)} \cdot \hat{\mathbf{t}}) + \frac{1}{1-\nu} (\mathbf{b}^{(1)} \times \hat{\mathbf{t}}) \cdot (\mathbf{b}^{(2)} \times \hat{\mathbf{t}}) \right] \\
& \times \left[\frac{1}{L} \int_{-L/2}^{L/2} dx_3^{(2)} \int_{-L/2}^{L/2} \frac{dx_3^{(1)}}{[p^2 + (x_3^{(1)} - x_3^{(2)})^2]^{1/2}} \right] \\
& - \frac{1}{(1-\nu)} [(\mathbf{b}^{(1)} \times \hat{\mathbf{t}}) \cdot \mathbf{p}][(\mathbf{b}^{(2)} \times \hat{\mathbf{t}}) \cdot \mathbf{p}] \\
& \times \left[\frac{1}{L} \int_{-L/2}^{L/2} dx_3^{(2)} \int_{-L/2}^{L/2} \frac{dx_3^{(1)}}{[p^2 + (x_3^{(1)} - x_3^{(2)})^2]^{3/2}} \right] \Bigg\},
\end{aligned} \tag{16.85}
$$

where the limits of integration, $\pm L/2$, will be increased without limit. The first double line integral (with $p \ll L$) corresponds to the integral in Eq. (12.248), and the bracketed term containing this integral is therefore given by $[-2\ln p + 2\ln(2L/e)]$ (see Eq. (12.251)). However, the L dependent part of this result may be dropped since it must be associated with end effects.[1] The bracketed term containing the second double line integral is readily integrated and is given by $2/p^2$. Putting these results into Eq. (16.85), and adding a constant term proportional to $-\ln p_\circ$, then produces Eq. (16.37).[2]

16.2 Show that the interaction energy between the two coaxial circular dislocations in Fig. 16.9 in an isotropic system is given by

$$
\begin{aligned}
W_{\text{int}}^{(1)/(2)} = \frac{\mu b^2 a^2}{2(1-\nu)} \int_0^{2\pi} \Bigg\{ & \left[\frac{1}{R} - \frac{a^2(1-\cos\phi)^2}{R^3} \right] \cos\phi \\
& + \frac{a^2(1-\cos\phi)\sin^2\phi}{R^3} \Bigg\} d\phi.
\end{aligned} \tag{16.86}
$$

Solution. Start with Eq. (16.34), which, for this problem, reduces to

$$
W_{\text{int}}^{(1)/(2)} = \frac{\mu}{4\pi(1-\nu)} \oint_{C^{(1)}} \oint_{C^{(2)}} (\mathbf{b} \times d\mathbf{s}^{(1)}) \cdot \underline{\mathbf{T}} \cdot (\mathbf{b} \times d\mathbf{s}^{(2)}). \tag{16.87}
$$

[1] Ignoring end effects, the interaction energy for two parallel infinitely long dislocations should be independent of *L* and a function of *q* only.

[2] Adding this term is permissible, since it merely shifts the reference level of the interaction energy by an arbitrary constant.

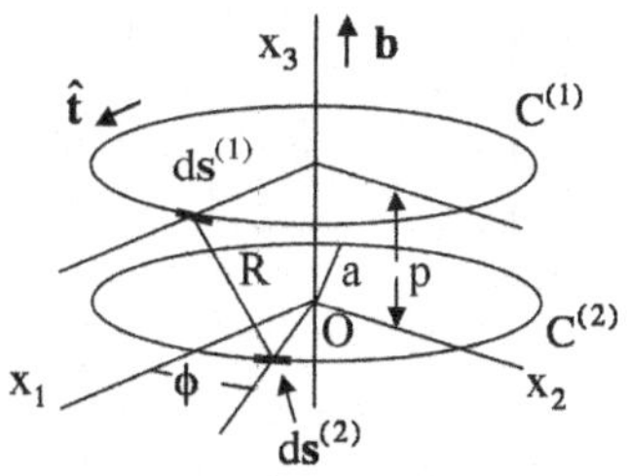

Fig. 16.9 Two circular coaxial dislocation loops with the same $\mathbf{b}$ and the same $\hat{\mathbf{t}}$ sense.

If Eq. (16.87) is integrated first around $C^{(2)}$, the contribution of this integration to every $ds^{(1)}$ element in the second integration around $C^{(1)}$ is the same because of the circular symmetry. Therefore, holding $ds^{(1)}$ in the first integration conveniently at $\mathbf{ds}^{(1)} = \hat{\mathbf{e}}_2 ds^{(1)}$, and with $\mathbf{ds}^{(2)} = (-\hat{\mathbf{e}}_1 \sin\phi + \hat{\mathbf{e}}_2 \cos\phi) ds^{(2)}$, $\mathbf{b} = (0, 0, b)$, and

$$T_{ij} = \frac{\partial^2 R}{\partial x_i^{(1)} \partial x_j^{(1)}} = \frac{\delta_{ij}}{R} - \frac{(x_i^{(1)} - x_i^{(2)})(x_j^{(1)} - x_j^{(2)})}{R}, \quad (16.88)$$

where

$$\begin{aligned} x_1^{(1)} &= a \quad x_2^{(1)} = 0 \quad x_3^{(1)} = p \\ x_1^{(2)} &= a\cos\phi \quad x_2^{(2)} = a\sin\phi \quad x_3^{(2)} = 0 \\ R &= (4a^2 \sin^2(\phi/2) + p^2)^{1/2} \end{aligned} \quad (16.89)$$

Eq. (16.87) takes the form

$$\begin{aligned} W_{int}^{(1)/(2)} &= \frac{\mu b^2}{4\pi(1-\nu)} \oint_{C^{(1)}} ds^{(1)} \oint_{C^{(2)}} (T_{11}\cos\phi + T_{12}\sin\phi) ds^{(2)} \\ &= \frac{\mu b^2 a^2}{2(1-\nu)} \int_0^{2\pi} \left\{ \left[\frac{1}{R} - \frac{a^2(1-\cos\phi)^2}{R^3} \right] \cos\phi \right. \\ &\quad \left. + \frac{a^2(1-\cos\phi)\sin^2\phi}{R^3} \right\} d\phi. \end{aligned} \quad (16.90)$$

Hirth and Lothe (1982) show that the integrals in Eq. (16.90) are elliptic integrals that take relatively simple asymptotic forms when $p \ll a$ or $p \gg a$.

16.3 (a) Use Eq. (16.49) to determine the total force exerted on a straight screw dislocation segment, AB, by an orthogonal straight screw seg-

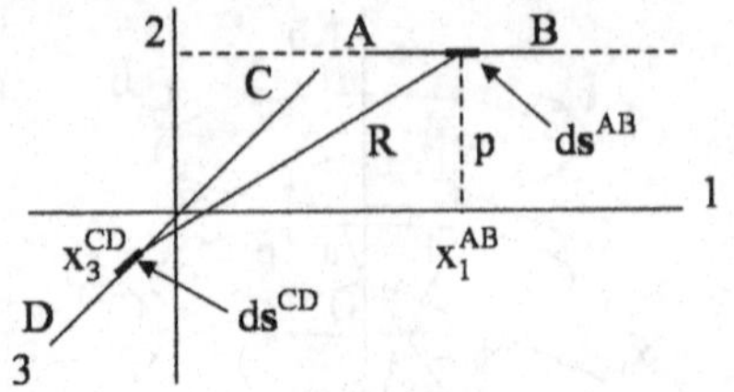

Fig. 16.10 Geometry for determining force exerted on segment AB by segment CD.

ment, CD, as in Fig. 16.10, in an isotropic system. Find the force when CD $\to \infty$, and then verify the result directly by using the Peach–Koehler force equation and the known stress field of an infinitely long screw dislocation.

Solution. Using the Cartesian coordinate system shown in Fig.16.10, $\mathbf{b}^{AB} = (b^{AB}, 0, 0)$, $\hat{\mathbf{t}}^{AB} = (1,0,0)$, $\mathbf{b}^{CD} = (0,0,b^{CD})$ and $\hat{\mathbf{t}}^{CD} = (0,0,1)$. The last three terms in Eq. (16.49) therefore vanish, and with $R = [(x_1^{AB})^2 + p^2 + (x_3^{CD})^2]^{1/2}$, $\mathbf{ds}^{AB} = \hat{\mathbf{t}}^{AB} dx_1^{AB}$ and $\mathbf{ds}^{CD} = \hat{\mathbf{t}}^{CD} dx_3^{CD}$, the first term yields

$$\begin{aligned}\mathbf{F} &= -\frac{\mu b^{AB} b^{CD}}{4\pi} \int_{x_1^A}^{x_1^B} \int_{x_3^C}^{x_3^D} \left[-\hat{\mathbf{e}}_2 \cdot \nabla \left(\frac{1}{R} \right) \right] dx_1^{AB} dx_3^{CD} \hat{\mathbf{e}}_2 \\ &= -\frac{\mu b^{AB} b^{CD}}{4\pi} \int_{x_1^A}^{x_1^B} \int_{x_3^C}^{x_3^D} \frac{p\, dx_1^{AB} dx_3^{CD}}{[(x_1^{AB})^2 + p^2 + (x_3^{CD})^2]^{3/2}} \hat{\mathbf{e}}_2 \\ &= -\frac{\mu b^{AB} b^{CD} p}{4\pi} \int_{x_1^A}^{x_1^B} \left[\frac{x_3^{CD}}{[(x_1^{AB})^2 + p^2 + (x_3^{CD})^2]^{1/2}} \right]_{x_3^C}^{x_3^D} dx_1^{AB} \hat{\mathbf{e}}_2.\end{aligned} \tag{16.91}$$

When $x_3^C \to -\infty$, and $x_3^D \to \infty$, the force given by Eq. (16.91) becomes

$$\mathbf{F} = -\frac{\mu b^{AB} b^{CD} p}{2\pi} \int_{x_1^A}^{x_1^B} \frac{dx_1^{AB}}{p^2 + (x_1^{AB})^2} \hat{\mathbf{e}}_2. \tag{16.92}$$

Next, the force is obtained by the alternative method of using the Peach–Koehler force equation and the known stress field of an infinitely long screw dislocation. By employing Eq. (13.11), and the components of the CD screw dislocation stress field given by Eq. (12.55),

$$\mathbf{F} = \int_{x_1^A}^{x_1^B} (\mathbf{d} \times \hat{\mathbf{t}}^{AB}) dx_1^{AB} = b^{AB} \int_{x_1^A}^{x_1^B} \sigma_{13}^{CD} dx_1^{AB} \hat{\mathbf{e}}_2$$
$$= -\frac{\mu b^{AB} b^{CD} p}{2\pi} \int_{x_1^A}^{x_1^B} \frac{dx_1^{AB}}{p^2 + (x_1^{AB})^2} \hat{\mathbf{e}}_2 \qquad (16.93)$$

in agreement with Eq. (16.92).

16.4 Consider two infinitely long straight parallel edge dislocations in an isotropic system and determine the force per unit length that one exerts on the other by direct use of Eq. (16.49). Then show that this force can be determined equally well by differentiating the interaction energy between them with respect to their relative coordinates.

Solution. Start with finite segments of length L, with segment 1 along the line $(p, 0, x_3^{(1)})$ and segment 2 along the line $(0, 0, x_3^{(2)})$ in a $(\hat{\mathbf{e}}_1, \hat{\mathbf{e}}_2, \hat{\mathbf{e}}_3)$ coordinate system. With $\mathbf{b}^{(1)} = \mathbf{b}^{(2)} = (b, 0, 0)$, $\hat{\mathbf{t}}^{(1)} = \hat{\mathbf{t}}^{(2)} = (0, 0, 1)$, $d\mathbf{s}^{(1)} = \hat{\mathbf{t}}^{(1)} dx_3^{(1)}$ and $d\mathbf{s}^{(2)} = \hat{\mathbf{t}}^{(2)} dx_3^{(2)}$, the first two terms of Eq. (16.49) vanish, and after integrating over the segments, the force is given by

$$\mathbf{F} = -\frac{\mu}{4\pi(1-\nu)} \int_{-L/2}^{L/2} \int_{-L/2}^{L/2} \left\{ \left[(\mathbf{b}^{(2)} \times d\mathbf{s}^{(2)}) \cdot \nabla \right] \left[d\mathbf{s}^{(1)} \times \mathbf{b}^{(1)} \underline{\mathbf{T}} \right] \right.$$
$$\left. - \left[(\mathbf{b}^{(2)} \times d\mathbf{s}^{(2)}) \cdot \nabla(\nabla^2 R) \right] \left[d\mathbf{s}^{(1)} \times \mathbf{b}^{(1)} \right] \right\}, \qquad (16.94)$$

where

$$\left[(\mathbf{b}^{(2)} \times d\mathbf{s}^{(2)}) \cdot \nabla \right] \left[d\mathbf{s}^{(1)} \times \mathbf{b}^{(1)} \underline{\mathbf{T}} \right] = b^2 dx_3^{(1)} dx_3^{(2)} \left(\frac{\partial T_{12}}{\partial x_2^{(1)}} \hat{\mathbf{e}}_1 - \frac{\partial T_{11}}{\partial x_2^{(1)}} \hat{\mathbf{e}}_2 \right)$$
$$= -\frac{b^2 p dx_3^{(1)} dx_3^{(2)}}{R^3} \hat{\mathbf{e}}_1 \qquad (16.95)$$

and

$$\left[(\mathbf{b}^{(2)} \times d\mathbf{s}^{(2)}) \cdot \nabla(\nabla^2 R) \right] = \left(b dx_3^{(2)} \hat{\mathbf{e}}_2 \right)$$
$$\cdot \left\{ \frac{2}{R^3} \left[p\hat{\mathbf{e}}_1 + \left(x_3^{(1)} - x_3^{(2)} \right) \hat{\mathbf{e}}_3 \right\} = 0. \qquad (16.96)$$

Substituting these relationships, Eq. (16.94) then takes the form

$$\mathbf{F} = \frac{\mu b^2 p}{4\pi(1-\nu)} \int_{-L/2}^{L/2} dx_3^{(2)} \int_{-L/2}^{L/2} \frac{dx_3^{(1)}}{[p^2 + (x_3^{(1)} - x_3^{(2)})^2]^{3/2}} \hat{\mathbf{e}}_1. \tag{16.97}$$

After integrating with respect to $x_3^{(1)}$ and letting $L \to \infty$ in the first integration,

$$\mathbf{F} = \frac{\mu b^2}{2\pi(1-\nu)p} \int_{-L/2}^{L/2} dx_3^{(2)} \hat{\mathbf{e}}_1. \tag{16.98}$$

According to Eq. (16.98), the force per unit length is therefore constant at the value

$$\boldsymbol{f} = \frac{\mathbf{F}}{L} = \frac{\mu b^2}{2\pi(1-\nu)p} \hat{\mathbf{e}}_1. \tag{16.99}$$

Alternatively, according to Eq. (16.37), the interaction energy between the two dislocations per unit length is

$$\frac{W_{int}^{(1)/(2)}}{L} = -\frac{\mu b^2}{2\pi(1-\nu)} \ln \frac{p}{p_\circ} \tag{16.100}$$

since $\mathbf{p} = (p, 0, 0)$. The force per unit length between the dislocations is then

$$\begin{aligned} \boldsymbol{f} = \frac{\mathbf{F}}{L} &= -\frac{\partial}{\partial p}\left(\frac{W_{int}^{1/2}}{L}\right)\hat{\mathbf{e}}_1 = -\frac{\partial}{\partial p}\left(-\frac{\mu b^2}{2\pi(1-\nu)} \ln \frac{p}{p_\circ}\right)\hat{\mathbf{e}}_1 \\ &= \frac{\mu b^2}{2\pi(1-\nu)p}\hat{\mathbf{e}}_1 \end{aligned} \tag{16.101}$$

in agreement with Eq. (16.99).

16.5 (a) Determine the force that a cubically symmetric point defect, located at the origin, exerts on an identical point defect, located at $\mathbf{x}$, in an isotropic system as described in Sec. 16.2.2. Neglect effects beyond those due to octopoles. (b) Show that in the high symmetry directions (1, 0, 0), (1, 1, 0) and (1, 1, 1) the force is a central force, i.e., is parallel to $\mathbf{x}$. (c) Despite the fact that the force is not a central force in all directions, show that the interaction energy between such a point defect at the origin and a uniform distribution of identical defects on the surface of a sphere centered on the origin vanishes.

Solution. (a) Substituting Eq. (16.9) into (5.40), the force is given by

$$F_\alpha(\mathbf{x}) = -\frac{\partial E_{\text{int}}}{\partial x_\alpha} = -Ax^{-6} l_\alpha \left[3 + 4l_\alpha^2 - 9\left(l_1^4 + l_2^4 + l_3^4\right)\right], \qquad (16.102)$$

where A = constant.

(b) Straightforward substitution then shows that

$$\text{when } \mathbf{x} \text{ is parallel with } (1,0,0), \mathbf{F}_{100} = 2Ax^{-6}(1,0,0),$$

$$\text{when } \mathbf{x} \text{ is parallel with } (1,1,0), \mathbf{F}_{110} = -\frac{Ax^{-6}}{2\sqrt{2}}(1,1,0), \qquad (16.103)$$

$$\text{when } \mathbf{x} \text{ is parallel with } (1,1,1), \mathbf{F}_{111} = -4\frac{Ax^{-6}}{3\sqrt{3}}(1,1,1).$$

(c) Using Eq. (16.9), and assuming a unit sphere with N defects on its surface, the total interaction energy between these defects and the defect at the origin, E_{int}, is given by

$$E_{\text{int}} = A \oiint_{\text{N defects}} \left(-\frac{3}{5} + l_1^4 + l_2^4 + l_3^4\right) dS, \qquad (16.104)$$

where the integral is over the N defects on the surface. Using the spherical coordinate system of Fig. 1Ab, $l_1 = \sin\phi\cos\theta$, and the number of defects on the surface area element dS is $NdS/(4\pi)$ where $dS = \sin\phi d\phi d\theta$. Therefore, substituting these quantities into Eq. (16.104), and taking advantage of the spherical symmetry of the problem,

$$E_{\text{int}} = A\left[-\frac{3}{5}N + 4\int_0^{\pi/2}\int_0^{\pi} 3\sin^4\phi\cos^4\theta \frac{N\sin\phi d\phi d\theta}{4\pi}\right]$$

$$= AN\left[-\frac{3}{5} + \frac{3}{\pi}\int_0^{\pi/2}\sin^5\phi d\phi \int_0^{\pi}\cos^4\theta d\theta\right]. \qquad (16.105)$$

Finally, carrying out the integrations in Eq. (16.105),

$$E_{\text{int}} = AN\left[-\frac{3}{5} + \frac{3}{\pi}\left(\frac{8}{15}\right)\left(\frac{3\pi}{8}\right)\right] = 0. \qquad (16.106)$$

16.6 Consider the relatively long-range dipole-dipole interaction between two tetragonal point defects, D1 and D2, each mimicked by a double force as in Fig. 10.3. The medium is isotropic, and $\nu = 1/3$. (a) If D1 lies on the $\hat{\mathbf{e}}_1$ axis at the distance q, and D2 is at the origin, and their tetragonal axes are parallel to $\hat{\mathbf{e}}_3$, show that the central force between them is repulsive. (b) On the other hand, if D1 is rotated so that its tetragonal axis lies along $\hat{\mathbf{e}}_1$, and the axes of the two defects are therefore orthogonal, show that the central force becomes attractive.

Solution. (a) In the parallel case, where $P_{ij}^{D1} = P_{ij}^{D2} = P\delta_{3i}\delta_{3j}$, the interaction energy is given by the first term in Eq. (16.6), i.e.,

$$E_{int}^{D1/D2}(para) = \frac{P^2}{16\pi\mu(1-\nu)}\frac{\partial^2 G_{33}}{\partial x_3^2} = \frac{P^2}{16\pi\mu(1-\nu)} \times \left[\frac{(4\nu-1)}{x^3} - \frac{6(1+2\nu)x_3^2}{x^5} + \frac{15x_3^4}{x^7}\right]. \quad (16.107)$$

after using Eq. (4.110). Since $x_1 = x = q$ and $x_2 = x_3 = 0$, the central force (along $\hat{\mathbf{e}}_1$) is given by

$$F_1^{D1/D2}(para) = -\frac{\partial E_{int}^{D1/D2}(para)}{\partial x_1} = \frac{P^2}{16\pi\mu(1-\nu)}\left[\frac{3(4\nu-1)}{q^4}\right] \quad (16.108)$$

which is positive, and therefore repulsive.

(b) In the corresponding orthogonal case, where $P_{ij}^{D1} = P\delta_{1i}\delta_{1j}$ and $P_{ij}^{D2} = P\delta_{3i}\delta_{3j}$, the interaction energy becomes

$$E_{int}^{D1/D2}(orth) = \frac{P^2}{16\pi\mu(1-\nu)}\frac{\partial^2 G_{13}}{\partial x_1 \partial x_3} = \frac{P^2}{16\pi\mu(1-\nu)}\left[\frac{1}{x^3} - \frac{3(x_1^2+x_3^2)}{x^5} + \frac{15x_1^2x_3^3}{x^7}\right] \quad (16.109)$$

and the central force is then

$$F_1^{D1/D2}(orth) = -\frac{\partial E_{int}^{D1/D2}(orth)}{\partial x_1} = -\frac{P^2}{16\pi\mu(1-\nu)}\left[\frac{6}{q^4}\right] \quad (16.110)$$

which is negative, and therefore, attractive.

16.7 Equation (16.65), for the interaction energy between a point defect represented by a force multipole located at **x** and a dislocation at the origin, has been obtained by a rather tortuous procedure beginning with the definition of the interaction energy given by Eq. (5.15) and then involving a sequence of equations including Eqs. (5.16),

(11.1) and (16.1). Show that Eq. (16.65) can be derived directly from Eq. (5.15) in a very direct manner.

Solution. Using the nomenclature employed in Eq. (16.65), Eq. (5.15) takes the form

$$E_{int}^{D/DIS} = W^{D+DIS} - W^{D} - W^{DIS} + \Phi^{D+DIS} - \Phi^{D} \quad (16.111)$$

since $\Phi^{DIS} = 0$. Then, since $W_{int}^{D/DIS} = W^{D+DIS} - W^{D} - W^{DIS} = 0$ by virtue of Eq. (5.25),

$$E_{int}^{D/DIS} = \Phi^{D+DIS} - \Phi^{D} = \Delta\Phi. \quad (16.112)$$

The forces associated with the force multipole in the D + DIS system are displaced relative to their positions in the D system because of the displacement field of the dislocation, and, by virtue of Eq. (5.3), we can therefore write Eq. (16.65) directly, i.e.,

$$\begin{aligned} E_{int}^{D/DIS} &= \Phi^{D+DIS} - \Phi^{D} = \Delta\Phi \\ &= -\sum_{q} u_j^{DIS}(\mathbf{x} + \mathbf{s}^{D(q)})F_j^{D(q)}. \end{aligned} \quad (16.113)$$

16.8 In Sec. 12.5.1.1 it is argued that if the dislocation loop in Fig. 12.1 is created by the cut and displacement method on the surface, Σ, in the presence of the force, $\mathbf{F}$, then the work, $\Delta\mathcal{W}$, performed during this process by the tractions on Σ caused by the stress field of $\mathbf{F}$, must be equal to the change of potential energy, $\Delta\Phi$, of the force caused by the displacement field of the dislocation (see Eq. (12.64)). In the previous exercise (Exercise 16.7) it is shown that the interaction energy between a dislocation and a defect force multipole is given by

$$E_{int}^{D/DIS} = \Delta\Phi = \sum\nolimits_i \Delta\Phi_i, \quad (16.114)$$

where the sum is over all forces in the multipole. We should therefore expect (see Eq. (12.64)) that the interaction energy should just as well be given by

$$E_{int}^{D/DIS} = \Delta\mathcal{W} = \sum\nolimits_i \Delta\mathcal{W}_i. \quad (16.115)$$

Verify, by means of a detailed calculation, that this is indeed the case for the dislocation and multipole illustrated in Fig. 16.11. Use the force dipole approximation.

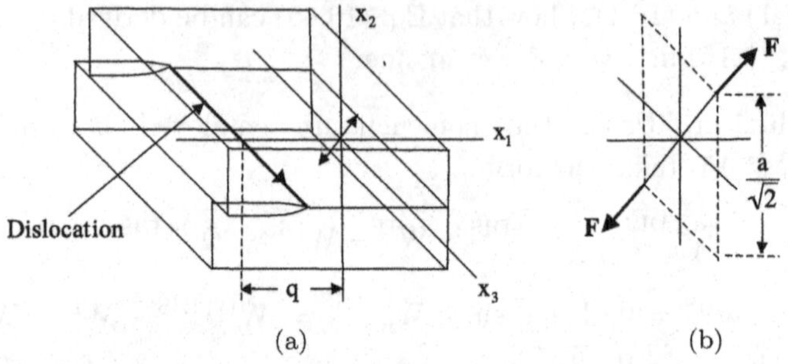

Fig. 16.11 (a) Straight screw dislocation lying along $x_1 = -q$ with $x_2 = x_3 = 0$, and $\mathbf{b} = (0, 0, b)$ and $\hat{\mathbf{t}} = (0, 0, 1)$, in the presence of a force dipole located at origin of $(\hat{\mathbf{e}}_1, \hat{\mathbf{e}}_2, \hat{\mathbf{e}}_3)$ coordinate system. (b) Enlarged view of dipole corresponding to a double force with its axis along $(0, 1, 1)$.

Solution. First, we evaluate $E_{int}^{D/DIS}$ by employing Eq. (16.114). Here, we can use Eq. (16.66), i.e.,

$$E_{int}^{D/DIS} = -P_{ij}^{D} \frac{\partial u_j^{DIS}}{\partial x_i} \tag{16.116}$$

to determine $E_{int}^{D/DIS}$ since it is based on the change of potential energy of the forces in the multipole due to the displacement field of the dislocation. To determine the required dipole tensor in the coordinate system of Fig. 16.11, P_{ij}^{D}, we refer to Eq. (10.14) and employ the tensor transformation law given by Eq. (2.23) to obtain

$$P_{ij}^{D} = \frac{1}{\sqrt{2}} \begin{bmatrix} \sqrt{2} & 0 & 0 \\ 0 & 1 & -1 \\ 0 & 1 & 1 \end{bmatrix} \begin{bmatrix} 0 & 0 & 0 \\ 0 & aF & 0 \\ 0 & 0 & 0 \end{bmatrix} \frac{1}{\sqrt{2}} \begin{bmatrix} \sqrt{2} & 0 & 0 \\ 0 & 1 & 1 \\ 0 & -1 & 1 \end{bmatrix}$$

$$= \frac{aF}{2} \begin{bmatrix} 0 & 0 & 0 \\ 0 & 1 & 1 \\ 0 & 1 & 1 \end{bmatrix}. \tag{16.117}$$

The required derivative of the dislocation displacement at $\mathbf{x} = (q, 0, 0)$ is obtained by use of Eq. (12.274) in the form

$$\frac{\partial u_3^{DIS}}{\partial x_2} = \frac{b}{2\pi} \left(\frac{x_1}{x_1^2 + x_2^2} \right) = \frac{b}{2\pi q}. \tag{16.118}$$

Then, by substituting these results into Eq. (16.116),

$$E_{\text{int}}^{\text{D/DIS}} = \Delta\Phi = -\frac{\text{aFb}}{4\pi\text{q}}. \tag{16.119}$$

Next, we evaluate $E_{\text{int}}^{\text{D/DIS}}$ by employing Eq. (16.115). Here, after invoking the direction of $\hat{\mathbf{n}}$ and the Σ cut and displacement rules (p. 325), the traction in the $\hat{\mathbf{e}}_3$ direction imposed on the negative side of the Σ cut (where $\hat{\mathbf{n}} = (0, -1, 0)$) by the stress field of the dipole during the creation of the dislocation is $T_3^{\text{D}} = -\sigma_{23}^{\text{D}}$. The work performed over the entire cut during the displacement, $\mathbf{b}$, is then

$$\begin{aligned}\Delta\mathcal{W} &= \int\!\!\int_{\Sigma} \mathbf{b}\cdot\mathbf{T}^{\text{D}}\text{dS} = \int\!\!\int_{\Sigma} \text{b}T_3^{\text{D}}\text{dS} \\ &= -\text{bLim}_{\text{L}\to\infty}\int_{-\text{L}}^{-\text{q}}\text{dx}_1\int_{-\text{L}}^{\text{L}}\sigma_{23}^{\text{D}}\text{dx}_3.\end{aligned} \tag{16.120}$$

The stress, σ_{23}^{D}, can be found by first determining the strain $\varepsilon_{23}^{\text{D}}$ in the form

$$\begin{aligned}\varepsilon_{23}^{\text{D}} &= \frac{1}{2}\left(\frac{\partial u_2^{\text{D}}}{\partial x_3} + \frac{\partial u_3^{\text{D}}}{\partial x_2}\right) = -\frac{\text{aF}}{4}\left[\left(\frac{\partial^2}{\partial x_3^2} + \frac{\partial^2}{\partial x_2 \partial x_3}\right)G_{22}\right. \\ &\quad \left. + \left(\frac{\partial^2}{\partial x_2^2} + 2\frac{\partial^2}{\partial x_2\partial x_3} + \frac{\partial^2}{\partial x_3^2}\right)G_{23} + \left(\frac{\partial^2}{\partial x_2^2} + \frac{\partial^2}{\partial x_2 \partial x_3}\right)G_{33}\right]\end{aligned} \tag{16.121}$$

by the use of Eq. (10.10). Then, by substituting Eq. (4.110) into Eq. (16.121) and employing Eq. (2.123),

$$\sigma_{23}^{\text{D}} = \frac{\text{aF}}{8\pi(1-\nu)}\left(\frac{1-2\nu}{x^3} + \frac{3\nu x_3^2}{x^5}\right). \tag{16.122}$$

Finally, substituting Eq. (16.122) into Eq. (16.120) and performing the integration,

$$\begin{aligned}E_{\text{int}}^{\text{D/DIS}} &= \Delta\mathcal{W} = -\frac{\text{aFb}}{8\pi(1-\nu)}\text{Lim}_{\text{L}\to\infty} \\ &\quad \times\int_{-\text{L}}^{-\text{q}}\text{dx}_1\int_{-\text{L}}^{\text{L}}\left(\frac{1-2\nu}{x^3} + \frac{3\nu x_3^2}{x^5}\right)\text{dx}_3 = -\frac{\text{aFb}}{4\pi\text{q}}\end{aligned} \tag{16.123}$$

in agreement with Eq. (16.119). See Exercise 16.9 for yet another method to determine $E_{int}^{D/DIS}$.

16.9 In Exercise 16.8 an explicit expression for the interaction energy between the defect force dipole and screw dislocation in Fig. 16.11, $E_{int}^{D/DIS}$, is determined by two methods employing Eqs. (16.114) and (16.66) in the first and then Eq. (16.115) in the second. Now, show that it can also be obtained by a third method involving the simple and direct use of Eq. (16.65).

Solution. By virtue of Eq. (12.274), Eq. (16.65) takes the form

$$E_{int}^{D/DIS} = -F_3(u_3^{DIS} - u_3^{DIS'}) = -\frac{F}{\sqrt{2}}\frac{b}{2\pi}(\theta - \theta'), \quad (16.124)$$

where the primed and non-primed quantities refer to opposite ends of the force dipole. However, the angular difference $(\theta - \theta')$ is relatively small, and, therefore

$$(\theta - \theta') = \frac{a}{q\sqrt{2}}. \quad (16.125)$$

Substitution of Eq. (16.125) into (16.124) then yields the desired result.

Chapter 17

Defect Self-interactions and Self-forces

17.1 Introduction

The previous chapter dealt with interactions and forces between separate defects. However, as already pointed out in the introduction to Ch. 5, individual defects which are extended in at least one dimension, i.e., inclusions, dislocations and interfaces, generally experience a self-force. The elastic strain energy of such a defect varies with its shape and a force therefore generally exists urging the defect to change its shape (configuration) in a direction that decreases its energy.

The chapter begins with an analysis of the self-force acting at a point on a smoothly curved dislocation loop. This self-force has often been interpreted as arising from a dislocation *line tension*, defined as an energy per unit dislocation length, and this concept is examined. Following this, a treatment is given of the self-force acting at a point on a straight finite dislocation segment which is part of a segmented loop. Finally, the self-force acting at a point on the surface of an inclusion is analyzed.

17.2 Self-force Experienced by a Smoothly Curved Dislocation

17.2.1 *Circular planar loop*

We adopt the analysis of Barnett (1976) and Gavazza and Barnett (1976) and start with the relatively simple case of finding the self-force acting at a point on a circular planar dislocation loop, C, of radius R. The force is found by virtually displacing the loop by expanding it radially by δR, determining the change in the loop energy and then employing Eq. (5.38).

Having these results, the self-force exerted at any point along a smoothly curved planar loop of arbitrary shape is found.

17.2.1.1 *Loop self-stress*

To determine the change in loop energy, and ultimately the self-force on the loop, it will be necessary to have an expression for the loop self-stress in the near vicinity of the loop circumference, i.e., the dislocation core. We, therefore, begin by finding the self-stresses at field points that are opposite each other just inside and outside the circumference. These can be obtained by integrating Brown's formula, i.e., Eq. (12.115),

$$\sigma_{ij}(\mathbf{x}) = \frac{1}{2}\oint_C \frac{1}{|\mathbf{x}-\mathbf{x}'|}\left[\Sigma_{ij}(\theta) + \frac{d^2\Sigma_{ij}(\theta)}{d\theta^2}\right] d\theta. \tag{17.1}$$

To obtain the contribution of each segment of the loop lying at the angle α to the reference line to the integral in Eq. (17.1), we employ the geometry of Fig. 17.1 (see caption for the details). The parameter, k, defined by

$$k = 1 - \frac{\varepsilon}{R} \tag{17.2}$$

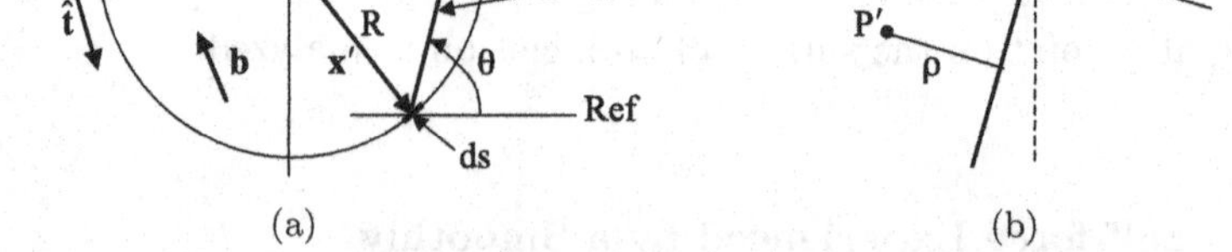

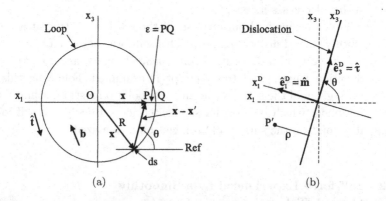

Fig. 17.1 (a) Geometry used to find the self-stress of a circular planar loop, with Burgers vector $\mathbf{b} = (b_1, b_2, b_3)$, at a field point P inside the loop at a distance ε from the circumference. All quantities are referred to a $(\hat{\mathbf{e}}_1, \hat{\mathbf{e}}_2, \hat{\mathbf{e}}_3)$ crystal coordinate system with coordinate distances measured by (x_1, x_2, x_3). (b) Infinitely long straight dislocation, with a Burgers vector $\mathbf{b}^D$ identical to that of the loop, lying along the $\hat{\mathbf{e}}_3^D$ axis of a second crystal coordinate system, $(\mathbf{e}_1^D, \mathbf{e}_2^D, \mathbf{e}_3^D)$, which is rotated relative to the $(\hat{\mathbf{e}}_1, \hat{\mathbf{e}}_2, \hat{\mathbf{e}}_3)$ system in (a) by the angle $-(90-\theta)$ around their common $\hat{\mathbf{e}}_2 = \hat{\mathbf{e}}_2^D$ axes. The dislocation is therefore parallel to $\mathbf{x} - \mathbf{x}'$ and lies at the angle θ with respect to the reference line. A third coordinate system, i.e., the $(\hat{\mathbf{m}}, \hat{\mathbf{n}}, \hat{\boldsymbol{\tau}})$ system of the integral formalism, is also present and is oriented so that $\hat{\boldsymbol{\tau}} = \hat{\mathbf{e}}_3^D$ and $\hat{\mathbf{m}} = \hat{\mathbf{e}}_1^D$.

is next introduced to obtain an expression for $|\mathbf{x} - \mathbf{x}'|$ in Eq. (17.1). Using the law of cosines for the triangle bounded by $\mathbf{x}$, $\mathbf{x}'$ and $|\mathbf{x} - \mathbf{x}'|$,

$$|\mathbf{x} - \mathbf{x}'|^2 - 2\mathrm{Rk}|\mathbf{x} - \mathbf{x}'|\cos\theta - \mathrm{R}^2(1 - \mathrm{k}^2) = 0 \tag{17.3}$$

and therefore

$$\frac{1}{|\mathbf{x} - \mathbf{x}'|} = \frac{1}{\mathrm{R}} \frac{[(1 - \mathrm{k}^2 \sin^2\theta)^{1/2} - \mathrm{k}\cos\theta]}{1 - \mathrm{k}^2}. \tag{17.4}$$

Then, substituting Eq. (17.4) into Eq. (17.1)

$$\sigma_{\mathrm{ij}}(\mathrm{R} - \varepsilon) = \frac{1}{2\mathrm{R}(1 - \mathrm{k}^2)} \int_0^{2\pi} \left[\Sigma_{\mathrm{ij}}(\theta) + \frac{\mathrm{d}^2\Sigma_{\mathrm{ij}}(\theta)}{\mathrm{d}\theta^2}\right] \times [(1 - \mathrm{k}^2 \sin^2\theta)^{1/2} - \mathrm{k}\cos\theta]\mathrm{d}\theta. \tag{17.5}$$

The terms in Eq. (17.5) containing the factor $\mathrm{k}\cos\theta$ vanish in the integration because of the periodicity of $\cos\theta$ and the $\Sigma_{\mathrm{ij}}(\theta)$ function which (see Exercise 17.1) has the property

$$\Sigma_{\mathrm{ij}}(\theta + \pi) = \Sigma_{\mathrm{ij}}(\theta). \tag{17.6}$$

The term containing the factor $\mathrm{d}^2\Sigma_{\mathrm{ij}}(\theta)/\mathrm{d}\theta^2$ in Eq. (17.5) can be integrated by parts twice so that

$$\int_0^{2\pi} \frac{\mathrm{d}^2\Sigma_{\mathrm{ij}}(\theta)}{\mathrm{d}\theta^2}(1 - \mathrm{k}^2\sin^2\theta)^{1/2}\mathrm{d}\theta = -\mathrm{k}^2 \int_0^{2\pi} \frac{(\cos^2\theta - \sin^2\theta + \mathrm{k}^2\sin^4\theta)}{(1 - \mathrm{k}^2\sin^2\theta)^{3/2}}\Sigma_{\mathrm{ij}}(\theta)\mathrm{d}\theta \tag{17.7}$$

and, by substituting these results into Eq. (17.5), and again invoking the periodicity of the integrand,

$$\sigma_{\mathrm{ij}}(\mathrm{R} - \varepsilon) = \frac{1}{2\mathrm{R}} \int_0^{2\pi} \frac{\Sigma_{\mathrm{ij}}(\theta)}{(1 - \mathrm{k}^2\sin^2\theta)^{3/2}}\mathrm{d}\theta = \frac{1}{2\mathrm{R}}\left[\int_0^{\pi} \frac{\Sigma_{\mathrm{ij}}(\theta)}{(1 - \mathrm{k}^2\sin^2\theta)^{3/2}} + \int_{\pi}^{2\pi} \frac{\Sigma_{\mathrm{ij}}(\theta)}{(1 - \mathrm{k}^2\sin^2\theta)^{3/2}}\right]\mathrm{d}\theta$$

$$
\begin{aligned}
&= \frac{1}{R}\int_0^{\pi} \frac{\Sigma_{ij}(\theta)}{(1-k^2\sin^2\theta)^{3/2}} d\theta \\
&= \frac{1}{R}\left[\int_0^{\pi/2} \frac{\Sigma_{ij}(\theta)}{(1-k^2\sin^2\theta)^{3/2}} + \int_{\pi/2}^{\pi} \frac{\Sigma_{ij}(\theta)}{(1-k^2\sin^2\theta)^{3/2}}\right] d\theta.
\end{aligned} \tag{17.8}
$$

The range of the integration can be reduced further by making the change of variable $\theta \to -\theta'$ in Eq. (17.8) to obtain,

$$
\begin{aligned}
\sigma_{ij}(R-\varepsilon) &= \frac{1}{R}\left[\int_0^{\pi/2} \frac{\Sigma_{ij}(\theta)}{(1-k^2\sin^2\theta)^{3/2}} d\theta \right. \\
&\quad \left. - \int_{-\pi/2}^{-\pi} \frac{\Sigma_{ij}(-\theta')}{[1-k^2\sin^2(-\theta')]^{3/2}} d\theta' \right] \\
&= \left[\frac{1}{R}\int_0^{\pi/2} \frac{\Sigma_{ij}(\theta)}{(1-k^2\sin^2\theta)^{3/2}} d\theta - \int_{\pi/2}^{0} \frac{\Sigma_{ij}(-\theta')}{(1-k^2\sin^2\theta')^{3/2}} d\theta'\right] \\
&= \frac{1}{R}\left[\int_0^{\pi/2} \frac{\Sigma_{ij}(\theta)}{(1-k^2\sin^2\theta)^{3/2}} - \int_{\pi/2}^{0} \frac{\Sigma_{ij}(-\theta)}{(1-k^2\sin^2\theta)^{3/2}}\right] d\theta \\
&= \frac{1}{R}\int_0^{\pi/2} \frac{[\Sigma_{ij}(\theta)+\Sigma_{ij}(-\theta)]}{(1-k^2\sin^2\theta)^{3/2}} d\theta.
\end{aligned} \tag{17.9}
$$

Since we are interested in the case where $k \cong 1$ (or, alternatively, $\varepsilon/R \ll 1$), Eq. (17.9) can be integrated to a good approximation by noting that, since $k \cong 1$, the integrand becomes strongly divergent when $\theta \to \pi/2$. The great majority of the integral must therefore accumulate in this region. This can be captured by replacing the functions in the integrand by their Taylor expansions in the vicinity of $\theta = \pi/2$ given by

$$
\begin{aligned}
\Sigma_{ij}(\theta) &= \Sigma_{ij}(\pi/2) + \left[\frac{d\Sigma_{ij}(\theta)}{d\theta}\right]_{\pi/2} (\theta - \pi/2) \\
&\quad + \frac{1}{2}\left[\frac{d^2\Sigma_{ij}(\theta)}{d\theta^2}\right]_{\pi/2} (\theta - \pi/2)^2 + \cdots
\end{aligned}
$$

$$\Sigma_{ij}(-\theta) = \Sigma_{ij}(-\pi/2) + \left[\frac{d\Sigma_{ij}(\theta)}{d\theta}\right]_{-\pi/2} (-\theta + \pi/2)$$
$$+\frac{1}{2}\left[\frac{d^2\Sigma_{ij}(\theta)}{d\theta^2}\right]_{-\pi/2} (-\theta + \pi/2)^2 + \cdots$$
$$(\theta - \pi/2) = \sin(\theta - \pi/2) = \cos\theta \qquad (17.10)$$

so that, by employing Eqs. (17.10) and (17.6), the quantity $[\Sigma_{ij}(\theta)+\Sigma_{ij}(-\theta)]$ in the vicinity of $\theta = \pi/2$ is given by

$$\Sigma_{ij}(\theta) + \Sigma_{ij}(-\theta) = 2\Sigma_{ij}(\pi/2) + \left[\frac{d^2\Sigma_{ij}(\theta)}{d\theta^2}\right]_{\pi/2} \cos^2\theta. \qquad (17.11)$$

Then substituting this result into Eq. (17.9),

$$\sigma_{ij}(R - \varepsilon) = \frac{2}{R}\Sigma_{ij}(\pi/2) \int_0^{\pi/2} \frac{d\theta}{(1 - k^2 \sin^2\theta)^{3/2}}$$
$$+ \frac{1}{R}\left[\frac{d^2\Sigma_{ij}(\theta)}{d\theta^2}\right]_{\pi/2} \int_0^{\pi/2} \frac{\cos^2\theta d\theta}{(1 - k^2 \sin^2\theta)^{3/2}}. \qquad (17.12)$$

The integrals in Eq. (17.12) are known integrals which are expressed in terms of the complete elliptic integrals, E(k) and K(k) (Gradshteyn and Ryzhik 1980). By substituting these integrals into Eq. (17.12),

$$\sigma_{ij}(R - \varepsilon) = \frac{2}{R}\Sigma_{ij}(\pi/2)\frac{E(k)}{(1 - k^2)}$$
$$- \frac{1}{R}\left|\frac{d^2\Sigma_{ij}(\theta)}{d\theta^2}\right|_{\pi/2} \frac{[E(k) - K(k)]}{k^2}. \qquad (17.13)$$

However, since $k \cong 1$, we may employ first order expansions for these integrals in the region where $k \cong 1$ (or, alternatively, $\varepsilon/R \ll 1$) which, according to Cayley (1985), have the forms

$$\begin{aligned} E(k) &= 1 + \frac{\varepsilon}{2R}\left[\ln\left(\frac{8R}{\varepsilon}\right) - 1\right], \\ K(k) &= \frac{1}{2}\ln\left(\frac{8R}{\varepsilon}\right) + \frac{\varepsilon}{4R}\left[\ln\left(\frac{8R}{\varepsilon}\right) - 2\right]. \end{aligned} \tag{17.14}$$

Therefore, by substituting Eq. (17.14) into Eq. (17.13), and then invoking the condition $\varepsilon/R \ll 1$,

$$\begin{aligned} \sigma_{ij}(R - \varepsilon) = \frac{\Sigma_{ij}(\pi/2)}{\varepsilon} - \frac{1}{R}\Bigg\{&\left(\frac{d^2\Sigma_{ij}(\theta)}{d\theta^2}\right)_{\pi/2} \\ &- \frac{1}{2}\left[\Sigma_{ij}(\pi/2) + \left(\frac{d^2\Sigma_{ij}(\theta)}{d\theta^2}\right)_{\pi/2}\right]\ln\frac{8R}{\varepsilon}\Bigg\}. \end{aligned} \tag{17.15}$$

Comparison of Eq. (17.15) with Eq. (12.107) shows that the leading term in Eq. (17.15), which is by far the largest term, corresponds to the stress at P (Fig. 17.1a) produced by a straight and infinitely long dislocation tangent to the loop at Q as might be expected at a point very close to the loop circumference.

When the field point is outside the loop, Barnett (1976) and Gavazza and Barnett (1976) have shown, by means of a generally similar method, that Eq. (17.15) again holds but with the sign of the first term reversed.[1] Therefore, the complete result is given by

$$\begin{aligned} \sigma_{ij}(R \pm \varepsilon) = \mp\frac{\Sigma_{ij}(\pi/2)}{\varepsilon} - \frac{1}{R}\Bigg\{&\left(\frac{d^2\Sigma_{ij}(\theta)}{d\theta^2}\right)_{\pi/2} \\ &- \frac{1}{2}\left[\Sigma_{ij}(\pi/2) + \left(\frac{d^2\Sigma_{ij}(\theta)}{d\theta^2}\right)_{\pi/2}\right]\ln\frac{8R}{\varepsilon}\Bigg\}. \end{aligned} \tag{17.16}$$

[1] This is a reflection of the asymmetry of the stress field around an infinitely long straight dislocation [see Eq. (12.58)].

Furthermore, Eq. (17.16) refers to the special case where the tangent vector to the loop at Q (Fig. 17.1) lies at the angle $\pi/2$ measured from the reference line. It can therefore be generalized to apply to any point on the loop (where the tangent vector lies at the angle α as in Fig. 12.14) by writing it in the form

$$\sigma_{ij}(R \pm \varepsilon) = \mp\frac{\Sigma_{ij}(\alpha)}{\varepsilon} - \frac{1}{R}\left\{\left(\frac{d^2\Sigma_{ij}(\alpha)}{d\alpha^2}\right) - \frac{1}{2}\left[\Sigma_{ij}(\alpha) + \left(\frac{d^2\Sigma_{ij}(\alpha)}{d\alpha^2}\right)\right]\ln\frac{8R}{\varepsilon}\right\}. \qquad (17.17)$$

17.2.1.2 *Loop energy*

Having the loop self-stress directly adjacent to the loop core in hand, we continue to follow Gavazza and Barnett (1976) and next determine the loop self-energy prior to its virtual displacement. To avoid the usual complication posed by the elastic singularity at the loop core (see Sec. 12.3.2 for the straight dislocation case) the core region is enclosed by a tube of small radius, ε, and area $\mathcal{S}^{TUBE}$ as indicated in Fig. 17.2, and the loop self-energy is approximated by the strain energy, W, in the region, $\mathcal{V}$, outside the tube.[2] Using Eq. (2.133) and the divergence theorem, this energy can

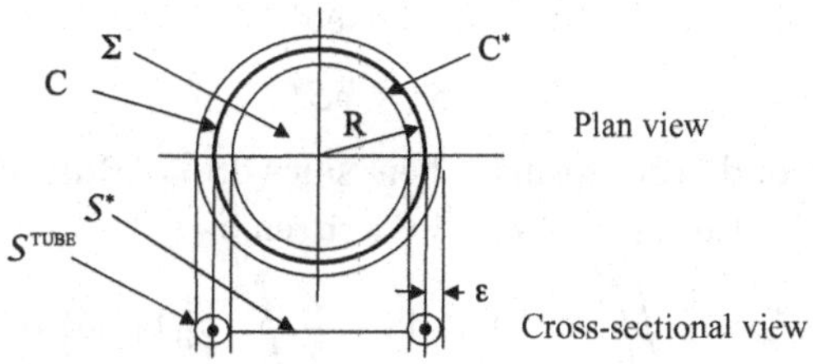

Fig. 17.2 Circular planar dislocation loop, C, of radius R, enclosed by tube of small radius, ε. Σ is the planar surface bounded by C and corresponds to the cut used to create loop C. $\mathcal{S}^*$ is the planar surface bounded by C^*, i.e., the inner intersection of the tube with the Σ surface.

[2]As seen below, the results obtained with this assumption have only a relatively weak logarithmic dependence upon the magnitude of the radius of the tube, ε, which is of the order of b.

be written as

$$\begin{aligned} W &= \frac{1}{2}\iiint_{\mathcal{V}} \sigma_{ij}\varepsilon_{ij}dV = W^* + W^{TUBE} \\ W^* &= \frac{1}{2}\iint_{\mathcal{S}^*} \sigma^C_{ij} b_i \hat{n}_j dS \\ W^{TUBE} &= \frac{1}{2}\oiint_{\mathcal{S}^{TUBE}} \sigma^C_{ij} u^C_i \hat{n}_j dS \end{aligned} \tag{17.18}$$

where W^* and W^{TUBE} are the strain energies arising from the work done on the $\mathcal{S}^*$ and the $\mathcal{S}^{TUBE}$ surfaces, respectively, during the production of the loop C, and σ^C_{ij} and u^C_i are the corresponding stresses and displacements.

17.2.1.3 *Change in loop energy due to virtual displacement*

Change in W^*

Now, displace the loop virtually by expanding it radially by $\delta R(\alpha)$, and distinguish the final loop quantities by a prime. The change in the W^* energy is then

$$\begin{aligned} \delta W^* = W'^* - W^* &= \frac{1}{2}\iint_{\mathcal{S}^*+\delta\mathcal{S}^*} (\sigma^C_{ij} + \delta\sigma^C_{ij}) b_i \hat{n}_j dS - \frac{1}{2}\iint_{\mathcal{S}^*} \sigma^C_{ij} b_i \hat{n}_j dS \\ &= \frac{1}{2}\iint_{\mathcal{S}^*} \delta\sigma^C_{ij} b_i \hat{n}_j dS + \frac{1}{2}\iint_{\delta\mathcal{S}^*} \sigma^C_{ij} b_i \hat{n}_j dS \end{aligned} \tag{17.19}$$

after using the relationships

$$\begin{aligned} \sigma^{C'}_{ij} &= \sigma^C_{ij} + \delta\sigma^C_{ij} \\ \mathcal{S}'^* &= \mathcal{S}^* + \delta\mathcal{S}^* \end{aligned} \tag{17.20}$$

and dropping second order terms. Then, since $dS = \delta R ds$, where s measures distance along C^*, Eq. (17.15) can be written as

$$\delta W^* = \frac{1}{2}\iint_{\mathcal{S}^*} \delta\sigma^C_{ij} b_i \hat{n}_j dS + \frac{1}{2}\oint_{C^*} \sigma^C_{ij} b_i \hat{n}_j \delta R ds. \tag{17.21}$$

However, as illustrated in Fig. 17.3 it is seen that the difference between the final loop and the initial loop due to the displacement, is equivalent to the loop (C′ − C) (see caption). The stress, $\delta\sigma^C_{ij}$, in the above equations must therefore correspond to the stress produced by the loop (C′ − C). Inspection of Eq. (17.21) shows that the first term is of the form of half the interaction energy expected between the loop (C′ − C) in Fig. 17.3 and a dislocation loop produced by a cut and displacement over the area $\mathcal{S}^*$ bounded by C^* if it is imagined that this dislocation, termed the C^* loop,

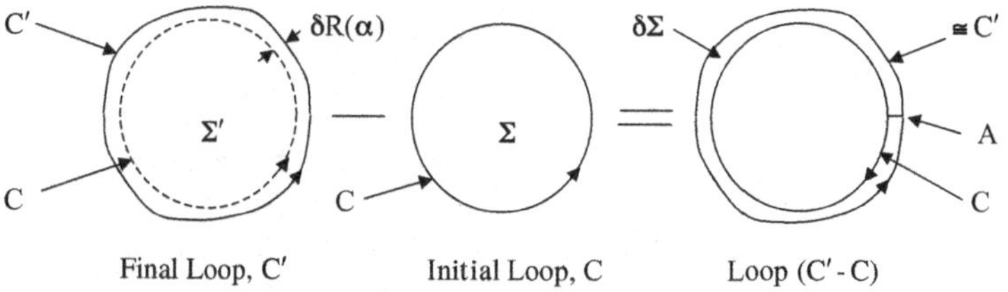

Fig. 17.3 On the left side of the above "equation" the initial loop, C, is subtracted from the displaced loop, C′, (which is shown relative to the dashed initial loop). On the right is a loop labeled (C′ − C), of area $\delta\Sigma$, that is essentially equivalent to the difference between the two loops on the left, since it is composed of an outer arc essentially equivalent to C′, an inner arc essentially equivalent to C (but of opposite sense), and two short transverse segments at A (of length δR but opposite sense) that are infinitesimally close together and therefore effectively annihilate one another.

is produced in the presence of the stress field of the (C′ − C) loop. However, the same interaction energy can be obtained by imagining that the (C′ − C) loop is created in the stress field of the C* loop, and we therefore must have

$$\iint_{S^*} \delta\sigma_{ij}^{C} b_i \hat{n}_j dS = \iint_{\delta\Sigma} \sigma_{ij}^{C^*} b_i \hat{n}_j dS = \oint_C \sigma_{ij}^{C^*} b_i \hat{n}_j \delta R ds. \tag{17.22}$$

Then, substitution of this result into Eq. (17.21) produces the symmetrical result

$$\delta W^* = \frac{1}{2} \oint_C \sigma_{ij}^{C^*} b_i \hat{n}_j \delta R ds + \frac{1}{2} \oint_{C^*} \sigma_{ij}^{C} b_i \hat{n}_j \delta R ds. \tag{17.23}$$

However, by the use of Eq. (17.17), Eq. (17.23) for the energy change, δW^*, can be put into a simpler form. Let ds and ds* be differential arc lengths on C and C*, respectively, so that

$$\frac{ds^*}{ds} = \frac{R^* d\theta}{R d\theta} = \frac{R - \varepsilon}{R} = 1 - \frac{\varepsilon}{R}. \tag{17.24}$$

Then, by substituting $\sigma_{ij}(R + \varepsilon)$ and $\sigma_{ij}(R - \varepsilon)$ from Eq. (17.17) for σ_{ij}^{C*} and σ_{ij}^{C}, respectively, into Eq. (17.23), and also employing Eq. (17.24),

$$\delta W^* = \frac{1}{2R} \oint_C \left\{ -\Sigma_{ij}(\alpha) - 2\frac{\partial^2 \Sigma_{ij}(\alpha)}{\partial \alpha^2} \right. \\ \left. + \left(\Sigma_{ij}(\alpha) + \frac{\partial^2 \Sigma_{ij}(\alpha)}{\partial \alpha^2} \right) \ln \frac{8R}{\varepsilon} \right\} b_i \hat{n}_j \delta R ds. \tag{17.25}$$

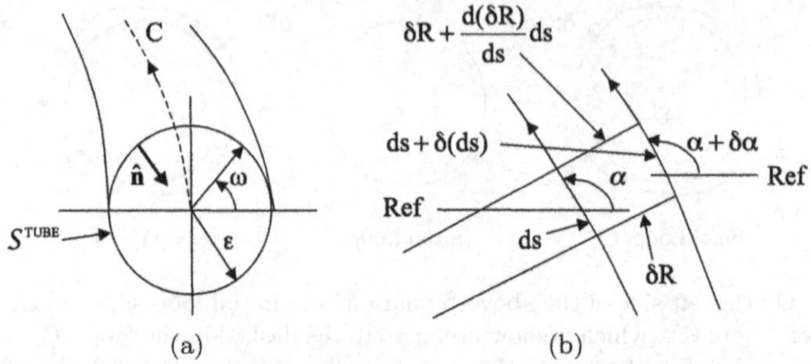

Fig. 17.4 (a) Parameters and geometry used to integrate over the tube surface via Eq. (17.26). (b) Geometry used to deduce Eqs. (17.30) and (17.32).

Change in W^{TUBE}

The work W^{TUBE}, given by Eq. (17.18), can be further expressed as

$$W^{TUBE} = \frac{1}{2}\oiint_{S^{TUBE}} \sigma^C_{ij} u^C_i \hat{n}_j dS = \frac{1}{2}\oiint_{S^{TUBE}} \sigma^C_{ij} u^C_i \hat{n}_j \varepsilon ds d\omega \quad (17.26)$$

where the parameters and geometry used for the integration over the tube surface are shown in Fig. 17.4a. As pointed out by Bullough and Foreman (1964) and Gavazza and Barnett (1976), the stresses and displacements at the tube surface, S^{TUBE}, in the integral in Eq. (17.26) can be well approximated by using the stresses and displacements, i.e., σ^∞_{ij} and u^∞_i, produced at the radial distance ε from an infinitely long straight dislocation running tangent to the tube at ds.[3] Therefore, adopting this approximation, and recognizing that, since ε is constant, the inner integral is a function of α only, we can write

$$W^{TUBE} = \oint_C F(\alpha) ds, \quad (17.27)$$

where $F(\alpha)$ is given by

$$F(\alpha) = \frac{1}{2}\oiint_{S^{CYL}} \sigma^\infty_{ij} u^\infty_i \hat{n}_j \varepsilon d\omega. \quad (17.28)$$

A determination of $F(\alpha)$ as given by Eq. (17.28), for a circular edge-type dislocation loop, is carried out in Exercise 17.5.

[3]Note that this corresponds to the leading term in Eq. (17.17) for the elastic field of the loop in Fig. 17.1a at the field points $R \pm \varepsilon$, which is by far the dominant term in the equation. Furthermore, the contribution of the tube force to the total force is relatively minor thus reducing the impact of the approximation.

Now, if the loop is radially displaced by $\delta R(\alpha)$,

$$\delta W^{TUBE} = \oint_C F(\alpha)\delta(ds) + \oint_C \delta F(\alpha) ds. \tag{17.29}$$

But, from Fig. 17.4b,

$$\begin{aligned} \delta(ds) &= (R + \delta R)\phi - R\phi = \delta R\phi = \delta R\frac{ds}{R}, \\ \delta\alpha &= -\left[\delta R + \frac{d(\delta R)}{ds}ds - \delta R\right] \Big/ ds = -\frac{d(\delta R)}{ds}, \end{aligned} \tag{17.30}$$

and substituting these results along with $\delta F(\alpha) = [dF(\alpha)/d\alpha]\delta\alpha$ into Eq. (17.29),

$$\delta W^{TUBE} = \oint_C F(\alpha)\frac{\delta R}{R} ds - \oint_C \frac{dF(\alpha)}{d\alpha}\frac{d(\delta R)}{ds} ds. \tag{17.31}$$

Then, by integrating by parts and using $d\alpha = ds/R$ obtained from Fig. 17.4b,

$$\delta W^{TUBE} = \oint_C \left[F(\alpha) + \frac{d^2F(\alpha)}{d\alpha^2}\right]\frac{\delta R}{R} ds. \tag{17.32}$$

17.2.1.4 *Self-force*

We first determine the self-force associated with the change in the energy, δW^*. For this purpose, Eq. (5.38) can be written in the integral form

$$\delta W^* = -\oint_C f^* \delta R ds, \tag{17.33}$$

where f^* is the self-force (per unit dislocation arc length). Therefore, upon comparing Eqs. (17.33) and (17.25),

$$f^*(\alpha) = \frac{1}{2R}\left\{\Sigma_{ij}(\alpha) + 2\frac{d^2\Sigma_{ij}(\alpha)}{d\alpha^2} - \left(\Sigma_{ij}(\alpha) + \frac{d^2\Sigma_{ij}(\alpha)}{d\alpha^2}\right)\ln\frac{8R}{\varepsilon}\right\} b_i\hat{n}_j. \tag{17.34}$$

Similarly, for the force associated with δW^{TUBE},

$$\delta W^{TUBE} = -\oint_C f^{TUBE}\delta R ds, \tag{17.35}$$

and upon comparing this with Eq. (17.32),

$$f^{\mathrm{TUBE}}(\alpha) = -\frac{1}{\mathrm{R}}\left(\mathrm{F}(\alpha) + \frac{\mathrm{d}^2\mathrm{F}(\alpha)}{\mathrm{d}\alpha^2}\right). \tag{17.36}$$

The total self-force acting at a point on a circular loop of radius R is therefore finally

$$\begin{aligned} f(\alpha) &= f^*(\alpha) + f^{\mathrm{TUBE}}(\alpha) \\ &= \frac{1}{\mathrm{R}}\left\{\left[\frac{\Sigma_{\mathrm{ij}}(\alpha)}{2} + \frac{\mathrm{d}^2\Sigma_{\mathrm{ij}}(\alpha)}{\mathrm{d}\alpha^2} - \frac{1}{2}\left(\Sigma_{\mathrm{ij}}(\alpha) + \frac{\mathrm{d}^2\Sigma_{\mathrm{ij}}(\alpha)}{\mathrm{d}\alpha^2}\right)\ln\frac{8\mathrm{R}}{\varepsilon}\right]\mathrm{b_i}\hat{\mathrm{n}}_\mathrm{j} \right.\\ &\quad \left. - \left(\mathrm{F}(\alpha) + \frac{\mathrm{d}^2\mathrm{F}(\alpha)}{\mathrm{d}\alpha^2}\right)\right\}, \end{aligned} \tag{17.37}$$

with $\mathrm{F}(\alpha)$ given by Eq. (17.28). The tube force, f^{TUBE}, is generally small relative to f^* for loops of common sizes and is frequently neglected in the literature. See Exercise 17.5 for the case of an edge type loop.

17.2.2 *General smoothly curved planar loop*

For a general smoothly curved planar loop, as in Fig. 17.5, Barnett (1976) has shown, by a rather lengthy analysis which will not be reproduced here, that near any point Q on the loop where the local radius of curvature is R, the in-plane stresses are given, with acceptable accuracy, by Eq. (17.17) referred to a circular dislocation loop corresponding to the osculating circle to the loop at Q as illustrated in Fig. 17.5.[4] The previous analyses and results therefore apply with R corresponding to the local radius of curvature.

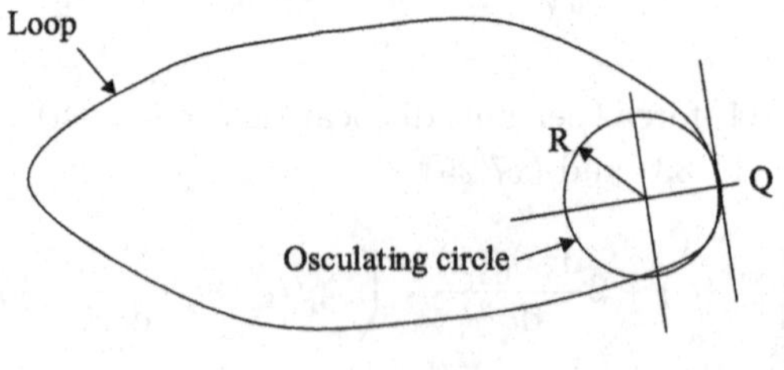

Fig. 17.5 Osculating circle to a point Q on a generally curved planar dislocation loop.

[4] An osculating circle to a point Q on a general smoothly curved line just "kisses" the line at Q and possesses the same radius of curvature, R, as the local radius of curvature of the line at Q.

17.2.3 *Some results for isotropic systems*

17.2.3.1 *Loop self-stress*

In the case of an isotropic system an exact analytical expression can be found for the quantity $[\Sigma_{ij}(\theta) + \Sigma_{ij}(-\theta)]$ in Eq. (17.9). This allows us to obtain, after integration, an expression for $\sigma_{ij}(P)$ in Fig. 17.1a that is free of the Taylor expansion approximation used to obtain Eq. (17.17). We start by determining the function $\Sigma_{ij}(\theta)$ by using Fig. 17.1b and Eqs. (12.107) and (12.24) to write

$$\Sigma^{D}_{mn} = \Sigma^{D}_{mn}|\mathbf{x}^{D}| = \frac{b^{D}_{s}C^{D}_{mnip}}{2\pi}\{-\hat{m}^{D}_{p}S^{D}_{is} + \hat{n}^{D}_{p}[(\hat{n}\hat{n})^{D}_{ik}]^{-1}[4\pi B^{D}_{ks} + (\hat{n}\hat{m})^{D}_{kr}S^{D}_{rs}]\}, \tag{17.38}$$

where the superscript D indicates that the various quantities are referred to the $(\mathbf{e}^{D}_{1}, \mathbf{e}^{D}_{2}, \mathbf{e}^{D}_{3})$ coordinate system, and the $\hat{\mathbf{m}}^{D}$ and $\hat{\mathbf{n}}^{D}$ vectors are as indicated by the figure. The field vector $\mathbf{x}^{D}$ is directed along $\hat{\mathbf{m}}^{D}$ as required by Eq. (12.24), and $|\mathbf{x}^{D}| = \rho$. Then, using Eq. (3.141) with $\hat{\mathbf{m}}^{D} = (1, 0, 0)$ and $\hat{\mathbf{n}}^{D} = (0, 1, 0)$,

$$[(\hat{n}\hat{n})^{D}]^{-1} = \frac{1}{\mu}\begin{bmatrix} 1 & 0 & 0 \\ 0 & \dfrac{1-2\nu}{2(1-\nu)} & 0 \\ 0 & 0 & 1 \end{bmatrix} \quad (\hat{n}\hat{m})^{D} = \mu\begin{bmatrix} 0 & 1 & 0 \\ \dfrac{2\nu}{1-2\nu} & 0 & 0 \\ 0 & 0 & 0 \end{bmatrix}. \tag{17.39}$$

With the use of the above relationships and Eqs. (3.147) and (2.120) we then obtain

$$[\Sigma^{D}] = \frac{\mu}{2\pi(1-\nu)}\begin{bmatrix} b^{D}_{2} & b^{D}_{1} & 0 \\ b^{D}_{1} & b^{D}_{2} & (1-\nu)b^{D}_{3} \\ 0 & (1-\nu)b^{D}_{3} & 2\nu b^{D}_{2} \end{bmatrix}. \tag{17.40}$$

However, we seek a final expression for $\Sigma_{ij}(\theta)$ in the $(\hat{\mathbf{e}}_1, \hat{\mathbf{e}}_2, \hat{\mathbf{e}}_3)$ system of Fig. 17.1a. Therefore, by making the necessary coordinate transformations using the relationships

$$[\Sigma] = [l]\,[\Sigma^{D}]\,[l]^{T} \quad [b^{D}] = [l]^{T}[b] \quad [l] = \begin{bmatrix} \sin\theta & 0 & -\cos\theta \\ 0 & 1 & 0 \\ \cos\theta & 0 & \sin\theta \end{bmatrix}, \tag{17.41}$$

obtained with the help of Eqs. (2.23) and (2.17), the components of $[\Sigma]$ are found to be

$$\Sigma_{ij}(\theta) = \frac{\mu}{2\pi(1-\nu)} \times \begin{bmatrix} b_2(\sin^2\theta + 2\nu\cos^2\theta) & b_1(1-\nu\cos^2\theta) + b_3\nu\sin\theta\cos\theta & b_2(1-2\nu)\sin\theta\cos\theta \\ b_1(1-\nu\cos^2\theta) + b_3\nu\sin\theta\cos\theta & b_2 & b_3(1-\nu\sin^2\theta) + b_1\nu\sin\theta\cos\theta \\ b_2(1-2\nu)\sin\theta\cos\theta & b_3(1-\nu\sin^2\theta) + b_1\nu\sin\theta\cos\theta & b_2(\cos^2\theta + 2\nu\sin^2\theta) \end{bmatrix}. \tag{17.42}$$

Then, having Eq. (17.42), the quantity $[\Sigma_{ij}(\theta) + \Sigma_{ij}(-\theta)]$ is easily obtained, and substituting this into Eq. (17.9) and integrating using the same procedure employed earlier to obtain Eq. (17.15) from Eq. (17.12), the stresses at P in Fig. 17.1a are given by

$$\begin{aligned}
\sigma_{11}(P) &= \frac{\mu b_2}{2\pi(1-\nu)}\left\{\frac{1}{\varepsilon} + \frac{1}{R}\left[2(1-2\nu) - \frac{(1-4\nu)}{2}\ln\frac{8R}{\varepsilon}\right]\right\}, \\
\sigma_{22}(P) &= \frac{\mu b_2}{2\pi(1-\nu)}\left\{\frac{1}{\varepsilon} + \frac{1}{R}\left[\frac{1}{2}\ln\frac{8R}{\varepsilon}\right]\right\}, \\
\sigma_{33}(P) &= \frac{\mu b_2 \nu}{\pi(1-\nu)}\left\{\frac{1}{\varepsilon} - \frac{1}{R}\left[\frac{(1-2\nu)}{\nu} + \frac{(1-\nu)}{2\nu}\ln\frac{8R}{\varepsilon}\right]\right\}, \\
\sigma_{12}(P) &= \frac{\mu b_1}{2\pi(1-\nu)}\left\{\frac{1}{\varepsilon} + \frac{1}{R}\left[2\nu + \frac{1}{2}(1-2\nu)\ln\frac{8R}{\varepsilon}\right]\right\}, \\
\sigma_{13}(P) &= 0, \\
\sigma_{23}(P) &= \frac{\mu b_3}{2\pi}\left\{\frac{1}{\varepsilon} - \frac{1}{R}\left[\frac{2\nu}{(1-\nu)} + \frac{(1+\nu)}{2(1-\nu)}\ln\frac{8R}{\varepsilon}\right]\right\}.
\end{aligned} \tag{17.43}$$

17.2.3.2 *Self-force*

Consider first the $f^*(\alpha)$ self-force given by Eq. (17.34) for the case of a pure edge loop with $\mathbf{b} = (0, b, 0)$ and $\hat{\mathbf{n}} = (0, 1, 0)$. Equation (17.34) then reduces to

$$f^*(\alpha) = \frac{1}{2R}\left\{\Sigma_{22}(\alpha) + 2\frac{\partial^2\Sigma_{22}(\alpha)}{\partial\alpha^2} - \left[\Sigma_{22}(\alpha) + \frac{\partial^2\Sigma_{22}(\alpha)}{\partial\alpha^2}\right]\ln\frac{8R}{\varepsilon}\right\}, \tag{17.44}$$

and, after substituting Eq. (17.42),

$$f^* = \frac{b}{2R}\Sigma_{22}\left[1 - \ln\frac{8R}{\varepsilon}\right] = -\frac{\mu b^2}{4\pi(1-\nu)R}\left[\ln\frac{8R}{\varepsilon} - 1\right]. \qquad (17.45)$$

The force is seen to be independent of α as expected, since the dislocation edge character is constant around the circumference.

The above result may be compared with the $f^*(\alpha)$ self-force obtained directly from the strain energy of the loop

$$W^* = \frac{\mu b^2}{2(1-\nu)}R\left[\ln\frac{8R}{\varepsilon} - 2\right] \qquad (17.46)$$

obtained in Exercise 17.3. Since the self-force is uniform around the loop, the self-force derived from a virtual radial loop displacement, δR, which we take here to be uniform around the circumference, must be given by

$$f^* = -\frac{1}{2\pi R}\frac{\delta W^*}{\delta R} = -\frac{\mu b^2}{4\pi(1-\nu)R}\left[\ln\frac{8R}{\varepsilon} - 1\right], \qquad (17.47)$$

which agrees with the result given by Eq. (17.45). In Exercise 17.2, similar agreement is found between the average self-force, $\langle f^*\rangle$, experienced by a pure shear loop determined by use of Eq. (17.34) and obtained directly from an expression for the loop strain energy, W^*, determined in Exercise 17.4.

An evaluation of the tube parameter, $F(\alpha)$, which is necessary for the determination of the f^{TUBE} self-force via Eq. (17.36), is carried out in Exercise 17.5.

17.3 Dislocation Line Tension

If the highly approximate assumption is made that a curved dislocation possesses a constant energy per unit length, regardless of its configuration, it

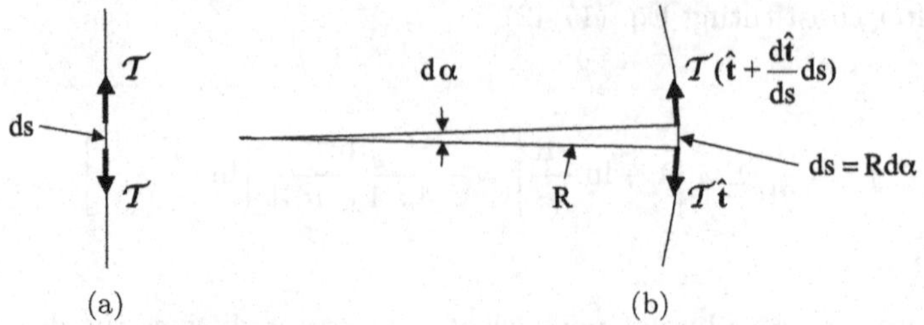

Fig. 17.6 (a) Line tension, $\mathcal{T}$, acting on an element, ds, of a straight dislocation. (b) Line tension acting on an element of a smoothly curved dislocation at a point where the local radius of curvature is R.

should behave like a flexible extensible line where every element is subjected to a constant tangential tensile force, or *line tension*, $\mathcal{T}$, as illustrated for a straight segment in Fig. 17.6a. If dW is the increase in dislocation energy that accompanies an increase in its length, dL, $\mathcal{T}$ must be given by

$$\mathcal{T} = \frac{dW}{dL}. \tag{17.48}$$

In the case of a smoothly curved dislocation it is readily shown that such a line tension would impose a self-force at every point in a direction towards the center of curvature which tends to straighten out the dislocation, thereby reducing its length and associated energy. Referring to Fig. 17.6, and writing the line tension force as $\mathcal{T}\hat{\mathbf{t}}$, the net vector self-force experienced by the differential segment ds of the curved dislocation in Fig. 17.6b is just

$$\begin{aligned} \boldsymbol{f}ds &= \mathcal{T}\left(\hat{\mathbf{t}} + \frac{d\hat{\mathbf{t}}}{ds}ds\right) - \mathcal{T}\hat{\mathbf{t}} = \mathcal{T}\frac{d\hat{\mathbf{t}}}{ds}ds, \\ \boldsymbol{f} &= \mathcal{T}\frac{d\hat{\mathbf{t}}}{ds} = \mathcal{T}\frac{d\hat{\mathbf{t}}}{Rd\alpha} = \mathcal{T}\frac{\hat{\mathbf{u}}d\alpha}{Rd\alpha} = \frac{\mathcal{T}}{R}\hat{\mathbf{u}}, \end{aligned} \tag{17.49}$$

where $\hat{\mathbf{u}}$ is a unit vector directed along the local radius of curvature, R, towards the center of curvature. Alternatively, the same result may be obtained by appealing to Fig. 17.4b. Here, the segment ds is virtually displaced by δR and, according to Eq. (17.30), is increased in length by $\delta(ds) = (\delta R/R)ds$. The increase in dislocation energy must then be $\mathcal{T}\delta(ds)$, and this must be equal to the work done against the self-force, i.e., $\boldsymbol{f}ds\delta R$.

Therefore,

$$f \mathrm{ds}\delta \mathrm{R} = \mathcal{T}\delta(\mathrm{ds}) = \mathcal{T}\frac{\delta \mathrm{R}}{\mathrm{R}}\mathrm{ds}, \qquad f = \frac{\mathcal{T}}{\mathcal{R}}, \tag{17.50}$$

in agreement with the magnitude of $\boldsymbol{f}$ given by Eq. (17.49).

According to the simple picture described above, the self-force at any point should merely be proportional to the local curvature, i.e., 1/R, via a constant line tension as in Eq. (17.50). However, examination of Eq. (17.37) shows that this is not the case, since the effective line tension, i.e., the quantity in curly brackets, is also dependent upon the local direction of the dislocation in the crystal via the angle α and exhibits an additional logarithmic dependence on R. Obviously, the simple constant line tension model is not valid. However, the effective line tension deduced from Eq. (17.37) is seen to be only relatively weakly dependent upon R via the logarithmic term, and, by neglecting the α dependence, the self-force is often crudely approximated in the literature by assuming a constant line tension and a self-force given by Eq. (17.49) with $\mathcal{T}$ given by Eq. (12.43). Further discussion of the problems associated with the concept of a dislocation line tension is given by Bacon, Barnett and Scattergood (1979b) and Hirth and Lothe (1982).

17.4 Self-force Experienced by Straight Dislocation Segment

We now consider the self-force acting at a point on a straight dislocation segment that is part of a segmented structure such as, for example, the planar polygonal loop shown in Fig. 17.7a. Such a loop, as indicated in the figure, can be regarded as an assemblage of five bi-angular dislocations of the type illustrated in Fig. 12.24 where the infinitely long straight segments associated the bi-angular dislocations would mutually annihilate, as is evident in Fig. 17.7a. The stress contributed by the segment AB at a field point P located at the normal distance ρ from it (Fig. 17.7b) therefore corresponds to the stress associated with the bi-angular dislocation containing it which is given by Eq. (12.182), i.e.,

$$\sigma_{\mathrm{ij}}^{\mathrm{AB}}(\mathrm{P}) = \frac{1}{2\rho}\left| -\cos(\theta-\alpha)\Sigma_{\mathrm{ij}}(\theta) + \sin(\theta-\alpha)\frac{\mathrm{d}\Sigma_{\mathrm{ij}}(\theta)}{\mathrm{d}\theta}\right|_{\theta^{\mathrm{A}}}^{\theta^{\mathrm{B}}}. \tag{17.51}$$

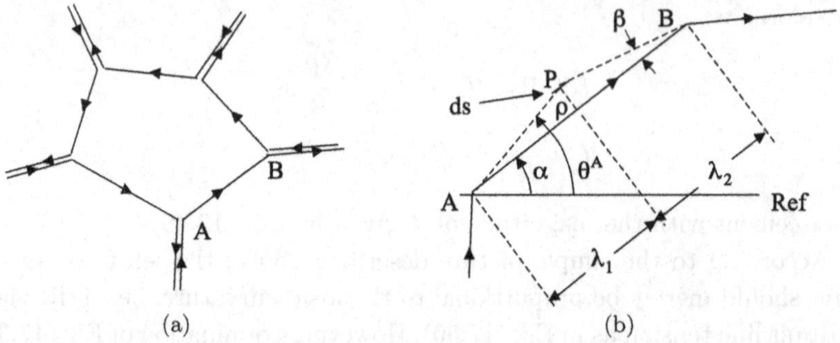

Fig. 17.7 (a) Planar polygonal dislocation loop corresponding to an assemblage of five bi-angular dislocations. (b) Enlarged view of segment AB. In addition, a segment, *ds*, of a parallel dislocation possessing the same Burgers vector as AB is present at a distance ρ from AB.

Following Hirth and Lothe (1982), the self-force on a differential element of AB can then be determined by first imagining that a differential dislocation segment, *ds*, of the same type as the segment AB is located at the field point P, as in Fig. 17.7b, and calculating the stress and the corresponding force imposed on it by the AB bi-angular dislocation. Then, by taking the limit of this force as $\rho \to 0$, the self-force experienced by a corresponding element of AB is obtained.

Expressions for the angles $(\theta^{\mathrm{A}} - \alpha)$ and $(\theta^{\mathrm{B}} - \alpha)$ required in Eq. (17.51) are obtained from the geometry in Fig. 17.7b. For $(\theta^{\mathrm{A}} - \alpha)$,

$$\sin(\theta^{\mathrm{A}} - \alpha) = \frac{\rho}{\sqrt{\rho^2 + \lambda_1^2}} \qquad \cos(\theta^{\mathrm{A}} - \alpha) = \frac{\lambda_1}{\sqrt{\rho^2 + \lambda_1^2}}, \tag{17.52}$$

while for $(\theta^{\mathrm{B}} - \alpha)$, since $\cos(\theta^{\mathrm{B}} - \alpha - \pi) = -\cos(\theta^{\mathrm{B}} - \alpha)$,

$$\sin(\theta^{\mathrm{B}} - \alpha) = \frac{-\rho}{\sqrt{\rho^2 + \lambda_2^2}} \qquad \cos(\theta^{\mathrm{B}} - \alpha) = \frac{-\lambda_2}{\sqrt{\rho^2 + \lambda_2^2}}. \tag{17.53}$$

Then, substituting these expressions into Eq. (17.51), the stress at P due to the AB segment is

$$\sigma_{\mathrm{ij}}^{\mathrm{AB}}(\mathrm{P}) = -\frac{1}{2}\left(\frac{1}{\sqrt{\rho^2 + \lambda_2^2}}\left[\frac{\mathrm{d}\Sigma_{\mathrm{ij}}(\theta)}{\mathrm{d}\theta}\right]_{\theta=\theta^{\mathrm{B}}} + \frac{1}{\sqrt{\rho^2 + \lambda_1^2}}\left[\frac{\mathrm{d}\Sigma_{\mathrm{ij}}(\theta)}{\mathrm{d}\theta}\right]_{\theta=\theta^{\mathrm{A}}}\right)$$
$$+\frac{1}{2\rho}\left(\frac{\lambda_2}{\sqrt{\rho^2 + \lambda_2^2}}\Sigma_{\mathrm{ij}}(\theta^{\mathrm{B}}) + \frac{\lambda_1}{\sqrt{\rho^2 + \lambda_1^2}}\Sigma_{\mathrm{ij}}(\theta^{\mathrm{A}})\right). \tag{17.54}$$

Also, if the AB segment is infinitely long, so that $\lambda_1 \to \infty$, $\lambda_2 \to \infty$, $\theta^A \to \alpha$, $\theta^B \to \alpha + \pi$, the stress at P given by Eq. (17.54) reduces, after use of Eq. (17.6), to

$$\sigma_{ij}^{AB\infty}(P) = \frac{1}{2\rho}\left[\Sigma_{ij}(\alpha + \pi) + \Sigma_{ij}(\alpha)\right] = \frac{\Sigma_{ij}(\alpha)}{\rho}, \tag{17.55}$$

in agreement with Eq. (12.107). Then, by use of the Peach–Koehler force equation, i.e., Eq. (13.9), the corresponding forces exerted on the segment *ds* located at P by these stresses are, respectively

$$dF_k^{ds/AB} = -e_{ksj} b_i \hat{t}_s \sigma_{ij}^{AB} ds \tag{17.56}$$

and

$$dF_k^{ds/AB\infty} = -e_{ksj} b_i \hat{t}_s \sigma_{ij}^{AB\infty} ds. \tag{17.57}$$

If we now subtract Eq. (17.57) from (17.56), and substitute Eqs. (17.54) and (17.55), and then take the limit of the result as $\rho \to 0$,

$$\begin{aligned}
&\lim_{\rho\to 0}(dF_k^{ds/AB} - dF_k^{ds/AB\infty}) \\
&= -\frac{1}{2}\lim_{\rho\to 0} e_{ksj} b_i \hat{t}_s ds \left\{ -\left(\frac{1}{\sqrt{\rho^2+\lambda_2^2}} \left[\frac{d\Sigma_{ij}(\theta)}{d\theta}\right]_{\theta=\theta^B} \right.\right. \\
&\left. + \frac{1}{\sqrt{\rho^2+\lambda_1^2}} \left[\frac{d\Sigma_{ij}(\theta)}{d\theta}\right]_{\theta=\theta^A} \right) \\
&+ \frac{1}{\rho}\left(\frac{\lambda_2}{\sqrt{\rho^2+\lambda_2^2}} \sigma_{ij}(\theta^B) + \frac{\lambda_1}{\sqrt{\rho^2+\lambda_1^2}} \Sigma_{ij}(\theta^A) \right) \\
&\left. - \frac{1}{\rho}\left[\Sigma_{ij}(\alpha+\pi) + \sigma_{ij}(\alpha)\right] \right\}.
\end{aligned} \tag{17.58}$$

However,

$$\lim_{\rho\to 0} dF_k^{ds/AB} = dF_k^{ABself} \tag{17.59}$$

where dF_k^{ABself} is the self-force experienced by a differential element on the finite segment AB, and

$$\lim_{\rho\to 0} dF_k^{ds/AB\infty} = dF_k^{AB\infty self} = 0, \tag{17.60}$$

where $dF_k^{AB\infty self}$ is the self-force experienced by a corresponding element on an infinitely long AB-type straight dislocation which is well known to vanish (Hirth and Lothe 1982). Therefore, substituting Eqs. (17.59) and (17.60) into Eq. (17.58), and taking the limit where $\theta^A \to \alpha$ and $\theta^B \to \alpha + \pi$, and using Eq. (17.6),

$$dF_k^{ABself} = \frac{1}{2} e_{ksj} b_i \hat{t}_s \left(\frac{1}{\lambda_1} + \frac{1}{\lambda_2} \right) \left[\frac{d\Sigma_{ij}(\theta)}{d\theta} \right]_{\theta=\alpha} ds. \tag{17.61}$$

The total force on the *ds* segment of AB in Fig. 17.7a is then the above self-force plus the forces imposed by the stress fields of the other four segments of the polygonal loop.

A corresponding expression for the self-force on the AB segment in an isotropic system is derived in Exercise 17.6.

17.5 Self-force Experienced by Inclusion

The local self-force (per unit area) acting normal to a point on the surface of an inclusion can be determined by employing the energy-momentum tensor force given by Eq. (15.6). Taking regions 1 and 2 in Fig. 15.1 to be the regions directly adjacent to the INC/M interface in the matrix and inclusion, respectively, the force is then given by

$$\mathcal{F} = (w^M - w^{INC}) - T_i \frac{\partial(u_i^{tot,M} - u_i^{tot,INC})}{\partial x_l} \hat{n}_l = (w^M - w^{INC})$$

$$-\sigma_{ij}^{INC} \hat{n}_j \frac{\partial(u_i^{tot,M} - u_i^{tot,INC})}{\partial x_l} \hat{n}_l, \tag{17.62}$$

where all quantities refer to regions directly adjacent to the INC/M interface and correspond to the self-stress field of the inclusion embedded in an infinite matrix.

For purposes of illustration we consider the simplest possible case, i.e., a homogeneous incoherent inclusion possessing a uniform transformation strain in the form of an ellipsoid of revolution in an elastically isotropic system. The inclusion possesses major axes *c* and *a* and lies in a Cartesian coordinate system with *c* parallel to x_3. Its shape and volume are therefore given by

$$\frac{x_1^2}{a^2} + \frac{x_2^2}{a^2} + \frac{x_3^2}{c^2} = 1, \quad V^{INC} = \frac{4}{3}\pi a^2 c = \text{constant}. \tag{17.63}$$

It will be assumed that is in a state of minimum elastic energy where, as described in Sec. 6.5, its incoherency allows it to relieve its transformation strain tensor, ε_{ij}^{T}, of any shear components while at the same time distributing the remaining cubical dilatation into normal transformation strains that minimize the strain energy and maintain hydrostatic stress throughout the inclusion. During this process the inclusion volume, including its cubical dilatation, must remain constant, since any relief would require long-range mass transport between the inclusion and sources/sinks present elsewhere in the matrix, a phenomenon not considered here.

Rather than writing out a complete expression for the force given by Eq. (17.62) we shall be content to derive suitable expressions for each of its terms. The strain energy density term, w^{INC}, can be determined by use of Eq. (2.136) which, in turn, requires the strains in the inclusion, ε_{ij}^{INC}. These consist only of normal strains which are present uniformly in the incoherent inclusion, and by use of Eqs. (6.1) and (6.96), are given by

$$\begin{aligned} \varepsilon_{11}^{INC} &= (S_{11}^{E} - 1)\varepsilon_{11}^{T} + S_{12}^{E}\varepsilon_{22}^{T} + S_{13}^{E}\varepsilon_{33}^{T}, \\ \varepsilon_{22}^{INC} &= (S_{12}^{E}\varepsilon_{11}^{T} + (S_{11}^{E} - 1)\varepsilon_{22}^{T} + S_{13}^{E}\varepsilon_{33}^{T}, \\ \varepsilon_{33}^{INC} &= (S_{31}^{E}\varepsilon_{11}^{T} + S_{31}^{E}\varepsilon_{22}^{T} + (S_{33}^{E} - 1)\varepsilon_{33}^{T}. \end{aligned} \tag{17.64}$$

Note that the cylindrical symmetry of the ellipsoid of revolution dictates that only five elements of the Eshelby tensor are independent, and that these have been taken to be S_{11}^{E}, S_{33}^{E}, S_{12}^{E}, S_{13}^{E} and S_{31}^{E}. The three normal transformation strains in Eq. (17.64), which serve to minimize the total elastic energy associated with the inclusion under the restraint that the cubical dilatation, e^{T}, is constant, can be found using the Lagrange multiplier method described in Sec. 6.5 and are given by[5]

$$\begin{aligned} \tilde{\varepsilon}_{11}^{T} = \tilde{\varepsilon}_{22}^{T} &= \frac{\begin{matrix}[-\nu(S_{11}^{E} + S_{12}^{E}) - (1-4\nu)S_{13}^{E} - (1-\nu)S_{31}^{E} \\ +(2-3\nu)S_{33}^{E} - 2(1-2\nu)]\end{matrix}}{2(1-2\nu)\left[S_{11}^{E} + S_{12}^{E} - 2(S_{13}^{E} + S_{31}^{E}) + 2S_{33}^{E} - 3\right]} e^{T}, \\ \tilde{\varepsilon}_{33}^{T} &= \frac{[(1-\nu)(S_{11}^{E} + S_{12}^{E}) - S_{13}^{E} - (1-3\nu)S_{31}^{E} - \nu S_{33}^{E} - (1-2\nu)]}{(1-2\nu)[S_{11}^{E} + S_{12}^{E} - 2(S_{13}^{E} + S_{31}^{E}) + 2S_{33}^{E} - 3]} e^{T}, \\ \tilde{\varepsilon}_{11}^{T} + \tilde{\varepsilon}_{22}^{T} + \tilde{\varepsilon}_{33}^{T} &= \tilde{e}^{T} = e^{T}. \end{aligned} \tag{17.65}$$

[5]It can be readily confirmed that use of Eqs. 17.64) and (17.65) produces the results for spherical, thin disk and needle inclusions given in Secs. 6.5.2, 6.5.3 and 6.5.4, respectively, when f = 1.

The w^{INC} term is then determined by substituting the normal transformation strains given by Eq. (17.65) into Eqs. (17.64) and (2.136).

The corresponding strain energy density in the matrix, w^M, required in Eq. (17.62), can be found by employing Eqs. (6.70) and (2.136). According to Eq. (6.1), $\varepsilon_{ij}^M = \varepsilon_{ij}^{C,M}$. Therefore, using Eq. (6.70), the strains in the matrix at the INC/M interface are given by

$$(\varepsilon_{il}^{M})_{INC/M} = (\varepsilon_{il}^{C,INC})_{INC/M} + \frac{1}{2\mu}\left[\frac{\sigma_{jk}^{T}}{(1-\nu)}\hat{n}_i\hat{n}_j\hat{n}_k\hat{n}_l - (\sigma_{ik}^{T}\hat{n}_k\hat{n}_l + \sigma_{lk}^{T}\hat{n}_i\hat{n}_k)\right]. \quad (17.66)$$

Here, the $(\varepsilon_{il}^{C,INC})_{INC/M}$ term is determined by use of Eq. (6.91), and the σ_{jk}^T stresses can be obtained by substituting the transformation strains given by Eq. (17.65) into Eq. (2.123). The w^M term is then finally obtained by substituting Eq. (17.66) into Eq. (2.136).

The σ_{ij}^{INC} stress term is determined by substituting Eq. (17.64), with the transformation strains given by Eq. (17.65) into Eq. (2.123). Also, by using Eq. (17.63) and standard methods (Hildebrand 1949), the normal vector, $\hat{\mathbf{n}}$, at a point on the inclusion surface can be written as

$$\hat{\mathbf{n}} = \frac{c^2x_1\hat{e}_1 + c^2x_2\hat{e}_2 + a^2x_3\hat{e}_3}{a[c^4 + (a^2 - c^2)x_3^2]^{1/2}}. \quad (17.67)$$

Finally, the quantity $(\partial u_i^{tot,M} - \partial u_i^{tot,INC})/\partial x_l$ in Eq. (17.62) can be determined using results obtained by Hill (1961) and Barnett (2015). According to Hill, it can be first written as

$$\left[\frac{\partial(u_i^{tot,M} - u_i^{tot,INC})}{\partial x_l}\right]_{INC/M} = \hat{\tau}_i\hat{n}_l, \quad (17.68)$$

where $\hat{\boldsymbol{\tau}}$ is a unit vector lying in the interface so that

$$\hat{\boldsymbol{\tau}} \cdot \hat{\mathbf{n}} = 0. \quad (17.69)$$

Then, following Barnett (2015), multiply Eq. (17.68) through by $\hat{\tau}_l$ so that

$$\hat{\tau}_l\left[\frac{\partial(u_i^{tot,M} - u_i^{tot,INC})}{\partial x_l}\right]_{INC/M} = \hat{\tau}_l\hat{n}_l\hat{\tau}_i = 0, \quad (17.70)$$

by virtue of Eq. (17.69). Next, a solution to Eq. (17.70) for the unknown derivative at the interface can be written in the form

$$\left[\frac{\partial(u_i^{tot,M} - u_i^{tot,INC})}{\partial x_l}\right]_{INC/M} = v_i \hat{n}_l, \tag{17.71}$$

where v_i is now the unknown quantity in the form of a vector to be determined. This vector can be found by first noting that use of Eq. (3.154) yields

$$(\sigma_{ij}^{M} - \sigma_{ij}^{INC})_{INC/M} = C_{ijkl}\left[\frac{\partial(u_k^{tot,M} - u_k^{tot,INC})}{\partial x_l}\right]_{INC/M} + \sigma_{ij}^{T}. \tag{17.72}$$

Then, after multiplying Eq. (17.72) throughout by $\hat{n}_i$,

$$\hat{n}_i(\sigma_{ij}^{M} - \sigma_{ij}^{INC})_{INC/M} = \hat{n}_i C_{ijkl} v_k \hat{n}_l + \sigma_{ij}^{T}\hat{n}_i \tag{17.73}$$

and defining $\mathbf{M}_{jk}$ by

$$M_{jk} \equiv C_{ijkl}\hat{n}_l\hat{n}_i \tag{17.74}$$

Eq. (17.73) takes the form

$$M_{jk}v_k + \sigma_{ij}^{T}\hat{n}_i = 0. \tag{17.75}$$

Then, solving Eq. (17.75) for v_i,

$$v_k = -M_{ij}^{-1}\sigma_{ij}^{T}\hat{n}_i \tag{17.76}$$

and, finally, substituting this result into Eq. (17.71),

$$\left[\frac{\partial(u_r^{tot,M} - u_r^{tot,INC})}{\partial x_l}\right]_{INC/M} = -M_{rj}^{-1}\sigma_{ij}^{T}\hat{n}_i\hat{n}_l \tag{17.77}$$

Suitable paths to expressions for all quantities in the self-force given by Eq. (17.62) have now been determined, and the analysis is therefore complete.

Exercises

17.1. Starting with Eq. (12.110), prove that the angular function, $\Sigma_{ij}(\theta)$, introduced by Eq. (12.107), has the periodic property expressed by

Eq. (17.6), i.e.,

$$\Sigma_{ij}(\theta + \pi) = \Sigma_{ij}(\theta). \tag{17.78}$$

Solution. Since, according to Eq. (12.110),

$$\Sigma_{ij}(\alpha) = \int_{\alpha}^{\alpha+\pi} \alpha_{ij}(\theta) \sin(\theta - \alpha) d\theta, \tag{17.79}$$

we must have

$$\begin{aligned} \Sigma_{ij}(\alpha + \pi) &= \int_{\alpha+\pi}^{\alpha+2\pi} \alpha_{ij}(\theta) \sin(\theta - \alpha - \pi) d\theta \\ &= -\int_{\alpha+\pi}^{\alpha+2\pi} \alpha_{ij}(\theta) \sin(\theta - \alpha) d\theta. \end{aligned} \tag{17.80}$$

However, according to Eq. (12.103), $\alpha_{ij}(\theta + \pi) = \alpha_{ij}(\theta)$, and since $(\theta - \alpha)$ ranges from 0 to π in the integrand of Eq. (17.79) and from π to 2π in the integrand of Eq. (17.80),

$$\int_{\alpha+\pi}^{\alpha+2\pi} \alpha_{ij}(\theta) \sin(\theta - \alpha) d\theta = -\int_{\alpha}^{\alpha+\pi} \alpha_{ij}(\theta) \sin(\theta - \alpha) d\theta. \tag{17.81}$$

Then, by substituting Eqs. (17.81) into (17.80), it is seen that

$$\Sigma_{ij}(\alpha) = \Sigma_{ij}(\alpha + \pi). \tag{17.82}$$

17.2. (a) Determine the average self-force, $\langle f^* \rangle$, experienced by a purely shear-type planar circular loop in an isotropic system. Note that in this case the Burgers vector component normal to the loop plane vanishes, and, in addition, the self-force, f^*, varies around the circumference. (b) Then, show that the result obtained in (a) is identical to the average self-force derived directly from the strain energy,

$$W^* = \frac{\mu b^2 (2 - \nu)}{4(1 - \nu)} R \left[\ln \frac{8R}{\varepsilon} - 2 \right], \tag{17.83}$$

associated with such a loop according to Chou and Eshelby (1962) and Exercise 17.4.

Solution. (a) Using Fig. 17.1, with $\mathbf{b} = \mathrm{b}(1,0,0)$ and $\hat{\mathbf{n}} = (0,1,0)$, Eq. (17.34) reduces to

$$f^*(\alpha) = \frac{\mathrm{b}}{2\mathrm{R}}\left\{\Sigma_{12}(\alpha) + 2\frac{\partial^2\Sigma_{12}(\alpha)}{\partial\alpha^2} - \left[\Sigma_{12}(\alpha) + \frac{\partial^2\Sigma_{12}(\alpha)}{\partial\alpha^2}\right]\ln\frac{8\mathrm{R}}{\varepsilon}\right\}. \quad (17.84)$$

Then, substituting Eq. (17.42) into Eq. (17.84),

$$f^*(\alpha) = \frac{\mu\mathrm{b}^2}{4\pi(1-\nu)\mathrm{R}}\left[(1 - 4\nu\sin^2\alpha + 3\nu\cos^2\alpha) - (1 + \nu\cos^2\alpha - 2\nu\sin^2\alpha)\ln\frac{8\mathrm{R}}{\varepsilon}\right]. \quad (17.85)$$

Next, using

$$\langle f^*(\alpha)\rangle = \frac{1}{2\pi}\int_0^{2\pi} f^*(\alpha)\mathrm{d}\alpha \quad (17.86)$$

and substituting Eq. (17.85) and performing the integration,

$$\langle f^*\rangle = -\frac{\mu\mathrm{b}^2(2-\nu)}{8\pi(1-\nu)\mathrm{R}}\ln\left[\frac{8\mathrm{R}}{\varepsilon} - 1\right]. \quad (17.87)$$

(b) The first equality in Eq. (17.47) also holds for the average self-force when the self-force varies around the loop and $\delta\mathrm{R}$ is taken to be uniform around the circumference. Therefore, substituting Eq. (17.83) into Eq. (17.47),

$$\langle f^*\rangle = -\frac{1}{2\pi\mathrm{R}}\frac{\delta\mathrm{W}^*}{\delta\mathrm{R}} = -\frac{\mu\mathrm{b}^2(2-\nu)}{8\pi(1-\nu)\mathrm{R}}\ln\left[\frac{8\mathrm{R}}{\varepsilon} - 1\right] \quad (17.88)$$

in agreement with Eq. (17.87).

17.3. Derive the expression for the strain energy, W^*, of a circular planar edge-type dislocation loop in an isotropic system given by Eq. (17.46).

Solution. Equation (17.18) referred to the $\hat{\mathbf{e}}'_{\mathrm{i}}$ coordinate system in Fig. 17.8, and with $\mathbf{b}' = (0,\mathrm{b},0)$ and $\hat{\mathbf{n}}' = (0,1,0)$, reduces to

$$\mathrm{W}^* = \frac{1}{2}\iint_{\mathcal{S}^*}\sigma'^{\mathrm{C}}_{\mathrm{ij}}\mathrm{b}'_{\mathrm{i}}\hat{\mathrm{n}}'_{\mathrm{j}}\mathrm{dS} = \frac{\mathrm{b}}{2}\iint_{\mathcal{S}^*}\sigma'^{\mathrm{C}}_{22}\mathrm{dS}. \quad (17.89)$$

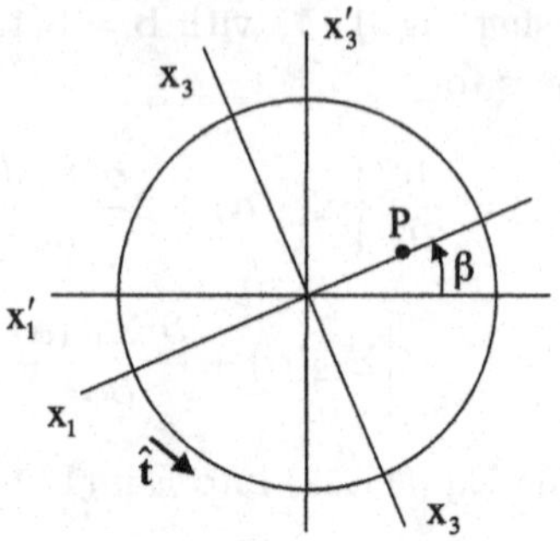

Fig. 17.8 Circular dislocation loop with $\hat{e}_i'$ coordinate system and also $\hat{e}_i$ coordinate system, which is rotated with respect to the $\hat{e}_i'$ system by the angle β.

The first step to obtain an expression for $\sigma_{22}^{\prime C}$ in the above integrand is to determine σ_{22}^{C} at P in the $\hat{e}_i$ coordinate system in Fig. 17.8. Using Eq. (17.9) for this purpose, the required expression for $\sum_{22}^{C}(\theta)$ is given by Eq. (17.42) in the form $\sigma_{22}^{C}(\theta) = \mu b/[2\pi(1-\nu)]$, and, therefore

$$\sigma_{22}^{C}(P) = \frac{\mu b}{\pi(1-\nu)R}\int_0^{\pi/2}\frac{d\theta}{(1-k'^2\sin^2\theta)^{3/2}} = \frac{\mu b}{\pi(1-\nu)R}\frac{E(k)}{(1-k^2)}. \tag{17.90}$$

Then, applying a standard transformation of coordinates using Eq. (2.23), the above stress, referred to the $\hat{e}_i'$ coordinate system in Fig. 17.8 is found to be unchanged, i.e.,

$$\sigma_{22}^{\prime C}(P) = \sigma_{22}^{C}(P). \tag{17.91}$$

Next, substituting Eqs. (17.91) and then (17.90) into Eq. (17.89),

$$W^* = \frac{\mu b^2}{2\pi(1-\nu)R}\iint_{\mathcal{S}^*}\frac{E(k)}{(1-k^2)}dS. \tag{17.92}$$

To integrate Eq. (17.92) it is convenient to introduce a new variable, ε', defined as the radial distance between a field point lying in the $\mathcal{S}^*$ surface and the loop circumference, so that over $\mathcal{S}^*$ it ranges between $\varepsilon' = R$ and $\varepsilon' = \varepsilon$, and, in addition, the corresponding variable $k' = 1-\varepsilon'/R$. Then, $dS = (R-\varepsilon')d\beta d(R-\varepsilon') = R^2k'dk'd\beta$,

and

$$W^* = \frac{\mu b^2}{2\pi(1-\nu)} R \int_0^{1-\varepsilon/R} \int_0^{2\pi} \frac{E(k')}{(1-k'^2)} k' dk' d\beta$$
$$= \frac{\mu b^2}{(1-\nu)} R |K(k') - E(k')|_0^{1-\varepsilon/R}, \qquad (17.93)$$

after making use of the integral tables given by Gradshteyn and Ryzhik (1980). Finally, substituting $E(0) = K(0) = \pi/2$, and

$$[K(1-\varepsilon/R) - E(1-\varepsilon/R)] = \frac{1}{2} \ln \frac{8R}{\varepsilon} - 1, \qquad (17.94)$$

obtained from Eq. (17.14) when $\varepsilon/R \ll 1$, we obtain

$$W^* = \frac{\mu b^2}{2(1-\nu)} R \left[\ln \frac{8R}{\varepsilon} - 2 \right] \qquad (17.95)$$

in agreement with Eq. (17.46).

17.4. Derive the expression for the strain energy, W^*, of a circular planar shear-type dislocation loop in an isotropic system given by Eq. (17.83).

Solution. The procedure is generally similar to that used in the solution for an edge-type loop in Exercise 17.3. Equation (17.18), referred to the $\hat{\mathbf{e}}'_i$ coordinate system in Fig. 17.8, and with $\mathbf{b}' = (b, 0, 0)$ and $\hat{\mathbf{n}}' = (0, 1, 0)$, reduces to

$$W^\bullet = \frac{1}{2} \iint_{S^\bullet} \sigma'^C_{ij} b'_i \hat{n}'_j dS = \frac{b}{2} \iint_{S^*} \sigma'^C_{12} dS. \qquad (17.96)$$

The first step to obtain an expression for σ'^C_{12} in the above integrand is to determine σ^C_{12} and σ^C_{23} at P in the $\hat{\mathbf{e}}_i$ coordinate system. Using Eq. (17.9) for this purpose, the required expressions for Σ^C_{12} and Σ^C_{23} are given by Eq. (17.42), with $\mathbf{b} = b(\cos\beta, 0, \sin\beta)$, in the forms

$$\Sigma^C_{12} = \frac{\mu b}{2\pi(1-\nu)} [(1 - \nu \cos^2\theta) \cos\beta + \nu \sin\theta \cos\theta \sin\beta],$$
$$\Sigma^C_{23} = \frac{\mu b}{2\pi(1-\nu)} [(1 - \nu \sin^2\theta) \sin\beta + \nu \sin\theta \cos\theta \cos\beta]. \qquad (17.97)$$

Then, substitution of Eq. (17.97) into Eq. (17.9) yields

$$\sigma_{12}^{C}(P) = \frac{\mu b \cos\beta}{\pi(1-\nu)R}\int_0^{\pi/2} \frac{(1-\nu\cos^2)d\theta}{(1-k^2\sin^2\theta)^{3/2}}$$

$$= \frac{\mu b \cos\beta}{\pi(1-\nu)R}\left[\frac{E(k)}{(1-k^2)} + \nu\frac{E(k)}{k^2} - \nu\frac{K(k)}{k^2}\right],$$

$$\sigma_{23}^{C}(P) = \frac{\mu b \sin\beta}{\pi(1-\nu)R}\int_0^{\pi/2} \frac{(1-\nu\sin^2)d\theta}{(1-k^2\sin^2\theta)^{3/2}}$$

$$= \frac{\mu b \sin\beta}{\pi(1-\nu)R}\left[\frac{E(k)}{(1-k^2)} - \nu\frac{E(k)}{k^2(1-k^2)} + \nu\frac{K(k)}{k^2}\right]. \quad (17.98)$$

Next, applying a standard transformation of coordinates using Eq. (2.23), the $\sigma_{12}'^{C}$ stress required in Eq. (17.96) is related to the above stresses by

$$\sigma_{12}'^{C} = \sigma_{12}^{C}\cos\beta + \sigma_{23}^{C}\sin\beta \quad (17.99)$$

Then, after substituting Eq. (17.98) into (17.99), and next putting the result into Eq. (17.96), and making the change of variable used in Exercise 17.3 to convert Eq. (17.92) into (17.93),

$$W^{\bullet} = \frac{\mu b^2}{2\pi(1-\nu)R}\int_0^{1-\varepsilon/R}\int_0^{2\pi} \times\left\{\cos^2\beta\left[\frac{E(k')}{(1-k'^2)} + \nu\frac{E(k')}{k'^2} - \nu\frac{K(k')}{k'^2}\right]\right.$$
$$\left. + \sin^2\beta\left[\frac{E(k')}{(1-k'^2)} - \nu\frac{E(k')}{k'^2(1-k'^2)} + \nu\frac{K(k')}{k'^2}\right]\right\} R^2 k' dk' d\beta. \quad (17.100)$$

Finally, integrating Eq. (17.100) by employing the same methods used to integrate Eq. (17.93) in Exercise 17.3,

$$W^{\bullet} = \frac{\mu b^2(2-\nu)}{2(1-\nu)}R\int_0^{1-\varepsilon/R}\frac{E(k')}{(1-k'^2)}k'dk'$$

$$= \frac{\mu b^2(2-\nu)}{4(1-\nu)}R\left[\ln\frac{8R}{\varepsilon} - 2\right], \quad (17.101)$$

in agreement with results obtained by Chou and Eshelby (1962).

17.5. Show that the tube force, f^{TUBE}, is generally small relative to the f^* force in the expression for the total force given by Eq. (17.37) for a circular edge-type dislocation loop in an isotropic system.

Solution The f^* force is already determined in the form of Eq. (17.45): f^{TUBE} is given by Eq. (17.36) in terms of the tube parameter, $\mathrm{F}(\alpha)$ which must, therefore, be evaluated first. For this purpose we employ Eq. (17.28), i.e.,

$$\mathrm{F}(\alpha) = \frac{1}{2}\iint_{\mathcal{S}^{\mathrm{CYL}}} \sigma_{\mathrm{ij}}^{\infty}\mathrm{u}_{\mathrm{i}}^{\infty}\hat{\mathrm{n}}_{\mathrm{j}}\varepsilon d\omega, \tag{17.102}$$

where $\sigma_{\mathrm{ij}}^{\infty}$ and $\mathrm{u}_{\mathrm{i}}^{\infty}$ are the stresses and displacements present at the cylindrical surface, $\mathcal{S}^{\mathrm{CYL}}$, which is co-axial to an infinitely long straight edge dislocation of the same type as the loop. The integration over this surface is complicated by the presence of the cut required to produce the loop (see Fig. 17.2), and the cut and additional geometry used to perform the integration are shown in detail in Fig. 17.9. In view of the discontinuity at the cut, the range of integration over ω is taken to be $-3\pi/2 \leq \omega \leq \pi/2$. Therefore, by substituting Eqs. (12.46) and (12.54) for $\sigma_{\mathrm{ij}}^{\infty}$ and $\mathrm{u}_{\mathrm{i}}^{\infty}$, respectively, along with $\hat{\mathbf{n}} = -(\cos\omega, \sin\omega, 0)$, into Eq. (17.102),

$$\mathrm{F}(\alpha) = \frac{1}{2}\int_{-3\pi/2}^{\pi/2} (\sigma_{11}^{\infty}\mathrm{u}_1^{\infty}\hat{\mathrm{n}}_1 + \sigma_{12}^{\infty}\mathrm{u}_2^{\infty}\hat{\mathrm{n}}_1 + \sigma_{22}^{\infty}\mathrm{u}_2^{\infty}\hat{\mathrm{n}}_2 + \sigma_{12}^{\infty}\mathrm{u}_1^{\infty}\hat{\mathrm{n}}_2)\varepsilon d\omega, \tag{17.103}$$

where

$$\sigma_{11}^{\infty}\mathrm{u}_1^{\infty}\hat{\mathrm{n}}_1\varepsilon = \mathrm{AB}\sin\phi\cos\phi(2\cos^2\phi + 1)(\phi + \mathrm{C}\sin\phi\cos\phi),$$

$$\sigma_{12}^{\infty}\mathrm{u}_2^{\infty}\hat{\mathrm{n}}_1\varepsilon = -\mathrm{A}\cos^2\phi(2\cos^2\phi - 1)(-\mathrm{D}\ln\varepsilon - \mathrm{E}\cos^2\phi + \mathrm{E}\sin^2\phi),$$

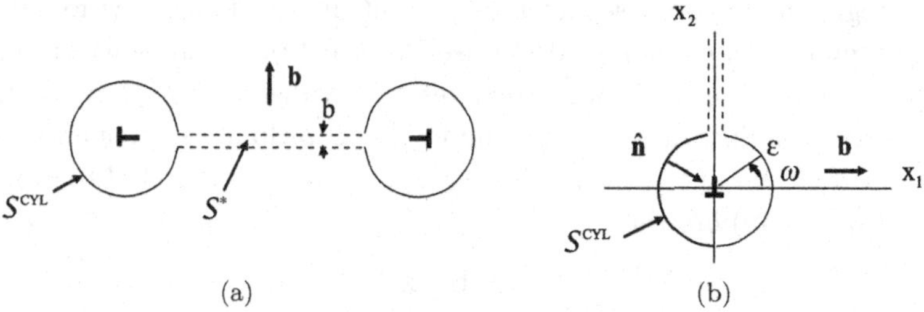

Fig. 17.9 Geometry used to integrate Eq. (17.102). (a) Cross-sectional view of loop showing cut in $\mathcal{S}^{\mathrm{CYL}}$. (b) Polar coordinate system and parameters used for the integration.

$$\sigma_{22}^{\infty} u_2^{\infty} \hat{n}_2 \varepsilon = -\mathrm{A} \sin^2\phi(2\cos^2\phi - 1)(-\mathrm{D}\ln\varepsilon - \mathrm{E}\cos^2\phi + \mathrm{E}\sin^2\phi),$$
$$\sigma_{12}^{\infty} u_1^{\infty} \hat{n}_2 \varepsilon = -\mathrm{AB}\sin\phi\cos\phi(2\cos^2\phi - 1)(\phi + \mathrm{C}\sin\phi\cos\phi),$$
$$\mathrm{A} = \frac{\mu b}{2\pi(1-\nu)} \quad \mathrm{B} = \frac{b}{2\pi} \quad \mathrm{C} = \frac{1}{2(1-\nu)}$$
$$\mathrm{D} = \frac{b(1-2\nu)}{4\pi(1-\nu)} \quad \mathrm{E} = \frac{b}{8\pi(1-\nu)} \tag{17.104}$$

Then, substituting Eq. (17.104) into (17.103) and performing the integrations,

$$\mathrm{F}(\alpha) = \frac{\mu b^2(3-2\nu)}{16\pi(1-\nu)^2} \tag{17.105}$$

in agreement with the result obtained by Gavazza and Barnett (1976).

Finally, substituting this result into Eq. (17.36),

$$f^{\mathrm{TUBE}} = -\frac{1}{\mathrm{R}}\mathrm{F}(\alpha) = -\frac{\mu b^2(3-2\nu)}{16\pi(1-\nu)^2\mathrm{R}} \tag{17.106}$$

and, with the use of Eq. (17.45),

$$f^{\mathrm{TUBE}}/f^* = \frac{3-2\nu}{4(1-\nu)[\ln(8\mathrm{R}/\varepsilon)-1]}. \tag{17.107}$$

Therefore, assuming $\nu = 1/3$, and $\varepsilon = 2\times10^{-10}$m, $f^{\mathrm{TUBE}}/f^* = 0.18$ and 0.12 when $\mathrm{R} = 100\times10^{-10}$m and 1000×10^{-10}m, respectively.

17.6. Use Eq. (16.49) to derive an expression for the self-force experienced by the segment AB in Fig. 17.7a in an isotropic system.

Solution. We employ the same general procedure which was used in Sec. 17.4 for the anisotropic case, in which the force exerted by the segment AB on a segment ds present at a distance ρ is found and then evaluated as $\rho \to 0$. To use Eq. (16.49), which generally yields the force imposed on segment $ds^{(1)}$ (on dislocation 1) by segment $ds^{(2)}$ (on dislocation 2), we employ the coordinate system and geometry shown in Fig. 17.10. The various quantities required to evaluate Eq. (16.49) are then

$$\begin{aligned}
&\mathbf{b}^{(1)} = \mathbf{b}^{(2)} = \mathbf{b} && \mathbf{x}^{(1)} = \hat{\mathbf{e}}_1\rho \\
&\mathbf{ds}^{(1)} = \hat{\mathbf{e}}_3 \mathrm{ds} && \mathbf{x}^{(2)} = \hat{\mathbf{e}}_3 s' \\
&\mathbf{ds}^{(2)} = \hat{\mathbf{e}}_3 \mathrm{ds}' && \mathbf{R} = \mathbf{x}^{(1)} - \mathbf{x}^{(2)}
\end{aligned} \tag{17.108}$$

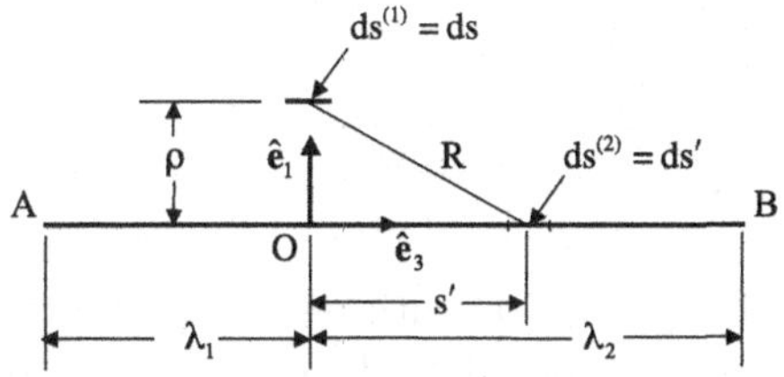

Fig. 17.10 Straight dislocation segment AB along with parallel differential segment ds present at the distance ρ.

The first term in Eq. (16.49) vanishes, and the various bracketed quantities in the three remaining terms then assume the following forms:

$$\{[\mathbf{b}^{(2)} \times \nabla(\nabla^2 R)] \times \mathbf{ds}^{(1)}\} = \frac{-2}{(\rho^2 + s'^2)^{3/2}}[(b_1 s' + b_3\rho)\hat{\mathbf{e}}_1 + b_2 s'\hat{\mathbf{e}}_2]ds,$$

$$[\mathbf{b}^{(1)} \cdot \mathbf{ds}^{(2)}] = b_3 ds',$$

$$[(\mathbf{b}^{(2)} \times \mathbf{ds}^{(2)}) \cdot \nabla] = \left(b_2\frac{\partial}{\partial x_1^{(1)}} - b_1\frac{\partial}{\partial x_2^{(1)}}\right) ds',$$

$$[\mathbf{ds}^{(1)} \times (\mathbf{b}^{(1)}\underline{\mathbf{T}})] = [(b_1 T_{11} + b_2 T_{12} + b_3 T_{13})\hat{\mathbf{e}}_2 - (b_1 T_{12} + b_2 T_{22} + b_3 T_{23})\hat{\mathbf{e}}_1]ds, \tag{17.109}$$

$$[(\mathbf{b}^{(2)} \times \mathbf{ds}^{(2)}) \cdot \nabla(\nabla^2 R)] = \frac{-2\rho b_2 ds'}{(\rho^2 + s'^2)^{3/2}},$$

$$[\mathbf{ds}^{(1)} \times \mathbf{b}^{(1)}] = (b_1\hat{\mathbf{e}}_2 - b_2\hat{\mathbf{e}}_1)ds.$$

Next, substituting these quantities into Eq. (16.49) and using the derivatives

$$\frac{\partial T_{ij}}{\partial x_1^{(1)}} = \frac{1}{R^3}\begin{bmatrix} -2\rho[1 - 3\rho^2/(2R^2)] & 0 & s'(1 - 3\rho^2/R^2) \\ 0 & 0 & 0 \\ s'(1 - 3\rho^2/R^2) & 0 & 3\rho s'^2/R^2 \end{bmatrix}$$

$$\frac{\partial T_{ij}}{\partial x_1^{(2)}} = \frac{1}{R^3}\begin{bmatrix} 0 & -\rho & 0 \\ -\rho & 0 & s' \\ 0 & s' & 0 \end{bmatrix} \tag{17.110}$$

obtained through use of Eq. (16.50), and then taking the limit of the result as $\rho \to 0$, the force exerted on ds by ds' is

$$\mathbf{dF}^{ds/ds'} = -\frac{\mu\nu b_3 ds}{4\pi(1-\nu)}(b_1\hat{\mathbf{e}}_1 + b_2\hat{\mathbf{e}}_2)\frac{ds'}{s'^2}. \tag{17.111}$$

Integration of Eq. (17.111) with respect to s′ over the length of AB will yield the force imposed on ds by the AB segment. However, this is equivalent to the self-force experienced by a differential segment of AB, and we therefore can write

$$\mathbf{dF}^{\mathrm{ABself}} = -\frac{\mu\nu b_3 ds}{4\pi(1-\nu)}(b_1\hat{\mathbf{e}}_1 + b_2\hat{\mathbf{e}}_2)\int_{-\lambda_1}^{\lambda_2}\frac{ds'}{s'^2} \tag{17.112}$$

using the notation of Eq. (17.56). To cope with the singularity in the above integrand we employ the same strategy used in Sec. 17.4 and use Eq. (17.112) to write the self-stress experienced by an element of an infinitely long AB segment, which is well known to vanish, as

$$\mathbf{dF}^{\mathrm{AB}\infty\mathrm{self}} = -\frac{\mu\nu b_3 ds}{4\pi(1-\nu)}(b_1\hat{\mathbf{e}}_1 + b_2\hat{\mathbf{e}}_2)\int_{-\infty}^{\infty}\frac{ds'}{s'^2} = 0. \tag{17.113}$$

Then, subtracting Eq. (17.113) from (17.112), and performing the integration, we obtain the final result for the force, i.e.,

$$\begin{aligned}
\mathbf{dF}^{\mathrm{ABself}} &= -\frac{\mu\nu b_3 ds}{4\pi(1-\nu)}(b_1\hat{\mathbf{e}}_1 + b_2\hat{\mathbf{e}}_2)\left(\int_{-\lambda_1}^{\lambda_2}\frac{ds'}{s'^2} - \int_{-\infty}^{\infty}\frac{ds'}{s'^2}\right) \\
&= \frac{\mu\nu b_3 ds}{4\pi(1-\nu)}(b_1\hat{\mathbf{e}}_1 + b_2\hat{\mathbf{e}}_2)\left(\int_{-\infty}^{-\lambda_1}\frac{ds'}{s'^2} + \int_{-\lambda_1}^{\lambda_2}\frac{ds'}{s'^2}\right. \\
&\quad \left. + \int_{\lambda_2}^{\infty}\frac{ds'}{s'^2} - \int_{-\lambda_1}^{\lambda_2}\frac{ds'}{s'^2}\right) ds \\
&= \frac{\mu\nu b_3 ds}{4\pi(1-\nu)}(b_1\hat{\mathbf{e}}_1 + b_2\hat{\mathbf{e}}_2)\left(\int_{-\infty}^{-\lambda_1}\frac{ds'}{s'^2} + \int_{\lambda_2}^{\infty}\frac{ds'}{s'^2}\right) ds \\
&= \frac{\mu b_3 \nu}{4\pi(1-\nu)}(b_1\hat{\mathbf{e}}_1 + b_2\hat{\mathbf{e}}_2)\left(\frac{1}{\lambda_1} + \frac{1}{\lambda_2}\right) ds
\end{aligned} \tag{17.114}$$

which is seen to be normal to the segment as must be the case.

Appendix A

Relationships Involving the ∇ Operator

A summary is presented of a number of relationships involving the del operator, ∇, in the cylindrical and spherical orthogonal curvilinear coordinate systems shown in Fig. A.1.

A.1 Cylindrical Orthogonal Curvilinear Coordinates

If $f = f(r, \theta, z)$, and $\mathbf{f} = \hat{\mathbf{e}}_r f_r + \hat{\mathbf{e}}_\theta f_\theta + \hat{\mathbf{e}}_z f_z$, where the unit base vectors $(\hat{\mathbf{e}}_r, \hat{\mathbf{e}}_\theta, \hat{\mathbf{e}}_z)$ are given in terms of Cartesian base vectors by Eq. (G.2),

$$\nabla f = \hat{\mathbf{e}}_r \frac{\partial f}{\partial r} + \hat{\mathbf{e}}_\theta \frac{1}{r}\frac{\partial f}{\partial \theta} + \hat{\mathbf{e}}_z \frac{\partial f}{\partial z}, \tag{A.1}$$

$$\nabla \cdot \mathbf{f} = \frac{1}{r}\frac{\partial}{\partial r}\left(r \frac{\partial f_r}{\partial r}\right) + \frac{1}{r}\frac{\partial f_\theta}{\partial \theta} + \frac{\partial f_z}{\partial z}, \tag{A.2}$$

$$\nabla^2 f = \frac{1}{r}\frac{\partial}{\partial r}\left(r \frac{\partial f}{\partial r}\right) + \frac{1}{r^2}\frac{\partial^2 f}{\partial \theta^2} + \frac{\partial^2 f}{\partial z^2}. \tag{A.3}$$

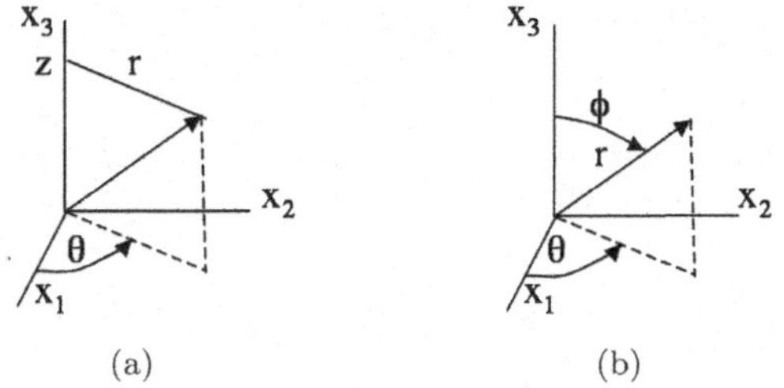

Fig. A.1 (a) Cylindrical (r, θ, z) coordinates. (b) Spherical (r, θ, ϕ) coordinates.

A.2 Spherical Orthogonal Curvilinear Coordinates

If $f = f(r, \theta, \phi)$, and $\mathbf{f} = \hat{\mathbf{e}}_r f_r + \hat{\mathbf{e}}_\theta f_\theta + \hat{\mathbf{e}}_\phi f_\phi$, where the unit base vectors $(\hat{\mathbf{e}}_r, \hat{\mathbf{e}}_\theta, \hat{\mathbf{e}}_\phi)$ are given in terms of Cartesian base vectors by Eq. (G.9),

$$\nabla f = \hat{\mathbf{e}}_r \frac{\partial f}{\partial r} + \hat{\mathbf{e}}_\phi \frac{1}{r}\frac{\partial f}{\partial \phi} + \hat{\mathbf{e}}_\theta \frac{1}{r \sin\theta}\frac{\partial f}{\partial \theta}, \tag{A.4}$$

$$\nabla \cdot \mathbf{f} = \frac{1}{r^2}\frac{\partial}{\partial r}\left(r^2 f_r\right) + \frac{1}{r\sin\phi}\frac{\partial}{\partial \phi}\left(f_\phi \sin\phi\right) + \frac{1}{r\sin\phi}\frac{\partial f_\theta}{\partial \theta}, \tag{A.5}$$

$$\nabla^2 f = \frac{1}{r^2}\frac{\partial}{\partial r}\left(r^2 \frac{\partial f}{\partial r}\right) + \frac{1}{r^2 \sin\phi}\frac{\partial}{\partial \phi}\left(\sin\phi \frac{\partial f}{\partial \phi}\right) + \frac{1}{r^2 \sin^2\phi}\frac{\partial^2 f}{\partial \theta^2}. \tag{A.6}$$

Appendix B

Integral Relationships

B.1 Divergence (Gauss's) Theorem

If $\mathbf{A}(\mathbf{x})$ is a vector field, and $\mathcal{V}$ is a region of volume enclosed by the surface $\mathcal{S}$, and $\hat{\mathbf{n}}$ is the positive unit vector normal to $\mathcal{S}$ (and therefore in the outward direction), as in Fig. B.1, the *divergence theorem* states that

$$\oiiint_{\mathcal{V}} \nabla \cdot \mathbf{A} \mathrm{dV} = \oiint_{\mathcal{S}} \mathbf{A} \cdot \hat{\mathbf{n}} \mathrm{dS} \tag{B.1}$$

or, alternatively,

$$\oiiint_{\mathcal{V}} \left(\frac{\partial \mathrm{A}_1}{\partial \mathrm{x}_1} + \frac{\partial \mathrm{A}_2}{\partial \mathrm{x}_2} + \frac{\partial \mathrm{A}_3}{\partial \mathrm{x}_3} \right) \mathrm{dV} = \oiint_{\mathcal{S}} (\mathrm{A}_1 \hat{\mathrm{n}}_1 + \mathrm{A}_2 \hat{\mathrm{n}}_2 + \mathrm{A}_3 \hat{\mathrm{n}}_3) \mathrm{dS}. \tag{B.2}$$

B.2 Stokes' Theorem

If $\mathbf{M}(\mathbf{x})$ is a vector field, *Stokes' theorem* is usually expressed as

$$\oint_{\mathrm{C}} \mathbf{M} \cdot \mathrm{d}\mathbf{x} = \iint_{\mathcal{S}} \nabla \times \mathbf{M} \cdot \hat{\mathbf{n}} \mathrm{dS} = \mathrm{e}_{\mathrm{ijk}} \iint_{\mathcal{S}} \frac{\partial \mathrm{M}_{\mathrm{k}}}{\partial \mathrm{x}_{\mathrm{j}}} \hat{\mathrm{n}}_{\mathrm{i}} \mathrm{dS}, \tag{B.3}$$

where the line integral is over the closed curve, C, in Fig. B.2 on which the surface, $\mathcal{S}$, terminates. The surface integral is over $\mathcal{S}$, and $\hat{\mathbf{n}}$ is the positive unit vector normal to $\mathcal{S}$, which is defined for this unclosed surface by the requirement that if C were shrunk down on $\mathcal{S}$ until it just traversed a circuit around $\hat{\mathbf{n}}$ in the direction of the tangent vector, $\hat{\mathbf{t}}$, the circuit would be clockwise when sighting along $\hat{\mathbf{n}}$.

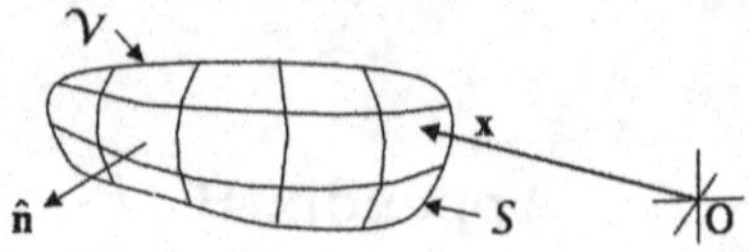

Fig. B.1 Geometry for divergence theorem.

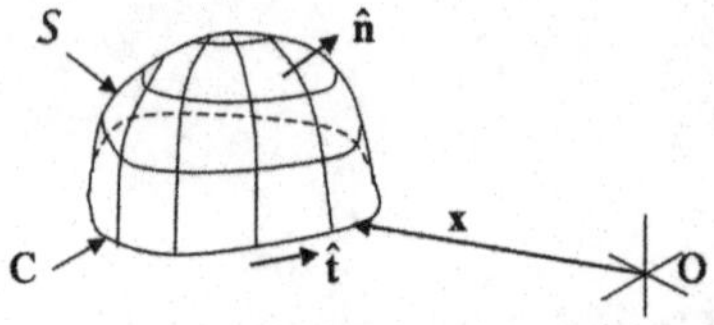

Fig. B.2 Geometry for Stokes' Theorem.

Substitution of the vector $\mathbf{M}(\mathbf{x}) = \phi(\mathbf{x})\hat{\mathbf{e}}_k$ into Eq. (B.3) yields the further form

$$\iint_{\mathcal{S}} \left[\frac{\partial\phi}{\partial x_j}\hat{n}_i - \frac{\partial\phi}{\partial x_i}\hat{n}_j\right] dS = e_{ijk} \oint_C \phi dx_k. \tag{B.4}$$

On the other hand, if $\underline{\mathbf{G}}(x)$ is a tensor field (Hetnarski and Ignaczak 2004),

$$\iint_{\mathcal{S}} [\text{curl}G]^T[\hat{n}]dS = \oint_C [G][dx], \tag{B.5}$$

where

$$(\text{curl}G)_{ij} = e_{ipq}\frac{\partial G_{jq}}{\partial x_p}. \tag{B.6}$$

B.3 Another form of Stokes' Theorem

If $g_i = g_i(\mathbf{x})$ is a scalar function of $\mathbf{x}$ in a body,

$$\iint_{\mathcal{S}} \left(\frac{\partial g_i}{\partial x_i}\delta_{jl} - \frac{\partial g_j}{\partial x_l}\right)\hat{n}_j dS = \oint_C e_{lij} g_i dx_j \tag{B.7}$$

as can be verified by applying Eq. (B.5) to the case where $\underline{\mathbf{G}}$ has the form $G_{lj} = e_{lij}g_i$. Then,

$$(\text{curl}G)^T_{lj}\hat{n}_j = e_{jpq}e_{lkq}\frac{\partial g_k}{\partial x_p}\hat{n}_j = e_{qjp}e_{qlk}\frac{\partial g_k}{\partial x_p}\hat{n}_j \tag{B.8}$$

and substitution of these quantities into Eq. (B.5), with the use of

$$e_{ijk}e_{imn} = \delta_{jm}\delta_{kn} - \delta_{jn}\delta_{km} \tag{B.9}$$

yields Eq. (B.7). However, if $\mathcal{S}$ is a closed surface and C is shrunk down to a point, we have

$$\oiint_{\mathcal{S}} \left(\frac{\partial g_i}{\partial x_i}\delta_{jl} - \frac{\partial g_j}{\partial x_l} \right) \hat{n}_j dS = \oiint_{\mathcal{S}} \left(\frac{\partial g_i}{\partial x_i} n_l - \frac{\partial g_j}{\partial x_l}\hat{n}_j \right) dS \tag{B.10}$$

Substitution of $g_i = \sigma^X_{ik}u^Y_k$, or alternatively $g_j = \sigma^X_{ji}u^Y_i$, into Eq. (B.10), then yields the further relationship

$$\oiint_{\mathcal{S}} \sigma^X_{ik}\frac{\partial u^Y_i}{\partial x_k}\delta_{jl}\hat{n}_j dS = \oiint_{\mathcal{S}} \left(\sigma^X_{ij}\frac{\partial u^Y_i}{\partial x_l} + u^Y_i\frac{\partial \sigma^X_{ij}}{\partial x_l} \right) \hat{n}_j dS. \tag{B.11}$$

Appendix C

The Tensor Product of Two Vectors

The tensor product of two vectors can be used to represent tensors as products of vectors. In many cases equations involving tensors can then be written in more compact vector forms.

The tensor product, $\underline{\mathbf{P}}$, of two vectors $\mathbf{a}$ and $\mathbf{b}$ is written as

$$\underline{\mathbf{P}} = \mathbf{a} \otimes \mathbf{b} \tag{C.1}$$

and possesses components given by

$$P_{ij} = a_i b_j \tag{C.2}$$

Therefore,

$$\underline{\mathbf{P}}\mathbf{u} = (\mathbf{a} \otimes \mathbf{b})\mathbf{u} = (\mathbf{b} \cdot \mathbf{u})\mathbf{a} \tag{C.3}$$

or, in matrix form

$$[P][u] = [a \otimes b][u] = \begin{bmatrix} a_1 b_1 & a_1 b_2 & a_1 b_3 \\ a_2 b_1 & a_2 b_2 & a_2 b_3 \\ a_3 b_1 & a_3 b_2 & a_3 b_3 \end{bmatrix} \begin{bmatrix} u_1 \\ u_2 \\ u_3 \end{bmatrix}$$

$$= [b_1 u_1 + b_2 u_2 + b_3 u_3] \begin{bmatrix} a_1 \\ a_2 \\ a_3 \end{bmatrix}. \tag{C.4}$$

Since

$$[\hat{e}_1 \otimes \hat{e}_1] = \begin{bmatrix} 1 & 0 & 0 \\ 0 & 0 & 0 \\ 0 & 0 & 0 \end{bmatrix} \quad [\hat{e}_2 \otimes \hat{e}_2] = \begin{bmatrix} 0 & 0 & 0 \\ 0 & 1 & 0 \\ 0 & 0 & 0 \end{bmatrix}$$

$$[\hat{e}_3 \otimes \hat{e}_3] = \begin{bmatrix} 0 & 0 & 0 \\ 0 & 0 & 0 \\ 0 & 0 & 1 \end{bmatrix} \tag{C.5}$$

the unitary tensor can be written in the matrix form

$$[\mathrm{I}] = [\hat{\mathrm{e}}_1 \otimes \hat{\mathrm{e}}_1] + [\hat{\mathrm{e}}_2 \otimes \hat{\mathrm{e}}_2] + [\hat{\mathrm{e}}_2 \otimes \hat{\mathrm{e}}_2] = \begin{bmatrix} 1 & 0 & 0 \\ 0 & 1 & 0 \\ 0 & 0 & 1 \end{bmatrix}. \tag{C.6}$$

Appendix D

Properties of the Delta Function

The one-dimensional delta function, $\delta(\mathrm{x}-\mathrm{x}_\circ)$ vanishes everywhere except at $\mathrm{x}=\mathrm{x}_\circ$, and has the property, e.g., Hassani (2000), that

$$\int_a^b f(x)\delta(x-x_\circ)dx = f(x_\circ). \quad (a < x_\circ < b) \tag{D.1}$$

Then, in three dimensions with vector arguments, the delta function appears as

$$\delta(\mathbf{x}) = \delta(x_1)\delta(x_2)\delta(x_3) \tag{D.2}$$

and

$$\iiint\!\!\!\!\!\!\!\!\!\!\!\!\bigcirc_{\mathcal{V}} f(\mathbf{x})\delta(\mathbf{x}-\mathbf{x}_\circ)dV = f(\mathbf{x}_\circ). \quad (\mathbf{x}_\circ \text{ inside } \mathcal{V}) \tag{D.3}$$

If $\mathbf{x}_\circ$ lies outside $\mathcal{V}$, the integral vanishes.

The Nth derivative of the delta function with respect to its argument, indicated here by the superscript (N), obeys the rule

$$\int_a^b \delta^{(N)}(x-x_\circ)f(x)dx = (-1)^N$$

$$\times \int_a^b \delta(x-x_\circ)f^{(N)}(x)dx = (-1)^N f^{(N)}(x_\circ). \quad (a < x_\circ < b) \tag{D.4}$$

Further properties, as given by Bacon, Barnett and Scattergood (1979b), are listed in Table D.1.

Table D.1 Further properties of the delta function, $\delta(x)$.

(a) $\delta(x-a)=\delta(a-x)$	(b) $\dfrac{d\delta(x)}{dx}=\dfrac{-d\delta(-x)}{dx}$	(c) $x\dfrac{d\delta(x)}{dx}=-\delta(x)$
(d) $\delta(ax)=\dfrac{\delta(x)}{\lvert a\rvert}$	(e) $\delta^{(N)}(ax)=\dfrac{\delta^{(N)}(x)}{\lvert a\rvert^{N+1}}$	(f) $\int_{-\infty}^{\infty}\delta(a-x)dx=1$
(g) $\delta(x)=\dfrac{dH(x)}{dx}$	(h) $\int_{-\infty}^{\infty}\dfrac{d\delta(a-cx)}{d(a-cx)}dx=-\dfrac{2\delta(c)}{a}$	

Still further useful relationships can be obtained from potential theory. The electrostatic potential, $v(\mathbf{x})$, due to a distribution of electrical charge density, $\rho(\mathbf{x})$, must satisfy Poisson's equation

$$\nabla^2 v(\mathbf{x}) = -4\pi A\rho(\mathbf{x}). \tag{D.5}$$

Therefore, inserting Eqs. (3.14) and $\rho(\mathbf{x}) = q\delta(\mathbf{x}-\mathbf{x}_\circ)$ into Eq. (D.5),

$$\delta(\mathbf{x}-\mathbf{x}_\circ) = -\frac{1}{4\pi}\nabla^2\frac{1}{|\mathbf{x}-\mathbf{x}_\circ|}. \tag{D.6}$$

Also, since $\nabla^2|\mathbf{x}-\mathbf{x}_\circ| = 2/|\mathbf{x}-\mathbf{x}_\circ|$,

$$\delta(\mathbf{x}-\mathbf{x}_\circ) = -\frac{1}{8\pi}\nabla^2\nabla^2|\mathbf{x}-\mathbf{x}_\circ|. \tag{D.7}$$

Appendix E

The Alternator Operator

The alternator operator, e_{ijk}, is conveniently expressed in the form

$$e_{ijk} = \hat{\mathbf{e}}_i \cdot (\hat{\mathbf{e}}_j \times \hat{\mathbf{e}}_k), \tag{E.1}$$

where, as usual, the $\hat{\mathbf{e}}_i$ are the base unit vectors of a Cartesian, right-handed, orthogonal coordinate system. Therefore, for example, it has the properties

$$\begin{aligned} e_{123} &= e_{231} = e_{312} = 1, \\ e_{132} &= e_{213} = e_{321} = -1, \\ e_{113} &= e_{221} = e_{233} = e_{122} = 0. \end{aligned} \tag{E.2}$$

Each time the operator is permutated by an exchange of nearest-neighbor indices, its sign is reversed. Hence,

$$e_{ijk} = -e_{ikj} = e_{kij} = -e_{kji}. \tag{E.3}$$

It also follows that

$$e_{ijk}e_{imn} = \delta_{jm}\delta_{kn} - \delta_{jn}\delta_{km}. \tag{E.4}$$

The operator is especially useful in expressing the vector product, since

$$\mathbf{a} \times \mathbf{b} = a_i b_j e_{ijk} \hat{\mathbf{e}}_k. \tag{E.5}$$

Appendix F

Fourier Transforms

The Fourier transform of a function, $\mathrm{g}(\mathbf{x})$, is often given (Sneddon 1951) as

$$\bar{\mathrm{g}}(\mathbf{k}) = \int_{-\infty}^{\infty}\int_{-\infty}^{\infty}\int_{-\infty}^{\infty} \mathrm{g}(\mathbf{x})\mathrm{e}^{\mathrm{i}\mathbf{k}\cdot\mathbf{x}}\mathrm{dx_1 dx_2 dx_3}. \tag{F.1}$$

The inverse transform is then

$$\mathrm{g}(\mathbf{x}) = \frac{1}{(2\pi)^3}\int_{-\infty}^{\infty}\int_{-\infty}^{\infty}\int_{-\infty}^{\infty} \bar{\mathrm{g}}(\mathbf{k})\mathrm{e}^{-\mathrm{i}\mathbf{k}\cdot\mathbf{x}}\mathrm{dk_1 dk_2 dk_3}. \tag{F.2}$$

Alternatively, the transform may be written in the form

$$\bar{\mathrm{h}}(\mathbf{k}) = \frac{1}{(2\pi)^3}\int_{-\infty}^{\infty}\int_{-\infty}^{\infty}\int_{-\infty}^{\infty} \mathrm{h}(\mathbf{x})\mathrm{e}^{-\mathrm{i}\mathbf{k}\cdot\mathbf{x}}\mathrm{dx_1 dx_2 dx_3} \tag{F.3}$$

and, in that case the inverse transform is

$$\mathrm{h}(\mathbf{x}) = \int_{-\infty}^{\infty}\int_{-\infty}^{\infty}\int_{-\infty}^{\infty} \bar{\mathrm{h}}(\mathbf{k})\mathrm{e}^{\mathrm{i}\mathbf{k}\cdot\mathbf{x}}\mathrm{dk_1 dk_2 dk_3}. \tag{F.4}$$

Appendix G

Equations from the Theory of Isotropic Elasticity

Presented below are selected equations from the theory of isotropic elasticity expressed in cylindrical and spherical coordinate systems (Sokolnikoff 1946).

G.1 Cylindrical Orthogonal Curvilinear Coordinates

A differential volume element and the components of the stress tensor are shown in Fig. G.1a. The displacement vector is written as

$$\mathbf{u} = \hat{\mathbf{e}}_r u_r + \hat{\mathbf{e}}_\theta u_\theta + \hat{\mathbf{e}}_z u_z, \tag{G.1}$$

where the unit base vectors $(\hat{\mathbf{e}}_r, \hat{\mathbf{e}}_\theta, \hat{\mathbf{e}}_z)$ are given in terms of Cartesian base vectors by

$$\begin{aligned} \hat{\mathbf{e}}_r &= \cos\theta\hat{\mathbf{e}}_1 + \sin\theta\hat{\mathbf{e}}_2, \\ \hat{\mathbf{e}}_\theta &= -\sin\theta\hat{\mathbf{e}}_1 + \cos\theta\hat{\mathbf{e}}_2, \\ \hat{\mathbf{e}}_z &= \hat{\mathbf{e}}_3. \end{aligned} \tag{G.2}$$

The strains are related to the displacements by

$$\begin{aligned} &\varepsilon_{rr} = \frac{\partial u_r}{\partial r}, \quad \varepsilon_{\theta\theta} = \frac{1}{r}\frac{\partial u_\theta}{\partial \theta} + \frac{u_r}{r}, \quad \varepsilon_{zz} = \frac{\partial u_z}{\partial z}, \\ &\varepsilon_{r\theta} = \frac{1}{2}\left(\frac{1}{r}\frac{\partial u_r}{\partial \theta} + \frac{\partial u_\theta}{\partial r} - \frac{u_\theta}{r}\right), \quad \varepsilon_{rz} = \frac{1}{2}\left(\frac{\partial u_z}{\partial r} + \frac{\partial u_r}{\partial z}\right), \\ &\varepsilon_{\theta z} = \frac{1}{2}\left(\frac{\partial u_\theta}{\partial z} + \frac{1}{r}\frac{\partial u_z}{\partial \theta}\right). \end{aligned} \tag{G.3}$$

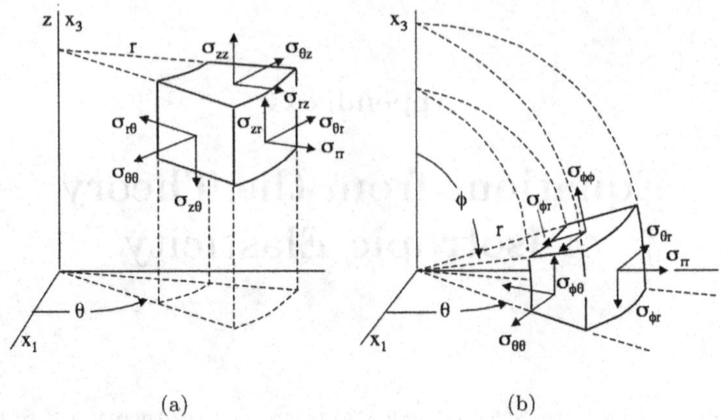

Fig. G.1 Differential volume elements and components of the stress tensor in: (a) cylindrical coordinates (r, θ, z); (b) spherical coordinates (r, θ, ϕ).

The relationships between stresses and strains, with $e = \varepsilon_{rr} + \varepsilon_{\theta\theta} + \varepsilon_{zz}$, are

$$\begin{aligned} &\sigma_{rr} = \lambda e + 2\mu\varepsilon_{rr} \quad \sigma_{\theta\theta} = \lambda e + 2\mu\varepsilon_{\theta\theta} \quad \sigma_{zz} = \lambda e + 2\mu\varepsilon_{zz} \\ &\sigma_{r\theta} = 2\mu\varepsilon_{r\theta} \quad \sigma_{rz} = 2\mu\varepsilon_{rz} \quad \sigma_{\theta z} = 2\mu\varepsilon_{\theta z} \end{aligned} \tag{G.4}$$

and the equations of equilibrium are

$$\begin{aligned} &\frac{\partial\sigma_{rr}}{\partial r} + \frac{1}{r}\frac{\partial\sigma_{r\theta}}{\partial\theta} + \frac{\partial\sigma_{rz}}{\partial z} + \frac{\sigma_{rr} - \sigma_{\theta\theta}}{r} + f_r = 0, \\ &\frac{\partial\sigma_{r\theta}}{\partial r} + \frac{1}{r}\frac{\partial\sigma_{\theta\theta}}{\partial\theta} + \frac{\partial\sigma_{\theta z}}{\partial z} + \frac{2}{r}\sigma_{r\theta} + f_\theta = 0, \\ &\frac{\partial\sigma_{rz}}{\partial r} + \frac{1}{r}\frac{\partial\sigma_{\theta z}}{\partial\theta} + \frac{\partial\sigma_{zz}}{\partial z} + \frac{1}{r}\sigma_{rz} + f_z = 0. \end{aligned} \tag{G.5}$$

The strains expressed in cylindrical coordinates are related to the strains expressed in Cartesian coordinates by the relations

$$\begin{aligned} &\varepsilon_{rr} = \varepsilon_{11}\cos^2\theta + \varepsilon_{22}\sin^2\theta + \varepsilon_{12}\sin 2\theta, && \varepsilon_{11} = \varepsilon_{rr}\cos^2\theta + \varepsilon_{\theta\theta}\sin^2\theta - \varepsilon_{r\theta}\sin 2\theta, \\ &\varepsilon_{\theta\theta} = \varepsilon_{11}\sin^2\theta + \varepsilon_{22}\cos^2\theta - \varepsilon_{12}\sin 2\theta, && \varepsilon_{22} = \varepsilon_{rr}\sin^2\theta + \varepsilon_{\theta\theta}\cos^2\theta + \varepsilon_{r\theta}\sin 2\theta, \\ &\varepsilon_{zz} = \varepsilon_{33}, && \varepsilon_{33} = \varepsilon_{zz}, \\ &\varepsilon_{r\theta} = (\varepsilon_{22} - \varepsilon_{11})\sin\theta\cos\theta + \varepsilon_{12}\cos 2\theta, && \varepsilon_{12} = (\varepsilon_{rr} - \varepsilon_{\theta\theta})\sin\theta\cos\theta + \varepsilon_{r\theta}\cos 2\theta, \\ &\varepsilon_{rz} = \varepsilon_{13}\cos\theta + \varepsilon_{23}\sin\theta, && \varepsilon_{13} = \varepsilon_{rz}\cos\theta - \varepsilon_{\theta z}\sin\theta, \\ &\varepsilon_{\theta z} = -\varepsilon_{13}\sin\theta + \varepsilon_{23}\cos\theta, && \varepsilon_{23} = \varepsilon_{rz}\sin\theta + \varepsilon_{\theta z}\cos\theta. \end{aligned}$$

Similarly, for the stresses,

$$\sigma_{rr} = \sigma_{11}\cos^2\theta + \sigma_{22}\sin^2\theta + \sigma_{12}\sin 2\theta, \qquad \sigma_{11} = \sigma_{rr}\cos^2\theta + \sigma_{\theta\theta}\sin^2\theta - \sigma_{r\theta}\sin 2\theta,$$

$$\sigma_{\theta\theta} = \sigma_{11}\sin^2\theta + \sigma_{22}\cos^2\theta - \sigma_{12}\sin 2\theta, \qquad \sigma_{22} = \sigma_{rr}\sin^2\theta + \sigma_{\theta\theta}\cos^2\theta + \sigma_{r\theta}\sin 2\theta,$$

$$\sigma_{zz} = \sigma_{33}, \qquad \sigma_{33} = \sigma_{zz},$$

$$\sigma_{r\theta} = (\sigma_{22} - \sigma_{11})\sin\theta\cos\theta + \sigma_{12}\cos 2\theta, \qquad \sigma_{12} = (\sigma_{rr} - \sigma_{\theta\theta})\sin\theta\cos\theta + \sigma_{r\theta}\cos 2\theta,$$

$$\sigma_{rz} = \sigma_{13}\cos\theta + \sigma_{23}\sin\theta, \qquad \sigma_{13} = \sigma_{rz}\cos\theta - \sigma_{\theta z}\sin\theta,$$

$$\sigma_{\theta z} = -\sigma_{13}\sin\theta + \sigma_{23}\cos\theta, \qquad \sigma_{23} = \sigma_{rz}\sin\theta + \sigma_{\theta z}\cos\theta.$$

G.2 Spherical Orthogonal Curvilinear Coordinates

A differential volume element and the components of the stress tensor are shown in Fig. G.1b. The displacement vector is written as

$$\mathbf{u} = \hat{\mathbf{e}}_r u_r + \hat{\mathbf{e}}_\theta u_\theta + \hat{\mathbf{e}}_\phi u_\phi, \tag{G.6}$$

where the unit base vectors $(\hat{\mathbf{e}}_r, \hat{\mathbf{e}}_\theta, \hat{\mathbf{e}}_\phi)$ are given in terms of Cartesian base vectors by

$$\begin{aligned} \hat{\mathbf{e}}_r &= \cos\theta\sin\phi\hat{\mathbf{e}}_1 + \sin\theta\sin\phi\hat{\mathbf{e}}_2 + \cos\phi\hat{\mathbf{e}}_3, \\ \hat{\mathbf{e}}_\theta &= -\sin\theta\hat{\mathbf{e}}_1 + \cos\theta\hat{\mathbf{e}}_2, \\ \hat{\mathbf{e}}_\phi &= \cos\theta\cos\phi\hat{\mathbf{e}}_1 + \sin\theta\cos\phi\hat{\mathbf{e}}_2 - \sin\phi\hat{\mathbf{e}}_3. \end{aligned} \tag{G.7}$$

The strains are related to the displacements by

$$\varepsilon_{rr} = \frac{\partial u_r}{\partial r}, \quad \varepsilon_{r\theta} = \frac{1}{2}\left(\frac{1}{r\sin\phi}\frac{\partial u_r}{\partial\theta} - \frac{u_\theta}{r} + \frac{\partial u_\theta}{\partial r}\right),$$

$$\varepsilon_{\theta\theta} = \frac{1}{r\sin\phi}\frac{\partial u_\theta}{\partial\theta} + \frac{u_r}{r} + u_\phi\frac{\cot\phi}{r},$$

$$\varepsilon_{r\phi} = \frac{1}{2}\left(\frac{1}{r}\frac{\partial u_r}{\partial\phi} - \frac{u_\phi}{r} + \frac{\partial u_\phi}{\partial r}\right),$$

$$\varepsilon_{\phi\phi} = \frac{1}{r}\frac{\partial u_\phi}{\partial\phi} + \frac{u_r}{r}, \quad \varepsilon_{\theta\phi} = \frac{1}{2}\left(\frac{1}{r}\frac{\partial u_\theta}{\partial\phi} - \frac{u_\theta\cot\phi}{r} + \frac{1}{r\sin\phi}\frac{\partial u_\phi}{\partial\theta}\right). \tag{G.8}$$

The relationships between stresses and strains, with $e = \varepsilon_{rr} + \varepsilon_{\theta\theta} + \varepsilon_{\phi\phi}$, are

$$\begin{gathered} \sigma_{rr} = \lambda e + 2\mu\varepsilon_{rr} \quad \sigma_{\theta\theta} = \lambda e + 2\mu\varepsilon_{\theta\theta} \quad \sigma_{\phi\phi} = \lambda e + 2\mu\varepsilon_{\phi\phi} \\ \sigma_{r\theta} = 2\mu\varepsilon_{r\theta} \quad \sigma_{r\phi} = 2\mu\varepsilon_{r\phi} \quad \sigma_{\theta\phi} = 2\mu\varepsilon_{\theta\phi} \end{gathered} \tag{G.9}$$

and the equations of equilibrium are

$$\frac{\partial\sigma_{rr}}{\partial r} + \frac{1}{r\sin\phi}\frac{\partial\sigma_{r\theta}}{\partial\theta} + \frac{1}{r}\frac{\partial\sigma_{r\phi}}{\partial\phi} + \frac{2\sigma_{rr} - \sigma_{\theta\theta} - \sigma_{\phi\phi} + \sigma_{r\phi}\cot\phi}{r} + f_r = 0,$$

$$\frac{\partial\sigma_{r\theta}}{\partial r} + \frac{1}{r\sin\phi}\frac{\partial\sigma_{\theta\theta}}{\partial\theta} + \frac{1}{r}\frac{\partial\sigma_{\theta\phi}}{\partial\phi} + \frac{3\sigma_{r\theta} + 2\sigma_{\theta\phi}\cot\phi}{r} + f_\theta = 0,$$

$$\frac{\partial\sigma_{r\phi}}{\partial r} + \frac{1}{r\sin\phi}\frac{\partial\sigma_{\theta\phi}}{\partial\theta} + \frac{1}{r}\frac{\partial\sigma_{\phi\phi}}{\partial\phi} + \frac{3\sigma_{r\phi} + (\sigma_{\phi\phi} - \sigma_{\theta\theta})\cot\phi}{r} + f_\phi = 0. \quad \text{(G.10)}$$

Appendix H

Components of the Eshelby Tensor in Isotropic System

Expressions given by Böhm, Fischer and Reisner (1997) for the non-zero components of the Eshelby tensor for ellipsoids of revolution in an isotropic system are listed below using the contracted matrix notation rules given by Eq. (2.88). The axis of revolution is along the x_3 axis, and $\alpha \equiv a_3/a_1 = a_3/a_2$.

$$
\begin{aligned}
S^{E}_{33} &= \frac{1}{2(1-\nu^{M})}\left[\frac{4\alpha^2-2}{\alpha^2-1} - 2\nu^{M} + \left(\frac{4\alpha^2-1}{1-\alpha^2} + 2\nu^{M}\right)g(\alpha)\right]\\
S^{E}_{11} = S^{E}_{22} &= \frac{1}{4(1-\nu^{M})}\left\{\frac{3\alpha^2}{2(\alpha^2-1)} + \left[\frac{4\alpha^2-13}{4(\alpha^2-1)} - 2\nu^{M}\right]g(\alpha)\right\}\\
S^{E}_{31} = S^{E}_{32} &= \frac{1}{2(1-\nu^{M})}\left\{\frac{\alpha^2}{1-\alpha^2} + 2\nu^{M} + \left[\frac{2\alpha^2+1}{2(\alpha^2-1)} - 2\nu^{M}\right]g(\alpha)\right\}\\
S^{E}_{13} = S^{E}_{23} &= \frac{1}{4(1-\nu^{M})}\left[\frac{2\alpha^2}{1-\alpha^2} + \left(\frac{2\alpha^2+1}{\alpha^2-1} - 2\nu^{M}\right)g(\alpha)\right] \qquad \text{(H.1)}\\
S^{E}_{12} = S^{E}_{21} &= \frac{1}{4(1-\nu^{M})}\left\{\frac{\alpha^2}{2(\alpha^2-1)} + \left[\frac{4\alpha^2-1}{4(1-\alpha^2)} + 2\nu^{M}\right]g(\alpha)\right\}\\
S^{E}_{44} = S^{E}_{55} &= \frac{1}{4(1-\nu^{M})}\left[\frac{2}{1-\alpha^2} - 2\nu^{M} + \frac{1}{2}\left(\frac{2\alpha^2+4}{\alpha^2-1} + 2\nu^{M}\right)g(\alpha)\right]\\
S^{E}_{66} &= \frac{1}{4(1-\nu^{M})}\left\{\frac{\alpha^2}{2(\alpha^2-1)} + \left[\frac{4\alpha^2-7}{4(\alpha^2-1)} - 2\nu^{M}\right]g(\alpha)\right\}
\end{aligned}
$$

with

$$
g(\alpha) = \frac{\alpha}{(\alpha^2-1)^{3/2}}\left[\alpha(\alpha^2-1)^{1/2} - \ln\left(\alpha + \sqrt{\alpha^2-1}\right)\right] \quad (\alpha > 1) \qquad \text{(H.2)}
$$

and

$$g(\alpha) = \frac{\alpha}{(1-\alpha^2)^{3/2}}\left[\cos^{-1}\alpha - \alpha(1-\alpha^2)^{1/2}\right]. \quad (\alpha < 1) \tag{H.3}$$

For a sphere ($\alpha = 1$),

$$S^E_{11} = S^E_{22} = S^E_{33} = \frac{7-5\nu^M}{15(1-\nu^M)},$$

$$S^E_{12} = S^E_{21} = S^E_{13} = S^E_{31} = S^E_{23} = S^E_{32} = \frac{5\nu^M - 1}{15(1-\nu^M)},$$

$$S^E_{44} = S^E_{55} = S^E_{66} = \frac{4-5\nu^M}{15(1-\nu^M)}. \tag{H.4}$$

For a needle ($\alpha \to \infty$),

$$S^E_{11} = S^E_{22} = \frac{5-4\nu^M}{8(1-\nu^M)}, \quad S^E_{12} = S^E_{21} = \frac{4\nu^M - 1}{8(1-\nu^M)},$$

$$S^E_{13} = S^E_{23} = \frac{\nu^M}{2(1-\nu^M)}, \quad S^E_{44} = S^E_{55} = \frac{1}{4},$$

$$S^E_{66} = \frac{3-4\nu^M}{8(1-\nu^M)}. \tag{H.5}$$

For a thin disk ($\alpha \to 0$),

$$S^E_{33} = 1, \quad S^E_{31} = S^E_{32} = \frac{\nu^M}{1-\nu^M}, \quad S^E_{44} = S^E_{55} = \frac{1}{2}. \tag{H.6}$$

Appendix I

Airy Stress Functions for Plane Strain

Tabulated below are various Airy stress functions, ψ, and the forms of their associated stresses, which apply when plane strain conditions exist (see Sec. 3.7).

Table I.1 Forms of the stresses associated with Airy stress functions, ψ, for plane strain. $\theta = \tan^{-1}(x_2/x_1)$, and $x = (x_1^2 + x_2^2)^{1/2}$.

ψ	σ_{11}	σ_{22}	σ_{12}
$x \ln x \sin\theta$	$\frac{x_2}{x^2} + \frac{2x_1^2 x_2}{x^4}$	$\frac{x_2}{x^2} - \frac{2x_1^2 x_2}{x^4}$	$-\frac{x_1}{x^2} + \frac{2x_1 x_2^2}{x^4}$
$x \ln x \cos\theta$	$\frac{x_1}{x^2} - \frac{2x_1 x_2^2}{x^4}$	$\frac{x_1}{x^2} + \frac{2x_1 x_2^2}{x^4}$	$-\frac{x_2}{x^2} + \frac{2x_1^2 x_2}{x^4}$
$x\theta \sin\theta$	$\frac{2x_1}{x^2} - \frac{2x_1 x_2^2}{x^4}$	$\frac{2x_1 x_2^2}{x^4}$	$\frac{2x_1^2 x_2}{x^4}$
$x\theta \cos\theta$	$-\frac{2x_1^2 x_2}{x^4}$	$-\frac{2x_2}{x^2} + \frac{2x_1^2 x_2}{x^4}$	$-\frac{2x_1 x_2^2}{x^4}$
$\ln x$	$-\frac{1}{x^2} + \frac{2x_1^2}{x^4}$	$\frac{1}{x^2} - \frac{2x_1^2}{x^4}$	$\frac{2x_1 x_2}{x^4}$
$\sin 2\theta$	$-\frac{12x_1 x_2}{x^4} + \frac{16x_1 x_2^3}{x^6}$	$-\frac{12x_1 x_2}{x^4} + \frac{16x_1^3 x_2}{x^6}$	$\frac{2}{x^2} - \frac{16x_1^2 x_2^2}{x^6}$
$\cos 2\theta$	$-\frac{4x_1^2}{x^4} + \frac{16x_1^2 x_2^2}{x^6}$	$\frac{4x_2^2}{x^4} - \frac{16x_1^2 x_2^2}{x^6}$	$\frac{8x_1 x_2}{x^4} - \frac{16x_1^3 x_2}{x^6}$
θ	$-\frac{2x_1 x_2}{x^4}$	$\frac{2x_1 x_2}{x^4}$	$-\frac{1}{x^2} + \frac{2x_1^2}{x^4}$

(*Continued*)

Table I.1 (*Continued*)

ψ	σ_{11}	σ_{22}	σ_{12}
$\dfrac{\sin\theta}{x}$	$\dfrac{2x_2}{x^4} - \dfrac{8x_1^2x_2}{x^6}$	$-\dfrac{2x_2}{x^4} + \dfrac{8x_1^2x_2}{x^6}$	$\dfrac{2x_1}{x^4} - \dfrac{8x_1x_2^2}{x^6}$
$\dfrac{\cos\theta}{x}$	$-\dfrac{2x_1}{x^4} + \dfrac{8x_1x_2^2}{x^6}$	$\dfrac{2x_1}{x^4} - \dfrac{8x_1x_2^2}{x^6}$	$\dfrac{2x_2}{x^4} - \dfrac{8x_1^2x_2}{x^6}$
$x^3\sin\theta$	$6x_2$	$2x_2$	$-2x_1$

Appendix J

Deviatoric Stress and Strain in Isotropic System

It is often helpful to split a general stress tensor, σ_{ij}, into a so-called *mean* part, i.e., $(\sigma_{mm}/3)\delta_{ij}$, and a *deviatoric* part, $\breve{\sigma}_{ij}$. The deviatoric part is then given by

$$\breve{\sigma}_{ij} = \sigma_{ij} - \frac{1}{3}\Theta\delta_{ij}$$

$$= \begin{bmatrix} \sigma_{11} - \dfrac{\sigma_{11}+\sigma_{22}+\sigma_{33}}{3} & \sigma_{12} & \sigma_{13} \\ \sigma_{12} & \sigma_{22} - \dfrac{\sigma_{11}+\sigma_{22}+\sigma_{33}}{3} & \sigma_{23} \\ \sigma_{13} & \sigma_{23} & \sigma_{33} - \dfrac{\sigma_{11}+\sigma_{22}+\sigma_{33}}{3} \end{bmatrix} \tag{J.1}$$

Similarly, the corresponding strain tensor can be split into mean and deviatoric parts so that

$$\breve{\varepsilon}_{ij} = \varepsilon_{ij} - \frac{1}{3}e\delta_{ij} \tag{J.2}$$

The deviatoric stresses and strains are coupled by the elastic constants in the usual way, and, since $\breve{e} = \breve{\Theta} = 0$,

$$\breve{\sigma}_{ij} = \lambda\breve{e}\delta_{ij} + 2\mu\breve{\varepsilon}_{ij} = 2\mu\breve{\varepsilon}_{ij} \quad \breve{\varepsilon}_{ij} = -\frac{\lambda\breve{\Theta}\delta_{ij}}{2\mu(3\lambda+2\mu)} + \frac{1}{2\mu}\breve{\sigma}_{ij} = \frac{1}{2\mu}\breve{\sigma}_{ij} \tag{J.3}$$

The deviatoric stresses and strains are therefore coupled by the simple relationship

$$\breve{\sigma}_{ij} = 2\mu\breve{\varepsilon}_{ij}. \tag{J.4}$$

References

Asaro, R. J. and Barnett, D. M. (1975). The non-uniform transformation strain problem for an anisotropic ellipsoidal inclusion. *J. Mech. Phys. Solids*, **23**, 77–83.

Asaro, R. J. and Barnett, D. M. (1976). Applications of the geometrical theorems for dislocations in anisotropic elastic media. *Nuclear Metall.*, **20**, 313–24.

Asaro, R. J. and Lubarda, V. A. (2006). *Mechanics of Solids and Materials*, Cambridge: Cambridge U. Press.

Balluffi, R. W., Allen, S. M. and Carter, W. C. (2005). *Kinetics of Materials*, Hoboken, NJ: Wiley.

Blin, J. (1955). Energie mutuelle de deux dislocations. *Acta Met.* **3**, 199–200.

Bacon, D. J., Barnett, D. M. and Scattergood, R. O. (1979a). On the anisotropic elastic field of a dislocation segment in three dimensions. *Phil. Mag. A*, **39**, 231–35.

Bacon, D. J., Barnett, D. M. and Scattergood, R. O. (1979b). Anisotropic continuum theory of lattice defects. *Prog. Mat. Sci.*, **23**, 51–262.

Bacon, D. J. and Groves, P. P. (1970). In *Fundamental Aspects of Dislocation Theory, National Bur. Standards Special Publ. 317,* **1**, ed. J. A. Simmons, R. deWit and R. Bullough. Washington, DC: U. S. Gov. Printing Office, pp. 35–45.

Barnett, D. M. (1971). On nucleation of coherent precipitates near edge dislocations. *Scripta Metall.*, **5**, 261–66.

Barnett, D. M. (1972). The precise evaluation of derivatives of the anisotropic elastic Green's function. *Phys. Stat. Sol. (b)*, **49**, 741–48.

Barnett, D. M. (1976). The singular nature of the self-stress field of a plane dislocation loop in an anisotropic elastic medium. *Phys. Stat. Sol. (a)*, **38**, 637–46.

Barnett, D. M. (1985). The displacement field of a triangular dislocation loop. *Phil. Mag. A*, **51**, 383–87.

Barnett, D. M. (2007). Unpublished research, Palo Alto: Stanford University.

Barnett, D. M. (2015). Private communication.

Barnett, D. M. and Balluffi, R. W. (2007). The displacement field of a triangular dislocation loop-a correction with commentary. *Phil. Mag. Lett.*, **87**, 943–45.

Barnett, D. M., Lee, J. K., Aaronson, H. I. and Russell, K. C. (1974). The strain energy of a coherent ellipsoidal precipitate. *Scripta Metall.*, **8**, 1447–50.

Barnett, D. M. and Lothe, J. (1974). An image force theorem for dislocations in anisotropic bicrystals. *J. Phys. F: Metal Phys.*, **4**, 1618–35.

Belov, A. Y. (1992). Dislocations emerging at planar boundaries. In *Elastic Strain Fields and Dislocation Mobility,* ed. V. L. Indenbom and J. Lothe. Amsterdam: North-Holland, pp. 391–446.

Bilby, B. A. (1990). John Douglas Eshelby. *Bio. Mem. Fellows Roy. Soc.*, **36**, 126–50.

Böhm, H. J., Fischer, F. D. and Reisner, G. (1997). Evaluation of elastic strain energy of spheroidal inclusions with uniform volumetric and shear eigenstrains. *Scripta Mater.*, **36**, 1053–59.

Boussinesq, J. (1885). *Applications des Potentiels.* Paris: Gauthier-Villars.

Burgers, J. M. (1939a). Some considerations on the fields of stress connected with dislocations in a regular crystal lattice I. *Proc. Kon. Nederl. Akad. Wetenschap.*, **42**, 293–325.

Burgers, J. M. (1939b). Some considerations on the fields of stress connected with dislocations in a regular crystal lattice II. *Proc. Kon. Nederl. Akad. Wetenschap.*, **42**, 378–99.

Byrd, P. F. and Friedman, M. D. (1954). *Handbook of Elliptical Integrals.* Berlin: Springer-Verlag.

Cayley, A. (1895). *An Elementary Treatise on Elliptic Functions, 2nd Edition.* London: George Bell and Sons.

Chalon, F. and Montheillet, F. (2003). The interaction of two spherical gas bubbles in an infinite elastic solid. *J. Appl. Mech.*, **70**, 789–98.

Chou, Y. T. and Eshelby, J. D. (1962). The energy and line tension of a dislocation in a hexagonal crystal. *J. Mech. Phys. Solids*, **10**, 27–34.

Chou, P. C. and Pagano, N. J. (1967). *Elasticity.* Princeton, NJ: Van Nostrand Co.

Christian, J. W. (1975). *The Theory of Transformations in Metals and Alloys.* Oxford: Pergamon Press.

Cochardt, A. W., Schoeck, G. and Wiedersich, H. (1955). Interaction between dislocations and interstitial atoms in body-centered cubic metals. *Acta Mater.*, **3**, 533–37.

Devincre, B. (1995). Three-dimensional stress field expressions for straight dislocation segments. *Solid State Comms.*, **93**, 875–78.

Devincre, B., Kubin, L. and Hoc, T. (2006). Physical analyses of crystal plasticity by DD simulations. *Scripta Mater.*, **54**, 741–46.

Devincre, B., Kubin, L. P., Lemarchand, C. and Madec, R. (2001). Mesoscopic simulations of plastic deformation. *Mat. Sci. and Eng. A*, **309–310**, 211–19.

deWit, R. (1960). The continuum theory of stationary dislocations. In *Solid State Physics*, **10**, ed. F. Seitz and D. Turnbull, New York: Academic Press, pp. 249–92.

deWit, R. (1967). Some relations for straight dislocations. *Phys. Stat. Sol.*, **20**, 567–73.

Dundurs, J. (1969). Elastic interactions of dislocations with inhomogeneities. In *Mathematical Theory of Dislocations*, ed. T. Mura. New York: Amer. Soc. Mech. Engrs., pp. 70–115.

Dundurs, J. and Sendeckyj, G. P. (1965). Behavior of an edge dislocation near a bimetallic interface. *J. Appl. Phys.* **36**, 3353–54.

Eringen, C. (2002). *Nonlocal Continuum Field Theories.* New York: Springer.

Eshelby, J. D. (1951). The force on an elastic singularity. *Phil. Trans. Roy. Soc. London,* **244**A, 87–112.

Eshelby, J. D. (1954). Distortion of a crystal by point imperfections. *J. Appl. Phys.*, **25**, 255–61.

Eshelby, J. D. (1955). The elastic interactions of point defects. *Acta Metall.*, **3**, 48790.

Eshelby, J. D. (1956). The continuum theory of lattice defects. In *Solid State Physics*, **3**, ed. F. Seitz and D. Turnbull. New York: Academic Press, pp. 79–144.

Eshelby, J. D. (1957). The determination of the elastic field of an ellipsoidal inclusion, and related problems. *Proc. Roy. Soc. A*, **241**, 376–96.

Eshelby, J. D. (1959). The elastic field outside an ellipsoidal inclusion. *Proc. Roy. Soc. A*, **252**, 561–69.

Eshelby, J. D. (1961). Elastic inclusions and inhomogeneities. In Vol. 2 of *Progress in Solid Mechanics*, ed. I. N. Sneddon and R. Hill. Amsterdam: North-Holland, pp. 87–140.

Eshelby, J. D. (1975). The elastic energy-momentum tensor. *J. Elasticity,* **5**, 321–35.

Eshelby, J. D. (1977). Interaction and diffusion of point defects. In *Vacancies '76*, ed. R. E. Smallman and J. E. Harris. London: The Metals Soc., pp. 3–10.

Eshelby, J. D. (1979). Boundary problems. In *Dislocations in Solids*, **1**, Chap. 3, ed. F. R. N. Nabarro. Amsterdam: North-Holland, pp. 167–221.

Eshelby, J. D. and Laub, T. (1967). Interpretation of terminating dislocations. *Can. J. Phys.*, **45**, 887–92.

Eshelby, J. D., Read, W. T. and Shockley, W. (1953). Anisotropic elasticity with applications to dislocation theory. *Acta Met.*, **1**, 251–59.

Eshelby, J. D. and Stroh, A. N. (1951). Dislocations in thin plates. *Phil. Mag.*, **42**, 1401–05.

Gavazza, S. D. and Barnett, D. M. (1975). The image force on a dislocation loop in a bounded elastic medium. *Scripta Metall.*, **9**, 1263–65.

Gavazza, S. D. and Barnett, D. M. (1976). Self-force on a planar dislocation loop in an anisotropic linear-elastic medium. *J. Mech. Phys. Solids*, **24**, 171–85.

Gel'fand, I. M., Graev, I. M. and Vilenkin, N. Ya (1966). Vol. 5 of *Generalized Functions.* New York: Academic Press.

Gel'fand, I. M. and Shilov, G. E. (1964). Vol. 1 of *Generalized functions.* New York: Academic Press.

Gradshteyn, I. S. and Ryzhik, I. M. (1980). *Tables of Integrals, Series and Products.* New York: Academic Press.

Groves, P. P. and Bacon, D. J. (1969). Elastic centers of strain and dislocations. *J. Appl. Phys.*, **40**, 4207–09.

Hardy, J. R. and Bullough, R. (1967). Point defect interactions in harmonic cubic lattices. *Phil. Mag.*, **15**, 237–46.

Hardy, J. R. and Bullough, R. (1968). Strain field interaction between vacancies in copper and aluminium. *Phil. Mag.* **17**, 833–42.

Hassani, S. (2000). *Mathematical Methods for Students of Physics and Related Fields*, Berlin: Springer.

Hehenkamp, T. (1994) Absolute vacancy concentrations in noble metals and some of their alloys. *J. Phys. Chem. Solids*, **55**, 907–15.

Hetnarski, R. B. and Ignaczak, J. (2004). *Mathematical Theory of Elasticity.* London: Taylor and Francis.

Hildebrand. F. B. (1949). *Advanced Calculus for Engineers.* New York: Prentice-Hall.

Hill, R. (1961). Discontinuity relations in the mechanics of solids. In *Progress in Solid Mechanics*, **2**, ed. I. N. Sneddon and R. Hill. Amsterdam: North-Holland, pp. 246–276.

Hirth, J. P., Barnett, D. M. and Lothe, J. (1979). Stress fields of dislocation arrays in bicrystals. *Phil. Mag. A*, **40**, 39–47.

Hirth, J. P. and Lothe, J. (1982). *Theory of Dislocations.* New York: John Wiley.

Indenbom, V. L. and Dubnova, G. N. (1967). Interactions of dislocations at nodes and equilibrium of dislocations. *Soviet Phys. Sol. State, USSR*, **9**, 915–19.

Indenbom, V. L. and Lothe, J., eds. (1992). *Elastic Strain Fields and Dislocation Mobility.* Amsterdam: North-Holland.

Indenbom, V. L. and Orlov, S. S. (1968). The general solutions for dislocations in an anisotropic medium. In *Proc. Kharkov Conf. on Dislocation Dynamics*, Physical-Technical Inst. of Low Temperatures, Kharkov: USSR Acad. Sci., pp. 406–417.

James, R. W. (1954). *The Optical Principles of the Diffraction of X-rays.* London: G. Bell and Sons.

Kato, M. and Fujii, T. (1994). Elastic state and orientation of plate-shaped inclusions. *Acta Metall. Mater.*, **42**, 2929–36.

Kato, M., Fujii, T. and Onaka, S. (1996a). Elastic states of inhomogeneous spheroidal inclusions. *Mat. Trans. Jap. Inst. Metals*, **37**, 314–18.

Kato, M., Fujii, T. and Onaka, S. (1996b). Elastic state and orientation of needle-shaped inclusions. *Acta Mater.*, **44**, 1263–69.

Kato, M., Fujii, T. and Onaka, S. (1996c). Elastic strain energies of sphere, plate and needle inclusions. *Mats. Sci. Eng. A*, **211**, 95–103.

Kellogg, O. D. (1929). *Foundations of Potential Theory.* Berlin: Springer.

Khraishi, T. A., Hirth, J. P. and Zbib, H. M. (2000). The stress field of a general circular Volterra dislocation loop: analytical and numerical approaches. *Phil. Mag. Lett.*, **80**, 95–105.

Khraishi, T. A., Hirth, J. P., Zbib, H. M. and Khaleel, M. A. (2000). The displacement, and strain-stress fields of a general circular Volterra dislocation loop. *Int. J. Eng. Sci.*, **38**, 251–66.

Kroupa, F. (1960). Circular edge dislocation loop. *Czech. J. Phys. B*, **10**, 284–93.

Kroupa, F. (1962). Continuous distribution of dislocation loops. *Czech. J. Phys. B*, **12**, 191–201.

Kroupa, F. (1966). Dislocation loops. In *Theory of Crystal Defects*, ed. B. Gruber. New York: Academic Press, pp. 275–316.

Leibfried. G. (1953). Versetzungen in anisotropem material. *Z. f. Physik*, **135**, 23–43.

Leibfried, G. and Breuer, N. (1978). *Point Defects in Metals I.* Springer Tracts in Modern Physics, 81. Berlin: Springer-Verlag.

Leipholz, H. (1975). *Theory of Elasticity.* Leyden: Noordhoff International Publishing.

Lekhnitskii, S. G. (1963). *Theory of Elasticity of an Anisotropic Body.* Moscow: Mir Publishers.

Lothe, J. (1982). Dislocations in anisotropic media: the interaction energy. *Phil. Mag. A*, **46**, 177–80.

Lothe, J. (1992a). Dislocations in continuous elastic media. In *Elastic Strain Fields and Dislocation Mobility*, ed. V. L. Indenbom and J. Lothe. Amsterdam: North-Holland, pp. 175–235.

Lothe, J. (1992b). Dislocations in anisotropic media. In *Elastic Strain Fields and Dislocation Mobility*, ed. V. L. Indenbom and J. Lothe. Amsterdam: North-Holland, pp. 269–328.

Lothe, J. (1992c). Dislocations interacting with surfaces, interfaces or cracks. In *Elastic Strain Fields and Dislocation Mobility*, ed. V. L. Indenbom and J. Lothe. Amsterdam: North-Holland, pp. 329–89.

Lothe, J., Indenbom, V. L. and Chamrov, V. A. (1982). Elastic field and self-force of dislocations emerging at the free surfaces of an anisotropic halfspace. *Phys. Stat. Sol. (b)*, **111**, 671–77.

Love, A. E. H. (1944). *A Treatise on the Mathematical Theory of Elasticity.* New York: Dover Publ.

MacMillan, W. D. (1930). *The Theory of the Potential.* New York: McGraw-Hill.

Meisner, M. J. and Kouris, D. A. (1995). Interaction of two elliptic inclusions. *Int. J. Solids Structures*, **32**, 451–66.

Mindlin, R. D. (1936a). Note on the Galerkin and Papkovitch stress functions, *Bull. Amer. Math. Soc.*, **42**, 373–76.

Mindlin, R. D. (1936b). Force at a point in the interior of a semi-infinite solid. *Physics*, **7**, 195–202.

Mindlin, R. D. (1953). Force at a point in the interior of a semi-infinite solid. In *Proc.of First Midwestern Conference on Solid Mechanics*, Urbana, IL: College of Eng. and Panel on fluid and Solid Mechanics. pp. 56–59.

Mindlin, R. D. and Cheng, D. H. (1950). Thermoelastic stress in the semi-infinite solid. *J. Appl. Phys.*, **21**, 931–33.

Morse, P. M. and Feshbach. H. (1953). *Methods of Theoretical Physics*, New York: McGraw-Hill.

Mura, T. (1963). Continuous distribution of moving dislocations, *Phil. Mag.*, **8**, 843–57.

Mura, T. (1987). *Micromechanics of Defects in Solids.* The Hague: Martinus Nijhoff.

Mura, T and Cheng, P. C. (1977). The elastic field outside an ellipsoidal inclusion. *J. Mech. Phys. Solids*, **44**, 591–94.

Muskhelishvili, N. I. (1953). *Some Basic Problems of the Mathematical Theory of Elasticity.* Groningen-Holland: Noordhoff.

Nabarro, F. R. N. (1940). The strains produced by precipitation in alloys. *Proc. Roy. Soc. A*, **175**, 519–38.

Nakahara, S. and Willis, J. R. (1973). Some remarks on interfacial dislocations. *J. Phys. F: Metal Phys.*, **3**, L249–54.

Nakahara, S., Wu, J. B. C. and Li, J. C. M. (1972). Dislocations in a welded interface between two isotropic media. *Mater. Sci. Eng.*, **10**, 291–96.

Neuber, H. (1944). *Kerbspannungslehre,* Ann Arbor, MI: J. W. Edwards.

Nishioka, K. and Lothe, J. (1972). Isotropic limiting behavior of the six-dimensional formalism of anisotropic dislocation theory and anisotropic Green's function theory. I. sum rules and their applications. *Phys. Stat. Sol. (b)*, **51**, 645–56.

Nowick, A. S. and Berry, B. S. (1972). *Anelastic Relaxation in Crystalline Solids.* New York: Academic Press.

Nowick, A. S. and Heller, W. R. (1963). Anelasticity and stress-induced ordering of point defects in crystals. *Advances in Physics*, **12**, 251–98.

Nye, J. F. (1953). Some geometrical relations in dislocated crystals. *Acta Metall.*, **1**, 153–62.

Nye, J. F. (1957). *Physical Properties of Crystals.* London: Oxford University Press.

Onaka, S. (2001). Averaged Eshelby tensor and elastic strain energy of a superspherical inclusion with uniform eigenstrains. *Phil. Mag. Lett.*, **81**, 265–72.

Onaka, S., Fujii, T. and Kato, M. (1995). The elastic strain energy of a coherent inclusion with deviatoric misfit strains. *Mech. Mats.*, **20**, 329–36.

Pan, E. and Yuan, F. G. (2000). Three-dimensional Green's functions in anisotropic bimaterials. *Internat. J. Solids and Structures*, **37**, 5329–51.

Papkovitch, P. F. (1932). Solution generale des equations differentielles fondamentales d'elasticite, exprimees par trois fonctions harmoniques. *Comptes Rendus, Acad. des Sciences, Paris*, **195**, 513–15.

Peach, M. and Koehler, J. S. (1950). The forces exerted on dislocations and the stress fields produced by them. *Phys. Rev.*, **80**, 436–39.

Poincare, H. (1899). *Theorie du Potentiel Newtonien*, Paris: Gauthier-Villars.

Pond, R. C., Ma, X., Chai, Y. W. and Hirth, J. P. (2007). Topological modelling of martensitic transformations. In *Dislocations in Solids*, **13**, Chap. 74, ed. F. R. N. Nabarro and J. P. Hirth. Amsterdam: Elsevier B. V., pp. 225–61.

Read, W. T. (1953). *Dislocations in Crystals.* New York: McGraw-Hill.

Read, W. T. and Shockley, W. (1950). Dislocation models of crystal grain boundaries. *Phys. Rev.*, **78**, 275–89.

Rice, J. R. (1968). A path independent integral and approximate analysis of strain concentration by notches and cracks. *J. Appl. Mech.*, **38**, 379–86.

Rietz, H. L., Reilly, J. F and Woods, R. (1936). *Plane and Spherical Trigonometry.* New York: MacMillan.

Rongved, L. (1955). Force interior to one of two joined semi-infinite solids. In: *Proceedings of the 2nd Midwestern Conference on Solid Mechanics*, Lafayette, IN: Research Series No. 129, Eng. Exp. Station, Purdue Univ., pp. 1–13.

Schilling, W. (1978). Self-interstitial atoms in metals. *J. Nucl. Mats.*, **69** & **70**, 465–89.

Seidman, D. N. and Balluffi, R. W. (1965). Sources of thermally generated vacancies in single-crystal and polycrystalline gold. *Phys. Rev. A*, **139**, 1824–40.

Shaibani, S. J. and Hazzledine, P. M. (1981). The displacement and stress fields of a general dislocation close to a free surface of an isotropic solid. *Phil. Mag. A*, **44**, 657–65.

Sharma, P. and Ganti, S. (2003). The size-dependent elastic state of inclusions in non-local elastic solids. *Phil. Mag. Lett.*, **83**, 745–54.

Siems, R. (1968) Mechanical interactions of point defects. *Phys. Stat Sol.*, **30**, 645–58.

Simmons, R. O. and Balluffi, R. W. (1960). Measurements of equilibrium vacancy concentrations in aluminum. *Phys. Rev.*, **117**, 52–61.

Smythe, W. R. (1950). *Static and Dynamic Electricity*, New York: McGraw-Hill.

Sneddon, I. N. (1951). *Fourier Transforms*, New York: McGraw-Hill.

Sokolnikoff, I. S. (1946). *Mathematical Theory of Elasticity*, New York: McGraw-Hill.

Sokolnikoff, I. S. and Redheffer, R. M. (1958) *Mathematics of Physics and Modern Engineering.* New York: McGraw-Hill.

Soutas-Little, R. W. (1999). *Elasticity.* Mineola, NY: Dover Publications.

Steeds, J. W. and Willis, J. R. (1979). Dislocations in anisotropic media. In *Dislocations in Solids*, **1**, Chap. 2, ed. F. R. N. Nabarro. Amsterdam: North-Holland, pp. 145–65.

Stroh, A. N. (1958). Dislocations and cracks in anisotropic elasticity. *Phil. Mag.*, **3**, 625–46.

Stroh, A. N. (1962). Steady state problems in anisotropic elasticity. *J. Math. Phys.*, **41**, 77–103.

Sutton, A. P. and Balluffi, R. W. (2006). *Interfaces in Crystalline Materials.* Oxford: Clarendon Press.

Teodosiu, C. (1982). *Elastic Models of Crystal Defects.* Berlin: Springer-Verlag.

Timoshenko, S. P. and Goodier, J. N. (1970). *Theory of Elasticity.* New York: McGraw-Hill.

Voigt, W. (1910). *Lehrbuch der Kristallphysik.* Leipzig: Teubner.

Volterra, V. (1907). Sur l'equilibre des carps elastiques multiplement connexes. *Annal. Sci. de l'Ecoles Norm. Super., Paris*, **24**, 401–517.

Weir, D. W., Hass, J. and Giordano, F. R. (2005). *Thomas' Calculus. Boston*: Addison Wesley.
Willis, J. R. (1970). Stress field produced by dislocations in anisotropic media. *Phil. Mag.*, **21**, 931–49.
Yoffe, E. H. (1960). The angular dislocation. *Phil. Mag.*, **5**, 161–75.
Yoffe, E. H. (1961). A dislocation at a free surface. *Phil Mag.*, **6**, 1147–55.
Zbib, H. M., Rhee, M. and Hirth, J. P. (1998). On plastic deformation and the dynamics of 3D dislocations. *Int. J. Mech. Sci.*, **40**, 113–27.

Index

Printed in the United States
By Bookmasters